Digital Signal Processing

Digital Signal Processing

Edited by
Lawrence R. Rabiner
Member, Technical Staff
Bell Telephone Laboratories, Inc.

Charles M. Rader
Assistant Leader of
Spacecraft Technology
M.I.T. Lincoln Laboratory

A volume in the IEEE PRESS Selected Reprint Series,
prepared under sponsorship of
the IEEE Audio and Electroacoustics Group.

IEEE PRESS

The Institute of Electrical and Electronics Engineers, Inc. New York

IEEE PRESS
445 Hoes Lane, P.O. Box 1331
Piscataway, New Jersey 08855-1331

ISBN 0-87942-018-9

IEEE Order Number: PP0018-2

Library of Congress Catalog Card Number 72-90358

Printed in the United States of America

Acknowledgement

The selection of papers in this monograph represents the work of several members of the IEEE Group on Audio and Electroacoustics Technical Committee on Digital Signal Processing, including, G. D. Bergland, Bell Laboratories, J. Cooley, IBM Corporation, H. Helms, Bell Laboratories, L. B. Jackson, Rockland Systems Corporation, J. Kaiser, Bell Laboratories, A. Oppenheim, Massachusetts Institute of Technology, L. R. Rabiner, Bell Laboratories, C. M. Rader, Massachusetts Institute of Technology, Lincoln Laboratories, R. W. Schafer, Bell Laboratories, K. Steiglitz, Princeton University, C. Weinstein, Massachusetts Institute of Technology, Lincoln Laboratories.

The contributions of the following individuals are also acknowledged: R. Kaenel, AMF Corporation, R. Singleton, Stanford University, T. Stockham, University of Utah, F. Van Veen, Teradyne, P. Welch, IBM Corporation.

Contents

Part 3—Effects of Finite Word Lengths

Introduction

As man becomes more certain of his control of physical things, an ever more important part of his work is the manipulation of the symbols he uses to describe and control these physical things. It is now common to consider a job essentially done when it has been completely planned; hence the great importance of the digital computer in our world. Even before this particular advance in electronic technology, one of the primary reasons for the importance of electronics was the ease with which electrical representations of information could be manipulated. The field of waveform manipulation, as in radio, radar, sonar, seismology, etc., is one of the keystones of electrical engineering.

During the decade of 1960–1970, an important new approach to waveform manipulation, or signal processing, came into importance. It became practical to represent information-bearing waveforms digitally, that is, symbolically, and to do signal processing on the digital representation of the waveform. Thus all the flexibility of digital computers became available for processing of waveforms, and this has had an effect on all the fields in which waveform processing is important. During the decade, it developed that one could manipulate waveforms digitally in ways that would be totally impractical with continuous representations. It was obvious at the start of the decade that one would expect one advantage in this respect, namely, that there is a freedom from the constraints of time when a waveform is in a digital memory. Continuous time waveforms occur and are gone. Delay lines provide one means of remembering waveforms but are cumbersome, expensive, and imperfect. The hardware to speed up, or slow down, a waveform by analog techniques is complicated. However, once a waveform, although transient, is in a computer memory, it may be considered to have any time base associated with it; it may even be restored to continuous form with a new time base, even a time reversal. At the beginning of 1960, a few workers were making use of the permanence of digital representations of waveforms to permit some digital processing. They were limited mainly by the time required for any but the simplest processing. It was understood at the beginning of the decade that a computer could do spectral analysis, but the computation time was so impractical that computer signal processing involving spectral analysis was most often done in conjunction with a bank of analog filters to perform the spectrum analysis. Digital filters, as known to most potential users of computers, were of the nonrecursive type, and could perform the most typical signal processing requirements only at the cost of considerable computing time. During the decade, the recursive filtering techniques were developed and became widely known, and computational algorithms for rapid Fourier transformation became available. As a result of the latter development, it is practical to do signal processing on waveforms in either the time or the frequency domains, something never practical with continuous systems. The Fourier transform became not just a theoretical description, but a tool. With this development, the field of digital signal processing grew from obscurity to importance, and it is likely to remain a tool of engineers in the years to come.

Digital signal processing is not just the processing of signals by computer. There are already a number of free standing devices, available as commercial products, that use digital processing techniques to achieve filters, signal generators, spectrum analyzers, correlation measurements, and displays. Meanwhile, the availability of a new set of computer techniques has greatly extended the scope of digital computers, so that they are now used for projects on which they could not have had any impact in the early sixties. For example, simulation of analog waveform processing schemes, such as modems, that would have required unaffordable amounts of computer time in 1960, is a commonly used method of system design today. As a result of the sort of flexibility offered by computers, experiments that could not be tried without them are

This work was supported by the U.S. Air Force.

now practical. What made this possible was the combination of three factors. One factor was the considerable advance in the availability of digital circuitry at low cost and capable of high-speed operation. The other two factors were the development of several efficient algorithms for filtering and spectral analysis, and significant new understandings of the theory of representing waveforms by discrete time series and modifying them in the time and frequency domains. By 1970, it was practical to attempt processing more than one hundred times as complex as was possible at the beginning of the decade.

The major event of the decade in digital signal processing was the publication, in April 1965, of a paper by J. W. Cooley and J. W. Tukey that disclosed an algorithm for computing Fourier series. The paper was concerned not with the computation of the Fourier series of continuous periodic waveforms, but of the so-called discrete Fourier transform. The former type of frequency domain representation mapped a waveform continuous in time into a sequence of complex numbers, generally an infinite sequence. The discrete Fourier transform, on the other hand, maps one finite sequence of numbers into another, and is therefore amenable to digital computation. The DFT has both a superficial similarity to the Fourier transform of continuous signal theory, and a rigorous set of analogous properties, such as a convolution theorem, a Parseval relationship, and a set of theorems concerning the transforms of odd or even, real or imaginary waveforms. The Cooley-Tukey paper demonstrated that there was an algebraic structure in the computation of discrete Fourier transforms that could be exploited to speed up the computation of such transforms by orders of magnitude for some practical examples that needed the DFT. The paper was published in *Mathematics of Computation*, and is included in this volume. Its publication was soon followed by useful expositions of the algorithm in several IEEE publications, some of which are also included here.

The publication of the Cooley-Tukey paper was not the first discovery of the exploitable structure of the DFT. It was, however, the first such publication to be widely noticed. One reason why the Cooley-Tukey paper was important, while the paper by I. J. Good a decade earlier was largely ignored (in spite of having described an essentially equivalent algorithm), was that by April 1965, a number of people were already involved in the use of computers to process waveforms. The work of James Kaiser had led to the ready adoption of a mathematical description widely known in control theory, the z-transform calculus, to sampled data representations of waveforms of interest in communication theory. Kaiser worked with recursive digital filters and showed how much of the well-developed theory of the design of filters made of resistors, capacitors, and inductors could be translated, with the aid of the z-transform, into straightforward digital filter design techniques. As with the Cooley-Tukey paper, many of the results we credit to Kaiser were exploited by previous authors, as he himself points out in his excellent 1966 survey of digital filtering in the book, *System Analysis by Digital Computer.* Kaiser, however, unified the field of digital filtering, making the necessary comparisons between recursive and nonrecursive filters, and considering the practical computational implications of the various designs. His work was applied rather quickly at Bell Telephone Laboratories, where digital filtering became an important tool in the computer simulation of speech compression systems.

Even earlier than this, some workers began to seriously consider the use of recursive digital filtering as a technique for filtering signals apart from a digital computer. Although this idea was impractical because of cost when first conceived, it is quite practical today, and digital filters are available today as off-the-shelf components.

Beginning with the publication of the Cooley-Tukey paper in 1965, the field of digital signal processing, that had up to then been advancing on the strength of a succession of borrowings from established fields, began to make advances on its own, based on a recognition on the part of several workers that there were some signal theory results to be obtained as a result of insights into the peculiar properties of sampled signals. That is to say, the attempt to approximate continuous systems ended and development of a theory of discrete signals began. The fast Fourier transform (FFT) of Cooley and Tukey may have stimulated this way of thinking for some workers. At any rate, the next fruit of the new thinking, published in the spring of 1966, was the method of computing convolutions and correlations by means of the fast Fourier transform. This involved the recognition that what was merely a means of proving theorems about continuous waveforms was for discrete waveforms a means of actually implementing processing, and because of the computational advantages of the FFT, it was actually an extremely attractive computational route. After the publication of the method of high-speed convolution and correlation, workers in the field routinely considered the frequency domain hospitable. They could operate on signals rather than using a mathematical device to describe what they were doing.

More recently, as workers in the broader field of representation of signals and systems have become more familiar and facile with the theory of digital signal processing, they have found that some of the concepts that are difficult to illustrate with continuous signals are considerably simpler when applied to sampled signals. As an example, a signal can be thought of as having minimum phase and maximum phase components. With continuous signals, this is of theoretical interest. Equations can be written that relate the maximum and minimum phase components. In digital processing theory, however, the two components can be computed and used as a pattern recognition cue, or in further processing. Theoretical questions, such as realizability, then become transparently simple. This is still another instance of where a theoretical tool for understanding continuous signal and system theory becomes a practical tool for manipulating digital representations of waveforms. From 1965 to the present, the digital signal processing field has been blessed with a variety of advances of this type, while the earlier phase of borrowing the results of previously developed fields has not completely ended.

Meanwhile, as the techniques of digital signal processing became more established and respected, applications developed. One of the earliest important applications of digital signal processing was in the filtering of Doppler-shifted radar returns. In one very common radar configuration, short pulses are transmitted, and the signal returning at a given delay after transmission is associated with a target at a corresponding distance. As successive pulses are transmitted, the returns from a given distance, or range, constitute a sampled waveform. Frequency components of this waveform due to Doppler shift are of compelling interest in many radar systems. Since the signal is already sampled in form, and since the identical processing function is very frequently required on the return from a large number of ranges, the use of digital filtering is seen to be reasonably well matched to the particular problem. There are a number of other radar problems for which a number of techniques of digital signal processing have proved invaluable. The most important technique is pulse compression, i.e., the ability to obtain the range resolution of a very short transmitted pulse with the energy content of a very long transmitted pulse.

Digital signal processing has been an important technique in a number of fields besides the radar and speech fields already mentioned. There have been applications in seismic exploration, analysis of vibration, analysis of biomedical signals, picture processing, reliable communications, and sonar. We can expect the applications to increase for some time as costs come down and as the techniques continue to become widely understood. This collection should help somewhat in that respect. This collection is one of a continuing series of efforts by the IEEE Group on Audio and Electroacoustics (G-AE) to provide the field of digital signal processing with the necessary framework for orderly and fruitful growth. This effort began halfway through the decade, when Dr. William Lang sparked the interest of the G-AE, one of many IEEE Professional Groups that could naturally have been expected to be interested in the new techniques. The G-AE began to welcome publications on signal processing, and even started to solicit them. The Group sponsored conference sessions with the aim of making the techniques more widely known, and sponsored workshops at which the leaders of the new field could exchange ideas. The G-AE Technical Committee on Digital Signal Processing, in addition to compiling this volume, is developing a document detailing the confusions in terminology that quite naturally abound in a field that was borrowed widely from other disciplines. That document should be available by late 1972.

This collection has been assembled and organized primarily to fill a need for access to original source material in the field. Each paper included covers at least one important aspect of digital signal processing not covered by other papers in the collection. For this reason, many excellent papers have been omitted as redundant. The Committee wishes to emphasize that papers included are not necessarily more meritorious than papers omitted, but only that we intended to make what we considered the most effective use of a limited number of pages. For the same reason, we have elected to entirely omit several subjects in order to cover others more fully; among these are random number generation, multidimensional digital signal processing, system simulation, and applications.

The papers we have chosen for this collection are grouped in three main categories: digital filters, fast Fourier transforms, and quantization effects. Under the digital filters category, the first ten papers deal with fundamental relationships between digital and analog signals, the representation of digital signals and systems, and the fundamental types of algorithms that can perform digital filtering. The next eleven papers deal with methods of design of digital filters. It is characteristic of the subject that design consists almost entirely of solving the approximation problem—an implementation usually follows quite directly from the design at that stage. Three of the eleven papers deal with design methods that naturally lead to recursive filters, although a more accurate characterization would be filters whose impulse response has infinite duration. The other eight papers deal with design techniques that naturally lead to filters with finite duration impulse response, typically realized by nonrecursive computation algorithms. Rounding out the section on digital filters are two papers on the implementation of such filters in hardware.

The next major subdivision of papers deals with the fast Fourier transform. Five papers deal with the history and fundamentals. The next seven papers deal with various programming considerations, structural organizations, and variations that permit flexibility. Some of these considerations have merit in hardware implementations, some in software, some offer maximum computational speed, some offer utmost simplicity, and some offer low cost. The various possibilities make for a challenging task of system design when a fast Fourier transform is called for.

The next five FFT papers deal with the application of the FFT in the computation of correlations, convolutions, and power spectra, while the last two FFT papers are concerned with special-purpose hardware systems for computing the DFT.

The last major subdivision of papers consists of those concerned with the problems associated with the fact that digital quantities are represented only to a finite (generally fixed) precision. Among the effects related to this fixed precision (or fixed imprecision) are these.

1) All signals considered are inaccurately represented and must be considered as corrupted by noise.
2) The results of computations must generally be rounded, adding futher noise. In recursive computations and in the FFT, this rounding noise is processed by a linear system, just as is the signal, although the linear system that processes the noise may differ in detail from that seen by the signal.
3) Finite precision representation of fixed constants in any desired process may result in a different process being implemented, and the extent of this difference must be assessed.
4) Overflow, the problem of handling signals out of the normal range of representation, can have strange effects, including instability of an otherwise stable filter.
5) In any conversion between digital and analog form, the practical realities of analog-to-digital and digital-to-analog conversion can have unsuspected effects on the useful fidelity of the representations.

The last group of papers deals with each of these effects, and they are an absolutely necessary part of the considerations needed for the successful use of digital signal processing techniques.

Today it is perhaps fair to suggest that some knowledge of the methods of representing and processing digital waveforms should be part of the stock in trade of most electrical engineers. We hope that this collection of papers will help to disseminate that knowledge.

Part 1
Digital Filters

The Equivalence of Digital and Analog Signal Processing*

K. Steiglitz

Department of Electrical Engineering, Princeton University, Princeton, New Jersey

A specific isomorphism is constructed via the transform domains between the analog signal space $L^2(-\infty, \infty)$ and the digital signal space l_2. It is then shown that the class of linear time-invariant realizable filters is invariant under this isomorphism, thus demonstrating that the theories of processing signals with such filters are identical in the digital and analog cases. This means that optimization problems involving linear time-invariant realizable filters and quadratic cost functions are equivalent in the discrete-time and the continuous-time cases, for both deterministic and random signals. Finally, applications to the approximation problem for digital filters are discussed.

LIST OF SYMBOLS

$f(t)$, $g(t)$	continuous-time signals
$F(j\omega)$, $G(j\omega)$	Fourier transforms of continuous-time signals
A	continuous-time filters, bounded linear transformations of $L^2(-\infty, \infty)$
$\{\mathbf{f}_n\}$, $\{\mathbf{g}_n\}$	discrete-time signals
$\mathbf{F}(z)$, $\mathbf{G}(z)$	z-transforms of discrete-time signals
$\mathbf{A}$	discrete-time filters, bounded linear transformations of l_2
μ	isomorphic mapping from $L^2(-\infty, \infty)$ to l_2
$\mathfrak{F}L^2(-\infty, \infty)$	space of Fourier transforms of functions in $L^2(-\infty, \infty)$
$\mathfrak{z}l_2$	space of z-transforms of sequences in l_2
$\lambda_n(t)$	nth Laguerre function

* This work is part of a thesis submitted in partial fulfillment of requirements for the degree of Doctor of Engineering Science at New York University, and was supported partly by the National Science Foundation and partly by the Air Force Office of Scientific Research under Contract No. AF 49-(638)-586 and Grant No. AF-AFOSR-62-321.

Reprinted with permission from *Inform. Contr.*, vol. 8, pp. 455–467, 1965.

I. INTRODUCTION

The parallel between linear time-invariant filtering theory in the continuous-time and the discrete-time cases is readily observed. The theory of the z-transform, developed in the 1950's for the analysis of sampled-data control systems, follows closely classical Fourier transform theory in the linear time-invariant case. In fact, it is common practice to develop in detail a particular result for a continuous-time problem and to pay less attention to the discrete-time case, with the assumption that the derivation in the discrete-time case follows the one for continuous-time signals without much change. Examples of this can be found in the fields of optimum linear filter and compensator design, system identification, and power spectrum measurement.

The main purpose of this paper is to show, by the construction of a specific isomorphism between signal spaces $L^2(-\infty, \infty)$ and l_2, that the theories of processing signals with linear time-invariant realizable filters are identical in the continuous-time and the discrete-time cases. This will imply the equivalence of many common optimization problems involving quadratic cost functions. In addition, the strong link that is developed between discrete-time and continuous-time filtering theory will enable the data analyst to carry over to the digital domain many of the concepts which have been important to the communications and control engineers over the years. In particular, all the approximation techniques developed for continuous-time filters become available for the design of digital filters.

In the engineering literature, the term *digital filter* is usually applied to a filter operating on samples of a continuous signal. In this paper, however, the term *digital filter* will be applied to any bounded linear operator on the signal space l_2, and these signals will not in general represent samples of a continuous signal. For example if $\{\mathbf{x}_n\}$ and $\{\mathbf{y}_n\}$ are two sequences, the recursive filter

$$\mathbf{y}_n = \mathbf{x}_n - 0.5\,\mathbf{y}_{n-1}$$

will represent a digital filter whether or not the $\mathbf{x}_n$ are samples of a continuous signal. The important property is that a digital computer can be used to implement the filtering operation; the term *numerical filter* might in fact be more appropriate.

II. PRELIMINARIES

The Hilbert space $L^2(-\infty, \infty)$ of complex valued, square integrable, Lebesgue measurable functions $f(t)$ will play the role of the space of continuous-time signals. The Hilbert space l_2 of double-ended sequences of complex numbers $\{\mathbf{f}_n\}_{n=-\infty}^{\infty}$ that are square summable will play the role of the space of discrete-time signals. A function in $L^2(-\infty, \infty)$ will be called an *analog* signal, and a sequence in l_2 will be called a *digital* signal. Similarly, a bounded linear transformation A of $L^2(-\infty, \infty)$ will be called an *analog filter*, and a bounded linear transformation A of l_2 will be called a *digital filter*. An analog filter A will be called *time-invariant* if

$$A: f(t) \to g(t), \qquad f(t), g(t) \in L^2(-\infty, \infty), \tag{1}$$

implies

$$A: f(t+\tau) \to g(t+\tau) \tag{2}$$

for every real number τ. Time-invariant analog filters can be represented by the convolution integral

$$g(t) = \int_{-\infty}^{\infty} f(\tau) a(t-\tau)\, d\tau, \tag{3}$$

where $a(t)$, the impulse response of the filter A, need not belong to $L^2(-\infty, \infty)$. Similarly, a digital filter A will be called *time-invariant* if

$$\mathrm{A}: \{\mathbf{f}_n\} \to \{\mathbf{g}_n\}, \qquad \{\mathbf{f}_n\}, \{\mathbf{g}_n\} \in l_2, \tag{4}$$

implies

$$\mathrm{A}: \{\mathbf{f}_{n+\nu}\} \to \{\mathbf{g}_{n+\nu}\} \tag{5}$$

for every integer ν. Time-invariant digital filters can be represented by the convolution summation

$$\{\mathbf{g}_n\} = \left\{ \sum_{i=-\infty}^{\infty} \mathbf{f}_i\, \mathbf{a}_{n-i} \right\}, \tag{6}$$

where the sequence $\{\mathbf{a}_n\}$, the impulse response of the filter A, need not belong to l_2.

Our program is to construct a specific isomorphism between the analog signal space and the digital signal space via their isomorphic transform domains. Hence, we now define the Fourier transform on the analog signal space, mapping $L^2(-\infty, \infty)$ to another space $L^2(-\infty, \infty)$ called the Fourier transform domain and denoted by $\mathfrak{F}L^2(-\infty, \infty)$. We need the following key results (Wiener, 1933; Titchmarsh, 1948):

THEOREM 1 (Plancherel). *If* $f(t) \in L^2(-\infty, \infty)$, *then*

$$F(s) = \underset{R\to\infty}{\mathrm{l.i.m.}} \int_{-R}^{R} f(t) e^{-st}\, dt \tag{7}$$

exists for $s = j\omega$, *and* $F(j\omega) \in L^2(-\infty, \infty)$. *Furthermore,*

$$(f, f) = \int_{-\infty}^{\infty} |f(t)|^2 \, dt = \frac{1}{2\pi j} \int_{-j\infty}^{j\infty} |F(s)|^2 \, ds, \tag{8}$$

and

$$f(t) = \underset{R \to \infty}{\text{l.i.m.}} \int_{-jR}^{jR} F(s) e^{st} \, ds. \tag{9}$$

Analytic extension of $F(j\omega)$ *to the rest of the s-plane (via (7) when it exists, for example) gives the two-sided Laplace transform.*

THEOREM 2 (Parseval). *If* $f(t), g(t) \in L^2(-\infty, \infty)$, *then*

$$(f, g) = \int_{-\infty}^{\infty} f(t) g^*(t) \, dt = \frac{1}{2\pi j} \int_{-j\infty}^{j\infty} F(s) G^*(t) \, ds. \tag{10}$$

The theory required for the analogous construction of a z-transform domain for digital signals is really no more than the theory of Fourier series. Consider the digital signal as a sequence of Fourier coefficients, and consider the periodic function with these Fourier coefficients as the z-transform evaluated on the unit circle in the z-plane. The Riesz-Fischer Theorem (Wiener, 1933) then reads:

THEOREM 3 (F. Riesz-Fischer). *If* $\{\mathbf{f}_n\} \in l_2$, *then*

$$\mathbf{F}(z) = \underset{N \to \infty}{\text{l.i.m.}} \sum_{n=-N}^{N} \mathbf{f}_n z^{-n} \tag{11}$$

exists for $z = e^{j\omega T}$, *and* $\mathbf{F}(e^{j\omega T}) \in L^2(0, 2\pi/T)$, *where* ω *is the independent variable of* $L^2(0, 2\pi/T)$, *and this* ω *is unrelated to the* ω *used in the s-plane. Furthermore,*

$$(\{\mathbf{f}_n\}, \{\mathbf{f}_n\}) = \sum_{n=-\infty}^{\infty} |\mathbf{f}_n|^2 = \frac{1}{2\pi j} \oint |\mathbf{F}(z)|^2 \frac{dz}{z}, \tag{12}$$

and

$$\mathbf{f}_n = \frac{1}{2\pi j} \oint \mathbf{F}(z) z^n \frac{dz}{z}, \tag{13}$$

where integrals in the z-plane are around the unit circle in the counter-clockwise direction.

As in the analog case, the analytic extension of $\mathbf{F}(e^{j\omega T})$ to the rest of the z-plane will coincide with the ordinary z-transform, which is usually defined only for digital signals of exponential order.

THEOREM 4 (Parseval). *If* $\{\mathbf{f}_n\}, \{\mathbf{g}_n\} \in l_2$, *then*

$$(\{\mathbf{f}_n\}, \{\mathbf{g}_n\}) = \sum_{n=-\infty}^{\infty} \mathbf{f}_n \mathbf{g}_n^* = \frac{1}{2\pi j} \oint \mathbf{F}(z)\mathbf{G}^*(z) \frac{dz}{z}. \qquad (14)$$

We denote the space $L^2(0, 2\pi/T)$ *of z-transforms of digital signals by* $_{\mathfrak{z}}l_2$.

III. A SPECIFIC ISOMORPHISM BETWEEN THE ANALOG AND DIGITAL SIGNAL SPACES

Intuitively, if we wish to connect the space of analog signals with the space of digital signals in such a way as to preserve the time-invariance and realizability of filters, we should somehow connect the $j\omega$-axis in the s-plane with the unit circle in the z-plane. The natural correspondence provided by the instantaneous sampling of analog signals matches e^{sT} with z, but is not one-to-one and hence cannot be an isomorphism. The next natural choice is the familiar bilinear transformation

$$s = \frac{z-1}{z+1}, \qquad z = \frac{1+s}{1-s}. \qquad (15)$$

There is an additional factor required so that the transformation will preserve inner products. Accordingly, the image $\{\mathbf{f}_n\} \in l_2$ corresponding to $f(t) \in L^2(-\infty, \infty)$ will be defined as the sequence with the z-transform

$$\mathbf{F}(z) = \frac{\sqrt{2}}{z+1} F\left(\frac{z-1}{z+1}\right). \qquad (16)$$

Thus the mapping $L^2(-\infty, \infty) \rightarrow l_2$ is defined by a chain which goes from $L^2(-\infty, \infty)$ to $\mathfrak{F}L^2(-\infty, \infty)$ to $_{\mathfrak{z}}l_2$ to l_2 as follows:

$$\mu: f(t) \rightarrow F(s) \rightarrow \frac{\sqrt{2}}{z+1} F\left(\frac{z-1}{z+1}\right) = \mathbf{F}(z) \rightarrow \{\mathbf{f}_n\}. \qquad (17)$$

The inverse mapping is easily defined, since each of these steps is uniquely reversible:

$$\mu^{-1}: \{\mathbf{f}_n\} \rightarrow \mathbf{F}(z) \rightarrow \frac{\sqrt{2}}{1-s} \mathbf{F}\left(\frac{1+s}{1-s}\right) = F(s) \rightarrow f(t). \qquad (18)$$

We then have

THEOREM 5. *The mapping*

$$\mu: L^2(-\infty, \infty) \rightarrow l_2$$

defined by (17) *and* (18) *is an isomorphism.*

Proof: μ is obviously linear and onto. To show that it preserves inner product, let $z = (1 + s)/(1 - s)$ in Parseval's relation (10), yielding

$$(f, g) = \frac{1}{2\pi j} \int_{-j\infty}^{j\infty} F(s)G^*(s)\, ds = \frac{1}{2\pi j} \oint \mathbf{F}(z)\mathbf{G}^*(z) \frac{dz}{z} \tag{19}$$
$$= (\{\mathbf{f}_n\}, \{\mathbf{g}_n\}).$$

We can show that μ is one-to-one in the following way: if $f \neq g$, then $(f - g, f - g) = (\{\mathbf{f}_n\} - \{\mathbf{g}_n\}, \{\mathbf{f}_n\} - \{\mathbf{g}_n\}) \neq 0$; which implies that $\{\mathbf{f}_n\} \neq \{\mathbf{g}_n\}$, and hence that μ is one-to-one.

We note here that under the isomorphisms μ and μ^{-1} signals with rational transforms are always matched with signals with rational transforms, a convenience when dealing with the many signals commonly encountered in engineering problems with transforms which are rational functions of s or z.

IV. THE ORTHONORMAL EXPANSION ATTACHED TO μ

The usual way of defining an isomorphism from $L^2(-\infty, \infty)$ to l_2 is to map an arbitrary function in $L^2(-\infty, \infty)$ to the sequence in l_2 of its coefficients in some orthonormal expansion. It comes as no surprise, then, that the isomorphism μ could have been so generated. This section will be devoted to finding this orthonormal expansion.

We start with the z-transform of the digital signal $\{\mathbf{f}_n\}$ which is the image under μ of an arbitrary analog signal $f(t)$:

$$\mathbf{F}(z) = \frac{\sqrt{2}}{z + 1} F\left(\frac{z - 1}{z + 1}\right) = \sum_{n=-\infty}^{\infty} \mathbf{f}_n z^{-n}. \tag{20}$$

By (13), the formula for the inverse z-transform, we have

$$\mathbf{f}_n = \frac{1}{2\pi j} \oint \frac{\sqrt{2}}{z + 1} F\left(\frac{z - 1}{z + 1}\right) z^n \frac{dz}{z}. \tag{21}$$

Letting $z = (1 + s)/(1 - s)$, this integral becomes

$$\mathbf{f}_n = \frac{1}{2\pi j} \int_{-j\infty}^{j\infty} F(s) \frac{\sqrt{2}}{1 + s} \left(\frac{1 + s}{1 - s}\right)^n ds. \tag{22}$$

By Parseval's relation (10) this can be written in terms of time functions as

$$\mathbf{f}_n = \int_{-\infty}^{\infty} f(t)\lambda_n(t)\, dt, \tag{23}$$

where the $\lambda_n(t)$ are given by the following inverse two-sided Laplace transform

$$\lambda_n(t) = \mathcal{L}^{-1}\left[\frac{\sqrt{2}}{1-s}\left(\frac{1-s}{1+s}\right)^n\right]. \tag{24}$$

We see immediately that, depending on whether $n > 0$ or $n \leqq 0$, $\lambda_n(t)$ vanishes for negative time or positive time. By manipulating a standard transform pair involving Laguerre polynomials we find:

$$\lambda_n(t) = \begin{cases} (-1)^{n-1}\sqrt{2}e^{-t}L_{n-1}(2t)u(t), & n = 1, 2, 3, \cdots \\ (-1)^{-n}\sqrt{2}e^{t}L_{-n}(-2t)u(-t), & n = 0, -1, -2, \cdots \end{cases} \tag{25}$$

where $u(t)$ is the Heaviside unit step function, and $L_n(t)$ is the Laguerre polynomial of degree n, defined by

$$L_n(t) = \frac{e^t}{n!}\frac{d^n}{dt^n}(t^n e^{-t}), \qquad n = 0, 1, 2, \cdots \tag{26}$$

The set of functions $\lambda_n(t)$, $n = 1, 2, 3, \cdots$, is a complete orthonormal set on $(0, \infty)$ and are called Laguerre functions. They have been employed by Lee (1931–2), Wiener (1949), and others for network synthesis; and are tabulated in Wiener (1949), and, with a slightly different normalization, in Head and Wilson (1956). The functions $\lambda_n(t)$, $n = 0, -1, -2, \cdots$, are similarly complete and orthonormal on $(-\infty, 0)$, so that the orthonormal expansion corresponding to (23) is

$$f(t) = \sum_{n=-\infty}^{\infty} \mathbf{f}_n \lambda_n(t). \tag{27}$$

We see that the values of the digital signal $\{\mathbf{f}_n\}$ for $n > 0$ correspond to the coefficients in the Laguerre expansion of $f(t)$ for positive t; and that the values of $\{\mathbf{f}_n\}$ for $n \leqq 0$ correspond to the coefficients in the Laguerre expansion of $f(t)$ for negative t.

V. THE INDUCED MAPPING FOR FILTERS

Thus far, we have explicitly defined four isomorphic Hilbert spaces as follows

$$\begin{array}{ccc} L^2(-\infty, \infty) & \text{———} & l_2 \\ | & & | \\ \mathfrak{F}L^2(-\infty, \infty) & \text{———} & {}_{\mathfrak{F}}l_2 = L^2(0, 2\pi/T) \end{array} \tag{28}$$

Therefore an analog or a digital filter as a bounded linear transformation has image transformations induced on the remaining three spaces.

A time-invariant analog filter A, defined by the convolution integral (3), has an image in $\mathfrak{F}L^2(-\infty, \infty)$, in l_2, and in $_{\mathfrak{z}}l_2$. Its image in $\mathfrak{F}L^2(-\infty, \infty)$ is multiplication by $A(s)$, the Fourier transform of $a(t)$. Its image A in l_2 can be found in the following way: let $\mathbf{x}$ be any digital signal. There corresponds to $\mathbf{x}$ a unique analog signal $\mu^{-1}(\mathbf{x})$. The result of operating on this analog signal by the analog filter A, $A\mu^{-1}(\mathbf{x})$, is also uniquely defined. This new analog signal can then be mapped by μ into a unique digital signal $\mu A\mu^{-1}(\mathbf{x})$, which we designate as the result of operating by A on $\mathbf{x}$. Thus we define A to be the composite operator

$$\mathrm{A} = \mu A \mu^{-1}(\ \ ). \tag{29}$$

To find the image of the analog filter A in $_{\mathfrak{z}}l_2$, notice that the Fourier transform of the analog signal Af is $A(s)F(s)$ and the z-transform of the digital signal μAf is

$$\frac{\sqrt{2}}{z+1} A\left(\frac{z-1}{z+1}\right) F\left(\frac{z-1}{z+1}\right). \tag{30}$$

Therefore, the image in $_{\mathfrak{z}}l_2$ of A and hence of A is multiplication by

$$\mathrm{A}(z) = A\left(\frac{z-1}{z+1}\right). \tag{31}$$

Similarly, a time-invariant digital filter A has an image in $_{\mathfrak{z}}l_2$ given by multiplication by $\mathrm{A}(z)$, the z-transform of the impulse response $\{\mathbf{a}_n\}$; an image in $L^2(-\infty, \infty)$ given by

$$A = \mu^{-1}\mathrm{A}\mu(\ \ ); \tag{32}$$

and an image in $\mathfrak{F}L^2(-\infty, \infty)$ given by multiplication by

$$A(s) = \mathrm{A}\left(\frac{1+s}{1-s}\right). \tag{33}$$

We have therefore proved

THEOREM 6. *The isomorphism μ always matches time-invariant analog filters A with time-invariant digital filters A. Furthermore,*

$$\mathrm{A}(z) = A\left(\frac{z-1}{z+1}\right), \tag{34}$$

and

$$A(s) = \text{A}\left(\frac{1+s}{1-s}\right). \tag{35}$$

Those time-invariant filters which are physically realizable in the sense that they are nonanticipatory are of great importance in many fields. A time-invariant analog filter A will be called *realizable* if $Af = 0$ for $t < 0$ whenever $f = 0$ for $t < 0$. Similarly, a time-invariant digital filter A will be called *realizable* if $\text{A}\{\mathbf{f}_n\} = 0$ for $n \leqq 0$ whenever $\{\mathbf{f}_n\} = 0$ for $n \leqq 0$. It is an important property of the mapping μ that it preserves the realizability of time-invariant filters. To see this, suppose first that A is a time-invariant realizable analog filter. Let $\{\mathbf{f}_n\}$ be any digital signal for which $\{\mathbf{f}_n\} = 0$ for $n \leqq 0$. Then its analog image $f(t)$ is such that $f(t) = 0$ for $t < 0$, by (27). Thus $Af = 0$ when t is negative, which implies that $\text{A}\{\mathbf{f}_n\} = 0$ for $n \leqq 0$, by (23). Hence A is a realizable digital filter. The same argument works the other way, and this establishes:

THEOREM 7. *The mapping μ always matches time-invariant realizable analog filters with time-invariant realizable digital filters.*

VI. OPTIMIZATION PROBLEMS FOR SYSTEMS WITH DETERMINISTIC SIGNALS

We are now in a position to see how some optimization problems can be solved simultaneously for analog and digital signals. Suppose, for example, that a certain one-sided analog $r(t)$ is corrupted by a known additive noise $n(t)$, and that we are required to filter out the noise with a stable, realizable time-invariant filter H whose Laplace transform is, say, $H(s)$. If we adopt a least integral-square-error criterion, we require that

$$\int_0^\infty [r - H(r+n)]^2\, dt = \text{min.} \tag{36}$$

As described by Chang (1961), this can be transformed by Parseval's relation to the requirement

$$\frac{1}{2\pi j}\int_{-j\infty}^{j\infty} [R - H(R+N)][R - H(R+N)]^*\, ds = \text{min.}, \tag{37}$$

where R, H, and N are functions of s, and $(\quad)^*$ means that s is replaced by $-s$. It can be shown, using an adaptation of the calculus of variations, that the realizable solution for $H(s)$, say $H_0(s)$, is given by

$$H_0(s) = \frac{1}{Y}\left[\frac{(R+N)^*R}{Y^*}\right]_{\mathrm{LHP}}, \tag{38}$$

where

$$YY^* = (R+N)(R+N)^* \tag{39}$$

Y has only left-half plane poles and zeros, and Y^* has only right-half plane poles and zeros. The notation $[\quad]_{\mathrm{LHP}}$ indicates that a partial fraction expansion is made and only the terms involving left-half plane poles are retained.

The fact that a least integral-square-error criterion is used means that the optimization criterion (36) can be expressed within the axiomatic framework of Hilbert space. Thus, in $L^2(-\infty, \infty)$, (36) becomes

$$\| r - H(r+n) \| = \min. \tag{40}$$

If we now apply the isomorphism μ to the signal $r - H(r+n)$, we have

$$\| r - H(r+n) \| = \| \mu[r - H(r+n)] \| = \| \mathbf{r} - \mathrm{H}(\mathbf{r} + \mathbf{n}) \|, \tag{41}$$

since μ preserves norm. Hence H_0 is the solution to the optimization problem

$$\| \mathbf{r} - \mathrm{H}(\mathbf{r} + \mathbf{n}) \| = \min. \tag{42}$$

Furthermore, since μ matches one-sided analog signals with one-sided digital signals and realizable time-invariant analog filters with realizable time-invariant digital filters, we see that H_0 is the solution to a digital problem that is completely analogous to the original analog problem. Thus

$$\mathrm{H}_0(z) = H_0\left(\frac{z-1}{z+1}\right) = \frac{z}{\mathbf{Q}}\left[\frac{(\mathbf{R}+\mathbf{N})^*\mathbf{R}}{z\mathbf{Q}^*}\right]_{\mathrm{in}}, \tag{43}$$

where

$$\mathbf{QQ}^* = (\mathbf{R}+\mathbf{N})(\mathbf{R}+\mathbf{N})^*. \tag{44}$$

Now $\mathbf{R}$, $\mathbf{N}$, and H_0 are functions of z; $(\quad)^*$ means that z is replaced by z^{-1}; $\mathbf{Q}$ and $\mathbf{Q}^*$ have poles and zeros inside and outside the unit circle respectively; and the notation $[\quad]_{\mathrm{in}}$ indicates that only terms in a partial fraction expansion with poles inside the unit circle have been retained.

In other optimization problems we may wish to minimize the norm of some error signal while keeping the norm of some other system signal within a certain range. In a feedback control system, for example, we may want to minimize the norm of the error with the constraint that the norm of the input to the plant be less than or equal to some prescribed number. Using Lagrange's method of undetermined multipliers, this problem can be reduced to the problem of minimizing a quantity of the form

$$\| e \|^2 + k \| i \|^2, \tag{45}$$

where e is an error signal, i is some energy limited signal, and both e and i depend on an undetermined filter H. Again, if $H_0(s)$ is the time-invariant realizable solution to such an analog problem, then $\mathrm{H}_0(z)$ is the time-invariant realizable solution to the analogous digital problem determined by the mapping μ.

More generally, we can state

THEOREM 8. *Let ν be an isomorphism between $L^2(-\infty, \infty)$ and l_2. Further, let the following optimization problem be posed in the analog signal space $L^2(-\infty, \infty)$: Find analog filters $H_1, H_2, \cdots, H_n$ which minimize some given function of some norms in a given analog signal transmission system and which are in a class of filters $\mathcal{K}$. Then if the class of filters $\mathcal{K}$ is invariant under ν, the corresponding digital problem is equivalent to the original analog problem in the sense that, whenever one can be solved, the other can be also. In particular, when ν is μ, $\mathcal{K}$ can be taken as the class of time-invariant filters or the class of time-invariant realizable filters. In this situation, the optimum filters are related by*

$$\mathrm{H}_i(z) = H_i\left(\frac{z-1}{z+1}\right), \qquad i = 1, 2, 3, \cdots, n. \tag{46}$$

VII. RANDOM SIGNALS AND STATISTICAL OPTIMIZATION PROBLEMS

While the consideration of systems with deterministic signals is important for many theoretical and practical reasons, it is often the case that the engineer knows only the statistical properties of the input and disturbing signals. For this reason the design of systems on a statistical basis has become increasingly important in recent years. The method of connecting continuous-time theory with discrete-time theory described above can be extended to the random case in a natural way if we restrict ourselves to random processes which are stationary with zero mean, ergodic, and have correlation functions of exponential order. For our purposes, such processes will be characterized by their second order properties. In the analog case these are the correlation function $\phi_{xy}(t)$ and its Fourier transform $\Phi_{xy}(s)$. In the digital case these are the correlation sequence $\boldsymbol{\phi}_{xy}(n)$ and its z-transform $\boldsymbol{\Phi}_{xy}(z)$.

We define the mapping μ for correlation functions in the following way, motivated by mapping the signals in the ensembles by the isomorphism μ for signals:

$$\mu\colon \phi_{xy}(t) \rightarrow \Phi_{xy}(s) \rightarrow \frac{2z}{(z+1)^2}\,\Phi_{xy}\left(\frac{z-1}{z+1}\right) = \boldsymbol{\Phi}_{xy}(z) \rightarrow \boldsymbol{\phi}_{xy}(n). \quad (47)$$

The inverse mapping is

$$\mu^{-1}\colon \boldsymbol{\phi}_{xy}(n) \rightarrow \boldsymbol{\Phi}_{xy}(z) \rightarrow \frac{2}{1-s^2}\,\boldsymbol{\Phi}_{xy}\left(\frac{1+s}{1-s}\right) = \Phi_{xy}(s) \rightarrow \phi_{xy}(t). \quad (48)$$

The important invariants under μ are the quantities

$$\phi_{xy}(0) = E[x(t)y(t)], \quad (49)$$

and

$$\boldsymbol{\phi}_{xy}(0) = E[\mathbf{x}_n\mathbf{y}_n], \quad (50)$$

which correspond to the inner products in the deterministic case. As before, time-invariant filters are matched with time-invariant filters, and time-invariant realizable filters are matched with time-invariant realizable filters. Hence, we have

THEOREM 9. *Let the following optimization problem be posed for random analog signals: Find analog filters* $H_1, H_2, \cdots, H_n$ *which minimize some given function of the mean-square values of some signals in an analog signal transmission system and which are in a class of filters* $\mathcal{K}$. *Then if* $\mathcal{K}$ *is the class of time-invariant filters, or the class of time-invariant realizable filters, the corresponding digital problem is equivalent to the original analog problem in the sense that, whenever one can be solved, the other can be also. If the correlation functions and power spectral densities are related by* μ, *the optimum filters are again related by* (46).

In summary, we have shown that in the time-invariant case the theory of processing analog signals and the theory of processing digital signals are the same.

VIII. THE APPROXIMATION PROBLEM FOR DIGITAL FILTERS

The mapping μ can be used to reduce the approximation problem for digital filters to that for analog filters (Steiglitz, 1962; Golden and Kaiser, 1964). Suppose that we wish to design a digital filter with a rational transform and a desired magnitude or phase characteristic as a function of ω, $-\pi/T \leqq \omega \leqq \pi/T$. For real frequencies the transformation μ relates the frequency axes by

$$\omega = \tan \omega T/2. \tag{51}$$

We can therefore transform the desired characteristic to a function of ω simply by stretching the abscissa according to (51). This new characteristic can be interpreted as the frequency characteristic of an analog filter, and we can approximate this with the rational analog filter $A(s)$. $\mathrm{A}(z) = A((z - 1)/(z + 1))$ will then be a rational function digital filter with the appropriate frequency characteristic. Many of the widely used approximation criteria, such as equal-ripple or maximal flatness, are preserved under this compression of the abscissa. Also, by Theorems 6 and 7, the time-invariant or the time-invariant realizable character of the approximant is preserved. Applications to the design of windows for digital spectrum measurement are discussed elsewhere (Steiglitz, 1963).

ACKNOWLEDGMENT

The author wishes to express his gratitude to Professor S. S. L. Chang for his many valuable comments during the course of this work.

RECEIVED: January 22, 1964

REFERENCES

CHANG, S. S. L. (1961), "Synthesis of Optimum Control Systems," Chaps. 2–6. McGraw-Hill, New York.

GOLDEN, R. M., AND KAISER, J. F. (1964), Design of wideband sampled-data filters. *Bell System Tech. J.* **43,** No. 4, Pt. 2, 1533–1546.

HEAD, J. W., AND WILSON, W. P. (1956), "Laguerre Functions: Tables and Properties." Monograph No. 183-R of the Institution of Electrical Engineers.

LEE, Y. W. (1931–2), Synthesis of electrical networks by means of the Fourier transforms of Laguerre's functions. *J. Math. Physi* **11,** 83–113.

STEIGLITZ, K. (1962), "The Approximation Problem for Digital Filters," Tech. Rept. no. 400–56 (Department of Electrical Engineering, New York University).

STEIGLITZ, K. (1963), "The General Theory of Digital Filters with Applications to Spectral Analysis," AFOSR Report No. 64-1664 (Eng. Sc.D. Dissertation, New York University, New York).

TITCHMARSH, E. C. (1948), "Introduction to the Theory of Fourier Integrals." Oxford Univ. Press, Oxford.

WIENER, N. (1933), "The Fourier Integral and Certain of Its Applications." Reprinted by Dover, New York.

WIENER, N. (1949), "Extrapolation, Interpolation and Smoothing of Stationary Time Series." Wiley, New York.

DESIGN METHODS FOR SAMPLED DATA FILTERS

J. F. KAISER
Bell Telephone Laboratories
Murray Hill, New Jersey

Introduction

The simulation of linear continuous filter networks and the filtering of data signals on digital computers require the use of sampled data filters or their difference equation equivalents. Several well known procedures for obtaining sampled data designs for continuous filters have been given in the literature but details outlining their limitations and possible ways to overcome them are still found wanting.

In this paper three methods are presented for the design of sampled data filters. It is assumed that the continuous filter for which a sampled approximation is sought is linear and is specified in terms of its frequency domain characteristics. The first two design methods require a rational function representation for the continuous filter while for the third design method this restriction is removed.

Further it is assumed that the frequency range over which the sampled filter characteristics are to approximate the continuous filter characteristics is to be as close to the Nyquist interval (minus to plus one-half the sampling frequency) as possible. If the system or filter is linear and if the filter input signals have been suitably bandlimited to the Nyquist interval then there theoretically is no reason to use a sampling rate any greater than twice the highest frequency of the input signal. In this paper a sampled filter will be termed "wideband" if the frequency range of useful approximation approaches the Nyquist limit. Thus the sampling frequency of a "wideband" filter will usually be from two to two and one-half times the highest data or signal frequency of interest. This is in somewhat sharp contrast to the sampled filters used in control system design where the sampling rate is usually chosen to lie between five and twelve or more times the effective control system bandwidth so as not to introduce excessive amounts of delay in the control loop.

The use of a minimal sampling rate will in general result in a minimum total amount of computation to be performed in any simulation or data processing operation unless, of course, the use of this sampling rate requires an inordinately

Reprinted from *Proc. 1st Annu. Allerton Conf. Circuit System Theory*, pp. 221-236, 1963.

complex sampled filter design. The relationship between sampled filter complexity as measured by the number of filter coefficients and the effective bandwidth of useful approximation is indicated for each design method.

The first design method employs the standard z-transformation with modifications where necessary. The use of the bilinear z-transformation with suitable scale modification constitutes the second method. A procedure utilizing a Fourier series expansion of the filter magnitude characteristics with truncation errors minimized by using spectral windows is the method advanced for the design of nonrecursive sampled data filters.

The Standard Z-Transform

The nature of the approximation of the standard z-transform[1] can be seen from the defining equations for the transform. Let H(s) be the transfer function (the Laplace transform of its impulse response) of a linear continuous filter. Then the standard z-transform of this transfer characteristic is defined by

$$H^*(s) = \sum_{n=-\infty}^{\infty} H(s+jn\omega_s) \tag{1}$$

or equivalently by

$$H^*(z) = T \sum_{n=0}^{\infty} h(nT)z^{-n} \tag{2}$$

where:

$s = \sigma + j\omega$

$\omega_s = \frac{2\pi}{T}$ = the radian sampling frequency

$z^{-1} = e^{-sT}$ the unit delay operator

$h(t)$ = impulse response of the continuous filter

The asymptotic behavior for large s is assumed to be

[1] See: Wilts, C. H., "Principles of Feedback Control," Addison-Wesley, Reading, Mass. 1960

$$\lim_{s \to j\infty} |H(s)| = \lim_{s \to j\infty} |1/(s/\omega_c)^n|, \quad n > 0 \qquad (3)$$

Thus, from (2), the z-transform is so defined that the impulse response of the sampled filter is identical to the sampled impulse response of the continuous filter. Further, $H^*(s)$ is periodic in ω of period ω_s. From (1) it follows that in the baseband $(-\omega_s/2 \leq \omega \leq \omega_s/2)$ the frequency response characteristics of the obtained sampled filter, $H^*(s)$, differ from those of the continuous filter, $H(s)$, the difference being the amount added or "folded" in through terms of the form $H(s+jn\omega_s)$, $n \neq 0$. If $H(s)$ is bandlimited to the baseband, i.e., $|H(s)| = 0$ for $|\omega| > \omega_s/2$, then there is no folding error and the frequency response of the sampled-data filter is identical to that of the continuous filter. Unfortunately, when $H(s)$ is a rational function of s, it is not bandlimited and therefore $H(s) \neq H^*(s)$ in the baseband.

The magnitude of the errors resulting from the folding is directly related to the high frequency asymptotic behavior of $H(s)$ as defined in Eq. 3. If n is large and $\omega_c << \omega_s/2$, then the folding errors will be small and the standard z-transform generally will yield a satisfactory sampled-data filter design. However, in wideband simulations, ω_c is usually an appreciable fraction of $\omega_s/2$. Furthermore, many continuous filter designs, such as elliptic filters or Chebyshev Type II filters, yield transfer functions in which n is no greater than 1. These two conditions, namely $\omega_c \approx \omega_s/2$ and $n = 1$, can create large folding errors in the frequency response characteristics of the filter obtained by using the standard z-transform and thus render an unusable result.

To reduce the possibility of error due to folding the $H(s)$ can be modified by adding in cascade a wideband low-pass filter $G(s)$, having an 'n' sufficiently large. This "guard" filter, $G(s)$, is usually chosen to have flat magnitude and linear phase characteristics in the frequency range where it is desired to have the $H^*(s)$ faithfully approximate $H(s)$. This may be obtained by using one of the standard low-pass all-pole filter forms such as the Butterworth or Chebyshev Type I and if necessary followed by an all-pass filter for phase equalization.

The procedure is to form $H_m(s)$ as

$$H_m(s) = G(s)\ H(s) \tag{4}$$

and then to apply the standard z-transformation to this $H_m(s)$.

With H(s) the quotient of a numerator polynomial N(s) of order n_N and a denominator polynomial D(s) of order n_D where $n_D > n_N$, application of the standard z-transformation requires first determining the zeros of D(s), then obtaining a partial fraction expansion of H(s) in these poles, and finally z-transforming each of the individual terms. From the following transform pair all the necessary pairs can be derived.

$$\frac{1}{s+a} \Longrightarrow \frac{T}{1-e^{-aT}z^{-1}} \tag{5}$$

Thus it is seen that H*(z) will be rational in z^{-1} and will have at most $2n_D$ coefficients. If in addition a "guard" filter of order n_G is used then the obtained $H^*_m(z)$ will be more complex now having at most $2(n_D+n_G)$ coefficients.

Another limiting feature of the standard z-transformation is the fact that the moments of h(t) do not remain invariant under z-transformation, i.e., in general.

$$\sum_{\ell=0}^{\infty} (\ell T)^n\ h(\ell T) \neq \int_0^{\infty} t^n h(t)dt \tag{6}$$

However, the numerator coefficients of H*(z) may usually be varied slightly so as to satisfy an equality for the first few moments without seriously affecting the overall frequency characteristics of H*(z).

The Bilinear Z-Transformation

To circumvent the "folding" problem of the standard z-transform, a transformation can be employed which will map the entire complex s plane into the horizontal strip in the s_1 plane bounded by the lines $s_1 = -j\omega_s/2$ and $s_1 = +j\omega_s/2$. Since a permissible H*(z) must also be periodic in ω of period ω_s this transformation must also cause H(s) to be mapped identically in each of the other horizontal strips bounded by the lines $s_1 = j(n-\frac{1}{2})\ \omega_s$ and $s_1 = j(n+\frac{1}{2})\ \omega_s$ where n is an integer.

A transformation having these properties is the bilinear z-transform or z-form which is defined as

$$s = \frac{2}{T} \tanh \left(\frac{s_1 T}{2}\right) \tag{7}$$

which becomes upon substituting

$$z^{-1} = e^{-s_1 T}$$

$$s = \frac{2}{T} \frac{(1-z^{-1})}{(1+z^{-1})} \tag{8}$$

Thus in terms of the z^{-1} plane this algebraic transformation (8) has the property of uniquely mapping the entire left half of the s plane into the exterior of the unit circle in the z^{-1} plane. Folding errors are eliminated since no folding occurs.

The z-form is applied simply by making the substitution indicated by (8) in the transfer characteristic H(s). Hence

$$H^*(z) = H(s) \Big|_{s = \frac{2}{T} \frac{(1-z^{-1})}{(1+z^{-1})}} \tag{9}$$

The price paid for this feature is the nonlinear warping imparted to the frequency scale as can be seen by setting $s_1 = j\omega_1$ and $s = j\omega$ in (7) to yield

$$\frac{\omega T}{2} = \tan \left(\frac{\omega_1 T}{2}\right) \tag{10}$$

The deviation from linearity of this relation is shown in Figure 1.

Because of this frequency warping aspect, the z-form is most useful in obtaining sampled-data filter approximations for continuous filters whose magnitude characteristics can be divided along the frequency scale into successive pass and stop bands where the gain or loss is essentially constant in each band. This type of frequency behavior is typical of many low-pass, bandpass and bandstop filters. Compensation can be made for the effect of warping by prewarping the continuous filter design in the opposite way such that upon applying the z-form the critical frequencies will be shifted back to the desired values.

On the other hand if the magnitude frequency characteristics are not essentially piecewise constant over the major portion of the Nyquist frequency interval, such as for a wideband differentiating filter, the frequency warping inherent in the z-form method can render an unsatisfactory filter design.

Thus the z-form has two primary advantages. First, the magnitude frequency characteristics of the continuous filter are carried directly over to the obtained sampled-data filter except for a warping of the frequency scale, i.e., continuous filters with equiripple magnitude characteristics transform into equiripple sampled filters with only the position of the maxima and minima being shifted in frequency a calcuable and compensatable amount. Secondly, application of the z-form is algebraic in form and can be applied equally well to the rational transfer characteristics of the continuous filter in either polynomial or factored form.

With respect to the moments of h(t) it can be shown that only the first two non-zero moments remain invariant under application of the bilinear z-transformation. As in the case of the standard z-transform, agreement can be forced for higher order moments if slight perturbations are made in the numerator coefficients of $H^*(z)$.

In comparing the z-form method with the standard z-transform method it is seen that both methods yield sampled-data transfer functions which are rational in z^{-1} and of the same degree namely that of the degree of the continuous filter. Further when the ratio of the moduli of the poles of the continuous filter to the sampling frequency approaches zero the two methods give identical results as one would expect. For obtaining "wideband" designs by use of the standard z-transformation guard filters must usually be used thus raising the order of the sampled filter. For obtaining "wideband" designs using the z-form the frequency warping must always be compensated for; although no increase in filter order is necessary this method is restricted to filters with essentially piecewise constant magnitude characteristics.

In those situations where a rational approximation is not known for the desired magnitude characteristics of

H(s) or where the magnitude specifications cover the entire Nyquist sampling interval ($|\omega| \leq \omega_s/2$) and where neither the standard z or bilinear z-transformations can be readily applied, the design procedure described in the next section may be used to advantage.

A Nonrecursive Filter Method

The method consists first in expanding the magnitude frequency characteristics desired of the continuous filter H(s) for the band $|\omega| < \omega_s/2$ in a Fourier series about zero frequency. The error resulting from using only a finite number of terms in this expansion is then reduced by applying a suitable weighting function to the expansion coefficients. Finally the individual cosine or sine terms in the expansion are written in terms of the unit delay operator, z^{-1}, by using the exponential forms for the trigonometric functions. The result is then a nonrecursive sampled-data filter.

Since the Fourier expansion is made about zero frequency, the choice of a sine or cosine series is made on the basis of the behavior of H(s) at very small frequencies. If $H(s) \approx Ks^m$ for small s then a cosine series is chosen when m is even and a sine series when m is odd. Thus

$$H(\omega) = \sum_{n=0}^{\infty} a_n \cos n\omega T, \quad m \text{ even} \tag{11a}$$

or

$$H(\omega) = \sum_{n=1}^{\infty} b_n \sin n\omega T, \quad m \text{ odd} \tag{11b}$$

but by definition

$$z^{-1} = e^{-j\omega T} \tag{12}$$

hence

$$H^*(z) = a_o + \tfrac{1}{2} \sum_{n=1}^{\infty} a_n (z^n + z^{-n}), \quad m \text{ even} \tag{13a}$$

and

$$H^*(z) = \tfrac{1}{2} \sum_{n=1}^{\infty} b_n (z^n - z^{-n}), \quad m \text{ odd} \tag{13b}$$

If the convergence of the Fourier series is not sufficiently rapid enough to make errors resulting from truncating the infinite series of (13) small then in order to yield a satisfactory approximation for (13) with a finite number of terms the coefficients a_n or b_n must be modified. This can be conveniently accomplished by multiplying the time response h(nT) by a time limited even function w(t), i.e., $w(t) = w(-t)$ and $w(t) = 0$ for $|t| > \tau$.

Multiplication in the time domain corresponds to convolution in the frequency domain. Thus, if the weighting function, w(t), is chosen such that the frequency content of its transform is concentrated primarily in the central lobe, i.e., $|\omega| \leq \omega_o$ then the effect of this convolution on the frequency characteristics of H*(z) will be to smooth out the sharp transititions or slope discontinuities in H*(z). This procedure is sketched in Fig. 2 where the desired filter is a wideband differentiator.

The design equations thus become

$$H_1^*(z) = a_o\, w(o) + \tfrac{1}{2} \sum_{n=1}^{N} [a_n w(nT)][z^n + z^{-n}], \quad m \text{ even} \qquad (14a)$$

or

$$H_1^*(z) = \tfrac{1}{2} \sum_{n=1}^{N} [b_n w(nT)][z^n - z^{-n}], \quad m \text{ odd} \qquad (14b)$$

where N is the greatest integer in (τ/T).

An especially simple weighting function which may be used is Hamming's[2] window function defined by

$$w_h(t) = \begin{cases} 0.54 + 0.46 \cos(\pi t/\tau), & |t| < \tau \\ 0 & , |t| > \tau \end{cases} \qquad (15)$$

For this function 99.96% of its energy lies in the band $|\omega| \leq 2\pi/\tau$ with the peak amplitude of the side lobes of $W_h(j\omega)$ being less than 1% of the peak.

A family of weighting functions with nearly optimum characteristics is given by the Fourier cosine transform pair

---

[2] Blackman, R. B. and J. W. Tukey, "The Measurement of Power Spectra," Dover Publications, Inc., New York, 1958, pp. 95-99.

$$w(t) = \begin{cases} \dfrac{I_o\left[\omega_a\sqrt{\tau^2-t^2}\,\right]}{I_o(\omega_a\tau)} & |t| < \tau \\ 0 & |t| > \tau \end{cases} \tag{16a}$$

$$W(j\omega) = \frac{2}{I_o(\omega_a\tau)} \; \frac{\sin \tau\sqrt{\omega^2-\omega_a^2}}{\sqrt{\omega^2-\omega_a^2}} \tag{16b}$$

where I_o is the modified Bessel function of the first kind and order zero. By varying the product $\omega_a\tau$ the energy in the central lobe can be changed. The usual range on values of $\omega_a\tau$ is $4 < \omega_a\tau < 9$ corresponding to a range of side lobe peak heights of 3.1% down to 0.047%. Fig. 3 shows w(t) and W(jω) for $\omega_a\tau = 6.0$. These functions can be shown to approximate closely the prolate spheroidal wave functions of order zero and whose bandlimiting properties[3] are well known.

By viewing this scheme for increasing the convergence of Fourier series at points removed from discontinuities, the relation between the number of terms required in the sampled filter H*(z) and the usable frequency range of the filter becomes clear. Referring to Fig. 2 again, the halving of ω_o and the concomitant doubling of τ yield a filter design having twice as many terms as before.

This nonrecursive filter design procedure is best applied to filters whose magnitude characteristics are fairly well behaved especially at very small frequencies and at frequencies approaching the Nyquist limit. For "wideband" filter designs this procedure is superior to the standard numerical analysis approach which usually consists of a Taylor series expansion about zero frequency of the desired magnitude characteristics the result being expressed in differences of increasing order.

Examples

To illustrate the range of the design methods put forth in this paper, three sample designs are briefly presented. The first consists of the design of a "wideband"

[3] D. Slepian, H. O. Pollak, and H. J. Landau, "Prolate Spheroidal Wave Functions, Fourier Analysis and Uncertainty," Parts I and II, Bell System Technical Journal, Vol. 40, No. 1, pp. 43-84, January 1961.

differentiating filter whose magnitude characteristics are to be within 0.1 db out to 80% of the Nyquist limit. Because the magnitude characteristic is not piecewise constant, the bilinear z-form may not be used satisfactorily. Instead the standard z-transform is used. A guard filter is necessary to reduce folding error at all frequencies to within specifications. The guard used is a 0.1 db ripple Chebyshev low-pass filter of ninth order chosen to give 50 db loss at 120% of the Nyquist limit. The magnitude response of the resulting sampled filter is shown in Fig. 4.

To design a sampled-data bandstop filter having equiripple passband magnitude characteristics and equiripple stopband loss the bilinear z-form is used with prewarping of the band edge frequencies being required. The result of the design is shown in Fig. 5. This result could not have been arrived at by the standard z-transform method.

To illustrate the nonrecursive filter design procedure, the "wideband" differentiating filter of the first example is to be obtained in nonrecursive form. To satisfy the 0.1 db ripple requirement $\omega_a \tau$ is set at 6.0 for the modified Bessel weighting function. The value of $\omega_a/2\pi$ is set at 1 kc to satisfy the bandwidth requirement. The resulting τ/T is then 9.55. Using these parameters the coefficients of the nonrecursive filter are then found. The frequency response of this filter is shown in Fig. 6. In comparing this design with that of Fig. 4 it is seen that both designs have approximately the same number of coefficients but that the nonrecursive filter has a constant phase slope equal to an integer number of samples and obtained at the price of the larger delay, 9T as compared to 3T for the recursive design.

Summary

In this paper the essential features of basically three different design methods for sampled data filters have been treated briefly. By discussion and through examples these methods have been shown to cover a wide variety of design problems. The choice of methods presented was not intended to be complete (as evidenced by the omission of least squares, orthogonal function, and other polynomial methods) but only representative of the range of procedures available.

Absent from the discussion has been mention of the details of the numerical analysis problems such as required coefficient accuracy, error control, and noise propagation problems which show themselves primarily in the higher order filter designs. Obtaining detailed insight into the computational aspects of these design methods is essential for their effective and intelligent use in realizing satisfactory simulations of continuous filter networks.

List of Figure Captions

Fig. 1 Nonlinear warping of the frequency scale in the bilinear z-transformation.

Fig. 2 Nonrecursive filter design procedure using weighting functions.

Fig. 3 A modified Bessel function of the first kind and zeroth order as a weighting function.

Fig. 4 The frequency response of a sampled data differentiating filter of the recursive type. The filter was obtained by the standard z-transform utilizing a cascade guard filter.

Fig. 5 Frequency response of a sampled data bandstop filter of the elliptic type obtained using the bilinear z-transformation with warping correction.

Fig. 6 Frequency response of a nonrecursive differentiating filter obtained using a modified Bessel function spectral window for truncation.

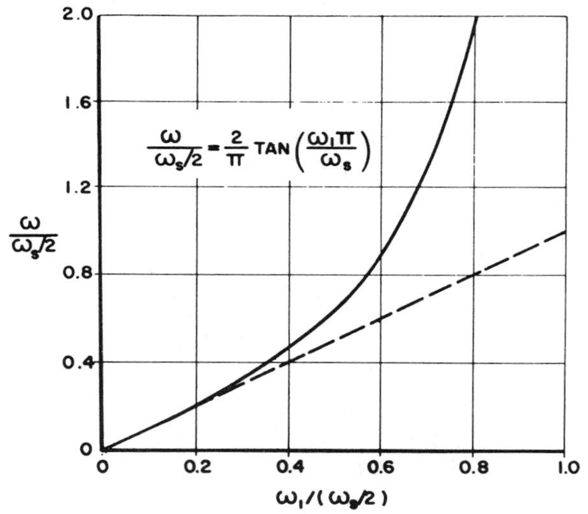

Fig.1

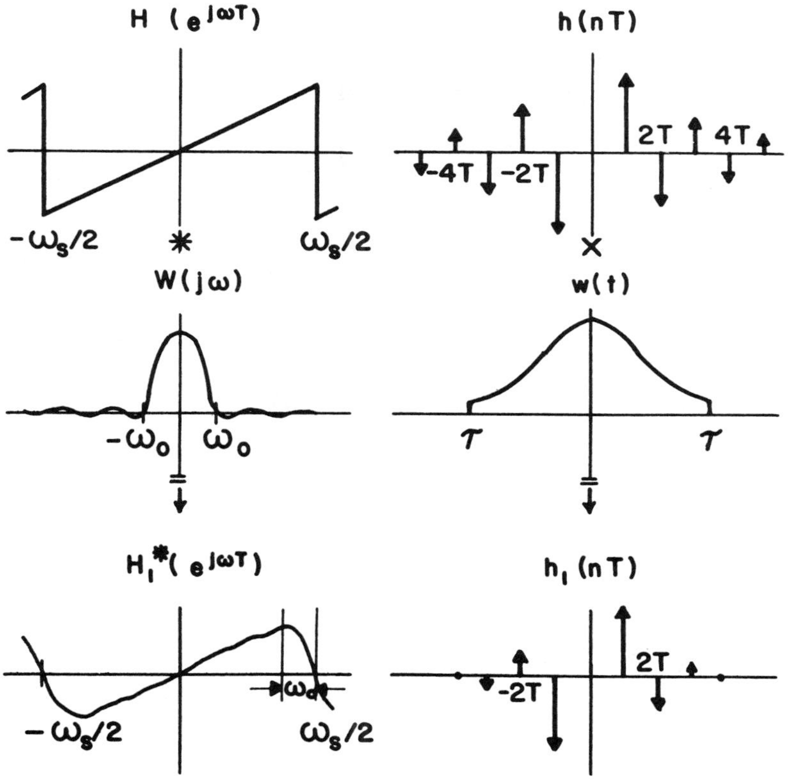
H (e^{jωT})
h(nT)
2T
4T
-4T
-2T
-ω_s/2
ω_s/2
W(jω)
w(t)
-ω_0
ω_0
τ
τ
H_1*(e^{jωT})
h_1(nT)
2T
-2T
ω_d
-ω_s/2
ω_s/2

Fig. 2

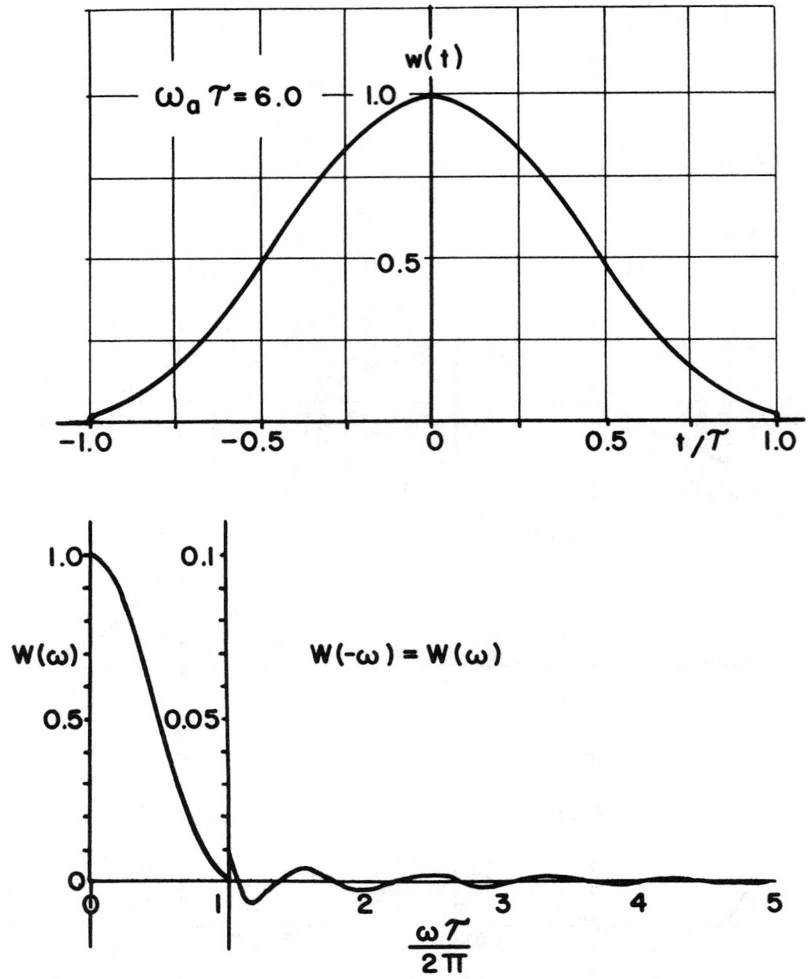

Fig. 3

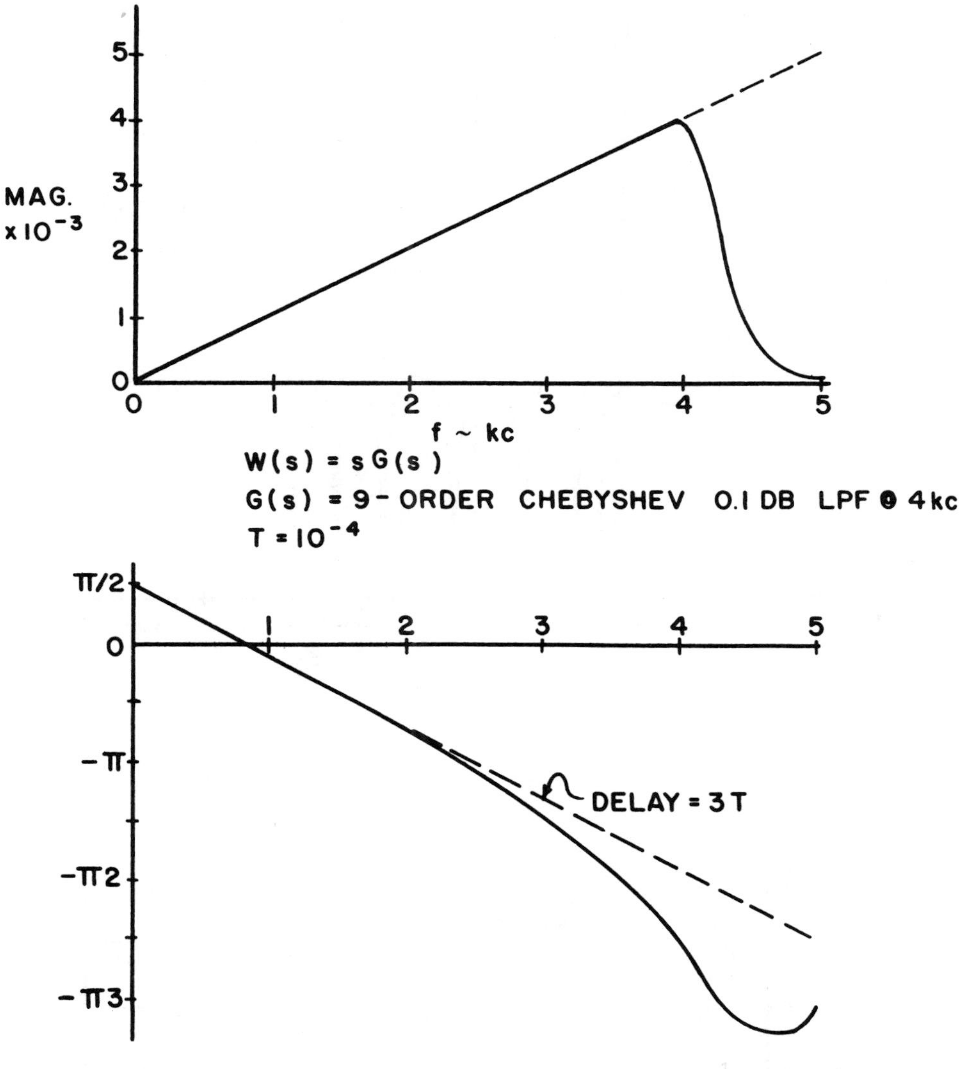
MAG.
x 10^-3
5
4
3
2
1
0
0
1
2
3
4
5
f ~ kc
W(s) = s G(s)
G(s) = 9 - ORDER CHEBYSHEV 0.1 DB LPF @ 4 kc
T = 10^-4
π/2
0
-π
-π2
-π3
1
2
3
4
5
DELAY = 3 T

Fig.4

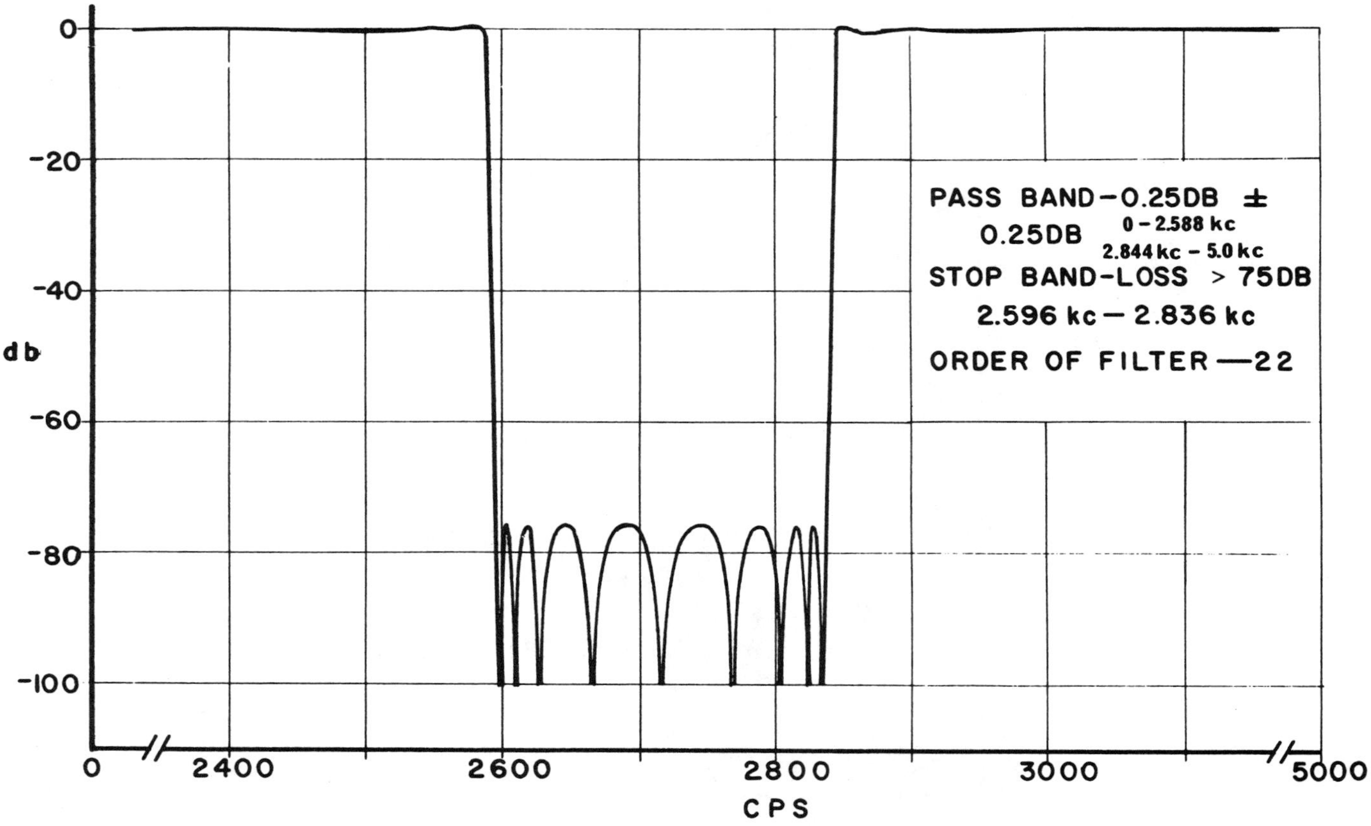
PASS BAND-0.25DB ±
0.25DB 0-2.588 kc
2.844 kc - 5.0 kc
STOP BAND-LOSS > 75DB
2.596 kc — 2.836 kc
ORDER OF FILTER —22
db
0
-20
-40
-60
-80
-100
0
2400
2600
2800
3000
5000
CPS

Fig. 5

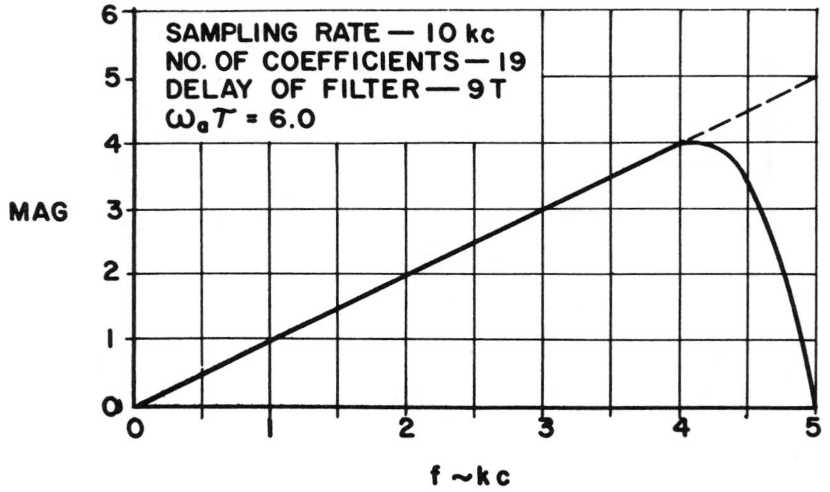
SAMPLING RATE — 10 kc
NO. OF COEFFICIENTS — 19
DELAY OF FILTER — 9T
$\omega_a T$ = 6.0
MAG
6
5
4
3
2
1
0
0
1
2
3
4
5
f ~ kc

Fig. 6

The Design of Digital Filters

A. J. GIBBS Ph.D.
A.P.O. Research Laboratories

The frequency-domain approximation problem for recursive digital filters is considered in this paper. The properties of the bilinear-transform design method are used (i) to show that to every well-defined digital approximation problem there is a corresponding well-defined analogue approximation problem, and (ii) that these corresponding problems are equivalent in the sense that one has a solution if and only if the other has a solution. In addition, the uniqueness of several important classes of digital filters is established.

The paper concludes with an illustration of the application of the bilinear-transform design method to group delay equalizer design. The successful use of the bilinear-transform method for this purpose, together with the successful use of it in designing filters to meet attenuation specifications implies that existinganalogue design methods—in fact, existing computer programs (with only minor modifications) — can be used to design filters which satisfy commonly encountered frequency-domain specifications.

1. Introduction

The increased interest in the design of computational algorithms using frequency-domain concepts was stimulated by the need to simulate complex analogue systems on digital computers. Several of the more important methods of obtaining digital representations of analogue systems are discussed in the companion paper (Ref. 109). One of these, the Bilinear-Transform method, is the natural method for solving the frequency-domain approximation problem for digital filters. The method consists of transforming the open left half-plane onto the interior of the unit circle, in a one-to-one fashion. Rational functions transform to rational functions, and because the transformation is onto and one-to-one, *to every digital filter represented by a rational function, there exists a unique analogue filter represented by a rational function, and vice versa.*

Furthermore, it is shown here that because the frequency-domain distortion inherent in the bilinear transformation is well defined, *to every well-defined frequency-domain specification for a digital filter there corresponds a well-defined specification for an analogue filter* Thus many of the frequency-domain methods for designing analogue filters are directly applicable to the design of digital filters.

In addition, it is known that the solutions of many important analogue approximation problems are unique, and in this paper, it is shown that this implies the solutions of the equally important corresponding digital approximation problems are unique—and thus, *irrespective of the method used to design such a digital filter, the result is identical to that obtained by transforming the corresponding analogue filter.*

The possibility of achieving physically realizable linear-phase filters is an interesting facet of the frequency-domain behaviour of nonrecursive digital filters. However, in order to achieve a sharp cutoff characteristic, a nonrecursive filter of high complexity is required, which is prohibitive in many real-time applications. Thus, in these cases, recursive digital filters must be used. It is known (Ref. 105) that recursive digital filters, of necessity, introduce phase distortion, and thus to achieve a linear phase, equalization must be employed.

The equivalence of the approximation problem for analogue and digital filters implies that phase equalization of digital filters can be achieved by designing a corresponding analogue equalizer and transforming the result. The applicability of using an existing method of phase equalization in this context is demonstrated in this paper, and thus *existing computer programs, with only minor modifications, can be employed in the solution of the corresponding approximation problems for analogue and digital filters.*

2. The Bilinear Transform

Steiglitz (27, 28) investigates in some detail the equivalence of certain aspects of digital and analogue signal processing, and he touches briefly on the equivalence of the approximation problems for analogue and digital filters. Central to this discussion are certain properties of the bilinear transformation:

$$s = \frac{z-1}{z+1} = \frac{1-z^{-1}}{1+z^{-1}}. \qquad (1)$$

This transformation has many significant properties (Ref. 106 p. 271); those of importance in this discussion are:

Reprinted with permission from *Aust. Telecommun. Res. J.*, vol. 4, pp. 29–34, 1970.

(a) It is a one-to-one mapping of the open left half-plane onto the open unit disc.
(b) It is a one-to-one mapping of the imaginary axis, including the point at infinity, onto the unit circle.
(c) It is a one-to-one mapping of the open right half-plane onto the complement, in the complex plane, of the closed unit disc.
(d) The inverse transformation

$$z = \frac{1+s}{1-s}$$

has analogous reciprocal properties.

Thus rational functions transform to rational functions, poles in the open left half-plane transform to poles in the open unit disc, and zeros on the imaginary axis and at infinity transform to zeros on the unit circle. The frequency-domain performance of the transformation is readily evaluated. Property (b) above and equation (1) show that

$$\omega = \tan \theta/2, \tag{2}$$

$$\theta = 2\tan^{-1}\omega \tag{3}$$

for $s = j\omega$ and $z = e^{j\theta}$.

3. The Corresponding Approximation Problems For Analogue And Digital Filters

The specifications for an analogue or digital filter usually constrain certain functions of the frequency-transfer function, namely, the real part, imaginary part, magnitude, phase, group delay or the phase delay. These quantities are defined as follows:

Definition: If G(s) is a rational function representing an analogue filter, and

$$G(j\omega) = \hat{R}(\omega) + j\hat{X}(\omega) = |G(j\omega)| \exp(j\hat{\beta}(\omega)),$$

then:

$$\begin{aligned}
&\text{real part of } G(j\omega) &&= \hat{R}(\omega),\\
&\text{imaginary part of } G(j\omega) &&= \hat{X}(\omega),\\
&\text{Magnitude of } G(j\omega) &&= |G(j\omega)|,\\
&\text{phase of } G(j\omega) &&= \hat{\beta}(\omega),\\
&\text{group delay of } G(j\omega) &&= \hat{\tau}(\omega)\\
& &&= -d\hat{\beta}(\omega)/d\omega,\\
&\text{phase delay of } G(j\omega) &&= \hat{\psi}(\omega)\\
& &&= -\hat{\beta}(\omega)/\omega;\ \omega \neq 0;\\
& &&= \hat{\tau}(\omega);\ \omega = 0.
\end{aligned}$$

If H(z) is a rational function representing a digital filter, and

$$H(e^{j\theta}) = R(\theta) + jX(\theta) = |H(e^{j\theta})| \exp(j\beta(\theta)),$$

then:

$$\begin{aligned}
&\text{real part of } H(e^{j\theta}) &&= R(\theta),\\
&\text{imaginary part of } H(e^{j\theta}) &&= X(\theta),\\
&\text{magnitude of } H(e^{j\theta}) &&= |H(e^{j\theta})|,\\
&\text{phase of } H(e^{j\theta}) &&= \beta(\theta),\\
&\text{normalized group delay}\\
&\text{of } H(e^{j\theta}) &&= \tau(\theta) = -d\beta(\theta)/d\theta,\\
&\text{normalized phase delay}\\
&\text{of } H(e^{j\theta}) &&= \psi(\theta) = -\beta(\theta)/\theta; \theta \neq 0,\\
& &&= \tau(\theta); \theta = 0.
\end{aligned}$$

To solve a digital approximation problem by solving a corresponding analogue approximation problem, the following results are needed.

Theorem 1 : If G(s) is a rational function and

$$H(z) = G\left(\frac{z-1}{z+1}\right)$$

then,

(a) $\hat{R}(\tan \theta/2) = R(\theta)$
(b) $\hat{X}(\tan \theta/2) = X(\theta)$
(c) $|G(\tanh j\theta/2)| = |H(e^{j\theta})|$
(d) $\hat{\beta}(\tan \theta/2) = \beta(\theta)$
(e) $\frac{\sec^2 \theta/2}{2} . \hat{\tau}(\tan \theta/2) = \tau(\theta)$
(f) $\frac{\tan \theta/2}{2} . \hat{\psi}(\tan \theta/2) = \psi(\theta)$

where the notation of the Definition above is used.

Proof: Because

$$G\left(\frac{z-1}{z+1}\right) = H(z)$$

it follows that $G(\tanh j\theta/2) = H(e^{j\theta})$.
Hence,

$$\hat{R}(\tan \theta/2) + j\hat{X}(\tan \theta/2) = R(\theta) + jX(\theta)$$

and

$$|G(\tanh j\theta/2)| \exp(j\hat{\beta}(\tan \theta/2)) = |H(e^{j\theta})| \exp(j\beta(\theta)),$$

thus proving (a), (b), (c) and (d).
Now,

$$\begin{aligned}
\frac{\sec^2 \theta/2}{2} . \hat{\tau}(\tan \theta/2) &= \frac{\sec^2 \theta/2}{2} . \frac{d\hat{\beta}(\tan \theta/2)}{d(\tan \theta/2}\\
&= \frac{\sec^2 \theta/2}{2} . \frac{d\beta(\theta)}{d\theta} . \frac{d\theta}{d(\tan \theta/2)}\\
&= \tau(\theta)
\end{aligned}$$

and,

$$\begin{aligned}
\frac{\tan \theta/2}{\theta} . \hat{\psi}(\tan \theta/2) &= \frac{\tan \theta/2}{\theta} \frac{\hat{\beta}(\tan \theta/2)}{\tan \theta/2}\\
&= \beta(\theta)/\theta\\
&= \psi(\theta),
\end{aligned}$$

thus proving (e) and (f).

The type of approximation problem considered here is: Find a rational function H(z), with real coefficients, poles in the open unit disc, and degree of numerator not greater than degree of denominator, with the property that some function, $f(\theta)$, of $H(e^{j\theta})$, meets a prescribed specification. The function $f(\theta)$ is either the real part $R(\theta)$, imaginary part $X(\theta)$, magnitude $|H(e^{j\theta})|$, phase $\beta(\theta)$, normalized group delay $\tau(\theta)$, or the normalized phase delay $\psi(\theta)$, and $f(\theta)$ meets a specification if

$$f(\theta)\begin{cases} = F_1(\theta) + E(\theta); & \theta \in A_1\\ \leqslant F_2(\theta) & ; \theta \in A_2\\ \geqslant F_3(\theta) & ; \theta \in A_3\end{cases} \tag{4}$$

where the F_i and E are prescribed functions of θ, and the A_i are in $[-\pi, \pi]$.

A typical example is that of group delay equalization: where $f(\theta) = \tau(\theta)$, $F_1(\theta)$ is an arbitrary constant minus the group delay distortion to be equalized, $E(\theta)$ represents the allowable equalization error, A_1 is the set in which equalization is required, and A_2, A_3 are null. Another example is that of filter design:

where $f(\theta) = |H(e^{j\theta})|$ (or $-\log|H(e^{j\theta})|$), A_1 is null, $F_2(\theta)$ is the allowable loss in the pass-band A_2, $F_3(\theta)$ represents the desired stop-band loss, and A_3 is the stop-band.

The results of Theorem 1, give the following:

Theorem 2: If G(s), a rational function with real coefficients, poles in the open left half-plane, and degree of numerator no greater than degree of denominator, has the property that:

$$\left.\begin{matrix} \hat{R}(\omega) \\ \hat{X}(\omega) \\ |G(j\omega)| \\ \hat{\beta}(\omega) \\ \frac{1+\omega^2}{2}\hat{\tau}(\omega) \\ \frac{\omega}{2\tan^{-1}\omega}\hat{\psi}(\omega) \end{matrix}\right\} \begin{matrix} = F_1(2\tan^{-1}\omega) + E(2\tan^{-1}\omega); \quad 2\tan^{-1}\omega \in A_1 \\ \leqslant F_2(2\tan^{-1}\omega); \quad 2\tan^{-1}\omega \in A_2 \\ \geqslant F_3(2\tan^{-1}\omega); \quad 2\tan^{-1}\omega \in A_3 \end{matrix} \qquad (5)$$

and if $H(z) = G\left(\frac{z-1}{z+1}\right)$, then, H(z) is a rational function with real coefficients, poles in the open unit disc and degree of numerator no greater than degree of denominator, which has the property that

$$\left.\begin{matrix} R(\theta) \\ X(\theta) \\ |H(e^{j\theta})| \\ \beta(\theta) \\ \tau(\theta) \\ \psi(\theta) \end{matrix}\right\} \begin{matrix} = F_1(\theta) + E(\theta); & \theta \in A_1 \\ \leqslant F_2(\theta) & ; \theta \in A_2 \\ \geqslant F_3(\theta) & ; \theta \in A_3 \end{matrix} \qquad (6)$$

where the notation of the Definition is used.

Proof: That H(z) has the stated properties follows from the fact that the bilinear transformation is a rational transformation with real coefficients, does not increase the degree of a rational function, and maps the open left half-plane onto the open unit disc. Because

$$G(\tanh j\theta/2) = G(j\tan\theta/2) = H(e^{j\theta}),$$

by using Theorem 1, and substituting $\omega = \tan\theta/2$ in equation 5, we have equation 6, thus proving the theorem.

It is seen that equations 6 represent the approximation problem for a digital filter; and if equations 5 are designated as the corresponding analogue approximation problem, the following important result follows:

Theorem 3 : The digital approximation problem can be solved if and only if the corresponding analogue approximation problem can be solved.

Proof: If the digital approximation problem can be solved by H(z), then because the bilinear transformation is onto, there exists a G(s) such that $G\left(\frac{z-1}{z+1}\right) = H(z)$: which implies, by Theorem 2, that the corresponding analogue approximation can be solved by G(s). The converse follows from Theorem 2.

Similar results are obtained by considering solving the analogue approximation problem in terms of a corresponding digital approximation problem, the only changes being that $2(\cos^2\theta/2)\,\tau(\theta)$ corresponds to $\hat{\tau}(\omega)$ and $\theta\cdot\psi(\theta)/(\tan\theta/2)$ corresponds to $\hat{\psi}(\omega)$. Thus we have the following result:

Theorem 4 : The corresponding approximation problems for digital and analogue filters are equivalent in the sense that one has a solution if and only if the other has a solution.

This result allows consideration of an approximation problem in whichever of the two domains is the more convenient, and of course it allows digital filters to be designed by using well established techniques for the approximation of analogue filters, if the corresponding relationships expressed by equations 5 are employed.

4. Implications of Uniqueness of Analogue Filter Approximations

In many important cases of analogue approximation it is known that the solution of a particular approximation problem is unique; and the fact that the bilinear transformation is one-to-one gives the following important result.

Theorem 5: The solution of a digital approximation is unique if the solution of the corresponding analogue approximation problem is unique.

Proof: Suppose $H_1(z)$ and $H_2(z)$ are solutions of a digital approximation problem for which the corresponding analogue approximation problem has a unique solution. Since the bilinear transformation is onto and one-to-one, there exists $G_1(s)$ and $G_2(s)$ which are solutions of the corresponding analogue approximation problem. The solution of this problem is unique: hence $G_1 = G_2$, which implies $H_1 = H_2$.

This result is particularly useful in those cases where an important aspect of the specification is invariant under the bilinear transformation. Such a case is that of filter design. Here the requirement is that an H(z) be found such that the minimum difference between $-\log|H(e^{j\theta})|$, (the attenuation), and the loss specification, is maximized: i.e., for a specified degree of H(z), the specification is exceeded by as great an amount as is possible. The corresponding analogue approximation problem is identical, because the property of minimum difference of attenuation is invariant under the bilinear transformation. It is known that the solution of this analogue approximation problem is one where all the minimum differences are equal (i.e., equi-error approximation) and further, that the solution is unique (Refs. 110–112). Thus, irrespective of the method of solution of the above stated digital approximation problem, the solution obtained is identical to that obtained by transforming the solution of the corresponding analogue approximation problem. Analogue filters with Chebychev pass-band and analogue elliptic filters have the properties that the minimum difference is maximized, and that they are unique; and thus all digital filters with Chebychev pass-band, and digital elliptic filters are transformations of their analogue counterparts. Thus direct synthesis of these types of digital filters (Refs. 86, 87) is superfluous.

The problem of group delay equalization also shows an invariance of an important aspect of the specification. Here, the problem is to design an equalizer such that the sum of the group delay distortion to be equalized, and of the group delay of the equalizer, is an arbitrary constant within a specified error, the magnitude of which varies with θ. The corresponding analogue approximation problem is to design an equalizer such that the sum of the corresponding group delay distortion and of $(1 + \omega^2)/2$ times the group delay of the analogue equalizer is an arbitrary constant within a corresponding error specification. Experience with analogue group delay equalization shows that the error specification is used most efficiently if the magnitude of the minimum differences between the total group delay (distortion plus equalizer), and the error envelope, are equal. Thus, if the corresponding analogue equalization is performed so that the magnitude of the minimum differences between the corresponding error envelope, and the sum of the corresponding group delay distortion and of $(1 + \omega^2)/2$ times the group delay of the analogue equalizer, are equal, the digital approximation has the property that the error specification is used most efficiently. The uniqueness of a solution of this nature is open to some little doubt. A solution of the conventional analogue group delay equalization problem (i.e., one in which the group delay of the equalizer is not weighted by $(1+\omega^2)/2$), which has the property that the magnitude of the minimum differences are equal is known to be unique (Ref. 104, p. 1854). Because the weighting term $(1 + \omega^2)/2$ is monotonic, it appears safe to conjecture that the weighting does not change this result; and assuming this, it may be concluded that the solution of the digital equalization problem is unique. This, of course, implies that irrespective of the method used to obtain an equal minimum-difference equalization, the result is the same as that obtained by transforming the solution of the corresponding analogue equalization problem. Similar remarks and conclusions can also be made for the problem of phase delay equalization.

5. Group Delay Equalization by Means of Digital Filters

In Theorem 2 the relationship between the group delay of an analogue filter and the normalized group delay of the corresponding digital filter (under the bilinear transformation) is presented, and it is shown that if the group delay $\hat{\tau}(\omega)$ of the analogue filter is weighted by $(1 + \omega^2)/2$, then this transforms to the normalized group delay $\tau(\theta)$ of the digital filter, i.e.,

$$\frac{1+\omega^2}{2}\cdot\hat{\tau}(\omega) \xrightarrow{\omega\,=\,\tan\theta/2} \tau(\theta).$$

The purpose of this section is to show that a presently available analogue equalization technique, with a simple modification, can be applied to the digital problem.

In a real-time application of digital phase equalization, the elements of the input sequence constitute samples of an analogue signal, and the group delay contributed by the digital filter is proportional to the sampling period, T. If normalized group delay and the dimensionless variable θ are used, the group delay distortion to be equalized must also be normalized. The following notation is adopted:

The physical angular frequency variable is denoted by $\bar{\omega}$, the sampling period is denoted by T, and thus $\theta = \bar{\omega}T$.

The group delay of the digital equalizer is denoted by $\bar{\tau}_E(\bar{\omega})$, and the group delay distortion is denoted by $\bar{\tau}_D(\bar{\omega})$. It is clear that

$$\bar{\tau}_E(\bar{\omega}) = -\frac{d\beta(\theta)}{d\,\bar{\omega}} = T\tau_E(\theta) \tag{7}$$

where $\tau_E(\theta)$ is the normalized group delay of the digital equalizer.

The angular frequency in the transformed analogue domain is denoted by ω, and $\omega = \tan\theta/2$.

The group delay equalization problem is to find a transfer function H(z) such that its group delay $\bar{\tau}_E(\bar{\omega})$, satisfies

$$\bar{\tau}_D(\bar{\omega}) + \bar{\tau}_E(\bar{\omega}) = \text{const.} + \bar{E}(\bar{\omega}), \tag{8}$$

where $\bar{E}(\bar{\omega})$ is the allowable equalization error. Using equation 7

$$\frac{1}{T}\bar{\tau}_D(\bar{\omega}) + \tau_E(\theta) = \text{const.} + \frac{1}{T}\bar{E}(\bar{\omega})$$

and thus

$$\tau_D(\theta) + \tau_E(\theta) = \text{const.} + E(\theta), \tag{9}$$

where $\tau_D(\theta)$ and $E(\theta)$ are the normalized group delay distortion and normalized error, respectively: i.e.,

$$\bar{\tau}_D(\bar{\omega}) = T\tau_D(\theta);\ \bar{E}(\bar{\omega}) = TE(\theta). \tag{10}$$

Using the bilinear transformation in equation 9

$$\hat{\tau}_D(\omega) + \frac{1+\omega^2}{2}\hat{\tau}_E(\omega) = \text{const.} + \hat{E}(\omega) \tag{11}$$

where $\hat{\tau}_D(\omega)$ and $\hat{E}(\omega)$ are, respectively, the transformed analogue group delay distortion and error specification, and $\hat{\tau}_E(\omega)$ is the group delay of the corresponding analogue equalizer with transfer function G(s), where

$$G\left(\frac{z-1}{z+1}\right) = H(z). \tag{12}$$

Thus the transfer function of the digital equalizer is found by designing an analogue equalizer satisfying equation 11 and using equation 12 to obtain H(z).

Analogue group delay equalization is performed by employing analogue all-pass transfer functions of the form

$$G(s) = \prod_{k=1}^{N}\frac{s^2 - 2a_k s + a_k{}^2 + b_k{}^2}{s^2 + 2a_k s + a_k{}^2 + b_k{}^2}, \tag{13}$$

the group delay of which is given by

$$\tau(\omega) = \sum_{k=1}^{N} 2a_k\left[\frac{1}{a_k{}^2 + (\omega - b_k)^2} + \frac{1}{a_k{}^2 + (\omega + b_k)^2}\right]. \tag{14}$$

The equalization problem is then one of finding N pairs, (a_k, b_k), such that equation 11 is satisfied, which, except for the weighting factor $(1 + \omega^2)/2$, is identical to the conventional analogue equalization problem. The equalization method to be used is that developed by the author (Refs. 113, 114), which allows equalization within tolerances which vary with frequency. Because the method is well-detailed in the literature, a concise account suffices here. The steps are as follows:

(a) Choose a set of (a_k, b_k) pairs which constitute a first approximation.

(b) Compute the minimum differences between the total group delay (distortion plus equalizer) and the allowable equalization error.

(c) Compute changes $(\Delta a_k, \Delta b_k)$ so as to improve the approximation, i.e., so that the magnitude of the minimum differences tend to be equal. This involves linearization of the problem using a truncated Taylor series involving first partial derivatives with respect to a_k and b_k.

(d) Repeat (a) to (c) with $(a_k + \Delta a_k, b_k + \Delta b_k)$ until all minimum differences are equal, within some prescribed tolerance.

A consideration of steps (a) to (d) indicates that the only change required for the digital problem is to include the weighting $(1 + \omega^2)/2$ in the calculation of the total group delay, step (a), and in the calculation of the partial derivatives, step (c). The method requires a first approximation; and experience shows that provided this has the required number of minimum differences, $2N + 2$, the solution converges. The required number of minimum differences is ensured by spacing the values of the b_k over the frequency range of interest, and using sufficiently small values of a_k, and employing this technique in the digital problem produces acceptable first approximations.

An estimate of the value of N can be obtained by considering the area, ((normalized group delay) times θ), which is required to *fill in* the distortion characteristic. Each weighted component of equation 14 contributes

$$\int_0^{\pi} \tau_E(\theta)d\theta$$
$$= \int_0^{\infty} \frac{2\tau_E(2\tan^{-1}\omega)}{1+\omega^2}\,d\omega = \int_0^{\infty} \hat{\tau}(\omega)d\omega = 2\pi, \qquad (15)$$

and thus if A is the area required to be added, an estimate of N is given by $A/2\pi$. The actual value required depends on the allowable equalization error and the steepness of the distortion near the band-edges, but the above estimate is useful. The value of N is therefore dependent on the sampling period T. Equation (10) shows that both the steepness of the distortion at the band edges and the allowable error are inversely proportional to T. Thus increasing the value of T decreases the steepness which tends to decrease N, but the decrease in the allowable error tends to increase N: and it appears difficult to make any general conclusion regarding this effect.

Illustrative Example: This example is typical of group delay equalization problems encountered in practice.

The group delay distortion chosen is that of a typical telephone channel (see Figure 1), and the equalization error is

$$|\bar{E}(\bar{\omega})| \leqslant \begin{cases} 180\ \mu s \text{ between 500 Hz and 600 Hz} \\ 190\ \mu s \text{ between 600 Hz and 1 k Hz} \\ 30\ \mu s \text{ between 1 kHz and 2·6 k Hz} \\ 180\ \mu s \text{ between 2·6 kHz and 2·8 k Hz.} \end{cases} \qquad (16)$$

In accordance with usual practice, linear interpolation is used between the specification breakpoints (Ref. 114). The sampling frequency chosen is 8 kHz: i.e., $T = 125\ \mu s$.

From Figure 1, the area to be added is approximately 4π giving an estimate for N of 2.

In fact $N = 3$ is required to perform the equalization, and the (a_k, b_k) pairs are (0·229129, 0·455670); (0·309712, 0·841496); (0·548503, 1·348887).

The equalization error between 1·0–2·6kHz is 29·5 μs, and graphs of the group delay of the equalizer and the total group delay are given in Figure 1.

If the same channel is equalized within the same tolerances, but with a sampling frequency of 7kHz, then $N = 3$ is also required, and the equalization error between 1·0–2·6 kHz is 22·8 μs. Thus, in this particular example the decrease in steepness of the distortion is the dominant effect.

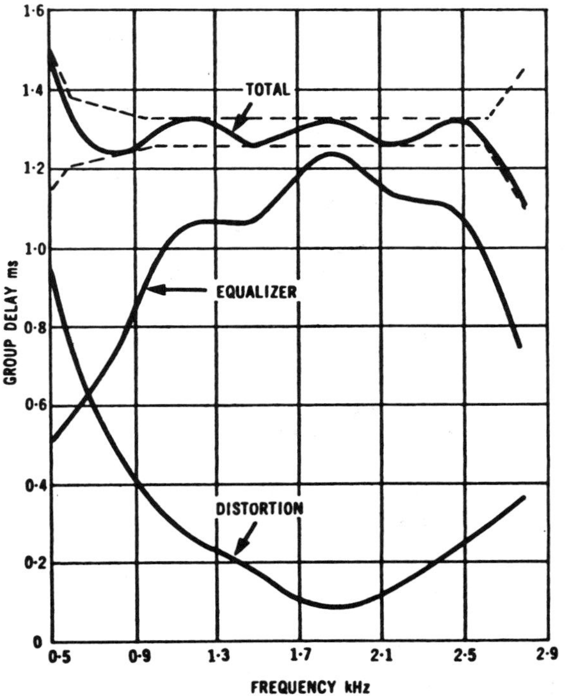

Fig. 1—Digital Group Delay Equalisation.

6. Summary

The use of the bilinear transform allows the digital approximation problem to be solved by solving a corresponding analogue problem; and thus well-established techniques can be brought to bear on the digital problem. If an attenuation characteristic is required, existing analogue techniques can be used, and if group delay or phase delay equalization is required, only a minor modification to existing analogue techniques is needed.

Thus, the conclusion which can be drawn is that for the type of frequency-domain approximation problem considered—which, in fact, is the one most commonly encountered in practice—the design of analogue filters is equivalent to the design of digital filters: and the possibility of designing certain types of analogue filters by considering the corresponding digital problem is a facet of analogue filter approximation which has yet to be utilized.

7. Acknowledgement

The author expresses his appreciation of the help and encouragement given by Professor T. J. Higgins of the Department of Electrical Engineering, University of Wisconsin.

8. References

1. Whittaker, E. T., and Robinson, G., "The Calculus of Observations", Blackie and Son, Ltd., Glasgow, 1929.
2. Jordan, C.. "Calculus of Finite Differences", Chelsea, New York. 1960. (Reprint of 1939 Edition).
3. Wiener. N.. "The Extrapolation, Interpolation and Smoothing of Stationary Time Series", John Wiley and Sons. Inc.. New York, 1949.
4. Blackman. R. B., and Tukey. J. W.. "The Measurement of Power Spectra". Dover Publications. Inc., New York. 1959.
5. Blackman. R. B.. "Smoothing and prediction of time series by cascaded simple averages". IRE Convention Record. 1960. 18. [2]. 21-24.
6. Blackman. R. B. "Linear Data Smoothing and Prediction Theory in Practice". Addison-Wesley Publishing Company. Inc.. Reading. Mass.. 1965.
7. James. H. M.. Nichols. N. B., and Phillips, R. S.. "Theory of Servomechanisms", Radiation Laboratory Series. 1947. vol. 25, McGraw-Hill, Inc.. New York. pp. 231-261.
8. Ragazzini. J. R.. and Franklin, G. F.. "Sampled-Data Control Systems". McGraw-Hill. Inc.. New York. 1958.
9. Jury. E. I.. "Sampled-Data Control Systems". John Wiley and Sons. Inc., New York, 1958.
10. Tou. J. T.. "Digital and Sampled-Data Control Systems". McGraw-Hill. Inc.. New York. 1959.
11. Wilts. C. H.. "Principles of Feedback Control". Addison-Wesley Publishing Company. Inc.. Reading. Mass.. 1960, pp. 195-220.
12. Mishkin. E.. and Braun. L.. Jr.. "Adaptive Control Systems". McGraw-Hill. Inc.. New York, 1961. pp. 119-183.
13. Kuo. B. C.. "Analysis and Synthesis of Sampled-Data Control Systems". Prentice-Hall. Inc.. Englewood Cliffs. N. J.. 1963.
14. Freeman. H.. "Discrete Time Systems". John Wiley and Sons. Inc.. New York, 1965.
15. DeRusso, P. M., Roy, R. J., and Close, C. M., "State Variables for Engineers", John Wiley and Sons, Inc.. New York. 1965. pp. 158-186.
16. Elgerd. O. I.. "Control System Theory", McGraw-Hill. Inc.. New York. 1967. pp. 371-411.
17. Jury, E. I.. "Theory and Application of the z—Transform Method", John Wiley and Sons, Inc., New York, 1964.
18. Bridgland, T. F., Jr.. "A linear algebraic formulation of sampled-data control", Journal of the SIAM. 1959. 7. [4]. 431-446.
19. Gold, B., "Experiment with speechlike phase in a spectrally flattened pitch-excited channel vocoder". Journal of the Acoustical Society of America, 1964. 36, [10], 1892-1894.
20. Baxter. D. C., "The digital simulation of transfer functions". Mechanical Engineering Report MK-13. National Research Council of Canada, Ottawa, Canada. 1964.
21. Baxter. D. C.. "The step response of digital simulations". Mechanical Engineering Report MK-14. National Research Council of Canada, Ottawa, Canada. 1965.
22. Baxter, D. C., "Digital simulation using approximate methods", Mechanical Engineering Report MK-15. National Research Council of Canada, Ottawa, Canada. 1965.
23. Rader. C. M.. "Study of vocoder filters by digital simulation". Journal of the Acoustical Society of America. 1964, 36, [6], 1023.
24. Golden. R. M.. "Digital computer simulation of communication systems using a block diagram compiler: BLODIB". Proceedings of the Third Annual Allerton Conference on Circuit and System Theory. Monticello, Ill.. 1965. pp. 690-707.
25. Jackson. L. B.. Kaiser. J. F.. and McDonald. H. S.. "Implementation of digital filters", IEEE Convention Record. 1968. 16. [2]. 213.
26. Linvill. W. K., "Sampled-data control systems studied through comparison of sampling with amplitude modulation", AIEE Transactions. 1951, 70, [II]. 1779-1788.
27. Steiglitz. K.. "The general theory of digital filters with applications to spectral analysis", Technical Report 400-99, Laboratory of Electroscience Research. New York University. New York. N. Y.. 1963.
28. Steiglitz. K.. "The equivalence of digital and analogue signal processing". Information and Control. 1965. 8. [5]. 455-467.
29. Zadeh. L. A., and Desoer. C. A.. "Linear System Theory", McGraw-Hill. Inc., New York. 1963, p. 418.
30. Bennet, W. R.. "Spectra of quantized signals", Bell Systems Technical Journal, 1948, 27. [3]. 446-472.
31. Widrow. B.. "A study of rough amplitude quantization by means of Nyquist sampling theory". IRE Transactions on Circuit Theory, 1956, CT-3, [4]. 266-276.
32. Widrow, B. "Statistical analysis of amplitude quantized sampled-data control systems". AIEE Transactions. 1961. 79. [II]. 555-568.
33. Wilkinson. J. H.. "Rounding Errors in Algebraic Processes". Prentice-Hall. Inc.. Englewood Cliffs, N. J.. 1963.
34. Knowles. J. B.. and Edwards. R.. "Effect of finite word-length computer in a sampled-data feedback system". Proceedings of the IEE. 1965. 112. [6]. 1197-1207.
35. Knowles. J. B.. and Edwards. R.. "Finite word-length effects in a multirate direct digital control system". Proceedings of the IEE, 1965. 112, [12]. 2376-2384.
36. Kaiser. J. F.. "Some practical considerations in the design of linear digital filters", Proceedings of the Third Allerton Conference on Circuit and System Theory. Monticello, Ill.. 1965. pp. 621-633.
37. Mantey. P. E.. "Digital-Computer implementation of linear systems", Technical Report No. SU-SEL-66-063. Stanford Electronics Laboratories. Stanford University. Stanford. Calif.. 1966.
38. Rader, C. M.. and Gold. B.. "Effects of parameter quantization of the poles of a digital filter". Proceedings of the IEEE, 1967. 55. [5]. 688-689.
39. Sandberg. I. W.. "Floating point roundoff in digital filter realizations". Proceedings of the Fifth Allerton Conference on Circuit and System Theory. Monticello. Ill.. 1967. pp. 1-3.
40. Knowles. J. B.. and Olcayto. E. M.. "Coefficient accuracy and digital filter response. "IEEE Transactions on on Circuit Theory, 1968, CT-15, [1]. 31-41.
41. Tustin. A.. "A method of analysing the behaviour of linear systems in terms of time series". Journal of the IEE. 1947, 94, [IIA]. 130-142.
42. Fleck, J. T. and Fryer, W. D.. "An exploration of numerical filtering techniques". CAL Report No. XA-869-P-1, Cornell Aeronautical Laboratory, Inc., Cornell University. Buffalo, N. Y.. 1953.
43. Salzer, J. M.. "Frequency analysis of digital computers operating in real time". Proceedings of the IRE. 1954. 42. [2]. 457-466.
44. Boxer, R., and Thaler. S., "A simplified method of solving linear and non-linear systems, "Proceedings of the IRE, 1956, 44, [1], 89-101.
45. Thaler, S. and Boxer, R., "An operational calculus of numerical analysis", IRE Convention Record, 1956. 4. [2]. 100-105.
46. Boxer, R., "A note on numerical transform calculus". Proceedings of the IRE, 1957, 45, [10]. 1401-1406.
47. Blum, M. "An extension of the minimum mean-

square prediction theory for sampled input signals". IRE Transactions on Information Theory, 1956, IT-2, [3], 176-184.
48. Martin, M. A., "Frequency domain applications in data processing". Technical Information Series No. 575D340. General Electric Co., Missile and Ordinance Systems Department. Philadelphia, Pa., 1957.
49. Lewis II, P. M., "Synthesis of sampled signal network", IRE Transactions on Circuit Theory. 1958, CT-5, [1], 74-77.
50. Martin, M. A., "Frequency domain applications in data processing". IRE Transactions on Space Electronics and Telemetry. 1959, SET-5, [1], 33-41.
51. Blum, M., "On exponential digital filters". Journal of the ACM, 1959, 6, [2], 283-304.
52. Welch, P. D., "A direct digital method of power spectrum estimation", IBM Journal of Research and Development. 1961, 5, [2], 141-156.
53. Ormsby, J. F. A., "Design of numerical filters with applications to missile data processing". Journal of the ACM. 1961, 8, [3], 440-466.
54. Trench, W. F., "A general class of discrete time invariant filters", Journal of the SIAM. 1961, 9, [3], 405-421.
55. Goodman, N. R., "Some comments on the spectral analysis of time series". Technometrics, 1961, 3, [2], 221-228.
56. Steiglitz, K., "The approximation problem for digital filters", Technical Report 400-56, College of Engineering, New York University, New York, N. Y., 1962.
57. Martin, M. A., "Digital filters for data processing". Technical Information Series No. 62SD484, General Electric Co., Missile and Space Division, Valley Forge, Pa., 1962.
58. Musa, J. D., "Discrete smoothing filters for correlated noise", Bell Systems Technical Journal, 1963, 42, [5], 2121-2151.
59. Kaiser, J. F. "Design methods for digital filters". Proceedings of the First Allerton Conference on Circuit and System Theory, Monticello, Ill., 1963, pp. 221-236.
60. Graham, R. J., "Determination and analysis of numerical smoothing weights". NASA Technical Report R-179. National Aeronautics and Space Administration, Washington, D. C., 1963.
61. Carney, R., "Design of digital notch filter with tracking requirements", IEEE Transactions on Space Electronics and Telemetry, 1963, SET-9, [4], 109-119.
62. Gibson, J. E., "Non-linear Automatic Control", McGraw-Hill, Inc., New York, 1963, pp. 147-159.
63. Golden, R. M., "Digital computer simulation of a sampled-data voice-excited vocoder". Journal of the Acoustical Society of America, 1963, 35, [9], 1358-1366.
64. Gauss, E. J., "Estimation of power spectral density by filters", Journal of the ACM, 1964, 11, [1], 98-103.
65. Golden, R. M., and Kaiser, J. F., "Design of wide band sampled-data filters". Bell System Technical Journal, 1964, 43, [4], part 2, 1533-1546.
66. Whittlesey, J. R. B., "A rapid method of digital filtering". Communications of the ACM, 1964, 7, [9], 552-556.
67. Mantey, P. E., "Convergent automatic synthesis procedures for sampled-data networks". Technical Report SU-SEL-64-112, Stanford Electronics Laboratories, Stanford University, Stanford, Calif., 1964.
68. Treitel, S., and Robinson, E. A., "The stability of digital filters", IEEE Transactions on Geoscience Electronics, 1964, GE-2, [1], 6-18.
69. Anders, E. B., Johnson, J. J., Lasaine, A. D., Spikes, P. W., and Taylo, J. T., "Digital filters", NASA Contractor Report CR-136, National Aeronautics and Space Administration, Washington, D. C., 1964.
70. Robertson, H. H., "Approximate design of digital filters", Technometrics, 1965, 7, [3], 387-403.
71. Kaiser, J. F., and Kuo, F., "System Analysis by Digital Computer", John Wiley and Sons, Inc., New York, 1966, 218-285.
72. Broome, P., "A frequency transformation for digital filters", Proceedings of the IEEE, 1966, 54, [2], 326-327.
73. Fleischer, P. E., "Digital realization of complex transfer functions", Simulation, 1966, 6, [3], 171-180.
74. Holtz, H., and Leondes, C. T., "The synthesis of recursive digital filters", Journal of the ACM, 1966, 13, [2], 262-280.
75. Roberts, P. D., "Digital filter with independently adjustable parameters for use with a general time-weighted mean-square error criterion", Electronics Letters, 1966, 2, [4], 150.
76. Roberts, P. D., "Self-adjusting orthogonal digital filter for identification of an unknown process", Electronics Letters, 1966, 2, [5], 187-188.
77. Halbertson, J. H., "Recursive, complex Fourier analysis for real-time applications", Proceedings of the IEEE, 1966, 54, [6], 903.
78. Treitel, S., and Robinson, E. A., "The design of high resolution digital filters", IEEE Transactions on Geoscience Electronics, 1966, GE-4, [1], 25-38.
79. Shaw, H., "Discrete analogues of continuous filters", Journal of the ACM, 1966, 13, [4], 600-604.
80. Negron, C. D., "Digital one-third octave spectral analysers", Journal of the ACM, 1966, 13, [4], 605-614.
81. Heyliger, G. E., "Design of numerical filters: Scanning functions and equal ripple approximations", Proceedings of the Fourth Allerton Conference on Circuit and System Theory, Monticello, Ill., 1966.
82. Rader, C. M., and Gold, B., "Digital filter design techniques in the frequency domain", Proceedings of the IEEE, 1967, 55, [2], 149-171.
83. Mantey, P. E. and Franklin, G., "Comments on 'Digital filter design techniques in the frequency domain' ", Proceedings of the IEEE, 1967, 55, [12], 2196.
84. Simpson, R. S., and Tranter, W. H., "Easily implemented real-time digital data filters", IEEE Region III Convention Record, 1967, 29, 549-562.
85. "The Fast Fourier Transform", IEEE Transactions on Audio and Electroacoustics, 1967, AU-15, [2].
86. Constantinides, A. G., "Synthesis of Chebychev digital filters", Electronics Letters, 1967, 3, [3], 124-126.
87. Constantinides, A. G., "Elliptic digital filters", Electronics Letters, 1967, 3, [6], 255-256.
88. Greaves, J. C., and Cadzow, J. A., "The optimal discrete filter corresponding to a given analogue filter", IEEE Transactions on Automatic Control, 1967, AC-12, [3], 304-307.
89. Vich, R., "Selective properties of digital filters obtained by convolution approximation", Electronics Letters, 1968, 4, [1], 1-2.
90. Golden, R. M., "Digital filter synthesis by sampled data transformation". Technical Information from Autonetics, Autonetic Division of North American Rockwell Corporation, Anaheim, Calif., 1968.
91. Constantinides, A. G., "Frequency transformations for digital filters", Electronics Letters, 1968, 4, [7], 115-116.
92. Constantinides, A. G., "Online digital filtering", Electronics Letters, 1968, 4, [12], 252-253.
93. Titchmarsh, E. C., "The Theory of Functions", Oxford University Press, London, 1964, pp. 411-415. (Reprint of 1932 edition).
94. Slepian, D., and Pollak, O. H., "Prolate spheroidal wave functions, Fourier analysis and uncertainty, I and II", Bell System Technical Journal, 1961, 40, [1], 43-84.
95. Schelkunoff, S. A., "A mathematical theory of linear arrays", Bell System Technical Journal, 1943, 22, [1], 80-107.
96. Cheng, D. K., and Ma, M. T., "A new mathematical approach for linear array analysis", IRE Transactions on Antennas and Propagation, 1960, AP-8, [e], 255-259.
97. Ma, M. T., "Application of Bernstein polynomials and interpolation theory to linear array synthesis", IEEE Transactions on Antennas and Propagation, 1964, AP-12, [6], 668-677.

98. Ma, M. T., "An application of the inverse z — transform theory to the synthesis of linear antenna arrays", IEEE Transactions on Antennas and Propagation, 1964, AP-12. [6]. 798-799.
99. Lorentz, G. G., "Bernstein Polynomials", Toronto University Press, Toronto, Canada, 1953.
100. Scarborough, J. B., "Numerical Mathematical Analysis", The John Hopkins Press, Baltimore, Md., 1955.
101. Lanczos, C., "Applied Analysis", Prentice-Hall, Inc., Englewood Cliffs, 1956, pp. 229-239.
102. Papoulis, A., "The Fourier Integral and its Applications", McGraw-Hill, Inc., New York. 1962, 201 and 209.
103. Guillemin, E. A., "Synthesis of Passive Networks". John Wiley and Sons, Inc., New York, 1957, 614-619.
104. Temes, G. C., and Calahan, D. A., "Computer aided network optimization: The state-of-the-art", Proceedings of the IEEE, 1967, 55, [11]. 1832-1863.
105. Gibbs, A. J., "On the Frequency Domain Responses of Causal Digital Filters", Ph.D. Thesis, University of Wisconsin, Madison, Wis., Chapter 4, 1969.
106. Rudin, W., "Real and Complex Analyses", McGraw-Hill, Inc., New York, 1966.
107. Tamarkin, J. D., "Theory of Fourier Series", Lectures delivered at Brown University during 1932-33, Providence, R.I., p. 271, 1933.
108. Gibbs, A. J., "The Design of Digital Filters". To be published in A.T.R., 4, [1].
109. Gibbs A. J. "An introduction to digital filters," Australian Telecommunications Research, 1969, 3, [2], 3-14.
110. Fujisawa, T. "Theory and procedure for optimization of low-pass attenuation characteristics," IEEE Transactions on Circuit Theory, 1964, CT-11, [4], 449–456.
111. Smith, B. R. and Temes, G. C. "An iterative approximation procedure for automatic filter synthesis," IEEE Transactions on Circuit Theory, 1965, CT-12, [1], 107–112.
112. Gibbs, A. J. "Optimum design of filter characteristic functions," IEEE Transactions on Circuit Theory, 1966, CT-13, [4], 445–447.
113. Gibbs, A. J. "The attainment of Tchebycheff approximations to prescribed phase delay and group delay characteristics with the aid of an electronic computer," Proc. IREE Aust., 1966, 27, [11], 312–320.
114. Gibbs, A. J. "Delay approximations within variable tolerances," Proc. IREE Aust., 27, [11], A18–A19, 1966.

Digital Filter Design Techniques in the Frequency Domain

CHARLES M. RADER, MEMBER, IEEE, AND BERNARD GOLD

Abstract—Digital filtering is the process of spectrum shaping using digital components as the basic elements. Increasing speed and decreasing size and cost of digital components make it likely that digital filtering, already used extensively in the computer simulation of analog filters, will perform, in real-time devices, the functions which are now performed almost exclusively by analog components. In this paper, using the z-transform calculus, several digital filter design techniques are reviewed, and new ones are presented. One technique can be used to design a digital filter whose impulse response is like that of a given analog filter; other techniques are suitable for the design of a digital filter meeting frequency response criteria. Another technique yields digital filters with linear phase, specified frequency response, and controlled impulse response duration. The effect of digital arithmetic on the behavior of digital filters is also considered.

Manuscript received August 24, 1965; final revision November 4, 1966.

The authors are with the Lincoln Laboratory, Massachusetts Institute of Technology, Lexington, Mass. (Operated with support from the U. S. Air Force.)

Section I

A. Introduction

DIGITAL FILTERING is the process of spectrum shaping using digital hardware as the basic building block. Thus the aims of digital filtering are the same as those of continuous filtering, but the physical realization

Reprinted from *Proc. IEEE*, vol. 55, pp. 149–171, Feb. 1967.

is different. Linear continuous filter theory is based on the mathematics of linear differential equations; linear digital filter theory is based on the mathematics of linear difference equations.

Our interest in digital filtering derives from recent work on computer simulation of speech communications systems [1]–[6] and from problems arising in the processing of geophysical data [7]. In addition, however, increasing speeds and decreasing costs of microelectronic digital circuitry make possible the conception of real-time digitized systems which perform filtering operations usually performed by analog hardware.

Linear digital filter theory is based on the well-known mathematics of linear difference equations with constant coefficients [19]. In addition, a large body of work exists on the subject of sampled data systems and numerical analysis [8], [13], [20], [21] and this work is of direct use in the design of digital filters. Recently [22], [23], [29], [30], a start has been made towards the development of design techniques from the point of view of frequency selectivity. The main intent of this paper is to review, extend, and make an initial attempt at unification of digital filter design from this same point of view. Thus, emphasis will be placed on the frequency selectivity of filters rather than on the questions of stability and time response which are usually of interest in control systems. However, it should be stressed that digital filters are mathematically equivalent to continuous systems with sampled data inputs and outputs. In the continuous case, the sampling interval T usually depends on the nature of the discrete input, such as a pulsed radar return. The constraint on the sampling interval of a digital filter could also be caused by the computation time needed to execute the difference equations.

The basic mathematical tool of digital filters is the z transform calculus, used by Hurewicz [8] to develop his theory of pulsed filters. We will first briefly review the z-transform method and then present a number of design techniques with illustrative examples.

Real-time digital filters have several advantages over continuous filters. A greater degree of accuracy can be attained in the digital filter realization. A greater variety of digital filters can be built, since certain realization problems (akin to negative elements) do not arise. No special components are needed to realize filters with time-varying coefficients. Aggregates of digital filters should be especially economic in the very low frequency band (0.01 to 1 Hz) where the size of analog components becomes appreciable.

In addition to the frequency-selective design techniques mentioned above, there has been some recent work [24], [25] of interest which pertains to the effects of quantization in a computer with finite word length. These quantization effects can either cause deviations from the original design criteria or create a noise at the filter output. We have reserved a separate section for the treatment of these effects, but otherwise, in the remainder of the paper, the variables and coefficients of the difference equations will be assumed unquantized.

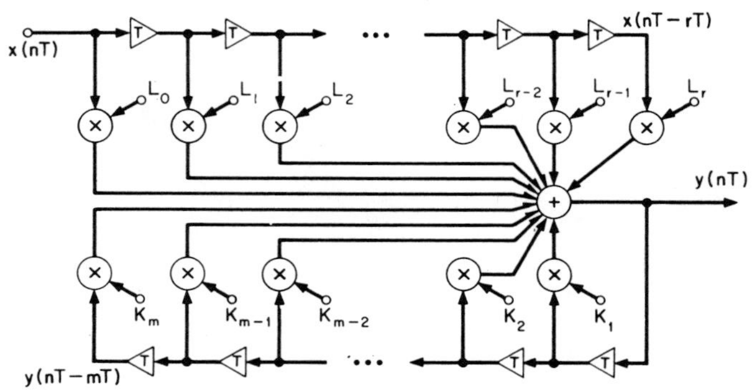

Fig. 1. Pictorial representation of mth-order linear difference equation.

B. Difference Equations of Digital Filters

An mth-order linear difference equation may be written as

$$y(nT) = \sum_{i=0}^{r} L_i x(nT - iT) - \sum_{i=1}^{m} K_i y(nT - iT). \quad (1)$$

The form of (1) emphasizes the iterative nature of the difference equation; given the m previous values of the output y and the $r + 1$ most recent values of the input x, the new output may be computed from (1). Physically, the input numbers are samples of a continuous waveform and real-time digital filtering consists of performing the iteration (1) for each arrival of a new input sample. Design of the filter consists of finding the constants K_i and L_i to fulfill a given filtering requirement. Real-time filtering implies that the execution time of the computer program for computing the right side of (1) is less than T, the sampling interval. Simulation programs in which the computations are not carried out in real time are understood to be simulations of the real-time digital filters and the original sampling interval T of the continuous input data remains the reference parameter. It should be realized that quantization as well as sampling is performed on the continuous data before entry into the program and, also, that the multiplications indicated in (1) contain an inherent round-off error caused by finite register lengths; but, in this section, these effects will be neglected.

A useful pictorial representation of (1) consisting of unit delays (of time T), adders, and multipliers is shown in Fig. 1.

Many practical designs utilize arrangements of simple (first- and second-order) difference equations. Arrangements can be serial, parallel, or combinations of both. For example, the two equations

$$\left.\begin{aligned} y_1(nT) &= K_{11}y_1(nT-T)+K_{12}y_1(nT-2T)+L_{11}x(nT) \\ y_2(nT) &= K_{21}y_2(nT-T)+K_{22}y_2(nT-2T)+L_{21}y_1(nT) \end{aligned}\right\} \quad (2)$$

constitute a serial arrangement whereby the output y_1 of the first equation is used as input to the second equation, as seen in Fig. 2. Similarly the equations

$$\left.\begin{aligned} y_1(nT) &= K_{11}y_1(nT-T)+K_{12}y_1(nT-2T)+L_{11}x(nT) \\ y_2(nT) &= K_{21}y_2(nT-T)+K_{22}y_2(nT-2T)+L_{21}x(nT) \\ y_3(nT) &= y_1(nT)+y_2(nT) \end{aligned}\right\} \quad (3)$$

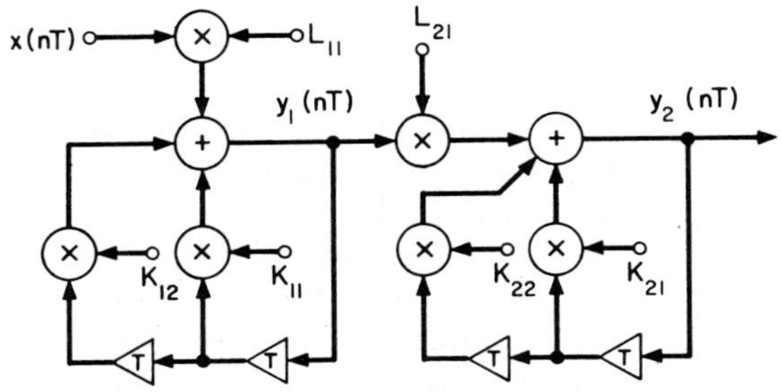

Fig. 2. Serial arrangement of difference equations.

shown in Fig. 3, constitute a parallel arrangement. The sets of equations (2) and (3) can be rewritten as single fourth-order difference equations; however, the significant properties of the digital filters are more readily understood via structures like Figs. 2 and 3.

Substantial literature [13] exists on the problems of presampling filtering of the input and ultimate filtering of the sampled output to convert it to a continuous wave. We shall restrict ourselves to studying only the properties of the samples.

C. The Z Transform

In this section, we will assume that the set of input and output samples $x(nT)$ and $y(nT)$ are zero for n less than zero.

The z transform of a sequence $x(nT)$ is defined as

$$X(z) = \sum_{n=0}^{\infty} x(nT)z^{-n}. \tag{4}$$

For many sequences, the infinite sum of (4) can be expressed in closed form. For example, the z transform of the sequence $x(nT)=0$ for $n<0$, $x(nT)=1$ for $n\geq 0$ is

$$X(z) = \sum_{n=0}^{\infty} z^{-n} = \frac{1}{1-z^{-1}}.$$

The transformation variable z is, in general, a complex variable and $X(z)$ is a function of a complex variable.

The z transform of a convergent sequence uniquely defines that sequence. By multiplying both sides of (4) by z^{n-1} and integrating around any closed curve in the z plane which encloses all the singular points (poles) of $X(z)$ and the origin, we obtain

$$x(nT) = \frac{1}{2\pi j}\oint X(z)z^{n-1}\,dz \tag{5}$$

where $j=\sqrt{-1}$. The inverse z transform (5) explicitly determines the sequence $x(nT)$ associated with a given $X(z)$. For sequences which converge to zero the unit circle can be the closed curve of integration.

D. Solution of Difference Equations by z-Transform Techniques

The z transform will now be used to obtain an explicit solution of the mth-order linear difference equation (1). First, rewrite (1) as

$$\sum_{i=0}^{m} K_i y(nT-iT) = \sum_{i=0}^{r} L_i x(nT-iT). \tag{6}$$

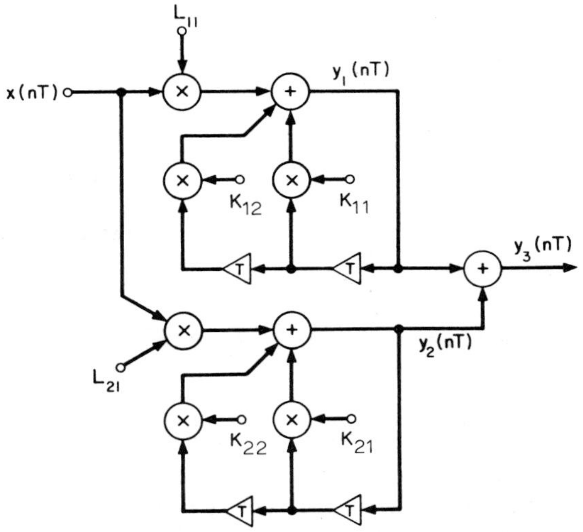

Fig. 3. Parallel arrangement of difference equations.

Next, take the z transform of (6), to obtain

$$\sum_{i=0}^{m} K_i \sum_{n=0}^{\infty} y(nT-iT)z^{-n} = \sum_{i=0}^{r} L_i \sum_{n=0}^{\infty} x(nT-iT)z^{-n}. \tag{7}$$

Recognizing that the z transform of a sequence delayed by i samples is equal to the z transform of the original sequence multiplied by z^{-i}, (7) reduces to

$$\left.\begin{aligned} &Y(z)\sum_{i=0}^{m} K_i z^{-i} = X(z)\sum_{i=0}^{r} L_i z^{-i} \\ \text{or}\quad & \\ &Y(z) = X(z)\frac{\sum_{i=0}^{r} L_i z^{-i}}{\sum_{i=0}^{m} K_i z^{-i}} = X(z)H(z) \end{aligned}\right\} \tag{8}$$

Thus $Y(z)$ is explicitly determined as the product of the input z transform and a system function $H(z)$ which is a rational fraction in z^{-1} and is a function of the constant coefficients in the original difference equation. By the inverse z transform (5), the output $y(nT)$ may be written as the inverse transform of $Y(z)$.

If we choose as the input sequence

$$x(nT) = \begin{cases} 1 & \text{for } n=0 \\ 0 & \phantom{\text{for }} n\neq 0 \end{cases},$$

then in (8) $X(z)=1$. Clearly the response to this input is the inverse z transform of $H(z)$. Thus the sequence $1, 0, 0, \cdots$ plays the part in digital filter theory which the unit impulse plays in continuous filter theory. We shall refer to the inverse z transform of $H(z)$ as the impulse response of a digital filter.

We can use (8) to derive a general representation of a digital filter which differs from that of Fig. 1. We can define an intermediate variable $W(z)$ corresponding to an intermediate sequence $w(nT)$, such that

$$W(z) = X(z) \frac{1}{\sum_{i=0}^{m} K_i z^{-i}}$$

$$Y(z) = W(z) \sum_{i=0}^{r} L_i z^{-i}. \tag{8a}$$

The equations of (8a) result in simultaneous difference equations

$$w(nT) = x(nT) - \sum_{i=1}^{m} K_i w(nT - iT)$$

$$y(nT) = \sum_{i=0}^{r} L_i w(nT - iT) \tag{8b}$$

which lead to the structure shown in Fig. 4. This structure requires fewer delays (less storage) and may, therefore, be preferred for some realizations.

The primary importance of the system function $H(z)$ for our purposes is in its interpretation as a frequency selective function. Let us assume that the input is a sampled complex exponential wave.

$$x(nT) = e^{jn\omega T}. \tag{9}$$

The solution to (6) for the input (9) is also a complex exponential wave which can be represented as

$$y(nT) = F(e^{j\omega T})e^{jn\omega T} \tag{10}$$

and, by substitution of (9) and (10) into (6), we quickly arrive at the result

$$F(e^{j\omega T}) = \frac{\sum_{i=0}^{r} L_i e^{-ji\omega T}}{\sum_{i=0}^{m} K_i e^{-ji\omega T}} = H(e^{j\omega T}). \tag{11}$$

The response to a sampled sinusoidal input can be readily found from (11).

E. *Representation and Geometric Interpretation of System Function*

We see that the system function $H(z)$ of (8) is interpretable as a frequency response function for values of z on the *unit circle* in the complex z plane. Note that the radian frequency ω is a continuous frequency so that the physical significance of the frequency response function is the same as that for continuous systems. Furthermore, the system function is a rational fraction whose numerator and denominator can be factored, so that $H(z)$ is uniquely defined, except for a constant multiplier, by the positions of its poles and zeros in the z plane, and its value for any point z determined directly by the distances of that point from the singularities. We are thus led directly to the geometric interpretation of Fig. 5 whereby the value of the frequency response function for any frequency ω is obtained by rotating by an angle ωT about the circle, measuring the distances to the zeros $R_1, R_2 \cdots$, the distances to the poles $P_1, P_2 \cdots$, and then forming the ratio, so that in the example of Fig. 5,

$$|H(e^{j\omega T})| = \frac{R_1 R_2}{P_1 P_2 P_3}. \tag{12}$$

$H(e^{j\omega T})$ also has associated with it a phase, which is given by

$$\phi = \phi_1 + \phi_2 - (\psi_1 + \psi_2 + \psi_3). \tag{13}$$

The geometric basis for digital filter design is thus identical in principle with the geometric basis for continuous filter design with the following single and important difference: in the continuous filter, frequency is measured along the imaginary axis in the complex s plane whereas, in the digital filter, frequency is measured along the circumference of the unit circle in the z plane.

F. *Several Examples*

First-order difference equation: The difference equation

$$y(nT) = Ky(nT - T) + x(nT) \tag{14}$$

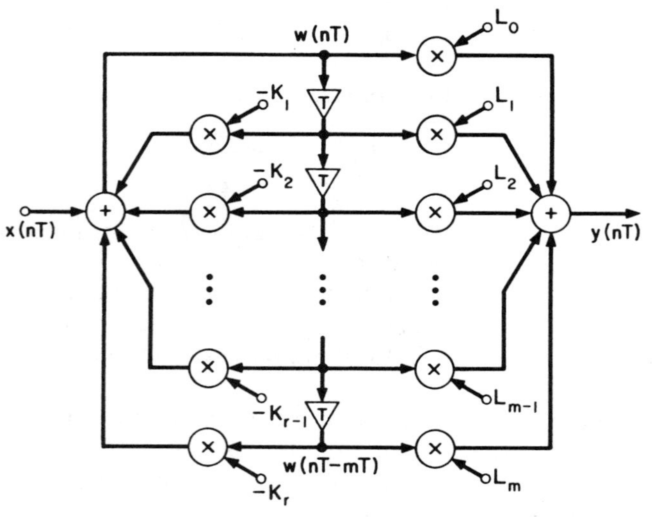

Fig. 4. Alternate representation of digital filter (drawn for $m=r$).

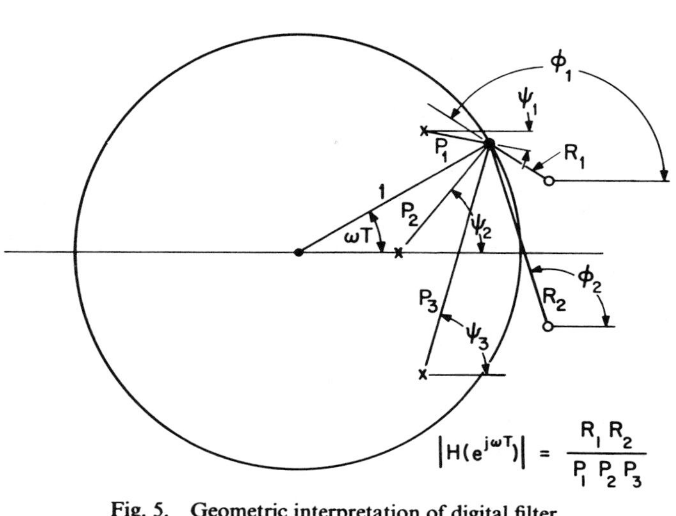

Fig. 5. Geometric interpretation of digital filter frequency response function.

has the system function

$$H(z) = \frac{1}{1 - Kz^{-1}}. \quad (15)$$

The frequency response function is then obtained by letting $z = {}^{j\omega T}$ in (15), which yields

$$|H(e^{j\omega T})| = \frac{1}{\sqrt{1 + K^2 - 2K\cos\omega T}}$$

and

$$\phi = \text{angle of } H(e^{j\omega T}) = -\tan^{-1}\frac{\sin\omega T}{\cos\omega T - K}. \quad (16)$$

Second-order difference equation: The difference equation

$$y(nT) = K_1 y(nT - T) + K_2 y(nT - 2T) + x(nT) \quad (17)$$

has the system function

$$H(z) = \frac{1}{1 - K_1 z^{-1} - K_2 z^{-2}} \quad (18)$$

and the resulting frequency response magnitude

$$|H(e^{j\omega T})| = \frac{1}{[(1 - K_1\cos\omega T - K_2\cos 2\omega T)^2 + (K_1\sin\omega T + K_2\sin 2\omega T)^2]^{\frac{1}{2}}} \quad (19)$$

and phase

$$\phi(e^{j\omega T}) = -\tan^{-1}\left[\frac{K_1\sin\omega T + K_2\sin 2\omega T}{1 - K_1\cos\omega T - K_2\cos 2\omega T}\right]. \quad (20)$$

The pole positions γ and β of $H(z)$ are given by

$$\left.\begin{aligned}\gamma &= \frac{K_1}{2} + \sqrt{\frac{K_1^2}{4} + K_2}\\ \beta &= \frac{K_1}{2} - \sqrt{\frac{K_1^2}{4} + K_2}\end{aligned}\right\}. \quad (21)$$

For K_2 negative and $|K_2| > (K_1^2/4)$, the poles become complex conjugate and, as they approach the unit circle, the well-known resonance effect is evidenced. Figure 6 is a plot of magnitude versus ωT for $T = 10^{-4}$, a resonant frequency of 1500 Hz and several values of damping. In the vicinity of the poles, the magnitude function behaves very much like a continuous resonator. Of course, the magnitude and phase must be periodic in ωT; this follows from the original premise of a sampled input and is embodied in the unit circle z plane representation. We state, without proof, the well-known facts that all poles of $H(z)$ must lie within the unit circle for stability and that poles and zeros must occur on the real axis or in conjugate complex pairs. Also, it should be clear from (8) that an mth-order difference equation has a z-transform system function containing m poles and r zeros, r being the number of unit delays of the input.

An approach to digital filtering, employing the relationships between linear, constant coefficient difference equations, z-plane representations, and frequency response has been summarized in this section. These digital filters can be called "recursive," since the output at any time depends on both input and previous values of the output. By contrast, nonrecursive filters are defined by equations wherein each output sample is an explicit function of only the input samples. This corresponds to a filter in which all the K_i's in Fig. 1 are set equal to zero. The chief advantage of the recursive technique has been in the smaller number of operations usually required to obtain a desired frequency response. Recent work [31] has shown that program execution time for nonrecursive filters can be substantially reduced by making use of fast Fourier transform techniques [38]. Nonrecursive filter design has been treated extensively and we will not discuss it further in this paper.

Section II

A. General Discussion of Digital Filter Design Techniques

Since much information is available on continuous filter design, a useful approach to digital filter design involves finding a set of difference equations having a system function $H(z)$ which significantly resembles known analog system functions. The work of Hurewicz [8] provides a technique for doing this in an "impulse invariant" way. By this we mean that the discrete responses to an impulse function (see Section I-D) of the derived digital filter will be the samples of the continuous impulse response of the given continuous filter.

Digital filters can be specified from a desired squared magnitude function using procedures akin to that of Butterworth and Chebyshev continuous filter design procedure.

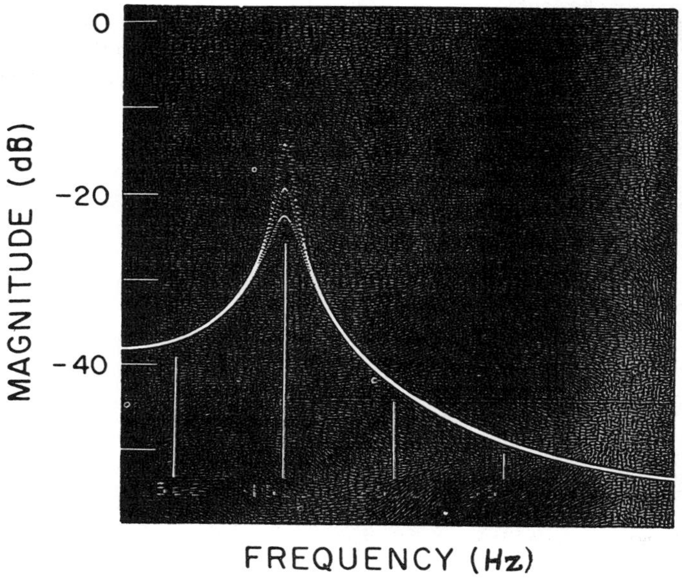

Fig. 6. Response of digital resonator for several values of damping.

This method will be described and realizability conditions discussed in the Appendix.

Another technique used by various workers [9]–[12], [26] uses conformal mapping to transform a digital filter design problem into a continuous filter design problem, for which a vast body of design literature exists. We shall refer to this technique as "bilinear transformation" although other transformations have sometimes been used.

Finally, a technique referred to as "frequency sampling" makes use of the special properties of an elemental digital filter. This elemental filter's frequency response curve resembles a sin x/x function and has a linear phase versus frequency characteristic. By suitably combining these elemental filters, a simple design technique for a large variety of digital filters can be developed.

Where the same filtering requirements can be adequately met by several different digital filters, the choice between them depends on the speed of execution of a computer program which performs the difference equation. An important factor in this speed is the number of multiplications (not counting multiplications by such factors as ± 1 or 2^n, which computers can do quickly). Some digital filters are able to meet essentially the same requirements as others with substantially fewer multiplications per output sample, and these are to be preferred. It is important to stress that speed of execution is the main limiting factor in the utilization of digital filtering methods.

B. Technique 1—Impulse Invariance

We first show that a digital filter with an impulse response equal to the sampled impulse response of a given continuous filter can be derived via the correspondence

$$Y(s) = \sum_{i=1}^{m} \frac{A_i}{s + s_i} \Rightarrow \sum_{i=1}^{m} \frac{A_i}{1 - e^{-s_i T} z^{-1}} = H(z). \tag{22}$$

The impulse response $k(t)$ of a continuous filter is defined as the inverse Laplace transform of its system function $Y(s)$, given in general form[1] by the left side of (22). Similarly, the impulse response $h(nT)$ of a digital filter is defined as the inverse z transform of its system function $H(z)$, which can be expressed generally by the right side of (22).

Thus

$$k(t) = \mathscr{L}^{-1}\left(\sum_{i=1}^{m} \frac{A_i}{s + s_i} \right) = \sum_{i=1}^{m} A_i e^{-s_i t}. \tag{23}$$

If we desire that

$$h(nT) = k(t), \qquad t = 0, T, 2T \cdots \tag{24}$$

$$h(nT) = \sum_{i=1}^{m} A_i e^{-s_i n T}. \tag{25}$$

Taking the z transform of (25),

$$H(z) = \sum_{n=0}^{\infty} h(nT) z^{-n} = \sum_{i=1}^{m} A_i \sum_{n=0}^{\infty} e^{-s_i n T} z^{-n}$$

$$= \sum_{i=1}^{m} \frac{A_i}{1 - e^{-s_i T} z^{-1}}. \tag{26}$$

Thus the condition (24) that the impulse response of the digital filter be equal to the sampled impulse response of a given continuous filter $Y(s)$ leads to a digital filter defined by (26), where all constants A_i and s_i have already been specified from $Y(s)$. Thus, by means of the correspondence (22), z transforms can be tabulated [13].

Example: The simple one-pole RC low-pass filter is transformed to a digital filter via the correspondence

$$\frac{a}{s + a} \Rightarrow \frac{a}{1 - e^{-aT} z^{-1}}. \tag{27}$$

This example may be used to illustrate two problems inherent in the design of digital filters by the impulse invariant method. One problem is analogous to the spectrum "folding" effect encountered when a time function is sampled. A digital filter can be "impulse invariant" only in a sampled sense. The frequency response of the digital filter may differ markedly from that of the corresponding analog filter if the latter has a significant frequency response above one half of the sampling frequency. Figure 8 shows the frequency responses of corresponding analog and digital filters of the form of (27) for various values of a and unity sampling rate.

The other problem encountered with the impulse invariant method is the variation of the gain of the filter with sampling rate. Even if aliasing effects can be neglected, the gain of an impulse invariant digital filter is proportional to sampling frequency, and for practical problems this gain may be 10^4 or more. It is important to be aware of the gain of a filter [from (11)], to compensate for it if necessary, and to protect against the possibility of overflow in the computer program.

System functions of various resonant circuits may be expanded by partial fractions, leading to the correspondences

$$\frac{s + a}{(s + a)^2 + b^2} \Rightarrow \frac{1 - e^{-aT}(\cos bT) z^{-1}}{1 - 2e^{-aT}(\cos bT) z^{-1} + e^{-2aT} z^{-2}} \tag{28}$$

$$\frac{b}{(s + a)^2 + b^2} \Rightarrow \frac{e^{-aT}(\sin bT) z^{-1}}{1 - 2e^{-aT}(\cos bT) z^{-1} + e^{-2aT} z^{-2}}. \tag{29}$$

C. Design of a Digital Lerner Filter which is "Impulse Invariant"

The Lerner filter [14] is defined by the continuous system function

$$Y(s) = \sum_{i=1}^{m} \frac{B_i(s + a)}{(s + a)^2 + b_i^2} \tag{30}$$

[1] To be completely general, (22) should also include terms of the form

$$\frac{A_i}{(s + s_i)^l}$$

which correspond to terms of the form

$$\left[\frac{(-1)^{l-1}}{(l-1)!} \frac{\partial^{l-1}}{\partial a^{l-1}} \left(\frac{A_i}{1 - e^{-aT} z^{-1}} \right) \right] \Bigg|_{a = s_i}$$

with

$$B_1 = \frac{1}{2}, \quad B_m = \frac{(-1)^{m+1}}{2},$$
$$B_i = (-1)^{i+1} \quad \text{for } i = 2, \cdots, m-1$$

and the pole positions $(-a \pm b_i)$ shown in Fig. 7(a). It has been shown that Lerner filters have a high degree of phase linearity and reasonably selective passbands.

From the correspondence (28), the z transform of the Lerner filter impulse response is

$$H(z) = \sum_{i=1}^{m} \frac{B_i(1 - e^{-aT}(\cos b_i T)z^{-1})}{1 - 2e^{-aT}(\cos b_i T)z^{-1} + e^{-2aT}z^{-2}}. \quad (31)$$

The excellent magnitude and phase characteristics of Lerner filters will, however, not be retained in digital realization unless the aliasing effect mentioned in the preceding section is not significant. This is the case when the bandwidth of each individual pole is small compared with the sampling frequency. The invariance of the impulse response may be all that is required in some situations, and this is, of course, not a function of the sampling frequency.

Figure 7(b) shows the digital realization of a fourth-order bandpass Lerner filter. Each of the four parallel suboutputs y_i is computed by the difference equation

$$y_i(nT) = e^{-aT}(\cos b_i T)[2y_i(nT-T) - x(nT-T)] - e^{-2aT}y_i(nT-2T) + x(nT) \quad i = 1, 2, 3, 4 \quad (32)$$

and the output $y(nT)$ is given by

$$y(nT) = \tfrac{1}{2}y_1(nT) - y_2(nT) + y_3(nT) - \tfrac{1}{2}y_4(nT). \quad (33)$$

D. Gain of Digital Resonators

The correspondences (28) and (29) define two digital resonators which are "impulse invariants" of given continuous resonators. In practice, digital resonators can be specified without reference to continuous resonators. Specification consists of placing the pair of complex conjugate poles and, in most cases, a single zero, from which the difference equation can be quickly derived. Since many digital filters are simple serial or parallel combinations of these resonators, it is important to understand their behavior.

The z transform of a resonator with poles at $z = re^{\pm j\omega_0 T}$ and a zero at q is

$$H(z) = \frac{1 - qz^{-1}}{1 - 2r\cos\omega_0 Tz^{-1} + r^2 z^{-2}}. \quad (34)$$

The magnitude versus frequency function for (34) is $|H(e^{j\omega T})|$ and can be written down, by inspection, from Fig. 5, using the law of cosines,

$$|H(e^{j\omega T})| = \left[\frac{1 + q^2 - 2q\cos\omega T}{[1 + r^2 - 2r\cos(\omega - \omega_0)T][1 + r^2 - 2r\cos(\omega + \omega_0)T]}\right]^{\frac{1}{2}}. \quad (35)$$

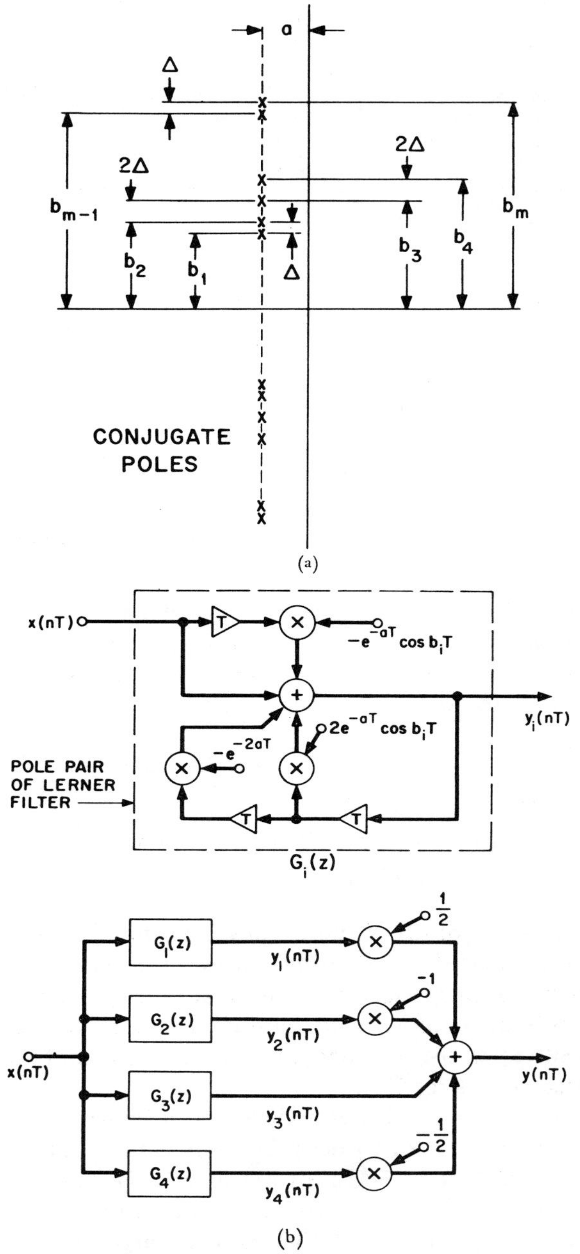

Fig. 7. (a) Pole positions for Lerner filter. (b) Configuration of fourth-order (eight-pole) digital Lerner filter [dotted box shows configuration of pole pair $G(z)$].

Case 1, $q = \cos\omega_0 T$: For values of r close to unity, the value of $|H(e^{j\omega T})|$ at the resonance ω_0 can be approximated by

$$|H(e^{j\omega_0 T})| = \frac{1}{2(1 - r)\sqrt{r}} \quad (36)$$

which is independent of ω_0. Thus, this choice of q makes possible the design of an equal gain bank of resonators (or filters composed of these resonators) covering a wide frequency range.

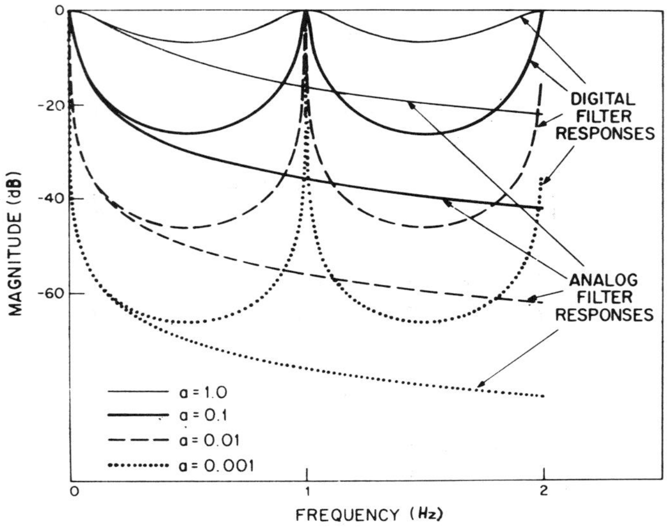

Fig. 8. Frequency responses of analog and digital one-pole filters.

For narrowband resonators in which r is close to unity, (36) shows clearly that the gain at resonance is usually appreciably greater than unity. Knowledge of filter gains is required for determination of appropriate register word lengths.

Case 2, $q = r \cos \omega_0 T$: For this case, the difference equation for (34) can be written

$$y(nT) = r \cos \omega_0 T[2y(nT - T) - x(nT - T)] + r^2 y(nT - 2T) + x(nT). \tag{37}$$

Execution of (37) requires only 2 multiplication instructions compared to 3 multiplications for general q, and this case is thus of special interest for real-time applications and when total computer running time is inordinately lengthy. Sensitivity of the resonant gain with the resonant frequency is greater than for Case 1.

Case 3, $q = 1$: This case yields zero gain for "dc," when $\omega = 0$, which is often desirable. The gain at resonance is highly sensitive to $\omega_0 T$; for r nearly 1, the resonant gain for $\omega_0 T = \pi/2$ is $\sqrt{2}$ times the resonant gain when $\omega_0 T$ is close to zero, provided that $\omega_0 T$ is large compared to $(1 - r)$.

Case 4, $q = 0$: In this case, the zero disappears. In the design of formant vocoders, it is desirable to design resonators without zeros, but with constant gain for $\omega = 0$, independent of ω_0. This is obtained from the digital system function

$$H(z) = \frac{1 - 2r \cos \omega_0 T + r^2}{1 - 2r(\cos \omega_0 T)z^{-1} + r^2 z^{-2}}. \tag{38}$$

E. Design of Digital Filters from Continuous Filters which have Zeros at Infinity

A large class of analog filters are defined by system functions of the form

$$Y(s) = \frac{1}{\prod_{i=1}^{m} (s + s_i)} \tag{39}$$

where the denominator is a product. Such filters have m poles at finite values of s and an mth-order zero for infinite s. Included in this category are Butterworth, Chebyshev, and Bessel filters.

In order to design impulse invariant digital filters based on (39), the procedure outlined in Section II-B can be used; $Y(s)$ is expanded in partial fractions, the A_i are found, and $H(z)$ is obtained from the correspondence (22). In general, this causes zeros to appear in $H(z)$, although there were no finite zeros in $Y(s)$. However, when the poles at $-s_i$ are close to the imaginary axis in the s plane, such that $e^{-s_i T}$ is close to unity, the zeros of $H(z)$ can be ignored and $H(z)$ may be approximated by

$$H(z) = \frac{1}{\prod_{i=1}^{m} (1 - e^{-s_i T} z^{-1})}. \tag{40}$$

In practice, digital bandpass filters several hundred cycles wide of the Bessel, Butterworth or Chebyshev-type have been successfully programmed for 10 000 cycle sampling rates, using the form (40).

Example, Three-Pole Butterworth Lowpass Filter: The system function of the continuous filter is

$$Y(s) = \frac{s_1 s_2 s_3}{(s + s_1)(s + s_2)(s + s_3)} \tag{41}$$

with

$$s_1 = \omega_c, \quad s_2 = \frac{1}{2}(1 + j\sqrt{3})\omega_c, \quad s_3 = \frac{1}{2}(1 - j\sqrt{3})\omega_c$$

ω_c being the cutoff frequency, defined by $|Y(j\omega_c)| = 0.707$. Expanding (41) into partial fractions and defining $\eta(\omega_c T)$ as

$$e^{-\omega_c T/2}\left[\cos \frac{1}{2}\sqrt{3}\,\omega_c T + \frac{1}{\sqrt{3}} \sin \frac{1}{2}\sqrt{3}\,\omega_c T\right],$$

the correspondence (22) leads to the z transform

$$H(z) = \omega_c \left\{ \frac{1}{1 - e^{-\omega_c T} z^{-1}} + \frac{-1 + \eta(\omega_c T) z^{-1}}{1 - 2e^{-\omega_c T/2} \cos\left(\frac{\sqrt{3}}{2}\omega_c T\right) z^{-1} + e^{-\omega_c T} z^{-2}} \right\} \tag{42}$$

which has the diagrammatic representation shown in Fig. 9.

If a cascade representation is desired, (42) can be rewritten to give

$$H(z) = \frac{Cz^{-1} + Dz^{-2}}{(1 - e^{-\omega_c T} z^{-1})\left(1 - 2e^{-\omega_c T/2}\left(\cos\frac{\sqrt{3}}{2}\,\omega_c T\right)z^{-1} + e^{-\omega_c T} z^{-2}\right)} \tag{43}$$

with

$$C = \omega_c\left[e^{-\omega_c T} + e^{-\omega_c T/2}\left(\frac{1}{\sqrt{3}}\sin\frac{\sqrt{3}}{2}\,\omega_c T - \cos\frac{\sqrt{3}}{2}\,\omega_c T\right)\right]$$

and

$$D = \omega_c\left[e^{-\omega_c T} - e^{-3\omega_c T/2}\left(\frac{1}{\sqrt{3}}\sin\frac{\sqrt{3}}{2}\,\omega_c T + \cos\frac{\sqrt{3}}{2}\,\omega_c T\right)\right].$$

$H(z)$ is seen to have two zeros, one at $z = 0$ and the other at $z = -(D/C)$. This zero is on the real axis and increases as $\omega_c T$ is increased from zero. Note that the denominator of (43) can be written down by inspection of the poles of the continuous filter, since an s-plane pole transforms directly into a z-plane pole via $z = e^{sT}$. Thus the zero of (43) at $-(D/C)$ causes the only error in the assumption that three s-plane resonators in cascade transform directly into three z-plane resonators in cascade.

For small values of $\omega_c T$, the zero of $H(z)$ is approximately at -1, and thus has but a small effect on the bandpass characteristics of the three-pole Butterworth filter. Figure 10 shows a plot of $-(D/C)$ versus $\omega_c T$. We see, for example, that the zero moves to -0.94 for $\omega_c T = 0.1$, which still introduces but slight distortion in the passband of the cascade approximation.

F. Review of Butterworth and Chebyshev Filters

In this section, we review briefly design procedures for continuous Butterworth and Chebyshev filters, which will be needed in the later discussion of digital filter design. More complete treatments are available in standard texts [15]–[17].

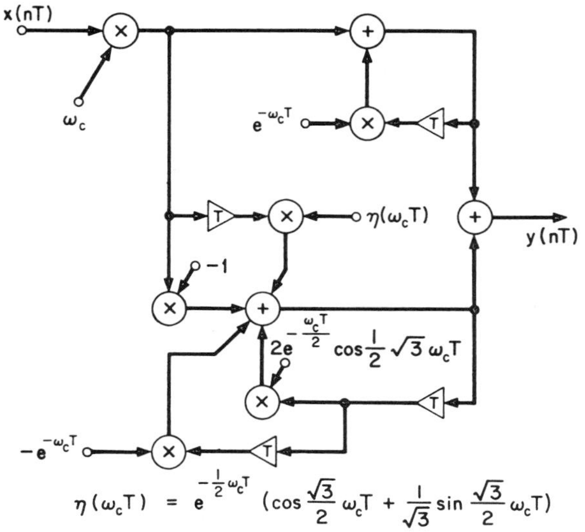

Fig. 9. Impulse invariant three-pole Butterworth lowpass filter.

The Butterworth filter can be specified by the relationship

$$|F(j\omega)|^2 = \frac{1}{1 + \left(\dfrac{\omega}{\omega_c}\right)^{2n}} \tag{44}$$

where ω_c is a cutoff frequency and $|F(j\omega)|^2$ is the squared magnitude of a filter transfer function. The poles of (44) lie equally spaced on a circle of radius ω_c in the s plane, as shown in Fig. 11. For n odd, there will be poles at angles of 0 and π; for n even, the first pole (beginning at angle 0) occurs at an angle of $\pi/2n$. It can be shown that the desired transfer function $F(s)$ is a rational function with unity numerator and denominator determined by the left-half plane poles of Fig. 11. Plots of $|F(j\omega)|$ of (44) for several

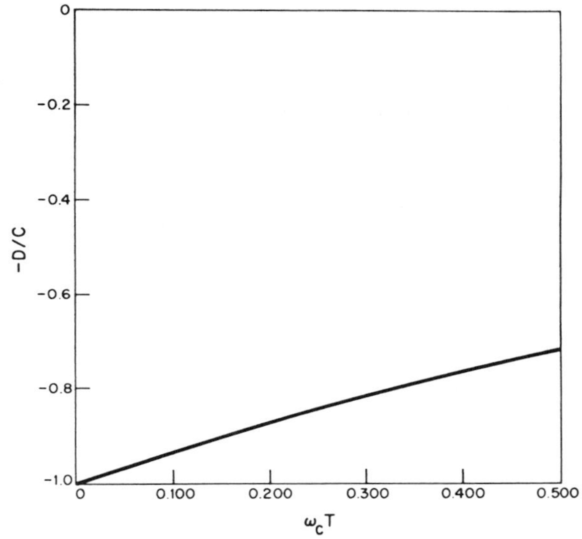

Fig. 10. Graph of D/C showing position of extra zero introduced by impulse invariant design technique.

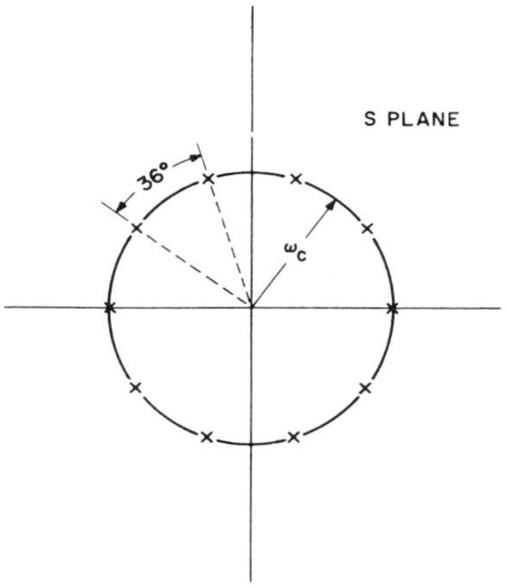

Fig. 11. Poles of Butterworth squared-magnitude function, (44), for $n = 5$. Poles are equally spaced with angular separation 36°. The system function $Y(s)$ is defined by the poles in left-half plane.

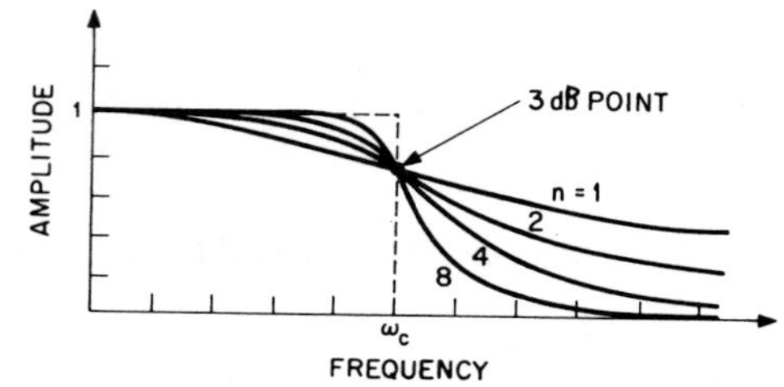

Fig. 12. Magnitude versus frequency of Butterworth filters.

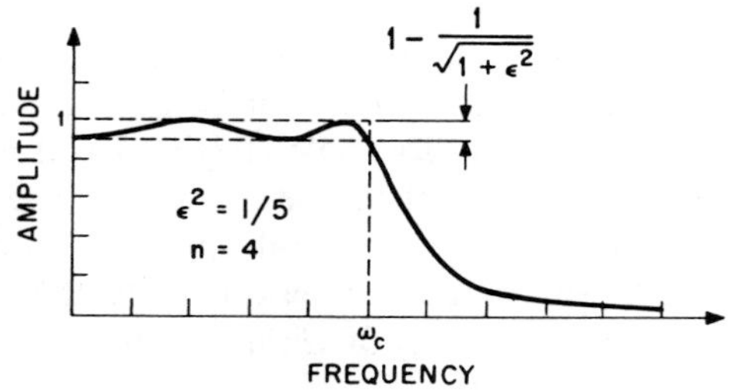

Fig. 13. Magnitude versus frequency of fourth-order Chebyshev lowpass filter.

values of n are shown in Fig. 12. From these plots the selectivity properties of the Butterworth filter become clear.

The Chebyshev filter is specified by

$$|F(j\omega)|^2 = \frac{1}{1 + \varepsilon^2 V_n^2\left(\frac{\omega}{\omega_c}\right)} \tag{45}$$

where $V_n(x)$ is a Chebyshev polynomial of order n which can be generated by the recursion formula

$$V_{n+1}(x) - 2xV_n(x) + V_{n-1}(x) = 0 \tag{46}$$

with

$$V_1(x) = x \quad \text{and} \quad V_2(x) = 2x^2 - 1.$$

The Chebyshev polynomial has the property of equal ripple over a given range, which, with added specification of ε, leads to a magnitude function of a form given by Fig. 13, an equal ripple in the passband and a monotonic decay in the stopband. The ripple amplitude δ is given by

$$\delta = 1 - \frac{1}{\sqrt{1+\varepsilon^2}}. \tag{47}$$

The poles of (45) lie on an ellipse which can be determined totally by specifying ε, n, and ω_c. Figure 14 shows this ellipse, the vertical and horizontal apexes being given by $b\omega_c$, $a\omega_c$, where

$$b, a = \frac{1}{2}\{(\sqrt{\varepsilon^{-2}+1} + \varepsilon^{-1})^{1/n} \pm (\sqrt{\varepsilon^{-2}+1} + \varepsilon^{-1})^{-1/n}\} \tag{48}$$

where the b is given for the plus sign and a for the minus sign. The poles on the ellipse may be geometrically related to the poles of two Butterworth circles of radius $a\omega_c$ and $b\omega_c$. The vertical position of the ellipse pole is equal to the vertical position of the pole on the large circle, while the horizontal position is that of the horizontal position of the small circle pole.

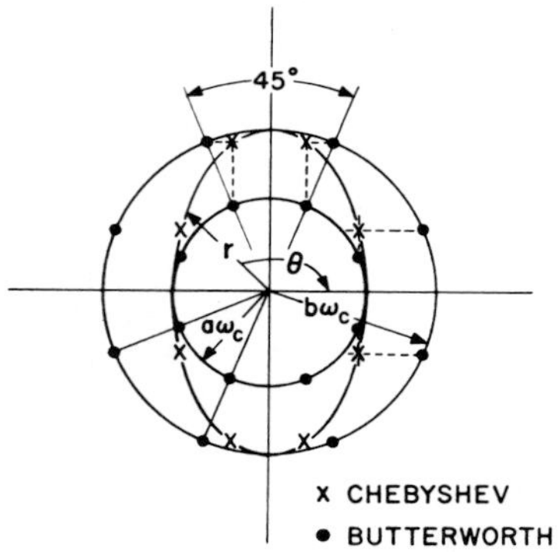

Fig. 14. Poles of Chebyshev squared-magnitude function for $n=4$.

G. Technique 2—Digital Filter Specification from Squared Magnitude Function

We have seen in Section II-F that the Butterworth and Chebyshev filters are specified by choosing suitably selective squared-magnitude functions such as in (44) and (45). The same procedure is possible for digital filters and is described in this section.

Having established in Section I-D that the digital filter system function $H(z)$ is a rational fraction in z^{-1}, it follows that $H(z)$ for z on the unit circle is a rational fraction of $e^{j\omega T}$. Thus, the squared magnitude $|H(e^{j\omega T})|^2$ can always be expressed as the ratio of two trigonometric functions of ωT [27].

Two examples of squared magnitude functions suitable for low-pass filtering are

$$|H(e^{j\omega T})|^2 = \frac{1}{1 + \dfrac{\tan^{2n}\dfrac{\omega T}{2}}{\tan^{2n}\dfrac{\omega_c T}{2}}} \tag{49a}$$

and

$$|H(e^{j\omega T})|^2 = \frac{1}{1 + \dfrac{\sin^{2n}\dfrac{\omega T}{2}}{\sin^{2n}\dfrac{\omega_c T}{2}}}. \tag{49b}$$

The equations are plotted in Fig. 15(a) and 15(b) for $\omega_c T = \pi/2$ and for several values of n. The curves obtained are similar to those of the Butterworth filter of (44) plotted in Fig. 12. The cutoff frequency ω_c plays the same role in both the continuous and digital cases.

Working with (49a), letting $z = e^{j\omega T}$, we obtain

$$|H(z)|^2 = \frac{\tan^{2n}\dfrac{\omega_c T}{2}}{\tan^{2n}\dfrac{\omega_c T}{2} + (-1)^n\left(\dfrac{z-1}{z+1}\right)^{2n}}. \tag{50}$$

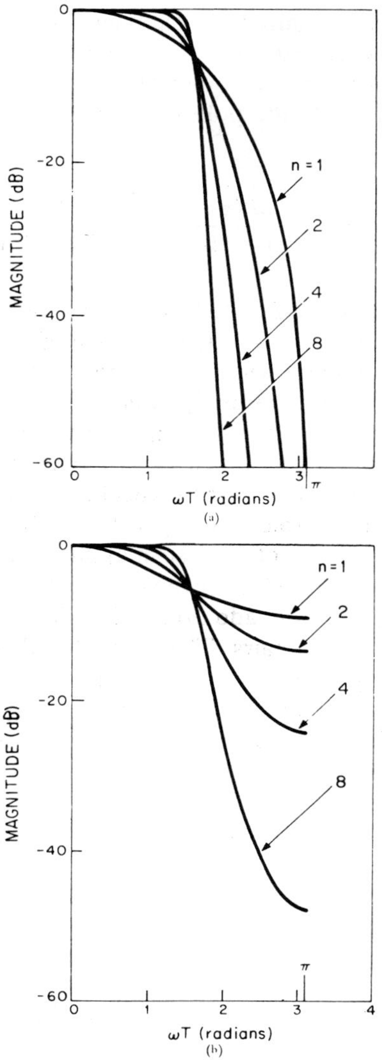

Fig. 15. Magnitude versus frequency of frequency invariant digitized Butterworth filters. (a) Form of (49a). (b) Form of (49b).

We see that (50) is a rational fraction in z which has a zero of order $2n$ at $z=-1$. The poles are found by substituting, in (50),

$$p = \frac{z-1}{z+1} \tag{51}$$

from which we can ascertain that the $2n$ poles of p are uniformly spaced around a circle of radius $\tan \omega_c T/2$. The poles of z are then readily found by the transformation inverse to (51), namely

$$z = \frac{1+p}{1-p}. \tag{52}$$

Letting $p=x+jy$ and $z=u+jv$, we find from (52) the component relations

$$u(x, y) = \frac{1-x^2-y^2}{(1-x)^2+y^2}; \quad v(x, y) = \frac{2y}{(1-x)^2+y^2}. \tag{53}$$

The circle containing the poles in the p plane satisfies the equation

$$x^2 + y^2 = \tan^2 \frac{\omega_c T}{2}. \tag{54}$$

From (53) and (54) it can be shown that the circle of (54) maps into a circle in the z plane centered at (u, v) with radius ρ

$$\left.\begin{aligned} u &= \frac{1+\tan^2 \frac{\omega_c T}{2}}{1-\tan^2 \frac{\omega_c T}{2}} = \sec \omega_c T, \quad v = 0 \\ \rho &= \frac{2\tan^2 \frac{\omega_c T}{2}}{1-\tan^2 \frac{\omega_c T}{2}} = \tan \omega_c T \end{aligned}\right\}. \tag{55}$$

For odd values of n, the $2n$ poles of p have the x and y coordinates

$$\left.\begin{aligned} x_m &= \tan \frac{\omega_c T}{2} \cos \frac{m\pi}{n} \\ y_m &= \tan \frac{\omega_c T}{2} \sin \frac{m\pi}{n} \end{aligned}\right\} m = 0, 1, \cdots, 2n-1. \tag{56}$$

For even values of n, the coordinates are

$$\left.\begin{aligned} x_m &= \tan \frac{\omega_c T}{2} \cos \frac{2m+1}{2n} \pi \\ y_m &= \tan \frac{\omega_c T}{2} \sin \frac{2m+1}{2n} \pi \end{aligned}\right\} m = 0, 1, \cdots, 2n-1. \tag{57}$$

From (53) and (56) the corresponding poles in the z plane are computed to be

$$\left.\begin{aligned} u_m &= \frac{1-\tan^2 \frac{\omega_c T}{2}}{1 - 2\tan \frac{\omega_c T}{2} \cos \frac{m\pi}{n} + \tan^2 \frac{\omega_c T}{2}} \\ v_m &= \frac{2\tan \frac{\omega_c T}{2} \sin \frac{m\pi}{n}}{1 - 2\tan \frac{\omega_c T}{2} \cos \frac{m\pi}{n} + \tan^2 \frac{\omega_c T}{2}} \end{aligned}\right\} m = 0, 1, \cdots, 2n-1. \tag{58}$$

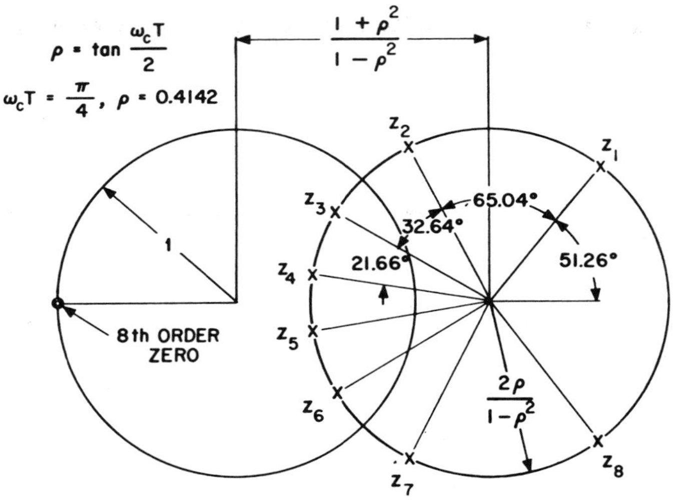

Fig. 16. Poles and zeros of squared-magnitude function for example of Section II-G.

Replacing

$$\frac{m\pi}{n} \text{ by } \frac{2m+1}{2n}\pi$$

yields equivalent formulas for n even.

Example: Find the poles and zeros of the squared magnitude function of a lowpass filter with 3 dB attenuation at 1250 Hz and with at least 20 dB attenuation at 2000 Hz. Let the sampling rate be 10 000 Hz.

The cutoff frequency of 1250 Hz corresponds to $\omega_c T = 45°$. The frequency 2000 Hz corresponds to $\omega T = 72°$. The squared-magnitude function (49a) becomes

$$|H(e^{j\omega T})|^2 = \frac{1}{1 + \dfrac{\tan^{2n}\dfrac{\omega T}{2}}{\tan^{2n}\dfrac{\pi}{8}}}. \tag{59}$$

The appropriate value of n in (59) is $n = 4$, obtained by setting $\omega T = 72°$ and $|H(e^{j\omega T})|^2$ to 0.01, thus satisfying the 20 dB attenuation condition. The eight poles in the p plane are found from (57). Equation (58) can now be used to find the z-plane poles shown in Fig. 16, which also shows the $2n$ zeros located at $z = -1$, which are directly derivable from (50). The squared-magnitude function is thus completely specified as pole-zero placements in the z plane.

Problem: Find the pole-zero description of the lowest order ($n=1$) filter of the form of (49b).

Let

$$\sin^2\frac{\omega_c T}{2} = F$$

$$|H(e^{j\omega T})|^2 = \frac{F}{F + \sin^2\dfrac{\omega T}{2}}$$

$$|H(z)|^2 = \frac{-4Fz}{(z-1)^2 - 4Fz}.$$

The squared-magnitude function has a zero at $z=0$, which can be ignored since it does not contribute to the magnitude response. There are two poles at $z = 1 + 2F \pm 2F\sqrt{1 + 1/F}$. To form $H(z)$, only the pole inside the unit circle is kept, giving

$$H(z) = \frac{2F(1 - \sqrt{T + 1/F})}{z - (1 + 2F - 2F\sqrt{T + 1/F})}.$$

For a given order n, (49a) always results in a sharper lowpass filter than (49b). The general case of (49b) is worked out in some detail in Holtz and Leondes [27].

In the Appendix, an analysis is made of the necessary relations between an assumed squared-magnitude function and the digital filter specified by that function. It is shown that, in order for a filter to be realizable, any pole inside the unit circle (for example z_4 in Fig. 16), must have a mate outside the unit circle of inverse magnitude and the same angle. Thus, if $z_4 = re^{j\psi}$, there must be a pole (in this case z_8) given by $(1/r)e^{j\psi}$. In addition, all poles must occur in complex conjugate pairs. Therefore, the digital filter derived from Fig. 16 has the conjugate poles z_4, z_5 and z_3, z_6.

The above argument holds for the zeros as well, including the special case of zeros on the real axis. In Fig. 16, all eight zeros occur at $z = -1$. The derived filter has four zeros at $z = -1$.

If the squared-magnitude function is given by

$$|H(e^{j\omega T})|^2 = \frac{1}{1 + \varepsilon^2 V_n^2\left[\dfrac{\tan\dfrac{\omega T}{2}}{\tan\dfrac{\omega_c T}{2}}\right]} \tag{60}$$

then it can be shown that the poles of $p = (z-1)/(z+1)$ lie on an ellipse in the p plane which has the same properties as the Chebyshev ellipse of Fig. 14. Using the notation of Section II-F and Fig. 14, the p-plane components can be written

$$x = a\tan\frac{\omega_c T}{2}\cos\theta$$
$$y = b\tan\frac{\omega_c T}{2}\sin\theta. \tag{61a}$$

Substituting (61a) into (53) yields

$$\left.\begin{aligned} u &= \frac{2\left(1 - a\tan\dfrac{\omega_c T}{2}\cos\theta\right)}{\left(1 - a\tan\dfrac{\omega_c T}{2}\cos\theta\right)^2 + b^2\tan^2\dfrac{\omega_c T}{2}\sin^2\theta} - 1 \\ v &= \frac{2b\tan\dfrac{\omega_c T}{2}\sin\theta}{\left(1 - a\tan\dfrac{\omega_c T}{2}\cos\theta\right)^2 + b^2\tan^2\dfrac{\omega_c T}{2}\sin^2\theta}\end{aligned}\right\}. \tag{61b}$$

Figure 17 shows the z-plane mapping for $a \tan \omega_c T/2 = 0.5$ and $b \tan \omega_c T/2 = 1$. The ellipse of Fig. 14 maps into the cardioid-like curve of Fig. 17 and the inner circle of Fig. 14 maps into the right-hand circle of Fig. 17. The outer circle of Fig. 14 maps into a circle of infinite radius, shown by the straight line of Fig. 17. The points shown on the mapped ellipse of Fig. 17 are the actual computed points from (61b).

H. Technique 3—Design of Digital Filters using Bilinear Transformation of Continuous Filter Functions

The previous section has shown how the poles and zeros of a digital filter with a suitable squared magnitude function can be found. This was completely analogous to the mathematics of analog filter design, which has reached a considerable degree of sophistication. For many digital filter design problems, the substitution of (51) transforms the digital filter design problem into a problem which can be recognized as identical to an already solved analog filter design problem. For example, (51) transformed the problem of finding the pole locations for (49a) and (60) into the already-solved problems of finding the pole locations of analog Butterworth and Chebyshev filters. This observation leads to another digital filter design approach, using design directly in the s plane.

Suppose we have a stable analog filter described by $H(s)$. Its frequency response is found by evaluating $H(s)$ at points on the imaginary axis of the s plane. If in the function $H(s)$, s is replaced by a rational function of z which maps the imaginary axis of the s plane onto the unit circle of the z plane, then the resulting $H'(z)$, evaluated along the unit circle, will take on the same set of values as $H(s)$ evaluated along the imaginary axis.

Of course, this does not mean the functions are the same, for the frequency scales are distorted relative to one another. This is illustrated by the simplest rational function[2] which maps the $j\omega$ axis onto the unit circle,

$$s \to \frac{z-1}{z+1}. \tag{62}$$

Let the analog frequency variable be ω_A, and let the digital frequency variable be $\omega_D T$. Then the functions $H(\omega_A)$, $H'(\omega_D T)$ take on the same values for

$$\omega_A = \tan \frac{\omega_D T}{2}. \tag{63}$$

Note that the transformation (62) leads to a ratio of polynomials in z. Since it maps the left half of the s plane onto the inside of the unit circle, we can be sure that it will always yield for $H'(z)$ a realizable, stable digital filter.

Equations (62) and (63) yield a technique for designing a digital filter by analog techniques. The procedure is as follows.

1) Note the critical frequencies and ranges (passband or stopband, maximum attenuation point, etc.) of the desired digital filter, and call them $\omega_{D_i} T$. Compute a new set of frequencies ω_{A_i} by

$$\omega_{A_i} = \tan \frac{\omega_{D_i} T}{2}. \tag{64}$$

2) Design a transfer function $H(s)$ with the properties of the digital filter at the new frequencies and ranges. There is, of course, no need to synthesize $H(s)$.

3) Replace s by $(z-1)/(z+1)$ in $H(s)$, and perform the algebra necessary to express the resulting $H'(z)$ as a ratio of polynomials—this yields the desired digital filter.

[2] This is like (51) except that here we are identifying the p plane with the s plane.

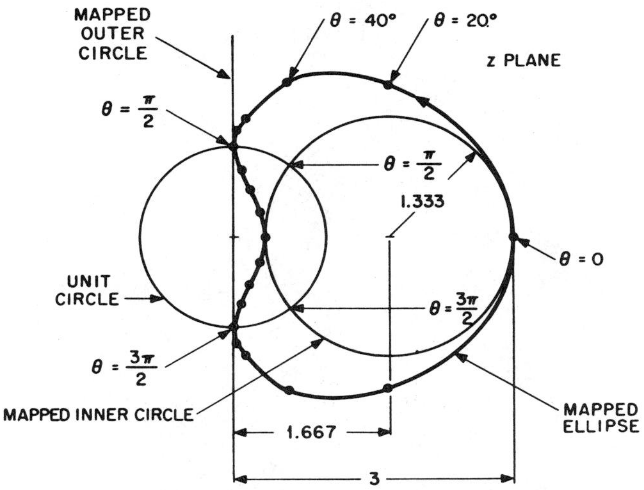

Fig. 17. Bilinear mapping of Chebyshev ellipse into z plane for $a \tan \omega_c T/2 = 0.5$, $b \tan \omega_c T/2 = 1$. Points shown are mapped from 20-degree increments of θ.

This technique is illustrated by the following example in which we design a digital filter for a 10 kHz sampling rate, which is flat to 3 dB in the passband of 0 to 1000 Hz, and which is more than 10 dB down at frequencies beyond 2000 Hz. The filter must be monotonic in passbands and stopbands.

From Fig. 12, we see that a Butterworth filter meets the above requirement in the analog domain. The critical frequencies are $\omega_{D_1} T = 2\pi \times 0.1$ and $\omega_{D_2} T = 2\pi \times 0.2$.

1) Compute ω_{A_1}, ω_{A_2}.

$$\omega_{A_1} = \tan \frac{2\pi \times 0.1}{2} = 0.3249$$

$$\omega_{A_2} = \tan \frac{2\pi \times 0.2}{2} = 0.7265.$$

2) We design a Butterworth filter with 3 dB point at $\omega_c = \omega_{A_1} = 0.3249$. $\omega_{A_2}/\omega_c = 2.235$. To find the order n we solve $1 + (2.235)^{2n} = 10$ and obtain $n = 2$. A second-order Butterworth filter with $\omega_c = 0.3249$ has poles at $s = 0.3249 \times (-0.707 \pm j \cdot 707) = -0.23 \pm j0.23$, and no zeros.

$$H(s) = \frac{s_1 s_2}{(s+s_1)(s+s_2)} = \frac{2 \times 0.23^2}{(s+0.23)^2 + (0.23)^2}$$

$$= \frac{0.1058}{s^2 + 0.46s + 0.1058}.$$

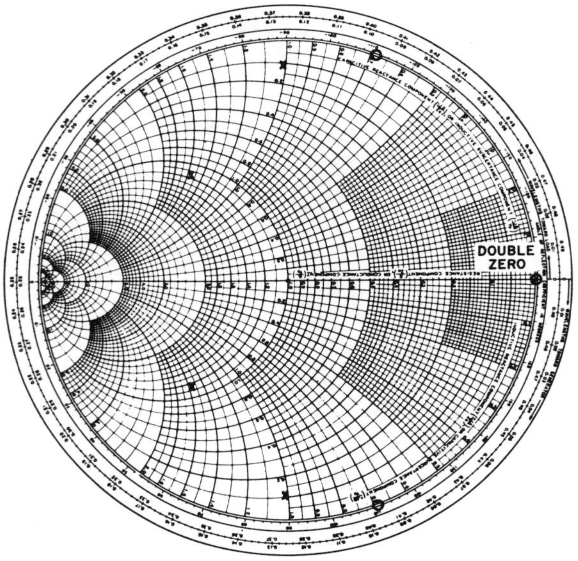

Fig. 18. Poles and zeros of elliptic high-pass filter, from Smith Chart.

3) We replace s by $(z-1)/(z+1)$, yielding $H'(z)$.

$$H'(z) = \frac{0.1058}{\left(\frac{z-1}{z+1}\right)^2 + 0.46\left(\frac{z-1}{z+1}\right) + 0.1058}$$

$$H'(z) = \frac{0.1058(z^2 + 2z + 1)}{1.5658z^2 - 1.7884z + 0.6458}$$

$$= (0.0675)\frac{z^2 + 2z + 1}{z^2 - 1.141z + 0.413}.$$

This is the required digital filter. It requires three multiplications per output point, or if the constant gain factor of 0.0675 can be ignored, only two multiplications.

Problem: Design a digital filter passing from 0 to 100 Hz with $\frac{1}{2}$ dB ripple, which falls off monotonically to at least -19 dB at 183 Hz. Use 1000 Hz sampling rate.

1) The critical frequencies 100 and 183 Hz are transformed to analog frequencies.

$$\omega_c = \tan\frac{2\pi \times 100}{2 \times 1000} = \tan 18^\circ = 0.32492$$

$$\omega_s = \tan\frac{2\pi \times 183}{2 \times 1000} = 0.6498 \approx 2\omega_c.$$

2) We now design an analog filter of the Chebyshev variety. One-half dB ripple corresponds to $\varepsilon^2 = 0.1220184$. To find the required order, we solve $1 + \varepsilon^2 V_n^2(\omega_s/\omega_c) = 10^{1.9}$. The lowest order n satisfying this relationship is $n = 3$.

A unity-bandwidth, $\frac{1}{2}$ dB ripple Chebyshev filter is

$$H_1(s) = \frac{\text{constant}}{s^3 + 1.252913s^2 + 1.5348954s + 0.7156938}.$$

Replacing s by s/ω_c yields

$$H(s) = \frac{0.0255842155}{s^3 + 0.4127346s^2 + 0.16656307s + 0.0255842155}$$

where the constant has been adjusted for unity gain at $s = 0$.

3) We replace s by $(z-1)/(z+1)$, giving for the digital filter desired, after multiplying numerator and denominator by $(z+1)^3$,

$$H'(z) = \frac{0.0255842155(z^3 + 3z^2 + 3z + 1)}{1.60488218z^3 - 3.169418z^2 + 2.44628634z - 0.728243953}$$

which is the required digital filter design.

It is worth noting a useful geometric interpretation of Step 3). Replacing s by $(z-1)/(z+1)$ is a mapping of points in the s plane onto points in the z plane. The following short table gives the correspondence of some critical points in the s and z planes

s plane	z plane
$0+j0$	$1+j0$
∞	$-1+j0$
$0+j1$	$0+j1$
$0-j1$	$0-j1$
$-1+j0$	$0+j0$
point on real axis	point on real axis
point on imaginary axis	point on unit circle
point on any line	point on circle passing through $-1+j0$

A very similar mapping has been found useful in some applications, and graph paper which performs the transformation can be purchased under the name *Smith Chart.* To perform the mapping of our application we take a conventional Smith Chart and rotate it 180°, giving a chart like Fig. 18.

The location of any point $-a+jb$ in the s plane (left half) is used to find the corresponding location in the z plane as follows.

1) Locate a, a point along the centerline. All points along the circle passing through this point correspond to points in the s plane with real part $= -a$.

2) Locate b, a point along the perimeter of the outer circle. For $b>0$ use the top semicircle; for $b<0$ use the lower semicircle. The circular arcs passing through this point correspond to points in the s plane with imaginary part b.

3) The intersection of the circle with the circular arc is the point in the z plane corresponding to $-a+jb$ in the s plane.

The Smith Chart is useful when the function in the s plane is known in terms of poles and zeros or poles and residues, especially when the z-plane representation is desired in that form.

Example: Design of a digital highpass filter with stopband (0–500 Hz) attenuation greater than 36 dB and passband (above 660 Hz) ripple of 1.25 dB. Sampling rate is 2.5 Hz.

1) The critical frequencies are transformed to analog, giving, as the stopband limit, 0.72654 rad/s, and as the passband limit, 1.09 rad/s.

2) The specifications are met with a fourth-order elliptic filter. The design of elliptic filters is covered in the literature [17] and only the result is given here. The poles are at $s = -1.1915812 \pm j1.5528835$, and -0.11078815

$\pm j1.09445605$, and the zeros are at $s=0$ (double), and $\pm j0.69117051$.

These are located on a Smith Chart in Fig. 18. Using a ruler and protractor, the zeros in the z plane are found to be at $z=1$ (double), and $e^{\pm j69.3^\circ}$ and the poles are found to be at $z=0.586e^{\pm j131.6^\circ}$ and $0.895e^{\pm j90.6^\circ}$. The function of z defining the filter is thus

$$H(z) = \frac{(z-1)^2(z^2 - 0.707z + 1)}{(z^2 + 0.777z + 0.3434)(z^2 + 0.01877z + 0.801)}.$$

A prime difficulty with the Smith Chart method, illustrated by this example, is that the poles cannot be located with great accuracy. In compensation, a good deal of insight into the operation of the digital filter is gained in the course of the design. Where greater accuracy is needed, the z-plane poles can be computed from the s-plane poles using

$$z = \frac{1+s}{1-s}$$

which is the inverse of (62).

The transformation $s \to (z-1)/(z+1)$ is not the only rational function of z which maps the imaginary axis of the s plane onto the unit circle. For example, another such transformation is

$$s \to \frac{z^2 - 2z\cos\psi_0 T + 1}{z^2 - 1}. \tag{65}$$

For this transformation, the imaginary axis of the s plane maps onto the top arc of the unit circle, and also onto the bottom arc. The origin of the s plane maps onto the two points $e^{\pm j\psi_0 T}$.

Thus, (65) maps a frequency function $H(\omega_A)$ into $H'(\omega_D T)$ with

$$\omega_A = \frac{\cos\psi_0 T - \cos\omega_D T}{\sin\omega_D T}. \tag{66}$$

This means that (65) transforms a lowpass $H(\omega_A)$ into a bandpass $H'(\omega_D T)$.[3] Equation (65) can thus be used to design bandpass digital filters as follows.

1) Decide on a center frequency $\psi_0 T$ for the digital filter.[4] This may be forced by the specifications or it may be available to simplify the choice of other parameters. Compute the critical frequencies of the desired analog filter from the critical frequencies in the specification of the digital filter using (66). Equation (66) will often yield negative frequencies, which is all right, since $H(\omega_A)$ will be an even function of ω_A.

2) Design an analog filter $H(s)$ with the translated specifications. It is likely that one or more of the specifications will be superfluous.

3) Replace s by

$$s = \frac{z^2 - 2z\cos\psi_0 T + 1}{z^2 - 1}$$

in $H(s)$, and perform the required algebra to manipulate the resulting $H'(z)$ into a ratio of polynomials; this is the required filter.

One example will probably suffice to illustrate the method.

Example: Design a digital bandpass filter for a 1000 Hz sampling rate, to pass 100 to 400 Hz with ripple-free attenuation between 0 and 3 dB. At 45 and 450 Hz, the filter must be at least 20 dB down, and must fall off monotonically beyond both frequencies.

1) $\psi_0 T$ can be chosen so the two analog 3 dB points are negatives of each other. If the digital 3 dB points are L_1 and L_2, we can see from (66) that $\psi_0 T$ should be chosen so that

$$\cos\psi_0 T = \frac{\cos\frac{1}{2}(L_1 + L_2)}{\cos\frac{1}{2}(L_1 + L_2)}. \tag{67}$$

Using (67) yields $\cos\psi_0 T = 0$ for this case. This means that the transformation from digital-to-analog critical frequencies becomes

$$\omega_A = -\cot\omega_D T.$$

The 3 dB points now translate to

$$\omega_{3\text{dB}} = \pm 1.3764.$$

The 20 dB points are not equal in magnitude. They are

$$\omega_{s_1} = -3.442$$
$$\omega_{s_2} = 3.078.$$

2) The problem is now to design a monotonic filter with 0 to 3 dB in the region $0 < \omega < 1.3764$. The filter must be 20 dB down by $\omega = 3.078$, which will automatically satisfy the requirement at $\omega = 3.442$. A Butterworth design seems to be called for, with $\omega_c = 1.3764$.

For $\omega/\omega_c = 2.23$ we demand 20 dB attenuation. We calculate the order n,

$$1 + (2.23)^{2n} = 100.$$

It is clear that $n=3$ will suffice. A unity-bandwidth Butterworth filter of order 3 is

$$H_1(s) = \frac{K}{s^3 + 2s^2 + 2s + 1}.$$

$H(s)$ is found by replacing s by $s/1.3764$,

$$H(s) = \frac{2.6075581}{s^3 + 2.7528s^2 + 3.788954s + 2.6075581}.$$

3) To find the required filter, we let $s = (z^2+1)/(z^2-1)$ in $H(s)$. Note that since this is going to yield a function of z^2, the resulting digital filter will be quite simple.

[3] In fact, (65) was found by preceding the bilinear transformation by the well-known bandpass transformation. Other interesting transformations may be found by combining the bilinear transformation with highpass or bandstop transformations.

[4] All that is ever really needed is $\cos\psi_0 T$.

$$H'(z) = C_1 \frac{z^6 - 3(z^4 - z^2) + 1}{z^6 + C_2 z^4 + C_3 z^2 + C_4}$$

where

$$C_1 = 0.256919688$$
$$C_2 = -0.577263586$$
$$C_3 = 0.421794133$$
$$C_4 = -0.0562997861.$$

The techniques of this section can also be used to design highpass and band-elimination filters.

I. Technique 4—Frequency Sampling Technique

The difference equation

$$y(nT) = x(nT) - x(nT - mT) \tag{68}$$

has the z transfer function $1 - z^{-m}$, which has m zeros equally spaced around the unit circle, at points

$$z_k = e^{j2\pi \frac{k}{m}}; \qquad k = 0, 1, \cdots, m-1. \tag{69}$$

If, in (68), the subtraction is replaced by an addition, the transfer function becomes $1 + z^{-m}$, for which the zeros are also equally spaced around the unit circle, at

$$z_k = e^{j2\pi((k+1/2)/m)}; \qquad k = 0, 1, \cdots, m-1. \tag{70}$$

The magnitude-versus-frequency curves of these filters repeat with period $2\pi/m$ radians, and the filters are often referred to as comb filters. These filters can be incorporated into an especially appealing type of design. Before presenting this design, let us examine the practical significance of (68). If successive values of $x(nT)$ are thought of as contents of registers in computer memory, it is clear that m registers must be set aside for buffer storage in order to execute (68). This is usually a substantial amount of memory compared to that needed for execution of second-order difference equations. Thus possible practical use of the filters we are about to describe is limited to systems where the necessary memory is available or to systems where many filters share a common input. It is important to note that the delays implied by (68) can be effected by digital delay lines, which are presently relatively cheap forms of memory.

A simple resonator can be placed in cascade with the comb filter.[5] Let the resonator be of the type considered in Section II-D, Case 4, except that the poles lie directly on the unit circle. Further let the angle ω_r of a resonator pole be such that the pole is coincident with a zero of the comb filter.

$$\omega_r = \begin{Bmatrix} \dfrac{2\pi k}{m} \\[2ex] \dfrac{2\pi(k+1/2)}{m} \end{Bmatrix} \text{for a comb filter of the} \begin{Bmatrix} \text{first} \\[2ex] \text{second} \end{Bmatrix} \text{type.} \tag{71}$$

The poles of the resonator cancel the kth zero of the comb filter, and its conjugate. In what follows, we shall refer to the resonator which is used to cancel the kth zero as the kth elemental filter.

The cascade of an elemental filter with a comb filter is a composite filter with the following properties.

1) The impulse response is of finite duration, mT.

2) The magnitude-versus-frequency response is

$$|H(e^{j\omega T})| = \left| \frac{\sin m\omega T/2}{\cos \omega T - \cos \omega_r T} \right| \tag{72}$$

which is zero at all the radian frequencies for which the comb filter is zero, except at ω_r. The magnitude at ω_r is $(m/2) \csc \omega_r T$.

3) The phase versus frequency is exactly linear except for discontinuities of π radians. These discontinuities occur where the magnitude response is zero.

4) The phase difference between two composite filters with resonant frequencies ω_r and ω_{r+1} is π for $\omega_r < \omega < \omega_{r+1}$ and is zero outside these bounds.

5) The amplitude of any composite filter is zero at the resonant frequencies of all the other composite filters.

As m is made large, the magnitude response of a cascaded filter becomes like

$$\left[\frac{\sin(\omega - \omega_r)T}{(\omega - \omega_r)T} + \frac{\sin(\omega + \omega_r)T}{(\omega + \omega_r)T} \right]$$

in shape. These properties suggest that any desired magnitude response could be obtained by adding together the weighted outputs of cascaded comb and elemental filters, just as any "bandlimited" time function can be formed from a weighted sum of delayed sin t/t functions. Let us examine this idea, which we call "frequency sampling," in some detail.

A sufficiently "narrowband" frequency response function (one for which the frequency response is a sufficiently smooth function of frequency) is sampled at equally spaced points, with radian frequencies

$$\omega_k = \frac{2\pi k}{m} \text{ or } \frac{2\pi(k+1/2)}{m}; \qquad k = 0, 1, \cdots, m-1$$

depending on which kind of comb filter will be used. Let the sample value of the amplitude at frequency $\omega_k T$ be W_k. An elemental filter of resonant frequency ω_k, cascaded with a comb filter of delay mT and a gain of $W_k \sin \omega_k T$ are used to provide an "elemental frequency response" of W_k at radian frequency $\omega_k T$ and zero at the other sampling frequencies.

Since the phases at resonance of the consecutive elemental filters differ by π, the gains of the odd numbered elemental filters are to be multiplied by -1. The desired input to the filter is applied to the comb filter, which is shared among all the elemental filters, followed by the gains (and sign changers for odd-numbered elemental filters). The outputs of all the elemental filters, with proper gains, are added together to give the desired filter output. The resulting filter has an impulse response of duration mT, a frequency response with linear phase, and an amplitude response which agrees with specifications at the sampling frequencies and

[5] Golden [2] attributes this idea to B. F. Logan.

connects the sampling points smoothly. The variety of filters which can be programmed with this technique is quite large.

There are some practical problems to be considered before the frequency sampling method is applied. For one thing, the resonant poles of an elemental filter cannot exactly cancel the zeros of a comb filter because of quantization. Thus it is wise to move both the zeros of the comb filter and the poles of the elemental filters slightly inside the unit circle, with a radius of something like $e^{-aT}=1-2^{-26}$.

We have successfully programmed filters with the poles and zeros at radii ranging from $1\text{–}2^{-12}$ to $1\text{–}2^{-27}$, with little change in the behavior of the filter.

Another change, while not necessary, is useful to keep in mind for bandpass filters. In the passband, it is common for the samples W_k to be equal. Thus it would be desirable if all the elemental filters had the same gains at resonance. If a zero is put at $(\cos \omega_k T)e^{-aT}$, the gain of each elemental in cascade with the comb filter becomes $m/2$. This has a slight effect on the magnitude and phase response, negligible for large m. The modified comb filter z transform thus becomes

$$H(z) = 1 - e^{-maT}z^{-m} \tag{73}$$

and the modified kth elemental filter becomes

$$H_k(z) = \frac{1 - e^{-aT}\cos \omega_k T z^{-1}}{1 - 2e^{-aT}\cos \omega_k T z^{-1} + e^{-2aT}z^{-2}}. \tag{74}$$

It is worth noting that the introduction of the additional zero does not require another multiplication since twice the numerator coefficient is also present in the denominator. The response of a filter of the form of (73) and (74) is shown in Fig. 19.

Example, Bank of Bandpass Filters: It was desired to design a bank of bandpass filters (with a common input), each 400 Hz wide, covering the band 300 Hz to 3100 Hz. The filters are to be as selective as possible but with minimum ringing time. A further requirement is that the contiguous filters cross at -3 dB of the midband gain. None of the standard designs had satisfactory selectivity combined with short ringing.

The filters chosen were frequency sampling filters, each composed of seven elemental filters. The consecutive zeros were 100 Hz apart. Since the sampling rate was 12.5 kHz, m was 125. The general form of such a filter has z transform

$$H(z) = (1 - e^{-maT}z^{-m})$$

$$\sum_{k=r}^{r+6} (-1)^k \frac{1 - e^{-aT}\cos \dfrac{2\pi k}{m} z^{-1}}{1 - 2e^{-aT}\cos \dfrac{2\pi k}{m} z^{-1} + e^{-2aT}z^{-2}} W_k. \tag{75}$$

The design of the filter is completely specified by choosing r and the set of W_k in (75). Since 3 dB crossovers 400 Hz apart were required, W_{r+1} and W_{r+5} were chosen to be 0.707. The three center terms had $W_k=1$. The end-term

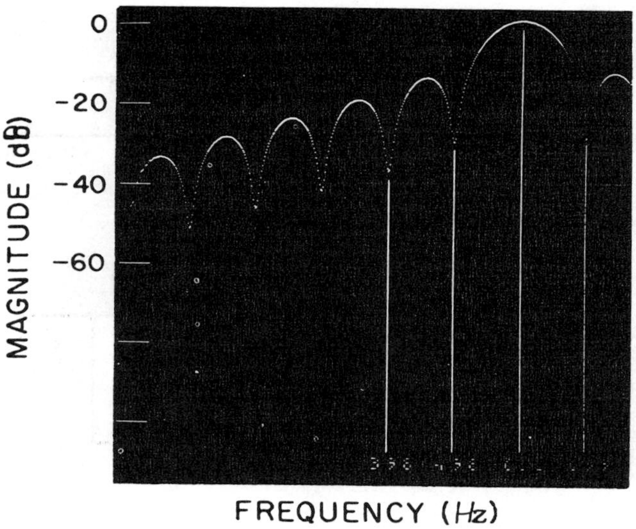

Fig. 19. Response of elemental filter of (74).

gains, W_r and W_{r+6} were found empirically to be 0.221 for satisfactory out-of-band rejection.

The design of the 300 to 700 Hz filter is illustrated in Fig. 20, and the experimental frequency responses of it and the next higher filter (700 to 1100 Hz) are shown in Fig. 21.

The frequency sampling technique is especially satisfactory for designing such banks of bandpass filters, since, if the same elemental filter is a component of several bandpass filters, it may be shared among them. In the above example only thirty-one elemental filters were needed for seven bandpass filters of seven elemental filters each.

Section III

Quantization Noise of Digital Filters

When a digital filter is realized with a digital arithmetic element, as is the case with a computer, additional considerations are necessary to describe the performance of the filter. There are three obvious degradations: 1) quantization of the coefficients of the difference equation, 2) quantization of the input, and 3) quantization of the results of computations. The first two effects, although important, are simple to understand. Quantization of the coefficients changes them to slightly different coefficients. This happens once and for all, resulting in a new, slightly different filter [24]. For very high Q filters this may be important; even instability may result. Quantization of the input is equivalent to adding a noiselike term which has been well described in the literature. We rely heavily on the results of Bennett [18] in treating this and similar effects of quantization in a statistical manner.

Quantization of results of computation is more complicated because these results are used in later computations, and thus the effect of the error on future computations must be understood [25]. Let us first note that quantization of the results cannot be avoided because after each iteration of the difference equation, the number of bits required for exact representation of the result increases by about as many bits as in the representation of the coefficients. Thus,

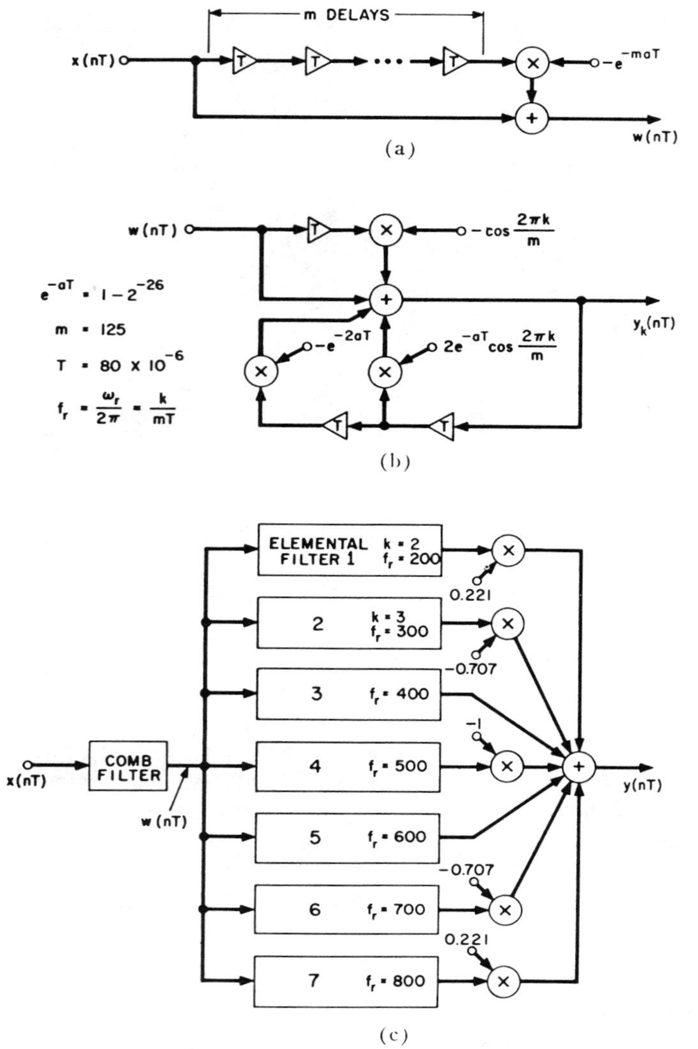

Fig. 20. Representation of frequency sampling filter with bandwidth from 300 to 700 Hz. (a) Comb filter. (b) Elemental filter. (c) Complete filter.

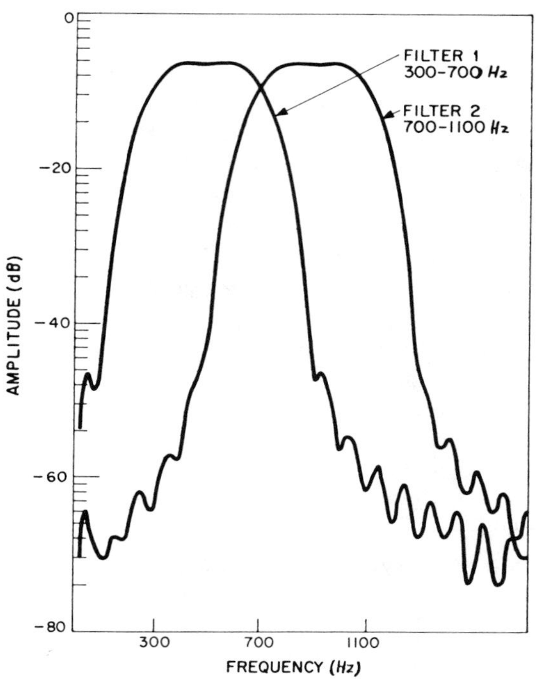

Fig. 21. Experimental amplitude versus frequency for contiguous bandpass filters using frequency sampling techniques.

without quantization, the number of bits required would grow without bound.

In the following, we define quantization as the replacement of the exact value of a quantity by the value of the nearest of a set of levels differing by steps of E_0. Thus, the maximum error introduced by a quantization is $E_0/2$, and, except for a few degenerate cases, the mean-squared error is $E_0^2/12$.

Figure 22 shows a representation of a digital filter in which the output is quantized before being used in further iterations.

We can write a "linear" difference equation describing this device if we represent the quantization error by $v(nT)$.

$$r(nT) = v(nT) + \sum_{i=0}^{N} a_i x(nT - iT) + \sum_{i=1}^{M} b_i r(nT - iT). \quad (76)$$

This equation has a z transform form

$$R(z) = V(z) = \sum_{i=0}^{N} a_i z^{-i} X(z) + \sum_{i=1}^{M} b_i z^{-i} R(z) \quad (77)$$

$$R(z) = \frac{V(z) + \sum_{i=0}^{N} a_i z^{-i} X(z)}{1 - \sum_{i=1}^{M} b_i z^{-i}}. \quad (78)$$

We notice that the z transform of the output consists of two terms, one of which is the desired output. The other term depends only on the poles of the filter and on the noise sequence $v(nT)$.

In what follows, assume that the input to the filter is of such level that the output with no quantization noise has a mean-squared value of unity. Assume that this output is large compared to E_0, so that the samples $v(nT)$ are independent and uniformly distributed from $-E_0/2$ to $E_0/2$. Let us call the denominator of the right side of (78) $Q(z)$.

Our aim is to determine the mean-squared value of the noise term of (78) in terms of E_0, the quantization step size.

For this determination, the two-sided z transform [13] of the autocorrelation function of a sequence is needed. If a one-sided sequence has z transform $A(z)$, then its autocorrelation function has a two-sided z transform $A(z)A(1/z)$. It follows that a sequence with z transform $V(z)/Q(z)$ has an autocorrelation function whose z transform is the ratio of the z transforms of the autocorrelation functions of $v(nT)$ and $q(nT)$.

We are really interested in a normalized autocorrelation function,

$$\lim_{k\to\infty} \frac{1}{k} \sum_{i=0}^{k-1} g(iT)g(nT + iT)$$

which, for $v(nT)$, is the sequence

$$A_v(nT) = 0, \cdots, 0, E_0^2/12, 0, \cdots, 0, \cdots$$

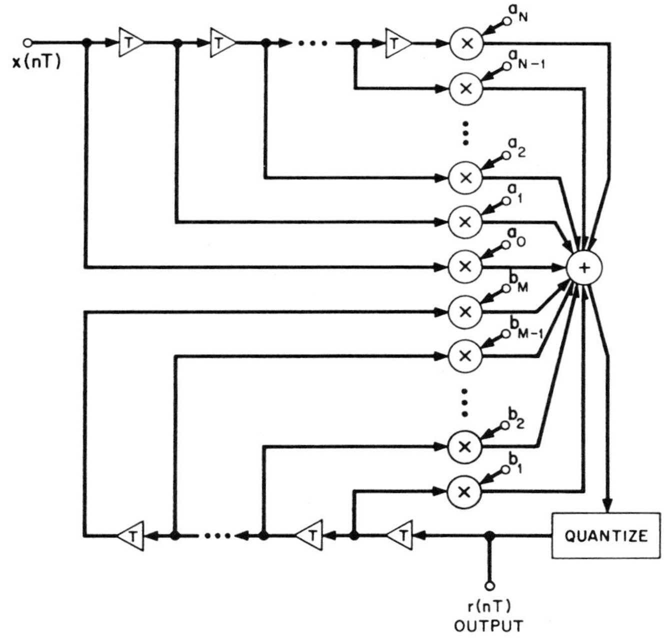

Fig. 22. Digital filter with quantization error.

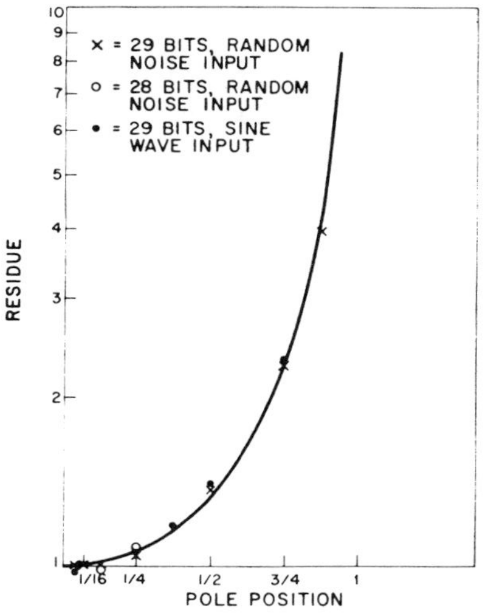

Fig. 23. Quantization noise for one-pole filter.

with z transform $A_v(z) = E_0^2/12$.

For the noise output, the normalized autocorrelation function has z transform

$$\frac{E_0^2/12}{Q(z)Q\left(\frac{1}{z}\right)}.$$

The mean-squared output noise (MSON) is, of course, given by the coefficient of z^0 in the z transform of the autocorrelation function of the output noise. For a two-sided z transform, this coefficient is given by

$$\text{MSON} = \frac{E_0^2}{12}\left(\sum \text{residues of } \frac{1}{zQ(z)Q\left(\frac{1}{z}\right)}\right) \tag{79}$$

where only the residues at poles inside the unit circle are to be included in the sum.

Example: Consider a one-pole filter with $H(z) = 1/(1-az^{-1})$. Then $1/zQ(z)Q(1/z)$ has a pole at $z=a$ and a residue of $1/(1-a^2)$ at that pole. The mean-squared output noise for a digital realization of this filter is proportional to this residue, which is plotted versus pole position in Fig. 23, with the constant of proportionality being $E_0^2/12$.

Two filters, one using 36-bit arithmetic and one using a variable number of bits fewer than 36, were simulated, with a common input to both filters, and with $H(z)$ as above. The 36-bit case was considered unquantized and the difference in the filter outputs was thus a good approximation to the quantization noise introduced by the less precise filter. The mean-squared value of this noise was measured for a random noise input with 28- and 29-bit arithmetic, and for a sine wave input at about 0.1 of the sampling frequency, using 29-bit arithmetic. The measured results, normalized with respect to $E_0^2/12$, are shown along with the theoretical result in Fig. 23.

We have arrived at a very fortunate expression since we

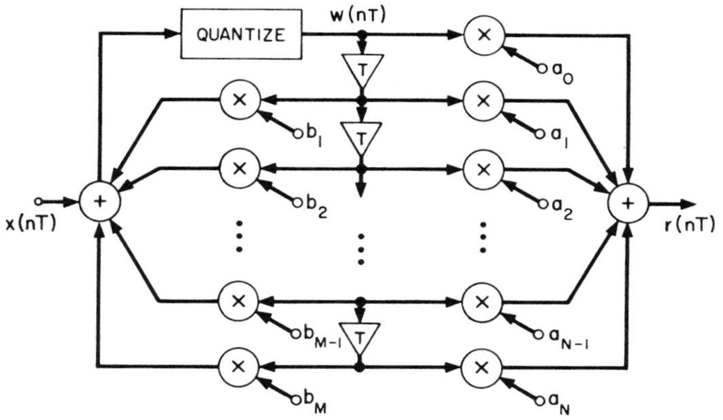

Fig. 24. Canonic representation of digital filter with quantization (drawn for $M = N$).

can quickly solve it for $(-\log_2 E_0)$ which is the number of bits of quantization needed below the output signal level, for a given amount of mean-squared output noise. Say we require that the mean-squared output noise be F dB below a unit output signal level. Then

$$F = 10\log_{10}\left\{\frac{E_0^2}{12}\left(\sum \text{residues}\right)\right\} \tag{80}$$

$$B_{-1} = -\log^2 E_0 = -0.166F - 1.79 + 1.66\log_{10}\left(\sum \text{residues of } \frac{1}{zQ(z)Q\left(\frac{1}{z}\right)}\right). \tag{81}$$

B_{-1} is the number of bits which must be retained below the unit signal level; an additional 3 to 5 bits ought to be retained above the unit signal level to protect against overflow.

The preceding description of quantization noise is valid for digital filters realized in the form of Fig. 1. If the realization is in the canonic form of Fig. 4, then quantization noise is added to $w(nT)$ rather than to $r(nT)$, since it is $w(nT)$ which is used in further computations. Figure 24 shows the digital arrangement of the canonic form. We write the two difference equations

$$w(nT) = x(nT) + \sum_{i=1}^{M} b_i w(nT - iT) + v(nT) \tag{82}$$

$$r(nT) = \sum_{i=0}^{N} a_i w(nT - iT). \tag{83}$$

These may be z transformed, giving

$$W(z) = (X(z) + V(z))/Q(z) \tag{84}$$

$$R(z) = P(z)W(z) \tag{85}$$

where

$$P(z) = \sum_{i=0}^{N} a_i z^{-i}.$$

Combining these into a single equation

$$R(z) = X(z)\frac{P(z)}{Q(z)} + V(z)\frac{P(z)}{Q(z)} \tag{86}$$

we see that in the canonic representation the quantization noise is filtered by both the poles and the zeros of the filter. If desired, the procedures leading to (79), (80), and (81) could be repeated, giving as the mean-squared output noise

$$\text{MSON} = \frac{E_0^2}{12}\sum \text{residues of}\left(\frac{P(z)P\left(\frac{1}{z}\right)}{zQ(z)Q\left(\frac{1}{z}\right)}\right) \tag{87}$$

and giving the number of bits required below a unit signal level as

$$B_{-1} = -0.166F - 1.79 + 1.66\log_{10}\left(\sum \text{residues of}\frac{P(z)P\left(\frac{1}{z}\right)}{zQ(z)Q\left(\frac{1}{z}\right)}\right). \tag{88}$$

From (79) and (87), explicit formulas for the mean-square error were derived for both the "direct" and "canonic" realization of a digital filter with z transform given by the right side of (28). The results are, for the direct realization

$$\text{MSON} = \frac{E_0^2}{12}\left\{\frac{1 + r^2}{(1 - r^2)(r^4 + 1 - 2r^2\cos 2bT)}\right\} \tag{89}$$

and for the canonic realization

$$\text{MSON} = \frac{E_0^2}{12}\left[\frac{1}{1 - r^2} - r^2\sin^2 bT\left[\frac{1 + r^2}{(1 - r^2)(r^4 + 1 - 2r^2\cos 2bT)}\right]\right]. \tag{90}$$

In both (89) and (90), r is equal to the term e^{-aT} of (28). If r is close to unity, so that it is possible to let $r = 1 - \varepsilon$, ε being a very small number, the right sides of (89) and (90) reduce, respectively, to $E_0^2/(48\varepsilon\sin^2 bT)$ and $E_0^2/48\varepsilon$. Thus, for low values of the radian resonance frequency bT, the canonic form yields substantially lower noise.

Both (89) and (90) have been verified experimentally, using the technique already described for the one-pole case.

When one attempts to apply (79) and (87) to systems for which the poles fall on the unit circle, the result obtained is an infinite mean-squared output noise. Recalling that (79) and (87) are steady-state formulas, this result is reasonable and tells us, for example, that digital computer programs for generating sine and cosine waves will eventually yield very noisy outputs. Our primary concern, however, is with the time variation of the noise statistics rather than the steady-state result, since one is interested in knowing how long the program will run before the results become unusable. We shall first derive a time-varying mean-squared output noise for a generalized digital filter $H(z)$ [impulse response $h(nT)$] and then apply this result to estimate the length of life of a sine-cosine generator.

Letting the input to $H(z)$ be $e(nT)$ and the output be $f(nT)$, the convolutional relation between the two is

$$f(nT) = \sum_{m=0}^{n} h(mT)e(nT - mT). \tag{91}$$

Squaring both sides of (91), taking the statistical means of the squares, and assuming that successive samples of $e(nT)$ are independent with zero mean and mean-square value, $E_0^2/12$ results in

$$\text{MSON} = E[f^2(nT)] = \frac{E_0^2}{12}\sum_{m=0}^{n} h^2(mT). \tag{92}$$

Equation (92) allows the computation of the MSON as a function of nT. Letting n approach infinity will cause (92) to converge to (79) or (87), depending on the realization.

Let us apply (92) to the sine-cosine generator represented by the pair of equations

$$y(nT + T) = \cos bTy(nT) + \sin bTx(nT) + e_1(nT)$$
$$x(nT + T) = -\sin bTy(nT) + \cos bTx(nT) + e_2(nT) \tag{93}$$

with initial conditions $x(o) = 1$, $y(o) = 0$. The noises $e_1(nT)$ and $e_2(nT)$ are assumed to be the roundoff caused by the combined effect of the two multiplications in the top and bottom equation of (93), respectively. If we further assume that $e_1(nT)$ and $e_2(nT)$ are independent random variables, each with a zero mean and a mean-square of $E_0^2/12$, we obtain, after some manipulation,

$$\text{MSON} = \frac{E_0^2}{12}n. \tag{94}$$

Equation (94) holds for either output $x(nT)$ or $y(nT)$, and shows that the noise increases linearly with the number of iterations of the difference equations. Thus, for example, a million iterations would result in a standard deviation of the noise of about 10 bits. Assuming that one iteration is performed in 50 microseconds, several hours would pass before the outputs would be noisy, if, for example, a 36-bit word length were used. Thus, it is clear that for most practical problems, (93) is perfectly acceptable for generating sines and cosines.

If an alternate iteration

$$y(nT+2T) = 2\cos bTy(nT+T) - y(nT),$$
$$y(o) = 1, \qquad y(T) = \cos bT \tag{95}$$

is used to generate a cosine wave, then application of (92) causes the mean-square output noise to be of the form $n/\sin^2 bT$ and again we have the phenomenon that a different programming technique will lead to a much larger noise for low frequencies.

Conclusions

The z-transform calculus is the mathematical basis for the analysis and design of digital filters. Such filters are best understood by emphasizing the relations between the difference equations, the pictorial representation or block diagram, the z-plane geometry, and the frequency response function.

Several formal bases for digital filter design have been presented. For the cases of design techniques using either the standard z transformation or the bilinear transformation, advantage is taken of available material in the field of analog filter design. In these cases, it must be kept in mind that the digital filter can never be exactly the same as the corresponding analog filter, although the two may be made sufficiently equivalent in many practical situations by choosing the appropriate sampling rate.

Digital filter design based on either assumed squared-magnitude functions or assumed frequency response curves can be realized by the techniques demonstrated in this paper. The latter technique, called "frequency sampling" has the added advantage of yielding linear phase and involves little computation.

The two crucial design parameters for digital filters are the sampling rate and the register lengths. The former is best determined by consideration of the design techniques as enumerated in the previous paragraph. The latter must be derived primarily from filter gain and quantization noise considerations. When quantization error is small compared to register length, a useful and simple model of the noise generated by quantization of the signals is derivable, which in many practical cases leads to specification of required word lengths.

Appendix

Realizability of Digital Filters with Certain Squared-Magnitude Functions

In this Appendix, we shall find it useful to define polynomials of the form

$$a_0 z^k + a_1 z^{k-1} + \cdots + a_1 z + a_0,$$

such that the coefficient of z^r is equal to the coefficient of z^{k-r}, as mirror-image polynomials of order k, and to summarize some of the properties of the polynomials.

1) A mirror-image polynomial (MIP) of order k has roots which occur in reciprocal pairs for k even.

Proof: Let z be replaced by $1/z$ in the original MIP and equate it to zero.

$$a_0 z^{-k} + a_1 z^{-k+1} + a_1 z^{-1} + a_0 = 0.$$

If we multiply by z^k, which introduces no new roots to the equation, we see that the roots of the polynomial in $1/z$ are the same as the roots of the polynomial in z. Thus, the roots either occur in reciprocal pairs, or are self-reciprocals. For k odd, one of the roots cannot be part of a reciprocal pair. This root is its own reciprocal, and since it must be real, it is either $+1$ or -1. We will generally not be interested in odd-order MIP's.

2) The sum of MIP's of the same order is an MIP of that order.

3) An MIP of order k plus z^r times an MIP of order $(k-2r)$ is an MIP of order k.

4) The product of MIP's is an MIP.

5) A polynomial whose roots occur in reciprocal pairs is an MIP. Step 5) is proved by multiplying together the factors which are reciprocals, noting that each product equals an MIP, and applying step 4).

We shall use the above properties of MIP's to prove the existence of digital filters with certain squared-magnitude functions. Suppose a squared-magnitude function $F(\omega T)$ is given. We can always replace ωT by $-j \log z$, equivalent to $z = e^{j\omega T}$, to give a function $G(z)$ which is equal to $F(\omega T)$ when evaluated along the unit circle. For our purposes $G(z)$ must be a rational function of z with real coefficients. We can guarantee this if, in the squared-magnitude function, ωT only appears as part of the following expressions,

$$\sin^2\frac{\omega T}{2}, \cos^2\frac{\omega T}{2}, \tan^2\frac{\omega T}{2}, \cot^2\frac{\omega T}{2}, \sec^2\frac{\omega T}{2}, \csc^2\frac{\omega T}{2},$$

and if these expressions are combined only by addition, subtraction, multiplication, and division. This is because the squared trigonometric functions of $\omega T/2$ yield rational functions of z, and rational functions form a field with multiplication and addition. The correspondence of squared trigonometric functions of $\omega T/2$ with rational functions of z is given in Table I.

TABLE I

$\sin^2\frac{\omega T}{2} \rightarrow$	$\frac{(z-1)^2}{-4z}$
$\cos^2\frac{\omega T}{2} \rightarrow$	$\frac{(z+1)^2}{4z}$
$\tan^2\frac{\omega T}{2} \rightarrow$	$\frac{-(z-1)^2}{(z+1)^2}$
$\cot^2\frac{\omega T}{2} \rightarrow$	$\frac{-(z+1)^2}{(z-1)^2}$
$\sec^2\frac{\omega T}{2} \rightarrow$	$\frac{4z}{(z+1)^2}$
$\csc^2\frac{\omega T}{2} \rightarrow$	$\frac{-4z}{(z-1)^2}$

Now let us assume that we have been given an $F(\omega T)$ with the preceding restrictions, found $G(z)$, and found all the poles and zeros of $G(z)$. To find a transfer function $H(z)$ with squared-magnitude function $F(\omega T)$, we should like to use the following procedure.

1) Discard any poles or zeros at the origin.

2) Put the remaining poles in one-to-one correspondence with each other in such a way that for each pair of poles the distance from one pole to any point on the unit circle is in constant ratio to the distance from that point on the unit circle to the other pole of the pair.

3) Repeat step 2) for the zeros.

4) Discard one of each corresponding pair of poles and zeros, making sure that the retained poles are all within the unit circle.

5) Form $H(z)$ from the retained poles and zeros only.

This procedure works only if steps 2) and 3) are possible. The magnitude of $H(z)$ is proportional to the product of the distances from the unit circle to retained zeros divided by the product of distances to the retained poles, and thus is proportional to the square root of $F(\omega T)$. Note that the singularities of $G(z)$ at the origin contribute nothing to a magnitude function since the distance from the origin to the unit circle is always one.

Steps 2) and 3) are possible if the poles (and zeros) occur in pairs whose locations are equal or, more commonly, conjugate reciprocals of each other, $re^{j\theta}$ and $1/re^{j\theta}$. Since the singularities of $G(z)$ occur in conjugate pairs, it is enough to require the roots to occur in equal or reciprocal pairs. By the following proof, we show that the ratio of distances from any point on the unit circle to two singularities which are conjugate reciprocals is constant. *Proof:* Let the two fixed singularities be $re^{j\theta}$ and $1/re^{j\theta}$. By the theorem of Appollonius, the locus of points, the ratio of whose distance to two fixed points is constant, is a circle whose center is collinear with the two fixed points. Consider the ratios at the two points $e^{j\theta}$ and $e^{j(\theta+\pi)}$. The ratios are equal to r at both points, and the only circle passing through both points whose center is collinear with the two fixed points is the unit circle. Thus, the unit circle is the circle of Appollonius, and the ratio of the distance to the singularities is constant.

The corresponding proof for equal singularities is trivial.

If after discarding poles or zeros at the origin the numerator and denominator of $G(z)$ become MIP's, it is possible by following the five-step procedure given above to find $H(z)$ with squared-magnitude function $F(\omega T)$.[6] We shall

now enumerate a few types of $F(\omega T)$ which yield MIP's for numerator and denominator of $G(z)$.

Consider $F(\omega T)$ a rational function of $\sin^2 \omega T/2$. Then $G(z)$, by substitution from Table I, is of the form

$$G(z) = \frac{a_0 \left[\dfrac{(z-1)^2}{-4z}\right]^n + a_1 \left[\dfrac{(z-1)^2}{-4z}\right]^{n-1} + \cdots}{b_0 \left[\dfrac{(z-1)^2}{-4z}\right]^m + b_1 \left[\dfrac{(z-1)^2}{-4z}\right]^{m-1} + \cdots}$$

which can be put in the form

$$G(z) = \frac{(a_0(z-1)^2)^n + a_1((z-1)^2)^{n-1}(-4z) + a_2((z-1)^2)^{n-2}(-4z)^2 + \cdots)(-4z)^{m-n}}{b_0((z-1)^2)^m + b_1((z-1)^2)^{m-1}(-4z) + b_2((z-1)^2)^{m-2}(-4z)^2 + \cdots}.$$

The $(-4z)^{m-n}$ contributes only poles or zeros at the origin and is discarded. Note that powers of $(z-1)^2$ are MIP's by property 5), and thus the entire numerator of $G(z)$ is an MIP by property 3). The same reasoning holds for the denominator.

Exactly similar reasoning shows that rational functions of $\cos^2 \omega T/2$ also yield realizable digital filters. Rational functions of $\tan^2 \omega T/2$ yield $G(z)$ of the form

$$G(z) = \frac{a_0(z-1)^{2n} + a_1(z-1)^{2(n-1)}(z+1)^2 + \cdots + a_n(z+1)^{2n}}{b_0(z-1)^{2m} + b_1(z-1)^{2(m-1)}(z+1)^2 + \cdots + b_m(z+1)^{2m}} (z+1)^{2(m+n)}.$$

Here we note by repeated application of property 4) that again both numerator and denominator are MIP's and therefore a digital filter exists with the desired squared-magnitude function. Again exactly similar reasoning suffices to extend the proof to rational functions of $\cot^2 \omega T/2$, $\sec^2 \omega T/2$ and $\csc^2 \omega T/2$.

Similar discussions show that rational functions of sums of products or ratios of the squares of the trig functions of Table I, i.e.,

$$\left(\sin^2 \frac{\omega T}{2} + \tan^4 \frac{\omega T}{2}\right)$$

also yield $G(z)$ which is a ratio of MIP's. Further extensions can be made, but we shall be content here with these few examples and techniques.

ACKNOWLEDGMENT

Much of the inspiration for this work came from M. J. Levin, formerly of M.I.T. Lincoln Laboratory, who was tragically killed in an automobile accident early in 1965. Dr. Levin's expert knowledge of z transform theory benefited the authors greatly when, as novices, they realized that efficient computer simulation of filters depended on this theory. Later he taught a seminar on the subject which unified the existing information. If he had lived, the present paper would certainly have been partly his work.

The authors are also indebted to J. Craig of the Lincoln

---

[6] This is not strictly correct if the MIP's have simple roots on the unit circle, which are then their own conjugate reciprocals. Subsequent statements should be qualified by taking this possibility into account. However, this is a limiting case, unlikely to occur.

Laboratory who brought many other points to their attention, notably the role of the bilinear transformation in digital filter theory.

References

[1] C. M. Rader, "Study of vocoder filters by computer simulation," *J. Acoust Soc. Am.*, vol. 36, p. 1023(A), 1964.
[2] R. M. Golden, "Digital computer simulation of a sampled-data voice excited vocoder," *J. Acoust. Soc. Am.*, vol. 35, pp. 1358–1366, 1963.
[3] B. Gold and C. M. Rader, "Bandpass compressor: a new type of speech compression device," *J. Acoust. Soc. Am.*, vol. 36, p. 1215, 1964.
[4] C. M. Rader, "Vector pitch detection," *J. Acoust. Soc. Am.*, vol. 36, p. 1963(L), 1964.
[5] ——, "Speech compression simulation compiler," *J. Acoust. Soc. Am.* (A), June 1965.
[6] B. Gold, "Experiment with speechlike phase in a spectrally flattened pitch-excited channel vocoder," *J. Acoust. Soc. Am.*, vol. 36, pp. 1892–1894, 1964.
[7] E. J. Kelly and M. J. Levin, "Signal parameter estimation for seismometer arrays," M.I.T. Lincoln Lab., Lexington, Mass., Tech. Rept. 339, January 8, 1964.
[8] H. M. James, N. B. Nichols, and R. S. Phillips, *Theory of Servo-Mechanisms*, M.I.T. Radiation Lab. Ser. 25, 1947, ch. 5, pp. 231–261.
[9] W. D. White and A. E. Ruvin, "Recent advances in the synthesis of comb filters," *1957 IRE Nat'l Conv. Rec.*, pt. 5, pp. 186–199.
[10] P. M. Lewis, "Synthesis of sampled-signal networks," *IRE Trans. on Circuit Theory*, March 1950.
[11] R. M. Golden and J. F. Kaiser, "Design of wideband sampled-data filters," *Bell Sys. Tech. J.*, vol. 48, July 1964.
[12] J. Craig, unpublished notes, 1963.
[13] J. R. Ragazzini and G. F. Franklin, *Sampled-Data Control Systems*. New York: McGraw-Hill, 1958.
[14] R. M. Lerner, "Band-pass filters with linear phase," *Proc. IEEE*, vol. 52, pp. 249–268, March 1964.
[15] E. A. Guilleman, *Synthesis of Passive Networks*. New York: Wiley, 1957.
[16] L. Weinberg, *Network Analysis and Synthesis*. New York: McGraw-Hill, 1962.
[17] J. E. Storer, *Passive Network Synthesis*. New York: McGraw-Hill, 1957.
[18] W. R. Bennett, "Spectra of quantized signals," *Bell Sys. Tech. J.*, vol. 27, pp. 446–472, July 1948.
[19] L. M. Milne-Thomson, *The Calculus of Finite Differences*. London: Macmillan, 1951.
[20] J. G. Truxal, *Automatic Feedback Control System Synthesis*. New York: McGraw-Hill, 1955.
[21] R. B. Blackman, *Linear Data-Smoothing and Prediction in Theory and Practice*. Reading, Mass.: Addison-Wesley, 1965.
[22] J. F. Kaiser, "Design methods for sampled data filters," *1963 Proc. 1st Allerton Conf.*, pp. 221–236.
[23] Notes from seminar given by M. J. Levin, 1963.
[24] J. F. Kaiser, "Some practical considerations in the realization of linear digital filters," *1965 Proc. 3rd Allerton Conf.*, pp. 621–633.
[25] J. B. Knowles and R. Edwards, "Effect of a finite-word-length computer in a sampled-data feedback system," *Proc. IEE (London)*, vol. 112, pp. 1197–1207, June 1965.
[26] K. Steiglitz, "The approximation problem for digital filters," Dept. of Elec. Engrg., New York University, New York, N. Y., Tech. Rept. 400–56, March 1962.
[27] H. Holtz and C. T. Leondes, "The synthesis of recursive digital filters," *J. ACM*, vol. 13, pp. 262–280, April 1966.
[28] B. Gold and C. M. Rader, "Effects of quantization noise in digital filters," *1966 Proc. Spring Joint Computer Conf.*, pp. 213–219.
[29] R. Carney, "Design of a digital notch filter with tracking requirements," *IEEE Trans. on Space Electronics and Telemetry*, vol. SET-9, pp. 109–114, December 1963.
[30] R. Boxer and S. Thaler, "A simplified method of solving linear and nonlinear systems," *Proc. IRE*, vol. 44, pp. 89–101, January 1956.
[31] T. G. Stockham, Jr., "High speed convolution and correlation," *1966 Proc. Spring Joint Computer Conf.*, pp. 229–233.
[32] J. W. Cooley and J. W. Tukey, "An algorithm for the machine calculation of complex Fourier series," *Math. Comput.*, vol. 19, pp. 297–301, April 1965.
[33] *System Analysis by Digital Computer*, J. F. Kaiser and F. Kuo, Eds. New York: Wiley, 1966, pp. 218–277.

Nonlinear Filtering of Multiplied and Convolved Signals

ALAN V. OPPENHEIM, MEMBER, IEEE, RONALD W. SCHAFER, MEMBER, IEEE, AND THOMAS G. STOCKHAM, JR., MEMBER, IEEE

Invited Paper

Abstract—**An approach to some nonlinear filtering problems through a generalized notion of superposition has proven useful. In this paper this approach is investigated for the nonlinear filtering of signals which can be expressed as products or as convolutions of components. The applications of this approach in audio dynamic range compression and expansion, image enhancement with applications to bandwidth reduction, echo removal, and speech waveform processing are presented.**

I. Introduction

IN THIS PAPER, a class of nonlinear filters is discussed. This class is based on an approach to the problem of synthesizing nonlinear systems from the same point of view as that used for linear system design and analysis. Specifically, there are many classes of nonlinear systems which obey a principle of superposition. This property can be exploited in much the same way as it is in characterizing linear systems.

The general theoretical structure for characterizing nonlinear systems in this way has been formulated and studied in detail by Oppenheim [1]–[3]. While the framework which this structure provides is quite broad, it has so far been pursued in depth for two specific cases: the synthesis of nonlinear filters for signals which can be expressed as a product of components and the synthesis of nonlinear filters for signals which can be expressed as a convolution of components.

The first part of the paper is directed toward a brief explanation of the notion of superposition as it applies to problems in nonlinear filtering. This explanation is followed by a detailed discussion of the analytical framework for the specific cases of the filtering of multiplied signals and the filtering of convolved signals. Following this analysis, the discussion is directed toward the applications of the theory which have thus far been pursued.

Four applications are presented, two involving multiplicative filtering and two involving convolutional filtering or deconvolution. The multiplicative applications, as developed by Stockham, involve audio dynamic range compression and expansion and image enhancement with applications to bandwidth reduction. The deconvolution examples involve echo removal and speech analysis as pursued by Schafer and Oppenheim, respectively. All four applications have progressed to the point where working models have been realized through computer simulation, and one to the point where specially designed hardware has been installed as part of an unrelated system.

The work was originally inspired to a large extent by the ideas and attitudes of Dr. M. V. Cerillo, and many readers will undoubtedly recognize the flavor of his thinking in some of the applications presented in the following.

II. Generalized Linear Filtering

When considering the problem of filtering signals that have been added, we often focus our attention on the use of a linear system. While this constraint does not always lead to a "best" choice for the filter, it has the advantage of analytical convenience. This analytical convenience is almost a direct result of the principle of superposition that linear systems satisfy. In contrast, when determining a filtering procedure to separate signals that have been nonadditively combined, such as through multiplication or convolution, it is usually more difficult, and in many cases less meaningful to use a linear system. However, we can imagine generalizing the notion of linear filtering in such a way that it encompasses this broader class of problems. Specifically, let us consider two signals $s_1(t)$ and $s_2(t)$ that have been combined according to some rule which we denote by $\circ$, so that the resulting signal $s(t)$ to be processed can be expressed as

$$s(t) = s_1(t) \circ s_2(t).$$

Let ϕ represent the transformation for the filter. Then in generalizing the notion of linear filtering, we require that ϕ have the property that

$$\phi[s_1(t) \circ s_2(t)] = \phi[s_1(t)] \circ \phi[s_2(t)]. \qquad (1)$$

Manuscript received April 5, 1968; revised June 5, 1968. *This invited paper is one of a series planned on topics of general interest.—The Editor.*

A. V. Oppenheim is with the Department of Electrical Engineering and the Research Laboratory of Electronics (supported in part by the Joint Services Electronics Program [Contract DA 28-043-AMC-02536(E)]), Massachusetts Institute of Technology, Cambridge, Mass. He is now on a leave of absence at M.I.T. Lincoln Laboratory (operated with support from the U. S. Air Force and the U. S. Advanced Research Project Agency) where a portion of the work was performed.

R. W. Schafer was formerly with the M.I.T. Research Laboratory of Electronics. A portion of the work reported here was submitted to the M.I.T. Department of Electrical Engineering in partial fulfillment of the requirements for the Ph.D. degree. He is now with the Bell Telephone Laboratories, Inc., Murray Hill, N. J.

T. G. Stockham, Jr., was formerly with the M.I.T. Lincoln Laboratory. He is now with the Department of Computer Sciences, University of Utah, Salt Lake City, Utah.

Reprinted from *Proc. IEEE*, vol. 56, pp. 1264–1291, Aug. 1968.

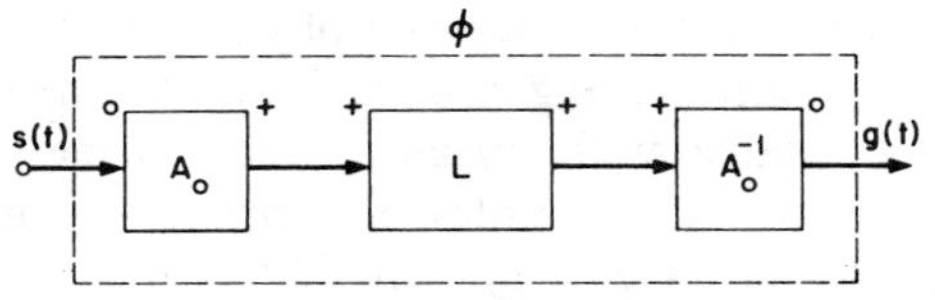

Fig. 1. The canonic representation for a homomorphic filter.

The formalism for representing systems having this property lies in interpreting the system inputs as vectors in a vector space with the rule ○ corresponding to vector addition, and the system transformation ϕ as an algebraically linear transformation on that space [1]. We must therefore restrict the operation ○ so that it satisfies the algebraic postulates of vector addition and associate with the set of inputs a rule for combining inputs with scalars, which we will call scalar multiplication and denote by :. To generalize the notion of linear filtering, then, we require that the class of systems, in addition to satisfying (1), also have the property that

$$\phi[c:s(t)] = c:\phi[s(t)]. \tag{2}$$

When the rule ○ corresponds to addition of the functions and the rule : corresponds to the product of the input with the scalar, then (1) and (2) reduce to the principle of superposition as it applies to linear systems. Systems in the class satisfying (1) and (2) have been referred to as homomorphic systems, emphasizing their interpretation as algebraically linear transformations between vector spaces.

The primary advantage in the restriction of the class of filters through (1) and (2) lies in the canonic representation for systems having this property. It has been shown [1] that if the system inputs constitute a vector space with the operations ○ and : corresponding to vector addition and scalar multiplication, then ϕ is representable as a cascade of three systems as shown in Fig. 1. The first system, $A_\circ$ in this representation, has the property that

$$A_\circ[s_1(t) \circ s_2(t)] = A_\circ[s_1(t)] + A_\circ[s_2(t)] \tag{3}$$

and

$$A_\circ[c:s(t)] = cA_\circ[s(t)]. \tag{4}$$

Furthermore, $A_\circ$ is characteristic of the class in the sense that it depends only on the operations ○ and : and not on the details of the system ϕ, and consequently is conveniently referred to as the *characteristic system*. The system L is a linear system and the system $A_\circ^{-1}$ is the inverse of the system $A_\circ$, i.e.,

$$A_\circ^{-1}\{A_\circ[s(t)]\} = s(t). \tag{5}$$

On the basis of this canonic representation we observe that generalized linear filtering corresponds to transforming the original problem to one in which the components are added, and after linear filtering, transforming the result back to the original space of inputs. Thus, once the characteristic system for the class has been determined, the problem reduces to a linear filtering problem.

III. Homomorphic Filtering of Multiplied Signals

One of the simplest examples of a rule of superposition satisfying the conditions above is that of ordinary multiplication. Of further interest is the fact that there exist several practical situations in which it is especially convenient to consider waveforms as products rather than as sums. Examples include problems involving fading channels, amplitude modulation, automatic gain control, audio dynamic range compression or expansion, and image processing. In these situations it is common to find two signals, one varying slowly and the other rapidly, combined as a product. In addition, it is frequently desirable to modify one signal and not the other or to process each according to separate objectives.

The product rule satisfies the algebraic postulates of vector addition. The companion rule for scalar multiplication is that of taking a signal to a scalar power. In terms of the symbols used earlier we have[1]

$$s_1(t) \circ s_2(t) = s_1(t) \cdot s_2(t)$$

and

$$c:s(t) = [s(t)]^c.$$

Equations (1) and (2) then become

$$\phi[s_1(t) \cdot s_2(t)] = \phi[s_1(t)] \cdot \phi[s_2(t)] \tag{6}$$

and

$$\phi\{[s(t)]^c\} = \{\phi[s(t)]\}^c. \tag{7}$$

Following the pattern of Fig. 1 we may construct Fig. 2 and, in analogy with (3), (4), and (5), we require that P have the property that

$$P[s_1(t) \cdot s_2(t)] = P[s_1(t)] + P[s_2(t)] \tag{8}$$

$$P[\{s(t)\}^c] = cP[s(t)] \tag{9}$$

and

$$P^{-1}\{P[s(t)]\} = s(t). \tag{10}$$

If we limit our consideration to include only positive real signals $s(t)$, and therefore real scalars c, the characteristic system P may be chosen as the ordinary logarithm function. It follows that P^{-1} is the corresponding exponential function. With this information this class of homomorphic systems can be represented more explicitly as in Fig. 3. An example in which we encounter only positive real signals is to be found in image processing in which the signals are formed of incoherent light. The physics of the situation guarantees the absence of negative or nonreal light intensities, while practical considerations almost certainly preclude zero intensity.

In the event that the signals to be processed cannot be restricted as above we may consider complex signals $s(t)$

[1] In these arguments we will assume that the signals involved are functions of time t. However, it is important for the reader to realize that there is nothing in the arguments to be presented which prevents the consideration of signals which are functions of space, frequency, or any other parameter. Neither is there any restriction to the consideration of one-dimensional signals.

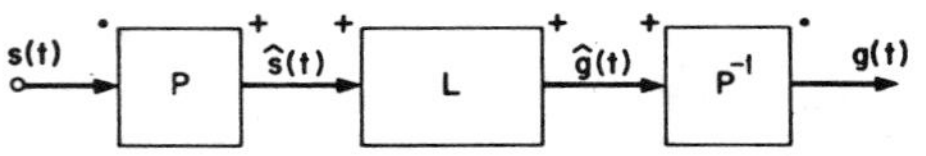

Fig. 2. The canonic representation for a multiplicative filter.

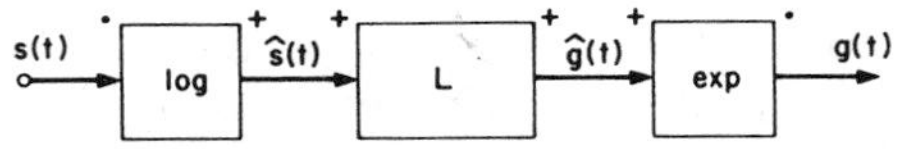

Fig. 3. The filter of Fig. 2 with P and P^{-1} specified as the logarithm and exponential transformations, respectively.

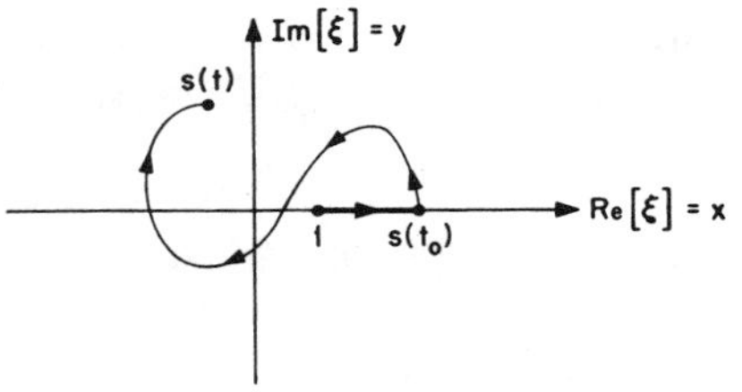

Fig. 4. A typical path of integration for defining the complex logarithm.

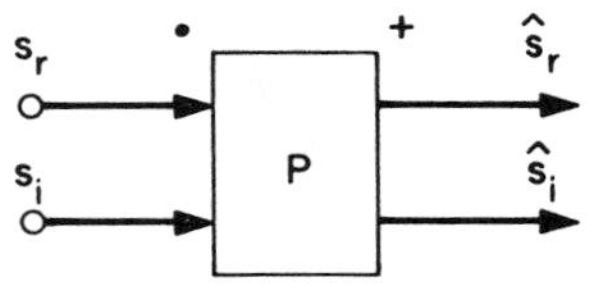

Fig. 5. The system P of Fig. 2 represented for complex signals.

and either real or complex scalars c. If we attempt to employ the complex logarithm function as the characteristic system in this situation, we encounter the immediate dilemma that the output $\hat{s}(t)$ of that system is not unique. The standard artifice of invoking the principal value of the complex logarithm cannot be used in this case, because the principal value of the logarithm of a product of complex signals is not always the sum of the principal values corresponding to the individual complex signals, violating (8).

There are restrictions which can be placed upon complex input signals such that a satisfactory characteristic system P closely related to the complex logarithm can be found. These restrictions require that complex inputs be continuous nonzero functions which attain a positive real value $s(t_0)$ at some prescribed instant of time t_0. In the case that complex scalars c are to be considered, $s(t_0)$ must be unity. The operation P is then taken to be

$$P[s(t)] = \int_1^{s(t)} \frac{d\xi}{\xi} = \hat{s}(t) \tag{11}$$

where the path of integration from the point $\xi = 1$ to the point $\xi = s(t)$ is constrained to be a straight line on the positive real axis from the point $\xi = +1$ to the point $\xi = s(t_0)$ followed thence to the point $\xi = s(t)$ via the continuous curve traced by $s(t)$ in the interval between t_0 and t. A typical path is shown in Fig. 4. The uniqueness of this transformation is assured by the fact that the path of integration is specified completely by the constraint placed upon it. While (11) is often used to define the complex logarithm function, only the end points of the contour are specified in that definition and thus multiples of $j2\pi$ may be added or subtracted by introducing arbitrary encirclements of the pole of unit residue at the origin of the ξ-plane. With our interpretation of the complex logarithm we may write $\hat{s}(t) = \log s(t)$. The inverse system P^{-1} is the complex exponential function.

Equation (11) serves as a formal definition for the transformation P, but requires further practical interpretation. Typically $\hat{s}(t)$ and $s(t)$ are realized as pairs of real signals:

$$s(t) = s_r(t) + js_i(t) \tag{12}$$

$$\hat{s}(t) = \hat{s}_r(t) + j\hat{s}_i(t). \tag{13}$$

These relationships are shown diagrammatically in Fig. 5. We now wish to establish an explicit relationship between the input and output signal pairs. Let us consider $\hat{s}_r(t)$ first. There is never any ambiguity about the real part of a complex logarithm. It is always the real logarithm of the magnitude of the complex argument. Thus we have

$$\begin{aligned} \hat{s}_r(t) &= \log|s(t)| = \log\sqrt{s_r^2(t) + s_i^2(t)} \\ &= (1/2)\log\left[s_r^2(t) + s_i^2(t)\right]. \end{aligned} \tag{14}$$

Next we consider $\hat{s}_i(t)$. Except for an ambiguity of multiples of 2π the imaginary part of the complex logarithm of a complex number is proportional to the angle of the complex number. The provisions we have made for resolving the ambiguity require more than a knowledge of $s_r(t)$ and $s_i(t)$ at a single instant. A complete history in the interval t_0 to t must be employed in the determination. We may accomplish this by noting that, from (11),

$$\frac{d\hat{s}_r(t)}{dt} + j\frac{d\hat{s}_i(t)}{dt} = \frac{d}{dt}\left[\int_1^{s(t)} \frac{d\xi}{\xi}\right] = \frac{1}{s(t)}\frac{ds(t)}{dt} \tag{15}$$

so that

$$\frac{d}{dt}\hat{s}_i(t) = \frac{s_r^2(t)}{s_r^2(t) + s_i^2(t)}\frac{d}{dt}\left[\frac{s_i(t)}{s_r(t)}\right] \tag{16}$$

and

$$\hat{s}_i(t_0) = 0.$$

Thus we can construct an expression for $\hat{s}_i(t)$ in integral form as

$$\hat{s}_i(t) = \int_{t_0}^{t} \frac{s_r^2(t)}{s_r^2(t) + s_i^2(t)}\frac{d}{dt}\left[\frac{s_i(t)}{s_r(t)}\right]dt \tag{17}$$

which becomes

$$\hat{s}_i(t) = \int_{t_0}^{t} \frac{1}{|s(t)|^2}\left[s_r(t)\frac{ds_i(t)}{dt} - s_i(t)\frac{ds_r(t)}{dt}\right]dt. \tag{18}$$

The inverse characteristic system P^{-1} is diagramed in Fig. 6. Explicit expressions for its input-output relations are

$$g_r(t) = e^{\hat{g}_r(t)}\cos\hat{g}_i(t) \tag{19}$$

$$g_i(t) = e^{\hat{g}_r(t)}\sin\hat{g}_i(t). \tag{20}$$

The canonic form for a multiplicative homomorphic system employing complex signals is depicted in Fig. 7. There is a

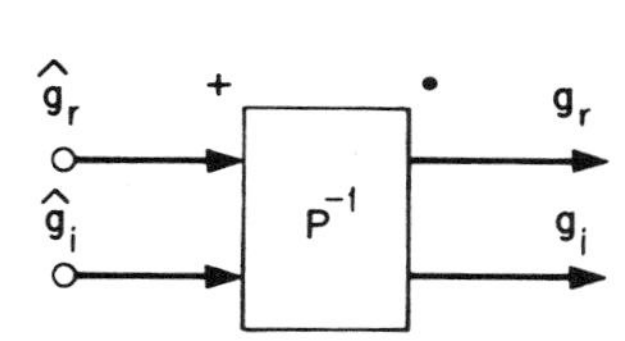

Fig. 6. The system P^{-1} of Fig. 2 represented for complex signals.

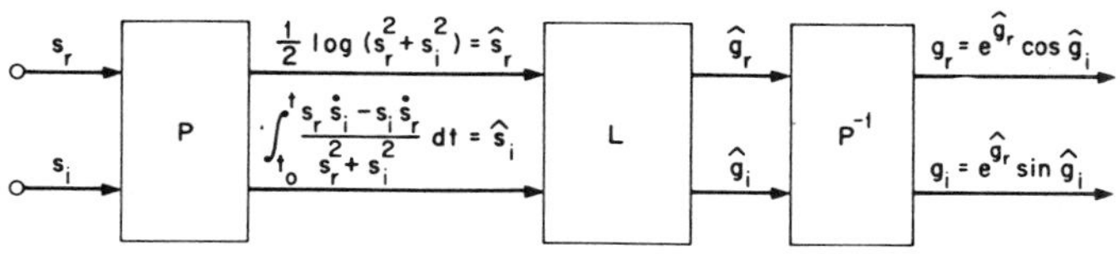

Fig. 7. The canonic representation for a multiplicative filter employing complex signals. $\dot{s}_i$ and $\dot{s}_r$ denote the time derivatives of s_i and s_r.

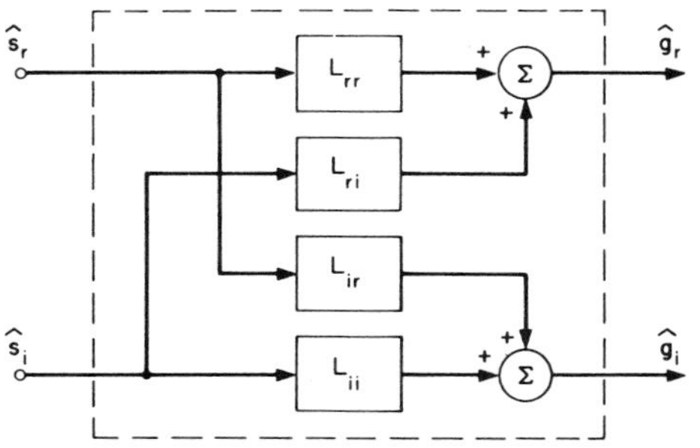

Fig. 8. The general topology for the system L of Fig. 7.

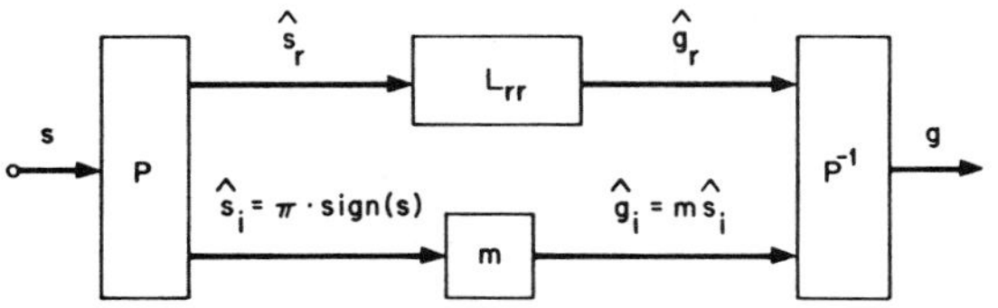

Fig. 9. The multiplicative filter of Fig. 7 specialized to employ real bipolar signals.

residual question concerning the form of the linear system L. If we restrict ourselves to real scalars c, then we place a different set of restrictions upon L than if we allow complex scalars. A completely general topology for the system L is presented in Fig. 8 in which the small internal systems are linear with real signals. In the case of real scalars c, the system L must obey superposition only when the input $s(t)$ is multiplied by a real scalar. Under these circumstances there are no restrictions on the four real linear systems. However, in the case of complex scalars the system L must obey superposition when the input $s(t)$ is multiplied by a complex scalar. This requires that

$$L_{rr} = L_{ii} \tag{21}$$

and

$$L_{ri} = -L_{ir} \tag{22}$$

which restricts the systems L that may be employed.

One of the important practical applications of the above ideas involves signals which are real and bipolar, that is, are sometimes positive and sometimes negative. Formally such signals do not fit within the above framework since the condition that the signals be nonzero and continuous simultaneously cannot be met. Consequently, if we want to apply the notion of homomorphic filtering to this case we must modify the bipolar signals so that they are complex. For example, we may treat the signals as real until they become smaller than some very small value at which time they assume complex values of fixed magnitude ε and varying angle. Another possible method involves adding a very small constant imaginary value to the signal, thus forcing it to be nonzero. All such methods are basically similar in nature and require that complex signals be considered. Formalizing this idea is difficult and seems to offer no real advantage. For the particular application to be discussed, where only one of the two multiplied signals is bipolar, no difficulties arise if we require that $L_{ir} = L_{ri} = 0$ and $L_{ii} = m$ (an integer). In this case, the system of Fig. 7 is reduced to that of Fig. 9.

IV. Homomorphic Filtering of Convolved Signals

There are many waveforms of interest which can be considered as a convolution of component signals which we wish to separate. Often, for example, a waveform is corrupted through reverberation, that is, the introduction of echos, which we would like to remove. In speech processing, it is often of interest to isolate the effects of vocal tract impulse response and excitation, which at least on a short-time basis can be considered to have been convolved to form the speech waveform [4]. Another example lies in the separation of probability density functions which have been convolved by the addition of independent random processes.

A common approach to deconvolution is the technique of inverse filtering. In this case the unwanted components of the signal to be processed are removed by filtering with a linear system whose system function is the reciprocal of the Fourier transform of these components. Clearly this method is reasonable only for those situations in which we have a detailed model or description of the components to be removed. This approach has been successful, for example, in recovering the excitation function from the speech waveform since accurate models of the vocal tract have been developed [5]. Inverse filtering is analogous to removing the effect of noise in the additive case (i.e., signal plus noise) by subtraction. If the noise is known exactly except for a few parameters then we might reasonably expect to recover the signal by subtracting the noise from the sum. In many cases, however, we do not have available detailed information about the unwanted components of the signal, and consequently this method of subtraction in the additive case or inverse filtering in the convolutional case is no longer feasible.

In applying the notion of generalized linear filtering to the separation of convolved signals, we must first determine the characteristic system for this class of filters. While we may formulate the results either in terms of continuous or discrete (sampled) inputs, the processing to be described is most easily realized on a digital computer. Consequently, the discussion will be phrased in terms of discrete time series. Thus, we consider a sequence $s(n)$ consisting of the discrete convolution of two sequences $s_1(n)$ and $s_2(n)$ so that

$$s(n) = \sum_{k=-\infty}^{+\infty} s_1(k)s_2(n-k)$$

or

$$s(n) = s_1(n) \otimes s_2(n) \tag{23}$$

where $\otimes$ denotes a discrete convolution. The canonic form for the class of filters is represented symbolically in Fig. 10 where D is the characteristic system for the class and has the property that

$$D[s_1(n) \otimes s_2(n)] = \hat{s}_1(n) + \hat{s}_2(n) \tag{24}$$

where $\hat{s}_1(n)$ and $\hat{s}_2(n)$ are the responses of D for inputs $s_1(n)$ and $s_2(n)$, respectively.

Let $S(z)$ and $\hat{S}(z)$ denote the two-sided z-transforms of $s(n)$ and $\hat{s}(n)$, respectively, so that

$$S(z) = \sum_{n=-\infty}^{+\infty} s(n)z^{-n} \tag{25a}$$

$$\hat{S}(z) = \sum_{n=-\infty}^{+\infty} \hat{s}(n)z^{-n} \tag{25b}$$

$$s(n) = \frac{1}{2\pi j}\oint_{C_1} S(z)z^{n-1}\,dz \tag{25c}$$

and

$$\hat{s}(n) = \frac{1}{2\pi j}\oint_{C_2} \hat{S}(z)z^{n-1}\,dz \tag{25d}$$

with c_1 and c_2 closed counterclockwise contours of integration in the z-plane. It will be assumed for notational convenience that c_1 and c_2 are always taken to be the unit circle $z=e^{j\omega}$. While this is somewhat restrictive the results obtained are easily modified to incorporate the general case.

It follows from (23) and the properties of the z-transform that

$$S(z) = S_1(z)S_2(z) \tag{26}$$

where $S_1(z)$ and $S_2(z)$ are the z-transforms of $s_1(n)$ and $s_2(n)$, respectively. Hence, from the results of the previous section, applied here to functions of frequency, we may relate $S(z)$ and $\hat{S}(z)$ through a suitably defined logarithmic transformation.

Let us require that both $S(z)$ and $\hat{S}(z)$ be analytic functions with no singularities on the unit circle. Letting

$$S(e^{j\omega}) = S_R(e^{j\omega}) + jS_I(e^{j\omega}) \tag{27a}$$

and

$$\hat{S}(e^{j\omega}) = \hat{S}_R(e^{j\omega}) + j\hat{S}_I(e^{j\omega}), \tag{27b}$$

we then require that S_R, $\hat{S}_R$, S_I, and $\hat{S}_I$ be continuous functions of ω. Since the z-transform is a periodic function of ω with period 2π we require in addition that $\hat{S}_R$ and $\hat{S}_I$ be

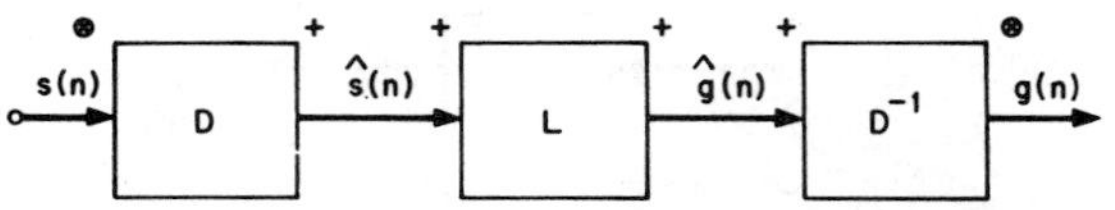

Fig. 10. The canonic representation for a deconvolution filter.

periodic functions of ω. Furthermore, we may impose the constraint that $s(n)$ and $\hat{s}(n)$ be real functions so that S_R and $\hat{S}_R$ are even functions of ω and S_I and $\hat{S}_I$ are odd functions of ω. Then from (14) and (16) we define

$$\hat{S}_R = \log|S| \tag{28}$$

and

$$\frac{d\hat{S}_I}{d\omega} = \frac{S_R^2}{S_I^2 + S_R^2}\frac{d}{d\omega}\left[\frac{S_I}{S_R}\right] \tag{29}$$

with

$$|\hat{S}_I(e^{j\omega})|_{\omega=0} = 0.$$

Thus the imaginary part of $\hat{S}$ is interpreted to be the angle of S considered as a continuous, odd, periodic function of ω. The response of the system D then corresponds to the inverse transform of the complex logarithm of the transform.

A similar transformation was introduced by Bogert, Healy, and Tukey in which the power spectrum of the logarithm of the power spectrum was proposed as a means for detecting echoes [6]. The result of this set of operations was termed the cepstrum. It is clear that $\hat{s}(n)$ bears a strong relationship to the cepstrum with the primary differences being embodied in the use of the *complex* Fourier transform and *complex* logarithm [7]. To emphasize the relationship while maintaining the distinction, it has been convenient to refer to $\hat{s}(n)$ as the complex cepstrum.

Properties of the Complex Cepstrum

While (28) and (29) define the complex cepstrum, it is possible to reformulate the relationship between $s(n)$ and $\hat{s}(n)$ in several ways which place more in evidence the properties of the complex cepstrum. From (28) and (29)

$$\hat{s}(n) = \frac{1}{2\pi j}\oint_C \log[S(z)]z^{n-1}\,dz. \tag{30}$$

Since the contour of integration is the unit circle and we have defined $\log[S(z)]$ so that it is a single-valued function, we may rewrite (30) as

$$\hat{s}(n) = \frac{1}{2\pi}\int_{-\pi}^{\pi} \log[S(e^{j\omega})]e^{j\omega n}\,d\omega.$$

Integrating by parts and using the fact that $S_I(e^{j\omega})$ is restricted to be a continuous, odd, periodic function of ω, we obtain the result that

$$\hat{s}(n) = \begin{cases} -\dfrac{1}{2\pi jn}\oint z\dfrac{1}{S(z)}\dfrac{dS(z)}{dz}z^{n-1}\,dz & n \neq 0 \\ \dfrac{1}{2\pi}\displaystyle\int_{-\pi}^{\pi}\log|S(e^{j\omega})|\,d\omega & n = 0. \end{cases} \tag{31}$$

An example of a class of functions $S(z)$ satisfying the requirement that both $S(z)$ and $\hat{S}(z)$ be analytic is the class of the form

$$S(z) = |K| \frac{\prod_{i=1}^{M_0} (1 - a_i z^{-1}) \prod_{i=1}^{M_1} (1 - b_i z)}{\prod_{i=1}^{P_0} (1 - c_i z^{-1}) \prod_{i=1}^{P_1} (1 - d_i z)} \tag{32}$$

where the a_i and c_i are the zeros and poles inside the unit circle and $(1/b_i)$ and $(1/d_i)$ are the zeros and poles outside the unit circle. For this class of examples, we note that the poles of the integrand in (31) occur at the poles and zeros of $S(z)$. Consequently, $\hat{s}(n)$ will be composed of a sum of exponentials divided by n.

Equation (31) can be rewritten in a somewhat different form by noting that, from (15),

$$\frac{d\hat{S}(z)}{dz} = \frac{1}{S(z)} \frac{dS(z)}{dz}$$

or

$$S(z) \frac{d\hat{S}(z)}{dz} = \frac{dS(z)}{dz}. \tag{33}$$

Using the fact that $d\hat{S}(z)/dz$ is the z-transform of $-n\hat{s}(n)$, and $dS(z)/dz$ is the z-transform of $-ns(n)$, the inverse z-transform of (33) is

$$[n\hat{s}(n)] \otimes s(n) = ns(n)$$

or

$$\sum_{k=-\infty}^{+\infty} \frac{k}{n} \hat{s}(k) s(n-k) = s(n) \qquad n \neq 0. \tag{34}$$

In general, this is an implicit relation between $s(n)$ and $\hat{s}(n)$ and cannot be computed. However, if it is assumed that $s(n)$ and $\hat{s}(n)$ are zero for n negative and that $s(0) \neq 0$, then (34) becomes

$$\hat{s}(n) = \begin{cases} \dfrac{s(n)}{s(0)} - \displaystyle\sum_{k=0}^{n-1} \frac{k}{n} \hat{s}(k) \frac{s(n-k)}{s(0)} & n \neq 0 \\ \log s(0) & n = 0. \end{cases} \tag{35}$$

For this case, the inverse of the characteristic system can be easily obtained by solving (35) for $\hat{s}(n)$ in terms of $s(n)$ with the result that

$$s(n) = \begin{cases} s(0)\hat{s}(n) + \dfrac{1}{n} \displaystyle\sum_{k=0}^{n-1} k\hat{s}(k)s(n-k) & n \neq 0 \\ e^{\hat{s}(0)} & n = 0. \end{cases} \tag{36}$$

The Complex Cepstrum of Minimum Phase Sequences

For a function $f(t)$ which is zero for $t<0$, the real and imaginary parts of its Fourier transform are related through the Hilbert transform. This relationship is derived by noting that $f(t)$ is uniquely expressible in terms of its even part [8].

For certain classes of functions, the magnitude and phase of the Fourier transform are also related through the Hilbert transform and such functions are generally referred to as minimum phase functions. The relationship between magnitude and phase is derived by treating the log magnitude and phase of the Fourier transform as the real part and imaginary part, respectively, of a new Fourier transform. If the time function associated with this new Fourier transform is zero for $t<0$, then its real and imaginary parts are related. An entirely parallel set of statements can be made for discrete sequences, with the Fourier transform replaced by the z-transform evaluated on the unit circle.

From the above discussion we note that a minimum phase sequence is one for which the complex cepstrum $\hat{s}(n)$ is zero for $n<0$, which is the same condition imposed in deriving (35). Thus we may conclude that for input sequences which are minimum phase, the input and output of the system D are related by the recursion relation of (35) and the input and output of the system D^{-1} are related by the recursion relation of (36). These recursion relations do not necessarily offer a computational advantage. However, they are conceptually important. In particular, they bring to light the fact that for minimum phase inputs, the transformation D is a realizable transformation, i.e., the response $\hat{s}(n)$ for $n=n_0$ is dependent only on samples of the input for $n \leq n_0$. Similarly, the inverse transformation D^{-1} is realizable for input sequences $\hat{s}(n)$ which are zero for $n<0$. From this we may conclude that for minimum phase input sequences the class of homomorphic filters defined by the canonic form of Fig. 10 is realizable in the sense that the output depends only on previous values of the input if the linear filter is also realizable.

An analogous discussion can be carried out for sequences whose complex cepstrum is zero for $n>0$. Such sequences, which have no minimum phase components, could appropriately be called maximum phase sequences. For these cases a relation similar to (35) can be derived, in which values of the complex cepstrum depend only on future rather than past values of the input. It should be remarked that any sequence can always be expressed as the convolution of a minimum phase sequence and a maximum phase sequence, i.e.,

$$s(n) = s_1(n) \otimes s_2(n).$$

The portion of $\hat{s}(n)$ for $n>0$ represents the contribution from the minimum phase component, and the portion for $n<0$ represents the contribution from the maximum phase component.

As a result of these considerations an interesting and perhaps useful result emerges. Consider a time-limited sequence $s(n)$ which contains $(N+1)$ samples. Let us choose the origin and polarity of the waveform so that $S(z)$ can be expressed in the form of (32). Now $s(n)$ can be expressed as the convolution of a minimum phase sequence $s_1(n)$ and a maximum phase sequence $s_2(n)$ where $s_1(n)$ and $s_2(n)$ are time-limited so that

$$\begin{aligned} s_1(n) &\neq 0 \quad 0 \leq n \leq N_1 \\ &= 0 \quad \text{otherwise} \end{aligned}$$

and

$$s_2(n) \neq 0 \quad -N_2 \leq n \leq 0$$
$$= 0 \quad \text{otherwise}$$

where

$$N_1 + N_2 = N.$$

The complex cepstrum of $s_1(n)$ is in general not time-limited. However, $\hat{s}_1(n)$ is zero for $n<0$ and $\hat{s}_2(n)$ is zero for $n\geq 0$. Thus, from (34),

$$s_1(n) = \begin{cases} e^{\hat{s}_1(0)} & n = 0 \\ \hat{s}_1(n)s_1(0) + \sum_{k=0}^{n-1} \left(\frac{k}{n}\right)\hat{s}_1(n)s_1(n-k) & n > 0 \end{cases}$$

and

$$s_2(n) = \begin{cases} 1 & n = 0 \\ \hat{s}_2(n) + \sum_{k=n+1}^{0} \left(\frac{k}{n}\right)\hat{s}_2(k)s_2(n-k) & n < 0. \end{cases}$$

Consequently, (N_1+1) values of $\hat{s}_1(n)$ are needed to recover $s_1(n)$ and N_2 values of $\hat{s}_2(n)$ are needed to recover $s_2(n)$, so that (N_1+N_2+1) values of the complex cepstrum are needed to obtain the (N_1+N_2+1) values of $s(n)$.

Sequences with Rational z-Transforms

Thus far, we have restricted the input sequences to be such that $S(z)$ and $\hat{S}(z)$ are analytic and for these cases, the logarithm of the z-transform on the unit circle was defined such that the imaginary part of the logarithm was a continuous, odd, periodic function of ω. It was remarked that this included all sequences with z-transforms of the form of (32). It is reasonable to assume that most input sequences of interest can be represented at least approximately by z-transforms which are rational, of the form

$$S(z) = Kz^r \frac{\prod_{i=1}^{M_0} (1 - a_i z^{-1}) \prod_{i=1}^{M_1} (1 - b_i z)}{\prod_{i=1}^{P_0} (1 - c_i z^{-1}) \prod_{i=1}^{P_1} (1 - d_i z)}. \tag{37}$$

Equation (37) differs from (32) in the inclusion of a term z^r representing a delay or advance of the sequence and removal of the absolute value on the multiplying constant so that $S(z)$ is no longer required to be positive for $z=1$ ($\omega=0$).

While it is possible to generate a formal structure which would include this more general case, it offers no real advantage. Specifically, if we consider the problem at hand, namely, carrying out a separation of convolved signals, we would not expect to be able to determine, and most likely would not be interested in determining how much of the constant K, including its sign, was contributed by each. Similarly we could not expect to be able to determine how much of the net advance or delay r was contributed by each. In summary, we can expect to be generally interested in the shape of the components and not their amplitudes or time origin.

If we are willing to permit this flexibility, then we can measure the algebraic sign of K and the value of r separately and then alter the input (or its transform) so that the z-transform is in the form of (32).

Computation of the Complex Cepstrum

On the basis of the previous discussion, for general input sequences the computation of the complex cepstrum requires a computation of the Fourier transform of the input. Thus practical considerations require that the input $s(n)$ contain only a finite number of points, that is, be time-limited, and that the transform be computed only at discrete frequencies. Thus, in an implementation of the transformation D, we replace the z-transform and its inverse by the discrete Fourier transform pair (DFT) defined as

$$F(k) = \sum_{n=0}^{N-1} f(n)W^{nk}$$

and

$$f(n) = \frac{1}{N}\sum_{k=0}^{N-1} F(k)W^{-nk}$$

where

$$W = e^{-2\pi j/N}.$$

Thus, the complex cepstrum computed by use of the DFT is given by

$$\begin{aligned} \hat{s}_d(n) &= \frac{1}{N}\sum_{k=0}^{N-1} [\log S(k)]W^{-nk} \\ &= \frac{1}{N}\sum_{k=0}^{N-1} [\log |S(k)| + j\theta(k)]W^{-nk} \end{aligned} \tag{38}$$

with

$$S(k) = \sum_{n=0}^{N-1} s(n)W^{nk}.$$

It is straightforward to verify that $\hat{s}_d(n)$ is an aliased version of $\hat{s}(n)$, i.e.,

$$\hat{s}_d(n) = \sum_{a=-\infty}^{+\infty} \hat{s}(aN + n).$$

The effect of the aliasing depends on the value chosen for the rate at which the spectrum is sampled, or equivalently the value of N. In many cases this is not a severe problem since relatively fast and efficient means for computing the discrete Fourier transform for large N have recently been developed [9].

The phase curve $\theta(k)$ can be computed by first computing the phase modulo 2π and then "unwrapping" it to satisfy the requirement that it be continuous and odd. Simple algorithms for doing this are easily generated, provided that the frequency spacing of adjacent points is sufficiently small.

As an alternative to computing the complex cepstrum by means of (38), we may obtain $\hat{s}(n)$ by forming the ratio of the derivative of the spectrum and the spectrum, as suggested by (31). In particular, since samples of the derivative of the spectrum, denoted by $\tilde{S}(k)$, can be obtained by

$$\tilde{S}(k) = -j \sum_{n=0}^{N-1} nf(n)W^{nk}$$

we obtain $\hat{s}_d(n)$ as

$$\hat{s}_d(n) = -\frac{1}{jn}\frac{1}{N}\sum_{k=0}^{N-1}\frac{\tilde{S}(k)}{S(k)}W^{-nk}. \tag{39}$$

The complex cepstrum computed on the basis of (39) differs somewhat from that computed from (38). The difference can be expressed by observing that $n\hat{s}_d(n)$ as represented by (39) is an aliased replica of $n\hat{s}(n)$, i.e.,

$$n\hat{s}_d(n) = \sum_{a=-\infty}^{+\infty} (aN + n)\hat{s}(aN + n)$$

or

$$\hat{s}_d(n) = \frac{1}{n}\sum_{a=-\infty}^{+\infty} (aN + n)\hat{s}(aN + n). \tag{40}$$

We note that in general the effect of the aliasing introduced by the use of (39) is more severe than that introduced by (38). On the other hand, use of (38) requires the explicit computation of the unwrapped phase curve, whereas use of (39) does not.

V. Applications of Homomorphic Filtering

In the preceding paragraphs we have discussed the analytical aspects of homomorphic filtering in general, and multiplicative and convolutional filtering in particular. We now wish to deemphasize the theory and concern ourselves with specific practical applications. The following discussions serve two distinct purposes. The first is to disclose a specific technology which has emerged as a direct result of the theory. The second is to lend to the theory a set of examples which hopefully will serve to clarify concepts and to foster further investigation.

The Multiplicative Processing of Audio Signals

The first application of multiplicative filtering to be discussed involves the processing of audio signals [10]. We are all familiar with the idea of analyzing audio waveforms as sums of harmonic oscillations. However, for the purposes of this discussion we conceive of analyzing audio waveforms as a product instead of a sum. Specifically, we are motivated by the obvious fact that audio signals bear a resemblance to amplitude-modulated waves. They are similar because each grows larger and smaller at a relatively slow rate while dancing around at some other relatively fast rate. In the case of audio the fast motion is irregular and varied. In the case of AM it is neither. In this respect, they are not similar. There are factors in the manufacture of audible signals which are certainly multiplicative in nature. A person modulates his voice both consciously and unconsciously. Musicians play loud and soft passages. Sounds form and die away as their energy is absorbed.

With this motivation let us analyze an audio signal as a product of two components. Let the first be an envelope $e(t)$ which is slowly varying but always positive. Let the

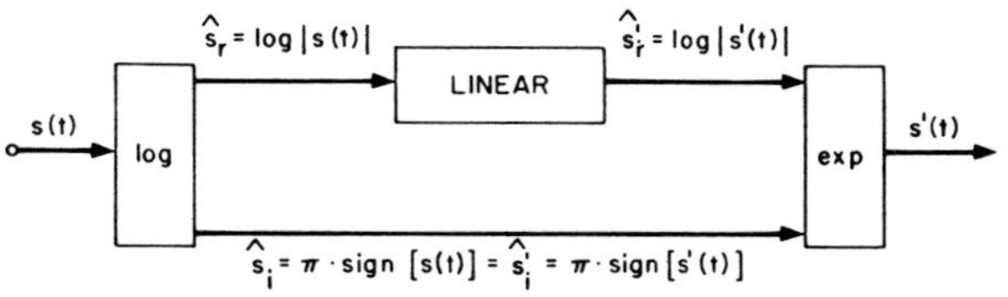

Fig. 11. A multiplicative filter for audio processing.

second be a carrier or vibration $v(t)$ which is rapidly varying and bipolar. If we call our audio signal $s(t)$, we obtain

$$s(t) = e(t)\cdot v(t). \tag{41}$$

Furthermore, let us process this signal with a multiplicative homomorphic filter such that the response $s'(t)$ will be given by

$$s'(t) = e'(t)\cdot v'(t) \tag{42}$$

where $e'(t)$ and $v'(t)$ are the responses for $e(t)$ and $v(t)$ acting separately. Fig. 11 shows this situation in accordance with our previous discussions concerning real bipolar signals and multiplicative filters. We have set $m=1$ since we wish to preserve the sign information embodied in $v(t)$. Notice that since $e(t)$ is always positive

$$\hat{e}_r(t) = \log|e(t)| = \log e(t) \tag{43}$$

and

$$\hat{e}_i(t) = 0 \tag{44}$$

such that

$$\hat{s}_i(t) = \hat{v}_i(t). \tag{45}$$

Furthermore, since

$$\hat{s}_i'(t) = \hat{s}_i(t) \tag{46}$$

it follows that

$$\hat{e}_i'(t) = 0 \tag{47}$$

which implies that $e'(t)$ is always positive as well. More explicitly

$$\hat{s}_i'(t) = \hat{v}_i'(t) \tag{48}$$

and

$$\hat{e}_r'(t) = \log|e'(t)| = \log e'(t). \tag{49}$$

With these equations in mind, we can reconstruct Fig. 11 as in Fig. 12.

If $\log e(t)$ and $\log|v(t)|$ were to possess frequency components occupying separate frequency bands, linear systems could be designed to perform different processing tasks upon each. While it is almost certainly true that any preconceived definition of $e(t)$ and $v(t)$ would not result in this condition, an extremely interesting situation is revealed through the examination of the spectra of $\log|s(t)|$ for typical audio signals. An example of such a spectrum is shown in Fig. 13. This curve is derived from a computer calculation of the periodogram of the log magnitude of seven seconds of speech waveform. We see that above a certain critical fre-

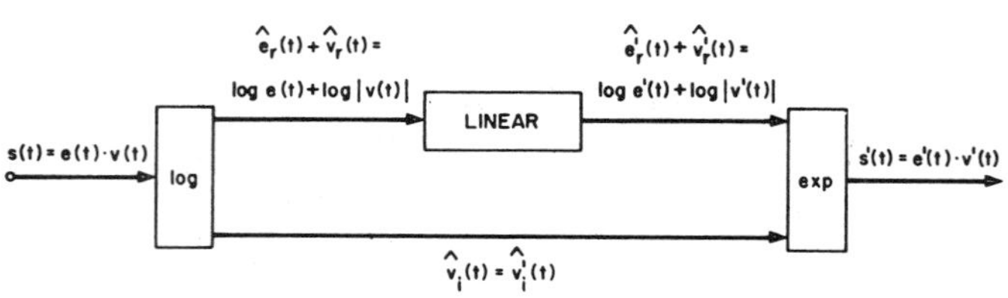

Fig. 12. The filter of Fig. 11 with component signals shown explicitly.

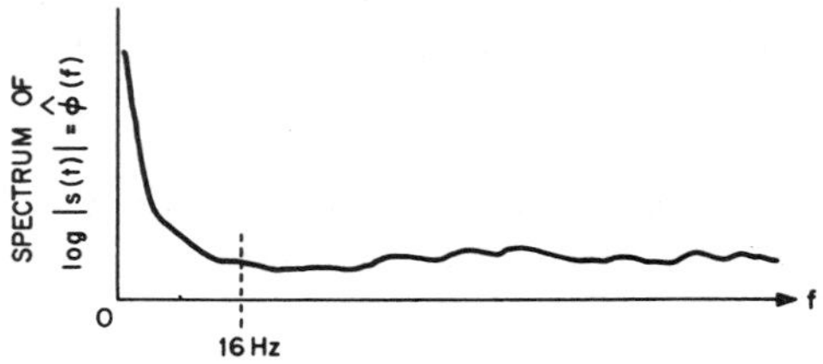

Fig. 13. A typical spectrum of log $|s(t)|$ for audio signals.

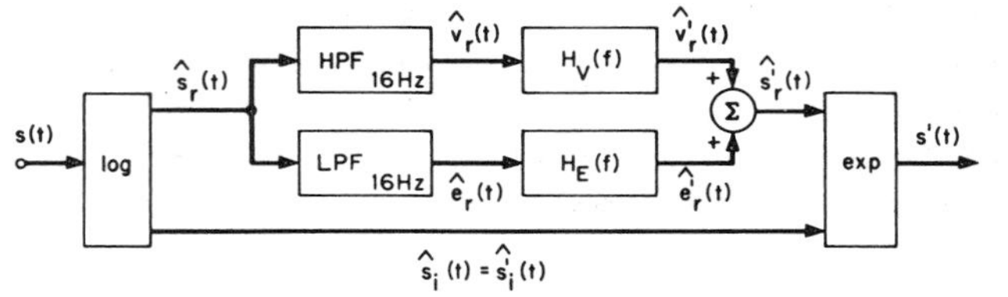

Fig. 14. A multiplicative filter which processes $\hat{E}(f)$ and $\hat{V}(f)$ separately.

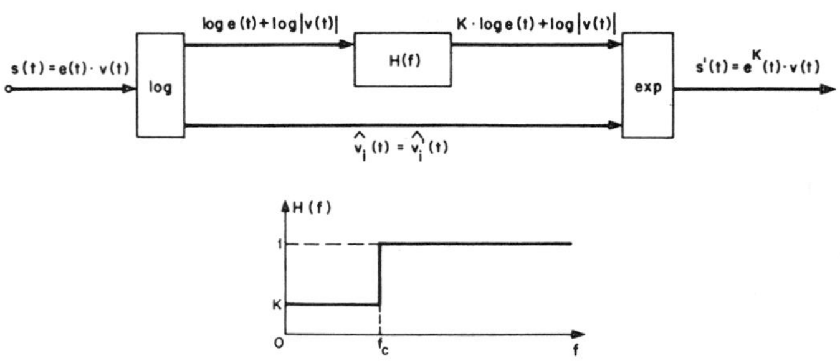

Fig. 15. The system of Fig. 14 with $H_V(f)=1$ and $H_E(f)=K$.

quency near 16 Hz the spectrum is nominally constant, but below that frequency it grows rapidly with decreasing frequency. This behavior suggests that this spectrum can be broken into two components, one more or less constant at all frequencies and the other large at low frequencies but decreasing rapidly with increasing frequency. Although these components overlap each other it is clear that the character of their spectral behavior is sufficiently different to permit effective partial separation by the methods of linear filtering.

For the sake of simplicity, however, let us assume that $\hat{\phi}(f)$ can be broken into two parts occupying distinct frequency bands. Let the first part, called $\hat{E}(f)$, be constituted from all components of $\hat{\phi}(f)$ below 16 Hz. Let the second part, called $\hat{V}(f)$, be constituted from all components of $\hat{\phi}(f)$ above 16 Hz. For any specific audio signal we will associate $\hat{E}(f)$ with the envelope signal and $\hat{V}(f)$ with the vibration signal.

A multiplicative filter which processes each of these components independently is shown in Fig. 14. The high-pass and low-pass filters serve to isolate the component signals of $\hat{s}_r(t)$ and the separate different linear filters $H_V(f)$ and $H_E(f)$ operate upon each independently.

An extremely interesting subclass of filters is generated by considering $H_V(f)=1$. This class leaves $v(t)$ entirely unaffected while operating only on the envelope $e(t)$. A simple specific example is formed by choosing $H_E(f)=K$. For this choice we can reconstitute the filter as shown in Fig. 15 where we find a single frequency response specification for the linear system. Since this system has a gain of unity for vibration signals, they are unaffected by the filter. However, for envelope signals the gain is K, and so for them the system is a power law device with exponent K. This arrangement results in a response $s'(t)$ as given by

$$s'(t) = e'(t)v'(t) = e^K(t) \cdot v(t). \tag{50}$$

If $K<1$, the envelope function is subject to rooting action, thus reducing the dynamic range of the composite signal $s(t)$. If $K>1$, this dynamic range is increased. In this way this multiplicative filter acts as a volume compressor or expander.

If $K=0$ the envelope signal response $e'(t)$ is reduced to unity and $s'(t)=v(t)$. This situation is similar to that obtained with automatic gain control circuits, because the amplitude of response is not dependent upon the amplitude of excitation.

A system based upon the diagram of Fig. 15 for effecting the modification of dynamic range has been simulated on the TX-2 computer at the M.I.T. Lincoln Laboratory and has been constructed for audio signals and employed in practice with remarkable success. Some interesting practical considerations arise in this respect which are worth mentioning here.

The first has to do with the realization of the system function $H(f)$ described in Fig. 15. The ideal filter $H(f)$ can be approximated in practice by a lumped parameter system $L(f)$. Whether employing few or many degrees of freedom, if $|L(0)|=K$ and $|L(\infty)|=1$ the basic operating characteristics discussed above can be realized. This is so because the need for a sharp transition characteristic is not implied by Fig. 13. The reader may ask about the effect of the phase characteristics of $L(f)$ upon this situation. Phase will have negligible effect upon system performance as long as it approaches zero as frequency becomes infinite. This condition assures that there is no delay for the high-frequency components of log $|v(t)|$, which is sufficient for a proper reconstruction of the axis crossing behavior of the component $v(t)$.

Another important practical consideration is that the bandwidth required for transmitting and processing log $|s(t)|$ is considerably wider than that required for $s(t)$ itself. This fact is best appreciated by reference to Fig. 16 which shows a typical $s(t)$ and log $|s(t)|$. Notice that as $s(t)$ passes through zero, log $|s(t)|$ attempts to become negatively infinite. In mathematical terms, when $s(t)$ possesses a zero, log $|s(t)|$ possesses a logarithmic pole. These logarithmic poles must be reproduced fairly well if the zero crossing behavior of $s'(t)$ is to resemble that of $s(t)$. In practice the bandwidth

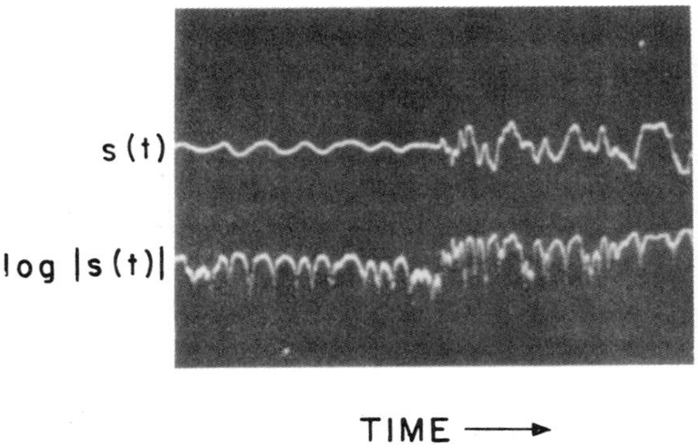

Fig. 16. A typical $s(t)$ and log $|s(t)|$ as measured in the laboratory

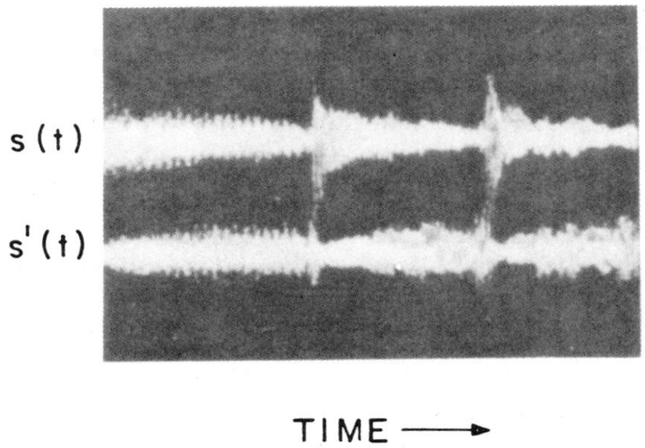

Fig. 17. A typical $s(t)$ and its supercompressed counterpart.

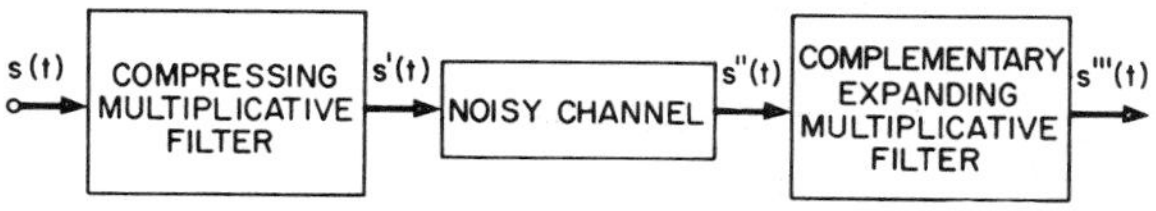

Fig. 18. A compression-expansion system employing a pair of complementary multiplicative filters.

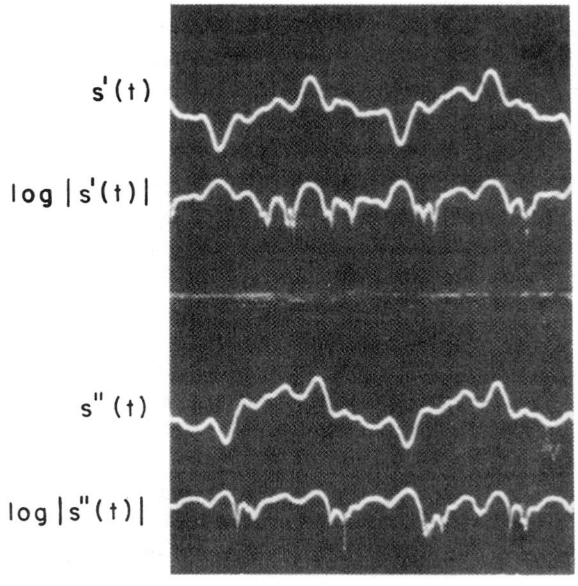

Fig. 19. $s(t)$ and log $|s(t)|$ before and after envelope distortion due to an ac coupled channel.

required for audio is a few kilohertz. One hundred or one thousand times this bandwidth might be required to process log $|s(t)|$ satisfactorily depending upon the degree of precision required. This performance is easily achieved with present technology.

An unusual situation arises if K is made negative. A form of supercompression results in which the role of loud and soft are reversed. For $K=-1$ very loud sounds become barely audible while barely audible sounds become very loud. Fig. 17 shows some typical signals.

It is possible to construct a simple filter $L(f)$ which causes compression and which also has a simple inverse filter $L^{-1}(f)$ which causes exactly complementary expansion. In this way a new type of compression-expansion system can be constructed. Fig. 18 shows such a system in which a noisy channel is to be upgraded for audio use by pre-compression and post-expansion [11]. If the channel is assumed perfect, which of course it is not, then the compressor and expander, being exactly inverse systems, operate as an exactly compensating pair and the received signal $s'''(t) \equiv s(t)$. For only mildly degraded channels this situation is closely approximated. This statement has been demonstrated empirically in the laboratory in applications involving data recording channels. In these situations it was discovered that the channel imperfection which most seriously affects performance is phase shift at low audio frequencies. For dc coupled channels this is never a problem, however. The level of degradation is mild in typical ac coupled situations but provides an occasional serious problem in applications requiring maximum quality. The mechanism of the difficulty is easily described in terms of a hypothetical $s(t)$ composed of two harmonically related sine waves, the amplitude and phase of which are adjusted to yield a waveform which is small for a large percentage of time. The log of this signal will have a relatively small average value as a result of this property. If the two sine waves are shifted in relative phase by 30° or so, $s(t)$ would no longer be small for such a large percentage of time and thus log $|s(t)|$ would have a larger average value. In this way channel phase shift at low frequencies can cause mild envelope distortion in terms of the operation of a multiplicative compressor. See Fig. 19.[2]

The Multiplicative Processing of Images

The second application of multiplicative filtering to be discussed involves the processing of images. This application is motivated very directly, because image formation is predominantly a multiplicative process. This statement applies equally to natural and photographic images [13]. In a natural scene, the illumination and reflectance of objects are combined by multiplication to form observable brightness. The illumination and reflectance in a scene vary independently from object to object and from point to point, thus forming a brightness image. In terms of its projection onto

[2] A channel imperfection which might at first seem troublesome is additive channel noise. It has been verified both theoretically [12] and experimentally that additive noise at moderate levels does not have a major effect on the performance of the type of systems which we are discussing.

the retina this image forms a two-dimensional spatial signal as expressed by

$$I_{x,y} = i_{x,y} \cdot r_{x,y} > 0 \tag{51}$$

where $I_{x,y}$ is the image, $i_{x,y}$ is its illumination component, and $r_{x,y}$ is its reflectance component.[3]

The first step in the production of a photographic scene is usually the manufacture of a negative transparency. The name negative is correct only in a multiplicative sense because it is really an *inverse* image as expressed by

$$N_{x,y} = \frac{1}{I_{x,y}} = I_{x,y}^{-1} = i_{x,y}^{-1} \cdot r_{x,y}^{-1} \tag{52}$$

where $N_{x,y}$ is the negative image. If two such negatives are superimposed by placing the transparencies one on top of the other,[4] a third negative image is formed. That combined image is the *product* of the two components as given by

$$^{(3)}N_{x,y} = {}^{(1)}N_{x,y} \cdot {}^{(2)}N_{x,y} = \frac{1}{^{(1)}I_{x,y} \cdot {}^{(2)}I_{x,y}}. \tag{53}$$

Thus if we wish to process images using a homomorphic system that combines its signals according to the law of image formation, that system must obey superposition multiplicatively.

The image processor thus formulated is depicted in Fig. 20 in accordance with our previous multiplicative discussions concerning real positive signals. It has a response image

$$I'_{x,y} = i'_{x,y} \cdot r'_{x,y} > 0 \tag{54}$$

which is the product of the separately processed illumination and reflectance components. This formulation places in evidence three important properties of multiplicative image processors. Regardless of the specific process invoked by the system, the response $I'_{x,y}$ is always a physically meaningful image in the sense that it cannot contain points of negative brightness. The effect that the process has upon the appearance of objects in a scene is independent of the light falling upon those objects, whether bright, dim, or variable. Similarly, the effect that the process has upon the apparent light falling upon objects is independent of the nature of those objects.

If $\log i_{x,y}$ and $\log r_{x,y}$ were to possess frequency components occupying separate frequency areas,[5] the image processor could be designed to perform different processing tasks upon the illumination and reflectance components of an image. In practice the illumination and reflectance components of typical images behave in a manner similar to the envelope and vibration functions of the audio application. Illumination generally varies slowly, while reflection is sometimes static and sometimes dynamic, because objects vary in texture and size and almost always have well-defined

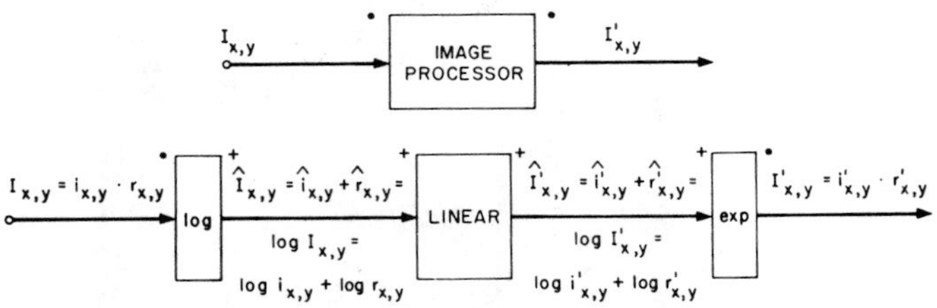

Fig. 20. A multiplicative filter for image processing.

edges. Thus only partially independent processing is possible.

A quantitative measure of the actual spectral content of typical logged images was obtained through computer analysis of the four scenes presented in Fig. 21. The two-dimensional periodograms for the various $\log I_{x,y}$ were computed. They are shown in Fig. 22 where relative brightness represents the magnitude of the periodograms on a decibel scale. In all cases the lowest frequency components are very dominant. To produce a broader view of spectral content the log images were processed by a whitening filter and the periodograms reevaluated. The results of this alternative process are presented in Fig. 23. The two-dimensional frequency characteristics of the whitening filter are shown in Fig. 24. Although the whitening process was only approximate the corresponding periodograms clearly show the high-frequency components of the log images.

The logarithms of all four tested scenes were characterized by an extreme peak in low-frequency energy content and, with minor exceptions, had similar whitened spectra. This two-dimensional data is reminiscent of the one-dimensional spectrum computed for typical log audio and shown in Fig. 13. Again there is an implicit suggestion of two spectral components, one a low-frequency peak, the other a middle- and high-frequency plateau. While it is probably incorrect to associate the peak solely with physical illumination and the plateau entirely with object reflectance, an approximate association of this type has proved most useful in effecting designs involving partially independent processing of the corresponding image components. Before discussing such designs let us explore some of the simpler aspects and uses of the image processor.

If the linear component of the image processor is chosen as a simple amplifier or attenuator with gain γ, the image processor becomes a power law device. The output is given in terms of the input by

$$I'_{x,y} = I^{\gamma}_{x,y}. \tag{55}$$

The parameter gamma is well known to photographers who, by selecting from a variety of photographic materials and using shorter or longer development times for them, control its numerical value. For negative photographic materials $\gamma < 0$ and so they can be thought of as multiplicative inverters.

The general situation calls for the linear component of the image processor to possess a gain which is a function of frequency in two dimensions. Using script letters to represent

[3] We preclude zero image values on practical grounds.

[4] Such superpositions are a prevalent practice in professional photography, expecially color lithography.

[5] Recall that for images the frequency domain is two-dimensional.

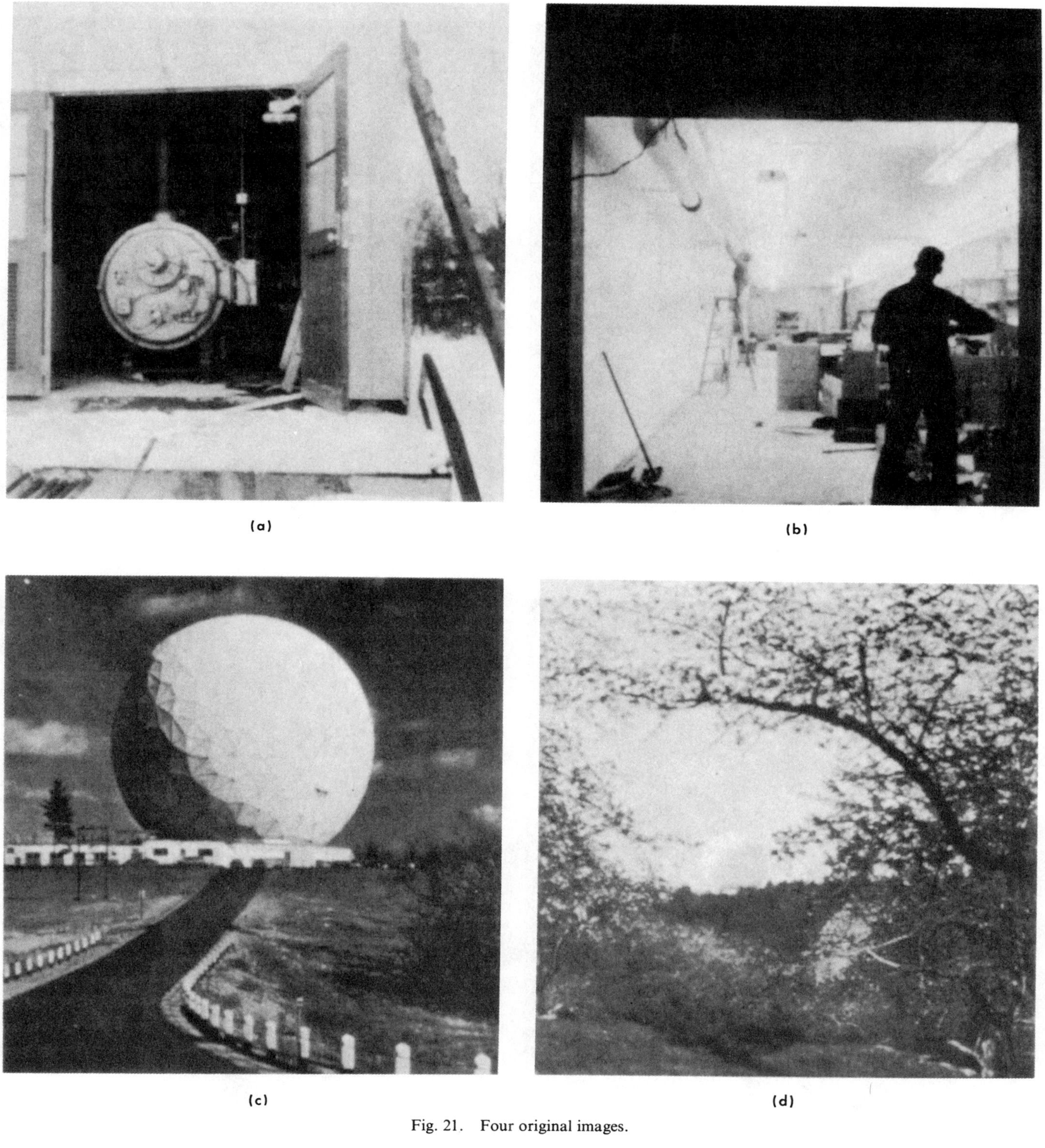

(a) (b) (c) (d)

Fig. 21. Four original images.

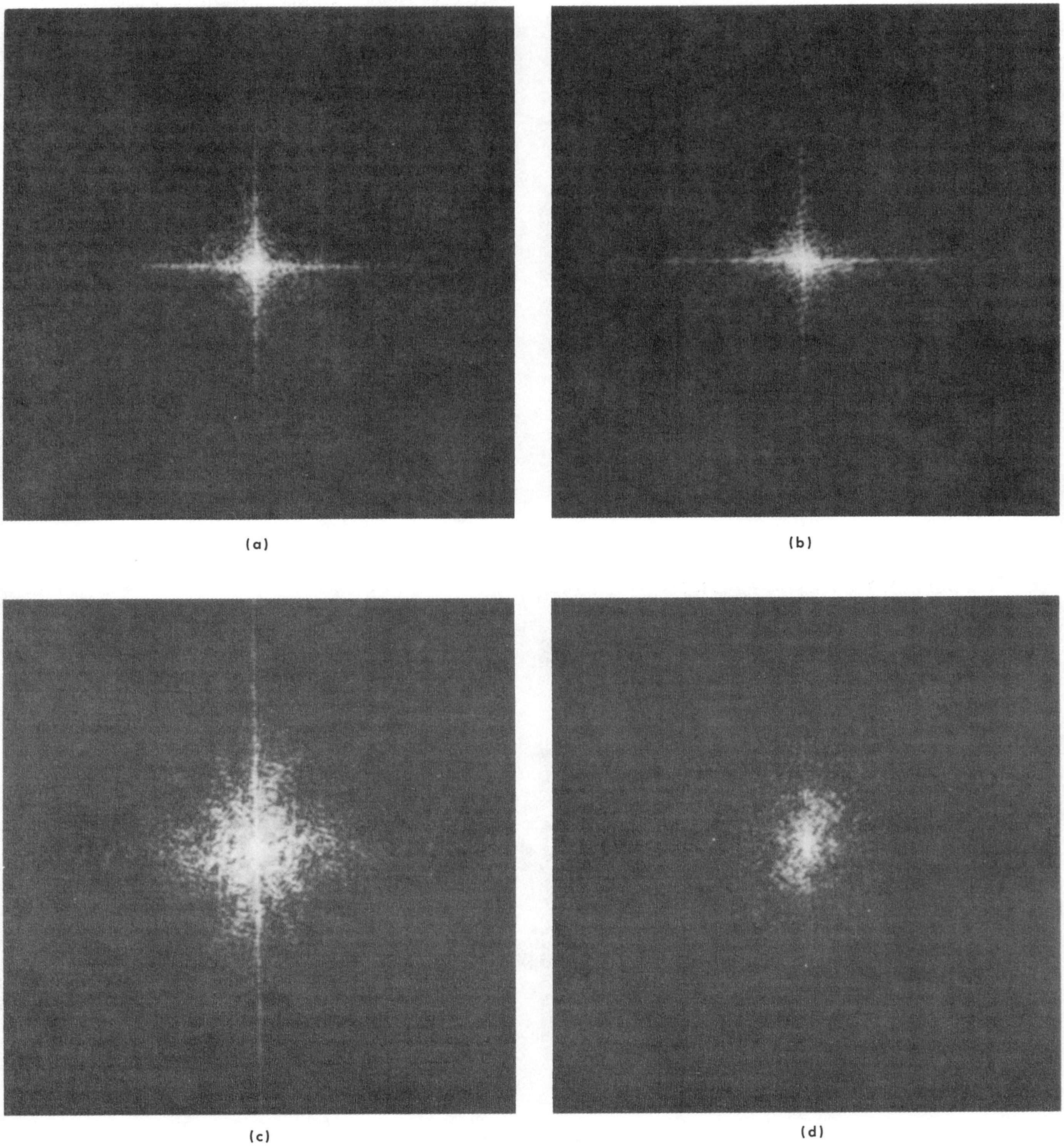

Fig. 22. Log periodograms for the log images corresponding to Fig. 21.

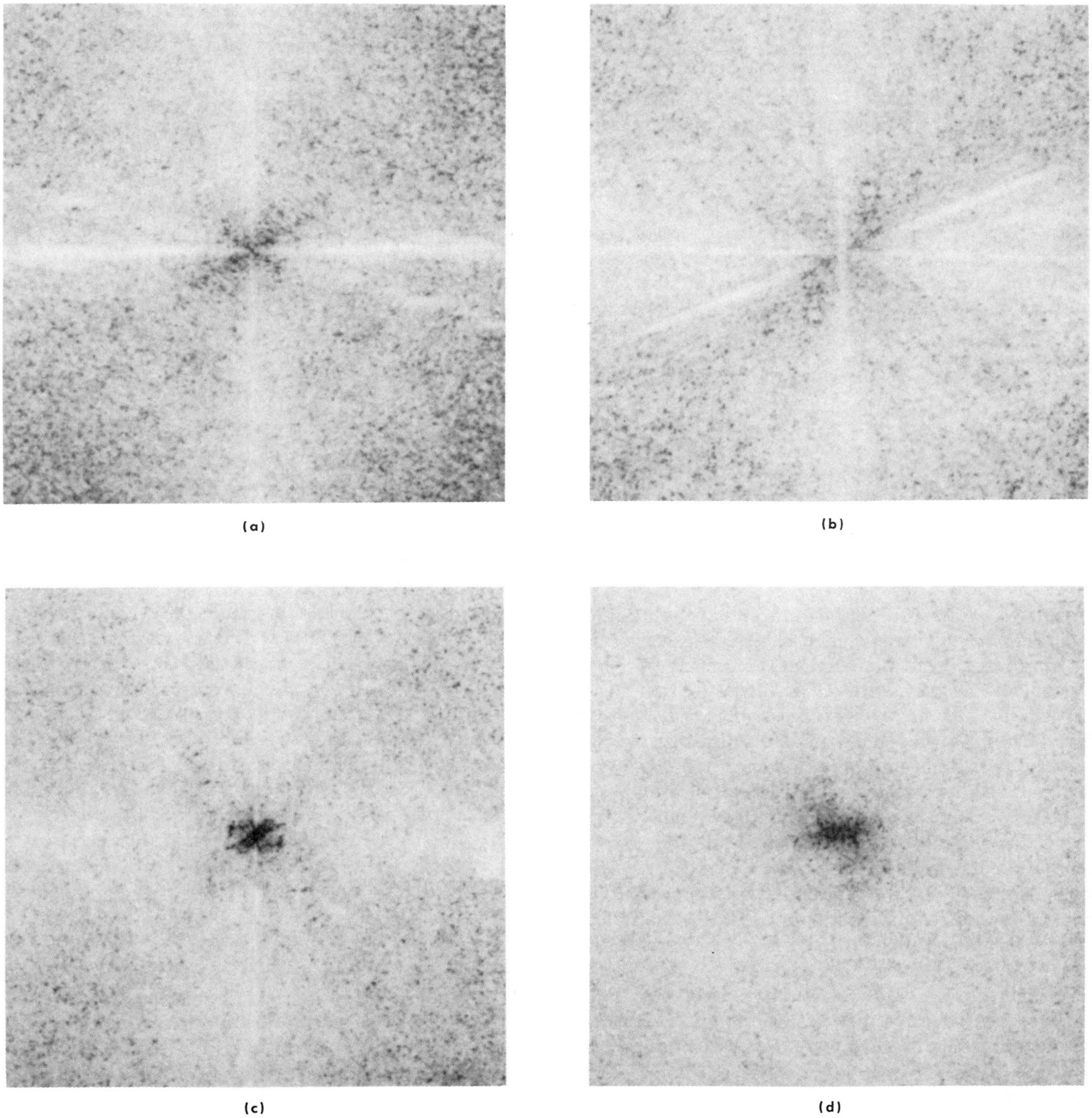

Fig. 23. Log periodograms for the whitened log images corresponding to Fig. 21.

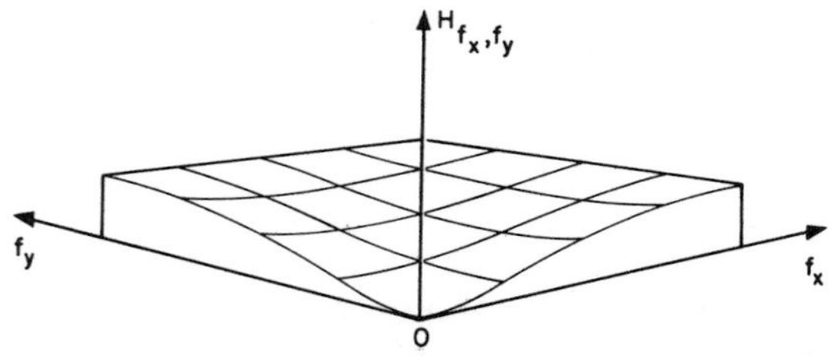

Fig. 24. The whitening filter frequency response. One quadrant shown.

two-dimensional Fourier transforms, and X and Y to represent the frequency variables corresponding to x and y, we can then write

$$\hat{\mathscr{I}}'_{X,Y} = \hat{\mathscr{I}}_{X,Y} \cdot \gamma_{X,Y}.$$

Under these circumstances the image processor can be thought of as having a frequency-sensitive gamma in the sense that it exhibits a different power law behavior for each sinusoidal component of the logarithm of the input image. If we write the input image as a product of components having sinusoidal logarithms, we obtain the following double product:

$$I_{x,y} = \prod_X \prod_Y P_{X,Y;x,y}. \tag{56}$$

The output image is then given by a double product in which each of these components is raised to the appropriate power:

$$I'_{x,y} = \prod_X \prod_Y (P_{X,Y;x,y})^{\gamma_{X,Y}}. \tag{57}$$

A problem common to all forms of image technology is that of excessive dynamic range. Scenes with excessive light-to-dark ratios are usually handled by cramming them to fit the available medium with the result that highlights lose their bright luster and lowlights are without detail. The four scenes of Fig. 21 have been treated in this manner. In an alternate approach gamma may be selected as less than unity so that the image has a reduced ratio and may be reproduced without exceeding the limited dynamic range of practical media. If carried to extremes, this gives the image a muddy or washed-out appearance. Fig. 25(a) shows the original scene of Fig. 21(a) after it has been processed by an image processor with a gamma of one-half. For some applications, gammas greater than one are used in an attempt to give scenes more sharpness about the edges of objects. For example, Fig. 25(b) shows the scene of Fig. 21(a) processed by a gamma of two. A common consequence is that dynamic range capacities are exceeded even more than before.

While the reduction of dynamic range and the enhancement of edge sharpness seem to be conflicting objectives, it is possible to achieve both simultaneously by employing a multiplicative image processor having a linear component with a frequency-dependent gain. To explain this we must return to our previous discussions concerning the partially independent processing of illumination and reflectance components.

The large dynamic range encountered in natural images is contributed to mostly by large variations in illumination $i_{x,y}$, which we recall contains primarily low frequencies in its logarithm. The edges of objects, on the other hand, contribute only to the reflectance component $r_{x,y}$ of a scene, indeed, primarily to the high frequencies of its logarithm. It follows that if one desires to maintain normal contrast for the details of an image, but demands a reduction in dynamic range, the gain of the linear component should be unity for high frequencies and less than unity at low frequencies. This situation is identical to that considered for the audio compressor.

Fig. 25(c) presents the image of Fig. 21(a) processed with the system of Fig. 20 in which the linear system had the frequency response depicted in Fig. 26. This filter was chosen to be spatially isotropic and so a one-dimensional plot of frequency response is sufficient for its unique specification. It also follows that the phase shift of the linear filter was zero at all frequencies. Notice that at low frequencies the gain of the filter was one-half, while at high frequencies it was unity. In Fig. 25(c) the areas which were dark in the original scene have been made far more visible as if illuminated with auxiliary lights, but without disturbing the rendition of the brightly lighted areas.[6] This effect and the effect of sharpening through the use of gammas larger than one are obtained simultaneously in the picture of Fig. 27. In this case, the frequency response was as given in Fig. 28. Again the filter had isotropic properties, so a one-dimensional frequency response curve was sufficient to specify it uniquely, and again the phase was zero. Notice that while the gain at low frequencies was maintained at one-half, the gain at high frequencies was increased to two. Fig. 27 is a kind of blend of Fig. 25(a) and (b) in which the best properties of both have been retained and the worst properties of both greatly reduced. Relying upon our approximate analysis which assigns the lowest-frequency components of an image to the illumination and the higher-frequency components to the reflectance, we can see that according to our definition of partially independent processing the illumination component has been treated by a gamma equal to one-half while the reflectance components have been treated by a gamma equal to two. This situation is summarized as follows:

$$I'_{x,y} = i'_{x,y} \cdot r'_{x,y} \approx i^{0.5}_{x,y} \cdot r^2_{x,y}. \tag{58}$$

Since the large brightness ratios in a subject are usually produced by large variations in illumination, the fact that $I'_{x,y}$ contains the square root of the original illumination explains why the brightness ratio is reduced. On the other hand, the reflectance component has the same [Fig. 25(c)] or greater (Fig. 27) variability than in the original and thus details are preserved or enhanced. As has been stated, (58) is only an approximate equation. In fact, the illumination which formed the scene certainly contains some high-frequency components while the reflectance function certainly

[6] Results similar to this can be obtained by means of the classical photographic technique of unsharp masking [14] or by use of a logetronics printer [15].

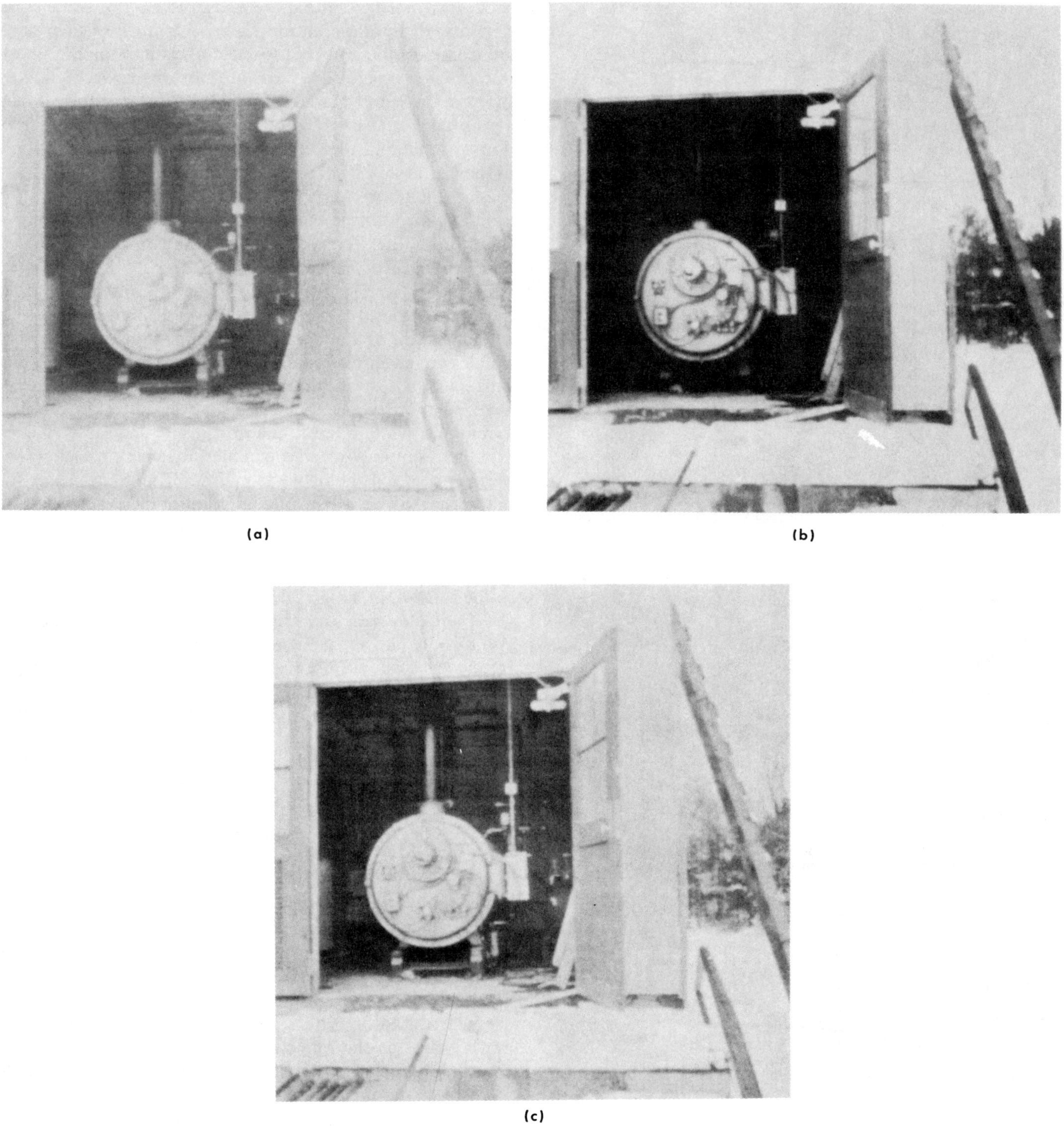

Fig. 25. The image of Fig. 21(a) processed using (a) $\gamma = 1/2$, (b) $\gamma = 2$, (c) a frequency-dependent γ with low-frequency attenuation.

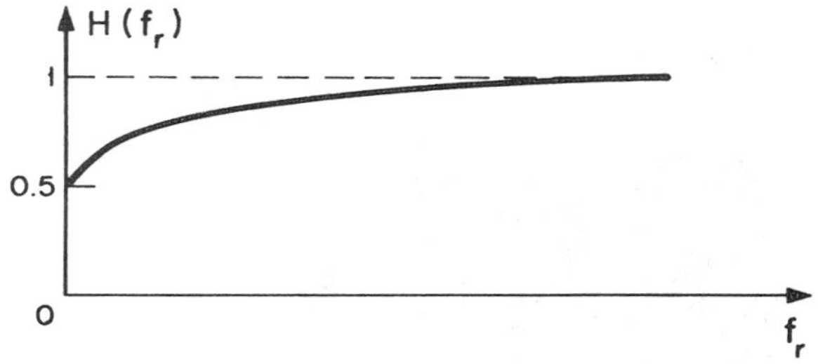

Fig. 26. The radial cross section of the multiplicative frequency response used to produce Fig. 25(c).

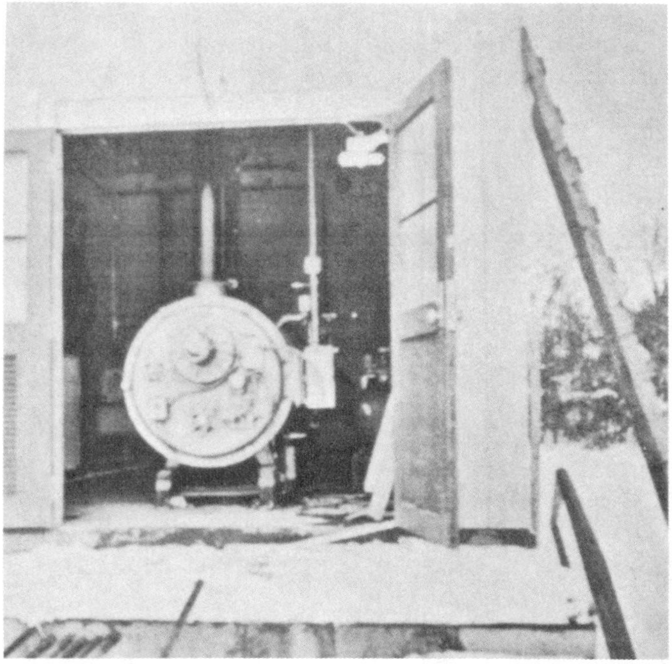

Fig. 27. The image of Fig. 21(a) processed using a frequency-dependent γ with low-frequency attenuation and high-frequency amplification.

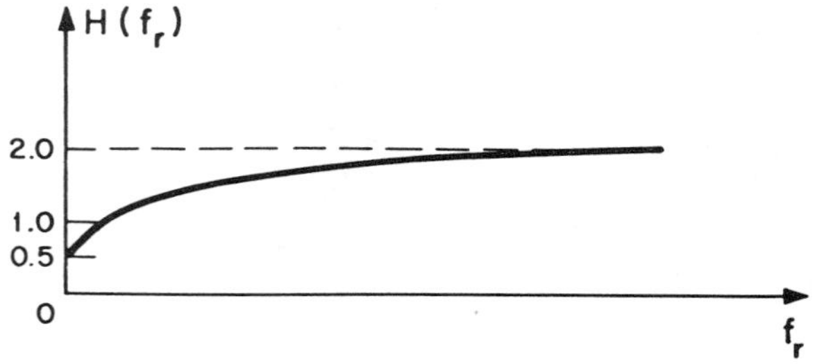

Fig. 28. The radial cross section of the multiplicative frequency response used to produce Fig. 27.

contains some low-frequency components. A communications engineer would say that there is cross talk between these two components of the original image. Thus, in the processing used to obtain the image of Fig. 27, some components of the illumination function have been increased and some components of the reflectance function have been decreased. Subjectively, these are not obtrusive. However, they are there and can be seen most easily, especially if they are pointed out. Specifically, in Fig. 27 the glow around the doorway making the building look whiter than it really is in the vicinity of the blackened room, the intense brightness level of the door dampener, and the white ring around the boiler-shaped object inside the room are typical artifacts of this type. Others can be seen in Fig. 29(a), (b), and (c), which bear the same relationship to Fig. 21(b), (c), and (d), respectively, as Fig. 27 does to Fig. 21(a). In spite of these approximations, the use of these methods in the processing of images has obvious practical implications wherever dynamic range is limited and the preservation of details is important. It has the additional advantage of obeying a law of superposition which facilitates analysis through classical techniques while operating according to the same rules of combination that form the original subject information. The degree to which the artifacts in Fig. 27 and Fig. 29(a), (b), and (c) are visible is a strong function of the exact nature in which the frequency response of Fig. 28 makes its transition from one-half at low frequencies to two at high frequencies. Fig. 30(a) and (b) were obtained from the image of Fig. 21(b) in the same manner as was Fig. 29(a), with the exception of the frequency response shape used. These frequency responses are shown in Fig. 31(a) and (b), respectively. Both frequency responses are characterized by a rather rapid transition between the high-frequency asymptote of two and the low-frequency asymptote of one-half. They differ, however, in that the transition occurs in the case of Fig. 31(a) at a relatively high frequency and in Fig. 31(b) at a relatively low frequency. The halo and flaring effects of Fig. 30(a) and (b) are more objectionable by far than those of Fig. 27. The frequency response of Fig. 28 is a compromise. That of Fig. 31(a) favors flare along large objects at the expense of halo around small ones. That of Fig. 31(b) favors halo around small ones at the expense of flare along large.

The compromise characteristics of the frequency response of Fig. 28 were arrived at through an attempt to find the frequency response which would treat large and small objects equally or nearly so. Since the two-dimensional Fourier spectrum of an object contracts as the object grows in size and spreads as the object shrinks in size, a frequency response characteristic which is somewhat invariant to changes in frequency scale would meet the objective. A characteristic possessing some invariance is the logarithmic frequency function. This invariance property is described by

$$\log Af = \log A + \log f. \tag{59}$$

If we think of the parameter A as a scale factor on an image of standard size, then we see that, except for an additive constant, the frequency response which an image en-

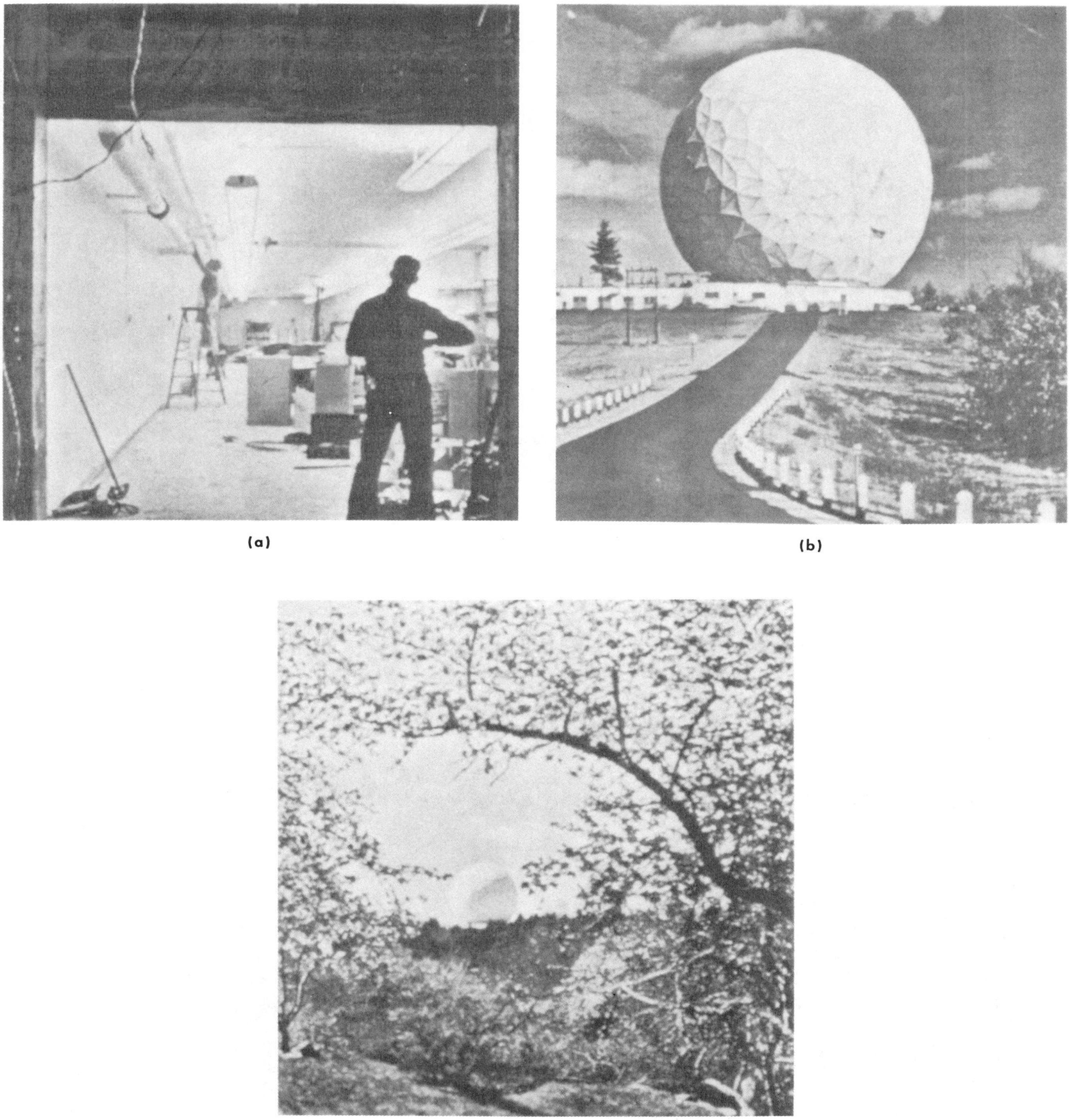

(a) (b) (c)

Fig. 29. The remaining images of Fig. 21 processed as in Fig. 27.

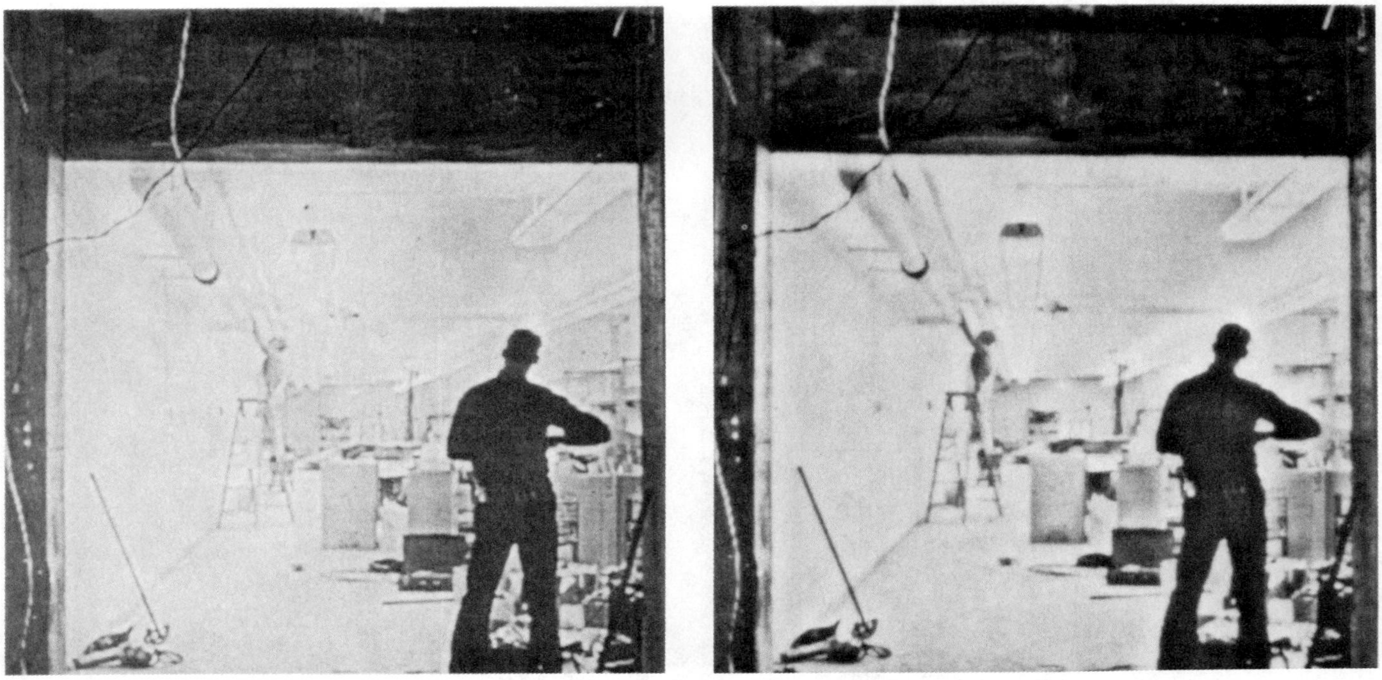

(a) (b)

Fig. 30. Images processed using abruptly changing frequency characteristics.

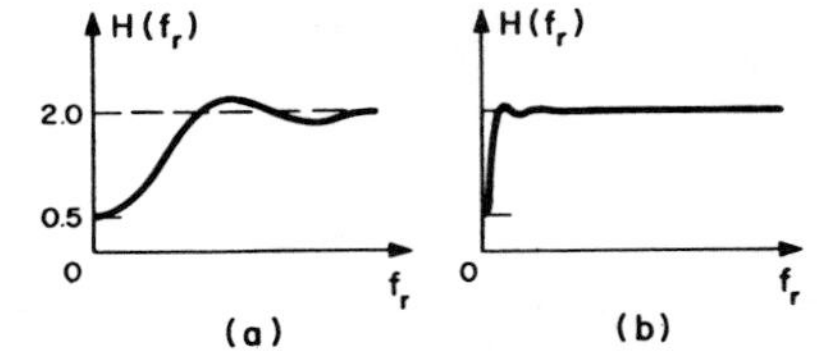

(a) (b)

Fig. 31. The multiplicative frequency responses used to produce the images of Fig. 30 from the image of Fig. 21(b).

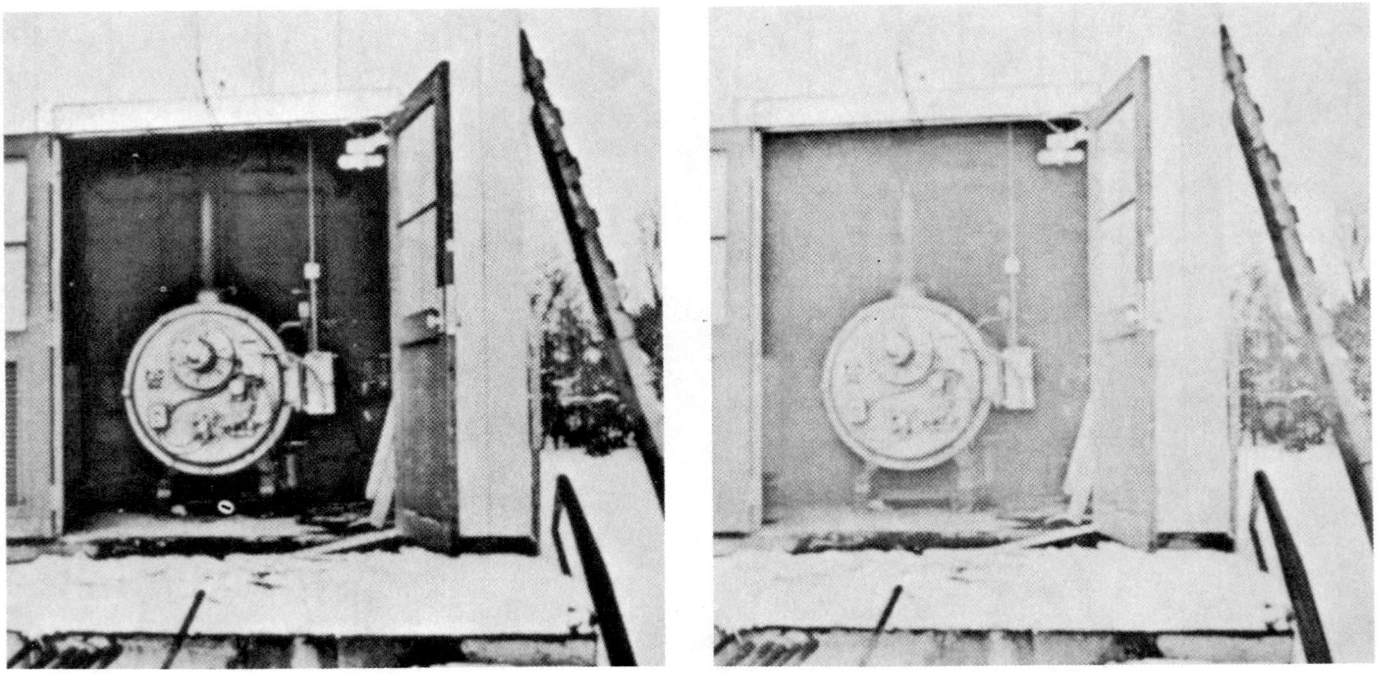

(a) Biased for best appearance. (b) Average value of image restored.

Fig. 32. Two versions of the image of Fig. 21(a) processed as in Fig. 27 but using an ordinary linear filter rather than a multiplicative filter.

counters after magnification or reduction by the scale factor A is the same as that which it encounters at its standard size. The frequency response of Fig. 28 is such a logarithmic characteristic to a first approximation. Only for frequencies extremely close to zero is the logarithmic variation altered to provide an asymptote to the value of one-half rather than minus infinity.

At this point it is quite reasonable to ask what effect ordinary linear filtering would have upon the images presented here and to draw a comparison in effectiveness. To this end we present Fig. 32(a) and (b). Fig. 32(a) represents the original scene of Fig. 21(a) processed using the frequency response of Fig. 28 in an ordinary linear filtering process and biased for best appearance. The results are instantly striking, but careful examination reveals some severe drawbacks. The most serious of these is the black halo around the inside of the doorway in which all detail is lost. While the visibility inside the room has been increased the improvement falls short of that obtained multiplicatively in Fig. 27. Finally, there are many places in the scene that are much darker than they should be. Most of these problems are associated with the fact that the linear processing employed produces negative brightness values in the processed image which due to the half-wave rectification of photographic processing appear as black. The addition of minimum bias sufficient to eliminate negative brightness results in a washed-out image of such poor quality that we do not show it here. Fig. 32(b) is a compromise in this respect in which bias sufficient to restore the average value of the original image brightness is used.

An effective approach to the problem of television bandwidth reduction [16], [17] centers around the idea of separating images into a relatively narrow-bandwidth low-frequency component and the complementary high-frequency component and preserving only a small fraction of the information in the high-frequency component in the form of edge contours. At the receiver, the edge contours are used in an attempt to re-create the high-frequency component. The results are combined with the low-frequency component to produce an approximation of the original image. Standard practice has been to consider the low- and high-frequency components of the image in the additive sense in keeping with the established traditions of signal processing. If these components are taken in the multiplicative sense, the results can be considerably enhanced. There are two reasons for this. When the components are taken in the additive sense, poorly illuminated edges may remain undetected during the process of bandwidth reduction. In the multiplicative case, all edges are represented equally by the high-frequency component since variations in illumination have been more or less separated out with the low-frequency component. Since any bandwidth reduction process will introduce errors in the image and practice indicates that these errors occur more or less uniformly throughout the picture, then in the additive case poorly illuminated portions of the image will be dominated by error and thus rendered totally useless. If the process is carried out multiplicatively, constant errors are made in terms of the logarithm of the picture and thus represent proportional errors in the final image. In this way, brightly illuminated areas and dimly illuminated areas are treated equivalently.

Fig. 33 represents an image as it appears in the various stages of bandwidth reduction when the process is carried out using both the traditional ideas of additive superposition and those of multiplicative superposition. The low-frequency images of Fig. 33(a) and (b) are subjectively nearly indistinguishable and, for all intents and purposes, appear much the same as a defocused photograph. The edge contours of Fig. 33(c) and (d) are markedly different, however. These images were produced by differentiating the original images in two dimensions and clipping the results into three levels as dictated by suitably adjusted thresholds. Notice that for linear processing there is less edge contour information in the regions corresponding to the dark areas of the original. The artificial-highs images of Fig. 33(e) and (f) which were produced directly from the edge contours by restoration filters can be compared similarly. The reduced-bandwidth recreated images of Fig. 33(g) and (h) differ much as would be expected from the statements made above. In the most brightly illuminated areas, the preservation of details is more faithfully carried out by the linear process. In the most dimly illuminated areas, however, the preservation of details is more faithfully carried out by the multiplicative process. Histograms of brightness for typical scenes are heavily skewed towards black. Similar histograms of the logarithm of these scenes reveal more or less rectangular distributions of log brightness placing equal weight on bright and dark areas. This fact further favors the use of the multiplicative scheme. Since the eye is sensitive more nearly to brightness ratios than to absolute brightness, it is not surprising that percentage errors are to be preferred on a subjective basis. A drawback to the multiplicative process is that the bandwidth-reduced edge contour image of Fig. 33(c) contains more information than its counterpart of Fig. 33(d) and thus, all else being equal, the use of the multiplicative scheme can result in smaller bandwidth reductions. This fact is suggested by the first-order entropies of the actual numerical samples of Fig 33(c) and (d) which are 1.077 and 0.945 bits per picture element, respectively.

The images presented here were all processed digitally by the TX-2 computer at the M.I.T. Lincoln Laboratory. The analog signal from a low-noise rotating scanner was fed to a twelve-bit analog-to-digital converter and the resulting numbers stored in the computer memory. Tests have shown that these numbers contain ten bits of significance. Each image was represented by a square array, 340 samples on a side. Before deposition in a permanent image library, the twelve-bit samples were converted to logarithms, the most significant nine bits of which were maintained.

The linear processing was performed through the use of high-speed convolution methods [18] applied in two dimensions. The two-dimensional isotropic convolution kernels were determined through the Hankel transforms [17] of the defining frequency characteristics of Figs. 26, 28, and 31 truncated to possess nonzero values inside circles with diameters of about 80 picture elements. Each convolu-

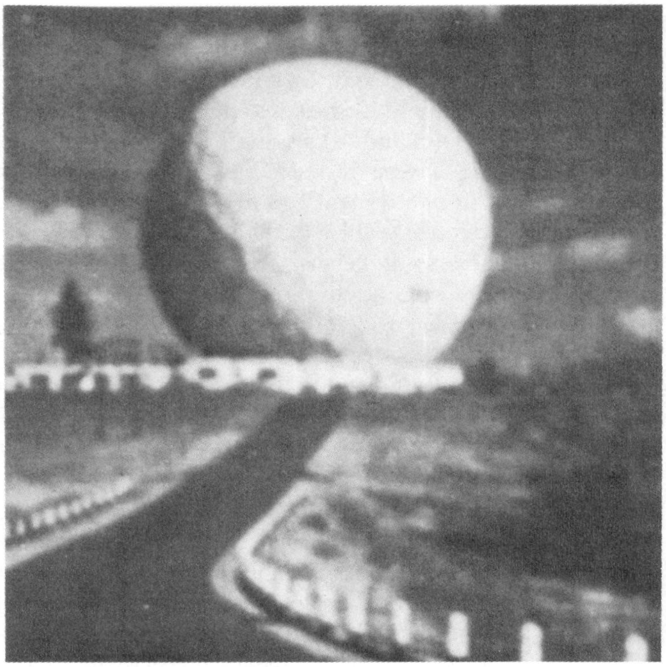

(a) Low-pass, multiplicative.

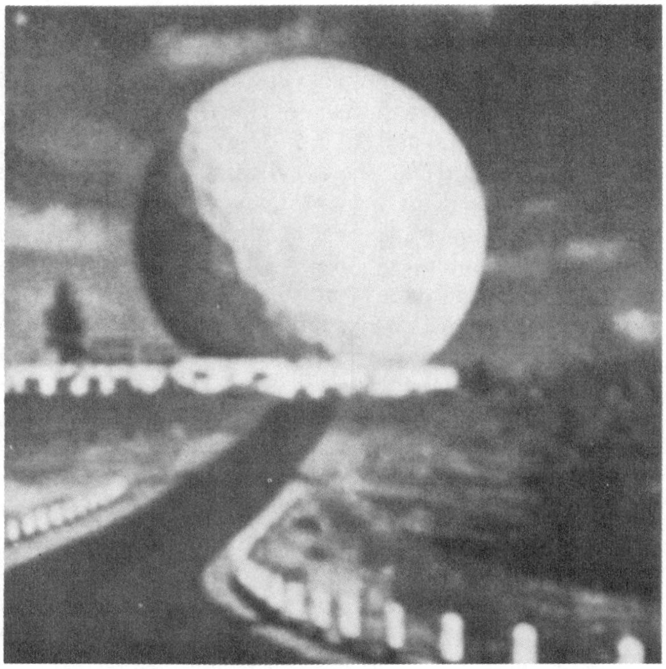

(b) Low-pass, linear.

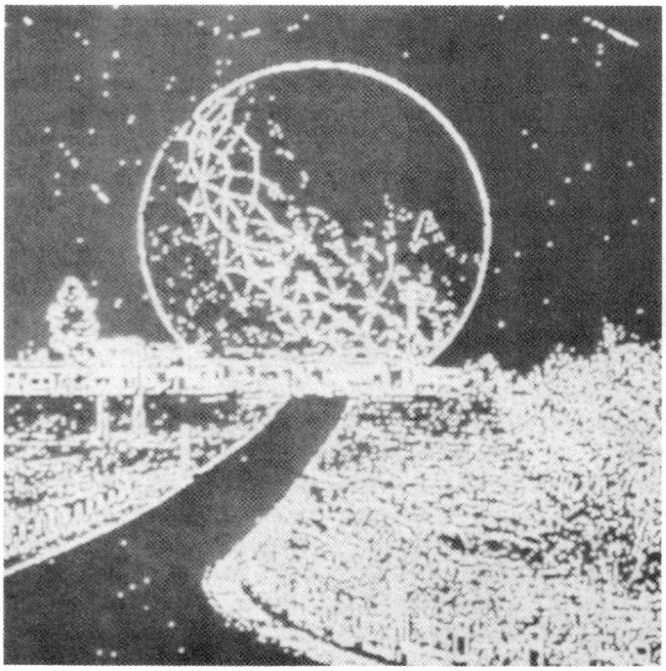

(c) Edge contours, multiplicative.

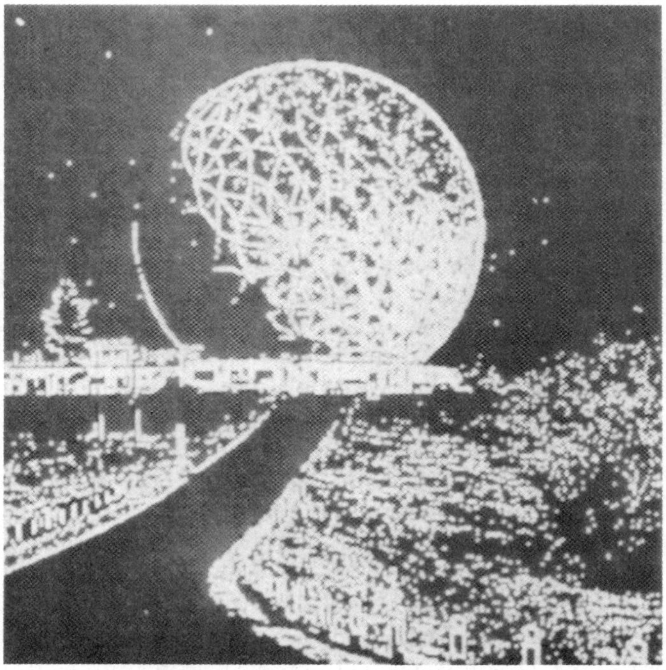

(d) Edge contours, linear.

Fig. 33. The image of Fig. 21(c) in various stages of bandwidth compression.

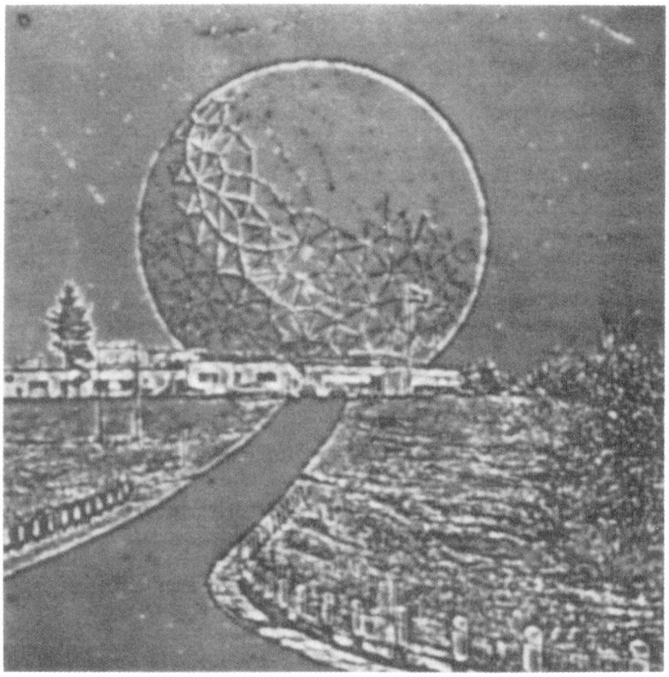

(e) Artificial highs, multiplicative.

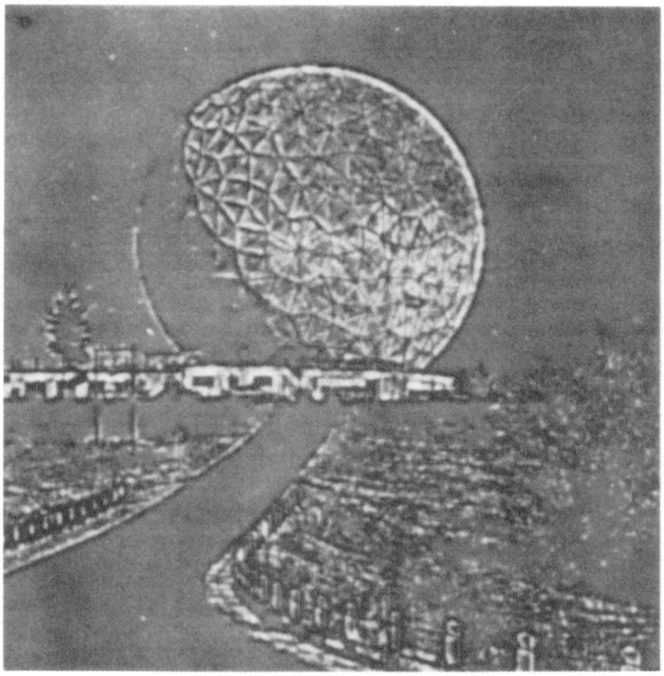

(f) Artificial highs, linear.

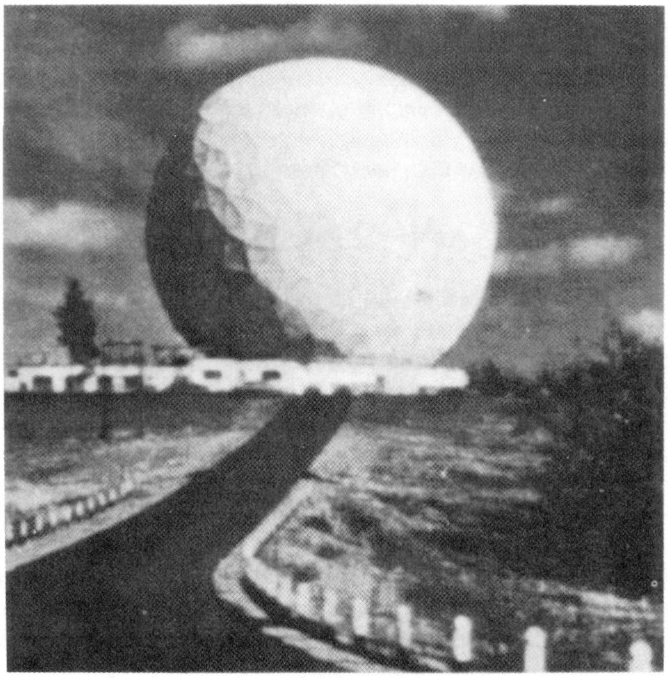

(g) Recreated, multiplicative.

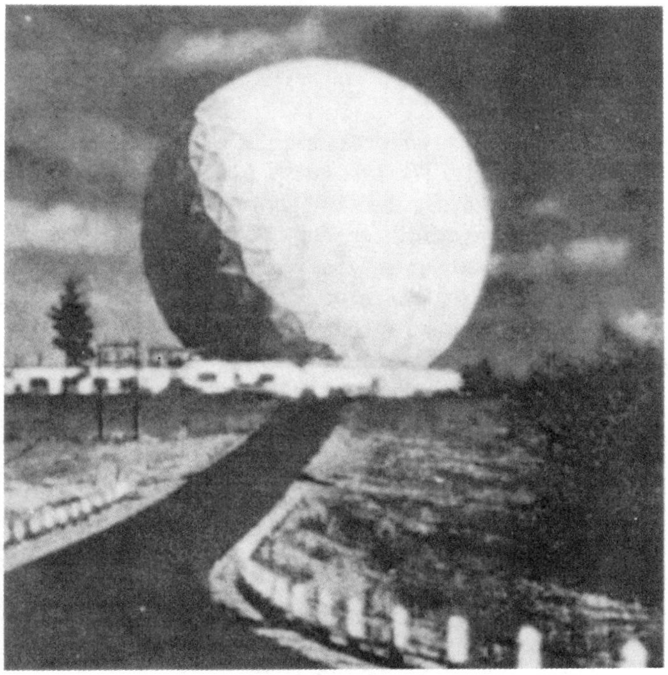

(h) Recreated, linear.

Fig. 33 (*Cont'd*).

tion required about 13 minutes.

Photographs of processed images were made by exponentiating the filtered image logarithms and controlling a vector-drawing computer graphics display [19] with the resulting eight-bit numerical values. The control was arranged to vary the velocity of scan while holding the cathode ray intensity at a standard constant level, thus avoiding the complications of beam and phosphor nonlinearity. The time to display an output image was between ten and fifteen seconds depending on average scene brightness, thus permitting real-time viewing in a darkened room as well as ordinary photographic recording through time exposure. For the latter, special digital compensation curves were used to straighten the nonlinear photographic characteristics of the films employed, thus resulting in much improved image quality.

Homomorphic Filtering of Echoed Signals [20], [21]

In many areas of application, signals are transmitted or recorded in a reverberant environment, i.e., one which introduces echoes. Reverberation arises, for example, in audio recording, in multipath communication, and in radar and sonar detection. In many cases we wish to remove the distortion represented by the echoes, or to recover the echo structure as a means of probing and characterizing the channel.

A simple model for the distortion introduced by reverberation is a convolution of the original waveform with a train of weighted samples, i.e.,

$$x(n) = s(n) \otimes p(n)$$

$$p(n) = \sum_{k=0}^{\infty} \alpha_k \delta(n - n_k)$$

where $s(n)$ and $x(n)$ are the original and distorted waveforms, respectively. The analysis presented previously suggests that in applying the notion of homomorphic deconvolution to separating the echo pattern and the original waveform, we determine the complex logarithm of the z-transform of $x(n)$, the distorted waveform, and then look for a property of each of the components that permits their separation by means of linear filtering. To help focus the approach let us first consider the case of a simple echo, i.e.,

$$p(n) = \delta(n) + \alpha\delta(n - n_0)$$

so that

$$x(n) = s(n) \otimes [\delta(n) + \alpha\delta(n - n_0)]. \tag{60}$$

The z-transform $X(z)$ evaluated on the unit circle is

$$X(e^{j\omega}) = S(e^{j\omega})[1 + \alpha e^{-j\omega n_0}].$$

We observe that the contribution due to the echo is a periodic function of ω with period $2\pi/n_0$. Furthermore, its repetition rate increases as n_0 increases so that longer echo times are manifested by more rapid fluctuations in the spectrum. Since the logarithm of a periodic waveform remains periodic with the same repetition rate, the echo is represented in the log spectrum as an *additive* periodic component.

The character of the log spectrum of $p(n)$ suggests the possibility that we may separate the contributions of $s(n)$ and $p(n)$ by removing the variations in the log spectrum which occur at repetition rates which are multiples of $2\pi/n_0$. Thus the linear filtering would convolve the complex log spectrum with a kernel designed to remove the periodic components.

Since convolution in frequency corresponds to a multiplication in time, we may view this linear filtering as a multiplication of the complex cepstrum by a fixed weighting. Specifically, we observe that periodic variations in the log spectrum contribute to the complex cepstrum only at values of n which are multiples of the echo time n_0, in precisely the same way that a periodic time function with period T has spectral components at only those frequencies which are multiples of $1/T$. Then the "comb" filtering suggested above will correspond to multiplying the complex cepstrum by a weighting which is unity except in regions that are multiples of the echo time n_0, in which case the weighting is zero. This class of filters is depicted in Fig. 34. Clearly the notion of comb filtering can only be successful if we have approximate information about the echo time. If we do not have this information, and if the complex cepstrum of $s(n)$ is concentrated near $n=0$, then we can replace the idea of comb filtering with that of "low-time" filtering, that is, weighting the complex cepstrum by unity near the origin and zero otherwise. In terms of the log spectrum this filtering corresponds to associating the slow variations with $s(n)$ and the rapid variations with the echo pattern $p(n)$. An alternative is to use an adaptive procedure whereby the parameters of a comb filter are based on a measurement of the echo time. Such a measurement can be based on the fact that a peak is expected to occur in the complex cepstrum at multiples of the echo time. If we return to the more general case of multiple echoes, we recognize that as long as the echoes are equally spaced so that

$$p(n) = \sum_{k=0}^{\infty} \alpha_k \delta(n - kn_0)$$

the approach is essentially identical to the case of a single echo. If the echoes are not equally spaced the situation becomes more complex since we can no longer localize the effect of the echo pattern in the complex cepstrum.

In principle the approach taken to echo removal by means of homomorphic filtering is based on the reasoning presented above. However, the above discussion assumes that we have available the entire waveform $x(n)$ and are able to compute its Fourier transform. The more typical situation in practice is that the waveform $s(n)$ is indefinite in duration and consequently it is impractical to compute the Fourier transform of the entire waveform. In addition, there are situations in which the reverberation times vary slowly. Thus we are led to considering echo removal in which a short-time analysis of the waveform is more appropriate. In this case the waveform is processed in pieces and the results fitted together to obtain the output. To illustrate how this can be done, let us again consider a simple echo as in (60) and for which $s(n)$ is a waveform of indefinite dura-

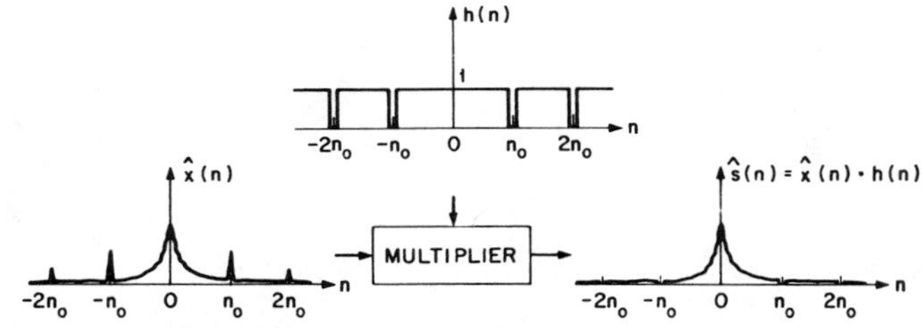

Fig. 34. A linear time-varying "comb" filter for removing the component in the complex cepstrum due to a simple echo.

tion. The situation is depicted in Fig. 35(a) where $s(n)$ is represented by the solid curve and $\alpha s(n-n_0)$ is represented by the dotted curve. Let us consider segments of $x(n)$ consisting of L samples. To facilitate our discussion, we define for $\xi=0, L, 2L, \cdots$,

$$x(\xi, n) = x(\xi + n) \quad 0 \le n < L$$
$$= 0 \quad \text{otherwise.}$$

From Fig. 35(a) we see that, in general, a particular segment of the input can be expressed in the form

$$x(\xi, n) = s(\xi, n) + \alpha s(\xi, n - n_0) + e(\xi, n)$$

where

$$s(\xi, n) \neq 0 \quad 0 \le n < L$$
$$= 0 \quad \text{otherwise}$$

and

$$e(\xi, n) \neq 0 \quad 0 \le n < n_0 \quad \text{and} \quad L \le n < L + n_0$$
$$= 0 \quad \text{otherwise.}$$

The term $e(\xi, n)$ accounts for the overlap of the echo from the previous segment at the beginning of the segment and the overlap into the next segment at the end of the segment. Since it is the amount by which $s(\xi, n)$ fails to have the desired form, $e(\xi, n)$ is referred to as the error in the segment.

The nature of the errors is depicted in Fig. 35(b) for three consecutive segments of the input of Fig. 35(a). It can be seen that the error at the end of a segment is the negative of the error at the beginning of the next segment.

If we take the z-transform of a segment of the input, we obtain

$$\begin{aligned} X(\xi, z) &= \sum_{n=0}^{L-1} x(\xi, n) z^{-n} \\ &= S(\xi, z)(1 + \alpha z^{-n_0}) + E(\xi, z) \qquad (61) \\ &= \left[S(\xi, z) + \frac{E(\xi, z)}{1 + \alpha z^{-n_0}} \right] (1 + \alpha z^{-n_0}) \end{aligned}$$

where $S(\xi, z)$ is given by

$$S(\xi, z) = \sum_{n=0}^{L-1} s(\xi, n) z^{-n}.$$

We note that $X(\xi, z)$ is not simply a product as it would be if the entire waveforms were transformed. It is true, however, that if

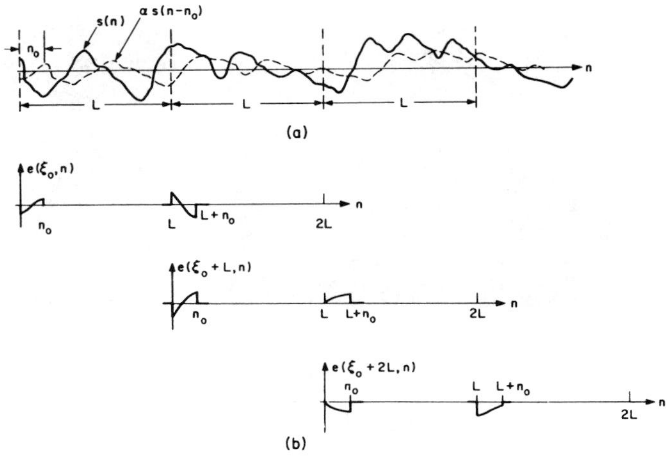

Fig. 35. (a) A signal $s(n)$ (solid line) and delayed scaled replica (dashed line). Each of the depicted segments L is to be represented separately as the sum of a component signal, an echo of this signal, and an error. (b) The error term in this representation depicted for each segment.

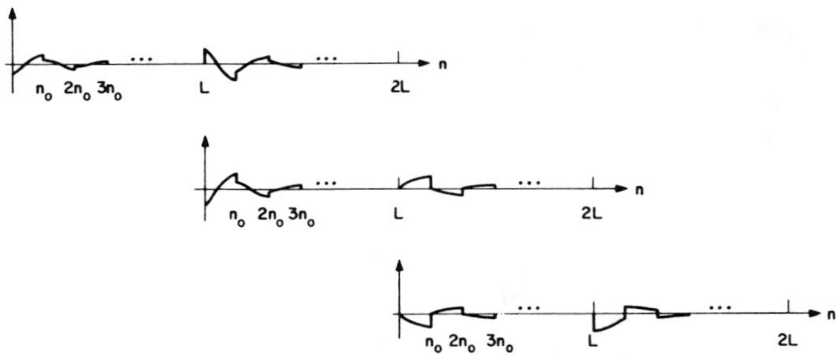

Fig. 36. The output due to the error terms of Fig. 35(b). Successive lines represent the response due to the error term for adjacent segments of length L in Fig. 35(a). Note that the error at the end of one segment is the negative of the error at the beginning of the next segment.

$$L \gg n_0$$

then there will still be impulses in the complex cepstrum at $n_0, 2n_0, \cdots$. Removing these impulses is equivalent to removing the factor $(1+\alpha z^{-n_0})$ in (61) so that operating with the inverse characteristic system on the complex cepstrum with the impulses removed yields an output whose z-transform is

$$Y(\xi, z) = S(\xi, z) + \frac{E(\xi, z)}{1 + \alpha z^{-n_0}}.$$

If $|\alpha|<1$, then the corresponding output sequence is

$$y(\xi, n) = s(\xi, n) + \sum_{k=0}^{\infty} (-\alpha)^n e(\xi, n - k n_0).$$

Thus the output consists of the desired output segment $s(\xi, n)$ plus an error term. This error term is effectively the error in the input segment passed through a linear system whose system function is the reciprocal of the z-transform of the impulse train which represents the echoes.

The error in the output for the three consecutive segments of Fig. 35 is depicted in Fig. 36. The figure suggests

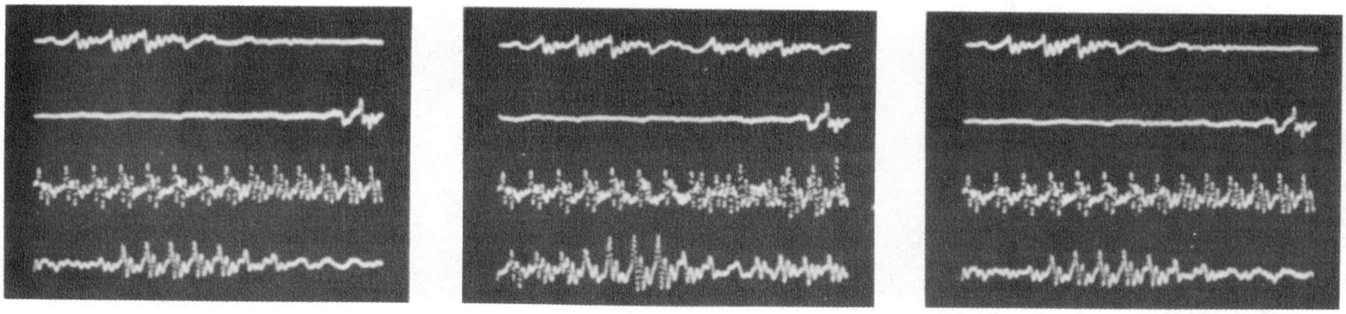

(a) (b) (c)

Fig. 37. An example of homomorphic echo removal. (a) 410 ms of speech sampled at 10 kHz with the four traces from top to bottom representing contiguous segments of 102.5 ms. (b) The speech sample of (a) with a 50-ms echo. (c) The speech sample of (b) processed to remove the echo.

how the output segments can be put together. If we simply add the appropriately delayed output segments, the result will be the desired output signal $s(n)$. An alternate procedure is suggested by the fact that if $L \gg n_0$, the error in the output decays as n gets large. Thus there may be large portions of each output segment which are relatively free of error. In this case we can process overlapping segments of length L, and save only part of each output segment. If we choose segments such that the portions saved are contiguous parts of the waveform, these pieces can simply be placed end-to-end to produce the output.

This approach to echo removal has been carried out on the TX-2 computer at the M.I.T. Lincoln Laboratory using speech as the original waveform with echoes artificially introduced. Informal listening tests indicate that echoes can be removed to the extent that they are inaudible with only minor degradation of the speech. An illustration of typical waveforms obtained is shown in Fig. 37. Fig. 37(a) represents 410 ms of speech sampled at 10 kHz with the four traces from top to bottom representing contiguous segments of 102.5 ms. This waveform with a 50-ms echo is shown in Fig. 37(b). The result of carrying out the processing which has been described above is shown in Fig. 37(c), indicating that the echo has, to a large extent, been removed.

Deconvolution of Speech [22], [23]

During voiced sounds such as vowels, the speech waveform may be considered as the result of periodic puffs of air released by the vocal cords exciting an acoustic cavity, the vocal tract [4]. Thus a simple and often useful model of the speech waveform consists of the convolution of three components, representing pitch, the shape of the vocal cord or glottal excitation, and the configuration of the vocal tract. Many systems for compressing the bandwidth of speech and for carrying out automatic speech recognition have as the basic strategy the separate isolation and characterization of the vocal tract excitation and the vocal tract impulse response. Thus many speech processing systems are directed in part toward carrying out a deconvolution of the speech waveform.

As in the previous example of echo removal, the speech waveform is a continuing signal and therefore must be processed on a short-time basis. Thus we consider a portion $s(t)$ of the speech waveform as viewed through a time-limited window $w(t)$. Although the vocal tract configuration changes with time we will choose the duration of the window to be sufficiently short so that we can assume that, over this duration, the shapes of the vocal tract impulse response and the glottal pulse are constant. Then, if we denote by $p(t)$ a train of ideal impulses whose timing corresponds to the occurrence of the pulses released by the vocal cords, by $g(t)$ the shape of the glottal pulse, and by $v(t)$ the impulse response of the vocal cavity, we express $s(t)$ approximately as

$$s(t) = [p(t) \otimes g(t) \otimes v(t)]w(t). \tag{62}$$

Furthermore, if $w(t)$ is smooth over the effective duration of the glottal pulse and the vocal tract impulse response, then we can approximate (62) as

$$s(t) = [p(t)w(t)] \otimes g(t) \otimes v(t). \tag{63}$$

Thus, if a smooth window is used to weight the speech waveform we can consider the weighted aperiodic function as a convolution of weighted pitch, glottal pulse, and vocal tract impulse response.

In keeping with the previous discussion we wish to phrase our remarks in terms of samples of $s(t)$, which we denote by $s(n)$. Assuming that we can replace the continuous convolution of (63) by a discrete convolution of samples of each of the component terms, we write that

$$s(n) = [p(n)w(n)] \otimes g(n) \otimes v(n)$$

or

$$s(n) = p_1(n) \otimes g(n) \otimes v(n) \tag{64}$$

where $w(n)$, $g(n)$, and $v(n)$ are samples of $w(t)$, $g(t)$, and $v(t)$, respectively, and $p_1(n)$ is a train of unit samples weighted with the window $w(n)$.

The vocal tract impulse response $v(n)$ is often modeled as the response of a cascade of damped resonators so that its z-transform is

$$V(z) = \frac{K}{\prod_{i=1}^{M}(1 - a_i z^{-1})(1 - a_i^* z^{-1})} \qquad |a_i| < 1.$$

In this case, $v(n)$ is minimum phase, and it follows from (31) that $\hat{v}(n)$, the complex cepstrum of $v(n)$, is of the form

$$\hat{v}(n) = \begin{cases} \sum_{i=1}^{M} \frac{|a_i|^n}{n} \cos \omega_i n & n > 0 \\ 0 & n < 0 \end{cases}$$

where

$$a_i = |a_i| e^{j\omega_i}.$$

Thus, $\hat{v}(n)$ decays as $1/n$ and therefore tends to have its major contribution near the origin for $n>0$.

An accurate analytical representation of the glottal pulse $g(n)$ is not known and consequently it is difficult to make any specific statements regarding the characteristics of its complex cepstrum $\hat{g}(n)$. However, we can expect in general that $g(n)$ is nonminimum phase [24]. Expressing $g(n)$ as the convolution of a minimum phase sequence $g_1(n)$ and a maximum phase sequence $g_2(n)$, we will assume that $\hat{g}_1(n)$, which is zero for $n < 0$, and $\hat{g}_2(n)$, which is zero for $n > 0$, both have an effective duration which is less than a pitch period.

The complex cepstrum of the train of weighted unit samples representing pitch is, as we have seen previously, a train of weighted unit samples with the same spacing. Thus, we can diagrammatically represent the complex cepstrum as in Fig. 38.

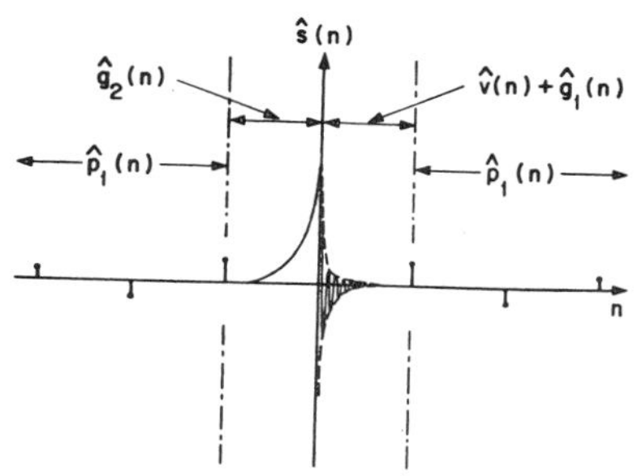

Fig. 38. An illustration of the characteristics of the complex cepstrum for speech.

The components of $s(n)$ due to pitch and to the combined effects of vocal tract and glottal pulse tend to provide their primary contributions in non-overlapping time intervals. The degree of separation will of course depend to some extent on the pitch, with more separation for low-pitched male voices than for high-pitched female voices. Experience has indicated, however, that except in cases of very high pitch a good separation of these components occurs. To illustrate, consider the example of Fig. 39. Fig. 39(a) shows a portion of the vowel "ah" as in "father," with a male speaker, and Fig. 39(b) shows the complex cepstrum. Based on the previous discussion, we can recover the term $p_1(n)$ of (64) by multiplying the complex cepstrum by zero in the vicinity of the origin (with a time-width of, say, 8 ms) and by unity elsewhere. Alternatively, to recover $v(n)\otimes g(n)$ we would multiply the complex cepstrum by unity in the vicinity of the origin and by zero elsewhere. After this weighting, the result is transformed by means of the system D^{-1}. In Fig. 39(c) is shown the result of attempting to recover the weighted train of pitch pulses $p_1(n)$. Pulses with the correct spacing are clearly evident.[7] In Fig. 39(d) the result of retaining only the low-time portion of $\hat{s}(n)$, corresponding to attempting to recover $[v(n)\otimes g(n)]$, is shown. To verify that the pulse of Fig. 39(d) can be considered as a convolutional speech component the speech was resynthesized by convolving this pulse with a train of unit samples whose spacing was chosen to be a pitch period as measured from the waveform of Fig. 39(c). The resynthesized speech is shown in Fig. 39(e) and should be compared with the original speech of Fig. 39(a).

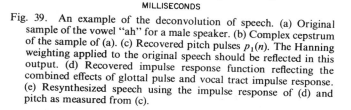

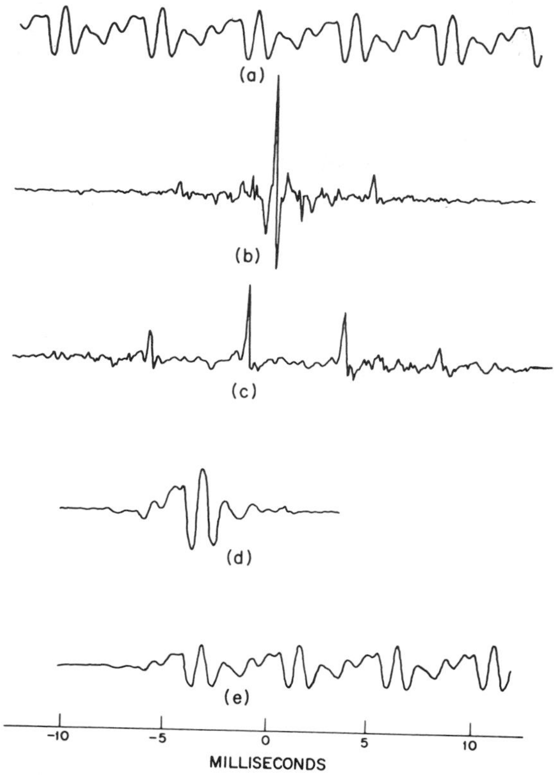

Fig. 39. An example of the deconvolution of speech. (a) Original sample of the vowel "ah" for a male speaker. (b) Complex cepstrum of the sample of (a). (c) Recovered pitch pulses $p_1(n)$. The Hanning weighting applied to the original speech should be reflected in this output. (d) Recovered impulse response function reflecting the combined effects of glottal pulse and vocal tract impulse response. (e) Resynthesized speech using the impulse response of (d) and pitch as measured from (c).

From the diagram of Fig. 38 it is clear that we could not expect to separate the glottal pulse $g(n)$ and vocal tract impulse response $v(n)$ by simple weighting of the complex cepstrum, although we might expect to recover the maximum phase part of the glottal pulse. This has been tried in a few cases to verify the idea. An example of the type of pulse obtained is shown in Fig. 40. However, the value of recovering only the maximum phase portion of the glottal pulse is not clear.

[7] Pitch detection based directly on a measurement of the location of a peak in the cepstrum (as defined by Bogert *et al.* [6]) has been successfully demonstrated by Noll [25].

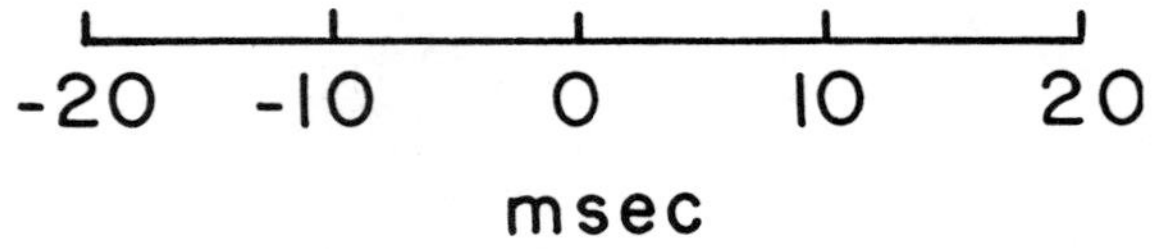

Fig. 40. Pulse obtained by retaining only those values in the complex cepstrum near the origin for $n<0$.

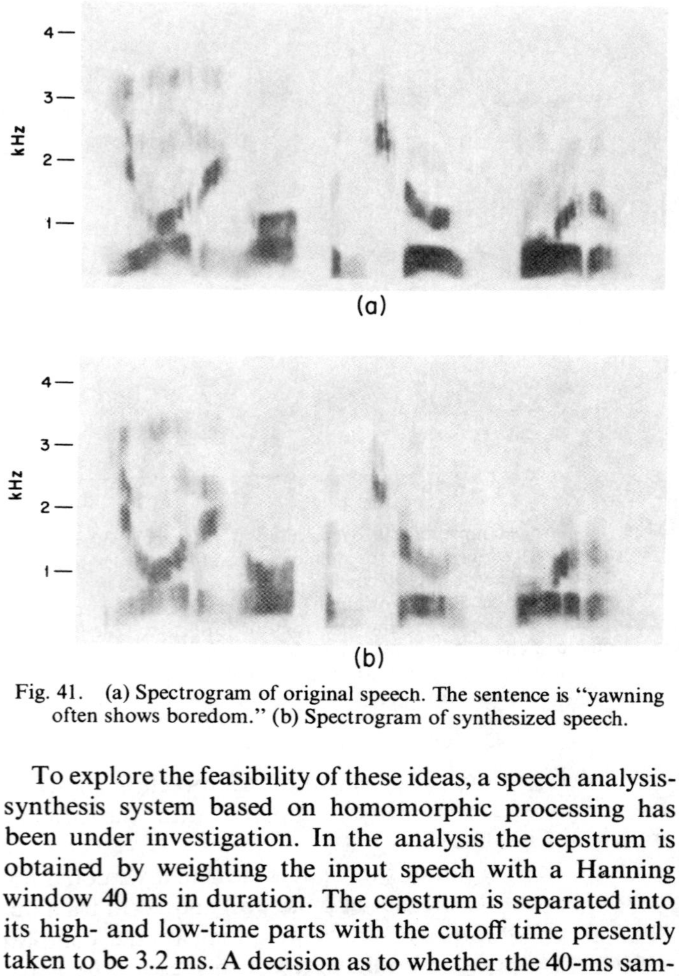

Fig. 41. (a) Spectrogram of original speech. The sentence is "yawning often shows boredom." (b) Spectrogram of synthesized speech.

To explore the feasibility of these ideas, a speech analysis-synthesis system based on homomorphic processing has been under investigation. In the analysis the cepstrum is obtained by weighting the input speech with a Hanning window 40 ms in duration. The cepstrum is separated into its high- and low-time parts with the cutoff time presently taken to be 3.2 ms. A decision as to whether the 40-ms sample is voiced or unvoiced and a measurement of the pitch frequency if voiced is made from the high-time portion. Thus, the synthesizer receives the low-time cepstral values, a voiced-unvoiced decision, and a pitch frequency measurement updated at 10-ms intervals. In the synthesizer, the pitch and voiced-unvoiced information is converted to an excitation function consisting of impulses during voicing and noise during unvoicing. The low-time cepstral values are converted to an impulse response function which is then convolved with the excitation function to form the synthetic speech. Informal listening tests indicate that the synthetic speech is of high quality and natural sounding. In Fig. 41 are shown spectrograms of a sentence before and after processing.[8]

VI. Conclusions

The applications which we have presented represent an attempt to apply the point of view provided by the theory of homomorphic systems to problems of general practical interest. The audio compression-expansion system has been realized economically in analog hardware and its success has led to its use in equipment in which dynamic range compression was required. As we have already stated, the other applications at present have been simulated on general-purpose digital computers.

The deconvolution of speech is being pursued as an approach to obtaining a high-quality speech bandwidth compression system which can be implemented in digital hardware with present technology. The homomorphic processing of images has immediate practical possibilities for non-real-time applications in which a small computer with a graphics facility is available. The removal of echoes by homomorphic deconvolution appears applicable to problems in which processing can be carried out on a large digital computer.

The full potential of these and other applications relies to a large extent on the advances which we can expect in system implementation. In particular, large-scale integration (both analog and digital) most certainly will play a vital role in providing an efficient and inexpensive realization of real-time processing of the kind which we have been discussing.

It is natural to ask if there are other areas of application for homomorphic filtering. In this respect we offer the following topics which we feel deserve consideration.

Applications of multiplicative filtering which may have potential are compensators for channel fading, systems for simultaneous amplitude and phase modulation and detection, automatic gain controls for other than audio application, ac and dc power regulators, and radar signal processing.

Some problems in convolutional filtering which may be promising involve the restoration of images blurred in an unknown manner [26], the suppression of multipath distortion, the enhancement of sonar signals, seismographic exploration, the sharpening of bioelectric signals blurred by propagation through tissue, the analysis of probability density functions, the measurement of auditorium acoustics,

[8] The speech analysis-synthesis system was simulated on the M.I.T. Lincoln Laboratory 1219 speech facility. Other parts of the work described in this section were carried out on the M.I.T. Lincoln Laboratory TX-2 computer and the PDP-1 computer facility operated by the Department of Electrical Engineering and the Research Laboratory of Electronics at M.I.T.

and the separation of antenna pattern and target impulse response in radar and sonar detection.

During the course of the research which resulted in this paper, the thinking of the authors was consistently influenced by some speculative ideas involving vision and hearing. While these ideas are presently being studied and no specific conclusions have been reached, we feel that the paper would be incomplete without mentioning them.

The presence of a logarithmic response in vision and hearing has been accepted for some time. Even more readily evident, and mechanized through the process of neural interaction, is the means for linear filtering [4], [27], [28]. This combination is so suggestive of both forms of homomorphic filtering which we have been discussing that questions concerning a possible relationship arose early during the research. Specifically, it seems reasonable to inquire whether to some approximation the processes of vision and hearing can be modeled as homomorphic systems directed toward a separation of multiplied components in the case of vision and a separation of convolved components in the case of hearing.[9] While the question as to whether a variety of psychophysical data can be modeled in these terms is purely speculative at present, some preliminary investigations have been encouraging. If it can in fact be verified that such models are reasonable, the resulting point of view may have a bearing on the design of communication systems for which the final receiver is the human eye or the human ear.

References

[1] A. V. Oppenheim, "Superposition in a class of non-linear systems," Research Lab. of Electronics, M.I.T., Cambridge, Mass., Tech. Rept. 432, March 31, 1965.

[2] ——, "Optimum homomorphic filters," Research Lab. of Electronics, M.I.T., Cambridge, Mass., Quart. Progr. Rept. 77, pp. 248–260, April 15, 1965.

[3] References [1] and [2] are summarized in A. V. Oppenheim, "Generalized superposition," *Information and Control*, vol. 11, pp. 528–536, November 1967.

[4] J. L. Flanagan, *Speech Analysis, Synthesis and Perception*. New York: Academic Press, 1965.

[5] R. L. Miller, "Nature of the vocal cord wave," *J. Acoust. Soc. Am.*, vol. 31, pp. 667–677, June 1959.

[6] B. Bogert, M. Healy, and J. Tukey, "The quefrency alanysis of time series for echoes," in *Proc. Symp. on Time Series Analysis*, M. Rosenblatt, Ed. New York, Wiley, 1963, ch. 15, pp. 209–243.

[7] A. V. Oppenheim, "Non-linear filtering of convolved signals," Research Lab. of Electronics, M.I.T., Cambridge, Mass., Quart. Progr. Rept. 80, pp. 168–175, January 15, 1966.

[8] E. A. Guillemin, *Theory of Linear Physical Systems*. New York: Wiley, 1963, ch. 18.

[9] J. W. Cooley and J. W. Tukey, "An algorithm for the machine calculation of complex Fourier series," *Math. of Comput.*, vol. 19, pp. 297–301, April 1965.

[10] T. G. Stockham, Jr., "The application of generalized linearity to automatic gain control," *IEEE Trans. Audio and Electroacoustics*, vol. AU-16, pp. 267–270, June 1968.

[11] C. W. Carter, Jr., A. C. Dickieson, and D. Mitchell, "Application of companders to telephone circuits," *Trans. AIEE*, vol. 65, pp. 1079–1086, December 1946.

[12] M. Medress, "Noise analysis of a homomorphic automatic volume control system," S.M. thesis, Dept. of Elec. Engrg., M.I.T., Cambridge, Mass., January 1968.

[13] C. B. Neblette, *Photography—Its Materials and Processes*. New York: Van Nostrand, 1962, chs. 20, 21, and 22.

[14] *Masking Color Transparencies*, Kodak Graphic Arts Data Book. Rochester, N. Y.: Eastman Kodak Co., 1960.

[15] E. G. St. John and D. R. Craig, "Logetronography," *Am. J. Roentgenol., Radium Therapy Nucl. Med.*, vol. 75, July 1957.

[16] W. F. Schreiber, "The mathematical foundation of the synthetic highs system," Research Lab. of Electronics, M.I.T., Cambridge, Mass., Quart. Progr. Rept. 68, p. 140, January 1963.

[17] D. N. Graham, "Image transmission by two-dimensional contour coding," *Proc. IEEE*, vol. 55, pp. 336–346, March 1967.

[18] T. G. Stockham, Jr., "High-speed convolution and correlation," *Spring Joint Computer Conf., AFIPS Proc.*, vol. 28, pp. 229–233, 1966.

[19] L. G. Roberts, "Conic display generator using multiplying digital-analog converters," *IEEE Trans. Electronic Computers* (*Short Notes*), vol. EC-16, pp. 369–370, June 1967.

[20] R. W. Schafer, "Echo removal by generalized linear filtering," *NEREM Record*, pp. 118–119, 1967.

[21] ——, "Echo removal by discrete generalized linear filtering," Ph.D. thesis, Dept. of Elec. Engrg., M.I.T., Cambridge, Mass., February 1968.

[22] A. V. Oppenheim, "Deconvolution of speech" (abstract), *J. Acoust. Soc. Am.*, vol. 41, p. 1595, 1967.

[23] A. V. Oppenheim and R. W. Schafer, "Homomorphic analysis of speech," *IEEE Trans. Audio and Electroacoustics*, vol. AU-16, pp. 221–226, June 1968.

[24] M. V. Mathews, J. E. Miller, and E. E. David., Jr., "Pitch synchronous analysis of voiced sounds," *J. Acoust. Soc. Am.*, vol. 33, pp. 179–186, February 1961.

[25] A. M. Noll, "Cepstrum pitch determination," *J. Acoust. Soc. Am.*, vol. 41, pp. 293–309, February 1967.

[26] C. M. Rader (private communication).

[27] S. S. Stevens, *Handbook of Experimental Psychology*. New York: Wiley, 1951, chs. 23–27.

[28] F. Ratcliff, *Mach Bands: Quantitative Studies on Neural Networks in the Retina*. San Francisco: Holden-Day, 1965.

[29] R. B. Marimont, "Linearity and the Mach phenomenon," *J. Opt. Soc. Am.*, vol. 53, pp. 400–401, March 1963.

[30] ——, "Model for visual response to contrast," *J. Opt. Soc. Am.*, vol. 52, pp. 800–806, July 1962.

[31] W. H. Huggins, "A phase principle for complex-frequency analysis and its implications in auditory theory," *J. Acoust. Soc. Am.*, vol. 24, pp. 582–589, November 1952.

[9] Although from a different point of view and with a different motivation, ideas suggestive of this have appeared before. Marimont [29], [30] suggests a model for vision similar to the canonic form of Fig. 2. With regard to hearing, Huggins [31] discusses the notion that hearing is a process of deconvolution, although the mechanism which he proposes is different from that discussed here.

THEORY AND IMPLEMENTATION OF THE DISCRETE HILBERT TRANSFORM

B. Gold, A. V. Oppenheim, C. M. Rader
Lincoln Laboratory, Massachusetts Institute of Technology, Lexington, Massachusetts

The Hilbert transform has traditionally played an important part in the theory and practice of signal processing operations in continuous system theory because of its relevance to such problems as envelope detection and demodulation, as well as its use in relating the real and imaginary components, and the magnitude and phase components of spectra. The Hilbert transform plays a similar role in digital signal processing. In this paper, the Hilbert transform relations, as they apply to sequences and their z-transforms, and also as they apply to sequences and their discrete Fourier transforms, will be discussed. These relations are identical only in the limit as the number of data samples taken in the discrete Fourier transforms becomes infinite.

The implementation of the Hilbert transform operation as applied to sequences usually takes the form of digital linear networks with constant coefficients, either recursive or nonrecursive, which approximate an all-pass network with 90° phase shift, or two-output digital networks which have a 90° phase difference over a wide range of frequencies. Means of implementing such phase shifting and phase splitting networks are presented.

1. INTRODUCTION

Hilbert transforms have played a useful role in signal and network theory and have also been of practical importance in various signal processing systems. Analytic signals, bandpass sampling, minimum phase networks and much of spectral analysis theory is based on Hilbert transform relations. Systems for performing Hilbert transform operations have proved useful in diverse fields such as radar moving target indicators, analytic signal rooting [1], measurement of the voice fundamental frequency [2, 3], envelope detection, and generation of the phase of a spectrum given its amplitude [4, 5, 6].

The present paper is a survey of Hilbert transform relations in digital systems, and of the design of linear digital systems for performing the Hilbert transform of an input signal. These subjects have been treated in the published literature for continuous signals and systems [7]. In this paper we present a treatment of the subject for digital systems. We will first present various Hilbert transform relationships followed by several design techniques for Hilbert transformers and a few examples and applications.

2. CONVOLUTION THEOREMS

In this section some notation is introduced and some well-known convolution theorems are quoted; the interested reader can find proofs of these and other

Reprinted with permission from *Proc. Symp. Comput. Process Commun.*, pp. 235–250.

theorems of z-transform theory in various books [8, 9]. Let $x(n)$ be a stable infinite sequence and define the z-transform of $x(n)$ as

$$X(z) = \sum_{n=-\infty}^{\infty} x(n)\, z^{-n} \ .$$

Given two such sequences $x(n)$ and $h(n)$ and their corresponding z-transforms $X(z)$ and $H(z)$, then, if $Y(z) = X(z)H(z)$, we have the convolution theorem

$$y(n) = \sum_{m=-\infty}^{m=\infty} x(n-m)\, h(m) = \sum_{m=-\infty}^{m=\infty} x(m)\, h(n-m) \ . \tag{1}$$

Similarly, if $y(n) = x(n)\, h(n)$, we have the complex convolution theorem

$$Y(z) = \frac{1}{2\pi j} \oint X(v)\, H(z/v)\, v^{-1}\, dv = \frac{1}{2\pi j} \oint X(z/v)\, H(v)\, v^{-1}\, dv \ , \tag{2}$$

where v is the complex variable of integration and the integration path chosen is the unit circle, taken counterclockwise.

The spectrum of a signal is defined as the value of its z-transform on the unit circle in the z-plane. Thus, the spectrum of $x(n)$ can be written as $X(e^{j\theta})$ where θ is the angle of the vector from the origin to a point on the unit circle.

If $x(n)$ is a sequence of finite length N, then it can be represented by its discrete Fourier transform (DFT). Denoting the DFT values by $X(k)$, we have

$$X(k) = \sum_{n=0}^{N-1} x(n)\, W^{-nk} \ ,$$

$$x(n) = \frac{1}{N} \sum_{k=0}^{N-1} X(k)\, W^{nk} \ , \tag{3}$$

with

$$W = e^{j2\pi/N} \ .$$

The convolution theorems for these finite sequences specify that if $Y(k) = H(k)\, X(k)$, then

$$y(n) = \sum_{m=0}^{N-1} x(m)\, h\big(((n-m))\big) = \sum_{m=0}^{N-1} x\big(((n-m))\big)\, h(m) \ , \tag{4}$$

and if $y(n) = x(n)\, h(n)$, then

$$Y(k) = \frac{1}{N} \sum_{\ell=0}^{N-1} X(\ell)\, H\big(((k-\ell))\big) = \frac{1}{N} \sum_{\ell=0}^{N-1} X\big(((k-\ell))\big)\, H(\ell) \ , \tag{5}$$

where the double parenthesis around the expressions $k - \ell$ and $n - m$ refer to these expressions modulo N; i.e., $((x))$ = the unique integer $x + kN$, satisfying $0 \le x + kN \le N - 1$.

Finally we define an infinite sequence $x(n)$ as "causal" if $x(n) = 0$ for $n < 0$. A finite duration sequence of length N is "causal" if $x(n)$ is zero in the latter half of the period $0, 1, \cdots, N-1$, i.e., for $n > \frac{N}{2}$.

3. HILBERT TRANSFORM RELATIONS FOR CAUSAL SIGNALS

The z-transform $X(z)$ of the impulse response $x(n)$ of a linear stable causal digital system is analytic outside the unit circle. Under these conditions, interesting relations between components of the complex function $X(z)$ can be derived, these relations being a consequence of the Cauchy integral theorem [10]. For example, $X(z)$ can be explicitly found outside the unit circle given either the real or imaginary components of $X(z)$ on the unit circle. These relations also hold on the unit circle; if we write

$$X(e^{j\theta}) = R(e^{j\theta}) + jI(e^{j\theta}) \ , \tag{6}$$

where $R(e^{j\theta})$ and $I(e^{j\theta})$ are the real and imaginary parts respectively of $X(e^{j\theta})$, then $X(e^{j\theta})$ can be explicitly found in terms of $R(e^{j\theta})$ or in terms of $I(e^{j\theta})$ and therefore $R(e^{j\theta})$ and $I(e^{j\theta})$ can be expressed as functions of each other. These various integral relationships will be referred to as Hilbert transform relations between components of $X(z)$.

First, we will derive an expression for $X(z)$ outside (not on) the unit circle given $R(e^{j\theta})$ (on the unit circle), beginning with the physically appealing concept of causality. A causal sequence can always be reconstructed from its even part, defined as

$$x_e(n) = \tfrac{1}{2}\,[x(n) + x(-n)] \ . \tag{7}$$

Since $x(n)$ is causal, it can be written

$$\left.\begin{aligned} x(n) &= 2\,x_e(n)\,s(n) \\ \text{where}\quad s(n) &= 1 \quad \text{for} \quad n > 0 \\ &= 0 \quad \text{for} \quad n < 0 \\ &= \tfrac{1}{2} \quad \text{for} \quad n = 0 \end{aligned}\right\} . \tag{8}$$

Now, consider $X(z)$ outside the unit circle, that is, for $z = re^{j\theta}$ with $r > 1$. Then

$$X(re^{j\theta}) = \sum_{n=-\infty}^{\infty} x(n)\,r^{-n}\,e^{-jn\theta} = 2 \sum_{n=-\infty}^{\infty} x_e(n)\,s(n)\,r^{-n}\,e^{-jn\theta} \ . \tag{9}$$

But on the unit circle, the z-transform of $x_e(n)$ is $R(e^{j\theta})$ and the z-transform of the sequence $s(n)r^{-n}$ is given by $\dfrac{1 + r^{-1}z^{-1}}{1 - r^{-1}z^{-1}}$. Thus, using the complex convolution theorem (2), we can rewrite (9) as

$$X(z)\big|_{z=re^{j\theta}} = \frac{1}{\pi j} \oint \frac{R(z/v)(v + r^{-1})}{v(v - r^{-1})}\, dv \ . \tag{10}$$

(In this and subsequent contour integrals, the contour of integration is always taken to be the unit circle).

Equation (10) expresses $X(z)$ outside the unit circle in terms of its real part on the unit circle. Equation (10) was written as a contour integral to stress the fact that in the physically most interesting case when $R(z)$ is a rational fraction, evaluation of (10) is most easily performed by contour integration using residues.

Similarly, we may construct $X(re^{j\theta})$ from $I(e^{j\theta})$ by noting that $x(n) = 2x_0(n)\, s(n) + x(0)\,\delta(n)$ where $x_0(n)$ denotes the odd part of $x(n)$ and $\delta(n)$ is the unit pulse, defined as unity for $n = 0$ and zero elsewhere. The result obtained is

$$X(z)\big|_{z=re^{j\theta}} = \frac{1}{\pi} \oint \frac{I(z/v)(v + r^{-1})}{v(v - r^{-1})}\, dv + x(0) \ . \tag{11}$$

Now, Equations (10) and (11) also hold in the limit as $r \longrightarrow 1$, provided care is taken to evaluate the integral correctly in the presence of a pole on the unit circle. This can be done formally by changing the integrals in (10) and (11) to the Cauchy principal values of these integrals, where the latter is defined as

$$\left.\begin{aligned} \frac{1}{2\pi j}\, P\oint \frac{f(z)}{z - z_0}\, dz &= f(z_0) && \text{for} \quad |z_0| < 1 \\ &= 0 && \text{for} \quad |z_0| > 1 \\ &= \tfrac{1}{2} f(z_0) && \text{for} \quad |z_0| = 1 \end{aligned}\right\} \ . \tag{12}$$

From (10), (11) and (12), it is a simple matter to construct explicit relations between $R(e^{j\theta})$ and $I(e^{j\theta})$. Alternately, these results could have been derived by appealing directly to Fig. 1, which shows the explicit relations between the real and imaginary part of a causal function to be

$$\left.\begin{aligned} x_0(n) &= \lim_{\alpha \to 1}\ x_e(n)\, w_1(n) \\ x_e(n) &= \lim_{\alpha \to 1}\ [x_0(n)\, w_1(n)] + \delta(n)\, x(0) \end{aligned}\right\} \ . \tag{13}$$

Figure 2 shows the ring of convergence for the z-transform $W_1(z)$ of $w_1(n)$ and Fig. 3 shows the poles and zeros of $W_1(z)$.

$$jI(z)\big|_{z=e^{j\theta}} = \frac{1}{2\pi j}\, P\oint \frac{R(z/v)(v+1)}{v(v-1)}\, dv \ . \tag{14}$$

$$R(z)\big|_{z=e^{j\theta}} = \frac{1}{2\pi}\, P\oint \frac{I(z/v)(v+1)}{v(v-1)}\, dv \ . \tag{15}$$

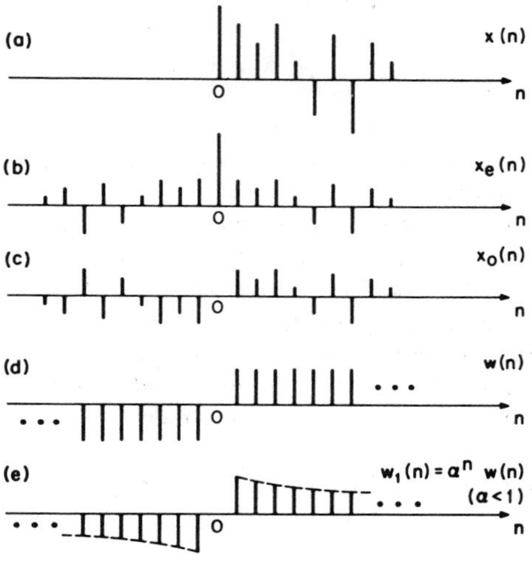

Figure 1.

By setting $z = e^{j\theta}$ and $v = e^{j\varphi}$, we change the contour integrals (14) and (15) to line integrals, yielding

$$\left.\begin{aligned} -I(e^{j\theta}) &= \frac{1}{2\pi}\, P \int_0^{2\pi} R(e^{j(\varphi-\theta)}) \cot(\varphi/2)\, d\varphi \\ R(e^{j\theta}) &= \frac{1}{2\pi}\, P \int_0^{2\pi} I(e^{j(\theta-\varphi)}) \cot(\varphi/2)\, d\varphi + x(0) \end{aligned}\right\} . \tag{16}$$

Similar results can be obtained for the real and imaginary parts of the discrete

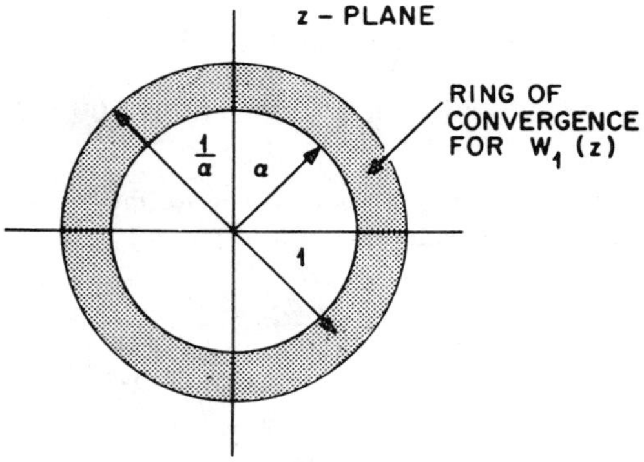

Figure 2.

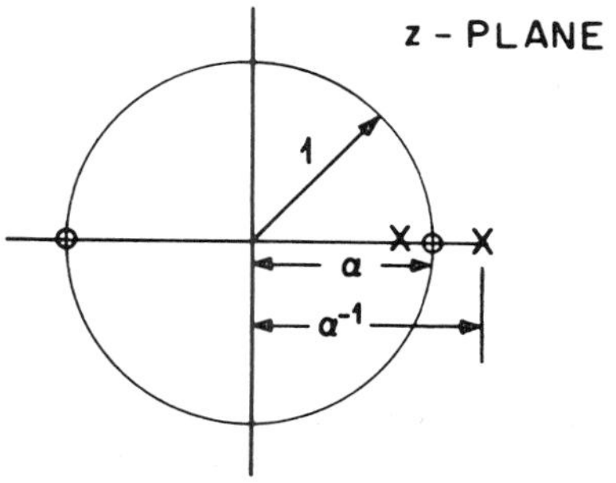

$$W_1(z) = \frac{z^2 - 1}{(z-\alpha)(z-\alpha^{-1})}$$

Figure 3.

Fourier transform of a finite duration sequence provided that the sequence is causal in the sense that, if the sequence $x(n)$ is considered to be of duration N, then $x(n) = 0$ for $n > \frac{N}{2}$. Defining the even and odd parts of $x(n)$ as

$$x_e(n) = \tfrac{1}{2}(x[((n))] + x[((-n))]) \quad ,$$

and

$$x_0(n) = \tfrac{1}{2}(x[((n))] - x[((-n))]) \quad ,$$

it follows (for N even) that

$$x(n) = x_e(n)\, f(n) \quad ,$$

and

$$x(n) = x_0(n)\, f(n) + x(0)\,\delta(n) + x\left(\frac{N}{2}\right)\delta\left(n - \frac{N}{2}\right) \quad ,$$

where

$$f(n) = \begin{cases} 1 & n = 0, \frac{N}{2} \\ 2 & n = 1, 2, \cdots, \frac{N}{2} - 1 \\ 0 & n = \frac{N}{2} + 1, \frac{N}{2} + 2, \cdots, N - 1 \end{cases}$$

From these relations, we can then derive that the real and imaginary parts, $R(k)$ and $I(k)$, of the DFT of $x(n)$ are related by

$$jI(k) = \frac{1}{N} \sum_{r=0}^{N-1} R(r)\, F[((k - r))] \quad , \qquad (17)$$

$$R(k) = \frac{1}{N} \sum_{r=0}^{N-1} jI(r)\, F[((k-r))] + x(0) + (-1)^k\, x\left(\frac{N}{2}\right) . \tag{18}$$

where $F(k)$ is given by

$$F(k) = \begin{cases} -j \cot \dfrac{\pi k}{N} & k \text{ odd} \\ 0 & k \text{ even} \end{cases}$$

Note that (17) and (18) are circular convolutions which can be numerically evaluated by fast Fourier transform methods. Similar but not identical relations can also be derived if N is odd.

If instead of working with the z-transform of a sequence, we choose to work with the logarithm of the z-transform, then comparable Hilbert transform relations can be derived between the log magnitude of the spectrum and its phase. However, certain theoretical restrictions arise, due to the fact that a) the logarithm of zero diverges and b) the definition of phase is ambiguous. However, the derivative of the phase (with respect to z) is not ambiguous; this leads to relationships based on the definition

$$D(z) = \frac{-z}{X(z)} \frac{dX(z)}{dz} . \tag{19}$$

The imaginary part of $D(e^{j\theta})$ is the derivative of $\log |D(e^{j\theta})|$ and its real part is the negative of the derivative of the phase of $D(e^{j\theta})$. If the inverse z-transform of $D(z)$ is causal, relations similar to our previous real-imaginary relations can be derived. For example, results analogous to (16)

$$F = \frac{|X(e^{j\theta})|'}{|X(e^{j\theta})|} = \frac{1}{2\pi} P \int_0^{2\pi} \Psi'(e^{j(\varphi-\theta)}) \cot(\varphi/2)\, d\varphi \tag{20}$$

$$\Psi'(e^{j\theta}) = -\frac{1}{2\pi} P \int_0^{2\pi} F(e^{j(\varphi-\theta)}) \cot(\varphi/2)\, d\varphi , \tag{21}$$

where $|X|$ is the magnitude of the spectrum and Ψ its phase and the primes denote differentiation with respect to θ.

If we impose the condition that Ψ is an odd function, then it must be zero for $\varphi = 0$ and (20) and (21) may be integrated to give

$$\log |X(e^{j\theta})| = \frac{1}{2\pi} P \int_0^{2\pi} \Psi(e^{j(\varphi-\theta)}) \cot(\varphi/2)\, d\varphi , \tag{22}$$

$$\Psi(e^{j\theta}) = -\frac{1}{2\pi} P \int_0^{2\pi} \log |X(e^{j(\varphi-\theta)})| \cot(\varphi/2)\, d\varphi . \tag{23}$$

The requirement that the inverse z-transform of $D(z)$ be zero for $n < 0$ imposes a restriction on the pole and zero locations of $X(z)$. Since the poles of $D(z)$ occur whenever there are either poles or zeros of $X(z)$ and since the inverse transform of $D(z)$ is zero for $n < 0$ only if the poles of $D(z)$ are all within the unit circle, it follows that both poles and zeros of $X(z)$ must be within the unit circle in order for Equations (20) through (23) to be valid. This is the well-known minimum phase condition [11].

It is also possible to relate the log magnitude and the phase of the DFT by analogous relations provided that the inverse DFT of the logarithm of the DFT is causal. The difficulty in applying this notion is that the logarithm of $X(k)$ is ambiguous since $X(k)$ is complex. For the previous case of the z-transform, this ambiguity was resolved in effect by considering the phase to be a continuous, odd, periodic function; this definition of the phase cannot be applied in this case. Nevertheless, it has been useful computationally for constructing a phase function from the log magnitude of a DFT by computing the inverse DFT of the log magnitude, multiplying by the function $f(n)$ and then transforming back [4, 5]. The real part of the result is the log magnitude as before and the imaginary part is an approximation to the phase.

4. HILBERT TRANSFORM RELATIONS BETWEEN REAL SIGNALS, AND A FEW APPLICATIONS

The relations of Section 3 were derived via the complex convolution theorem (2) and the requirement of causality. By interchanging time and frequency and using the convolution theorem (1), further relations can be found which are of practical and theoretical interest. One way of obtaining such relations is by the introduction of the "ideal" Hilbert transformer which has a spectrum defined as having the value $+j$ for $0 < \varphi < \pi$ and $-j$ for $\pi < \varphi < 2\pi$, or equivalently, a spectrum with flat magnitude vs. frequency and having a phase of $\pm\pi/2$. Thus, a Hilbert transformer is a (non-realizable) linear network with this transfer function and, as shown in Fig. 4a, the output of the network is the Hilbert transform of

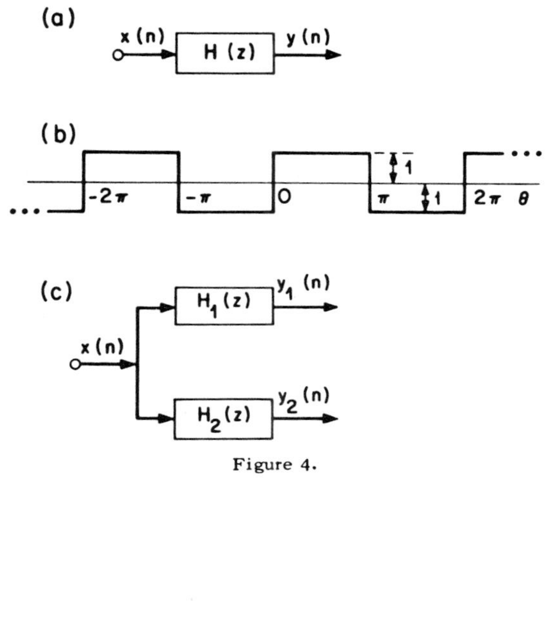

Figure 4.

the input. Hilbert transform relations can also be realized by having two all-pass networks having a phase difference of $\pi/2$, as shown in Fig. 4c; such a configuration is useful for synthesis of realizable approximate Hilbert transformers.

The unit pulse response of a Hilbert transformer can be derived by evaluating its inverse z-transform. Thus

$$h(n) = \frac{1}{2\pi j} \oint H(z)\, z^{n-1}\, dz \ ,$$

$$h(n) = \frac{1}{2\pi} \left\{ \int_0^{\pi} j\, e^{j\theta} d\theta - \int_{\pi}^{2\pi} j\, e^{j\theta}\, d\theta \right\} ,$$

$$\left.\begin{aligned} h(n) &= \frac{1 - e^{j\pi n}}{\pi n} && \text{for} \quad n \neq 0 \\ &= 0 && \text{for} \quad n = 0 \end{aligned}\right\} . \tag{24}$$

From (1) the input-output relations can be written down;

$$y(n) = \frac{1}{\pi} \sum_{\substack{m=-\infty \\ m \neq n}}^{\infty} \frac{x(m)\,[1 - e^{j\pi(n-m)}]}{(n-m)} \ . \tag{25}$$

Equation (25) can be inverted by noting that $X(z) = H^*(z)\, Y(z)$ (where $H^*(z)$ is the complex conjugate of $H(z)$); this yields

$$x(n) = -\frac{1}{\pi} \sum_{\substack{m=-\infty \\ m \neq n}}^{\infty} \frac{y(m)\,[1 - e^{j\pi(n-m)}]}{n-m} \ . \tag{26}$$

Thus (25) and (26) can be said to be a Hilbert transform signal pair. The graph of $\frac{1}{\pi}\, h(n)$ is shown in Figure 5.

The complex signal $s(n) = x(n) + j\, y(n)$ (where $x(n)$ and $y(n)$ are a Hilbert

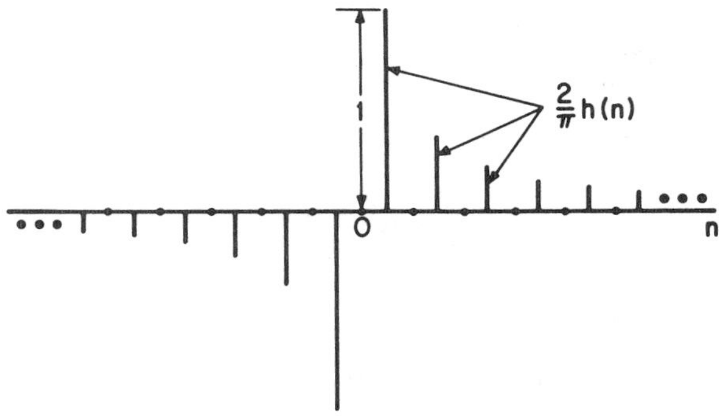

Figure 5.

transform pair) has been called the analytic signal and has the useful property that its spectrum is zero along the bottom half of the unit circle. One application of the analytic signal is to the bandpass sampling problem. Consider the problem of sampling a real signal having the spectrum of Figure 6a. If this signal is passed through the phase splitter of Fig. 4c, the resulting analytic signal has the spectrum shown in Fig. 6b, and thus can be sampled at the rate $1/B$. To reconstruct the original signal requires that the samples be applied to the unity gain bandpass filter shown in Figure 6c. The real part of the filtered signal corresponds to the original signal.

Another application of Hilbert transformers is to help create a bandpass spectrum which is arithmetically symmetric about an arbitrary center frequency. Effectively, the ability to do this allows us to design bandpass filters which are linear translations in frequency of prototype low pass filters, thus avoiding the distortions inherent in the standard low pass-bandpass transformations. Figure 7a illustrates a symmetric low pass. When a conventional transformation is applied, the non-symmetric bandpass of Fig. 7b results. Symmetry may be attained with the filter of Fig. 7c; however, we note that the output of such a filter is a sequence of complex numbers and, also, that by merely taking the real part we must introduce the complex conjugate pole, thus destroying the symmetry. To maintain symmetry of a real output over the range 0 through π can be accomplished by the configuration of Fig. 8, where $H_1(z)$ and $H_2(z)$ are all pass phase splitters such as shown in Figure 4. If only the real part of the signal is desired then a single phase splitter (rather than two) is needed. A filter satisfying the

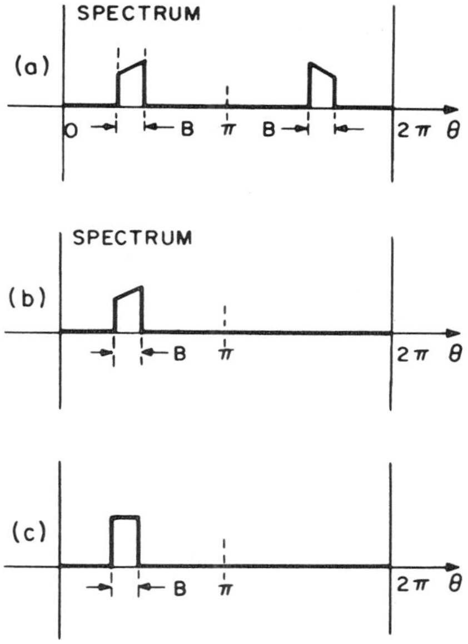

Figure 6.

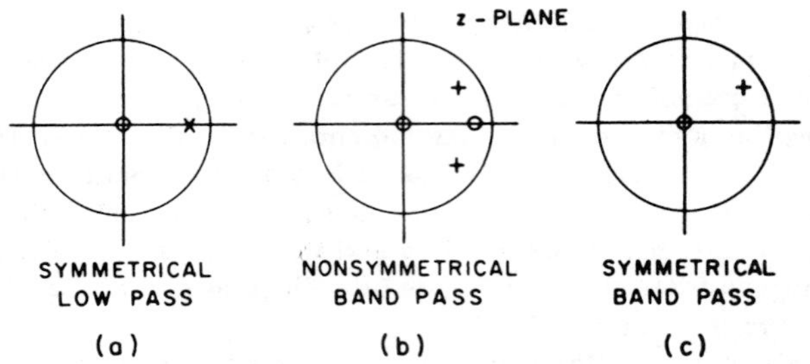

Figure 7.

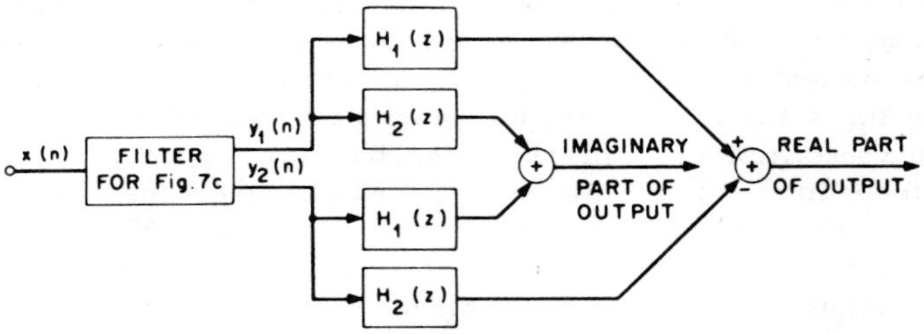

Figure 8.

pole-zero pattern of Fig. 7c is easily made and well-known and has usually been referred to as the coupled form [12, 9].

5. DIGITAL HILBERT TRANSFORM NETWORKS

A. Recursive Networks

Analog phase splitting networks have been extensively analyzed and synthesized [13, 9]. Since the desired networks are all-pass with constant phase difference over a frequency band, it is feasible to use the bilinear transformation [14, 15, 9] to carry analog designs into the digital domain. The resulting networks are all-pass, so that each pole at, say, $z = a$, has a matching zero at $z = 1/a$. An equiripple approximation to a constant 90° phase difference is obtained by the use of Jacobian elliptic functions [16], with the added advantage that all the poles and zeros lie on the real axis.

Let the two networks comprising the phase splitter be $H_1(z)$ and $H_2(z)$.

To synthesize the all pass networks $H_1(z)$ and $H_2(z)$ in an efficient manner, we note that the first order difference equation

$$y(n) = x(n-1) + a[y(n-1) - x(n)] \quad , \tag{27}$$

corresponds to the digital network

$$H(z) = \frac{z^{-1} - a}{1 - az^{-1}} \quad . \tag{28}$$

This shows that an all pass network with a pole at $z = a$ and a zero at $z = 1/a$

can be synthesized with a single multiply in a first order difference equation. In Fig. 9 a complete digital 90° phase splitter is shown which meets the requirements that the phase difference deviates from 90° in an equiripple manner by ±1° in the range 10° through 120° along the unit circle. From (28) we see that the coefficients in Fig. 9 are equal to the pole positions. The nomenclature of Fig. 9 is the following: the box z^{-1} signifies a unit delay, the plus signifies addition and a number-arrow combination signifies both direction of data flow and multiplication by the number. Arrows without numbers signify only data flow.

Now it is well known [9, 12, 17, 18, 19, 20, 21, 22] that because of finite register length, the performance of the actual filter deviates somewhat from that of the design. These effects can be categorized as follows:

a) Quantization of the input signal
b) Roundoff noise caused by the multiplications (fixed point arithmetic is assumed)
c) Deadband effect
d) A fixed deviation in the filter characteristic caused by inexact coefficients.

The analysis of these effects is simplified because the networks $H_1(z)$ and $H_2(z)$ are all-pass. Thus, signal to noise ratios caused by (a) are the same at the output of the networks as at the input. Item (b) can be analyzed for $H_1(z)$ or $H_2(z)$ by inserting noise generators at all adder nodes following multiplications. But each noise is then filtered by a cascade combination of the pole of the section in which the noise is generated, and an all-pass network. The well-known formula for the output variance of a network which has been subjected to a white noise input with uniform probability density of amplitude is given by

$$(E_0^2/12) \sum_{n=0}^{\infty} h^2(n) ,$$

where E_0 is a quantum step and $h(n)$ is the network impulse response. For a single pole at $z = a$, $h(n) = a^n$; assuming m independent noise generators and m poles at $a_1, a_2, \cdots, a_m$ causes a total variance

$$\sigma^2 = E_0^2/12 \sum_{i=1}^{m} \frac{1}{1 - a_i^2} . \tag{29}$$

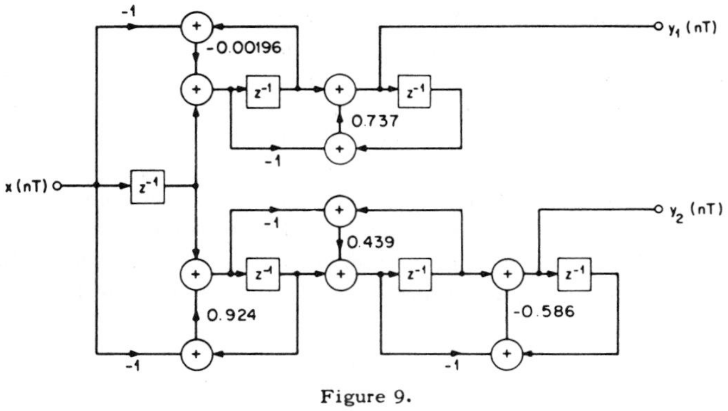

Figure 9.

We see that only values of a_l near unity cause much noise. Thus, for our numerical design example, only about 1 bit of noise is generated. Item (c) can be analyzed by similar considerations but it is probably not important for band-pass phase splitters anyway, since it is only an effect when the input is a constant.

For small errors in coefficients, item (d) can be analyzed in a manner similar to that of reference [12]. The realization chosen in Fig. 9 guarantees that even though a given coefficient is in error, the poles and zeros of the networks remain reciprocals so that only the phase response of the network can be effected. Let the phase response due to a pole zero pair at a, $1/a$ be $\Psi(\varphi, a)$. The phase error for a coefficient error Δa is approximated by

$$\Delta\Psi \approx \frac{\partial\Psi}{\partial a}\,\Delta a + \frac{1}{2}\frac{\partial^2\Psi}{\partial a^2}\,(\Delta a)^2 + \cdots .$$

It is easily shown, using (28), that

$$\Psi = 2\tan^{-1}\left(\frac{\sin\varphi}{a - \cos\varphi}\right),$$

$$\frac{\partial\Psi}{\partial a} = \frac{-2\sin\varphi}{a^2 - 2a\cos\varphi + 1},$$

$$\frac{1}{2}\frac{\partial^2\Psi}{\partial a^2} = \frac{2\sin\varphi(a - \cos\varphi)}{(a^2 - 2a\cos\varphi + 1)^2}.$$

Therefore a good approximation to the error in phase, is given by the early terms of the series

$$\Delta\Psi \approx \frac{-2\sin\varphi}{a^2 - 2a\cos\varphi + 1}\,\Delta a + \frac{2\sin\varphi(a - \cos\varphi)}{(a^2 - 2a\cos\varphi + 1)^2}\,(\Delta a)^2 + \cdots .$$

Using this approximation for each of the poles one can estimate how many bits are necessary to keep the phase error within a given tolerance. Of course, once a coefficient has been specified, the phase difference can be computed precisely.

B. Non-Recursive Networks

The phase splitting network we have described above is called recursive because, in the computation (27), a new output depends on a previous output. Recursive networks always have poles and unit pulse response of infinite duration. By contrast, a nonrecursive network has only zeros and a finite duration unit pulse response. As we have shown, perfectly all pass, recursive phase splitters with equiripple phase characteristics can be constructed. Such criteria cannot, in general, be met by nonrecursive networks. However, useful constructions are certainly possible. We have studied one method of nonrecursive design which is based on sampling in the frequency domain [19]. The sampling formula which relates the z-transform of a finite duration sequence of length N to the values H_k of its DFT is given by

$$H(z) = \frac{1 - z^{-N}}{N}\sum_{k=0}^{N-1}\frac{H_k}{1 - z^{-1}W^k}, \qquad W = e^{-j\frac{2\pi}{N}}. \tag{30}$$

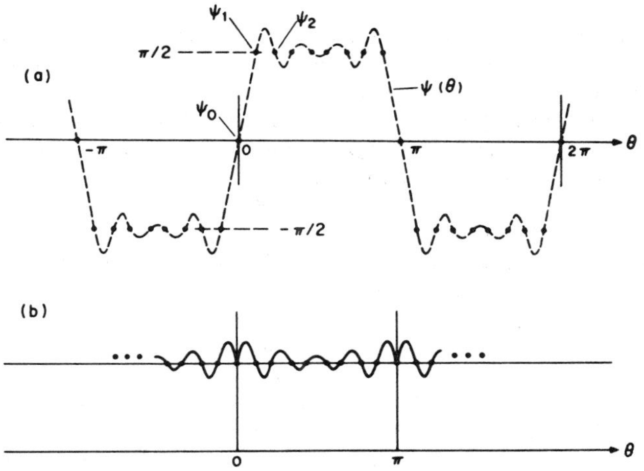

Figure 10.

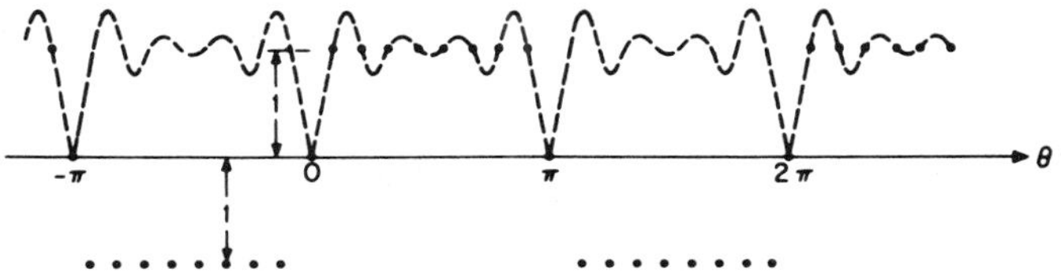

Figure 11.

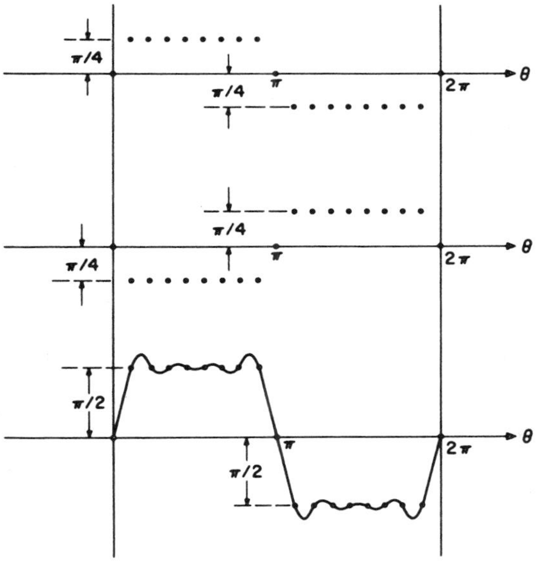

Figure 12.

Thus, an approach to the design of a nonrecursive Hilbert transformer is to specify the phase and magnitude of the network at the N equally spaced points around the unit circle, that is, the N values of H_k. If the unit pulse response of $H(z)$ is constrained to be real, then we must specify that H_k has even magnitude and odd phase. For example, if we specify the magnitude of H_k to be unity for all k and the phase to be $\pi/2$ in the first half of the period and $-\pi/2$ in the second half, as in Fig. 10, then an interpolated spectrum results, such as shown by the dotted curve of Figure 10.

Exact 90° phase can be attained at every frequency, by specifying that the H_k be purely imaginary. However, for real unit pulse response, such a phase shifter must have a magnitude characteristic which passes through zero at 0, π, 2π etc., as shown in Figure 11. Thus, an ideal phase can be attained by further degrading the all pass property of the network near 0 and π.

Nonrecursive networks can also be arranged as phase splitters, which has the advantage that the two components of the resulting analytic signal can have the same delay. It has been found experimentally that the arrangement indicated in Fig. 12, whereby each arm of the splitter has nominal phase of $\pm\pi/4$ leads to a low value of ripple of the interpolated phase difference. Further reduction of ripple is possible at the expense of bandwidth, by specifying intermediate values of H_k in the transition region when the sign of the phase is changing.

ACKNOWLEDGMENT

This work was sponsored by the U.S. Air Force.

REFERENCES

[1] M. R. Schroeder, J. L. Flanagan, E. A. Lundry, "Bandwidth Compression of Speech by Analytic-Signal Rooting," *Proc. IEEE*, 55, No. 3, 396–401, (March, 1967).

[2] R. M. Lerner, "A Method of Speech Compression," Ph.D. Thesis, (1957).

[3] C. M. Rader, "Vector Pitch Detection," *J. Acoust. Soc. Am.* L, 36, (1963).

[4] A. V. Oppenheim, R. W. Schafer, T. G. Stockham, Jr., *Proc. IEEE*, 56, No. 8, 1264–1291, (August, 1968).

[5] A. V. Oppenheim, "Speech Analysis-Synthesis System Based on Homomorphic Filtering," *J. Acoust. Soc. Am.*, 45, No. 2, 458–465, (1969).

[6] D. J. Sakrison, W. T. Ford, J. H. Hearne, "The z-Transform of a Realizable Time Function," *IEEE Trans. on Geoscience Electronics*, GE-5, No. 2, 33–41, (Sept., 1967).

[7] A. Papoulis, *The Fourier Integral and Its Application*, (New York, New York: McGraw-Hill, 1962).

[8] E. I. Jury, *Theory and Application of the z-Transform Method*, (New York, New York: John Wiley, 1964).

[9] B. Gold, C. M. Rader, *Digital Processing of Signals*, (New York, New York: McGraw-Hill, Spring, 1969).

[10] P. M. Morse, H. Feshback, *Methods of Theoretical Physics*, (New York, New York: McGraw-Hill, 1953).

[11] H. W. Bode, *Network Analysis and Feedback Amplifier Design*, (New York, New York: D. Van Nostrand Company, 1945).

[12] C. M. Rader, B. Gold, "Effects of Parameter Quantization on the Poles of a Digital Filter," *Proc. IEEE*, 688, (May, 1967).

[13] J. E. Storer, "Passive Network Synthesis," (New York, New York: McGraw-Hill, 1957).

[14] R. M. Golden, J. F. Kaiser, "Design of Wideband Sampled-Data Filters," *Bell Sys. Tech. J.*, 48, (July, 1964).

[15] C. M. Rader, B. Gold, "Digital Filter Design Techniques in the Frequency Domain," *Proc. IEEE*, 55, No. 2, 149–171, (February, 1967).

[16] E. T. Whittaker, G. N. Watson, "Modern Analysis," (Cambridge, England: Cambridge University Press, 1952, 4th Edition).

[17] J. F. Kaiser, "Some Practical Considerations in the Realization of Linear Digital Filters," *Proc.* 3rd *Allerton Conf.*, 621–633, (1956).

[18] J. B. Knowles, R. Edwards, "Effect of a Finite-Wordlength Computer in a Sampled-Data Feedback System," *Proc. IEE (London)*, 112, 1197–1207, (June, 1965).

[19] B. Gold, C. M. Rader, "Effects of Quantization Noise in Digital Filters," *Proc. Spring Joint Computer Conf.*, 213–219, (1966).

[20] J. B. Knowles, E. M. Olcayto, "Coefficient Accuracy and Digital Filter Response," *IEEE Trans. Circuit Theory*, CT-15, 31–41, (March, 1968).

[21] T. Kaneoko, B. Liu, "Round-off Error of Floating Point Digital Filters," 6th *Ann. Allerton Conf. on Circuit and System Theory*, (Oct. 2–4, 1968).

[22] C. Weinstein, A. V. Oppenheim, "A Comparison of Roundoff Noise in Floating Point and Fixed Point Digital Filter Realization," (to be published).

[23] B. Gold, K. L. Jordan, Jr., "Linear Programming Procedure for Designing Finite Duration Impulse Response Filters" (to be published) in *IEEE Trans. on Audio and Electroacoustics*.

Spectral transformations for digital filters

A. G. Constantinides, B.Sc.(Eng.), Ph.D.

Indexing term: Digital filters

Abstract

The paper describes certain general transformations for digital filters in the frequency domain. The term digital filter is used to denote a processing unit operating on a sampled waveform, so that the input, output and intermediate signals are only defined at discrete intervals of time; the signals may be either p.a.m. or p.c.m. The transformations discussed operate on a lowpass-digital-filter prototype to give either another lowpass or a highpass, bandpass or band-elimination characteristic. The transformations are carried out by mapping the lowpass complex variable z^{-1} [where $z^{-1} = \exp(-j\omega T)$ and T is the time interval between samples] by functions of the form

$$e^{j\theta} \prod_{i=1}^{n} \frac{z^{-1} - \alpha_i}{1 - \alpha_i^* z^{-1}}$$

known as unit functions.

List of principal symbols

a, b = real constant coefficients
f = frequency, Hz
F_s = sampling frequency, Hz
$G(\,)$ = pulse-transfer function of a digital filter
$g(\,)$ = unit function; spectral transformation
T = sampling period ($T = 1/F_s$)
z = unit advance, $z = \exp(j\omega T)$
z^{-1} = unit delay, $z = \exp(-j\omega T)$
α = constant parameter
α^* = complex conjugate of α
β = cutoff (angular) frequency of lowpass-digital-filter prototype
γ = constant coefficients
θ = angle of rotation
ξ = real constant coefficient
ω = angular frequency
Ω_s = angular sampling frequency, $\Omega_s = 2\pi F_s$

1 Introduction

This paper describes certain general transformations for digital filters in the frequency domain. With the exception of the restricted forms of the transformations published elsewhere,[1,2] spectral transformations for digital filters have been virtually nonexistent. A notable move towards establishing a set of such transformations has been made by Broome,[3] but the result of that study was rather unsatisfactory, in that certain distortion errors were introduced into the amplitude characteristics.

The general transformations proposed in this paper are compared with the restricted special cases of References 1 and 2, and Broome's translation formula[3] is critically examined.

Paper 6206 E, first received 15th January and in revised form 17th April 1970
Dr. Constantinides is with the UK Post Office Research Department, Brook Road, Dollis Hill, London NW2, England

The study included in this paper is completely theoretical, and the various results on digital-filter theory are taken as assumptions. An excellent introduction into the field of digital filters is given in Reference 6.

2 Assumptions

The following assumptions will be made:

(i) The inputs, outputs and intermediate signals in the systems that we shall be concerned with are sampled waveforms; sampling is carried out at constant frequency with the sampling-theorem condition satisfied.
(ii) The implementation of the filtering devices is carried out in a discrete form; i.e. processing of signals is either in p.c.m. or p.a.m. form.
(iii) Quantisation effects and other 'noise' errors are negligible. (These effects are pertinent to the operation of the devices, and will have no bearing on the present theoretical study.)
(iv) The transfer function of a digital filter (called a pulse-transfer function) is a real rational function in z^{-1} (for a proof see Reference 7.), where $z^{-1} = \cos\omega T - j\sin\omega T$, and T is the sampling period.
(v) The digital filters that are dealt with in this paper are stable;[7] i.e. they have their poles situated outside the unit circle $|z^{-1}| = 1$.

3 Projection of amplitude characteristic on a cylindrical surface

Here we give a simple representation of the amplitude characteristic of a given pulse-transfer function, on a cylindrical surface. This representation will be helpful in later Sections when we shall be dealing with the spectral transformations.

The representation is effected as follows. If z^{-1} is replaced by $\cos\omega T - j\sin\omega T$ in a given pulse-transfer function, the amplitude characteristic can be earily evaluated. To a particular frequency ω, there corresponds a definite point

Reprinted with permission from *Proc. Inst. Elec. Eng.*, vol. 117, pp. 1585–1590, Aug. 1970.

ωT on the unit circle at which the amplitude of the given function is calculated. Conceptually, we can represent the amplitude characteristic on the surface of a cylinder, normal to the z^{-1} plane, having the unit circle as base and a height of unity (i.e. normalised amplitude response). Then, for every point ωT, a height is erected on the generator of the cylinder equal in magnitude to the value of the amplitude characteristic. On joining these points on the surface of the cylinder for all arcs ωT (i.e. 2π radians), we have produced the amplitude characteristic. There is another advantage of this representation, in that the periodic amplitude characteristic has been reduced to one period only. Such a representation of the

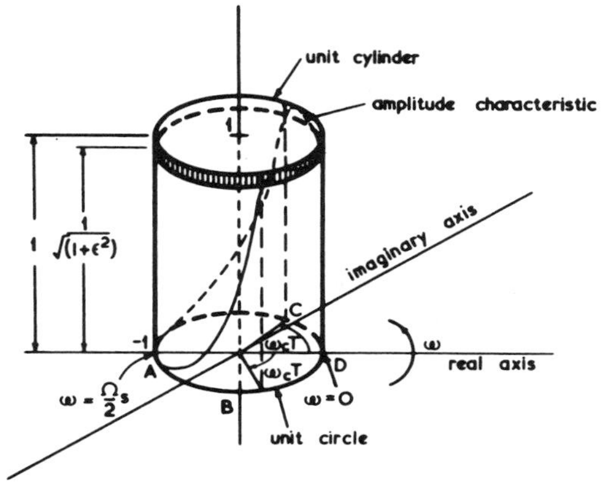

Fig. 1
Representation of 2nd-order Chebÿshev amplitude characteristic on cylindrical surface

amplitude characteristic is shown in Fig. 1 for a 2nd-order Chebÿshev lowpass filter, and it will be used later on an illustrating device.

4 Spectral-transformation problem

Statement of the problem: Given a pulse-transfer function of a lowpass digital filter $G(z^{-1})$, it is required to obtain the unique transformations

$$z^{-1} \rightarrow g(z^{-1})$$

for the following types of amplitude transformations:

(i) lowpass–lowpass
(ii) lowpass–highpass
(iii) lowpass–bandpass
(iv) lowpass–band-elimination

where the type of the amplitude characteristic is preserved.

Some preliminary points can be made concerning the general class of required transformations $g(z^{-1})$. Since $g(z^{-1})$ is to replace the variable z^{-1}, and, since the resulting pulse-transfer function must be real and rational in z^{-1}, and given that the original pulse-transfer function $G(z^{-1})$ is real and rational in z^{-1}, it follows that $g(z^{-1})$ must be a real rational function in z^{-1}.

Furthermore, the values assumed by the function $g(z^{-1})$ must correspond to those of z^{-1}, and no other values must be introduced. The above correspondence need not occur at the same point.

From stability considerations, the mappings must be such that the regions of stability and instability are preserved; i.e. the inside of the unit circle must map into the inside of a domain Γ, whereas the outside must map to the outside of the domain.

For the domain Γ, let

$$g(z^{-1}) = p(\omega)e^{j\phi(\omega)} \text{ for } z^{-1} = e^{-j\omega T}$$

where $p(\omega)$ is the amplitude and $\phi(\omega)$ is the phase of $g(z^{-1})$. If stability regions are to be preserved,

$$\text{for } |z^{-1}| \lesseqgtr 1 \qquad |g(z^{-1})| \lesseqgtr 1$$

Hence $p(\omega) = 1$. This means that the unit circle is mapped onto itself. In view of the above conditions, and from the maximum-modulus theorem, it follows that the mapping $g(z^{-1})$ [where $g(z^{-1})$ is not necessarily a real function] must be given by

$$g(z^{-1}) = e^{j\theta} \prod_{i=1}^{n} \frac{z^{-1} - \alpha_i}{1 - \alpha_i^* z^{-1}} \quad . \quad . \quad . \quad . \quad . \quad . \quad (1)$$

where $|\alpha_i| < 1$ and α_i^* is the complex conjugate of α_i.

The functions given by eqn. 1 are called 'unit functions'.

Since $g(z^{-1})$ is required to be real, it follows that the zeros α_i must occur in complex-conjugate pairs, and that

$$e^{j\theta} = \pm 1$$

(For digital filters with complex coefficients,[5] such restrictions on α_i and θ are removed.); i.e. the general spectral transformation for digital filters with real coefficients is given by

$$g(z^{-1}) = \pm \prod_{i=1}^{n} \frac{z^{-1} - \alpha_i}{1 - \alpha_i^* z^{-1}} \frac{z^{-1} - \alpha_i^*}{1 - \alpha_i z^{-1}} \quad . \quad . \quad . \quad (2)$$

where the order is now $2n$, and α_i are not necessarily all complex.

In eqn. 1, there are two parameters that need further attention if the zeros α_i are to be set aside for the moment.

First, it was mentioned earlier that the angle θ must be such that $e^{j\theta} = \pm 1$. Therefore θ must be a multiple of π. The significance of the angle can be best described as a rotation angle on the unit circle. Thus the only admissible rotations are multiples of π. The rotational nature of θ will be explained and used later.

A further parameter that needs to be considered is the order of the unit function m when $e^{j\theta} = \pm 1$. This order m dictates the nature of the unit-function mapping.

Since the modulus of $g(z^{-1})$ is unity for z^{-1} on the unit circle, the functional relationship between z^{-1} and $g(z^{-1})$ can be taken as a mapping of the unit circle onto itself. Furthermore, there will be m values of z^{-1} for every value of $g(z^{-1})$; or, conversely, for every rotation of z^{-1} on the unit circle around the origin, the function $g(z^{-1})$ rotates m times on the same unit circle around the origin. This means that a function $f(z^{-1})$ of z^{-1} assumes all its values m times for one revolution of z^{-1} when z^{-1} is replaced by $g(z^{-1})$.

The usual requirements for lowpass, highpass, bandpass and band-elimination characteristics, where one passband is required in the region $0 < \omega < \Omega_s/2$, can be translated in terms of the mapping functions as shown in Table 1.

Table 1

REQUIREMENTS FOR FOUR TYPES OF FILTER

Required characteristic	Number of passbands in relation to a lowpass prototype	Order
Lowpass	1	1
Highpass . . .	1	1
Bandpass . . .	2*	2
Band-elimination . .	2*	2

* Corresponding to the positive and the negative lowpass passbands

From this Table it can be seen that, for simple lowpass–lowpass, lowpass–highpass, lowpass–bandpass and lowpass–band-elimination transformations, we need only consider powers of maximum order two; for multiple bands in the Nyquist interval, however, the appropriate order must be chosen in conjunction with the general transformation of eqn. 2.

5 Lowpass–lowpass transformation

The lowpass–lowpass transformation is required to keep the amplitude characteristic unaltered with the points A and D shown in Fig. 1 fixed, but with the arc ABD (ACD) compressed or expanded along the circumference of the circle to result in a different cutoff frequency.

It is quite obvious that such a transformation is not a linear one for the frequency ω, as is the case for the lowpass–lowpass transformation in R, L, C and M filters where s is replaced by ks ($k \geqslant 0$ and constant).

Mathematically, it is required to map a unit circle into

itself in a 1:1 correspondence, with the points A and D invariant.

Let the mapping be denoted by

$$z^{-1} \rightarrow g(z^{-1})$$

so that, for one complete rotation of z^{-1} around the unit circle, $g(z^{-1})$ also takes one complete rotation.

Furthermore,

$$\left.\begin{aligned} g(1) &= 1 \\ g(-1) &= -1 \end{aligned}\right\} \quad . \quad . \quad . \quad . \quad . \quad . \quad . \quad . \quad . \quad . \quad (3)$$

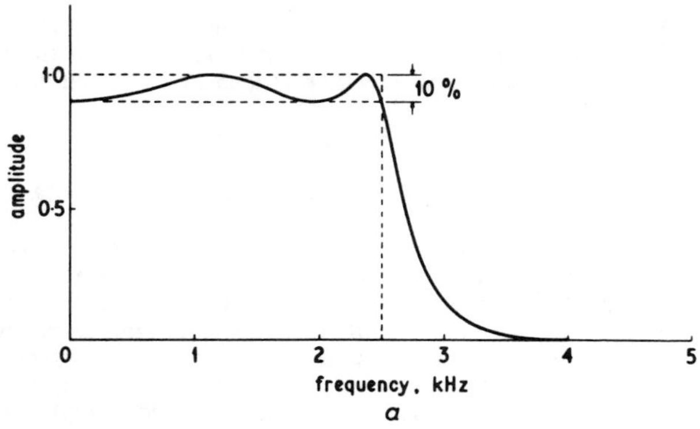

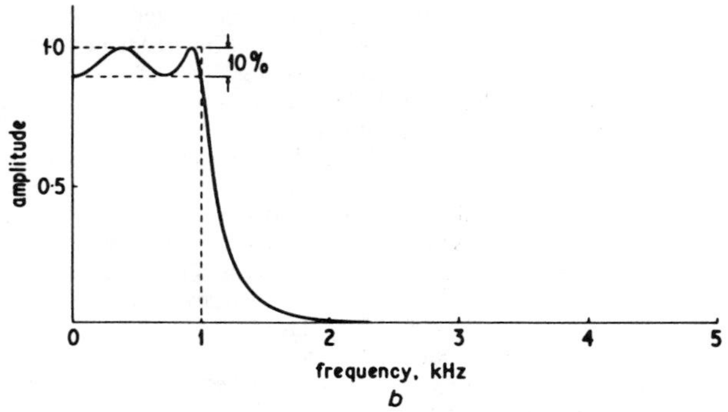

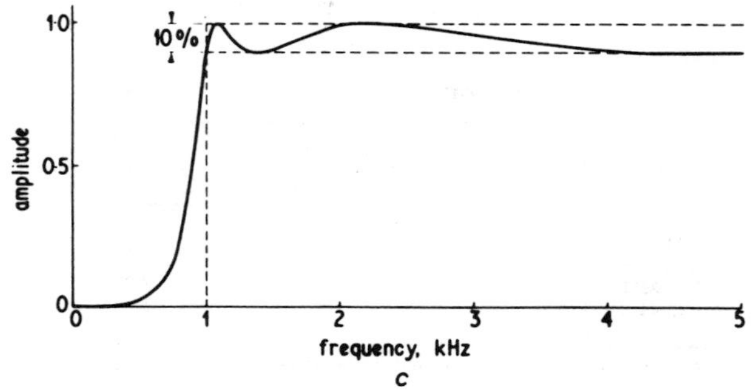

and, since there is to be a 1:1 correspondence between z^{-1} and $g(z^{-1})$, it follows from eqn. 2 that

$$g(z^{-1}) = \frac{z^{-1} - \alpha}{1 - \alpha z^{-1}} \quad . \quad . \quad . \quad . \quad . \quad . \quad . \quad . \quad . \quad (4)$$

where $|\alpha| < 1$ and is real.

Let the cutoff frequency of the lowpass prototype digital filter be β radians per second and that of the resulting lowpass digital filter be ω_c radians per second. Then the following relationship must hold:

$$\exp(-j\omega_c T) = \frac{\exp(-j\beta T) - \alpha}{1 - \alpha \exp(-j\beta T)}$$

from which we obtain

$$\alpha = \frac{\sin\left(\frac{\beta - \omega_c}{2}\right)T}{\sin\left(\frac{\beta + \omega_c}{2}\right)T} \quad . \quad . \quad . \quad . \quad . \quad . \quad . \quad . \quad . \quad (5)$$

The transformation given by eqn. 4 has no effect on the amplitude characteristic except for the change in the position of the frequencies on the unit circle. Thus, for negative α (i.e. $\omega_c > \beta$), the arc BDC of Fig. 1 is compressed on the circle, and, since A is invariant, the arc BAC is stretched. For positive α (i.e. $\omega_c < \beta$) the opposite effect occurs.

Fig. 2*a* shows the amplitude characteristic of a lowpass digital filter whose cutoff frequency is 2·5 kHz for a sampling frequency of 10 kHz. If $\omega_c/2\pi$ of eqn. 5 is chosen to be 1 kHz, as a result of transformation (eqn. 4) the amplitude characteristic is as shown in Fig. 2*b*.

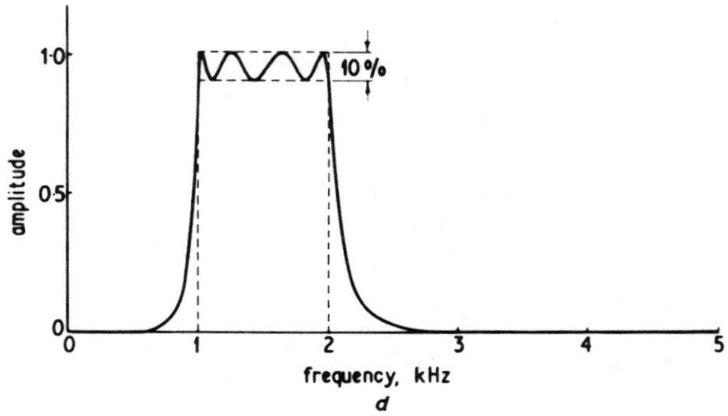

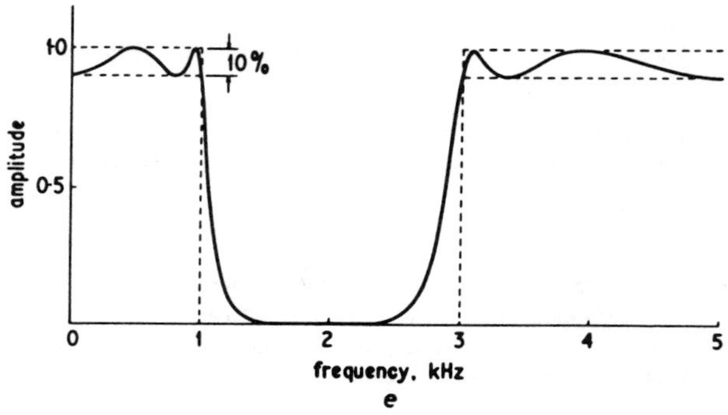

Fig. 2
Amplitude characteristics from spectral transformations
a 4th-order lowpass Chebyshev characteristic
$f_c = 2 \cdot 5\,\text{kHz}, F_s = 10\,\text{kHz}$
b Lowpass characteristic, $f_c = 1\,\text{kHz}$, resulting from lowpass-lowpass transformation (prototype filter shown in Fig. 2*a*)
c Highpass characteristic, $f_c = 1\,\text{kHz}$, resulting from lowpass-highpass transformation (prototype filter shown in Fig. 2*a*)
d Bandpass characteristic, $f_1 = 1\,\text{kHz}$, $f_2 = 2\,\text{kHz}$, resulting from a lowpass-bandpass transformation (prototype filter shown in Fig. 2*a*)
e Band-elimination characteristic, $f_1 = 1\,\text{kHz}$, $f_2 = 3\,\text{kHz}$, resulting from a lowpass-band-elimination transformation (prototype filter shown in Fig. 2*a*)

6 Lowpass–highpass transformation

Consider the lowpass characteristic of a digital filter as shown in Fig. 1. Now imagine that we have the freedom to rotate the cylinder at another angle relative to the positive direction of the real axis.

It can easily be seen that, when the cylinder has been rotated by 180°, we obtain a highpass amplitude characteristic (i.e. the passband and stopband of the lowpass filter have interchanged their relative positions) so that, if the cutoff frequency of the lowpass filter is β and that of the resulting highpass is ω_c,

$$\omega_c T + \beta T = \pi \text{ (angle of rotation)}$$

or

$$\omega_c = \frac{\Omega_s}{2} - \beta \quad . \quad . \quad . \quad . \quad . \quad . \quad . \quad . \quad (6)$$

where

$$\Omega_s = \frac{2\pi}{T}$$

The above method of achieving the required transformation

is rather restrictive, because it means that one has to replace z^{-1} in the lowpass-filter pulse-transfer function by $-z^{-1}$ where the two cutoff frequencies are restricted to satisfy the relationship given by eqn. 6.

In view of the lowpass–lowpass transformation given by eqn. 4, the above difficulty can be resolved by first applying the lowpass–lowpass transformation to obtain the particular lowpass cutoff frequency which, in conjunction with the required highpass cutoff frequency, satisfies eqn. 6, and then replacing z^{-1} by $-z^{-1}$ (i.e. rotate by 180°). These two steps, however, can be combined into one by replacing z^{-1} by $-z^{-1}$ in eqn. 4.

Thus the required transformation is given by

$$g(z^{-1}) = -\left(\frac{z^{-1}+\alpha}{1+\alpha z^{-1}}\right) \quad \ldots \quad (7)$$

It is quite simple to deduce the above transformation, however, from the general form of the unit function. The arguments proposed for the lowpass case are equally applicable here, except that we must take the angle of rotation θ of eqn. 2 to be 180°, which gives us the negative sign; and since there is to be a 1:1 correspondence between the lowpass prototype and the resulting highpass digital filter, we immediately obtain eqn. 7.

If the cutoff frequency of the highpass filter is ω_c and that of the prototype lowpass digital filter β, similarly to eqn. 5 we have

$$\alpha = -\frac{\cos\left(\frac{\beta-\omega_c}{2}\right)T}{\cos\left(\frac{\beta+\omega_c}{2}\right)T} \quad \ldots \quad (8)$$

We can state the above result formally in the form of a theorem.

Theorem 1: Given the pulse-transfer function $G(z^{-1})$ of a lowpass digital filter of cutoff frequency β, the pulse-transfer function of a highpass digital filter having the same type of amplitude characteristics as $G(z^{-1})$, and the cutoff frequency ω_c is given by replacing z^{-1} of $G(z^{-1})$ by the unit function of eqn. 7, where α is given by eqn. 8.

Fig. 2*c* shows the highpass amplitude characteristic that resulted from the lowpass one of Fig. 2*a*. In this case $\beta/2\pi$ = 2·5kHz, and $\omega_c/2\pi$ = 1kHz.

7 Lowpass–bandpass and lowpass–band-elimination transformations

The order of the unit function in this case will be two, as was shown earlier. The required mapping has the effect of reproducing the cylindrical surface of Fig. 1 twice on another cylindrical surface in such a way that the resulting two surfaces are symmetrical about a diameter of the base.

This cylindrical surface can be rotated in relation to the point $z^{-1} = 1$ on the circumference of the unit circle about its axis at any angle. This is equivalent to saying that θ (the angle of rotation) can take any value, but it was seen earlier that θ must be a multiple of π for real functions $g(z^{-1})$. It can easily be seen that, if $\theta = 0$, we have the band-elimination transformation, whereas for $\theta = \pi$ (or odd multiples of π) we have the bandpass transformation. From these conceptual considerations, let us return to the mathematical descriptions.

Consider the mapping requirements for the lowpass–bandpass transformation given in Table 2.

Table 2

LOWPASS–BANDPASS REQUIREMENTS

Lowpass frequency	Lowpass variable z^{-1}	Bandpass frequency	Bandpass variable
0	1	ω_0	$\exp(-j\omega_0 T)$
$-\beta$	$\exp(j\beta T)$	ω_1	$\exp(-j\omega_1 T)$
$+\beta$	$\exp(-j\beta T)$	ω_2	$\exp(-j\omega_2 T)$
$\Omega_s/2$	-1	$\Omega_s/2$	-1
		0	$+1$

Since the order of the required transformation is 2, we can write

$$g(z^{-1}) = \pm\frac{z^{-2}+\alpha_1 z^{-1}+\alpha_2}{\alpha_2 z^{-2}+\alpha_1 z^{-1}+1} \quad \ldots \quad (9)$$

By imposing the requirements of Table 2 on eqn. 9, we obtain

$$g(z^{-1}) = -\left(\frac{z^{-2}-\frac{2\alpha k}{k+1}z^{-1}+\frac{k-1}{k+1}}{\frac{k-1}{k+1}z^{-2}-\frac{2\alpha k}{k+1}z^{-1}+1}\right) \quad (10)$$

$$\text{where } \alpha = \cos\omega_0 T = \frac{\cos\left(\frac{\omega_2+\omega_1}{2}\right)T}{\cos\left(\frac{\omega_2-\omega_1}{2}\right)T} \quad \ldots \quad (11)$$

$$k = \cot\left(\frac{\omega_2-\omega_1}{2}\right)T\tan\frac{\beta T}{2} \quad \ldots \quad (12)$$

We therefore have the following theorem.

Theorem 2: Given the pulse-transfer function $G(z^{-1})$ of a lowpass digital filter of cutoff frequency β, the pulse-transfer function of a bandpass digital filter having the same type of amplitude characteristic as $G(z^{-1})$, centre frequency ω_0, and upper and lower cutoff frequencies ω_2 and ω_1, is given by replacing z^{-1} of $G(z^{-1})$ by the unit function of eqn. 10, where α and k are given by eqns. 11 and 12, respectively.

The transformation given by eqn. 10 can take different forms depending on the values of α and k.

In particular, if $k = 1$,

$$g(z^{-1}) = -\frac{z^{-1}(z^{-1}-\alpha)}{1-\alpha z^{-1}} \quad \ldots \quad (13)$$

and from eqn. 12 we have, with $k = 1$,

$$\omega_2 = \omega_1 = \beta \quad \ldots \quad (14)$$

The form of the transformation given by eqn. 13, along with the relationship of eqn. 14, has been given elsewhere.[1] This form is restricted in its application, because one has to use a lowpass filter of a particular cutoff frequency β given by eqn. 14 before applying eqn. 13. With the unrestricted form of eqn. 10 there is no particular difficulty, since the real coefficient k can be adjusted to satisfy eqn. 12.

Another form that the transformation can take corresponds to the case when $\alpha = 0$ and $k = 1$. This form is particularly important, since the resulting bandpass digital filter has an amplitude characteristic which is arithmetically symmetrical about the centre frequency. In this case,

$$g(z^{-1}) = -z^{-2} \quad \ldots \quad (15)$$

$$\left.\begin{aligned} \text{and} \quad \omega_1 &= \frac{\Omega_s}{4}-\frac{\beta}{2} \\ \omega_2 &= \frac{\Omega_s}{4}+\frac{\beta}{2} \\ \text{with} \quad \omega_0 &= \frac{\Omega_s}{4} \end{aligned}\right\} \quad \ldots \quad (16)$$

Fig. 2*d* shows the bandpass amplitude characteristic when $\omega_1/2\pi$ = 1kHz and $\omega_2/2\pi$ = 2kHz derived from the lowpass characteristic of Fig. 2*a*.

Now consider the lowpass–band-elimination transformation. The requirements for this case are tabulated in Table 3.

Table 3

LOWPASS–BAND-ELIMINATION REQUIREMENTS

Lowpass frequency	Lowpass variable z^{-1}	Band-elimination frequency	Band-elimination variable
0	$+1$	0	$+1$
		$\Omega_s/2$	-1
$-\beta$	$\exp(j\beta T)$	ω_2	$\exp(-j\omega_2 T)$
$+\beta$	$\exp(-j\beta T)$	ω_1	$\exp(-j\omega_1 T)$
$\Omega_s/2$	-1	ω_0	$\exp(-j\omega_0 T)$

Table 4

SPECTRAL TRANSFORMATIONS FROM A LOWPASS-DIGITAL-FILTER PROTOTYPE OF CUTOFF FREQUENCY β

Filter type	Transformation	Associated design formulas
Lowpass	$\dfrac{z^{-1}-\alpha}{1-\alpha z^{-1}}$	$\alpha = \dfrac{\sin\left(\dfrac{\beta-\omega_c}{2}\right)T}{\sin\left(\dfrac{\beta+\omega_c}{2}\right)T}$
Highpass	$\dfrac{z^{-1}+\alpha}{1+\alpha z^{-1}}$	$\alpha = -\dfrac{\cos\left(\dfrac{\beta-\omega_c}{2}\right)T}{\cos\left(\dfrac{\beta+\omega_c}{2}\right)T}$
Bandpass	$-\left(\dfrac{z^{-2}-\dfrac{2\alpha k}{k+1}z^{-1}+\dfrac{k-1}{k+1}}{\dfrac{k-1}{k+1}z^{-2}-\dfrac{2\alpha k}{k+1}z^{-1}+1}\right)$	$\alpha = \cos\omega_0 T = \dfrac{\cos\left(\dfrac{\omega_2+\omega_1}{2}\right)T}{\cos\left(\dfrac{\omega_2-\omega_1}{2}\right)T}$ $k = \cot\left(\dfrac{\omega_2-\omega_1}{2}\right)T_0 \tan\dfrac{\beta T}{2}$
Band-elimination	$\left(\dfrac{z^{-2}-\dfrac{2\alpha}{1+k}z^{-1}+\dfrac{1-k}{1+k}}{\dfrac{1-k}{1+k}z^{-2}-\dfrac{2\alpha}{1+k}z^{-1}+1}\right)$	$\alpha = \dfrac{\cos\left(\dfrac{\omega_2+\omega_1}{2}\right)T}{\cos\left(\dfrac{\omega_2-\omega_1}{2}\right)T} = \cos\omega_0 T$ $k = \tan\left(\dfrac{\omega_2-\omega_1}{2}\right)T\tan\dfrac{\beta T}{2}$

On imposing the constraints given in Table 3 on the unit function of eqn. 9, we obtain

$$g(z^{-1}) = \left(\frac{z^{-2}-\dfrac{2\alpha}{k+1}z^{-1}+\dfrac{1-k}{1+k}}{\dfrac{1-k}{1+k}z^{-2}-\dfrac{2\alpha}{1+k}z^{-1}+1}\right) \quad . \quad (17)$$

$$\text{where } \alpha = \cos\omega_0 T = \frac{\cos\left(\dfrac{\omega_2+\omega_1}{2}\right)T}{\cos\left(\dfrac{\omega_2-\omega_1}{2}\right)T} \quad . \; . \; . \; . \quad (18)$$

$$k = \tan\left(\frac{\omega_2-\omega_1}{2}\right)T\tan\frac{\beta T}{2} \quad . \; . \; . \; . \; . \quad (19)$$

From the above result, we can formulate a theorem similar to theorem 2.

Theorem 3: Given the pulse-transfer function $G(z^{-1})$ of a lowpass digital filter of cutoff frequency β, the pulse-transfer function of a bandpass digital filter having the same type of amplitude characteristic as $G(z^{-1})$, centre frequency ω_0, and upper and lower cutoff frequencies ω_2 and ω_1, is given by replacing z^{-1} of $G(z^{-1})$ by the unit function of eqn. 17, where α and k are given by eqns. 18 and 19, respectively.

There are again the following two restricted forms of the transformation. When $k = 1$, we have

$$g(z^{-1}) = \frac{z^{-1}(z^{-1}-\alpha)}{1-\alpha z^{-1}} \quad . \; . \; . \; . \; . \; . \; . \quad (20)$$

and, from eqn. 19,

$$\omega_2 - \omega_1 = \frac{\Omega_s}{2} - \beta \quad . \; . \; . \; . \; . \; . \; . \quad (21)$$

The above restricted form of eqn. 20, along with the relationship of eqn. 21, has been given elsewhere,[1] and the restrictions are precisely those stated for the equivalent bandpass case.

Now, when $\alpha = 0$ in eqn. 20, we have the arithmetically symmetrical band-elimination case, where

$$g(z^{-1}) = -z^{-2} \quad . \; . \; . \; . \; . \; . \; . \; . \; . \; . \quad (22)$$

$$\left.\begin{aligned} \omega_1 &= \frac{\beta}{2} \\ \omega_2 &= \frac{\Omega_s}{2} - \frac{\beta}{2} \\ \text{and} \quad \omega_0 &= \frac{\Omega_s}{4} \end{aligned}\right\} \quad . \; . \; . \; . \; . \; . \; . \; . \quad (23)$$

Fig. 2*e* shows the band-elimination amplitude characteristic when $\omega_1/2\pi = 1$ kHz and $\omega_2/2\pi = 3$ kHz, derived from the lowpass characteristic of Fig. 2*a*. All the transformations are summarised in Table 4.

8 Comparison of transformations to Broome's translation formula

Broome[3] gave the following formula as a means of transforming lowpass pulse-transfer functions into highpass, bandpass and band-elimination pulse-transfer functions.

Let $G(z^{-1})$ be the pulse-transfer function of a lowpass digital filter. Then

$$H(z^{-1}) = G\{z^{-1}\exp(-j\omega_0 T)\} + G\{z^{-1}\exp(+j\omega_0 T)\}$$

is a translated version of $G(z^{-1})$. Several comments have been produced by Kaiser in Reference 6.

Basically, the above formula represents two rotations on the z^{-1} plane, one in the clockwise and the other in the counter-clockwise direction. The sum of the two rotations of functions of the complex variable z^{-1} is a real rational function z^{-1} if the rotations are at equal angles. Because the pulse-transfer function resulting from the above formula is the sum of two individual pulse-transfer functions rotated to the angles $\pm\omega_0 T$, it will exhibit what Broome[3] and Kaiser[6] term a 'distortion error' which results from the tails of the image of $G(z^{-1})$ centred at $-\omega_0$. It is suggested in these References that a good lowpass-filter design will reduce the distortion error to a very small level.

No distortion will occur when the lowpass pulse-transfer function is rotated by $\pm 180°$ or multiples thereof, because, when $\omega_0 T = \pm 180°$,

$$\exp(j\omega_0 T) = -1$$

and hence

$$H(z^{-1}) = 2G(-z^{-1})$$

which is recognised as the restricted case of the general lowpass–highpass transformation of eqn. 7.

If $\omega_0 T = 0$, the pulse-transfer function remains unaltered except for a multiplier of value 2. This is not, however, a lowpass–lowpass transformation, because no change in cutoff frequency occurs.

Thus the Broome translation formula is only satisfactory under very special conditions where the distortion errors are small and when $\omega_0 T = 180°$, so that the special case of the lowpass–highpass transformation is obtained, and, in this restricted sense, the translation formula is exact.

The general transformations given in this paper, however, are therefore superior to the above translation formula, since they not only introduce no distortion errors but they are also unrestricted in their application, be it for wideband or narrowband digital filters.

9 Conclusions

We have given in this paper the complete set of transformations on the z^{-1} plane necessary to transform a given lowpass-digital-filter pulse-transfer function into a pulse-transfer function having the same type of amplitude characteristic and belonging to one of the following classes:

(i) lowpass
(ii) highpass
(iii) bandpass
(iv) band-elimination.

Furthermore, along with the transformations, the necessary design formulas were given which relate the critical frequencies of the required digital filter to the cutoff frequency of the lowpass-digital-filter prototype.

The transformations are quite general, and, their restricted forms that have been published elsewhere[1,2] are easily obtainable as indicated.

Proofs of the most important theorems are given in Appendix 12.

The spectral transformations possess the following important and useful features:

(*a*) They are allpass functions, and hence, because of their inherent form, the multipliers in the digital implementation can be reduced by a factor of 2 by multiplexing.

(*b*) Since the application of the transformations involves a substitution of z^{-1} in a lowpass-digital-filter transfer function by a unit function, it follows that the structure of the lowpass digital filter remains unaltered (i.e. the adders and multipliers are the same), and only an extra elementary transfer function, representing the transformation, is inserted in place of z^{-1}.

(*c*) As a consequence of property (*b*) above, certain forms of the transformations (e.g. eqns. 13 and 20) that have one variable parameter exhibit simple characteristics where the centre frequency can be varied over the Nyquist range but the bandwidth is kept constant.

10 Acknowledgments

The author wishes to thank the authorities of Northampton College of Advanced Technology, London (The City University, London) for financial support, and the Senior Director of Development, UK Post Office Telecommunications Headquarters, for facilities provided.

11 References

1 CONSTANTINIDES, A. G.: 'Frequency transformations for digital filters', *Electron. Lett.*, 1967, **3**, pp. 487–489
2 CONSTANTINIDES, A. G.: 'Frequency transformations for digital filters', *ibid.*, 1968, **4**, pp. 115–116
3 BROOME, P.: 'A frequency transformation for numerical filters', *Proc. Inst. Elect. Electron. Engrs.*, 1966, **54**, pp. 326–327
4 JURY, E. I.: 'Theory and application of the *z*-transform method' (Wiley, 1964)
5 CRYSTAL, T. H.: 'The design and applications of digital filters with complex coefficients', *IEEE Trans.*, 1968, **AU-16**, pp. 330–335
6 KAISER, J. F.: 'Digital filters' *in* KUO, F. F., and KAISER, J. F. (Eds.): 'System analysis by digital computer' (Wiley, 1966), pp. 218–253
7 CONSTANTINIDES, A. G.: 'Synthesis of recursive digital filters from prescribed amplitude characteristics'. Ph.D. thesis, University of London, 1968

12 Appendix

Proof of theorem 2: Rewriting eqn. 9 in a different form, we have

$$g(z^{-1}) = e^{j\theta} \frac{z^{-2} + \gamma_1 z^{-1} + \gamma_2}{\gamma_2 z^{-2} + \gamma_1 z^{-1} + 1}$$

From Table 2 we see that $g(1) = -1$.

Hence, by using this constraint on the above form of eqn. 9, we obtain $e^{j\theta} = -1$.

Furthermore, since $g\{\exp(-j\omega_0 T)\} = -1$, we have

$$(\gamma_2 + 1) + 2\gamma_1 \exp(-j\omega_0 T) + (\gamma_2 + 1)\exp(-j2\omega_0 T) = 0$$

and hence $(1 + \gamma_2)\cos\omega_0 T + \gamma_1 = 0$.

Let $\cos\omega_0 T = \alpha$, so that

$$g(z^{-1}) = -\frac{z^{-2} - \alpha(1 + \gamma_2)z^{-2} + \gamma_2}{\gamma_2 z^{-2} - \alpha(1 + \gamma_2)z^{-2} + 1}$$

and also let $\gamma_2 = \gamma$. Hence we can write the above equation in the form

$$g(z^{-1}) = -\left\{\frac{z^{-1}\left(\frac{z^{-1} - \alpha}{1 - \alpha z^{-1}}\right) + \gamma}{1 + \gamma z^{-1}\left(\frac{z^{-1} - \alpha}{1 - \alpha z^{-1}}\right)}\right\}$$

and, since $\alpha = \cos\omega_0 T$, $|\alpha| \leqslant 1$, so that the function $z^{-1}\left(\frac{z^{-1} - \alpha}{1 - \alpha z^{-1}}\right)$ is a unit function.

Furthermore $|\gamma| \leqslant 1$, since this quantity represents the product of two zeros of the unit function which are inside the unit circle $|z^{-1}| = 1$. Hence the quantity $-\left(\frac{z^{-1} + \gamma}{1 + \gamma z^{-1}}\right)$ is a unit function.

Let $E_1 = -\left(\frac{z^{-1} + \gamma}{1 + \gamma z^{-1}}\right)$ and $E_2 = z^{-1}\left(\frac{z^{-1} - \alpha}{1 - \alpha z^{-1}}\right)$

so that $g(z^{-1}) = E_1(E_2)$.

Now let us form $\frac{1 - g(z^{-1})}{1 + g(z^{-1})} = \frac{1 + \gamma E_2 + \gamma + E_2}{1 + \gamma E_2 - \gamma - E_2}$

i.e.

$$\frac{1 - g(z^{-1})}{1 + g(z^{-1})} = \frac{1 + \gamma}{1 - \gamma}\,\frac{1 + E_2}{1 - E_2}$$

$$= \frac{1 + \gamma}{1 - \gamma}\,\frac{z^{-2} - 2\alpha z^{-1} + 1}{1 - z^{-2}}$$

When $\omega = \omega_1$ in the bandpass characteristic, the function $g(z^{-1})$ corresponds to $\exp(j\beta T)$, and, when $\omega = \omega_2$, $g(z^{-1}) = \exp(-j\beta T)$. Hence

$$-\tan\frac{\beta T}{2} = \frac{\gamma + 1}{\gamma - 1}\,\frac{\cos\omega_1 T - \alpha}{\sin\omega_1 T}$$

and

$$\tan\frac{\beta T}{2} = \frac{\gamma + 1}{\gamma - 1}\,\frac{\cos\omega_2 T - \alpha}{\sin\omega_2 T}$$

Therefore

$$\alpha = \frac{\cos\left(\frac{\omega_2 + \omega_1}{2}\right)T}{\cos\left(\frac{\omega_2 - \omega_1}{2}\right)T}$$

Let $\frac{\gamma + 1}{\gamma - 1} = -k$, so that

$$k = \cot\left(\frac{\omega_2 - \omega_1}{2}\right)T\,\tan\frac{\beta T}{2}$$

and

$$\gamma = \frac{k - 1}{k + 1}$$

and the transformation becomes

$$g(z^{-1}) = -\left(\frac{z^{-2} - \frac{2\alpha k}{k + 1}z^{-1} + \frac{k - 1}{k + 1}}{\frac{k - 1}{k + 1}z^{-2} - \frac{2\alpha k}{k + 1}z^{-1} + 1}\right)$$

Hence the theorem is proved. The proof for theorem 3 is similar to the above.

A Note on Digital Filter Synthesis

Abstract—**It is commonly assumed that digital filters with both poles and zeros in the complex z-plane can be synthesized using only recursive techniques while filters with zeros alone can be synthesized by either direct convolution or via the discrete Fourier transform (DFT). In this letter it is shown that no such restrictions hold and that both types of filters (those with zeros alone or those with both poles and zeros) can be synthesized using any of the three methods, namely, recursion, DFT, or direct convolution.**

I. Introduction

A digital filter can be synthesized either by direct convolution, by linear recursive equations, or by the use of the discrete Fourier transform (DFT), usually via the fast Fourier transform (FFT). Kaiser[1] has used the terms "recursive" and "nonrecursive" to distinguish between filters that are defined by an impulse response of finite duration (nonrecursive) and those defined by an impulse response of infinite duration. The former type contains only zeros while the latter has both poles and zeros in the complex z-plane. It is well known that the design methods for filters with only zeros differ markedly from the design methods for filters with zeros and poles.

The term "recursive" is descriptive of the filter synthesis procedure whereby a new output sample can be computed as a linear combination of previous output samples as well as the latest and previous input samples. Similarly, "nonrecursive" describes a computation for which the new output sample is a linear function of only the input samples. Stockham[2] has shown that filters with zeros alone can be synthesized using the FFT algorithm.

The purpose of this letter is to show how 1) filters with only zeros can be synthesized recursively and how 2) filters with both poles and zeros can be synthesized using either direct convolution or the FFT. As a result, it becomes desirable to define our terms more carefully, and we propose that the terms "recursive," "direct convolution," and "FFT" be used to specify only the synthesis procedure. To distinguish between filters with zeros alone and those with poles and zeros, perhaps the terms "finite impulse response" (FIR) and "infinite impulse response" (IIR) could be used.

Manuscript received June 24, 1968.

[1] J. F. Kaiser, in *System Analysis by Digital Computer*, J. F. Kaiser and F. Kuo, Eds. New York: Wiley, 1966, pp. 218–277.

[2] T. G. Stockham, Jr., "High speed correlation and convolution," presented at the Spring Joint Computer Conf., 1966.

Reprinted from *Proc. IEEE*, vol. 56, pp. 1717–1718, Oct. 1968.

II. Recursive Synthesis of Filters with Zeros Alone

The z-transform of an FIR filter having impulse response (of duration NT) $h(nT)$ is given by

$$H(z) = \sum_{n=0}^{N-1} h(nT)z^{-n} \tag{1}$$

where T is the sampling interval.

Since $h(nT)$ is of finite duration, it has a DFT given by

$$H_k = \sum_{n=0}^{N-1} h(nT)e^{-j(2\pi/N)nk}, \qquad k = 0, 1, 2, \cdots N-1 \tag{2}$$

where the set of H_k are precisely the values of $H(z)$ at equally spaced points on the unit circle in the z-plane. The impulse response $h(nT)$ can then be expressed as the inverse DFT of the samples H_k and if this inverse DFT is substituted into (1), with the order of the resulting double summation inverted, we easily arrive at the result

$$H(z) = \frac{1 - z^{-N}}{N} \sum_{k=0}^{N-1} \frac{H_k}{1 - e^{j(2\pi k/N)}z^{-1}}. \tag{3}$$

Equation (3) can be physically represented by a comb filter with transfer function $1 - z^{-N}$ cascaded with a parallel arrangement of first-order recursive equations. In general, the coefficients of these equations $(e^{j(2\pi k/N)})$ are complex, but if in (3) we specify that H_{N-k} and H_k are complex conjugates, then combinations of second-order recursive networks with real coefficients can be derived. An interesting special case of (3) is the case $\theta_k = k\pi$ and $M_k = 1$, where $H_k = M_k e^{j\theta_k}$. This leads to a bandpass filter with linear phase, which has been called "frequency sampling" by Rader and Gold.[3] If two such filters are designed so that the frequency response of one is a linear translation of the other, then it is easy to show that the two outputs have a constant phase difference over the passbands common to both filters.

[3] C. M. Rader and B. Gold, "Digital filter design techniques in the frequency domain," *Proc. IEEE*, vol. 55, pp. 149–171, February 1968.

III. Direct Convolution and FFT Synthesis of Filters with Poles and Zeros

We consider the defining equation of an IIR filter to be

$$\sum_{i=0}^{m} K_i y(nT - iT) = \sum_{i=0}^{r} L_i x(nT - iT) \tag{4}$$

where the set of $x(nT)$ are the input samples, $y(nT)$ are the output samples, and K_i and L_i are constant coefficients. Taking the z-transform of (4) and including the initial condition yields

$$Y(z) = \frac{X(z)R(z)}{D(z)} + \frac{F(z) + G(z)}{D(z)} \tag{5}$$

where $X(z)$ is the z-transform of $x(nT)$, $Y(z)$ is the z-transform of $y(nT)$, $R(z)$ and $D(z)$ are given by

$$R(z) = \sum_{i=0}^{r} L_i z^{-i}, \qquad D(z) = \sum_{i=0}^{m} K_i z^{-i}, \tag{6}$$

and $F(z)$ and $G(z)$ are z-transforms which depend solely on the initial conditions and are given by

$$\left.\begin{aligned} F(z) &= \sum_{j=0}^{m-1} a_j z^{-j} \quad \text{with} \quad a_j = \sum_{i=0}^{m-j-1} K_{j-i+1} y(-T - iT) \\ G(z) &= \sum_{j=0}^{r-1} b_j z^{-j} \quad \text{with} \quad b_j = \sum_{i=0}^{r-j-1} L_{j-i+1} x(-T - iT). \end{aligned}\right\} \tag{7}$$

If we define $h_1(nT)$ to be the inverse z-transform of $R(z)/D(z)$, $h_2(nT)$ to be the inverse-transform of $1/D(z)$, $f(nT)$ to be the inverse z-transform of $F(z)$, and $g(nT)$ to be the inverse z-transform of $G(z)$, then the solution $y(nT)$ can be written as

$$y(nT) = \sum_{I=0}^{n} h_1(iT)\, x(nT - iT) + \sum_{i=0}^{m} h_2(iT) f(nT - iT) + \sum_{i=0}^{r} h_2(iT) g(nT - iT). \tag{8}$$

Equation (8) shows that the output can be written as the sum of three finite convolutions for any section of length $n+1$. We assume that the indices for that section are 0, 1, 2, $\cdots$, n. Thus, we imagine that each new section resets the origin to the beginning of that section, the contribution from previous sections being determined solely by the initial conditions and the system function. As can be seen in (5), the output for any contiguous section of samples can be viewed as the sum of an output determined entirely by the input signal and another output determined entirely by the initial conditions of the system.

Since a finite convolution can always be computed via DFT, it follows that (8) can be synthesized using either direct convolution or FFT.

IV. Conclusion

Sections II and III demonstrate the claim made in the Introduction, namely, that filters with zeros alone can be synthesized recursively while filters with zeros and poles can be synthesized by direct convolution or FFT. Thus, any of the known digital filter synthesis procedures can, in fact, be used to synthesize either type of filter.

B. Gold
K. L. Jordan, Jr.
M.I.T. Lincoln Lab.[4]
Lexington, Mass. 02173

[4] Operated with support from the U. S. Air Force.

Digital Filtering Via Block Recursion

HERBERT B. VOELCKER, Member, IEEE
University of Rochester
River Campus Station
Rochester, N. Y.

E. EUGENE HARTQUIST, Associate Member, IEEE
Ohio State University
Columbus, Ohio

Abstract

Digital filters which have poles in their transfer functions are usually implemented by various direct feedback arrangements. Such filters can also be synthesized in a form amenable to implementation via the FFT. The key feature of this procedure is block feedback through a special finite-response filter. Although the procedure appears to offer few immediate practical advantages, its asymptotic properties are interesting. It also provides a useful conceptual link between recursive and nonrecursive filtering techniques.

Manuscript received January 9, 1970; revised February 13, 1970.

This work was supported in part by NSF Grant GK-10, 861.

Portions of this work were submitted in partial fulfillment of the requirements for the B.S. degree in Electrical Engineering to the University of Rochester, Rochester, N. Y.

Introduction

A linear stationary digital filter can be described by the difference equation [1], [2]

$$y(nT) = \sum_{i=0}^{r} L_i x(nT - iT) - \sum_{i=1}^{m} K_i y(nT - iT). \quad (1)$$

The corresponding transfer or system function is found by z-transforming (1):

$$Y(z) = H(z)X(z) \quad (2)$$

where

$$H(z) = \frac{R(z)}{D(z)} \leftrightarrow h(nT) \quad (3a)$$

$$R(z) = \sum_{i=0}^{r} L_i z^{-i} \quad (3b)$$

$$D(z) = 1 + \sum_{i=1}^{m} K_i z^{-i}. \quad (3c)$$

The parameters m and r specify, respectively, the numbers of z-plane poles and zeros of the filter. The poles must be within the unit circle for stability.[1] The pulse response $h(nT)$ is of infinite duration if $H(z)$ has one or more net poles (i.e., poles not cancelled by zeros). Such filters are usually called recursive because the output y is defined on itself. If $H(z)$ has no net poles, the filter is nonrecursive and $h(nT)$ is of finite duration.

Implementation of (1) is straightforward. In principle, one need only connect special-purpose digital elements in the manner prescribed by (1), or program (1) directly for a general-purpose computer. In practice, one usually synthesizes complicated filters from series or parallel combinations of simple filters in order to minimize certain noise and stability problems [2]. It is important to note that with any such direct implementation, the amount of arithmetic per output sample grows linearly with $m+r$, the complexity or number of degrees of freedom of the filter.

Nonrecursive (zeros only) filters, however, are ostensibly special in a computational sense. Because their pulse responses are finite, they can be implemented "indirectly" via sequential overlapped finite convolutions. The finite convolutions, in turn, can be implemented via the FFT [2], [3], with the result that the amount of arithmetic per output sample need grow only as $\log(r)$. The computational savings can be impressive when r is large.

This last point leads one to ask: Can *recursive* filters somehow be implemented via sequential finite convolutions? An affirmative answer might imply useful computational savings through exploitation of FFT algorithms. Also, convolutional implementations might

[1] Rationalization of (3a) causes R and D to deposit, respectively, r poles and m zeros at the origin in the z-plane. These are, in a sense, "bogus" poles and zeros which we shall usually ignore.

Reprinted from *IEEE Trans. Audio Electroacoust.*, vol. AU-18, pp. 169–176, June 1970.

alleviate some of the noise and stability problems which complicate recursive filter design.

Gold and Jordan [4] have already provided an affirmative but relatively opaque answer to the above question. They argued, in essence, that (1) can be synthesized in terms of finite convolutions if initial conditions calculated from past results are introduced whenever a new finite convolution is begun. These initial conditions correspond physically to the data stored in a direct implementation of (1) at the instant when one convolution is terminated and another is begun. Unfortunately the Gold–Jordan solution is awkward to program directly, and it provides little insight into its own efficient parameterization and intrinsic efficiency. Chanoux [5] contributed a procedure for implementing the Gold–Jordan solution, but it led him to draw conclusions which appear misleading in the light of results presented here.[2]

We shall approach the question from a somewhat different viewpoint, and we shall provide a solution whose form, we believe, affords more insight. Our starting point is an immediate simplification. Henceforth, we shall deal exclusively with zero-free filters $H_p(z)$ where

$$H_p(z) = \frac{1}{D(z)} \cdot \tag{4}$$

This simplification is justified because the zeros-only feedforward portion $R(z)$ of our original filter can be viewed as a separate finite-impulse-response unit cascaded with $H_p(z)$. Also, we shall concentrate on schemes for implementing H_p via "block processing," in order to seek the computational economies of scale associated with the FFT and similar "block" algorithms.

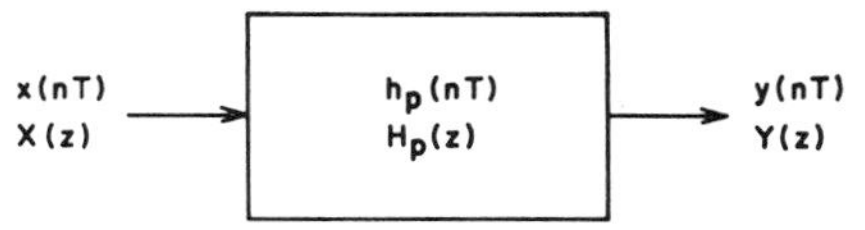

Fig. 1. Poles-only recursive filter.

Fig. 2. Block-recursive filter.

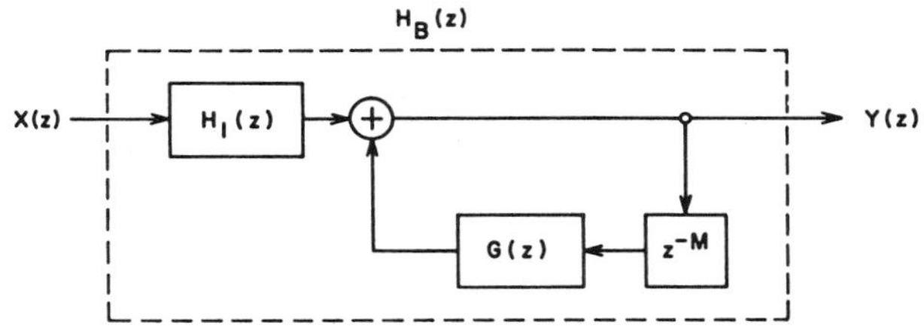

Recursive Block Processing

Let $x(nT)$, $y(nT)$, and $h_p(nT)$ in Fig. 1 be partitioned into sequences of equal size blocks of M samples where

$$M \geq m + 1. \tag{5}$$

(This rather arbitrary lower bound on M is a matter of temporary convenience.) Thus

$$\begin{aligned} h_1 &= \{h_p(0), h_p(T), \cdots, h_p(MT - T)\} \\ h_2 &= \{h_p(MT), \cdots \qquad\qquad\qquad\} \\ \cdot\ \cdot & \qquad\qquad \cdot\ \cdot\ \cdot \\ x_1 &= \{x(0), x(T), \cdots, x(MT - T)\} \\ \cdot\ \cdot & \qquad\qquad \cdot\ \cdot\ \cdot \text{ etc.} \end{aligned} \tag{6}$$

By the convolution theorem,

$$y_j = \sum_{i=1}^{j} \sum_{k=1}^{j} x_i * h_k. \tag{7}$$

We know how to compute fixed block-size convolutions efficiently. If $h_p(nT)$ were of finite extent, examination of (7) would show that the maximum number of block convolutions that must be calculated per output block is constant. But $h_p(nT)$ is not finite, and (7) suggests, rather ominously, that the number of block convolutions per output block grows linearly with the output block index. Thus a block feedforward approach is unrewarding.

One suspects, however, that recursive procedures may be applicable because of the "predictability" of the filter characteristics in successive blocks. (For example, one might guess—correctly—that the first h-block contains enough information to generate the other h-blocks.) We shall take a clue from Gold and Jordan [4] and recast (7) in the form

$$\begin{aligned} &(\text{current } y\text{-block}) \\ &\qquad = (\text{current } x\text{-block}) * (\text{fixed } h\text{-block}) \\ &\qquad\quad + \text{fun (previous } y\text{-block(s))}. \end{aligned} \tag{8}$$

Equation (7) suggests that the "fixed h-block" in (8) is $h_1 \leftrightarrow H_1(z)$. Let us hypothesize that the "fun" in (8) involves only a convolution of the immediately preceding y-block with a fixed function $g \leftrightarrow G(z)$. This hypothesis leads directly to the filter $H_B(z)$ shown in Fig. 2, wherein z^{-M} is a *block delay operator.*

The key feature of Fig. 2 is block-delayed feedback. This scheme will become potentially attractive only if we can show that a G exists that will make H_B equivalent to H_p, and that this G has a *finite* pulse response. Both conditions will be established in the next section. A subsequent section will show that Fig. 2 is relatively easy to implement via finite convolutions calculated through the FFT, because the block delay feature provides a "time gap" which permits the data entering the feedback loop to be "batched" for block processing.

Analysis

We shall outline the analytical approach before plunging into details. By inspection of Fig. 2,

[2] The authors thank Prof. A. Oppenheim for making available a copy of Chanoux's thesis.

$$H_B(z) = \frac{H_1(z)}{1 - z^{-M}G(z)}. \tag{9}$$

It is shown below that $H_1(z)$ can be written

$$H_1(z) = \frac{1 - z^{-M}Q(z)}{D(z)} \tag{10}$$

where $Q(z)$ is a polynomial in z^{-1} of order $m-1$. Given this fact, one need only equate $G(z)$ and $Q(z)$ in (9) and (10) to obtain

$$H_B(z)\big|_{G=Q} = \frac{1 - z^{-M}Q(z)}{D(z)} \cdot \frac{1}{1 - z^{-M}G(z)} \tag{11a}$$

$$= H_p(z) \tag{11b}$$

which is the desired result. The heart of the argument clearly resides in (10), and it is this equation that we shall now establish.

Let $H_p(z)$ be expanded in partial fractions:

$$H_p(z) = \frac{1}{\prod_{i=1}^{m}(1 - z_i/z)} = \sum_{i=1}^{m} \frac{a_i}{1 - z_i/z} \tag{12}$$

where the $\{z_i\}$ are the poles of $H_p(z)$, i.e., the zeros of $D(z)$, and the $\{a_i\}$ are constants.[3] Construct the finite-response filter $H_1(z)$ by truncating the inverse transform of each component of (12) at the Mth term. This can be done by inspection, and yields

$$H_1(z) = \sum_{i=1}^{m} \frac{a_i}{1 - z_i/z}[1 - (z_i/z)^M] \tag{13a}$$

$$= H_p(z) - z^{-M} \sum_{i=1}^{m} \frac{a_i z_i^M}{1 - z_i/z} \tag{13b}$$

$$= \frac{1 - z^{-M}Q(z)}{D(z)} \tag{13c}$$

where

$$Q(z) = \sum_{i=1}^{m} \frac{a_i z_i^M D(z)}{1 - z_i/z}. \tag{14}$$

Equation (13c) justifies the form (10), and (14) plus (3c) show that $Q(z)$ is a finite polynomial in z^{-1} of order $m-1$. The desired result, $H_B = H_p$, accrues per (11) by equating $G(z)$ in Fig. 2 to the $Q(z)$ specified by (14).

Equation (14) is sometimes inconvenient, and we shall derive an alternative form. By definition of the z-transform and h_1,

$$H_1(z) = 1 + \sum_{i=1}^{M-1} h_p(iT)z^{-i}. \tag{15}$$

Also,

$$H_1(z) = \frac{H_1(z)D(z)}{D(z)}, \tag{16}$$

and by comparing (13c) with (16),

$$1 - z^{-M}Q(z) = H_1(z)D(z). \tag{17}$$

Thus

$$Q(z) = z^M[1 - H_1(z)D(z)]. \tag{18}$$

This is a serendipitous result. The $H_1(z)D(z)$ product in (18) connotes a convolution of $h_1(nT)$ and $d(nT)$, and z^M is merely an advance operator. Thus $q(nT)$ can be evaluated algorithmically; see the last paragraph in the main body of the paper. An explicit equation for $q(nT)$ can be obtained either by inverse-transforming (18) directly, or by substituting (3c) and (15) into (18) and examining the coefficients of the powers of z^{-1}. One form of the result is

$$q(nT) = -\sum_{j=M+n-m}^{M-1} h_p(jT)K_{M+n-j}, \quad 0 \le n \le m-1. \tag{19}$$

Equation (19) specifies q (hence g, by (11)) in terms of the coefficients of $D(z)$ and the last m samples of the truncated pulse response h_1.

We have established that the procedure embodied in Fig. 2 is theoretically tenable: a stable poles-only filter can be synthesized, at least in principle, via two finite-response filters $H_1(z)$ and $G(z)$, plus a block-delay operator. One might interpret this result from two different viewpoints.

First, Gold and Jordan showed the necessity of introducing initial conditions when using sequential block processing. The feedback loop in Fig. 2 provides these. Indeed, the first term on the right of Gold and Jordan's (8) is our feedforward term, while their second term is our feedback contribution. (Their third term accounts for zeros in $H(z)$ which we agreed earlier to treat separately.)

Alternatively, one might argue that the nonrecursive H_1 in Fig. 2 cannot contain poles, but that its zeros must be related intimately to the poles of H_p because $h_1(nT)$ and $h_p(nT)$ are identical for $0 \le n \le M-1$. Equations (13c) and (17) specify the relationship: the numerator of (13c) can be factored into a pole-cancelling term and another (H_1) term. In this context, the feedback loop in Fig. 2 provides both the H_p poles and excess poles to cancel the zeros of H_1.

A Simple Example

Let $H_p(z)$ be a second-order ($m=2$) filter:

$$H_p(z) = \frac{1}{1 + k_1/z + k_2/z^2} \tag{20}$$

and let the block size M be 3. The H_1 filter derived from (20) is

$$H_1(z) = 1 - k_1/z + (k_1^2 - k_2)/z^2. \tag{21}$$

The G filter is specified via (11) and (19):

$$G(z) = (2k_1k_2 - k_1^3) + (k_2^2 - k_1^2k_2)/z. \tag{22}$$

[3] Poles of order higher than unity require that the argument be modified in a relatively obvious manner.

The block recursive filter is

$$H_B(z) = \frac{H_1(z)}{1 - z^{-3}G(z)}, \tag{23}$$

which can be shown to be equal to (20) by long division.

Stability

The preceding arguments imply that a block recursive filter H_B is stable if it is made precisely equivalent to a stable filter H_p. Suppose, however, that precise equivalence cannot be guaranteed because of small errors. Is the filter still stable?

Consider, for example,

$$\hat{H}_B(z)\,|_{G=Q} = \frac{H_1(z) + \delta H_1(z)}{1 - z^{-M}G(z)} \tag{24}$$

wherein the denominator is presumed exact and the numerator contains small errors, δH_1. We may write, using (17),

$$\hat{H}_B = \frac{H_1 + \delta H_1}{H_1 D}. \tag{25}$$

Clearly, $\hat{H}_B$ has the desired poles of H_p, plus additional poles at the zeros of H_1 which are *not* cancelled exactly. These additional poles must lie within the unit circle if $\hat{H}_B$ is to be stable.

Thus we have the following central question: Are the zeros which characterize H_1 wholly interior, i.e., within the unit circle? (This question is not merely academic, as anyone who experiments too casually with block recursion will discover.)

It is easy to show that the conditions $\{$all m poles of H_p interior, $M \geq m+1\}$ are *not* sufficient to insure that H_1 has wholly interior zeros. One need only consider the preceding specific example (20): the H_1 of (21) can have either interior or exterior zeros, depending on the specific interior locations of the two poles of H_p.

We can also show, however, that the zeros of H_1 must be wholly interior if M, the length of H_1, is made sufficiently large. To do this, let the tilde ($\tilde{}$) denote the result of the mapping $z^{-1} \rightarrow z$, i.e., $\tilde{f}(z) = f(z^{-1})$, and define $\tilde{A}(z)$ in (26) via (17).

$$\tilde{A}(z) = 1 - z^M\tilde{Q}(z) \tag{26a}$$

$$= \tilde{H}_1(z)\tilde{D}(z). \tag{26b}$$

Note that $\tilde{A}$ is a polynomial in z of order $M+m-1$, and that all zeros of $\tilde{D}$ are exterior. All zeros of $\tilde{H}_1$ will be exterior—hence, all zeros of H_1 will be interior—if and only if all zeros of $\tilde{A}$ are exterior.

Clearly, $\tilde{A}$ is analytic within and on the unit circle. The constant "1" in (26a) has no interior zeros. Hence, by Rouché's theorem,[4] $\tilde{A}$ will have no interior zeros if, on the unit circle,

$$|z^M\tilde{Q}(z)| < 1, \qquad z = e^{j\theta}. \tag{27}$$

From (14),

$$z^M\tilde{Q}(z)\,|_{z=e^{j\theta}} = e^{jM\theta}\sum_{i=1}^{m}\frac{a_i z_i^M \tilde{D}(e^{j\theta})}{1 - z_i e^{j\theta}}. \tag{28}$$

Now $|z_i| < 1$ for all i because the $\{z_i\}$ are the poles of the stable filter H_p. Thus, for any value of M greater than some critical value, (27) must be satisfied. (Equation (27) may obtain, of course, for certain values of M less than the critical value.) Thus we have shown that the zeros of H_1 will be interior, and that the block recursive filter $\hat{H}_B$ of (24) will be stable, if the length M of h_1 is sufficiently large.

The length constraint associated with stability has important practical implications, as we shall see later when we discuss computational efficiency. Fig. 3 is meant to provide some physical insight into the nature of the constraint. We assert that the $H_1(z)$ on the left of Fig. 3 is certain to have exterior zeros, whereas the $H_1(z)$ on the right is more likely to have wholly interior zeros. (The assertions are based on a test for exterior zeros mentioned briefly in a later section and presented in Appendix II.) The main qualitative point to keep in mind is that h_1 should encompass the major lobe of the pulse response; only the "tail" can be deleted. Stated differently, M will be relatively small if the poles of $H_p(z)$ are not near the unit circle (see (28)). This implies that the effective bandwidth of $H_p(e^{j\omega T})$ should be a significant fraction of the folding frequency $1/2T$.

Unfortunately, the authors can offer little additional guidance regarding M. Limited tests on two classes of filters—Gaussian, of the form $H_p(z) = (1 - a/z)^{-m}$, and low-order Butterworth—indicated that M is a sensitive function of pole placement and filter order.[5] (The preceding sentence contravenes one of Chanoux's conclusions [5]. Neither Gold and Jordan nor Chanoux considered stability, perhaps because the particular form of their solutions tends to mask the implicitly recursive nature of the problem.)

Two sensitivity problems which are variants of the stability problem might be mentioned in passing. Let us assume that H_1 has wholly interior zeros. Then $\hat{H}_B$ of (24) is stable, but the stability of the more generally perturbed $\hat{\hat{H}}_B$ of (29) is harder to assess. If

$$\hat{\hat{H}}_B = \frac{H_1 + \delta H_1}{1 - z^{-M}(G + \delta G)} \tag{29}$$

is stable, however, its derivations from the desired H_p characteristics will take two forms. There will be mistuning associated with pole perturbations, and there will also be an f-domain ripple component of the type one encounters in ordinary (nonrecursive) finite convolutions.

[4] Rouché's theorem [6]: "If $f(z)$ and $g(z)$ are analytic inside and on a closed contour C, and $|g(z)| < |f(z)|$ on C, then $f(z)$ and $f(z)+g(z)$ have the same number of zeros inside C."

[5] Some sample results for the Gaussian filter with $a=0.5$: $m=3$ requires $M \geq 6$; $m=7$ requires $M \geq 22$; $m=10$ requires $M \geq 35$; $m=15$ requires $M \geq 59$. Note that the bandwidth varies inversely with m. We suspect that this filter is unusually demanding.

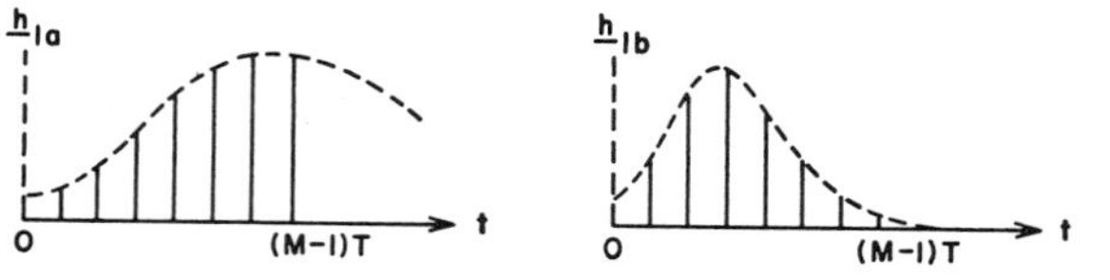

Fig. 3. Truncated pulse responses.

Implementation and Efficiency

The following discussion of the implementation of Fig. 2 presumes sequential block processing, i.e., "simultaneous" evaluation of all points of a finite convolution, such as occurs when a convolution is evaluated via the DFT. Thus, when the noncyclic convolution of two M-size blocks is to be found, no values of the convolution become available until both M-blocks are available in full; similarly, no values are available until all values are available, and the number of values is $2M-$ (or $2M$) rather than M.

Assume initially (see Fig. 2) that h_1 is of length $M \geq m+1$, and that $M-m$ zeros are appended to g so that it can also be considered to be of length M. Let the input and output x and y be partitioned into M-size blocks. Then a workable procedure is as shown in Fig. 4.

Each input x-block is convolved noncyclically with h_1 to produce $2M$ values.[6] A sequence of output y-blocks, each of size M, is generated by combining the first M values of each h_1 convolution with other feedforward and feedback terms, as shown in Fig. 4. As each y-block becomes available, it is convolved noncyclically with g to generate $2M$ feedback values.[6] Appendix I contains a FORTRAN-like program for implementing Fig. 4. (The logic of the program and Fig. 4 may be clearer if the reader initially ignores the feedback loop, i.e., sets g to zero. Fig. 4 and the program then become embodiments of a well-known procedure for on-going or "running" filtering with the nonrecursive filter h_1.)

As a further aid to understanding, imagine that the input to the filter is but a single pulse in the first position. Now expand (9):

$$H_B(z) = H_1(z) + z^{-M}G(z)H_1(z) + z^{-2M}G^2(z)H_1(z) + \cdots \tag{9}$$

Next, construct a table as follows, recalling that H_1 and G are polynomials in z^{-1} of order no higher than $M-1$.

Powers of z^{-1}	Contributing Terms from (9)
z^0 to $z^{-(M-1)}$	$H_1(z)$
z^{-M} to $z^{-(2M-1)}$	$z^{-M}G(z)H_1(z)$
z^{-2M} to $z^{-(3M-1)}$	$z^{-M}G(z)H_1(z)+z^{-2M}G^2(z)H_1(z)$
z^{-3M} to $z^{-(4M-1)}$	$z^{-2M}G^2(z)H_1(z)+z^{-3M}\cdots$
z^{-4M} to $z^{-(5M-1)}$	$z^{-2M}G^2(z)H_1(z)+z^{-3M}\cdots$
$\cdots\cdots$	$\cdots\cdots$

[6] A noncyclic convolution is insured by appending M zeros to each of the two constituent M-size blocks being convolved: see the definitions of $H1$, G, and Load (R,S,I) in Appendix I.

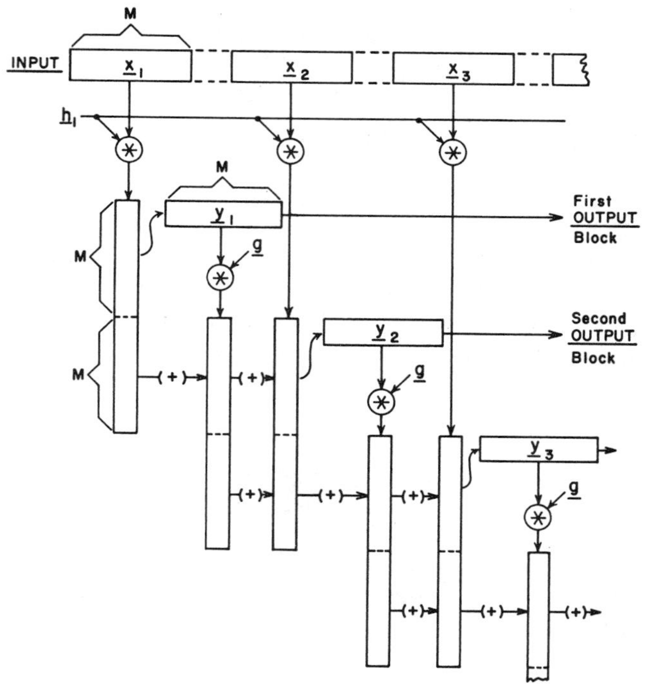

Fig. 4. Implementation of Fig. 2.

Finally, observe that the table reflects the structure of Fig. 4: each row in the table corresponds to an output y-block. This type of argument can be expanded into an (algebraically tedious) proof of the validity of the Fig. 4–Appendix I procedure, if such a proof is deemed necessary.

Let us estimate roughly the computational efficiency of the foregoing procedure on the assumptions that M is large, and that the convolutions are done via the FFT. Define

$$\text{Calc}\ (Y, M) = \text{number of operations per output } M\text{-block.} \tag{30}$$

Then, by inspection of Fig. 4 and/or Appendix I,

$$\text{Calc}\ (Y, M) \sim 4 \cdot \text{Calc (DFT of size } 2M) \tag{31a}$$

$$\sim 8M \log_2 2M \text{ cops} \tag{31b}$$

$$\sim 32M \log_2 2M \text{ ops.} \tag{31c}$$

(For our purposes, a complex operation (cop) can be defined as the multiplication of a pair of complex numbers and the addition of a pair of complex numbers. A real operation (op) is defined analogously for real-valued numbers. One cop is roughly equivalent to four ops.)

Now calculate the average number of ops per output sample.

$$\text{Calc}\ (Y, 1) = \frac{1}{M}\text{Calc}\ (Y, M) \tag{32a}$$

$$\sim 32 \log_2 2M \text{ ops.} \tag{32b}$$

Finally, recall that a conventional implementation of H_p, i.e., one not based on block recursion, requires about m ops per output sample. Thus we can define

$$\rho(BR) = \text{relative efficiency of block recursion}$$

$$= \frac{m}{\text{Calc }(Y,\,1)} \tag{33a}$$

$$\sim \frac{m}{32 \log_2 2M}. \tag{33b}$$

This result is interesting in a theoretical sense because it implies that the relative efficiency of block recursion increases monotonically with the complexity of the filter —*provided* that M grows (with m) at a slower than exponential rate. Unfortunately, it is less interesting in an immediately practical sense. Suppose that $M \approx m$; (33b) implies that the efficiency of block recursion exceeds that of conventional implementations only when the filter has more than about 300 poles! If one makes the less optimistic assumption that $M \approx m^{3/2}$, then the filter must have about 500 poles for block recursion to be more efficient.

One can argue, however, that (33b) is overly conservative because the constant "32" can be reduced by exploiting various efficiency-enhancing "tricks" (e.g., packing real data in complex arrays) which FFT practitioners have developed. Also, if M is significantly larger than m, one can improve the efficiency still more by implementing the feedback-loop calculations with several m-sized convolutions, rather than one M-sized convolution. (Several such variations of the "details" of the logic in Fig. 4 and Appendix I are permissible.) Enhancements of these types change the constants in (33b) but not the general form of the equation. Readers will have observed, no doubt, that the form (33b) also describes the relative efficiency of FFT-based implementations of nonrecursive filters.

Four additional points concerning implementation and efficiency warrant comment. First, suppose that one decides to implement the finite convolutions directly rather than through the FFT. Are computational savings possible for sufficiently large m? One need consider only the feedback loop in Fig. 2 to see that the answer must be negative. The G-network is necessarily of order $m-1$, and will require about m ops per sample if implemented directly. Thus, any computational savings must accrue from algorithms for fast processing of *blocks* of data. The merit of the block-recursive organization is that it permits one to deploy such algorithms.

Second, let us relate, through reorganization, block recursion to direct implementations. Imagine, merely for convenience, that M is shorter than m. Specifically, let $m = rM$, r an integer. $G(z)$ can be decomposed into M-size segments, each of order $M-1$.

$$G(z) = G_1(z) + z^{-M}G_2(z) + \cdots + z^{-M(r-1)}G_r(z). \tag{34}$$

Now redraw Fig. 2, as illustrated in Fig. 5 for $r=3$. It is easy to chart from Fig. 5 an augmented flow diagram similar to Fig. 4, and to calculate the computational

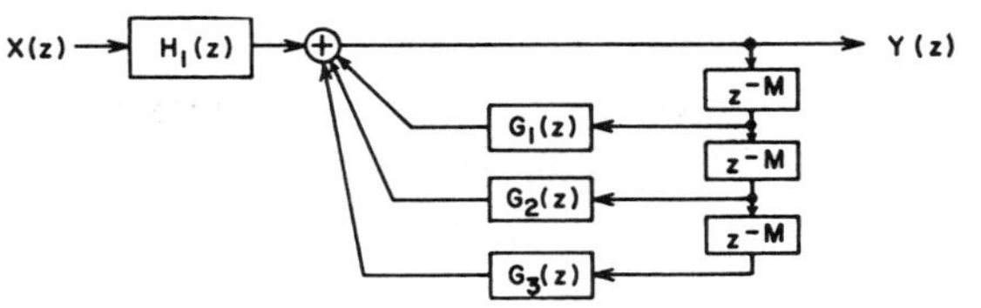

Fig. 5. Decomposition of G to facilitate M-size blocking; $r=3$.

efficiency. As $M \to 1$ and $r \to m$, the procedure proves to be retrogressive. Indeed, for $M=1$ and $r=m$, the Fig. 5 organization is a direct implementation! (Note that the decomposition of $G(z)$ per (34) into, say, N-size segments does not specifically require that h_1 be short. Note also that $H_1(z)$ can be segmented if desired.)

Third, nothing has been said about the internal noise and accuracy characteristics of block recursion. The authors can show that block recursion is at least comparable to conventional implementations in respect to noise when m is large and integer arithmetic is used, but they are not yet prepared to claim significant advantages. (Chanoux [5] came to a different conclusion for the special case $m=2$.) This area warrants considerably more study.

Finally, two closely related questions concerning extensions of the block-recursive organization arise. Might it be advantageous to incorporate the zeros of the original filter $H(z)$ directly into the block-recursive filter? Might it be advantageous to implement the main lobe of h_p in a regular nonrecursive filter, and to implement the tail via a separate, parallel, strictly m-block-size, block-recursive filter? Our studies imply negative answers for both questions.

An Overview

Let us assess block recursion from a conservative viewpoint. Suppose that one wishes to implement a complicated low-pass recursive filter via the FFT. Under accepted techniques, one first chooses a block size large enough to accomodate the "significant" portion of the true pulse response; the tail is amputated at the point at which it becomes consistently small. This finite response corresponds, of course, to H_1 in Fig. 2. Then, under accepted techniques, one modifies the finite pulse response via a multiplicative weighting function to control "picket fence effects" (ripple and relatively slow stopband attenuation). The source of these effects is precisely the finite aperature of the pulse response. In essence, the filter has insufficient memory to average-out progressively higher frequencies with uniformly increasing effectiveness.

What is the role of block recursion in such a context? It is, simply, to extend the response by producing, via feedback, the amputated tail. It thus obviates the need for window functions at a relatively small incremental cost.

In summary, it appears that block recursion does not offer (in its present state of development) dramatic, practically realizable computational savings. It holds scope for further work, however, and may prove attractive after additional development.

Some Practical Details

Readers who wish to experiment with block recursion will find that programming the procedure itself is not difficult. Finding the constituent h_1 and g functions, however, is slightly less obvious. We shall briefly discuss some applicable methods.

Usually the mth order polynomial $D(z)=H_p^{-1}(z)$ is known. The first goal is to find h_1, the truncated pulse response. This can be split into two tasks: finding the pulse response $h_p(nT)=Z^{-1}[H_p(z)]$, and then truncating h_p to the length M needed to insure stability.

Sometimes $h_p(nT)$ can be found by analysis, e.g., by inversion of a partial fraction expansion, by binomial expansion (for the Gaussian filter cited earlier), etc. For most filters, however, manual analysis becomes very tedious when m is large, and machine methods are almost mandatory. One such method is to deploy an algebraic language akin to FORMAC to automate the analytical tedium. Another is to deploy a form of the long-division algorithm. Yet another is to invoke the FFT to calculate the inverse transform of $H_p(e^{j\omega T})$. This last method seems to give good results if the FFT block size is sufficiently large and if H_p is well sited in the f-domain block. Still another method, which is suitable when m is not too large, is to obtain $h_p(nT)$ from a conventional recursive implementation.

Once $h_p(nT)$ is known, generation of h_1 by truncation is merely a matter of testing progressively longer blocks of samples from h_p to find the minimal block which has no exterior zeros. This is a readily machinable process for which a simple subroutine is provided in Appendix II.

When $h_1(nT)$ is known, $g(nT)$ can be found easily from (17), viz.,

$$H_1(z)D(z) = 1 - z^{-M}G(z)\,|_{Q=G}. \qquad (17)$$

One need only convolve $h_1(nT)$ and $d(nT)$, either directly or through the FFT. The sample values of $g(nT)$ are the negative of the values of the convolution at lags M, $M+1$, $\cdots$, $M+m-1$.

Appendix I

A Program for Block-Recursive Filtering

Definitions of Variables

- X, Y: large arrays for the input and output; possibly a real-time source and a real-time sink.
- $X1$, $X2$, $Y1$, $Y2$: four working arrays, each of size $2M$, initialized to zero.
- $H1$: $2M$-size array containing the DFT of {the M values of h_1 plus M zero}.
- G: $2M$-size array containing the DFT of {the m values of g plus $2M-m$ zeros}.
- J: a pointer used in indexing the X, Y arrays.

Subprogram Definitions

- Load(R, S, I): puts M values from S, starting at location I in S, into the first M places in R; loads second M places in R with zeros.
- Convolve(R, U): produces noncyclic convolution of R, u by DFTing R, forming the element-by-element product $R(j)\leftarrow R(j)U(j)$, then IDFTing R.
- Output(V, I, $Z1$, $Z2$, $Z3$, $Z4$): outputs, or loads into V, M values according to the formula:

$$V(j+I) = Z1(j) + Z2(j) + Z3(j+M) + Z4(j+M), \qquad j = 0, 1, \cdots, M-1.$$

Program

```
    J=0
10  Call Load(X1, X, J)
    Call Convolve(X1, H1)
    Call Output(Y, J, X1, Y2, X2, Y1)
    Call Load(Y1, Y, J)
    Call Convolve(Y1, G)
    J=J+M
    Call Load(X2, X, J)
    Call Convolve(X2, H1)
    Call Output(Y, J, X2, Y1, X1, Y2)
    Call Load(Y2, Y, J)
    Call Convolve(Y2, G)
    J=J+M
    IF(J<JSTOP) GO TO 10
```

Appendix II

A Test for Exterior Zeros

The following FORTRAN subroutine, SCHUR, is a partial implementation of the Schur test which underlies the Lehmer–Schur root-finding algorithm [7]. It detects the presence of interior zeros of polynomials of the form

$$f_n(z) = a_0 + a_1 z + \cdots + a_n z^n. \qquad (35)$$

Our application requires testing progressively longer sequences from

$$H_p(z) = h_p(0) + h_p(T)z^{-1} + h_p(2T)z^{-2} + \cdots \qquad (36)$$

for exterior zeros. Clearly, the Schur test is applicable directly because $z\rightarrow z^{-1}$ maps the interior of the unit circle into its exterior.

The subroutine is only a partial implementation because certain special cases are ignored. To use the routine, simply load an array with H_p coefficients and call SCHUR repeatedly on progressively longer sequences. The routine can be modified easily to cater for the omitted special cases, to return the number of successful T-tests on each call, etc. It can also be modified quite easily to detect zeros "near" the unit circle; this is important in protecting against instabilities of the type implied by (29). The user is responsible for determining the requisite arithmetic precision. (We found that FORTRAN double precision on a 360/65, which specifies numbers to an accuracy of about 17 decimal digits, began to give erratic results for polynomials whose orders ranged between 50 and 75.)

```
      SUBROUTINE SCHUR(X,N,IFLAG)
C   SCHUR PERFORMS A PARTIAL TEST ON THE (N-1)TH
C   ORDER POLYNOMIAL,
C   WHOSE COEFFICIENTS ARE REAL AND IN ARRAY X,
C   FOR ZEROS INSIDE THE UNIT CIRCLE.
C      IFLAG=0 : NO OBVIOUS ZEROS IN UNIT CIRCLE
C      IFLAG=1 : ONE OR MORE ZEROS IN UNIT CIRCLE
      DIMENSION X(N)
      DOUBLE PRECISION Y(128),Z(128)
      IF(X(1).EQ.O.)GO TO 40
      DO 5 I=1,N
      Y(I)=X(I)
    5 Y(I)=Y(I)/X(1)
      K=N+1
C   CALCULATE AND TEST REDUCED-POLYNOMIAL
C   T-FUNCTIONS
   10 K1=K
      K=K-1
      DO 15 I=1,K
   15 Z(I)=Y(I)-Y(K)*Y(K1-I)
      IF(Z(1).LT.O.)GO TO 40
      Y(1)=1.0
      DO 20 I=2,K
   20 Y(I)=Z(I)/Z(1)
      IF(K.GT.2)GO TO 10
      IFLAG=0
      RETURN
   40 IFLAG=1
      RETURN
      END
```

References

[1] C. M. Rader and B. Gold, "Digital filter design techniques in the frequency domain," *Proc. IEEE*, vol. 55, pp. 149–171, February 1967.
[2] B. Gold and C. M. Rader, *Digital Processing of Signals.* New York: McGraw-Hill, 1969.
[3] T. G. Stockham, Jr., "Highspeed convolution and correlation," *1966 Spring Joint Computer Conf., AFIPS Proc.*, vol. 28. Washington, D. C.: Thompson, 1966, pp. 229–233.
[4] B. Gold and K. L. Jordan, Jr., "A note on digital filter synthesis," *Proc. IEEE* (Letters), vol. 56, pp. 1717–1718, October 1968.
[5] D. Chanoux, "A method of digital filter synthesis," M.S. thesis, Massachusetts Institute of Technology, Cambridge, Mass., May 23, 1969.
[6] E. C. Titchmarsh, *Theory of Functions.* London, England: Oxford University Press, 1939.
[7] A. Ralston, *A First Course in Numerical Analysis.* New York: McGraw-Hill, 1965.

A General Theorem for Signal-flow Networks, with Applications

by Alfred Fettweis*

DK 621.372

A general theorem for signal-flow networks is derived which in many respects is similar to the difference form of Tellegen's theorem; it can be expressed for instantaneous signal values as well as in generalized form. As illustration, the theorem is used to obtain a simple proof for Mason's theorem on the transposition (flow reversal) of a signal-flow graph, and to derive properties of the transfer function sensitivity which are very similar to those known for conventional electrical networks.

Ein allgemeiner Satz für Signalflußnetzwerke mit Anwendungen

Ein allgemeiner Satz für Signalflußnetzwerke, ähnlich dem Tellegenschen Theorem, wird hergeleitet; dieser Satz kann sowohl unter Benützung der Augenblickswerte als auch in verallgemeinerter Form ausgedrückt werden. Zur Erläuterung der Anwendungsmöglichkeiten wird zunächst ein einfacher Beweis des Masonschen Satzes über die Transposition (Flußumkehr) eines Signalflußgraphen angegeben. Anschließend wird ein allgemeiner Ausdruck für die Empfindlichkeit der Übertragungsfunktion eines Signalflußnetzwerks gegen Bauelementeänderungen hergeleitet, der dem entsprechenden Ausdruck bei üblichen elektrischen Netzwerken sehr ähnlich ist.

1. Introduction

In the theory of conventional electrical networks, *Tellegen's theorem* plays an important role [1]—[6]. This is due largely to the amazingly simple and general way in which this theorem can be stated. Tellegen's theorem is a direct result of Kirchhoff's current and voltage laws.

Next to the conventional electrical networks, there exists a large variety of circuits which, at least in principle, are not based on Kirchhoff's laws but on a direct realization of signal-flow graphs; we shall designate circuits based on this principle by the general name of *signal-flow networks*. Among such networks, *digital filters* have lately become into particular prominence.

For signal-flow networks, Tellegen's theorem, clearly, does not apply. One may wonder, however, whether in this case a similar theorem cannot be stated. One may not expect this to be true for the simplest form of Tellegen's theorem since this latter theorem is related directly to power considerations. We shall show in this paper, however, that there exists for signal-flow networks a theorem which in a sense is similar to the difference form of Tellegen's theorem [3], [4] and which is also similarly general and powerful. It cannot be derived as an extension of Tellegen's theorem in the sense mentioned in chapter 8 of [4] since power considerations do not apply to signal-flow networks.

The usefulness of the new theorem will be illustrated first in Section 4 by extending Bordewijk's [2], [7] concepts of interreciprocity and transposition, giving in particular a very simple proof of a generalized form of Mason's theorem on the transposition (reversal) of a flow graph [8], [9], and then in Section 5 by showing how the theory on the sensitivity analysis and the related computer optimization of conventional electrical networks [5], [6], [10] can be extended to signal-flow networks.

* Prof. Dr. A. Fettweis, Lehrstuhl für Nachrichtentechnik der Universität, D-463 Bochum, Postfach 2148.

2. Signal-flow Networks

A usual signal-flow graph is composed of nodes and oriented branches [8]. All signals are, in general, either continuous or discrete functions of time. Let us designate the signals flowing into a given node by $x_1, \ldots, x_m$. The signals flowing out of this node are all the same; we may thus designate them all by the same letter, say y, and we have

$$y = x_1 + x_2 + \cdots + x_m . \quad (1)$$

The precise form of the relations between signals imposed by the branches is unimportant for our further considerations.

In addition to the nodes characterized by (1), it is customary to define source nodes and sink nodes [8]. As we shall see, it is more appropriate for our purpose, however, to consider sources and sinks as branches. A source of intensity $q = q(t)$ feeding into a node 1, can be represented as shown in Fig. 1a, while a sink fed from a node 1 can be represented as shown in Fig. 1b [11]—[13]. It is thus obvious, that source and sink can be represented by the common branch symbol shown in Fig. 1c which is characterized by the equation

$$x = q ,$$

thus corresponding to a source for $q \not\equiv 0$ and to a sink for $q \equiv 0$. Although a source branch can be connected between two arbitrary nodes just like any other branch, it is advantageous for our purpose to assume it to be connected as a self-loop (Fig. 1d).

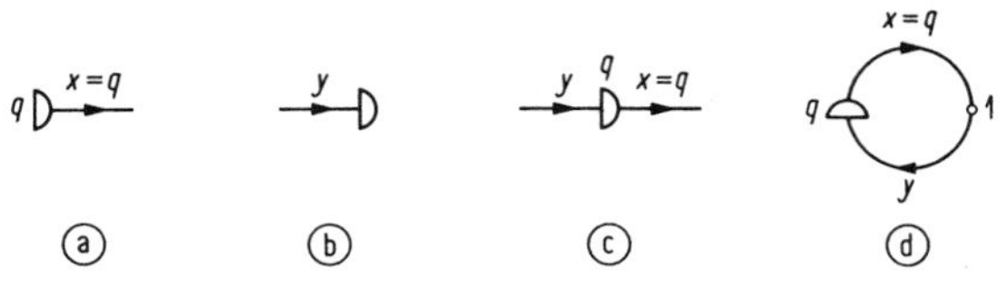

Fig. 1. (a) Representation of a source,
(b) representation of a sink,
(c) a branch representing a source for $q \not\equiv 0$ and a sink for $q \equiv 0$,
(d) a source branch connected as a self-loop.

Reprinted with permission from *Arch. Elek. Übertragung*, vol. 25, pp. 557–561, 1971.

Although branches and nodes suffice for the representation of most signal-flow networks occuring in practice, *building blocks* more general than branches are required, especially if the network is nonlinear. Such general building blocks, however, can also be useful for the representation of certain linear signal-flow networks (cf. the adaptors described in [11]—[13]).

A general building block, henceforth often simply called a *block*, has a certain number K of incoming signals $y_1, \ldots, y_K$ and a certain number L of outgoing signals $x_1, \ldots, x_L$ (Fig. 2a). In practice, we often have $K = L$. A branch, clearly, is a special case of a block. In order to avoid ambiguity, a general branch could be represented, not by a simple line, as is often done, but by the symbol of Fig. 2b.

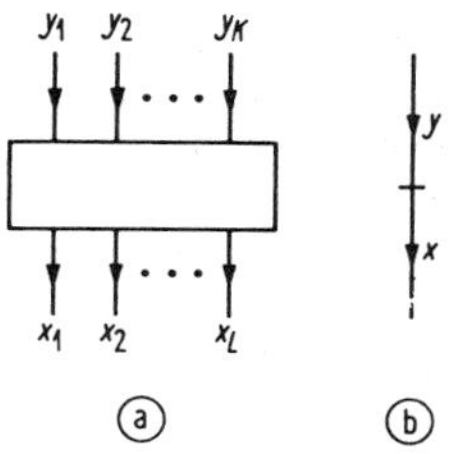

Fig. 2.
(a) A general building block having K input signals and L output signals,
(b) representation of a branch.

3. The Main Theorem

We now consider a general signal-flow network comprising N nodes numbered $n = 1, \ldots, N$, and B blocks numbered $b = 1, \ldots, B$. Let y_n be the value of the signals flowing out of node n, and let x_{bn} be the signal flowing from block b to node n; if there are several individual signals flowing from block b to node n, we designate by x_{bn} the sum of these individual signals, and if no signal flows from block b to node n, we write $x_{bn} = 0$. According to eq. (1), we may then write

$$y_n = \sum_{b=1}^{B} x_{bn} . \tag{2}$$

We consider now two distinct signal distributions compatible with the topology of the network (i. e. with the given number of blocks and nodes). As in case of Tellegen's theorem, the origin for the difference between these distributions may be quite different depending on the application to be considered. In accordance with the notation just defined, we designate the signals for the first of these distributions, as before, by x_{bn} and y_n, and those for the second by x'_{bn} and y'_n. We then have the following

Theorem 1 (main theorem):

The signals for two distinct distributions compatible with the topology of a given signal-flow network satisfy the general relation

$$\sum_{b=1}^{B} \sum_{n=1}^{N} (y_n x'_{bn} - y'_n x_{bn}) = 0 . \tag{3}$$

Proof: The proof of this theorem is immediate if we rewrite the left-hand side of eq. (3) in the form

$$\sum_n \left(y_n \sum_b x'_{bn} - y'_n \sum_b x_{bn} \right)$$

and make use of the expressions (2) and $y'_n = \sum_b x'_{bn}$.

The theorem just proved can be generalized in the same way as Tellegen's theorem [3], [4]. Thus, if X_{bn} and Y_n are derived through some operators from x_{bn} and y_n, respectively, and if we still have, similarly to eq. (2),

$$Y_n = \sum_{b=1}^{B} X_{bn} , \tag{4}$$

we can state the following

Theorem 2:

Consider two signal distributions compatible with the topology of the given signal-flow network and let the quantities just defined for these two distributions be designated by X_{bn} and Y_n and by X'_{bn} and Y'_n, respectively. We then have

$$\sum_{b=1}^{B} \sum_{n=1}^{N} (Y_n X'_{bn} - Y'_n X_{bn}) = 0 . \tag{5}$$

The proof of this theorem is the same as for theorem 1. For the applications to be discussed later, X_{bn} and Y_n may simply be considered to be the complex amplitudes corresponding to x_{bn} and y_n, respectively, for some steady-state conditions.

We now consider the important particular case of a *simple* signal-flow network, i.e. a signal-flow network in which the blocks are all branches and for which any two nodes are connected, for each of the two possible orientations, by one branch only; furthermore, each node may have a self-loop. In this case, a branch connecting a certain node, say m, to another node, say n, may also be designated by (m, n). The signal flowing from this branch to node n may similarly be designated by x_{mn}. These definitions include the possibility of a self-loop, in which case $m = n$.

Now, for a given node n, the summation over all b in eq. (3) needs only to be extended over those branches for which x_{bn} and x'_{bn} are not zero, i.e., in the present situation, only over the branches (m, n). It is thus clear that we can now write eq. (3) in the form

$$\sum_{m=1}^{N} \sum_{n=1}^{N} (y_n x'_{mn} - y'_n x_{mn}) = 0 . \tag{6}$$

Finally, interchanging the subscripts m and n in the double summation in eq. (6) when applied to the term $y'_n x_{mn}$, we obtain the

Corollary 1:

For a simple signal-flow network, eq. (3) in theorem 1 may be replaced by

$$\sum_{m=1}^{N} \sum_{n=1}^{N} (y_n x'_{mn} - y'_m x_{nm}) = 0$$

where both summations extend over all the nodes of the network.

This corollary can also easily be proved without reference to theorem 1 by making use of the expression

$$y_n = \sum_{m=1}^{N} x_{mn}$$

which, clearly, is now valid. In a similar way to corollary 1, we also have

Corollary 2:

In a simple signal-flow network, eq. (5) in theorem 2 may be replaced by

$$\sum_{m=1}^{N} \sum_{n=1}^{N} (Y_n X'_{mn} - Y'_m X_{nm}) = 0 \tag{7}$$

where both summations extend over the nodes of the network.

Note that these corollaries can be extended to cases where two nodes happen to be connected, in a given direction, by more than one branch. We then simply have to define x_{mn} to be the sum of the signals flowing into node n from branches originating at node m, etc.

4. Interreciprocity and Transposition

4.1. Reciprocal and interreciprocal signal-flow multipoles

A signal-flow network which is accessible from the outside at a certain number of its nodes and which is sourcefree, will be called a *signal-flow multipole*. In particular, if the number of accessible nodes is M, we can call it more precisely a *signal-flow M-pole*. We shall henceforth assume this multipole to be linear and we shall analyze it at steady-state conditions.

The accessible nodes will be numbered $m = 1, 2, \ldots, M$. At the accessible node m, we may assume an input signal X_m to be flowing from the outside into this node and an output signal Y_m to be flowing from this node towards the outside, as shown in Fig. 3. A general way of feeding the M-pole of Fig. 3a would be to add to each accessible node a source in self-loop connection as shown in Fig. 3b.

Consider now two arbitrary signal distributions in the multipole of Fig. 3a, with X_m and Y_m being the signal values for the first of these distributions and X'_m and Y'_m those for the second. In accordance with the definition used for conventional electrical networks, we may say that the signal-flow multipole under consideration is *reciprocal* if we have

$$\sum_{m=1}^{M} (X_m Y'_m - X'_m Y_m) = 0 \tag{8}$$

for any possible two signal distributions. In particular, if the M-pole under consideration is described by equations of the form

$$Y_m = \sum_{n=1}^{M} S_{mn} X_n\,, \quad m = 1, \ldots, M \tag{9}$$

where S_{mn} is the transfer function from port n to port m, condition (8) is equivalent to

$$S_{mn} = S_{nm}\,. \tag{10}$$

Indeed, substituting in eq. (8) first eq. (9) and then this same equation with Y_m and X_m replaced by Y'_m and X'_n, respectively, and interchanging the subscripts m and n in one of the two resulting double summations, we obtain

$$\sum_{m} \sum_{n} (S_{mn} - S_{nm}) X_m X'_n \tag{11}$$

which must be satisfied for all X_m and X_n.

We now consider two distinct signal-flow networks, which we assume to have the same number M of accessible nodes. Let the input and output signals X_m and Y_m refer to the first of these networks, and X'_m and Y'_m to the second. In accordance with the definition used by Bordewijk [2] for conventional electrical networks, we shall say that the signal-flow networks under consideration are *interreciprocal* if eq. (8) is fulfilled for all possible signal combinations X_m and Y_m and all possible signal combinations X'_m and Y'_m. In particular, if the networks are described by equations of the form (9) and

$$Y'_m = \sum_{n=1}^{M} S'_{mn} X'_n\,, \quad m = 1, \ldots, M\,, \tag{12}$$

respectively, substitution of eqs. (9) and (12) in eq. (8), followed by an appropriate interchange of indices, leads to the equivalent interreciprocity condition

$$S'_{mn} = S_{nm}\,. \tag{13}$$

4.2. Transposition

We first consider a simple signal-flow network as defined in Section 3 in relation with the corrolaries 1 and 2. For such a network, the operation of *transposition* (also called flow reversal) [8], [9], consists in reversing each individual branch. Thus, if in a given network a branch linking node m to node n is characterized by a branch transfer function H_{nm}, the transposed network will have a branch linking node n to node m with a transfer function H'_{mn} given by

$$H'_{mn} = H_{nm}\,. \tag{14}$$

If a branch contains a source of intensity Q, the transposed network will have the same source but in reverse direction; this will not produce any change if the sources are connected as self- loops, as we have assumed.

Consider next a general linear signal-flow network composed of blocks having arbitrary numbers of input and output signals. The operation of transposition then consists in transposing each individual block. In order to explain this latter operation, we consider an arbitrary block b; its steady-state equations can be written in the general form

$$X_{bn} = \sum_{m=1}^{N} H_{bnm} Y_m\,, \quad n = 1, \ldots, N\,. \tag{15}$$

The equations for the corresponding transposed block are then given by

$$X'_{bn} = \sum_{m=1}^{N} H'_{bnm} Y'_m\,, \quad n = 1, \ldots, N \tag{16}$$

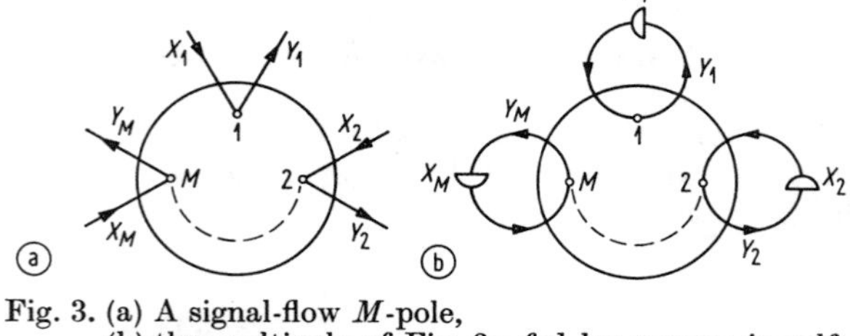

Fig. 3. (a) A signal-flow M-pole, (b) the multipole of Fig. 3a fed by sources in self-loop connection.

with $$H'_{bnm} = H_{bmn}. \qquad (17)$$

This definition encompasses the one given in the previous section.

We can now state the

Theorem 3:

A linear signal-flow multipole and its transpose are interreciprocal.

Proof: We consider first the contribution to the left-hand side of eq. (5) due to a sourcefree block. Substitution of eqs. (15) and (16) in the term for the corresponding value of b gives

$$\sum_{n=1}^{M}\sum_{m=1}^{M}(H'_{bnm}\,Y_n\,Y'_m - H_{bnm}\,Y'_n\,Y_m). \qquad (18)$$

Taking then into account eq. (17) and making an appropriate change of subscripts m and n, we see that the expression (18) ist actually equal to zero. Thus, as theorem 2 is always valid, eq. (5) reduces to

$$\sum_b{}'\sum_{n=1}^{N}(Y_n\,X'_{bn} - Y'_n\,X_{bn}) = 0 \qquad (19)$$

where the summation $\sum'$ has to be extended only over those branches which correspond to sources. As such branches are connected as self-loops, the quantities X_{bn} and X'_{bn} in eq. (19) are different from zero only if n is the node to which branch b is connected. This shows that with an appropriate change of notation, eq. (19) becomes equivalent to eq. (8).

For simple signal-flow networks (in the sense defined in Section 3) the proof of theorem 3 becomes particularly simple if we use corollary 2 rather than theorem 2, together with eq. (14) and the relations

$$X_{nm} = H_{nm}\,Y_m\,, \qquad X'_{mn} = H'_{mn}\,Y'_n\,.$$

Theorem 3 is then equivalent to Mason's theorem on the reversal of a signal-flow graph (see [8], section 4.18), since we have seen in subsection 4.1 that eq. (13) also corresponds to the interreciprocity condition. A proof of Mason's theorem under more restrictive conditions has recently been given by Jackson [9].

Note that the simplified proof indirectly also proves the general result. Indeed, the eqs. (15) of a building block can equivalently be obtained by branches having branch transfer functions H_{bnm}. This proves our statement since according to eq. (14), the flow reversal in these branches would be expressed by eq. (17) which is the equation defining the transposition of the block under consideration.

5. Sensitivity Analysis

In this section, we consider the sensitivity of a transfer function of a signal-flow network to changes in one of its building blocks. According to the remark made on the end of the last section, blocks can always be decomposed into branches. Hence, it is sufficient to consider changes in blocks which consist of one branch only.

We may consider without restriction a twopole whose nodes 1 and 2 serve as input and output, respectively (Fig. 4a). We assume that the block liable to change is a branch of transfer function Z

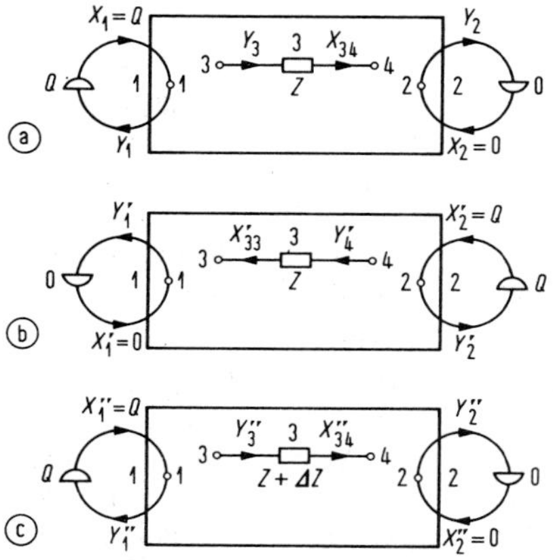

Fig. 4. (a) A signal-flow network whose node 3 is connected to node 4 by a branch of transfer function Z,
(b) the transpose of the network shown in (a),
(c) the network derived from (a) by changing Z into $Z + \Delta Z$.

linking node 3 to node 4. We consider furthermore the transpose of this network (Fig. 4b) as well as the network derived from the original one by changing Z into $Z + \Delta Z$ (Fig. 4c). We number the sink (source) branches at the input and output by $b = 1$ and $b = 2$, respectively. For these branches, we use the simplified notation X_1 rather than X_{11}, etc., as was also done in Section 4. The branch containing Z will be numbered $b = 3$.

Let us compare the networks of Fig. 4b and c. Applying theorem 2 of Section 3, we may write

$$\sum_b W_b = 0 \qquad (20)$$

where $$W_b = \sum_n (Y''_n\,X'_{bn} - Y'_n\,X''_{bn})\,. \qquad (21)$$

Furthermore, we have $W_b = 0$ for $b > 3$; this follows from the reasoning used in the proof of theorem 3 for reducing eq. (5) to eq. (19) if we note that the blocks in Fig. 4b and c with $b > 3$ are not source (sink) branches but are derived from one another by transposition. We thus can write

$$W_1 + W_2 + W_3 = 0 \qquad (22)$$

where we have, using the simplified notation mentioned above,

$$W_1 = Y''_1\,X'_1 - Y'_1\,X''_1 = -\,Y'_1\,Q\,, \qquad (23)$$

$$W_2 = Y''_2\,X'_2 - Y'_2\,X''_2 = Y''_2\,Q\,, \qquad (24)$$

$$W_3 = Y''_3\,X'_{33} - Y'_3\,X''_{33} + Y''_4\,X'_{34} - Y'_4\,X''_{34} = -\,Y''_3\,Y'_4\,\Delta Z\,, \qquad (25)$$

the second expressions for W_1, W_2, and W_3 being obtained by noting that

$$X'_2 = X''_1 = Q\,, \quad X'_1 = X''_2 = X'_{34} = X''_{33} = 0\,, \qquad (26)$$

and $$X'_{33} = Z\,Y'_4\,, \quad X''_{34} = (Z + \Delta Z)\,Y''_3\,.$$

The last two equalities in eq. (26) follow from the fact that there is no flow from branch 3 to node 4 in Fig. 4b, and to node 3 in Fig. 4c.

We now define the transfer functions

$$S_{21} = Y_2/Q\,, \quad S_{31} = Y_3/Q\,, \quad S'_{42} = Y'_4/Q\,. \qquad (27)$$

The change of the input-output transfer function

S_{21} due to the change of Z into $Z + \Delta Z$ is given by

$$\Delta S_{21} = (Y_2'' - Y_2)/Q\,. \qquad (28)$$

Thus, substituting eqs. (23) to (25) in eq. (22), taking into account eqs. (27) and (28) as well as the equality $Y_1' = Y_2$, which follows from the interreciprocity between the networks of Fig. 4a and b, and letting ΔZ go to zero, which also implies that Y_3'' approaches Y_3, we obtain finally

$$\partial S_{21}/\partial Z = S_{31} S_{42}'\,. \qquad (29)$$

Expression (29) is completely analogous to the corresponding expression for conventional electrical networks [5], [10]. As S_{31} and S_{42}' are transfer functions from the input and output nodes, respectively, we conclude in particular that it is also possible for signal-flow networks to determine the sensitivities of S_{21} with respect to all network elements by only analyzing once the original network and once its transpose.

The result just mentioned shows that the frequency-domain optimization of constant signal-flow networks can be carried out by the same simple method as known for conventional electrical networks [5], [10]. The extension to the time-domain optimization of much more general classes of networks, which DIRECTOR and ROHRER have described [6], is also just as possible for signal-flow networks. We shall not go into the details of the corresponding analysis, but simply derive the basic expression for the instantaneous quantities on which such an analysis must be based.

In order to do this, note first that eq. (3) is valid for the instantaneous signal values in the networks of Fig. 4a and b. Furthermore, the expression for the instantaneous quantities corresponding to eqs. (20) and (21) is given by

$$\sum_b \sum_n (y_n'' x_{bn}' - y_n' x_{bn}'') = 0\,. \qquad (30)$$

Hence, subtracting eqs. (3) and (30), we obtain the desired expression

$$\sum_b \sum_n (x_{bn}' \Delta y_n - y_n' \Delta x_{bn}) = 0\,,$$

where $\Delta y_n = y_n'' - y_n$ and $\Delta x_{bn} = x_{bn}'' - x_{bn}\,.$

6. Conclusions

We have shown that for signal-flow networks a general theorem can be given which in many respects is similar to the difference form of Tellegen's theorem for conventional electrical networks. In order to show the usefulness of this theorem, we have first given a simple proof for Mason's theorem on the flow reversal (transposition) in a signal-flow graph [8]. As a second application, we have computed the sensitivity of an input-output transfer function to variations of an individual branch transfer function. We have seen that just as for conventional networks all sensitivities for a given input-output transfer function can be computed by analyzing once the given network and once its transpose, i.e. by altogether two analyses only. This result can be extended by means of the generalized-adjoint-network approach described for conventional electrical networks by DIRECTOR and ROHRER [6].

7. Complementary Remarks

After submitting this paper, a paper by SEVIORA and SABLATASH [14] has become available which deals with essentially the same problem. The results by these authors, however, are less general and less systematically formulated so that the publication of the present paper remains fully justified. Some of the differences are listed hereafter.

Firstly, SEVIORA and SABLATASH consider only digital filter networks, while the present paper deals with more general signal-flow networks, including nonlinear ones.

Secondly, SEVIORA and SABLATASH do not distinguish between two types of variables. Due to this, their results can not be formulated as neatly as for Tellegen's theorem for classical networks or as in the present paper. This is due to the fact that the customary terminology for digital filters has led SEVIORA and SABLATASH to consider branch nodes and summation nodes as two types of elements. These two types of nodes, however, can be considered to be special cases of general nodes in signal-flow networks. This is the point of view adopted in the present paper. It has enabled us very easily to introduce two distinct types of variables which now play the role of voltage and current in classical networks.

Thirdly, SEVIORA and SABLATASH do not discuss the extension of BORDEWIJK's [2], [7] concepts of interreciprocity and transposition to signal-flow networks.

(Received March 12th, 1971.)

References

[1] TELLEGEN, B. O. H., A general network theorem, with applications. Philips Res. Rep. **7** [1952], 259–269.

[2] BORDEWIJK, J. L., Inter-reciprocity applied to electrical networks. Appl. Sci. Res. **B6** [1956], 1–74.

[3] PENFIELD, P., Jr., SPENCE, R., and DUINKER, S., A generalized form of Tellegen's theorem. Transact. Inst. Elect. Electron. Engrs. CT-**17** [1970], 302–305.

[4] PENFIELD, P., Jr., SPENCE, R., and DUINKER, S., Tellegen's theorem and electrical networks. M. I. T. Press, Cambridge, Mass. 1970.

[5] DIRECTOR, S. W. and ROHRER, R. A. Automated network design — "The frequency-domain case". Transact. Inst. Elect. Electron. Engrs. CT-**16** [1969], 330–337.

[6] DIRECTOR, S. W. and ROHRER, R. A., The generalized adjoint network and network sensitivities. Transact. Inst. Elect. Electron. Engrs. CT-**16** [1969], 318–323.

[7] BORDEWIJK, J. L., Comments on "Automated network design and the interreciprocity concept". Transact. Inst. Elect. Electron. Engrs. CT-**18** [1971], 179.

[8] MASON, S. J. and ZIMMERMANN, H. J., Electronic circuits, signals, and systems. J. Wiley & Sons, New York 1960.

[9] JACKSON, L. B., On the interaction of roundoff noise and dynamic range in digital filters. Bell Syst. tech. J. **49** [1970], 159–184.

[10] PINEL, J. F. and BLOSTEIN, M. L., Computer techniques for the frequency analysis of linear electrical networks. Proc. Inst. Elect. Electron. Engrs. **55** [1967], 1810–1819.

[11] FETTWEIS, A., Entwurf von Digitalfiltern in Anlehnung an Verfahren der klassischen Netzwerktheorie. "NTG-Report" of the "NTG-Fachtagung 'Analyse und Synthese von Netzwerken'", pp. 17–20, Stuttgart, October 1970.

[12] FETTWEIS, A., Digital filter structures related to classical filter networks. AEÜ **25** [1971], 79–89.

[13] FETTWEIS, A., Some principles of designing digital filters imitating classical filter structures. Transact. Inst. Elect. Electron. Engrs. CT-**18** [1971], 314–316.

[14] SEVIORA, R. E. and SABLATASH, M., A Tellegen's theorem for digital filters. Transact. Inst. Elect. Electron. Engrs. CT-**18** [1971], 201–203.

Digital Filters with Equiripple or Minimax Responses

HOWARD D. HELMS, Senior Member, IEEE
Bell Telephone Laboratories, Inc.
Whippany, N. J. 07981

Abstract

Techniques for determining the coefficients of digital filters which have equiripple or minimax errors are reviewed and occasionally extended. These techniques include: 1) mapping to provide equiripple errors in recursive filters; 2) windows for making Fourier spectrum measurements with minimax leakage; 3) the simplex method of linear programming to provide minimax errors in a nonrecursive filter's time response to a known pulse or Fourier transform of its coefficients; 4) nonlinear programming to provide minimax errors for nominally any response and filter; and 5) an integer programming technique to provide minimax error despite quantizing the coefficients of a nonrecursive filter. Some sources of computer programs embodying these techniques are indicated.

I. Introduction

Some techniques by which the coefficients of a digital filter can be determined are reviewed and occasionally extended in this paper. A digital filter is defined to be a linear difference equation having constant coefficients and operating on equally spaced samples of a signal. This paper describes only those techniques for which the errors in the response of the filter are made to be either equiripple or minimax. These errors are defined as the difference between the actual response of the digital filter and an ideal response. The response is assumed to be either the time response or the frequency response. (Although only one or the other of these responses is considered, other responses or combination of responses can be made to have minimax errors, if a nonlinear programming technique described later is used.)

The minimax value of the error is the minimum (obtained by varying the filter's coefficients) of the maximum of the absolute value of the error in the filter's response. This maximum is evaluated by examining the error at those times or frequencies which adequately characterize this error. An adequate characterization can often be obtained by sampling the error sufficiently often. An equiripple filter has more than one maximum of the absolute value of this error, and all of these maximum errors (e.g., ripple) are equal. An equiripple filter may or may not have minimax error. Conversely, a filter having minimax errors may or may not have equiripple error. The only class of digital filters for which the errors are known always to be both equiripple and minimax are the nonrecursive filters having coefficients which are even-symmetric and are determined by the simplex method of Section IV.

These minimax or equiripple techniques to be surveyed provide satisfactory values of the filter's coefficients only when the error falls within certain lower and upper bounds determined by the needs of the system containing the digital filter. A violation of any of these bounds requires changing the filter, e.g., by increasing the number of its coefficients and reusing the technique until the error is acceptable.

At least two reasons for considering techniques which provide minimax or equiripple errors can be given. One of these is that such techniques, which are almost as easy to use as the methods for obtaining least square errors, recently have been discovered. Another reason is that many applications require approximating ideal characteristics with specified upper and lower limits on the error. This latter reason is valid even when an application requires that the response must lie between two values; the mean of these values can be regarded as an ideal reponse to be approximated with errors whose moduluses must not exceed half of the difference between these values.

Rounding off the filter's coefficients to a fixed number of bits is assumed to be done only after unrounded values of the coefficients have been determined. Although no method for selecting this number of bits is reviewed in this paper, a subsequent section applies integer linear pro-

Manuscript received April 13, 1970; revised August 20, 1970.

A preliminary version of this paper was presented at the International Seminar on Digital Processing of Analog Signals, Zürich, Switzerland, March 11–13, 1970 and also at the Arden House Workshop on Digital Filtering, January 14, 1970.

Reprinted from *IEEE Trans. Audio Electroacoust.*, vol. AU-19, pp. 87–94, Mar. 1971.

TECHNIQUE	APPLICABLE TO WHAT TYPE OF DIGITAL FILTER			TYPES OF PERFORMANCE WHOSE ERROR CAN BE SPECIFIED				
				TIME RESPONSE	FREQUENCY RESPONSE			
	RECURSIVE	NONRECURSIVE	REDUCED-RATE NONRECURSIVE (SPECTRUM ANALYSIS)		COMPLEX	AMPLITUDE	PHASE	DELAY
MAPPING	√					√		√
WINDOWS		√	√		√			
LINEAR PROGRAMMING		√	√	√	√			
NONLINEAR PROGRAMMING	√	√	√	√	√	√	√	√

Fig. 1. Applicability of techniques for designing digital filters.

gramming to determine the coefficients in the presence of a constraint on this number.

Each of the techniques reviewed here is characterized in two ways. One characterization is the type of digital filter to which a technique can be applied. The other characterization is the type of performance which can be constrained by a technique. These characterizations appear as the column headings in Fig. 1, which compares the applicability of the techniques shown in the left column. These characterizations are discussed in the next two paragraphs.

The types of digital filters which can be designed are termed recursive, nonrecursive, and reduced-rate nonrecursive. "Recursive" and "nonrecursive" are defined as usual [1]. The (nonstandard) term "reduced-rate nonrecursive filter" denotes the usual digital method for measuring a power spectrum. This method is equivalent to computing the output of a narrowband nonrecursive filter at a rate which is some submultiple of the sampling frequency. (A submultiple is adequate because of the narrowness of the bandwidth of the equivalent nonrecursive filter.)

The types of performance which can be constrained in these techniques are divided into time and frequency responses. Time response is explained in the section on linear programming. Frequency response is subdivided into the following responses: 1) complex, 2) amplitude, 3) phase, and 4) delay. (This terminology is not standard but is convenient and is used throughout this paper.) The complex response is simply the Fourier transform of the impulse response of the filter. The amplitude and phase responses, respectively, are the absolute value and phase (i.e., arctan of the ratio of the imaginary part to the real part) of this Fourier transform. The delay response is the derivative (with respect to frequency) of the phase response. ("Group delay" is the standard term for the delay response.)

The best known of these techniques for determining the coefficients of a digital filter is described next. Perhaps most specifications for digital filters can be satisfied by this method.

II. Mapping

The coefficients of a recursive digital filter which approximates a piecewise-constant frequency response can be determined by mapping a Laplace transform into a z transform [1], [2]. The bilinear mapping can be substituted into a predistorted Fourier transform expressed as a rational fraction; the coefficients of the powers of the unit delay operator z^{-1} in the resulting rational fraction are the coefficients of the recursive filter whose amplitude response can be equiripple (e.g., as in an elliptic or Chebyshev filter) [1]–[4] or whose delay response can be equiripple [5]–[7]. The equiripple amplitude response can be made to be either low-pass, bandpass, bandstop, or high-pass [1]–[8]. Computer programs embodying these techniques exist [9], [10].

The impulse-invariant transformation (also called the standard z transform) can produce a close approximation to an equiripple (Chebyshev) amplitude response [1], [2]. This transformation has the effect of mapping the real frequency ($j\omega$) axis in the s plane (i.e., the plane in which the Laplace transform is evaluated) onto itself. Further explanation of the above techniques is available in many excellent descriptions [1]–[10].

III. Window Techniques

Window techniques can be used to measure the Fourier transform or the power spectrum of the samples of a waveform. A sequence of equally spaced samples of the waveform is multiplied by a sequence of weights obtained from the window function (usually by sampling it). Then the discrete Fourier transform of this sequence of products is computed with the aid of the fast Fourier transform [2] [11]. This procedure estimates the Fourier transform of the input waveform at a set of equally spaced frequencies. The power spectrum is estimated by squaring the absolute values of the estimates of the Fourier transform. This procedure is repeated after shifting the input sequence. The results of several repetitions of this procedure may be averaged to reduce the random error to an acceptable amount [12]. The number of samples by which the data is shifted often roughly corresponds to the rms width of the window function [12]. Each of the outputs of the fast Fourier transform is a sample of the input data's Fourier transform convolved with the Fourier transform of the weights. If the Fourier transform of the weights is chosen to be the Dolph–Chebyshev function,

$$D(x) = \cos\left(P \cos^{-1}\left(\frac{\cos \pi x}{\cos \pi B/2}\right)\right), \tag{1}$$

where x is normalized frequency, $|x| \leq 1/2$, the sidelobes of the Fourier transform of any sine wave in the input waveform will be minimax [13]. In (1), P is 1 less than the number of weights and B is the normalized bandwidth [14]. It is sometimes possible to choose P large enough so that leakage from the sidelobes of any sinusoid produce smaller errors than those from any other source (e.g., quantization or the environmental noise accompanying the sinusoid) [14]. The fast Fourier transform algorithm can be used to accurately and efficiently calculate the weights from the Dolph–Chebyshev function [14], [15].

Window techniques can be used also to determine the coefficients of ordinary nonrecursive filters whose complex responses have bounded errors. The window technique (or the Fourier series technique) involves forming the product of 1) a sequence of weights and 2) a sequence of samples computed by the inverse Fourier transform (or the inverse discrete Fourier transform) of an ideal complex response [1], [14]. The sequence of products is used as the coefficients of the nonrecursive digital filter.

The Fourier transform of this sequence is equal to the convolution of 1) the ideal complex response and 2) the Fourier transform of the weights. Although the error in this Fourier transform (defining the error as the difference between this Fourier transform and the ideal complex response) is not minimax or even equiripple (although it is often close to equiripple), this technique has the advantage that the maximum of the absolute value of its errors can be predicted if the ideal of the complex response is composed only of piecewise constant stopbands and passbands. The equation

$$P = \frac{1.46}{B} \log_{10}(RY) \tag{2}$$

holding for weights obtained from the fast Fourier transform of the Dolph–Chebyshev function provides an upper bound to the number $P+1$ of coefficients required for a nonrecursive digital filter having a relative error of $1/R$ in the sidelobes and having an ideal complex response which is piecewise constant; this ideal is composed of passbands and stopbands separated by Y transition bands in which the error is unspecified [14].

This preliminary upper bound to the number of coefficients can be used to obtain a rough (e.g., a factor of 2 too high) estimate of the number of coefficients required in the method described in the next section. The window method itself is not as effective as is this next method.

IV. The Simplex Method of Linear Programming

The simplex method of linear programming can be used to determine the coefficients of the nonrecursive digital filter which minimizes the maximum of the error in the complex response (defined in the introductory section). Samples of this error can be represented as

$$E_k = \sum_{q=-P/2}^{P/2} h_q \exp(-2\pi j x_k q) - F(2Wx_k) \qquad k = 1, \cdots, N \tag{3}$$

where j is the square root of -1, h_q denotes the coefficients of the digital filter, P is even, $F(2Wx_k)$ denotes the ideal complex response at a normalized frequency denoted by x_k, $|x_k| \leq 1/2$, and the sampling rate is $2W$. The usually equidistant spacing of the discrete frequencies x_k must be close enough so that all significant features of the complex response are represented. Intuition suggests that a necessary condition for adequate representation is that N be larger than P; N is usually chosen to be several times larger than P (e.g., $N=2P$). Assuming that the ideal complex response has the usual complex conjugate symmetry, i.e., $F(2Wx)=F^*(-2Wx)$, the coefficients h_q consequently should be chosen to be real. The real and the imaginary parts of the error are required to be less than some set of upper limits. This requirement can be written

$$\begin{aligned} -V_k &\leq \text{Re}\{E_k\} \leq V_k, \\ -V_k &\leq \text{Im}\{E_k\} \leq V_k, \qquad k = 1, \cdots, N \end{aligned} \tag{4}$$

where the V_k specify the limits on the real and imaginary parts of the error E_k. The V_k are usually chosen to be large or (4) is omitted, at frequencies lying in the transition bands (between passbands and stopbands).

The simplex method can provide a numerical solution of the following problem [16]–[18]: determine the coefficients h_q to minimize Q in

$$\begin{aligned} -QV_k &\leq \text{Re}\{E_k\} \leq QV_k \\ -QV_k &\leq \text{Im}\{E_k\} \leq QV_k \qquad k = 1, \cdots, N. \end{aligned} \tag{5}$$

If the minimum value of Q is greater than unity, the constraints (4) are not satisfied. Increasing the number $P+1$ of coefficients and repeating the linear programming can make the minimum value of Q be less than unity, thereby satisfying (4). Successive trials can reveal the smallest value of $P+1$ for which Q is less than unity.

The simplex method of linear programming can be applied also to minimizing the errors in the time response. These errors are defined to be samples of the sidelobes of the output of a digital filter whose input consists of samples of a known waveform [19], [20]. Again these sidelobes lie on both sides of the mainlobe of the output of the digital filter. If the input to the digital filter consists of samples of some known waveform (denoted b_i,

$i = -M/2, \cdots, M/2$), these sidelobes can be minimized by properly choosing the coefficients of the digital filter. The output at the nominal center point of the mainlobe is constrained to be not less than unity. The sidelobe level is defined as the minimum value of Q satisfying the inequality

$$-Q \leq \sum_{q=-P/2}^{P/2} \sum_{i=-M/2}^{M/2} h_q b_i \operatorname{sinc}\left(\frac{k}{I} - q - i\right) \leq Q,$$

$$k = -\frac{N}{2}, \cdots, -\frac{L}{2}, \frac{L}{2}, \cdots, \frac{N}{2}$$

$$\sum_{i=-\min(P/2,M/2)}^{\min(P/2,M/2)} h_i b_{-i} \geq 1 \qquad (6)$$

where I is an integer which is at least as great as 2, $P+1$ is the number of coefficients, which are denoted h_q, sinc (t) is defined as usual as $\sin(\pi t)/(\pi t)$, and $L+1$ is the number of samples in the mainlobe.

The minimum sidelobe level Q can be determined by the simplex method of linear programming [16], [17], which determines the values of the coefficients h_q which minimize Q. Inequalities (6) are written under the assumption that the b_i are real numbers. If the b_i are complex numbers, as arise in sampling the in-phase and quadrature-phase components of bandpass signals, the real and imaginary parts of inequalities (6) can be constrained as in inequalities (4).

In inequalities (6), the cardinal series (i.e., the ordinary sampling theorem) has been used to interpolate the samples of the output of the digital filter. Such interpolation is exact provided that the samples b_i of the input waveform are as usual taken at a rate $2W$ which is greater than twice the highest frequency present in the input waveform. The cardinal series does not appear in an original application of linear programming to this problem [19]. This absence can be explained under the assumption that this application involved sampling the input waveform at a rate (i.e., $2W$ samples per second) which was several times larger than the highest frequency present in the input waveform. Since this assumption does not hold in the present case, it is necessary to oversample the output of the digital filter. Oversampling by at least a factor of $I=2$ is intuitively necessary to adequately measure the error in the output, which is regarded as being continuous. One of the reasons for regarding the output as continuous is that the output samples are usually passed through a digital-to-analog converter connected to a low-pass filter; the resulting continuous waveform approximates the interpolation by the cardinal series.

Many programs embodying the simplex method for solving linear programs exist [21], [22]. Although linear programming packages in any other form can be used to minimize Q in inequalities (5) or (6) [18, p. 123], the dual simplex form is as efficient as any [21]. Techniques exist which might halve the required memory by taking advantage of the obvious symmetry in inequalities (5) or (6) [23], [24].

The running time of a program embodying the simplex method averages to approximately P^3 times the time required to do an operation consisting of 1) a floating point multiplication, 2) a floating point subtraction, and 3) a branching [25]. If $P=180$ and if an operation requires 10 μs, approximately 1 minute of computer time is required. This requirement seems modest.

A reason for discussing nonrecursive filters is that the time required for using such filters (i.e., computing the outputs of such filters) is small if the fast convolution application of the fast Fourier transform is used [2], [26]. Although the fast convolution often permits the outputs of nonrecursive filters to be computed as rapidly as the outputs of recursive filters, the amount of computer memory required by a nonrecursive filter is often much greater than that required by a recursive filter.

Other methods for solving linear programs exist. One method obtains a minimax error by using a steepest descent search [27]. In this linear program, the separation between frequencies [corresponding to x_k in (3)] often needs to be ten or more times the reciprocal of the total delay [28]. This separation between frequencies would be expected to correspond to a value of $I=10$ in (6). Although this steepest descent method and the simplex method obtain minimax errors at *discrete* times and/or frequencies, it is possible to use a nonlinear programming technique to produce equiripple error over the continuous subintervals of frequencies [29].

The technique in Section II for determining the coefficients of recursive filters is limited in the types of responses which can be approximated. This limitation theoretically does not apply to the next technique.

V. Nonlinear Programming

Both nonrecursive and recursive filters can be designed by nonlinear programming. The nonlinear programming approach to be outlined is similar to one which has been followed by Athanassopoulos [30]–[32]. This technique resembles the linear programming technique presented previously. An error E_k is formed by taking the difference between any previously mentioned response of the digital filter and the ideal values of this response at the frequency x_k. It is also possible to define more than one error corresponding to more than one type of response (e.g., amplitude and delay responses) at a single frequency x_k.

The nonlinear programming technique begins with a choice for the type and form of the digital filter. The type can be either nonrecursive or recursive. If the latter, the z transform [1], [2] of this filter is chosen usually to be either of the following forms (where z^{-1} as usual denotes a delay by one sample):

$$H(z) = g \prod_{i=1}^{S} \frac{1 + a_i z^{-1} + b_i z^{-2}}{1 + c_i z^{-1} + d_i z^{-2}}$$

or

$$H(z) = g + \sum_{i=1}^{S} \frac{a_i + b_i z^{-1}}{1 + c_i z^{-1} + d_i z^{-2}}. \tag{7}$$

The upper and lower equations, respectively, represent cascade and parallel connections of elementary sections. The complex response of either of these recursive filters can be evaluated by setting z equal to $\exp(2\pi jx)$, where in this notation the frequency x has been normalized so that the time between samples is unity and $|x| \leq 1/2$.

The error is chosen to be a differentiable function of all parameters a_i, b_i, c_i, d_i, and g. Whereas the amplitude response is not differentiable whenever it is zero (as might be expected to occur in a stopband), the square of the amplitude response is everywhere differentiable. Thus the error is defined frequently as the difference between the actual and the ideal values of the square of the amplitude response, as in the equation

$$E_k = |H(\exp(2\pi j x_k))|^2 - |F(2Wx_k)|^2.$$

The set of discrete frequencies x_k is chosen as described in the discussion following (3). The errors E_k are required to satisfy inequality (8), which resembles inequalities (4):

$$-L_k \leq E_k \leq U_k, \qquad k = 1, \cdots, N \tag{8}$$

where L_k and U_k are positive and real. In some cases either the lower or upper limit on E_k will be absent (e.g., L_k will be absent if an ideal value of the amplitude response is zero).

The following functions are defined:

$$\left.\begin{aligned} G_k &\triangleq QU_k - E_k \\ H_k &\triangleq QL_k + E_k \end{aligned}\right\} \qquad \begin{aligned} &k = 1, \cdots, N \\ &Q > 0. \end{aligned} \tag{9}$$

G_k or H_k are omitted for those values of k for which no upper or lower bound is specified for the error E_k.

A penalty function such as

$$Q + \sum_{k=1}^{N} \frac{r}{G_k} + \sum_{k=1}^{N} \frac{r}{H_k} \tag{10}$$

is formed [33], [34]. Additional summations of r times the reciprocals of quantities identical to G_k and H_k can be added to constrain errors of other responses (e.g., phase or delay response) [32]. A suitable program is used to minimize the penalty function simultaneously with respect to Q and to all of the coefficients g, a_i, b_i, c_i, and d_i. This can be done by programs which achieve an unconstrained minimization. An example of such a program is the conjugate gradient method of Davidon, Fletcher, and Powell [35]. This algorithm is available in computer programs [36]. (These programs must be modified to include tests on G_k and H_k to make certain that their values remain positive. They can become negative if r is reduced by too large a factor.)

After the minimum of the penalty function (10) has been obtained simultaneously with respect to all variables, i.e., Q, g, a_i, b_i, c_i, and d_i, $i = 1, \cdots, S$, the whole minimization procedure is repeated after dividing r by some suitable factor (e.g., 2 or 5). This process is continued until Q becomes nearly constant. This stationarity occurs when the terms which include r as a factor are negligible compared to Q. Thus, the process of repeatedly reducing r and minimizing the resulting penalty function eventually obtains the minimum of the error factor Q while satisfying all of the constraining inequalities. If Q exceeds unity, none of the constraints $-L_k \leq E_k \leq U_k$ are satisfied, and the whole of the above procedure is repeated with a larger number of stages S (e.g., increment S by unity). If Q is less than or equal to unity, all of the constraints are satisfied and the program terminates.

The initial value of Q should be chosen large enough so that the initial values of the right sides of (9) are positive. The initial value of r can be chosen so that the terms in (10) which include r as a factor are initially of the same size as the initial value of Q. Instead of choosing the initial values of the coefficients g, a_i, b_i, c_i, and d_i, $i = 1, \cdots, S$, arbitrarily (but with all poles inside the unit circle), these inital values might be the coefficients of digital filter (e.g., an elliptic filter) having a somewhat similar frequency response [32].

The result of this application of nonlinear programming will be some minimum value of Q, but this minimum may not be the lowest possible minimum corresponding to a different set of coefficients. The cause of this is that the topology is not convex and the process might converge to a local minimum without converging to the global minimum [35, p. 343]. Even if the lowest possible minimum is not obtained, the achieved minimum value of Q may be less than unity, thereby meeting the specifications. Except for a similar technique involving the use of nonlinear programming to determine the poles and zeros of a continuous frequency response which meets the desired specifications after applying the bilinear z transform, this is the only known procedure which is capable of optimally choosing the coefficients of digital filters having almost any form and meeting almost arbitrary specifications. A similar technique has been used for minimizing the squared error [37].

Other nonlinear programming algorithms can be used to obtain the coefficients of a recursive filter having minimax errors [38]. Some comparisons between these methods have been made [38]. Although the penalty function method described above may require several times more computing time than the fastest methods [38], the amount of computer memory required is nearly as small as possible. (A faster method, GRG 69 [38], requires excessive amounts of computer memory.)

VI. Integer Linear Programming

Integer programming techniques for solving a linear program in which the variables are integers exist [16, ch. 26]. These techniques can be used to determine the coefficients of a nonrecursive digital filter whose coefficients are quantized, i.e., expressed in fixed point arithmetic with only a finite number of bits. These techniques determine quantized coefficients which provide minimax errors in the response. This response can be either the complex response or the time response, as defined in Section IV. The only difference between the problem treated in the present section and in Section IV is the assumption that the coefficients are quantized.

Programs embodying these techniques exist [21, Sects. 1.11, 3.5.6, 4.12, 4.13], [39]. The amount of computing time required is only slightly larger than the amount required for ordinary linear programming [39]. Although no reports of the use of linear integer programming to solve minimax problems involving quantized variables are known to exist, experience suggests that minimax problems do not belong to one troublesome class [39, p. 234].

The coefficients in some nonrecursive digital filters are not only quantized but are also limited to some maximum value. These upper bounds on the coefficients can be straightforwardly included in the constants supplied as inputs to the integer linear program.

Other methods of solving integer linear programming exist [40]. A generalization of integer linear programming is nonlinear integer programming, as for example in determining the quantized coefficients of a recursive filter whose reponse has minimax error [41], [42].

VII. Conclusions

Several techniques are available for determining the coefficients of digital filters. Each of these techniques offers unique advantages. It therefore seems important to understand all of these techniques.

Application of these techniques is made usually on a digital computer. The last three of these techniques (Sections IV–VI) are applied conveniently by a computer in a time-shared mode. This mode permits varying constraints, parameters, and initial values until a satisfactory compromise is achieved.

Acknowledgment

The author wishes to thank J. Athanassopoulos, L. R. Rabiner, and J. F. Kaiser of Bell Telephone Laboratories, Murray Hill, N. J., for making many helpful suggestions, including citations to relevant sources.

References

[1] J. F. Kaiser, *Systems Analysis by Digital Computer.* New York: Wiley, 1966, ch. 7.
[2] C. M. Rader and B. Gold, *Digital Processing of Signals.* New York: McGraw-Hill, 1969.
[3] E. Christian and E. Eisenmann, *Filter Design Tables and Design Graphs.* New York: Wiley, 1966.
[4] A. G. Constantinides, "Digital filters with equiripple passbands," *IEEE Trans. Circuit Theory*, vol. CT-16, pp. 535–538, November 1969.
[5] D. Helman, "Tchebycheff approximations for amplitude and delay with rational functions," *Proc. Symp. Modern Network Theory*, Polytechnic Institute of Brooklyn, Brooklyn, N. Y., pp. 385–402.
[6] T. A. Abele, "Transmission factors with Chebyshev-type approximations of constant group delay," *Arch. Elek. Übertragung*, vol. 16, pp. 9–18, January 1962.
[7] E. Ulbrich and H. Piloty, "The design of allpass, lowpass, and bandpass filters with Chebyshev-type approximations of constant group delay," *Arch. Elek. Übertragung*, vol. 14, pp. 451–467, October 1960.
[8] A. G. Constantinides, "Spectral transformations for digital filters," *Proc. Inst. Elec. Eng.*, vol. 117, pp. 1585–1590, August 1970.
[9] J. F. Kaiser, "Computer aided design of classical continuous system transfer functions," in *Proc. Hawaii Int. Conf. System Sciences*, B. K. Kinariwala and F. P. Kuo, Eds. Honolulu: University of Hawaii Press, 1968, pp. 197–200.
[10] R. Edwards and A. Bradley, "Design of digital filters by computer," *Int. J. Numer. Methods Eng.*, vol. 2, pp. 311–333, 1970.
[11] W. T. Cochran, J. W. Cooley, D. L. Favin, H. D. Helms, R. A. Kaenel, W. W. Lang, G. C. Maling, Jr., D. E. Nelson, C. M. Rader, and P. D. Welch, "What is the fast Fourier transform?" *IEEE Trans. Audio. Electroacoust.*, vol. AU-15, pp. 45–55, June 1967; see also *Proc. IEEE*, vol. 55, pp. 1664–1674, October 1967.
[12] P. D. Welch, "The use of the fast Fourier transform for the estimation of power spectra: A method based on time averaging over short, modified periodograms," *IEEE Trans. Audio Electroacoust.*, vol. AU-15, pp. 70–73, June 1967.
[13] C. L. Dolph, "A current distribution for broadside arrays which optimizes the relationship between beamwidth and side-lobe level," *Proc. IRE*, vol. 34, pp. 335–348, June 1946.
[14] H. D. Helms, "Nonrecursive digital filters: Design methods for achieving specifications on frequency response," *IEEE Trans. Audio Electroacoust.*, vol. AU-16, pp. 336–342, September 1968.
[15] R. J. Stegen, "Excitation coefficients and beamwidths of Tschebyscheff arrays," *Proc. IRE*, vol. 41, pp. 1671–1674, November 1953.
[16] G. B. Dantzig, *Linear Programming and Extensions.* Princeton, N. J.: Princeton University Press, 1963.
[17] S. I. Gass, *Linear Programming*, 2nd ed. New York: McGraw-Hill, 1964.
[18] P. Rabinowitz, "Applications of linear programming to numerical analysis," *SIAM Rev.*, vol. 10, pp. 121–159, 1968.
[19] R. K. Cavin, C. H. Ray, and V. T. Rhyne, "The design of optimal convolution filters via nonlinear programming," *IEEE Trans. Geosci. Electron.*, vol. GE-7, pp. 142–145, July 1969.
[20] H. Groginsky *et al.*, various internal reports at Raytheon on the use of linear programming to determine the coefficients of transversal filters having optimum resolution, 1961 and subsequent years.
[21] H. P. Kunzi, H. G. Tzschach, and C. A. Zehnder, *Numerical Methods of Mathematical Optimization.* New York: Academic Press, 1968, sects. 1.7, 3.5.2, 4.4, 4.5.
[22] IBM Mathematical Programming System/360 (*360A-CO-14X*) *Linear and Separable Programming*, 1968. Also, *Introduction to the LP/600 Linear Programming System* (*CPB-1141C*), General Electric Co., 1968.
[23] J. E. Kelley, Jr., "An application of linear programming to curve fitting," *J. Soc. Ind. Appl. Math.*, vol. 6, pp. 15–22, 1958.
[24] I. Barrodale and A. Young, "Algorithms for best L_1 and L_∞ approximation on a discrete set," *Numer. Math.*, vol. 8, pp. 295–306, 1966.
[25] P. Wolfe and L. Cutler, "Experiments in linear programming," in *Recent Advances in Mathematical Programming*, R. L. Graves and P. Wolfe, Eds. New York: McGraw-Hill, 1963.
[26] H. D. Helms, "Fast Fourier transform method of computing difference equations and simulating filters," *IEEE Trans. Audio Electroacoust.*, vol. AU-15, pp. 85–90, June 1967.
[27] L. R. Rabiner, B. Gold, and C. A. McGonegal, "An approach to the approximation problem for nonrecursive digital filters," *IEEE Trans. Audio Electroacoust.*, vol. AU-18, pp. 83–106, June 1970.
[28] L. R. Rabiner, private communication, October 21, 1970.
[29] O. Herrmann, "Design of nonrecursive digital filters with linear phase," *Electron. Lett.*, vol. 6, pp. 328–329, May 1970.
[30] J. A. Athanassopoulos and A. D. Waren, "Design of discrete-time systems by mathematical programming," *Proc. 1968*

Hawaii Int. Conf. Syst. Sci. Honolulu: University of Hawaii Press, 1968, pp. 224–227.

[31] ——, "Time-domain synthesis by nonlinear programming," *1968 Proc. 4th Allerton Conf.*, pp. 766–775.

[32] J. A. Athanassopoulos and J. F. Kaiser, "Constrained optimization problems in digital filter design," presented at the Arden House Workshop on Digital Filtering, January 12, 1970.

[33] A. V. Fiacco and G. P. McCormick, *Nonlinear Programming.* New York: Wiley, 1968.

[34] W. I. Zangwill, *Nonlinear Programming.* Englewood Cliffs, N.J.: Prentice-Hall, 1969.

[35] R. Fletcher, *Optimization.* New York: Academic Press, 1969, pp. 1–12.

[36] IBM System/360 Programming Manual, Scientific Subroutine Package (360A-CM-03X), Version III, as programs FMFP, DFMFP, FMCG, and DFMCG. (These are FORTRAN programs of the Davidon and Fletcher–Powell algorithms.)

[37] K. Steiglitz, "Computer-aided design of recursive digital filters," *IEEE Trans. Audio Electroacoust.*, vol. AU-18, pp. 123–129, June 1970.

[38] J. Abadie and J. Guigou, *Integer and Nonlinear Programming,* Appendix 3. Amsterdam: North-Holland; New York: Elsevier, 1970.

[39] M. L. Balinski, "Some general methods in integer programming," in *Nonlinear Programming,* J. Abadie, Ed. Amsterdam: North Holland, 1967, pp. 222–247.

[40] M. L. Balinski and K. Spielberg, "Methods for integer programming: Algebraic, combinatorial, and enumerative," in *Progress in Operations Research*, vol. 3, J. S. Aronofsky, Ed. New York: Wiley, 1969, pp. 195–292.

[41] E. Avenhaus and W. Schuessler, "On the approximation problem in the design of digital filters with limited wordlength," unpublished paper.

[42] E. Avenhaus, "Zum entwurf digitaler filter mit begrenzter worlange," referenced in [41].

Time Domain Design of Recursive Digital Filters

CHARLES S. BURRUS, Member, IEEE
THOMAS W. PARKS, Member, IEEE
Department of Electrical Engineering
Rice University
Houston, Tex.

Abstract

The problem of synthesis of recursive digital filters to give a desired pulse response over a specified interval is studied. Realizability conditions are stated and a linear design method is developed. Several design procedures that require only linear calculations are given for approximate realization of recursive filters. Finally, an error analysis of the techniques is made.

Manuscript received November 10, 1969.

This work was supported in part by NASA Grant NGL 44-006-033.

I. Introduction

Recursive digital filters have been recognized as a very efficient and powerful implementation of many signal-processing procedures. The design methods have for the most part used Z transform techniques and frequency-domain criteria borrowed from continuous-time filter theory [1], [2].

This paper is concerned with the problem of time-domain synthesis of recursive digital filters. The main objectives of this paper are the following. 1) Develop precise conditions and tests for exact realizations and straightforward methods for calculating the realizations. 2) Develop several optimal approximate synthesis methods that require only linear calculations. These are based on a partitioning technique first suggested by J. L. Shanks [3]. 3) Develop interpretations based on a matrix formulation of the problem which greatly aid the designer in calculating the filter structure. 4) Develop an error analysis which relates the techniques of this paper to the classical problem of determining optimal exponents [4].

The description of the recursive filter will be in terms of the Z transform transfer function written in the form

$$G(z) = \frac{a_0 + a_1 z^{-1} + \cdots a_{N-1} z^{-N+1}}{1 + b_1 z^{-1} + \cdots b_{M-1} z^{-M+1}} \tag{1}$$

and having as a unit pulse response

$$g(k) = g_0, g_1, g_2, \cdots \tag{2}$$

which is the inverse transform of $G(z)$. The time-domain synthesis problem is to find the $M+N-1$ coefficients a_i and b_i such that the $g(k)$ approximates in some sense a given function $h(k)$ over the range of $k=0, 1, \ldots K-1$. The choice of M, N, and a criterion for approximation are the subjects of this paper and the resulting coefficients a_i and b_i are considered a realization of the filter.

II. Exact Techniques

Theorem 1: If $K=M+N-1$ then at least one exact recursive realization exists and it can be found using a noniterative linear procedure.

Proof: To prove this theorem a solution will be found for the equations resulting from setting the inverse transform of $G(z)$ equal to the desired output $h(k)$:

$$\frac{a_0 + a_1 z^{-1} + \cdots a_{N-1} z^{-(N-1)}}{1 + b_1 z^{-1} + \cdots b_{M-1} z^{-(M-1)}} = h_0 + h_1 z^{-1} + h_2 z^{-2} + \cdots. \tag{3}$$

In terms of convolution of number sequences this becomes

$$\sum_{n=0}^{M-1} b_n h_{i-n} = \begin{cases} a_i & i = 0, 1, \cdots N-1 \\ 0 & i \geq N \end{cases} \tag{4}$$

or in matrix form

Reprinted from *IEEE Trans. Audio Electroacoust.*, vol. AU-18, pp. 137–141, June 1970.

$$\begin{bmatrix} a_0 \\ a_1 \\ \vdots \\ a_{N-1} \\ 0 \\ \vdots \\ 0 \end{bmatrix} = \begin{bmatrix} h_0 & 0 & 0 \cdots 0 \\ h_1 & h_0 & 0 & \vdots \\ h_2 & h_1 & h_0 & \vdots \\ \vdots \\ h_{K-1} & h_{K-2} & \cdots & h_{K-M} \end{bmatrix} \begin{bmatrix} 1 \\ b_1 \\ b_2 \\ \vdots \\ b_{M-1} \end{bmatrix}. \tag{5}$$

This equation is partitioned such that

$$\begin{bmatrix} a_0 \\ \vdots \\ a_{N-1} \\ \hline 0 \\ \vdots \\ 0 \end{bmatrix} = \begin{bmatrix} h_0 & 0 \cdots 0 \\ \vdots \\ h_{N-1} \\ \hline h_N \\ \vdots \\ h_{K-1} & \cdots & h_{K-M} \end{bmatrix} \begin{bmatrix} 1 \\ b_1 \\ \vdots \\ b_{M-1} \end{bmatrix} \tag{6}$$

$$\begin{bmatrix} \boldsymbol{a} \\ \hline 0 \end{bmatrix} = \begin{bmatrix} H_1 \\ \hline H_2 \end{bmatrix} [\boldsymbol{b}]. \tag{7}$$

If $K=M+N-1$, then H_2 has $M-1$ rows and M columns, and therefore maximum rank of $M-1$; this guarantees a nontrivial solution to the lower partition of (7). Since b_0 is normalized, the equation for the lower partition of (7) can be written

$$\boldsymbol{h}^1 = -H_3\boldsymbol{b}^* \tag{8}$$

where $\boldsymbol{h}^1$ is the first column of H_2, H_3 is the remaining $M-1$ by $M-1$ square matrix, and $\boldsymbol{b}^*$ is the $\boldsymbol{b}$ matrix with the b_0 term removed.

$$\boldsymbol{b}^* = \begin{bmatrix} b_1 \\ b_2 \\ \vdots \\ b_{M-1} \end{bmatrix}. \tag{9}$$

If H_3 is nonsingular, then the b_i coefficients are unique and are found by

$$\boldsymbol{b}^* = -H_3^{-1}\boldsymbol{h}^1 \quad \text{and} \quad b_0 = 1. \tag{10}$$

If H_3 is singular, there are many solutions to the equation formed by the lower partition of (7) and this situation will be considered as two cases. If H_3 is singular and has rank R, then since this means there are R independent columns in H_3, there exists a solution $\boldsymbol{b}$ to the lower partition of (7) with R nonzero elements and $M-1-R$ zero components. If $\boldsymbol{h}^1$ lies in the space spanned by the columns of H_3, then this will be a solution of (10). If $\boldsymbol{h}^1$ does not lie in the space spanned by the columns of H_3, then the solution requires that b_0 be zero and (10) will have to be reformulated with b^1 equal to one. The recognition of the zero coefficient is important in the reduction of complexity in implementation of a digital filter. Once the denominator coefficients b_i are determined, a direct multiplication from the upper partition of (7) gives the numerator coefficients a_i.

$$\boldsymbol{a} = H_1\boldsymbol{b}. \tag{11}$$

Comments

It is interesting to note that an exact realization is possible for any choice of M and N as long as $M+N-1=K$. In order to use this freedom to advantage, one must understand how the nature of the impulse response of a recursive filter depends on M and N. The denominator of (3) determines an $M-1$ order difference equation so that $g(k)$ defined by (2) will be the sum of $M-1$ geometric sequences (except for the case of repeated natural frequencies).

$$g(k) = \sum_{i=1}^{M-1} C_i\lambda_i^k \qquad k = 0, 1, \cdots, K-1.$$

For $N<M-1$ the $M-1$ coefficients C_i are determined with N degrees of freedom by the numerator coefficients. For $N>M-1$ all the $M-1$ coefficients C_i are determined plus the first $N-M-1$ terms of $g(k)$. The two extremes would be, on one hand, $N=0$ and $g(k)$ realized summing $K=M-1$ geometric sequences, and on the other hand, by $M=0$ and the filter would be a simple $K-1$ length convolution.

The difference in filters designed with different M and N but with the same $M+N$ will be in $g(k)$ for $k\geq K$. Another reason for considering various M, N combinations is that it may be possible to have an exact realization for $M+N-1<K$. This case may exist when H_3 is singular and is considered in the next theorem.

Theorem 2: If $N=M-1$, a recursive realization exists if and only if

$$h(k) = \sum_{i=0}^{P-1} \sum_{n=0}^{L_i-1} k^n C_{in}\lambda_i^k \tag{12}$$

where L_i= the multiplicity of the λ_i geometric sequence and

$$M - 1 = \sum_{i=0}^{P-1} L_i.$$

Note this result is independent of K.

Proof: For simplicity and clarity the proof will be given for the case where all λ_i are nonzero and distinct. In this case (12) becomes

$$h(k) = \sum_{i=0}^{M-2} C_i\lambda_i^k. \tag{13}$$

The lower partition of (7), H_2, has $K-N$ rows and M columns. The columns of H_2 are denoted $\boldsymbol{h}^i$

$$H_2 = \left[\ \boldsymbol{h}^1 \mid \boldsymbol{h}^2 \mid \cdots \mid \boldsymbol{h}^M \ \right] \tag{14}$$

and a set of vectors is defined by

$$\boldsymbol{\lambda}_i = \begin{bmatrix} 1 \\ \lambda_i \\ \lambda_i^2 \\ \cdot \\ \cdot \\ \cdot \\ \lambda_i^{K-N-1} \end{bmatrix}. \qquad (15)$$

In terms of the $\boldsymbol{\lambda}$ vectors the columns of H_2 become

$$\begin{aligned} \boldsymbol{h}^M &= C_0\boldsymbol{\lambda}_0 + C_1\boldsymbol{\lambda}_1 + \cdots C_{M-2}\boldsymbol{\lambda}_{M-2} \\ \boldsymbol{h}^{M-1} &= (C_0\lambda_0)\boldsymbol{\lambda}_0 + (C_1\lambda_1)\boldsymbol{\lambda}_1 + \cdots . \\ &\cdot \\ &\cdot \\ &\cdot \end{aligned} \qquad (16)$$

This shows that, because of the shifted nature of the columns of H_2, these columns all lie in the subspace spanned by the $\boldsymbol{\lambda}_i$ vectors having nonzero C_i. This means $\boldsymbol{h}^1$ is in the space spanned by the other columns of H_2 and thus a solution to the lower partitions of (6) is guaranteed. Multiplication of the upper partition gives the a_i coefficients. The necessary part of the proof is direct since the lower partition of (6) represents an $M-1$ order homogeneous difference equation and (12) is the general form for the solution of a $M-1$ order difference equation.

Comments

The basic problem in applying this theorem is the testing of $h(k)$ to find out if an exact realization is possible for a given M. The rank of H_2 is equal to the number of independent columns of H_2, and for the lower partition of (7) to have a unique solution, this rank must be equal to $M-1$. If the rank is larger than $M-1$, then no exact solution exists and a larger M must be tried. If the rank of H_2 is less than $M-1$, then the individual components of $\boldsymbol{b}$ corresponding to the dependent columns of H_3 can be set to zero. The number of numerator coefficients N must be equal to $M-1$ in general. Any fewer than this will result in zeros in the upper right-hand corner of H_2, destroying the validity of (16). If fewer numerator coefficients are possible, then some of the a_i terms will be zero when the upper partition multiplication of (7) is carried out, but the partition must be set for $N=M-1$.

If an examination of $h(k)$ indicates that the rank of H_2 is not equal to $M-1$ for reasonable values of M, and that $h(k)$ fails to satisfy condition (12) because of the values of the first few terms, then partitioning (5) lower than after $M-1$ rows will remove the first terms of $h(k)$ from the H_2 matrix. If this causes the rank of H_2 to be $M-1$, then an exact realization is possible but N is larger than $M-1$.

The application of Theorem 2 will depend on the ability to analyze a particular $h(k)$ and choose values for M and N. Unfortunately in many cases H_2 has full rank and an exact reduced order realization is impossible. A more practical approach is to find a realization that will approximate the desired response in some sense. This is done in the next section and is based on a reformulation of an approach suggested by Shanks [3].

III. Approximate Techniques

The Direct Procedure

If the problem is directly attacked by minimizing the square of $g(k)-h(k)$, a rather complicated set of nonlinear equations [4] results, requiring iterative methods for this solution. To avoid this, an error measure similar to that used in the least-square inverse filter problem [5] is used here, and the same matrix partitioning employed in the first two theorems is used to uncouple the calculations of the a_i and b_i. The exact equation (6) is modified to define the error terms:

$$\begin{bmatrix} a_0 \\ a_1 \\ \cdot \\ \cdot \\ a_{N-1} \\ --- \\ \epsilon_N \\ \cdot \\ \cdot \\ \epsilon_{K-1} \end{bmatrix} = \begin{bmatrix} h_0 & 0 & 0 \cdots 0 \\ h_1 & h_0 & 0 & \\ \cdot & & & \\ \cdot & & & \\ h_{N-1} & & & \\ ------- & & & \\ h_N & & & \\ \cdot & & & \\ \cdot & & & \\ h_{K-1} & & & h_{K-M} \end{bmatrix} \begin{bmatrix} 1 \\ b_1 \\ b_2 \\ \cdot \\ \cdot \\ b_{M-1} \end{bmatrix}. \qquad (17)$$

Using the same notation of (7),

$$\begin{bmatrix} \boldsymbol{a} \\ \hat{\boldsymbol{\varepsilon}} \end{bmatrix} = \begin{bmatrix} H_1 \\ H_2 \end{bmatrix} [\boldsymbol{b}]. \qquad (18)$$

The first problem considered is the following. Given the desired response $h(k)$ and the values for N and M, find $\boldsymbol{a}$ and $\boldsymbol{b}$ such that $\hat{\boldsymbol{\varepsilon}}^T\hat{\boldsymbol{\varepsilon}}$ is minimized. The $\hat{\varepsilon}_i$ terms can be viewed as the inner product of $\boldsymbol{b}$ with the ith row of H_2 and the desired $\boldsymbol{b}$ will be "as orthogonal as possible" to the rows of H_2. Using the same notation as in (8) gives

$$\hat{\boldsymbol{\varepsilon}} = H_3\boldsymbol{b}^* + \boldsymbol{h}^1. \qquad (19)$$

The $\boldsymbol{b}^*$ that minimizes $\hat{\boldsymbol{\varepsilon}}^T\hat{\boldsymbol{\varepsilon}}$ is the solution of the normal equations [6] of this problem which are

$$H_3^TH_3\boldsymbol{b}^* = -H_3^T\boldsymbol{h}^1. \qquad (20)$$

If the rank of H_3 (and therefore $H_3^TH_3$) equals $M-1$, then (20) has one and only one solution given by the so-called pseudo-inverse [6].

$$\boldsymbol{b}^* = -[H_3^TH_3]^{-1}H_3^T\boldsymbol{h}^1. \qquad (21)$$

If the rank of H_2 is M, then the rank of H_3 is $M-1$; no exact solution exists and the unique approximate solution is given by (21). If the rank of H_2 is $M-1$ and the rank of H_3 is also $M-1$, then a unique exact solution exists and it is also given by (21). If the rank of H_2 is $M-1$ and the rank of H_3 is $M-2$, then a unique exact solution exists but it requires $b_0=0$ and therefore cannot be found from (20) or (21). In this case and the case for the rank of H_2 less than $M-1$, the comments following Theorem 2 apply.

This procedure has resulted in a set of b_i coefficients that minimize the norm of the error defined in (18), and the procedure is independent of determination of the a_i.

After the b vector is found, the a vector is obtained by the multiplication indicated in the upper partition of (18).

The procedure has solved the original problem and required only straightforward linear operations. If the conditions of Theorems 1 or 2 are satisfied for a given $h(k)$ then this procedure yields the exact result and zero error. Although rather time consuming, it is often very informative to consider the norm of $\hat{\varepsilon}$ as a function of M and N. If there is an abrupt drop in the norm of $\hat{\varepsilon}$ as M and N are increased this gives information as to the nature of the signal or the source of that signal.

The Modified Procedure

To this point the only error considered is the $(K-N)\times 1$ vector $\hat{\varepsilon}$ defined in (18). To develop a different procedure it will be necessary to consider a more general definition of ϵ given by

$$\begin{bmatrix} a \\ 0 \end{bmatrix} + [\varepsilon] = Hb \tag{22}$$

where ϵ is now a $K\times 1$ vector allowing a more general operation than the upper partition of (7) allowed. It is also necessary to reformulate the matrix equation (22). This is

$$\begin{bmatrix} a \\ 0 \end{bmatrix} + \varepsilon = \begin{bmatrix} b_0 & 0 & 0 \cdots 0 \\ b_1 & b_0 & 0 & \\ \cdot & & & \cdot \\ \cdot & & & \cdot \\ b_{M-1} & & & \cdot \\ 0 & & & 0 \\ 0 & & \cdots & b_0 \end{bmatrix} \begin{bmatrix} h_0 \\ h_i \\ \cdot \\ \cdot \\ h_{K-1} \end{bmatrix} = Bh \tag{23}$$

where B is a $K\times K$ lower triangular matrix which is therefore nonsingular. Next, the measure of error mentioned earlier as the difference between the actual and desired output is defined:

$$e(k) = g(k) - h(k). \tag{24}$$

This error can be written as a $K\times 1$ vector e and becomes

$$e = g - h. \tag{25}$$

The filter equation in terms of the actual output g is

$$\begin{bmatrix} a \\ 0 \end{bmatrix} = Bg, \tag{26}$$

and in terms of the desired output h and the two measures of error is

$$\begin{bmatrix} a \\ 0 \end{bmatrix} + \varepsilon = Bh \quad \text{and} \quad \begin{bmatrix} a \\ 0 \end{bmatrix} = B[h + e]. \tag{27}$$

If b is found according to the method in (20), there are several methods for finding a. Since B is nonsingular, (27) becomes

$$h + e = \beta \begin{bmatrix} a \\ 0 \end{bmatrix}$$

where

$$\beta = B^{-1}. \tag{28}$$

Partitioning β gives

$$h + e = \left[\beta_1 \mid \beta_2 \right] \begin{bmatrix} a \\ 0 \end{bmatrix} = \beta_1 a + \beta_2 0 \tag{29}$$

where β_1 is $K\times N$ and β_2 is $K\times(K-N)$. Consider the following different requirements.

1) If it is desired to make the first N outputs exact, i.e., $g(k)=h(k)$ or $e(k)=0$ for $k=0, 1, \ldots N-1$, then the first N rows of (29) are solved for a. This is always possible since B, and therefore β, are lower triangular, and thus the square matrix formed from the first N rows of β_1 is also lower triangular and nonsingular. Examination of the operation shows this result is obtained by the much more straightforward method of multiplication in the upper partition of (18).

2) If it is desired to make any N outputs exact, then the corresponding N rows of β_1 are used to form a square matrix which, if nonsingular, is inverted to give a. In this case there is no guarantee of a solution.

3) If it is desired to choose a so that the norm of e is minimized, the resulting normal equations are

$$\beta_1^T \beta_1 a = \beta_1^T h. \tag{30}$$

The result of this procedure which first solves for b by minimizing the norm of the lower partition of ε called $\hat{\varepsilon}$, and then solves for a by using the remaining degrees of freedom to minimize the norm of e, is the same as that obtained by Shank's method [3].

4) A generalization which includes the above three cases can be obtained by minimizing the weighted norm of e, which is given by $e^T W e$ where W is a $K\times K$ positive definite or positive semidefinite real symmetric matrix. The resulting normal equations are

$$\beta_1^T W \beta_1 a = \beta_1^T W h. \tag{31}$$

If W is chosen with ones along the first N positions of the main diagonal and zeros elsewhere, then the solution of (31) is the same as that for case 1). If N ones are placed at the points corresponding to the desired exact output, then the solution to case 2 results if a solution exists. If W is the identity matrix, then it is the same problem as case 3.

5) A combination of cases 1 and 3 can be formulated by requiring that a certain number of terms L, less than N, in $e(k)$ be zero, and the $N-L$ remaining degrees of freedom be used to minimize the norm of e. This is a quadratic minimization problem with equality constraints which also results in linear equations to solve for a.

IV. Error Relations

Since the uncoupling of the calculations of a and b and the resulting linear equations depend on the measure of error ε that is defined in (18) and (22), it is important to relate this error to the more usual measure e as defined in (25). From (27) the relating equation is obtained:

$$\varepsilon = -Be \tag{32}$$

where B is the $K\times K$ lower triangular matrix defined in (23).

Several observations can be made from this relation. First, the norm of $\boldsymbol{\varepsilon}$ is zero if and only if the norm of $\boldsymbol{e}$ is zero. Second, if the first L terms of either error are zero, then the first L terms of the other error must also be zero. Notice the first procedure discussed necessarily has zero error in the first N output terms. A useful view is to consider a system with a finite convolution operator $b(k)$ that takes an input $e(k)$ and gives as its output $\epsilon(k)$. This point of view allows qualitative comparison of the errors for a specific problem.

V. Conclusions

This paper has studied the problem of designing a recursive digital filter which will have a desired impulse response $h(k)$.

Precise statements of filter realizability for a given number sequence $h(k)$ have been stated in the two theorems along with a straightforward design procedure that only involves solving linear equations. The matrix formulation allows interpretations that aid in the solution for the order of the numerator and denominator. Since the actual design of recursive filters involves some trial and error in choosing a structure, this ease of intuitive analysis is very useful.

The very general matrix formulation of the approximate realization problem presented results in several design procedures that only require the solution of linear equations. This formulation allows a precise statement of the criteria of optimality and ease in comparisons of the methods. The optimality criteria include combinations of exact interpolation and mean-square minimization. An error analysis has been made that gives an exact relation for the types of error considered.

Presently the authors are studying the application of the procedure to determining pole locations and to parameter identification.

References

[1] C. M. Rader and B. Gold, "Digital filter design in the frequency domain," *Proc. IEEE*, vol. 55, pp. 149–171, February 1967.

[2] F. F. Kuo and J. F. Kaiser, *System Analysis by Digital Computer*. New York: Wiley, 1966, ch. 7.

[3] J. L. Shanks, "Recursion filters for digital processing," *Geophysics*, vol. 32, pp. 33–51, February 1967.

[4] R. N. McDonough and W. H. Huggins, "Best least-squares representation of signals by exponentials," *IEEE Trans. Automatic Control*, vol. AC-13, pp. 408–412, August 1968.

[5] J. F. Claerbout and E. A. Robinson, "The error in least-squares inverse filtering," *Geophysics*, vol. 29, pp. 118–120, 1964.

[6] T. N. E. Greville, "The pseudo inverse of a rectangular or singular matrix and its application to the solution of systems of linear equations," *SIAM Rev.*. vol. 1, no. 1, January 1959.

Computer-Aided Design of Recursive Digital Filters

KENNETH STEIGLITZ, Member, IEEE
Department of Electrical Engineering
Princeton University
Princeton, N. J. 08540

Abstract

A practical method is described for designing recursive digital filters with arbitrary, prescribed magnitude characteristics. The method uses the Fletcher–Powell optimization algorithm to minimize a square-error criterion in the frequency domain. A strategy is described whereby stability and minimum-phase constraints are observed, while still using the unconstrained optimization algorithm. The cascade canonic form is used, so that the resultant filters can be realized accurately and simply. Design examples are given of low-pass, wide-band differentiator, linear discriminator, and vowel formant filters.

Manuscript received September 10, 1969.

This work was supported by the U.S. Army Research Office-Durham, under Contract DA HC04 69 C 0012.

I. Introduction

While the problem of choosing the coefficients of a nonrecursive digital filter to approximate a specified magnitude characteristic has been thoroughly explored, the corresponding problem for recursive digital filters remains open [1], [2]. Design procedures for recursive filters generally deal only with the piecewise constant case, and involve transformations of well known continuous-time filter designs, such as the Butterworth or Chebyshev. The purpose of this paper is to describe a practical method for choosing the coefficients of a recursive digital filter to meet arbitrary specifications of the magnitude characteristic.

The proposed method uses the optimization algorithm described by Fletcher and Powell [3] to minimize a square-error criterion in the frequency domain. This technique has been used to design continuous-time filters [4]. In order to deal with the realization problem in the continuous-time case, a network topology is usually fixed, and the optimization method must incorporate the constraints that the element values be nonnegative. These restrictions are not present for digital filters, since any coefficients can be used for realization. The resulting digital filter must be stable, however, and this imposes the constraint that the poles lie inside the unit circle in the z-plane. It will be shown how this constraint, and an additional minimum-phase constraint, can be observed, while still using the unconstrained minimization method of Fletcher and Powell.

II. Choice of Canonic Form

The first important question to be resolved is the choice of the canonic form of the digital filter. A general recursive filter can be assumed to have the transfer function

$$Y(z) = \frac{\sum_{k=1}^{K} a_k z^{-(k-1)}}{1 + \sum_{k=1}^{N} b_k z^{-k}} . \qquad (1)$$

This so-called direct form suffers from the following difficulties. First, if control is to be exercised over the pole locations, the denominator must be factored at certain stages in the optimization process. Second, the pole locations may be extremely sensitive functions of the coefficients b_k for high-order filters [1]. This means that the b_k must be found to very high precision, and that the error surface may be badly skewed. The cascade form

$$Y(z) = A \prod_{k=1}^{K} \frac{1 + a_k z^{-1} + b_k z^{-2}}{1 + c_k z^{-1} + d_k z^{-2}} \qquad (2)$$

avoids these difficulties, and has the additional advantage of yielding the realization shown in Fig. 1, which is known to be practical for high-order filters. This form also has the advantage of making the zeros easy to find, a feature not shared by a third possibility, the parallel form. For

Reprinted from *IEEE Trans. Audio Electroacoust.*, vol. AU-18, pp. 123–129, June 1970.

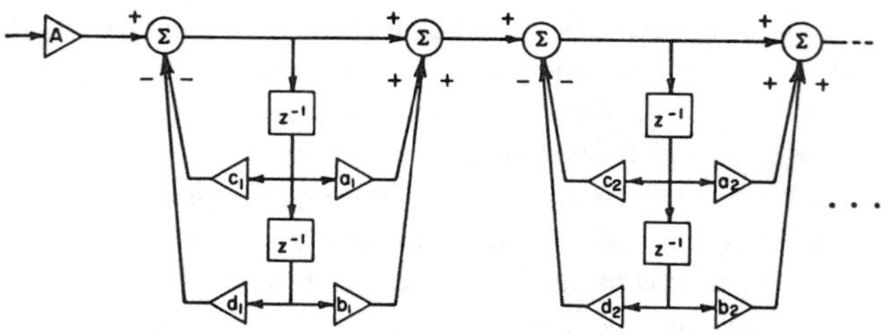

Fig. 1. Cascade realization corresponding to the canonic form.

these reaons the cascade canonic form will be assumed in what follows.

III. Statement of the Problem

Suppose now that the desired magnitude characteristic is prescribed at the discrete set of frequencies $W_1, \cdots, W_M$ where W_i is given in fractions of the Nyquist rate. These correspond to values of the variable z

$$z_i = e^{jW_i\pi} \qquad i = 1, \cdots, M. \tag{3}$$

Call the desired magnitude at these frequencies Y_i^d. Then the square-error in the frequency domain is

$$Q(\theta) = \sum_{i=1}^{M} (| Y(z_i) | - Y_i^d)^2 \tag{4}$$

where θ is the $(4K+1)$-vector of unknown coefficients

$$\theta = (a_1, b_1, c_1, d_1, a_2, b_2, c_2, d_2, \cdots, A)'. \tag{5}$$

The problem is to find a value of θ, say θ^*, such that for all θ

$$Q(\theta^*) \leq Q(\theta). \tag{6}$$

This square-error is a nonlinear function of the parameter vector θ, and an iterative method must be used to accomplish its minimization. Such numerical methods as are available seek a relative (local) minimum from a given starting point, and cannot in general be relied upon to find the global solution. Computational experience, gained by using different starting points for the same problem, often gives some indication of the likelihood that a given local solution is in fact global. In addition, a suboptimal value of θ can often provide a useful design.

IV. Elimination of *A* and Calculation of the Gradient

The method of Fletcher and Powell appears to be the most efficient and powerful nonlinear optimization method now available [3], [4]. It need not be described here, except to say that it performs a one-dimensional minimization at each cycle, along a direction determined by the gradient and an updated estimate of the Hessian. The double precision FORTRAN IV program DFMFP, supplied by IBM in the scientific subroutine package [5], was used without change. The Fletcher–Powell method requires the computation of the gradient of Q with respect to the parameter vector. This computation was performed using double precision complex arithmetic in FORTRAN IV.

The error function Q can be minimized analytically with respect to A for fixed a_i, b_i, c_i, d_i; and A need not be considered an unknown parameter. To eliminate A from Q, define the $4K$-dimensional parameter vector

$$\phi = (a_1, b_1, c_1, d_1, a_2, b_2, c_2, d_2, \cdots, d_K)' \tag{7}$$

and write

$$\begin{aligned} Y(z, A, \phi) &= A \prod_{k=1}^{K} \frac{1 + a_k z^{-1} + b_k z^{-2}}{1 + c_k z^{-1} + d_k z^{-2}} \\ &= AH(z, \phi). \end{aligned} \tag{8}$$

Then

$$Q(A, \phi) = \sum_{i=1}^{M} (| AH(z_i, \phi) | - Y_i^d)^2. \tag{9}$$

Differentiating with respect to $|A|$ and setting the result to zero yields the following optimum value of $|A|$, say $|A^*|$:

$$|A^*| = \frac{\sum_{i=1}^{M} | H(z_i, \phi) | Y_i^d}{\sum_{i=1}^{M} | H(z_i, \phi) |^2}. \tag{10}$$

The Fletcher–Powell method is then used to minimize the new error criterion

$$\hat{Q}(\phi) = Q(A^*, \phi). \tag{11}$$

Notice that the sign of A^* is immaterial, since it does not affect the magnitude characteristic. It will be taken positive. The gradient of $\hat{Q}$ with respect to ϕ can be computed as follows:

$$\frac{\partial \hat{Q}}{\partial \phi_n} = \frac{\partial Q(A^*, \phi)}{\partial \phi_n} + \frac{\partial Q(A^*, \phi)}{\partial A^*} \frac{\partial A^*}{\partial \phi_n} \qquad n = 1, \cdots, 4K. \tag{12}$$

The second term is zero, since A^* is chosen to minimize Q. Hence by (9),

$$\frac{\partial \hat{Q}}{\partial \phi_n} = 2A^* \sum_{i=1}^{M} (A^* | H(z_i, \phi) | - Y_i^d) \frac{\partial | H(z_i, \phi) |}{\partial \phi_n}. \tag{13}$$

Writing

$$| H(z_i, \phi) | = [H(z_i, \phi) \overline{H(z_i, \phi)}]^{1/2} \tag{14}$$

we have

$$\frac{\partial |H(z_i, \phi)|}{\partial \phi_n} = \frac{1}{|H(z_i, \phi)|} \cdot \mathrm{Re}\left\{\overline{H(z_i, \phi)} \frac{\partial H(z_i, \phi)}{\partial \phi_n}\right\} \tag{15}$$

which can be computed directly from (8) using complex arithmetic.

The subroutine which calculates $\hat{Q}(\phi)$ and grad $\hat{Q}(\phi)$, given ϕ, is summarized below.

1) Calculate

$$H_i = \prod_{k=1}^{K} \frac{1 + a_k z_i^{-1} + b_k z_i^{-2}}{1 + c_k z_i^{-1} + d_k z_i^{-2}}, \quad i = 1, \cdots, M. \tag{16}$$

2) Calculate

$$A^* = \frac{\sum_{i=1}^{M} |H_i| Y_i^d}{\sum_{i=1}^{M} |H_i|^2}. \tag{17}$$

3) Calculate

$$E_i = A^*|H_i| - Y_i^d, \quad i = 1, \cdots, M. \tag{18}$$

4) Calculate

$$\hat{Q} = \sum_{i=1}^{M} E_i^2. \tag{19}$$

5) Calculate

$$\begin{aligned} \frac{\partial |H_i|}{\partial a_k} &= \frac{1}{|H_i|} \mathrm{Re}\left\{\overline{H_i} \frac{\partial H_i}{\partial a_k}\right\} \\ &= \frac{1}{|H_i|} \mathrm{Re}\left\{\overline{H_i} H_i \frac{z_i^{-1}}{1 + a_k z_i^{-1} + b_k z_i^{-2}}\right\} \\ &= |H_i| \mathrm{Re}\left\{\frac{z_i^{-1}}{1 + a_k z_i^{-1} + b_k z_i^{-2}}\right\} \\ & k = 1, \cdots, K; \quad i = 1, \cdots, M \end{aligned} \tag{20}$$

and similarly,

$$\frac{\partial |H_i|}{\partial b_k} = |H_i| \mathrm{Re}\left\{\frac{z_i^{-2}}{1 + a_k z_i^{-1} + b_k z_i^{-2}}\right\} \tag{21}$$

$$\frac{\partial |H_i|}{\partial c_k} = -|H_i| \mathrm{Re}\left\{\frac{z_i^{-1}}{1 + c_k z_i^{-1} + d_k z_i^{-2}}\right\} \tag{22}$$

$$\frac{\partial |H_i|}{\partial d_k} = -|H_i| \mathrm{Re}\left\{\frac{z_i^{-2}}{1 + c_k z_i^{-1} + d_k z_i^{-2}}\right\}. \tag{23}$$

6) Calculate

$$\frac{\partial \hat{Q}}{\partial \phi_n} = 2A^* \sum_{i=1}^{M} E_i \frac{\partial |H_i|}{\partial \phi_n}, \quad n = 1, \cdots, 4K. \tag{24}$$

The elimination of A as an unknown parameter reduces by one the dimensionality of the search performed by the optimization program. An additional savings in computation time is achieved by computing the z_i once at the beginning of execution and storing them for later use.

V. Stability and Minimum-Phase Constraints

Suppose $Y(z)$ has a real pole at $z=\alpha$. Replacing this by a pole at $z=1/\alpha$ is equivalent to multiplying by the function

$$\frac{z - \alpha}{z - 1/\alpha} \tag{25}$$

which has magnitude $|\alpha|$ when z is on the unit circle. Hence the inversion of a real pole with respect to the unit circle does not affect the shape of the magnitude characteristic. Since the gain constant A is chosen optimally, $\hat{Q}(\phi)$ is not affected at all by such inversion. Similarly, $\hat{Q}(\phi)$ is unaffected by inversions of complex pairs of poles, or real or complex pairs of zeros. At convergence of the optimization program, poles and zeros will appear randomly inside or outside the unit circle, depending on the starting point of the iteration, and upon the course of the iteration itself. It will be taken as a design criterion that all the poles and zeros lie within the unit circle. The poles must do so in order that the filter be stable. The zeros lying inside the unit circle ensure that there is no excess phase.

VI. Final Strategy and Example 1: A Low-Pass Filter

At first thought, it would appear that the following procedure would yield an optimum transfer function with all its poles and zeros inside the unit circle.

1) Use of the optimization program to minimize $\hat{Q}(\phi)$ without constraining pole or zero locations.
2) At convergence, invert all poles or zeros outside the unit circle.

If the optimization program is started anew from the result of step 2, however, it is found that further reduction in $\hat{Q}(\phi)$ is sometimes possible. The following example will show how this can happen. Consider the specification of an ideal low-pass filter with cutoff frequency at one-tenth the Nyquist frequency:

$$\begin{aligned} &W = 0.00, 0.09\ (0.01); && Y^d = 1.0 \\ &W = 0.10; && Y^d = 0.5 \\ &W = 0.11, 0.20\ (0.01); && Y^d = 0.0 \\ &W = 0.2, \quad 1.0\ \ (0.1); && Y^d = 0.0. \end{aligned} \tag{26}$$

This specification weights frequencies below $W=0.2$ more heavily than those above. If the optimization for a one-section filter ($K=1$) is started at

$$\phi = (0., 0., 0., -0.25)' \tag{27}$$

convergence is obtained after 93 function evaluations. The resulting zeros and poles are given below and are plotted in Fig. 2(A):

$$\begin{aligned} &\text{zeros: } 0.67834430 \pm j\ 0.73474418 \\ &\text{poles: } 0.75677793, \qquad 1.3213916. \end{aligned} \tag{28}$$

The corresponding value of the error criterion is

$$\hat{Q} = 1.2611. \tag{29}$$

Notice that the two poles are very nearly inverses of each other. After inversion, 62 more function evaluations are required to produce convergence[1] to the following parameters (see Fig. 2(B)):

$$\begin{aligned} &\text{zeros: } 0.82191163 \ \pm j\ 0.56961501 \\ &\text{poles: } 0.89676390 \ \pm j\ 0.19181084 \\ &A = 0.11733978 \\ &\hat{Q} = 0.56731. \end{aligned} \tag{30}$$

The introduction of a complex pole pair is prevented in the first sequence of iterations by the fact that one pole is inside the unit circle and one is outside. After inversion, the two poles can split and become a complex pair, leading to the final minimum. The magnitude characteristic of the final filter is shown in Fig. 3.

The following algorithm was used to allow such convergence to take place.

1) Use the Fletcher–Powell optimization program until convergence takes place, or for a maximum of 25 cycles, and go to 2.
2) If any poles or zeros are outside the unit circle, invert them and go to 1. Otherwise, go to 3.
3) If convergence has not taken place, go to 1. Otherwise, go to 4.
4) Print out the final parameters and stop.

Fig. 3 also shows the resultant filter characteristic for $K=2$, corresponding to the following parameters at convergence:

$$\begin{aligned} &\text{zeros: } 0.92538461 \ \pm j\ 0.37902945 \\ &\qquad\quad 0.61137175 \ \pm j\ 0.79134343 \\ &\text{poles: } 0.93121838 \ \pm j\ 0.27718988 \\ &\qquad\quad 0.86454727 \ \pm j\ 0.13353860 \\ &A = 0.024867372 \\ &\hat{Q} = 0.033959. \end{aligned} \tag{31}$$

The final parameters obtained from the $K=1$ design were used as a starting point for the $K=2$ optimization, and 376 further function evaluations were required for convergence.

[1] The convergence criterion for this and all succeeding examples corresponds to the parameter EPS $= 10^{-5}$ in DFMFP.

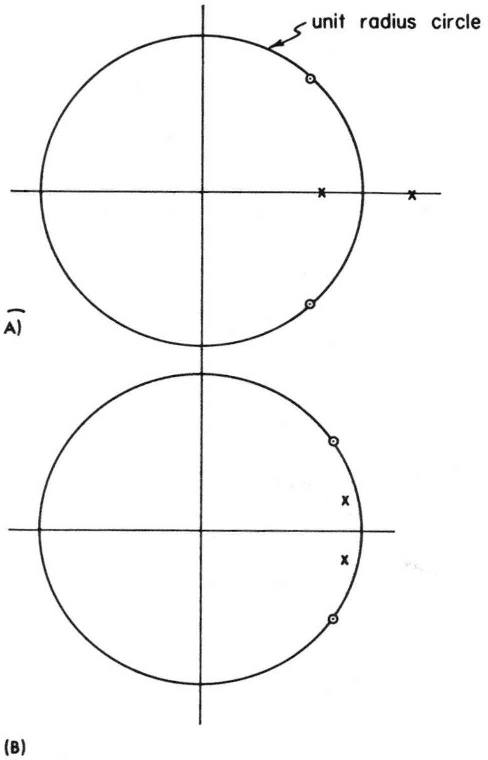

Fig. 2. Pole-zero configurations for the low-pass filter designs. (A) Intermediate local minimum. (B) Final minimum.

Fig. 3. Magnitude characteristic of the one- and two-section low-pass filters of Example 1.

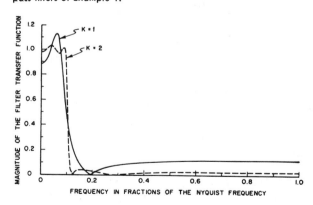

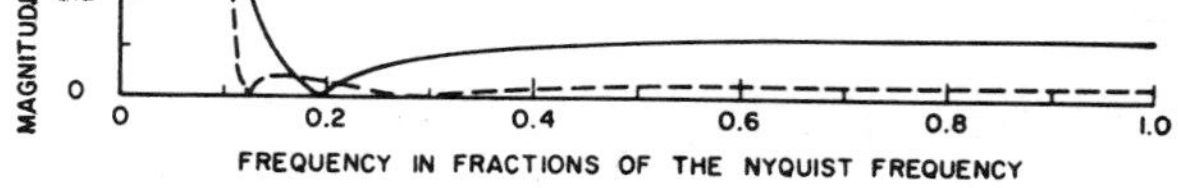

VII. Examples

Example 2: A Wide-Band Differentiator

Consider the following specification,

$$W = 0.0,\ 1.0\ (0.05); \qquad Y^d = W, \tag{32}$$

which represents a linear amplitude characteristic, and hence an ideal differentiating filter, ignoring consideration of the phase for the moment. The one-section design

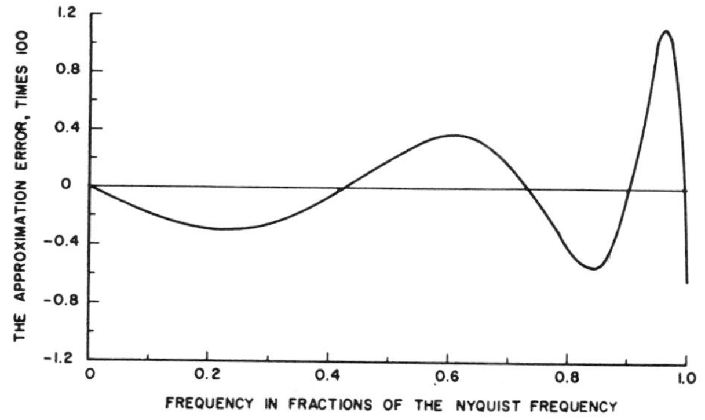

Fig. 4. The approximation error for the one-section wide-band differentiator filter.

Fig. 5. Phase characteristic of the one-section wide-band differentiator filter.

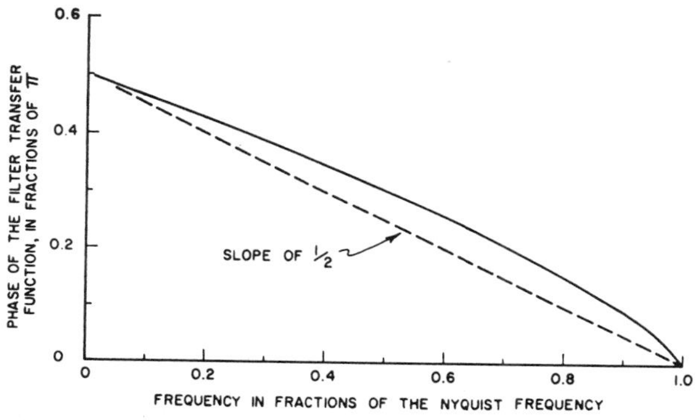

converged after 96 function evaluations to the following design:

$$\begin{aligned} \text{zeros:}\ & 1.0000000, \quad -0.67082621 \\ \text{poles:}\ & -0.14240300, \quad -0.71698670 \\ A &= 0.36637364 \\ \hat{Q} &= 2.7480 \times 10^{-4}. \end{aligned} \tag{33}$$

Fig. 4 shows the approximation error over the entire band of frequencies from zero to the Nyquist frequency. Of particular interest is the fact that the approximation is within about 1 percent of maximum over this entire range, in contrast with designs based on guard-band filters, which usually are accurate only up to about 80 percent of the Nyquist frequency (see [1]). Fig. 5 shows the phase characteristic, which approximates the phase of an ideal differentiator with an additional lag of one-half sampling period. Thus, this design introduces significantly less lag than designs reported in [1].

Starting with the one-section design above, 500 more function evaluations produced convergence to a more accurate two-section approximation, with $\hat{Q}=6.15\times 10^{-7}$. This two-section filter exhibited a similar characteristic ripple near the Nyquist frequency, and had almost the same phase characteristic. This points out the desirability of extending the method to include specifications on the phase characteristic.

Example 3: A Linear Discriminator

For the next example, consider the specification

$$W = 0.0,\ 1.0\ (0.05); \qquad Y^d = |1 - 2W| \tag{34}$$

which represents a linear discriminator with a zero at one-half the Nyquist frequency. After 40 function evaluations, the following one-section design was produced:

$$\begin{aligned} \text{zeros:}\ & 0.00000000 \pm j\ 1.00000000 \\ \text{poles:}\ & \pm 0.49614741 \\ A &= 0.35765018 \\ \hat{Q} &= 1.2299 \times 10^{-2}. \end{aligned} \tag{35}$$

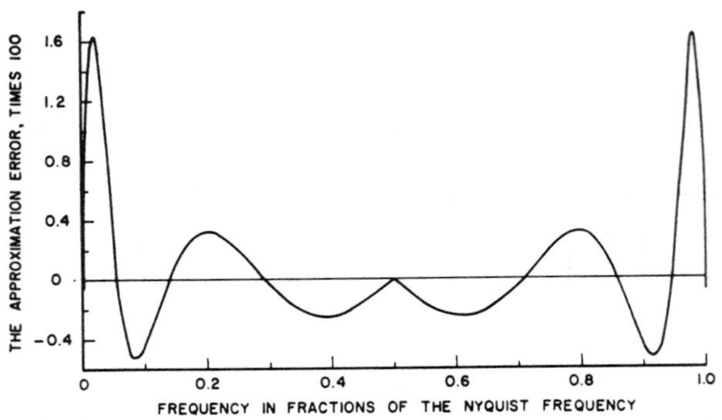

Fig. 6. The approximation error for the two-section linear discriminator filter.

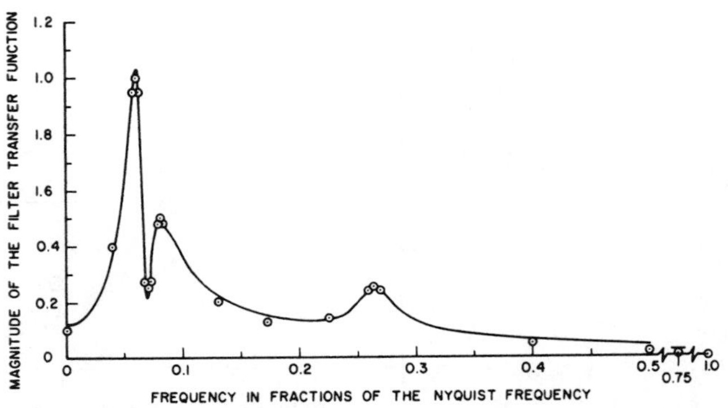

Fig. 8. Magnitude characteristic of the three-section vowel formant filter Circles indicate specification points.

Fig. 7. Phase characteristic of the two-section linear discriminator filter.

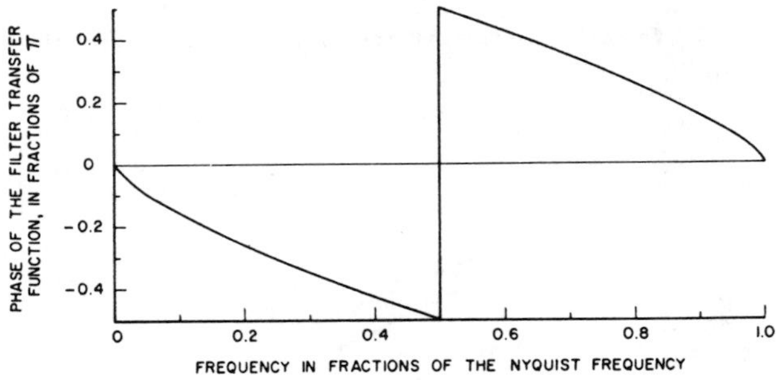

One-hundred and thirty more function evaluations produced convergence to the following two-section design:

$$\begin{aligned}
\text{zeros:}\ & 0.00000004 \ \pm j\ 0.99999931 \\
& 0.81492900, \ -\ 0.81492888 \\
\text{poles:}\ & 0.84492845, \ -\ 0.84492830 \\
& 0.37204922, \ -\ 0.37204934 \\
& A = 0.36676649 \\
& \hat{Q} = 1.0807 \times 10^{-4}.
\end{aligned} \tag{36}$$

As might be expected, the resulting pole-zero patterns are symmetric with respect to the imaginary axis, within the precision allowed by the convergence criterion. Figs. 6 and 7 show the approximation error and phase characteristic of the two-section filter.

Example 4: A Vowel Formant Filter

Fig. 8 shows the specification of a filter which is to have a magnitude characteristic corresponding to the formant for the vowel ɔ (as in "law") [6]. The principal requirements are taken to be that peaks occur at $W = 0.06$, 0.08, and 0.26; with values of 1.0, 0.5, and 0.25, respectively; and that troughs occur midway between these peaks, with values one-half the lower peak. This design problem is considerably more difficult than the previous ones, since it involves approximating a rather arbitrary and complex shape. An acceptably good design required three sections, 1809 function evaluations and 2 minutes 34 seconds of computation time on the IBM 360/65. The final design is also shown in Fig. 8 and corresponds to the following parameters:

$$\begin{aligned}
\text{zeros:}\ & 0.93470084, \ -\ 0.99966051 \\
& 0.62177384 \ \pm j\ 0.62464737 \\
& 0.97101465 \ \pm j\ 0.21383044 \\
\text{poles:}\ & 0.64343825 \ \pm j\ 0.70167849 \\
& 0.96361229 \ \pm j\ 0.19280318 \\
& 0.93515982 \ \pm j\ 0.20909432 \\
& A = 0.041075206 \\
& \hat{Q} = 9.4712 \times 10^{-3}.
\end{aligned} \tag{37}$$

VIII. Conclusions

A practical method has been described for designing recursive digital filters with arbitrary, prescribed magnitude characteristics. Examples have been given of such designs with 1, 2, and 3 cascade sections, corresponding to 5, 9, and 13 parameters. The most difficult of these designs takes about 2.5 minutes of computation time on the IBM 360/65 computer.

The important considerations in the development of this method have been: 1) a strategy for ensuring that the resulting filters are stable and minimum phase, 2) elimination of the gain factor A as an unknown parameter, 3) choice of the canonic form of the filter, and 4) choice of the unconstrained optimization algorithm.

Further work along these lines might take into account specifications on the phase characteristic, and arbitrary weighting of the errors at different specification points.

Acknowledgment

The author is indebted to the following for valuable discussions about this problem: Dr. G. C. Winham, Department of Music, Princeton University; C. M. Rader, Lincoln Laboratories; and Dr. J. F. Kaiser, Bell Telephone Laboratories.

References

[1] J. F. Kaiser, "Digital filters," in *System Analysis by Digital Filter*, F. F. Kuo and J. F. Kaiser, Eds. New York: Wiley, 1966, ch. 7.

[2] C. M. Rader and B. Gold, "Digital filter design techniques in the frequency domain," *Proc. IEEE*, vol. 55, pp. 149–171, February 1967.

[3] R. Fletcher and M. J. D. Powell, "A rapidly convergent descent method for minimization," *Computer J.*, vol. 6, no. 2, pp. 163–168, 1963.

[4] G. C. Temes and D. A. Calahan, "Computer-aided network design—the state of the art," *Proc. IEEE*, vol. 55, pp. 1832–1863, November 1967.

[5] "System/360 scientific subroutine package (360A-CM-03X), version III, programmer's manual," IBM Data Processing Division, White Plains, N. Y., Document H20-0205-3, 1968.

[6] R. K. Potter and J. C. Steinberg, "Toward the specification of speech," *J. Acoust. Soc. Am.*, vol. 22, pp. 807–820, November 1950.

Techniques for Designing Finite-Duration Impulse-Response Digital Filters

LAWRENCE R. RABINER, MEMBER, IEEE

Abstract—Several new techniques for designing finite-duration impulse-response digital filters have become available in the past few years. The motivation behind three design techniques that have been proposed are reviewed here, and the resulting designs are compared with respect to filter characteristics, ease of design, and methods of realization. The design techniques to be discussed include window, frequency-sampling, and equiripple designs.

I. Introduction

Interest in design techniques for finite-duration impulse-response (FIR) digital filters has been renewed in the past few years because of the application of powerful optimization algorithms to the design problem. Although closed-form solutions to the approximation problem cannot, in general, be obtained explicitly, iterative techniques can be made to yield optimum solutions. Two optimization techniques have been proposed recently [1]–[4] which, along with the classical window design method [5], [6], provide the user with several possibilities for approximating filters with arbitrary frequency-response characteristics. In this paper we will discuss the general theory behind windowing and two optimization techniques—frequency-sampling and equiripple designs—and then compare these methods with respect to several practical and theoretical criteria.

II. Terminology

Before discussing the design issues, it is important to distinguish the various types of digital filters that can be designed and to separate the way in which a filter is realized from the filter characteristics themselves. The following terms will be used throughout this paper.

1) *Finite-duration impulse-response (FIR)*: This term means that the duration of the filter impulse response h_n is finite; i.e.,

$$h_n = 0, \qquad n > N_1 < \infty$$
$$h_n = 0, \qquad n < N_2 > -\infty \tag{1}$$

and

$$N_1 > N_2.$$

2) *Infinite-duration impulse-response (IIR)*: This term means that the duration of the filter impulse response h_n is infinite; i.e., there exists no finite values of either N_1 or N_2 such that (1) is satisfied.

3) *Recursive realization*: This term describes the way a filter (either IIR or FIR) is realized. It means that the current filter output y_n is obtained explicitly in terms of past filter outputs $y_{n-1},\cdots$ as well as in terms of past and present filter inputs $x_n,x_{n-1},\cdots$. Thus the output of a recursive realization can be written as

$$y_n = F(y_{n-1},y_{n-2},\cdots,x_n,x_{n-1},\cdots). \tag{2}$$

4) *Nonrecursive realization*: This term means that the current filter output y_n is obtained explicitly in terms of only past and present inputs; i.e., previous outputs are not used to generate the current output. The representation on a nonrecursive realization can be written as

$$y_n = F(x_n,x_{n-1},\cdots). \tag{3}$$

The motivation behind this terminology is that it has been shown [7], [8] that FIR filters as well as IIR filters can be realized both nonrecursively and recursively. (It should be noted that, in general, recursive realizations of IIR filters and nonrecursive realizations of FIR filters are most efficient and are usually used.) Thus a term describing filter characteristics should be distinct from a term describing how the filter is realized. This is not how the terminology has traditionally been used.

III. Some Advantages of FIR Filters

Although our aim is to describe and compare design techniques for FIR filters, it is of interest to first discuss some reasons why FIR filters are of importance. These include the following.

1) FIR filters can easily be designed to approximate a prescribed magnitude frequency response to arbitrary accuracy with an exactly linear phase characteristic. In addition, FIR filters can approximate arbitrary frequency characteristics (both magnitude and phase), but IIR filters can also do this.

2) FIR filters can be realized efficiently both nonrecursively (using direct convolution, or high-speed convolution by using the fast Fourier transform (FFT) [9], [10]) and recursively (using a comb filter and a bank of resonators [11]).

3) An FIR filter realized nonrecursively is always stable. FIR filters realized nonrecursively contain only zeros in the finite z plane, and hence are always stable.

4) Quantization and roundoff problems inherent in recursive realizations of IIR filters are generally negligible in nonrecursive realizations of FIR filters.

Paper approved by the Data Communication Systems Committee of the IEEE Communication Technology Group for publication without oral presentation. Manuscript received October 2, 1970.
The author is with Bell Telephone Laboratories, Murray Hill, N. J. 07974.

Reprinted from *IEEE Trans. Commun. Technol.*, vol. COM-19, pp. 188–195, Apr. 1971.

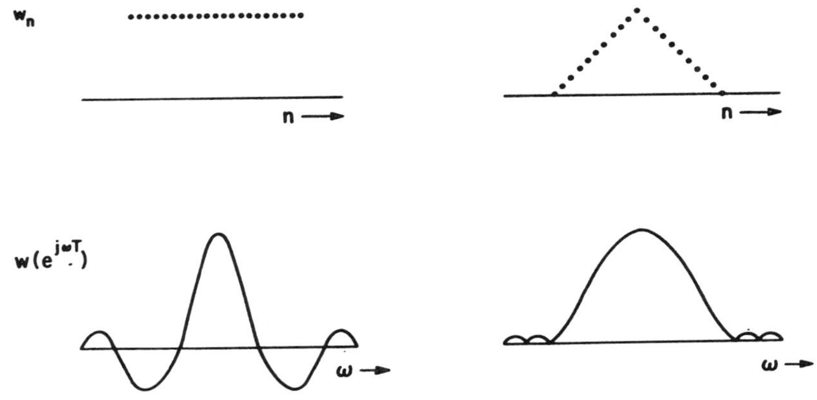

Fig. 1. Two examples of windows and their Fourier transforms.

5) The coefficient accuracy problems inherent in sharp cutoff IIR filters can often be made less severe for realizations of equally sharp FIR filters.

IV. Theory of Design

Since one of the most important reasons for designing FIR filters is that they can be designed with an exactly linear phase, we will restrict our discussion to this type. The general characteristics of the frequency response of a digital filter whose impulse-response coefficients are real are

$$H[\exp(j\omega T)] = H\left[\exp\left[j\left(\omega + n\frac{2\pi}{T}\right)T\right]\right], \quad n = 0, \pm 1, \pm 2, \cdots \quad (4)$$

$$|H[\exp(j\omega T)]| = |H[\exp(-j\omega T)]|, \quad 0 \le \omega \le \pi \quad (5)$$

$$\theta(\omega) = -\theta(-\omega), \quad 0 \le \omega \le \pi \quad (6)$$

where

$$H[\exp(j\omega T)] = |H[\exp(j\omega T)]| \exp[j\theta(\omega)] \quad (7)$$

where $H[\exp(j\omega T)]$ is the frequency response of the filter and T is the sampling period. Equation (4) shows the periodicity of sampled-data systems in frequency. Equations (5) and (6) show the symmetry of the magnitude function and asymmetry of the phase function for filters with real impulse responses. In the remainder of this section we will discuss three techniques for approximating the desired magnitude response characteristics, assuming linear phase, with an FIR filter.

A. Windowing

Since the frequency response of a digital system is periodic, it can be expanded in a Fourier series. The coefficients of this Fourier series are the filter impulse-response coefficients. Generally, there are an infinite number of nonzero Fourier-series coefficients. To obtain an FIR filter which approximates the original frequency response, the Fourier series must be truncated. Direct truncation of the series, however, leads to the well-known Gibbs phenomenon, i.e., a fixed percentage overshoot and ripple before and after an approximated discontinuity, making this technique unsatisfactory for approximating many standard types of filters. In order to control the convergence of the Fourier series a weighting function is used to modify the Fourier coefficients. This time-limited weighting function is called a window. Since the multiplication of Fourier coefficients by a window corresponds to convolving the original frequency response with the Fourier transform of the window, a design criterion for windows is to find a finite window whose Fourier transform has relatively small sidelobes. Fig. 1 shows two choices of windows and their respective frequency responses. The rectangular window at the top left corresponds to direct truncation of the Fourier series. The Fourier transform of the rectangular window, shown at the lower left, has a narrow center lobe, but has sidelobes which contain a large part of the total energy and which decay quite slowly. Another window, the triangular window, is shown in the upper right of Fig. 1. Its Fourier transform, shown at the lower right, has a main lobe twice the width of the rectangular window main lobe, but has much less energy in the sidelobes.

The search for windows to meet the criterion previously mentioned, i.e., a finite window with most of its energy in the main lobe of its Fourier transform, has led to several useful and in some sense optimum designs. The Hamming window, which is of the form:

$$w_n = 0.54 + 0.46 \cos\left(\frac{2\pi n}{N}\right), \quad \frac{-N}{2} \le n \le \frac{N}{2} \quad (8)$$

has 99.96 percent of its energy in its main lobe, with the peak amplitude of the sidelobes down more than 40 dB from the peak. The width of the main lobe is twice the width of the rectangular window's main lobe. The Blackman window, which is

$$w_n = 0.42 + 0.5 \cos\left(\frac{2\pi n}{N}\right) + 0.08 \cos\left(\frac{4\pi n}{N}\right), \quad \frac{-N}{2} \le n \le \frac{N}{2} \quad (9)$$

further reduces peak sidelobe ripple to less than 0.0001 of the main-lobe peak at the expense of a main lobe whose width is triple the width of the rectangular window main lobe. Optimum window designs have been proposed by Kaiser [5] and Helms [6]. The Kaiser window is an approximation to the prolate spheroidal wave functions whose band-limiting properties are well known [12]. By adjusting a parameter of the window, a tradeoff can be obtained between peak sidelobe ripple and the width of the main lobe. The main disadvantage of these windows is that one must compute Bessel functions to get the window coefficients.[1] Helms [6] has proposed the Dolph–Chebyshev window, which is optimum in the sense that the main-lobe width is as small as possible for a given peak ripple. The main disadvantage of this window is that inverse hyperbolic cosines must be evaluated to determine the window coefficients.

One disadvantage of the window design technique is that one must be able to compute Fourier-series coefficients for the periodic frequency response being approximated. Generally it is not trivial to determine closed-form expressions for these coefficients. The solution to this problem is found by approximately obtaining the Fourier-series coefficients as the discrete Fourier transform (DFT) of a sampled version of the continuous frequency response. By sampling the frequency response at a number of frequencies M much larger than the number of Fourier-series coefficients under the window N, one can obtain fairly good approximations to the first N Fourier coefficients. Fig. 2 presents a summary of the window design procedure. Fig. 2(a) shows the desired frequency response. Fig. 2(b) shows the same frequency response sampled at M equispaced frequencies, as well as a continuous interpolation in frequency. Through the use of the DFT formula [Fig. 2(c)] the M-point impulse response h_n is obtained and is shown below the formula. An N-point window w_n [Fig. 2(d)] with the Fourier transform $W(e^{j\omega T})$ [Fig. 2(e)] weights the impulse response to yield the N-point sequence a_n [Fig. 2(f)]. The sequence a_n is the filter impulse response and its Fourier transform [Fig. 2(g)] shows the final approximation to the desired response.

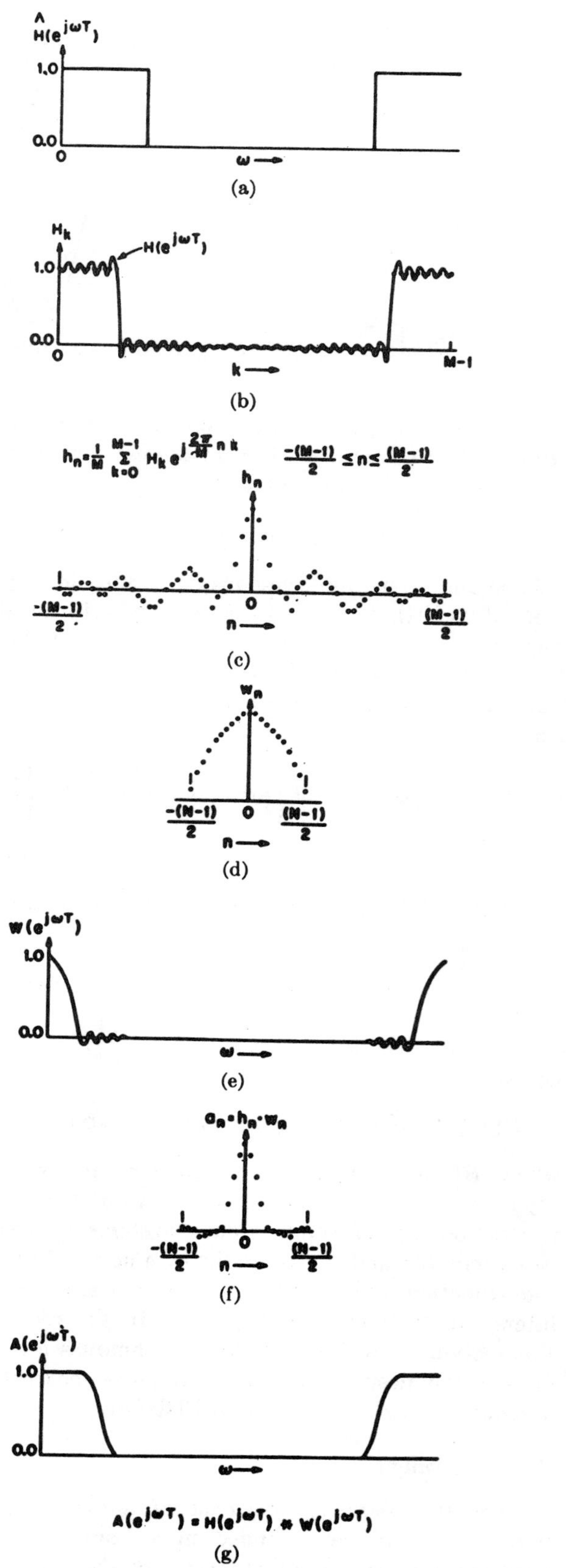

Fig. 2. Step by step realization of windowing. (a) Desired frequency response. (b) Sampled frequency response, $N \ll M$. (c) Impulse response obtained using DFT. (d) Typical window function. (e) Window frequency response. (f) Windowed impulse response. (g) Frequency response corresponding to windowed impulse response.

Frequency Sampling

A second technique for approximating a filter with given frequency-response specifications is to sample the desired frequency response at N equispaced frequencies, where N is the number of samples in the filter impulse response. By setting these frequency samples to be the DFT coefficients of the filter impulse response, one can derive an approximation to any desired continuous frequency response. For many types of filters, such as low-pass, bandpass, and high-pass filters, one can optimize the values of the frequency samples in transition bands to

[1] Kaiser has a simple 12-line Fortran 4 program which computes a power-series expansion (up to 25 terms) of the Bessel function $I_0(x)$.

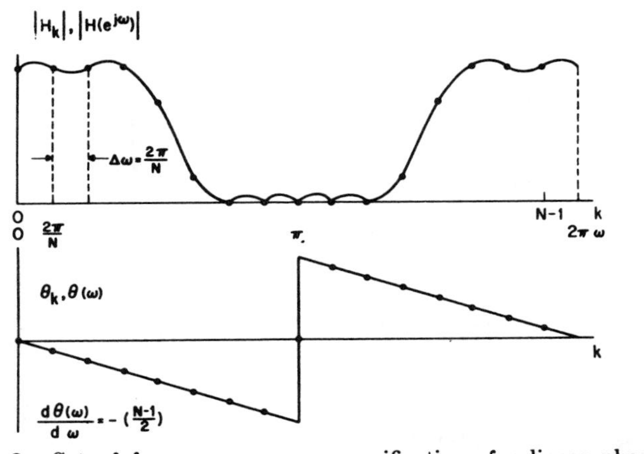

Fig. 3. Set of frequency-response specifications for linear phase filter. Solid curves show continuous frequency response and heavy dots show frequency samples.

optimize the filter design. The resulting filter designs will be shown to be quite efficient.

In Fig. 3 frequency-response specifications for a desired filter are shown (the solid curves) along with the frequency samples of this filter (the heavy points). The frequency samples are defined by the relation

$$H_k = |H_k| \exp(j\theta_k) = H(z)|_{z=\exp[j(2\pi/N)k]}, \quad k = 0,1,\cdots,N-1 \quad (10)$$

where $H(z)$ is the z transform of the FIR filter. By using the inverse discrete Fourier transform (IDFT), we can determine the filter impulse-response coefficients h_n in terms of the frequency samples as

$$h_n = \frac{1}{N}\sum_{k=0}^{N-1} H_k \exp\left(j\frac{2\pi}{N}nk\right), \quad n = 0,1,\cdots,N-1. \quad (11)$$

The z-transform relation gives

$$H(z) = \sum_{n=0}^{N-1} h_n z^{-n} = \frac{(1-z^{-N})}{N}\sum_{k=0}^{N-1}\frac{H_k}{1-z^{-1}\exp[j(2\pi/N)k]}. \quad (12)$$

Evaluation of (12) on the unit circle gives the continuous interpolation formula

$$H[\exp(j\omega T)] = \frac{\exp[-j(\omega NT/2)(1-1/N)]}{N} \cdot \sum_{k=0}^{N-1}\frac{H_k \exp[-j(\pi k/N)]\sin(\omega NT/2)}{\sin(\omega T/2 - \pi k/N)} \quad (13)$$

which can easily be evaluated using the FFT algorithm.

The basic ideas behind frequency sampling can be seen in (13). The filter frequency response is seen to be linearly related to the frequency samples, and thus linear optimization techniques can be used to optimally select values of several, or all, the frequency samples to give the best approximation to the desired filter. Furthermore, the interpolated frequency response is seen to consist of elementary functions of the form:

$$\frac{\sin(\omega NT/2)}{\sin(\omega T/2 - \theta)} \quad (14)$$

which, in the design of standard filters such as a low-pass filter, provide good ripple cancellation. Thus, by allowing variable frequency samples in a transition band between the in- and out-of-band regions, one can choose the frequency samples in this band to provide optimum ripple cancellation for either the out-of-band region, inband region, or a combination of the two. As the number of samples in the transition band increases, ever finer ripple cancellation is possible. Fig. 4 shows an example of a low-pass filter with an impulse-response duration of 256 samples, and 3 variable frequency samples in the transition band. Thirty-two inband frequency samples were set to 1.0, and the out-of-band samples set to 0.0. An optimization program determined the transition samples to minimize peak out-of-band ripple. As seen in Fig. 4(a), a peak out-of-band ripple of −88 dB was obtained. An expanded view of the frequency response of the filter is shown in Fig. 4(b).

Fig. 5 shows a frequency-sampling design for an ideal full-band differentiator. In this figure the filter impulse response (of 256-sample duration), the magnitude response, and the magnitude error are plotted. The peak magnitude error is less than 0.1 percent (the average error is considerably less than this) and there is no phase error at all.

It should be noted that the mathematical solution of the optimization problem is straightforward because of the linearity of the frequency response with respect to the unconstrained variables.

C. Equiripple Designs

A third technique for designing FIR filters solves a system of nonlinear equations to generate a filter with an equiripple approximation error [3], [4], [14]. In this method the unknown quantities are both the $(N+1)/2$ coefficients in the impulse response (assuming N odd, and a symmetrical impulse response) and a set of $(N-3)/2$ frequencies at which extrema of the approximation error occur. By writing constraint equations on the extrema and on the derivatives, a system of $(N-1)$ nonlinear equations in $(N-1)$ unknowns can be obtained. Standard nonlinear optimization techniques are used to solve these equations.

For simplicity, we assume the filter impulse response h_n, is symmetric and exists from $n = -(N-1)/2$ to $n = (N-1)/2$, where N is odd. Thus the filter frequency response can be written as

$$H[\exp(j\omega T)] = \sum_{n=-(N-1)/2}^{(N-1)/2} h_n \exp(-j\omega nT) = h_0 + \sum_{n=1}^{(N-1)/2} 2h_n \cos(\omega nT). \quad (15)$$

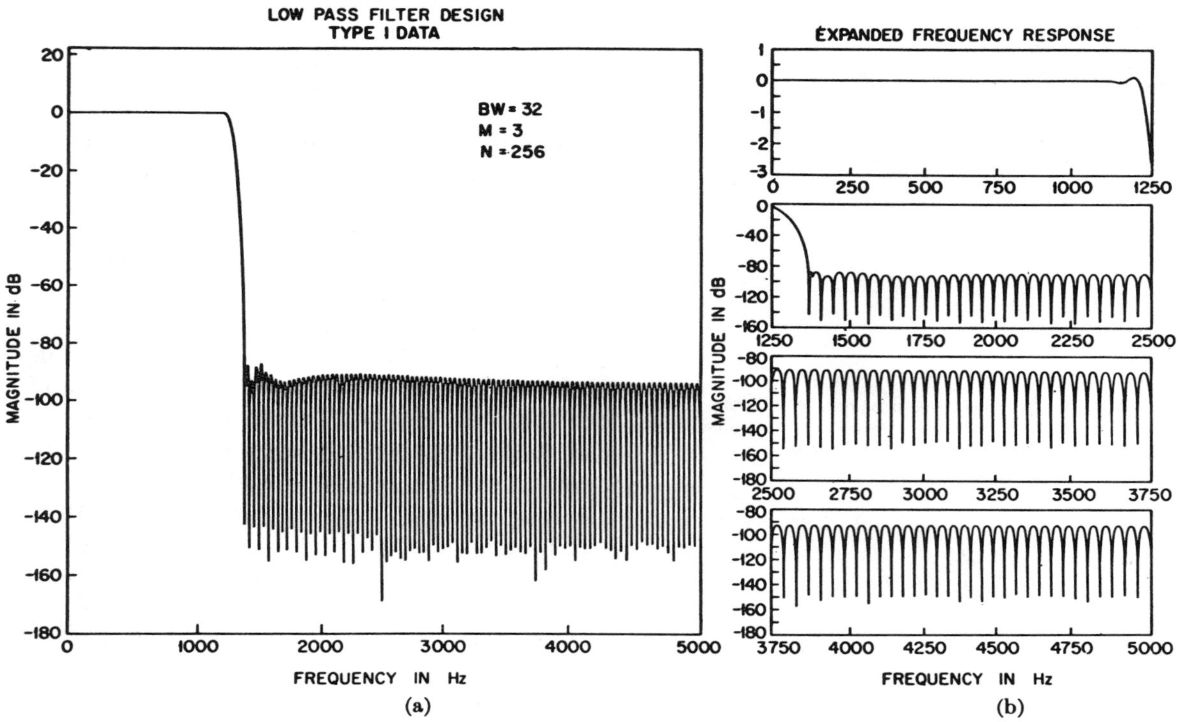

Fig. 4. (a) Low-pass filter with three transition samples and impulse-response duration of 256 samples, designed by frequency-sampling techniques. (b) Expanded view of filter frequency response.

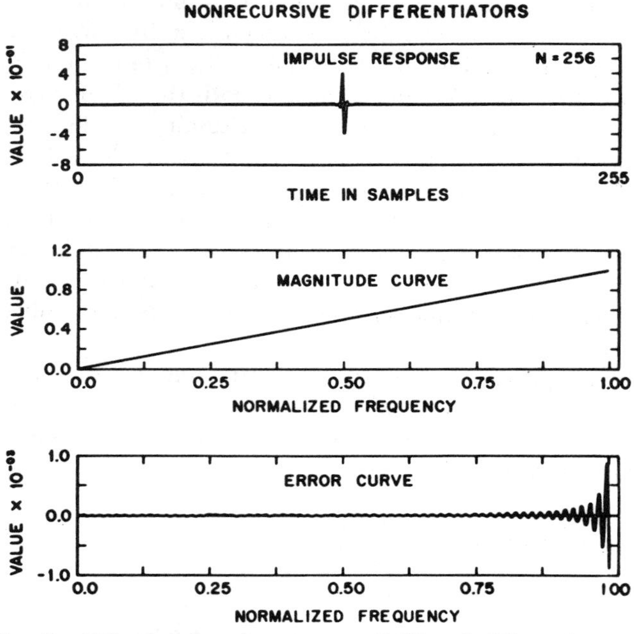

Fig. 5. 256-point impulse response full-band differentiator designed by frequency-sampling technique.

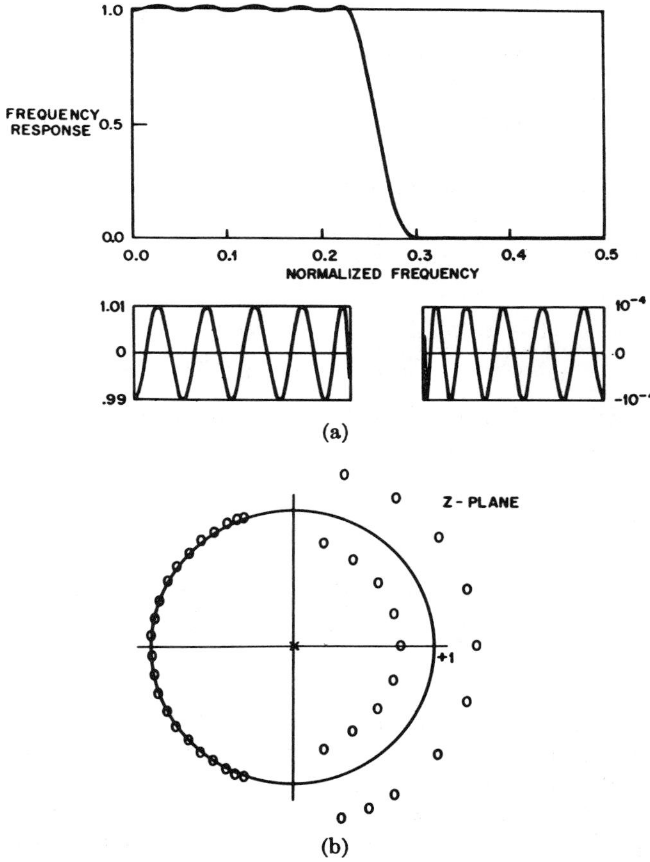

Fig. 6. (a) 41-point equiripple low-pass filter. (b) Plot of z-plane positions of zeros of filter.

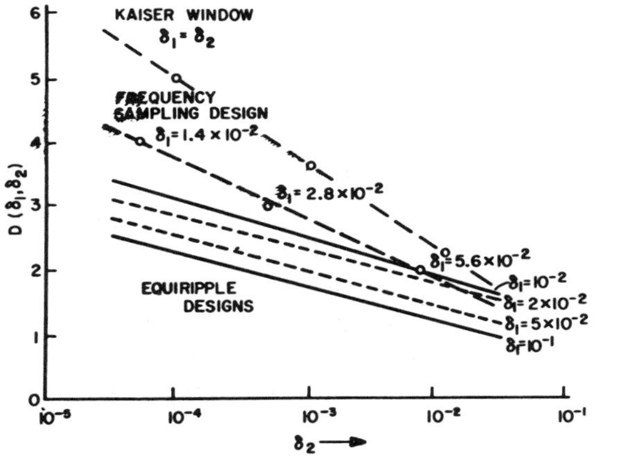

Fig. 7. Comparison of filter designs in terms of normalized width of transition band as function of out-of-band ripple δ_2.

The derivative of the frequency response can be evaluated as

$$H'[\exp(j\omega T)] = -\sum_{n=1}^{(N-1)/2} 2h_n \cdot nT \sin(\omega nT). \quad (16)$$

At each extremum of the approximation error, an equation can be written relating the frequency response to the desired value and the allowed error, i.e.,

$$H[\exp(j\omega_l T)] = H_d[\exp(j\omega_l T)] \pm \delta_l \quad (17)$$

where $H_d[\exp(j\omega T)]$ is the desired value at the frequency ω_l, δ_l is the allowable approximation error, and ω_l is an unknown frequency. A second equation can be written because the value of the frequency response at frequency ω_l is an extremum of the error, i.e.,

$$H'[\exp(j\omega_l T)] = 0. \quad (18)$$

A series of equations of the form of (17) and (18) can be solved using nonlinear optimization techniques to give the filter coefficients and the frequencies of the extrema. Fig. 6 shows typical results for the design of a low-pass filter with equiripple error both inband and out-of-band. The impulse-response duration is 41 samples long. The inband ripple in this case is 0.01, and the out-of-band ripple is 10^{-4}, or -80 dB. Fig. 6(b) shows the z-plane positions of the zeros of this filter. The out-of-band zeros are all on the unit circle, and the inband zeros lie in quadrouplets. For each zero at $z = r\exp(j\theta)$, there are corresponding zeros at $z = r\exp(-j\theta)$, and $z = (1/r)\exp(\pm j\theta)$. This geometric symmetry of the zeros, when they are off the unit circle, accounts for the linear phase response which is obtained.

V. Comparison of Design Techniques

There are many theoretical and practical ways of comparing different filter-design techniques. In this section we will compare the three methods discussed in terms of 1) the transition bandwidth for a standard low-pass filter, 2) the ease of design of new filters, and 3) the methods of realization.

A. Transition Bandwidth for Low-Pass Filters

One of the most basic ways of comparing low-pass filters is to compare the width of the transition band for different values of peak out-of-band and inband ripple. Fig. 7 shows a comparison of this type for the Kaiser window, frequency sampling, and equiripple designs. In this figure, the normalized width of the transition band as a function of the out-of-band ripple is plotted. The normalized width of the transition band is defined as follows[2]:

$$D(\delta_1,\delta_2) = NT[F_H(\delta_1,\delta_2) - F_L(\delta_1,\delta_2)] \quad (19)$$

where

$F_H(\delta_1,\delta_2)$	upper cutoff frequency in Hz
$F_L(\delta_1,\delta_2)$	lower cutoff frequency in Hz
δ_1	inband peak ripple
δ_2	out-of-band peak ripple.

In this figure the inband ripple δ_1 is shown as an additional parameter of the curves. The upper dashed curve shows the results for the Kaiser window, which, because of its time-limiting and band-limiting properties, tends to have the smallest normalized bandwidth. For an out-of-band ripple of -80 dB ($\delta_2 = 10^{-4}$) the normalized bandwidth is 5.01, whereas for a -40 dB ripple ($\delta_2 = 10^{-2}$) the normalized bandwidth is about 2.22. For these data points the inband peak ripple δ_1 is equal to δ_2.[3]

The middle dashed curve shows the results for frequency-sampling designs. For an out-of-band ripple of -86 dB, the normalized width of the transition band is 4, or about $\frac{3}{4}$ that of the Kaiser window. For values of δ_2 of -66 and -46 dB, the transition width decreases to 3 and 2 for these designs. The design tradeoff between the Kaiser window and equiripple designs is the larger values of δ_1 for the latter case versus the larger value of transition width for the former case.

The lowest set of curves show the results for equiripple designs for various values of δ_1. Since equiripple designs are optimum, i.e., they have the smallest width of transition band for fixed δ_1, δ_2, the data for this case falls below the data for the other two cases. The percentage difference in normalized bandwidth between equiripple designs and frequency-sampling designs, for fixed δ_1, δ_2, is only about 30 percent. Since the suboptimal frequency-sampling designs are not much less efficient than the optimal equiripple designs, there may be circumstances in which it is preferable to use the suboptimal design. We shall discuss such cases in the next section.

B. Ease of Design

An important issue to any filter designer or user is how easy it is to design a new filter to meet some particular

[2] The multiplication by NT in (19) makes $D(\delta_1,\delta_2)$ a dimensionless quantity, independent of NT.

[3] Low-pass filter designs, using the Kaiser window, have transition bandwidths that are within about 10 percent of the transition bandwidth of optimum equiripple filters.

specifications. For example, a user may desire an approximation to a frequency response which is not one for which the coefficients have already been precataloged. The following issues then arise.

1) What types of filters have been cataloged and can be referenced readily?

2) How easily can the design technique be applied to arbitrary magnitude and phase specifications?

3) Can filters with long impulse responses, i.e., large values of N, be designed to obtain sufficiently sharp transition ratios to meet the most stringent of design specifications?

In this section we shall try to answer these questions with respect to each of the three design techniques.

Since the window design technique is so straightforward to apply there is no need to catalog a large set of reference designs, and hence this has not been done. The window technique is readily amenable to approximating any set of filter specifications with no limitations on length of impulse response. The problems inherent in the window technique are the necessity to have Fourier-series coefficients of the desired periodic frequency response and the computation required for using optimum windows such as the Kaiser window or the Dolph–Chebychev window. The first problem is solved by the approximate procedure outlined in Fig. 2. Programs for computing Bessel functions for the Kaiser window, or inverse hyperbolic cosines for the Dolph–Chebychev window, are necessary. These are generally available as library subroutines on most computers. Thus the window technique tends to be relatively easy to use in the general design case.

In order to use the frequency-sampling technique, the user must program the linear optimization procedure, or use some available optimization program and adapt it to his specific needs. Once this has been accomplished, there is no problem designing filters with arbitrary magnitude and phase characteristics, or with any length impulse response. The linear nature of the problem guarantees convergence of the mathematical algorithms. For several standard filters there exists an extensive catalog of frequency-sampling designs [2], [13]. The types of filters included are low-pass and bandpass filters and wide-band (up to full-band) differentiators. From this catalog one could apply simple frequency transformations to obtain band-stop or high-pass filters [15].

In order to use the equiripple design techniques, the user must program the nonlinear optimization procedure, or use some available routine and again adapt it to his needs. The procedure can only design equiripple approximations to the magnitude response (assuming linear phase). It may be possible to modify it for arbitrary phase but this has not been done. The optimization technique used by Herrmann and Schuessler [3], [4] was capable of obtaining solutions only for small values of N (on the order of 40). Recent work by Herrmann [16] and Hofstetter *et al.* [14] have solved the mathematical problems and filters can be designed with large values of N. (A small catalog of equiripple designs for low-pass filters is available from Herrmann.[4])

C. Realization of FIR Filters

Any FIR filter can be realized nonrecursively by direct convolution or fast convolution, or recursively using a cascade of a comb filter and a bank of parallel resonators. To use a nonrecursive realization requires the coefficients of the filter impulse response h_n. For direct convolution the output y_n is determined explicitly from the inputs x_n as

$$y_n = \sum_{m=0}^{N-1} h_m x_{n-m}. \tag{20}$$

For fast convolution a block of output samples is obtained from a block of input samples by Fourier transforming the input, multiplying the transform by the Fourier transform of the impulse response, and inverse Fourier transforming the product. Details of this technique are explained in [9], [10].

The way in which an FIR filter may be realized recursively is seen from (11). Instead of the impulse-response coefficients, the DFT of the impulse response, or, as we have called them, the frequency samples, can be used to realize the filter as a cascade of a comb filter $(1 - z^{-N})$ and a parallel bank of complex resonators. The significance of this realization is that for frequency-sampling designs, in many cases several, if not the majority, of the frequency samples would be 0.0. Hence this realization can be much more efficient than nonrecursive realizations. For filters designed by the other techniques, all the DFT coefficients will be nonzero, in general, and this realization will be much less efficient. Furthermore, for frequency-sampling designs where the majority of the frequency samples are 0.0 or 1.0, the effects of quantization are much less severe in the recursive realization than in the nonrecursive realization.

VI. Conclusion

As in the design of IIR digital filters, there are now several techniques available for designing FIR filters. The choice of technique depends heavily on the decision whether to compromise accuracy of approximation, ease of design, or a method of realization with a fixed quantization accuracy.

Acknowledgment

The author is grateful for the insights he has gained into filter-design problems from discussions of this material with Dr. R. W. Schafer, Dr. J. F. Kaiser, Dr. B. Gold, Dr. H. Helms, Prof. A. V. Oppenheim, Prof. O. Herrmann, and Prof. H. W. Schuessler.

[4] O. Herrmann, Institut für Nachrichtentechnik, University of Erlangen–Nurenberg, 8520 Erlangen, Germany.

REFERENCES

[1] B. Gold and K. L. Jordan, "A direct search procedure for designing finite duration impulse response filters," *IEEE Trans. Audio Electroacoust.*, vol. AU-17, Mar. 1969, pp. 33–36.

[2] L. R. Rabiner, B. Gold, and C. A. McGonegal, "An approach to the approximation problem for nonrecursive digital filters," *IEEE Trans. Audio Electroacoust.*, vol. AU-18, June 1970, pp. 83–106.

[3] O. Herrmann and H. W. Schuessler, "On the design of selective nonrecursive digital filters," presented at the IEEE Arden House Workshop, Harriman, N. Y., Jan. 1970.

[4] O. Herrmann, "On the design on nonrecursive digital filters with linear phase," *Electron. Lett.*, 1970.

[5] J. F. Kaiser, "Digital filters," in *System Analysis by Digital Computer*, F. F. Kuo and J. F. Kaiser, Eds. New York: Wiley, 1966.

[6] H. D. Helms, "Nonrecursive digital filters: design methods for achieving specifications on frequency response," *IEEE Trans. Audio Electroacoust.*, vol. AU-16, Sept. 1968, pp. 336–342.

[7] B. Gold and K. L. Jordan, Jr., "A note on digital filter synthesis," *Proc. IEEE* (Lett.), vol. 56, Oct. 1968, pp. 1717–1718.

[8] H. B. Voelcker and E. E. Hartquist, "Digital filtering via block recursion," *IEEE Trans. Audio Electroacoust.*, vol. AU-18, June 1970, pp. 169–176.

[9] T. G. Stockham, Jr., "High-speed convolution and correlation with applications to digital filtering," in *Digital Processing of Signals*, B. Gold and C. M. Rader, Eds. New York: McGraw-Hill, 1969, ch. 7.

[10] H. D. Helms, "Fast Fourier transform method of computing difference equations and simulating filters," *IEEE Trans. Audio Electroacoust.*, vol. AU-15, June 1967, pp. 85–90.

[11] B. Gold and C. M. Rader, *Digital Processing of Signals*. New York: McGraw-Hill, 1969, pp. 78–86.

[12] D. Slepian and H. O. Pollak, "Prolate spheroidal wave functions, Fourier analysis and uncertainty—I and II," *Bell Syst. Tech. J.*, vol. 40, 1961, pp. 43–84.

[13] L. R. Rabiner and K. Steiglitz, "The design of wide-band recursive and nonrecursive digital differentiators," *IEEE Trans. Audio Electroacoust.*, vol. AU-18, June 1970, pp. 204–209.

[14] E. Hoffstetter, A. V. Oppenheim, and J. Siegel, paper on design techniques for equiripple filters, presented at Princeton Conf., Mar. 1971.

[15] A. G. Constantinides, "Spectral transformations for digital filters," *Proc. Inst. Elec. Eng.*, vol. 117, Aug. 1970, pp. 1585–1590.

[16] O. Herrmann, personal communication.

An Approach to the Approximation Problem for Nonrecursive Digital Filters

LAWRENCE R. RABINER, Member, IEEE
Bell Telephone Laboratories, Inc.
Murray Hill, N. J. 07974

BERNARD GOLD, Senior Member, IEEE
Lincoln Laboratory
Massachusetts Institute of Technology
Lexington, Mass.

C. A. McGONEGAL
Bell Telephone Laboratories, Inc.
Murray Hill, N. J. 07974

Abstract

A direct design procedure for nonrecursive digital filters, based primarily on the frequency-response characteristic of the desired filters, is presented. An optimization technique is used to minimize the maximum deviation of the synthesized filter from the ideal filter over some frequence range. Using this *frequency-sampling* technique, a wide variety of low-pass and bandpass filters have been designed, as well as several wide-band differentiators. Some experimental results on truncation of the filter coefficients are also presented. A brief discussion of the technique of nonuniform sampling is also included.

Manuscript received January 19, 1970.

Introduction

Nonrecursive digital filters have finite-duration impulse response and consequently contain no poles (only zeros) in the finite z-plane. The approximation problem is that of finding suitable approximations to various idealized filter transfer functions. A designer may be interested in approximating either the magnitude, or the phase, or both magnitude and phase of this ideal filter. A few examples of typical ideal filters are shown in Fig. 1(A) through (F). Fig. 1(A) shows an ideal low-pass filter while Fig. 1(B) through (D) show ideal high-pass, bandpass, and band-elimination filters. Fig. 1(E) shows the response of an ideal differentiator while Fig. 1(F) shows the phase response of an ideal Hilbert transformer which allows the two outputs to be in phase quadrature.

The approximation problem for recursive digital filters (having infinite-duration impulse response, and poles as well as zeros) has been treated extensively [1], [2]. Mathematically, in the recursive case the realizable approximation can be expressed as the ratio of two trigonometric polynomials, leading to filter designs based on classical analog filter theory. This leads, for example, to fairly sophisticated design techniques for Butterworth, Chebyshev, and elliptic filters to yield good magnitude response approximations. For nonrecursive digital filters the realizable approximations are trigonometric polynomials. Thus, the class of approximations is more constrained. The most widely used approach towards approximating the frequency domain filter characteristic is based on approximating the infinite-duration impulse response of the ideal filters by the finite-duration impulse response of the nonrecursive realization. The most significant result in this connection is the Gibbs phenomenon, illustrated in Fig. 2, which shows the resultant frequency response obtained when the "ideal" (infinite) impulse response corresponding to Fig. 1(A) is symmetrically truncated. As is well known, the amount of error or "overshoot" in the vicinity of the discontinuity does not diminish, even as the response is increased in duration. Recognition of this fact has prompted workers in the field to seek ways to decrease the ripple by decreasing the severity of the discontinuity. This can be accomplished by introducing a time-limited window function $w(n)$ having a z-transform $W(z)$. From the complex convolution theorem the z-transform of the product $h(n)\,w(n)$ is given by

$$F(z) = \frac{1}{2\pi j} \oint H(z/v) W(v) v^{-1} dv \qquad (1)$$

where $h(n)$ is the ideal impulse response and $H(z)$ is its z-transform. Thus, multiplying $h(n)$ by a window corresponds to smoothing the spectrum. Careful choice of a window can result in a frequency-response function with appreciably less in-band and out-of-band ripple, as can be seen by comparing Figs. 2 and 3.

Kaiser [1] has introduced a set of windows (which we

Reprinted from *IEEE Trans. Audio Electroacoust.*, vol. AU-18, pp. 83–106, June 1970.

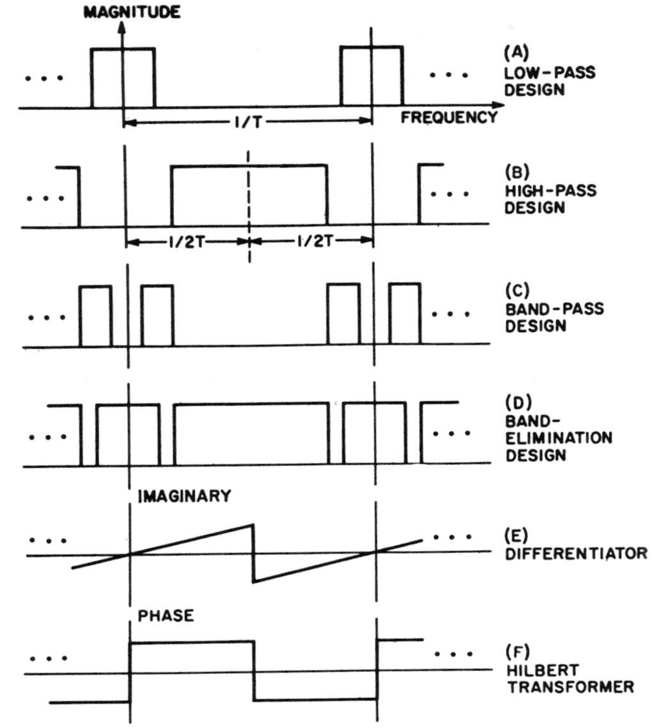

Fig. 1. Examples of typical ideal filters.

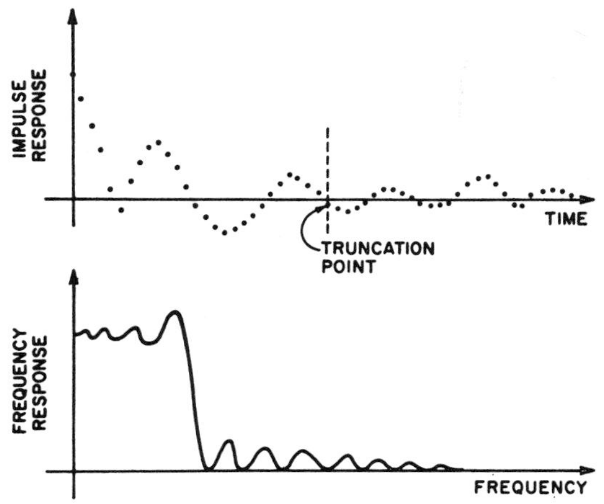

Fig. 2. An example of the effects of truncating the impulse response on the frequency response.

Fig. 3. Reduction of overshoot in the frequency response by windowing the truncated impulse response.

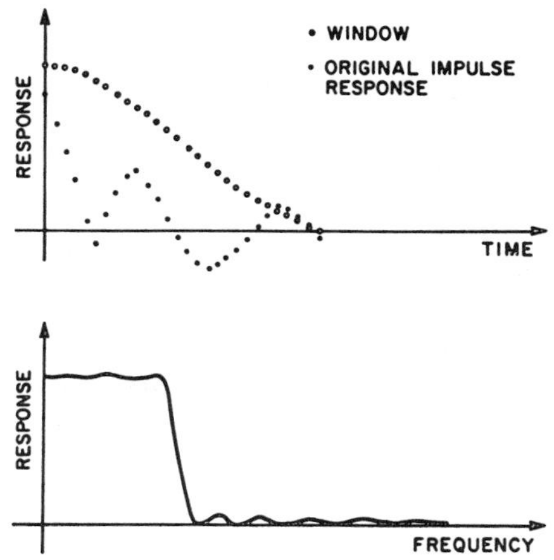

shall call Kaiser windows) which are very close to optimum. By adjusting a parameter of the window, the sidelobes can be diminished at the cost of increased transition bandwidth. Helms [3] recently proposed the Dolph–Chebyshev window because it has good spectral properties and because its parameters can be readily determined directly.

Window functions have also found great use in spectral analysis of random functions, but this subject will not be specifically discussed in this paper.

Design of nonrecursive filters from frequency-response specifications has been considered by Martin [4], who specified initial values of the frequency response at selected frequencies, leaving unspecified values of the frequency response in preselected transition bands. He then used a minimization procedure to solve for final values of the frequency response at equally spaced frequencies. The criterion used for the minimization was that the maximum deviation of the continuous frequency reponse from the ideal frequency response be minimized for both in-band and out-of-band frequencies. Martin obtained useful results for small values of N (the number of impulse-response samples) for the case of low-pass filters and for wide-band differentiators.

A recent paper by Gold and Jordan [5] introduced a somewhat different approach to the approximation problem for nonrecursive digital filters. In this approach the frequency response is specified exactly at N equispaced frequencies. If it is assumed that the number of frequency samples is equal to the number of samples in the impulse response, then the (continuous) frequency response is exactly determined. A simple example is shown in Fig. 4, where an ideal rectangular low-pass filter is sampled at equally spaced frequencies, resulting in a continuous frequency response with overshoot. (The transition band in Fig. 4 is the frequency range between the last in-band sample and the first out-of-band sample.) The impulse response corresponding to this frequency sampled filter is now no longer truncated, but rather *aliased* or *folded.* This fact should be well noted, as it serves to delineate sharply between this method (the sampling method) and the window method. It is not clear to us whether truncation or aliasing of an infinite impulse response is an intrinsically better procedure; however, this theoretical distinction makes it awkward to formulate the sampling method in terms of the window method. Our reasons for the rather extensive study of the sampling method to be presented in this paper are the following.

1) The designer trying to design filters to approximate a given ideal shape in the frequency domain need never

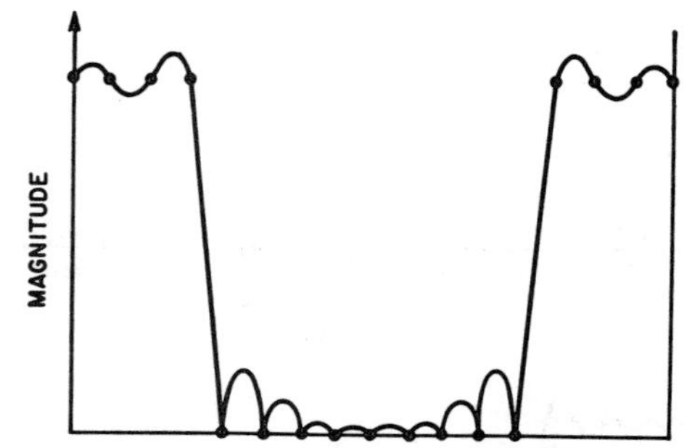

Fig. 4. The continuous frequency response of a filter derived from its frequency samples.

concern himself with an impulse response. This is intuitively appealing for filters with sufficiently long impulse responses (i.e., greater than about 30 samples), so that high-speed convolution using the fast Fourier transform is used for synthesis since the design results can be applied directly to yield the synthesis.

2) The sampling procedure is capable of being exploited to yield an "optimum" filter. As discussed above, the window technique results in a tradeoff between overshoot and transition bandwidth. By contrast, in the sampling technique, once the designer has chosen a transition bandwidth, he can, in a practical sense, calculate the best filter that will have such a transition bandwidth. As will be seen, this leads to quite efficient designs.

In the Gold and Jordan paper [5], results were obtained only for a few low-pass filters. Also, the computer optimization technique was semiautomatic, requiring an online interactive display oscilloscope. The computations needed for optimization were fairly lengthy and somewhat inaccurate.

In the present paper, the design method and optimization are treated more generally, and described in detail. The procedure has been fully automated and made computationally efficient. As a result, it has been possible to generate extensive design data, applicable in many cases to "cookbook" design. An analysis of low-pass and bandpass filters, as well as of wide-band differentiators, is presented. Numerical comparisons are made between the window and sampling methods for low-pass filters and differentiators. The effects of finite register length are discussed and a few results presented. Finally, the theory for a nonuniform sampling procedure is presented and a few numerical results are given.

Synthesis Techniques for Nonrecursive Filters

Before presenting the formalism of our design technique, it is worth discussing the filter synthesis question heuristically. We know of three useful ways of synthesizing a nonrecursive digital filter.

1) Direct Convolution: The impulse response of the filter is explicitly found and the filter is realized via the computation

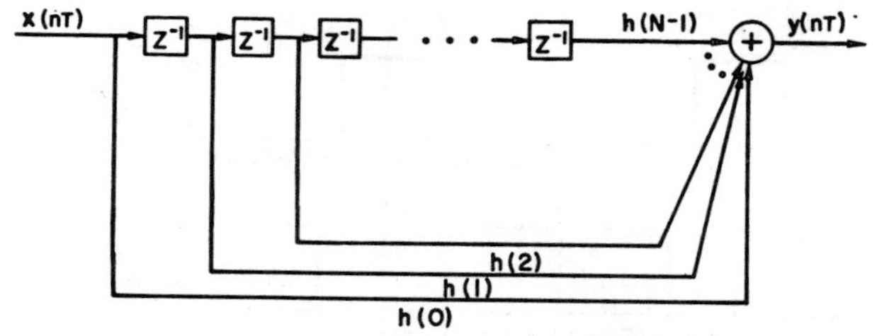

Fig. 5. Direct convolution realization of nonrecursive digital filter.

Fig. 6. Frequency sampling realization of nonrecursive digital filter.

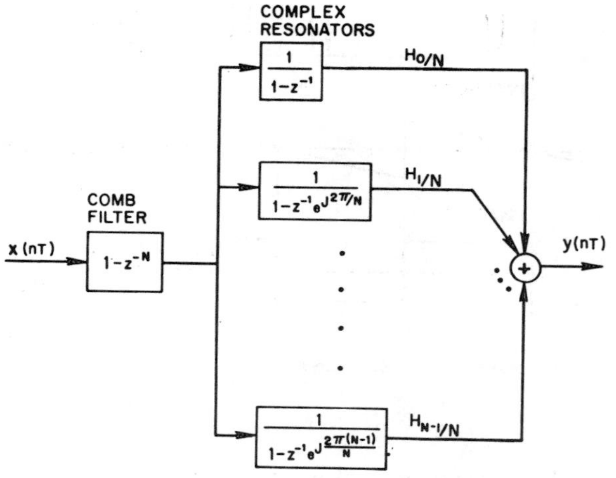

$$y(n) = \sum_{m=0}^{N-1} h(m)x(n-m) \tag{2}$$

where $h(m)$ represents the filter impulse response, $x(n)$ is the input sequence, and $y(n)$ is the output sequence. The realization of (2) is shown in Fig. 5. The limits in (2) imply that $h(m)$ is of duration N, so that $h(m)=0$ for $m \geq N$.

2) Fast Convolution: Here only values of the frequency response of the filter need to be explicitly found. First the discrete Fourier transform of $x(n)$ (suitably augmented with zero-valued time samples) is computed, then multiplied by samples of the filter frequency response, and then the product is inverse transformed to yield the output.

3) Frequency Sampling: Here the sampling theorem is specifically realized as a digital network [8]. As seen in Fig. 6, this network consists of a comb filter in cascade with a set of parallel complex exponential resonators, the outputs of which are suitably weighted and added to form the output.

Formulation of the Frequency Sampling Method of Filter Design

The sampling technique described in this paper can be applied to a finite set of samples of the z-transform of a filter evaluated anywhere in the z-plane. For the most part we will restrict ourselves to the case where the sample points are equally spaced around the unit circle, and the sample values represent values of the continuous frequency response of the filter. Later in this paper we will consider the more general case and, in particular, will examine the case of nonuniform frequency spacing of the samples.

For the case of uniformly spaced frequency samples the design procedure consists of a sequence of computations which can be summarized as follows.

1) Choose a set of frequencies at which the sampled frequency response is specified. The values of the sampled frequency response at some of these frequencies are generally left as parameters of the design problem. For the uniform frequency sampling considered here, the choice of a set of frequencies is merely the choice of a value for N, the number of impulse-response samples, and an initial frequency. Once N has been chosen, the frequency spacing between samples is $\Delta f = 1/NT$, where T is the sampling period. The choice of values of the frequency response at the sample frequencies is dictated by the ideal filter being approximated.

2) Obtain values of the continuous frequency response of the filter as a function of the filter parameters. The continuous frequency response can be determined as a function of the frequency samples, either as an explicit equation (i.e., the sampling theorem), or implicitly in terms of the fact Fourier transform algorithm (FFT) [9] or the chirp z-transform algorithm (CZT) [10].

3) Once the interpolated frequency response is obtained, a program automatically readjusts the filter parameters (the unspecified frequency samples) while searching for a minimum of some filter characteristic.

4) When the minimum has been obtained and verified, the final values of the free parameters are then used in the realization along with the fixed frequency samples.

There are a wide variety of filter problems where the designer requires a sharp cut-off amplitude characteristic and, preferably, a linear phase characteristic. For this reason, one of our aims was to obtain an interpolated frequency response which was pure real except for a linear phase shift. To achieve this goal requires careful consideration of the parameter N and the specific frequency positions of the samples. As a result, we found it useful to formulate the sampling theorem for four cases.

Case A: N even, frequency samples at

$$f_k = \frac{k}{NT}, \qquad k = 0, 1, 2, \cdots, N-1.$$

Case B: N even, frequency samples at

$$f_k = \frac{k+\frac{1}{2}}{NT}, \qquad k = 0, 1, 2, \cdots, N-1.$$

Case C: N odd, frequency samples at

$$f_k = \frac{k}{NT}, \qquad k = 0, 1, 2, \cdots, N-1.$$

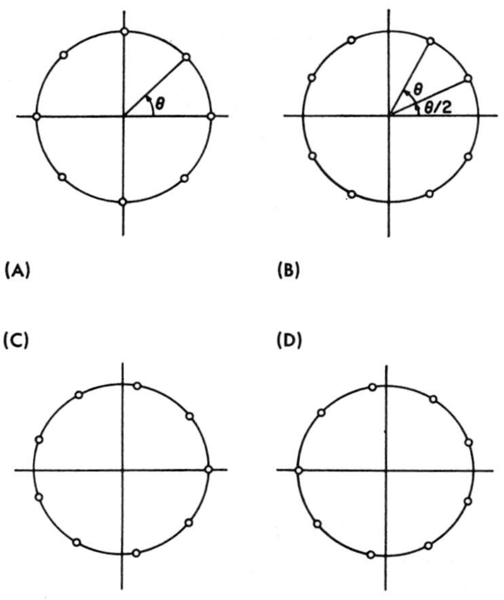

Fig. 7. The four possible orientations for uniformly spaced frequency samples. (A) and (C) show type-1 data whereas (B) and (D) show type-2 data.

Case D: N odd, frequency samples at

$$f_k = \frac{k+\frac{1}{2}}{NT}, \qquad k = 0, 1, 2, \cdots, N-1.$$

Fig. 7 illustrates these four cases, the circles representing the sampling points around the unit circle in the z-plane. For Cases A and B, N is 8, whereas for Cases C and D, N is 9. The data of Cases A and C will henceforth be referred to as type-1 data; whereas the data for Cases B and D will be referred to as type-2 data. The difference between the two types reflects the initial frequency at which the frequency response is sampled.

Derivation of Sampling Theorem for Case A

Given a finite-duration filter impulse response $h(0)$, $h(1), \cdots, h(N-1)$, the z-transform of this filter is

$$H(z) = \sum_{n=0}^{N-1} h(n) z^{-n}. \tag{3}$$

Since $h(n)$ is of finite duration, it can be represented in terms of its discrete Fourier transform (DFT) H_k, $k=0, 1, \cdots, N-1$, as follows:

$$h(n) = \frac{1}{N} \sum_{k=0}^{N-1} H_k e^{j2\pi kn/N} \tag{4}$$

where

$$H_k = H(z)\big|_{z=e^{j2\pi k/N}}. \tag{5}$$

Substituting (4) into (3) and interchanging sums, we observe that the sum over the n index can be evaluated in closed form so that

$$H(z) = \frac{1 - z^{-N}}{N} \sum_{k=0}^{N-1} \frac{H_k}{1 - z^{-1}e^{j2\pi k/N}}. \quad (6)$$

Evaluating (6) on the unit circle where $z = e^{j\omega T}$ leads to the interpolated frequency response

$$H(e^{j\omega T}) = \frac{\exp\left[-\frac{j\omega NT}{2}\left(1 - \frac{1}{N}\right)\right]}{N} \cdot \sum_{k=0}^{N-1} \frac{H_k e^{-j\pi k/N} \sin\left(\frac{\omega NT}{2}\right)}{\sin\left(\frac{\omega T}{2} - \frac{\pi k}{N}\right)}. \quad (7)$$

Let us now examine in detail the implications of (4) through (7). If the initial set of frequency samples H_k is chosen so that H_k is a real, symmetric sequence (i.e., $H_k = H_{N-k}$), then the interpolated frequency response cannot be pure real. A small oscillatory imaginary component of amplitude

$$A = \frac{1}{N} \sum_{k=0}^{N-1} H_k(-1)^k \quad (8)$$

will be part of the interpolated frequency response. In many cases the amplitude A is very small and can be tolerated. In other cases one is forced to look to other techniques for designing pure real nonrecursive filters.

One simple way of alleviating the problem of having an imaginary component (other than a linear phase shift) in the interpolated frequency response is suggested by (7). By making the substitution

$$H_k = G_k e^{j\pi k/N}, \quad (9)$$

the summation in (7) becomes pure real, and $H(e^{j\omega T})$ is real except for the linear phase-shift term outside the sum. A physical interpretation of the significance of the substitution of (9) can be obtained by examining the impulse response corresponding to this set of frequency samples. If the set G_k is chosen such that $G_{N/2} = 0$ and $G_k = -G_{N-k}$, then the impulse response $h(n)$ can be written as

$$h(n) = \frac{G_0}{N} + \sum_{k=1}^{(N/2)-1} \frac{2G_k}{N} \cos\left(\frac{\pi}{N}k + \frac{2\pi}{N}nk\right). \quad (10)$$

It is easily shown that $h(n)$ is a real sequence with the symmetry property

$$h(n) = h(N - 1 - n) \quad n = 0, 1, 2, \cdots, \frac{N}{2} - 1. \quad (11)$$

It should be noted that this is not the usual symmetry property of an N-point sequence. A typical impulse response is shown in Fig. 8 for the case $N = 16$. As seen in Fig. 8, the origin of symmetry of the impulse response lies midway between samples representing a delay of a noninteger number of samples. This half-sample delay can also be verified from (7) where the linear phase-shift term has a component equivalent to half a sample delay.

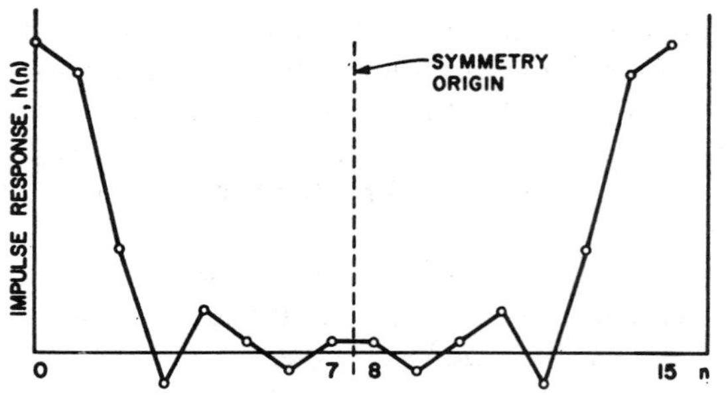

Fig. 8. A typical impulse response for Case A, showing the half-sample delay obtained for this case.

Sampling Theorem for Case C

The derivation of the sampling theorem for N odd is the same as for N even, leading to (7). However, for N odd, choosing the set of frequency samples H_k to be real and symmetric yields a real and symmetric impulse response whose origin of symmetry falls on a sampling point. Thus a pure real interpolated frequency response can be attained for this case. It is easy to show that the continuous frequency response is real by first deriving the impulse response and then computing the frequency response. The impulse response can be written as

$$h(n) = \frac{H_0}{N} + \sum_{k=1}^{(N-1)/2} \frac{2H_k}{N} \cos\left(\frac{2\pi}{N}nk\right). \quad (12)$$

The impulse response is a real and symmetric function with a unique peak at $n = 0$. By rotating the impulse response $(N-1)/2$ samples, i.e., replacing $h(n)$ by $h[(n-(N-1)/2) \bmod N]$, so that the peak occurs at $n = (N-1)/2$, and translating the entire impulse response by $(N-1)/2$ samples, the frequency response can be written as

$$H(e^{j\omega T}) = \sum_{n=-(N-1)/2}^{(N-1)/2} 2h(n) \cos(n\omega T) \quad (13)$$

which is purely real.

Summary of Computation Procedure for Cases A and C

The continuous frequency response can be computed directly from (7) for either N odd or even. However, our method of computation differs in that the FFT algorithm is used instead. We now present the detailed steps used to obtain the interpolated frequency response from the set of N frequency samples.

1) Given N, the designer must determine how fine an interpolation should be used. For the designs we investigated, where N varied from 15 to 256, we found that 16 N sample values of $H(e^{j\omega T})$ lead to reliable computations and results; i.e., 16 to 1 interpolation was used.

2) Given the set of N values of H_k, the FFT is used to compute $h(n)$, the inverse DFT of H_k. For both N odd and N even the set H_k which was used was real and symmetric; therefore $h(n)$ is real in all cases and symmetric for N odd.

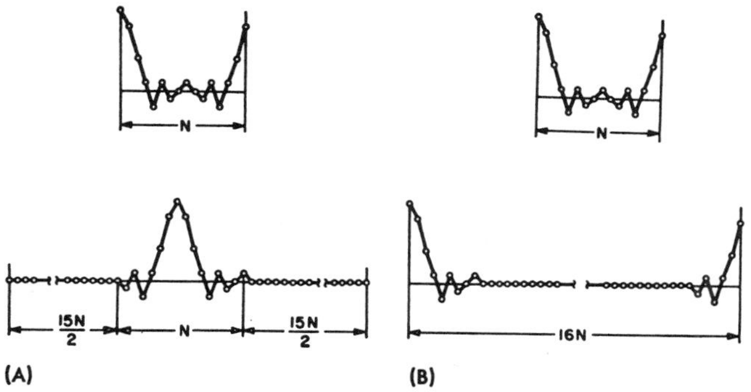

Fig. 9. Computational methods for obtaining the interpolated frequency response.

3) In order to obtain values of the interpolated frequency response one of two procedures is followed. Either a) $h(n)$ is rotated by $N/2$ samples (N even) or $[(N-1)/2]$ samples (N odd) to remove the sharp edges of the impulse response, and then 15 N zero-valued samples are symmetrically placed around the impulse response [as illustrated in Fig. 9(A)]; or b) $h(n)$ is split around the $(N/2)$nd sample value, and 15 N zero-valued samples are placed between the two pieces of the impulse response [as illustrated in Fig. 9(B)]. The zero-augmented sequences of Fig. 9(A) and (B) are transformed using the FFT to give the interpolated frequency responses. These two procedures can easily be shown to yield identical results, the differences being primarily computational ones.

Sampling Theorem for Case B

If the set of frequency samples is evaluated at $f_k = (k+\frac{1}{2})/NT$, $k=0, 1, \cdots, N-1$, and if this set is defined as F_k, then following a development similar to (4) through (7), we obtain

$$H(z) = \frac{1+z^{-N}}{(-N)} \sum_{k=0}^{N-1} \frac{F_k}{1 - z^{-1}\exp\left[\frac{j2\pi}{N}(k+\frac{1}{2})\right]}. \quad (14)$$

Evaluating (14) on the unit circle gives

$$H(e^{j\omega T}) = \exp\left[-\frac{j\omega NT}{2}\left(1-\frac{1}{N}\right)\right] \cdot \sum_{k=0}^{N-1} \frac{jF_k \cos\left(\frac{\omega NT}{2}\right)\exp\left[-\frac{j\pi}{N}(k+\frac{1}{2})\right]}{\sin\left(\frac{\omega T}{2} - \frac{\pi}{N}(k+\frac{1}{2})\right)}. \quad (15)$$

To perform the computation of (15) using the FFT requires a somewhat different procedure than for the previous Cases A and C. This is because in order to compute an inverse DFT using the FFT, it is assumed that the frequency position of the first sample is 0 Hz, whereas in Case B it is $1/(2NT)$ Hz. Therefore, the procedure is the following.

1) Shift the F_k by an angle of π/N clockwise thereby aligning the samples as required by the FFT.

2) Perform the FFT, obtaining a complex impulse response.

3) Either rotate the impulse response by $N/2$ samples and symmetrically augment with zero-valued samples, or split the impulse response at the center and fill in with 15 N zero-valued samples between the two halves of the impulse response.

4) Compute the 16 N point FFT to obtain an interpolated frequency response.

5) Rotate the frequency response data by an angle of π/N (8 samples for a 16 to 1 interpolation) counterclockwise, thereby compensating the original shift and producing the desired result.

The importance of data of Case B is that the interpolated frequency response, when the frequency samples form a real and symmetric set, is pure real. This can be proven from (15), but it is more easily shown to be true by examining the impulse response. For the conditions of Case B the symmetry of the frequency samples can be written

$$F_k = F_{N-1-k}. \quad (16)$$

Therefore the complex impulse response corresponding to step 2 above is

$$f(n) = \sum_{k=0}^{(N/2)-1} \frac{2F_k}{N} e^{-j\pi n/N} \cos\left(\frac{2\pi}{N} n(k+\tfrac{1}{2})\right) \quad (17)$$
$$n = 0, 1, 2, \cdots, N-1.$$

From (17) we see that the real part of $f(n)$ is symmetric, the imaginary part is antisymmetric, and $f(N/2)$ is identically zero. Therefore, the impulse response is technically of duration $(N-1)$ samples, although there are N independent frequency samples. It is, therefore, easy to find an axis of symmetry which coincides with a sample point. Thus the interpolated frequency response corresponding to step 4 above is real. For this case alone the original frequency samples, the true filter impulse response, and the interpolated frequency-response samples are all real.

Sampling Theorem for Case D

The development for Case D is identical to that for Case B. For the set F_k real and symmetric, the interpolated frequency response has a small and imaginary component similar to that of Case A discussed earlier. By making the set F_k complex, a real interpolated frequency response can be obtained as seen previously. Because of the similarity of this case to Case A no further discussion is necessary.

Rationale for Minimization Algorithm

There are several reasons why the different Cases A, B, C, and D are of interest. First, by inspection of (7) and (15), it is seen that when $H(e^{j\omega T})$ is real, it consists of a sum of elementary functions of the form

$$\sin\left(\frac{\omega NT}{2}\right) \Big/ \sin\left(\frac{\omega T}{2} - \theta\right). \tag{18}$$

In the design of, for example, a low-pass filter one would choose the frequency samples which occur in the passband to have value 1.0 and those which occur in the stopband to have value 0.0. The values of the frequency samples which occur in the transition band would be chosen according to some criterion. It is intuitively appealing to picture that the transition values found for any given optimum design produce functions of the form of (18) with ripples which cancel the ripples caused by the fixed samples. As the number of transition values is increased, it is easy to picture ever finer cancellation. Thus it is useful to obtain a real $H(e^{j\omega T})$.

Another reason for sampling at different frequencies (type-1 and type-2 data) arises when the designer chooses his bandwidth. If the frequency samples are to form an even function, then for Cases A and C, the bandwidth must contain an odd number of samples (the sample at frequency f_k is balanced by a sample at frequency f_{N-k}, except for the sample at $f=0$). Similarly for Cases B and D the bandwidth must contain an even number of frequency samples. Hence sampling at different frequencies provides additional flexibility to the designer. It also turns out that for small bandwidths (in terms of number of inband frequency samples) the sidelobe ripple cancellation is more efficient when the bandwidth is an even number of samples than when it is an odd number. Furthermore, as will be explained later, a convenient design for bandpass filters is based on rotation of low-pass prototypes. As such, the existence of data from all cases is of great value.

At this point, we now turn to a discussion of the optimization techniques which we have used.

The Minimization Algorithm

From (7) and (15) we observe that $H(e^{j\omega T})$ is a *linear* function of the samples H_k or F_k. In all of our problems most of the H_k or F_k will be preset, and the remaining few (the transition coefficients) will be varied until the maximum sidelobe is a minimum. Fig. 10 shows the typical specification for a low-pass filter. In this example, there are $2\,BW-1$ samples preset to 1.0, $2\,M$ transition samples, and the remaining samples are preset to 0.0. Symmetry considerations reduce the number of independent transition samples to M. Let us denote the transition coefficients by T_1, T_2, T_3, etc. Then Fig. 11 shows how $H(e^{j\omega_1 T})$, $H(e^{j\omega_2 T})$, etc. might vary with any one transition coefficient. All such variations are linear. It has been shown [11] that the upper envelope of these straight lines forms a convex function. (In Fig. 11 the upper envelope is drawn with heavy lines.) It has also been shown [12] that a convex function has a unique minimum (a local minimum is a global minimum). From this it follows that a procedure which searches for the minimum value of a

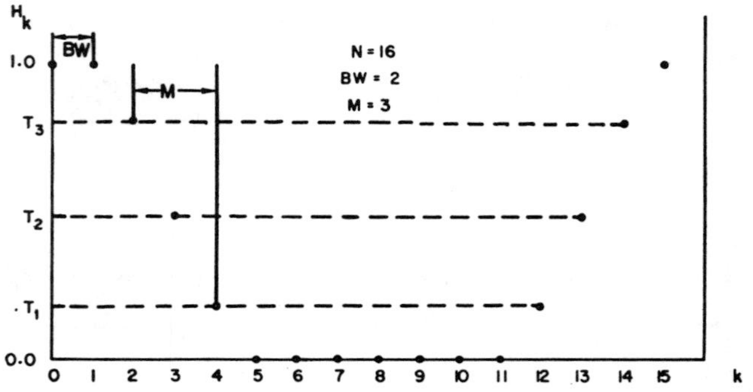

Fig. 10. Typical specifications for type-1 low-pass filters.

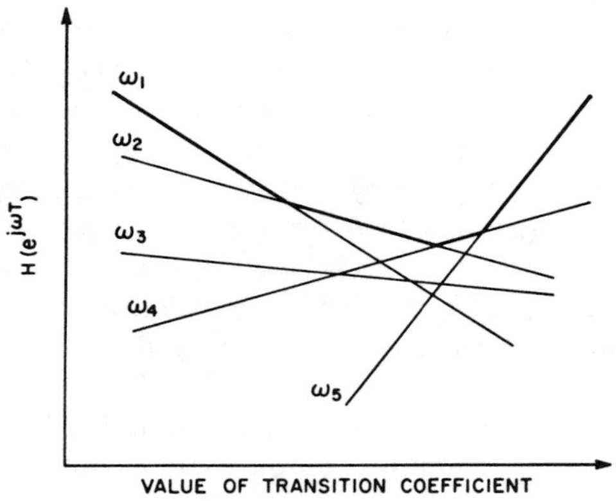

Fig. 11. Curves showing the linear variation of the frequency response versus transition coefficient at any given frequency.

maximum sidelobe must converge; i.e., the search will *not* result in a false minimum.

The above reasoning may be extended to more than one dimension. $H(e^{j\omega_1 T})$, $H(e^{j\omega_2 T})$, etc., can be plotted as a hyperline of the transition coefficients, T_1, T_2, etc. The upper envelope of the different hyperlines is a convex hypersurface and leads to the same result as before, namely, that a minimax search procedure as a function of T_1, T_2, etc., will converge.

The assurance of ultimate convergence does not necessarily mean that any given search procedure is feasible in terms of computer running time. Now is a good time to stress the discrete nature of our interpolation technique. This discreteness has two important effects. First, it makes it more or less impossible to locate and measure an *exact* minimax of the continuous function $H(e^{j\omega T})$. Experimentally, this is not bothersome; if a sufficient number of ω's are used, the computed result is within a fraction of a decibel of the exact answer. Second, the discreteness helps us by discretizing, in a sense, the convex hypersurface into a connected set of hyperlines. This is true because for small variations in T_1, T_2, etc., the (discrete) frequency position of the maximum sidelobe remains fixed. Thus, over this small variation, the *maximum* $H(e^{j\omega T})$ is a linear function of T_1, T_2, etc., which shows that the convex surface is really a connected set of hyperlines. When the frequency position of the maximum sidelobe changes, the slope of the resultant hyperline of steepest descent changes.

The above reasoning suggests the following search procedure.

1) Always begin with a one-dimensional search. For example, if it is desired to optimize over three transition coefficients, T_1, T_2, and T_3, begin by setting $T_3 = T_2 = 1$ and searching for the value of T_1 in the range 0.0 to 1.0 which yields a minimax. This value is labeled as point A in Fig. 12.

2) Now go to two dimensions. Let $T_2 = 1.0$ and the value of T_1 obtained from step 1 define a point on a two-dimensional line. To find another point, perturb T_2 slightly from its preset value of unity (to a slightly smaller value) and repeat the one-dimensional search, varying T_1, as before. This new two-dimensional point (point B in Fig. 12), along with the previous one, determines the appropriate straight line (the path of steepest descent) along which to do the full two-dimensional search.

3) A simple search is now made along the line found in step 2, yielding a minimum of $H(e^{j\omega T})$ (point C in Fig. 12). A new path of steepest descent is obtained by varying T_1 and keeping T_2 fixed at the value of point C, yielding point D; then perturbing T_2 slightly and again varying T_1 yielding point E. A simple search is made along the new line yielding a minimum at point F. If the difference between the values of $H(e^{j\omega T})$ at the minima of the searches along the lines of steepest descent (points C and F) is less than some prescribed threshold, the search is ended and point F is the two-dimensional solution. Otherwise the procedures of step 3 are iterated to yield refinements of the path of steepest descent until two consecutive searches yield minima whose difference satisfies the threshold condition. Practically it has been found that a two-dimensional search has always terminated within three iterations when the threshold is set to 0.1 dB.

4) Now go to three dimensions. Let $T_3 = 1.0$ and the two-dimensional result of step 3 define a point on a three-dimensional line. To find another point on the line, perturb T_3 slightly (to a smaller value) and repeat the two-dimensional search of steps 1 through 3. We now have two points on a three-dimensional line along which we can search for a minimum. At the minimum a new three-dimensional line of steepest descent is obtained and a new search is conducted. The search procedure is terminated when the difference in minima between two consecutive three-dimensional searches is less than a prescribed threshold.

Clearly the search procedure is more time consuming as the dimensionality increases; in fact, it is reasonable to expect that the search time is roughly an exponential function of the dimensionality. We have found experimentally that a four-dimensional search is attainable (within 300 seconds on a CDC-6600 computer), and that all searches have indeed converged.

A useful check on the convergence of the search can be made by examining sidelobes other than the minimax. In general, for M transition coefficients (M-dimensional search) there are $(M+1)$ *equal* minimax sidelobes. The proof of this assertion will not be given here, but philosophically it is similar to the proof given by Papoulis [13] in his discussion of elliptic filters.

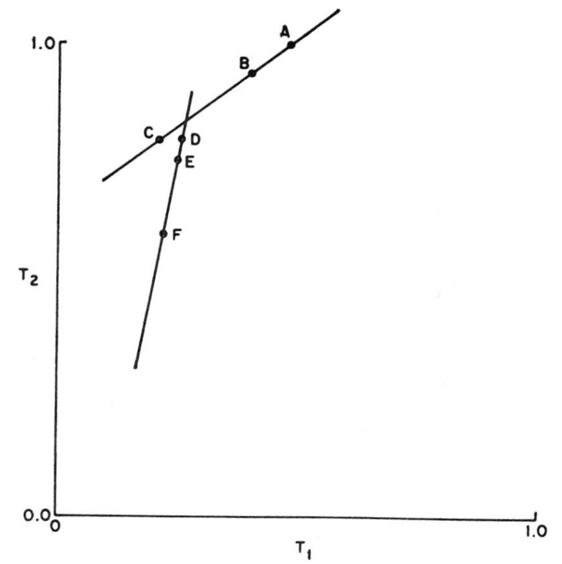

Fig. 12. Illustration of the path followed in a typical search for two optimum transition coefficients.

Note from (7), (15), and Fig. 10, that there are only M variable values of H_k or F_k; the remaining values are preset and remain fixed. This implies that during the course of a search (which may involve thousands of computations of (7) and (15) before convergence) increased computational efficiency results from separating (7) and (15) into two sums, namely, into those terms with the preset H_k or F_k and those terms with the variable H_k or F_k. The first sum may be evaluated once and stored in a table. The second sum consists of very few terms (one to four) and can either be rapidly computed for all values of the (discrete) interpolation for each step in the search or else broken into separate terms, each involving one transition coefficient, and also stored in tables. Using the second alternative, the (discrete) interpolation function is formed for the various values of T_1, T_2, etc., by multiplying the various tables by the appropriate transition coefficients and adding the results. This procedure is uneconomical of computer storage but exceedingly economical of computer running time.

Figs. 13 and 14 show typical examples of the results of a three-dimensional search for type-1 low-pass filters. Fig. 13(A) shows the entire frequency response with $N = 64$, $BW = 16$, and transition coefficients $T_1 = 0.030957$, $T_2 = 0.275570$, and $T_3 = 0.744348$. For this filter, ripple peaks 1, 2, 5, and 6 are equal within 0.54 dB. Fig. 13(B) shows an expanded view of the frequency response of Fig. 13(A). The first plot in Fig. 13(b) shows the nature of the in-band ripple. The greatly magnified vertical scale is in thousandths of a decibel. The ripple is very small near 0 frequency and increases steadily until the edge of the

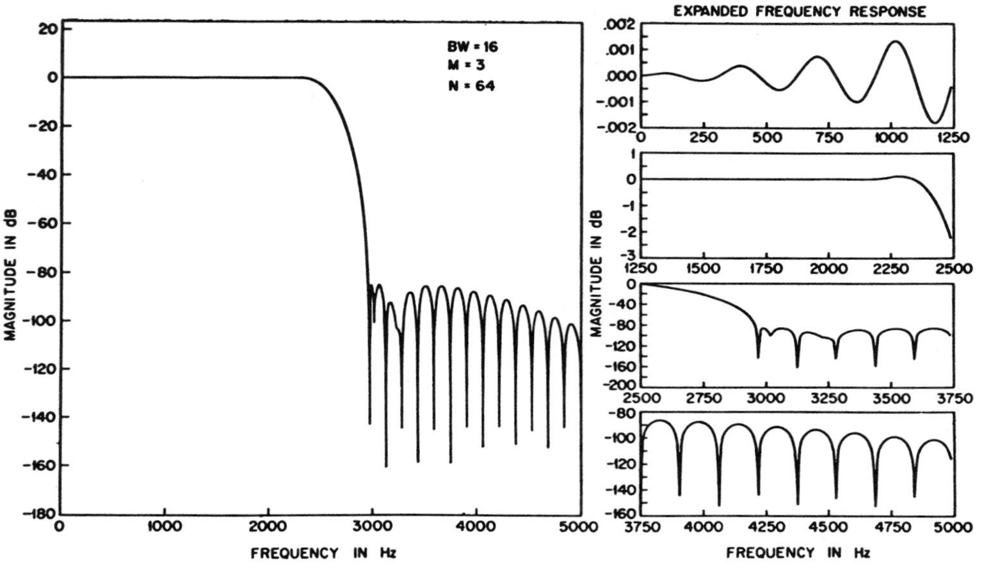

(A) (B)

Fig. 13. Interpolated frequency response for low-pass filter with three transition coefficients, for small value of N.

Fig. 14. Interpolated frequency response for low-pass filter with three transition coefficients, for large value of N.

(A) (B)

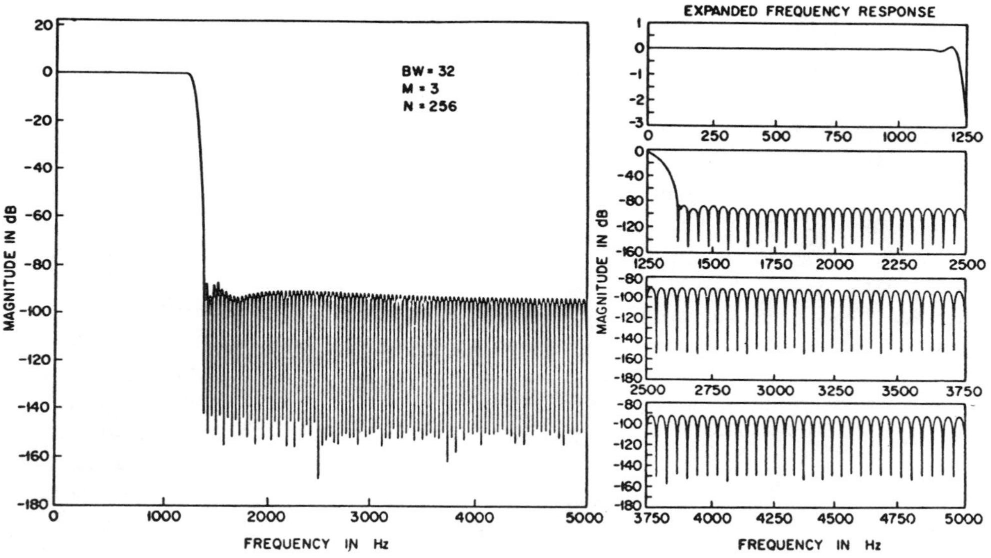

transition band at which point it reaches about 0.1 dB. The next three plots of Fig. 13(B) show the transition band and the out-of-band ripple. It is seen that the peak height of the out-of-band ripple decreases steadily as the frequency gets farther away from the edge of the transition band. This is due to the (sin ω/ω) type interpolation falloff from each of the nonzero values of H_k. Finally it is noted that the minimax solution of Fig. 13 is -85.01 dB.

Fig. 14 shows results for a larger value of N, this case being identical to the one discussed by Gold and Jordan [5]. For this set of data BW is 32, N is 256, and the transition coefficients are $T_1=0.025779$, $T_2=0.251635$, and $T_3=0.723071$. Fig. 14(A) shows the entire frequency response for this filter while (B) shows expanded horizontal and vertical scales. The minimax solution is -87.89 dB, and ripple peaks 1, 2, 4, and 5 are equal with 0.35 dB.

Results

Using the method explained in the previous sections, we have designed a large number of low-pass filters, bandpass filters, and wide-band differentiators. For low-pass filters we have considered type-1 and type-2 data for various values of N, BW, and M, as defined earlier in

TABLE I

Low-Pass Filter Design, One Transition Coefficient (Type-1 Data, N Even)

BW	Minimax	T_1
	$N=16$	
1	−39.75363827	0.42631836
2	−37.61346340	0.40397949
3	−36.57721567	0.39454346
4	−35.87249756	0.38916626
5	−35.31695461	0.38840332
6	−35.51951933	0.40155639
	$N=32$	
1	−42.24728918	0.42856445
2	−41.29370594	0.40773926
3	−41.03810358	0.39662476
4	−40.93496323	0.38925171
6	−40.85183477	0.37897949
8	−40.75032616	0.36990356
10	−40.54562140	0.35928955
12	−39.93450451	0.34487915
14	−38.91993237	0.34407349
	$N=64$	
1	−42.96059322	0.42882080
2	−42.30815172	0.40830689
3	−42.32423735	0.39807129
4	−42.43565893	0.39177246
5	−42.55461407	0.38742065
6	−42.66526604	0.38416748
10	−43.01104736	0.37609863
14	−43.28309965	0.37089233
18	−43.56508827	0.36605225
22	−43.96245098	0.35977783
26	−44.60516977	0.34813232
30	−43.81448936	0.29973144
	$N=128$	
1	−43.15302420	0.42889404
2	−42.59092569	0.40847778
3	−42.67634487	0.39838257
4	−42.84038544	0.39226685
5	−42.99805641	0.38812256
7	−43.25537014	0.38281250
10	−43.52547789	0.37826538
18	−43.93180990	0.37251587
26	−44.18097305	0.36941528
34	−44.40153408	0.36686401
42	−44.67161417	0.36394653
50	−45.17186594	0.35902100
58	−46.92415667	0.34273681
62	−49.46289873	0.28751221
	$N=256$	
1	−43.20187807	0.42891235
2	−42.66346216	0.40852051
3	−42.76789284	0.39846802
4	−42.94498634	0.39239502
5	−43.11306858	0.38829956
7	−43.38275146	0.38311157
10	−43.65783644	0.37877197
11	−43.72631788	0.37778931
18	−44.03770208	0.37368774
34	−44.34437370	0.37011490
50	−44.50268936	0.36840210
58	−44.56697893	0.36773071
66	−44.62925625	0.36708985
82	−44.76955509	0.36568604
98	−44.99645948	0.36352539
106	−45.22021151	0.36150513
114	−45.73309422	0.35722656
122	−47.75400352	0.34083862
123	−48.54518938	0.33485107
124	−50.05162334	0.32495117
125	−53.51222181	0.30643310
126	−55.34463978	0.28125000

TABLE II

Low-Pass Filter Design, Two Transition Coefficients (Type-1 Data, N Even)

BW	Minimax	T_1	T_2
	$N=16$		
1	−65.27693653	0.10703125	0.60559357
2	−62.85937929	0.12384644	0.62291631
3	−62.96594906	0.12827148	0.62855407
4	−66.03942485	0.12130127	0.61952704
5	−71.73997498	0.11066284	0.60979204
	$N=32$		
1	−67.37020397	0.09610596	0.59045212
2	−63.93104696	0.11263428	0.60560235
3	−62.49787903	0.11931763	0.61192546
5	−61.28204536	0.12541504	0.61824023
7	−60.82049131	0.12907715	0.62307031
9	−59.74928617	0.12068481	0.60685586
11	−62.48683357	0.13004150	0.62821502
13	−70.64571857	0.11017914	0.60670943
	$N=64$		
1	−70.26372528	0.09376831	0.58789222
2	−67.20729542	0.10411987	0.59421778
3	−65.80684280	0.10850220	0.59666158
4	−64.95227051	0.11038818	0.59730067
5	−64.42742348	0.11113281	0.59698496
9	−63.41714096	0.10936890	0.59088884
13	−62.72142410	0.10828857	0.58738641
17	−62.37051868	0.11031494	0.58968142
21	−62.04848146	0.11254273	0.59249461
25	−61.88074064	0.11994629	0.60564501
29	−70.05681992	0.10717773	0.59842159
	$N=128$		
1	−70.58992958	0.09445190	0.58900996
2	−68.62421608	0.10349731	0.59379058
3	−67.66701698	0.10701294	0.59506081
4	−66.95196629	0.10685425	0.59298926
6	−66.32718945	0.10596924	0.58953845
9	−66.01315498	0.10471191	0.58593906
17	−65.89422417	0.10288086	0.58097354
25	−65.92644215	0.10182495	0.57812308
33	−65.95577812	0.10096436	0.57576437
41	−65.97698021	0.10094604	0.57451694
49	−65.67919827	0.09865112	0.56927420
57	−64.61514568	0.09845581	0.56604486
61	−71.76589394	0.10496826	0.59452277
	$N=256$		
1	−70.65072060	0.09458618	0.58923281
2	−68.87253571	0.10375977	0.59425391
3	−68.10910606	0.10658569	0.59466635
4	−67.56728649	0.10690918	0.59322120
6	−67.11176777	0.10612793	0.58998313
9	−66.98584080	0.10502319	0.58671387
10	−66.96896267	0.10457764	0.58571956
17	−67.16303158	0.10323486	0.58217779
33	−67.59191895	0.10212250	0.57908094
49	−67.88601303	0.10168457	0.57769495
57	−67.98678684	0.10133057	0.57690264
65	−68.10634232	0.10109253	0.57627586
81	−68.26243591	0.09982910	0.57389784
97	−68.50218678	0.09773560	0.56999568
105	−68.69239998	0.09583130	0.56641188
113	−68.79911518	0.09375000	0.56186695
121	−68.76781750	0.08519897	0.54300842
122	−68.42778111	0.08584595	0.54292957
123	−68.08275509	0.08768921	0.54538271
124	−68.77609253	0.09545975	0.56272902
125	−72.14476395	0.09437866	0.56816792

TABLE III

Low-Pass Filter Design, Three Transition Coefficients (Type-1 Data, N Even

BW	Minimax	T_1	T_2	T_3
	$N=16$			
1	−96.63068199	0.01597290	0.19530278	0.67931499
2	−95.47730064	0.01951294	0.22385191	0.70432347
3	−98.72853756	0.01724854	0.21873067	0.70148393
4	−114.30107403	0.01259155	0.19831225	0.68572509
	$N=32$			
1	−93.11873436	0.01735230	0.20052231	0.68302930
2	−89.36560249	0.02354126	0.23959557	0.71593525
4	−86.97191620	0.02770996	0.26135787	0.73350248
6	−86.69376850	0.02871094	0.26670884	0.73796855
8	−88.41957283	0.02705688	0.26084303	0.73367810
10	−92.86338043	0.02340698	0.24691517	0.72345444
12	−115.13009739	0.01320190	0.20258948	0.68939742
	$N=64$			
1	−94.71326447	0.01544800	0.19125221	0.67535861
2	−89.93906212	0.02057495	0.22634942	0.70507613
3	−87.99901295	0.02438354	0.24519492	0.72204262
4	−87.48722935	0.02581177	0.25236063	0.72570913
8	−85.52755356	0.03010864	0.27213724	0.74181077
12	−85.35023785	0.02996826	0.27101526	0.74040381
16	−85.01383400	0.03095703	0.27556998	0.74434815
20	−85.68937778	0.02975464	0.27059980	0.74029023
24	−86.59522438	0.02882080	0.26751875	0.73850916
28	−115.01544189	0.01322021	0.20274639	0.68953717
	$N=128$			
1	−94.00015545	0.01566772	0.19122512	0.67492092
2	−92.48471928	0.01967163	0.22183722	0.70122715
3	−90.42631149	0.02271729	0.23824591	0.71450669
5	−88.68881035	0.02398071	0.24430176	0.71859482
8	−87.40693378	0.02510986	0.24892636	0.72166583
16	−86.15447140	0.02633057	0.25347440	0.72460093
24	−85.83375359	0.02739258	0.25816799	0.72843530
32	−85.56298637	0.02763062	0.25894149	0.72886454
40	−84.62227345	0.03014526	0.26996954	0.73824701
48	−85.57633972	0.02913208	0.26652967	0.73584086
56	−86.70246506	0.02863769	0.26648645	0.73743919
60	−114.79892921	0.01326904	0.20291216	0.68961047
	$N=256$			
1	−92.07104015	0.10647949	0.19387524	0.67664281
2	−93.63089275	0.01963501	0.22197911	0.70144920
3	−87.43575478	0.01908569	0.22085960	0.69990539
5	−89.86921692	0.02305298	0.24117076	0.71635813
8	−89.21122360	0.02479248	0.24843111	0.72164702
9	−88.34475231	0.02329712	0.24253562	0.71679420
16	−88.42712784	0.02444458	0.24629538	0.71900030
32	−87.89452744	0.02577896	0.25163493	0.72307099
48	−87.84068012	0.02421875	0.24359358	0.71550480
56	−86.96756554	0.02345581	0.23957232	0.71177494
64	−87.60656548	0.02396851	0.24199281	0.71380179
80	−87.11819744	0.02351685	0.23926844	0.71103085
96	−86.78892708	0.02435913	0.24219392	0.71293931
104	−85.55295181	0.02552490	0.24590992	0.71524908
112	−85.86081982	0.02607422	0.24926456	0.71857490
120	−88.45293331	0.02683105	0.25909273	0.73130690
121	−90.09288883	0.02561035	0.25523207	0.72916388
122	−93.23881817	0.02344360	0.24778644	0.72456966
123	−99.37811375	0.01946106	0.23070314	0.71099759
124	−113.12398720	0.01351929	0.20394843	0.69037794

TABLE IV

Low-Pass Filter Design, Four Transition Coefficients (Type-1 Data, N Even)

BW	Minimax	T_1	T_2	T_3	T_4
			$N=16$		
1	−127.30743676	0.00131836	0.03717696	0.25469056	0.71883166
			$N=128$		
16	−108.29668730	0.00606079	0.09324160	0.40820056	0.82096794

Fig. 10. For bandpass filters and differentiators we have considered type-1 data only.

Before proceeding to a discussion of the data, a few general remarks can be made about the results for low-pass filters.

1) For each filter there are three design parameters: N, M, and BW. A derived parameter, percentage bandwidth, defined as the ratio of in-band frequency bandwidth to half the sampling frequency, is of great value in visualizing the results.

2) For most cases the minimax lies between −40 and −50 dB for a single transition point, between −65 and −75 dB for two transition points, between −85 and −95 dB for three transition points, and about −105 dB for four transition points. To a rough approximation, adding a transition sample reduces the sidelobes by about 20 dB.

3) If the designer wants parameters that are not tabulated, he can find approximate values of the transition coefficients by linear interpolation of the tabulated values. Experimentally, we have found that the deviation of the result obtained by linear interpolation will be less than 6 dB from the optimum.

Low-Pass Filters

The data for type-1 low-pass filters, for N even, are tabulated in Tables I through IV. This set of data corresponds to Case A discussed previously with the frequency samples H_k constituting a real and symmetric set.

TABLE V

Low-Pass Filter Design, One Transition Coefficient (Type-1 Data, N Odd)

BW	Minimax	T_1
	$N=15$	
1	−42.30932283	0.43378296
2	−41.26299286	0.41793823
3	−41.25333786	0.41047363
4	−41.94907713	0.40405884
5	−44.37124538	0.39268189
6	−56.01416588	0.35766525
	$N=33$	
1	−43.03163004	0.42994995
2	−42.42527962	0.41042481
3	−42.40898275	0.40141601
4	−42.45948601	0.39641724
6	−42.52403450	0.39161377
8	−42.44085121	0.39039917
10	−42.11079407	0.39192505
12	−41.92705250	0.39420166
14	−44.69430351	0.38552246
15	−56.18293285	0.35360718
	$N=65$	
1	−43.16935968	0.42919312
2	−42.61945581	0.40903320
3	−42.70906305	0.39920654
4	−42.86997318	0.39335937
5	−43.01999664	0.38950806
6	−43.14578819	0.38679809
10	−43.44808340	0.38129272
14	−43.54684496	0.37946167
18	−43.48173618	0.37955322
22	−43.19538212	0.38162842
26	−42.44725609	0.38746948
30	−44.76228619	0.38417358
31	−59.21673775	0.35282745
	$N=125$	
1	−43.20501566	0.42899170
2	−42.66971111	0.40867310
3	−42.77438974	0.39868774
4	−42.95051050	0.39268189
6	−43.25854683	0.38579101
8	−43.47917461	0.38195801
10	−43.63750410	0.37954102
18	−43.95589399	0.37518311
26	−44.05913115	0.37384033
34	−44.05672455	0.37371826
42	−43.94708776	0.37470093
50	−43.58473492	0.37797851
58	−42.14925432	0.39086304
59	−42.60623264	0.39063110
60	−44.78062010	0.38383713
61	−56.22547865	0.35263062

TABLE VI

Low-Pass Filter Design, Two Transition Coefficients (Type-1 Data, N Odd)

BW	Minimax	T_1	T_2
	$N=15$		
1	−70.60540585	0.09500122	0.58995418
2	−69.26168156	0.10319824	0.59357118
3	−69.91973495	0.10083618	0.58594327
4	−75.51172256	0.08407593	0.55715312
5	−103.46078300	0.05180206	0.49917424
	$N=33$		
1	−70.60967541	0.09497070	0.58985167
2	−68.16726971	0.10585937	0.59743846
3	−67.13149548	0.10937500	0.59911696
5	−66.53917217	0.10965576	0.59674101
7	−67.23387909	0.10902100	0.59417456
9	−67.85412312	0.10502930	0.58771575
11	−69.08597469	0.10219727	0.58216391
13	−75.86953640	0.08137207	0.54712777
14	−104.04059029	0.05029373	0.49149549
	$N=65$		
1	−70.66014957	0.09472656	0.58945943
2	−68.89622307	0.10404663	0.59476127
3	−67.90234470	0.10720215	0.59577449
4	−67.24003792	0.10726929	0.59415763
5	−66.86065960	0.10689087	0.59253047
9	−66.27561188	0.10548706	0.58845983
13	−65.96417046	0.10466309	0.58660485
17	−66.16404629	0.10649414	0.58862042
21	−66.76456833	0.10701904	0.58894575
25	−68.13407993	0.10327148	0.58320831
29	−75.98313046	0.08069458	0.54500379
30	−104.92083740	0.04978485	0.48965181
	$N=125$		
1	−70.68010235	0.09464722	0.58933268
2	−68.94157696	0.10390015	0.59450024
3	−68.19352627	0.10682373	0.59508549
5	−67.34261131	0.10668945	0.59187505
7	−67.09767151	0.10587158	0.58921869
9	−67.05801296	0.10523682	0.58738706
17	−67.17504501	0.10372925	0.58358265
25	−67.22918987	0.10316772	0.58224835
33	−67.11609936	0.10303955	0.58198956
41	−66.71271324	0.10313721	0.58245499
49	−66.62364197	0.10561523	0.58629534
57	−69.28378487	0.10061646	0.57812192
58	−70.35782337	0.09663696	0.57121235
59	−75.94700718	0.08054886	0.54451285
60	−104.09012318	0.04991760	0.48963264

Values of minimax and transition coefficients are tabulated as functions of N and M.

The data for type-1 low-pass filters, for N odd, are tabulated in Tables V through VII. This set of data corresponds to Case C discussed previously.

The data for type-2 low-pass filters, for N even, are tabulated in Tables VIII through X. This set of data corresponds to Case B discussed previously. The data of these tables are shown graphically in Figs. 15 through 20. The horizontal axis for each of these figures is the percentage bandwidth defined (for type-2 data) as

$$\text{percentage bandwidth} = \frac{2BW - 1}{N}. \tag{19}$$

Figs. 15, 17, and 19 show the one-, two-, and three-dimensional minimax; Figs. 16, 18, and 20 show values of transition coefficients for one, two, and three transition coefficients.

The curves of minimax all show sharp drops for both large and small values of percentage bandwidth. The drop

TABLE VII

Low-Pass Filter Design, Three Transition Coefficients (Type-1 Data, N Odd)

BW	Minimax	T_1	T_2	T_3
		$N=15$		
1	−94.61166191	0.01455078	0.18457882	0.66897613
2	−104.99813080	0.01000977	0.17360713	0.65951526
3	−114.90719318	0.00873413	0.16397310	0.64711264
4	−157.29257584	0.00378799	0.12393963	0.60181154
		$N=33$		
1	−96.03734779	0.01373291	0.18448586	0.67025933
2	−86.96916771	0.01668701	0.20723432	0.68914992
4	−87.86485004	0.01990967	0.22577646	0.70374222
6	−93.33241367	0.02140503	0.23353566	0.70954533
8	−92.88408661	0.02062988	0.22815933	0.70362590
10	−99.76651382	0.01605835	0.20567451	0.68306885
12	−115.53731537	0.00855103	0.16082642	0.63829710
13	−160.64276695	0.00366821	0.12056040	0.59255887
		$N=65$		
1	−95.01340866	0.01552124	0.19101259	0.67492051
2	−93.53645134	0.01989136	0.22329262	0.70260139
3	−91.72289371	0.02214966	0.23609223	0.71288030
4	−89.08071899	0.02115479	0.23418217	0.71154775
8	−88.25607777	0.02576904	0.25203440	0.72436684
12	−88.23531914	0.02576904	0.25178881	0.72372888
16	−89.70906830	0.02368774	0.24385557	0.71742143
20	−84.97463989	0.01913452	0.22072406	0.69678377
24	−87.32207489	0.02396240	0.23899360	0.71008776
28	−115.10305405	0.00938721	0.16513107	0.64142790
29	−142.13555908	0.00352402	0.11881758	0.58955571
		$N=125$		
1	−94.76369476	0.01556396	0.19093541	0.67475127
2	−93.68562794	0.01968994	0.22232235	0.70177618
4	−88.66252041	0.02420654	0.24380589	0.71788831
6	−90.23277760	0.02402344	0.24543860	0.71966323
8	−87.43502617	0.02286987	0.23940384	0.71406829
16	−89.34566116	0.02507324	0.24955546	0.72206742
24	−87.72590733	0.02415771	0.24472233	0.71761565
32	−89.26216030	0.02522583	0.24994760	0.72208238
40	−88.83863926	0.02439575	0.24590832	0.71850440
48	−88.96858311	0.02440186	0.24453094	0.71649557
56	−93.85659409	0.01801758	0.21212543	0.68652023
57	−103.89656162	0.01030884	0.17459164	0.65248157
58	−110.04844761	0.00630951	0.14743829	0.62397790
59	−155.93431473	0.00361786	0.11937160	0.58988331

in minimax for the high percentage bandwidth is caused by the fact that very few ripples need to be canceled in the small out-of-band frequency range. The drop for the low percentage bandwidth is caused by the fact that there are very few contributions to the ripple; hence the small amount of ripple in the large out-of-band region is more perfectly canceled than for larger values of percentage bandwidth.

Before proceeding to the data on bandpass filters, two comments seem worthwhile.

1) As seen from Tables I through X or from Figs. 15 through 20, for a broad range of values of percentage bandwidth, values of minimax and transition coefficients do not change much, i.e., the curves tend to be flat topped.

2) For small values of bandwidth, ripple cancellation for type-2 filters is superior to ripple cancellation for

TABLE VIII

Low-Pass Filter Design, One Transition Coefficient (Type-2 Data, N Even)

BW	Minimax	T_1
	$N=16$	
1	−51.60668707	0.26674805
2	−47.48000240	0.32149048
3	−45.19746828	0.34810181
4	−44.32862616	0.36308594
5	−45.68347692	0.36661987
6	−56.63700199	0.34327393
	$N=32$	
1	−52.64991188	0.26073609
2	−49.39390278	0.30878296
3	−47.72596645	0.32984619
4	−46.68811989	0.34217529
6	−45.33436489	0.35704956
8	−44.30730963	0.36750488
10	−43.11168003	0.37810669
12	−42.97900438	0.38465576
14	−56.32780266	0.35030518
	$N=64$	
1	−52.90375662	0.25923462
2	−49.74046421	0.30603638
3	−48.38088989	0.32510986
4	−47.47863007	0.33595581
5	−46.88655186	0.34287720
6	−46.46230555	0.34774170
10	−45.46141434	0.35859375
14	−44.85988188	0.36470337
18	−44.34302616	0.36983643
22	−43.69835377	0.37586059
26	−42.45641375	0.38624268
30	−56.25024033	0.35200195
	$N=128$	
1	−52.96778202	0.25885620
2	−49.82771969	0.30534668
3	−48.51341629	0.32404785
4	−47.67455149	0.33443604
5	−47.11462021	0.34100952
7	−46.43420267	0.34880371
10	−45.88529110	0.35493774
18	−45.21660566	0.36182251
26	−44.87959814	0.36521607
34	−44.61497784	0.36784058
42	−44.32706451	0.37066040
50	−43.87646437	0.37500000
58	−42.30969715	0.38807373
62	−56.23294735	0.35241699
	$N=256$	
1	−52.98314095	0.25876465
2	−49.84995031	0.30517578
3	−48.54066038	0.32379761
4	−47.72320795	0.33405762
5	−47.17156410	0.34054565
7	−46.50723314	0.34817505
11	−45.86754894	0.35531616
18	−45.39137316	0.36023559
34	−44.99831438	0.36420288
50	−44.81827116	0.36600342
66	−44.68390656	0.36733399
82	−44.54332495	0.36871338
98	−44.33441591	0.37075806
114	−43.77572870	0.37611694
122	−42.27385855	0.38851929
123	−42.17442465	0.39050293
124	−42.62321901	0.39036255
125	−44.78945160	0.38365478
126	−56.23000479	0.35252228

TABLE IX

Low-Pass Filter Design, Two Transition Coefficients (Type-2 Data, *N* Even)

BW	Minimax	T_1	T_2
		$N=16$	
1	−77.26126766	0.05309448	0.41784180
2	−73.81026745	0.07175293	0.49369211
3	−73.02352142	0.07862549	0.51966134
4	−77.95156193	0.07042847	0.51158076
5	−105.23953247	0.04587402	0.46967784
		$N=32$	
1	−80.49464130	0.04725342	0.40357383
2	−73.92513466	0.07094727	0.49129255
3	−72.40863037	0.08012695	0.52153983
5	−70.95047379	0.08935547	0.54805908
7	−70.22383976	0.09403687	0.56031410
9	−69.94402790	0.09628906	0.56637987
11	−70.82423878	0.09323731	0.56226952
13	−104.85642624	0.04882812	0.48479068
		$N=64$	
1	−80.80974960	0.04658203	0.40168723
2	−75.11772251	0.06759644	0.48390015
3	−72.66662025	0.07886963	0.51850058
4	−71.85610867	0.08393555	0.53379876
5	−71.34401417	0.08721924	0.54311474
9	−70.32861614	0.09371948	0.56020256
13	−69.34809303	0.09761963	0.56903714
17	−68.06440258	0.10051880	0.57543691
21	−67.99149132	0.10289307	0.58007699
25	−69.32065105	0.10068359	0.57729656
29	−105.72862339	0.04923706	0.48767025
		$N=128$	
1	−80.89347839	0.04639893	0.40117195
2	−77.22580583	0.06295776	0.47399521
3	−73.43786240	0.07648926	0.51361278
4	−71.93675232	0.08345947	0.53266251
6	−71.10850430	0.08880615	0.54769675
9	−70.53600121	0.09255371	0.55752959
17	−69.95890045	0.09628906	0.56676912
25	−69.29977322	0.09834595	0.57137301
33	−68.75139713	0.10077515	0.57594641
41	−67.89687920	0.10183716	0.57863142
49	−66.76120186	0.10264282	0.58123560
57	−69.21525860	0.10157471	0.57946395
61	−104.57432938	0.04970703	0.48900685
		$N=256$	
1	−80.73009777	0.04661255	0.40166772
2	−77.22607231	0.06292725	0.47389125
3	−73.46642208	0.07637329	0.51333544
4	−71.95929623	0.08333740	0.53236954
6	−71.13770962	0.08861694	0.54725409
10	−70.46772671	0.09299927	0.55873035
17	−70.05028152	0.09572144	0.56546700
33	−69.59198570	0.09783325	0.57040469
49	−69.15290451	0.09915161	0.57313522
65	−68.87001705	0.10036621	0.57542419
81	−68.56850624	0.10156250	0.57766514
97	−67.91495895	0.10211182	0.57929190
113	−66.49593639	0.10307617	0.58240311
121	−69.16745663	0.10181885	0.58003297
122	−69.47800255	0.10039673	0.57756963
123	−70.38166904	0.09647827	0.57081524
124	−76.36186981	0.08026733	0.54389865
125	−104.06292439	0.04988327	0.48943934

TABLE X

Low-Pass Filter Design, Three Transition Coefficients (Type-2 Data, *N* Even)

BW	Minimax	T_1	T_2	T_3
		$N=16$		
1	−101.40284348	0.00737915	0.11050435	0.49920845
2	−104.84970379	0.00474243	0.11504207	0.54899404
3	−109.42771912	0.00368042	0.11363929	0.56198983
4	−129.92168427	0.00327759	0.10690445	0.58824614
		$N=32$		
1	−101.36804008	0.00750122	0.11125334	0.50005367
2	−99.62673664	0.01146851	0.15483406	0.59491490
4	−90.25220776	0.01828613	0.20058013	0.66114353
6	−90.54358006	0.01939697	0.21148202	0.67786618
8	−91.16841984	0.01898193	0.21139913	0.68027127
10	−106.64640045	0.01017456	0.16962101	0.64299580
12	−155.63167953	0.00352173	0.11656092	0.58328406
		$N=64$		
1	−108.25328732	0.00635376	0.10549329	0.49324515
2	−96.89522362	0.01149292	0.15247338	0.59059079
3	−93.50934601	0.01464844	0.17720349	0.63073228
4	−91.45420170	0.01686401	0.19265675	0.65259480
8	−89.21100140	0.02095337	0.22033124	0.68792394
12	−91.85765266	0.02094727	0.22664372	0.69828908
16	−91.86564636	0.02175903	0.23164135	0.70385697
20	−93.00161743	0.02034302	0.22588693	0.69939827
24	−90.38130093	0.02063599	0.22318429	0.69521828
28	−138.57364845	0.00465088	0.12720165	0.59773274
		$N=128$		
1	−109.42113113	0.00633545	0.10540650	0.49310051
2	−99.17228699	0.01088257	0.14983535	0.58804236
3	−94.18891430	0.01414185	0.17453122	0.62785978
5	−91.83029461	0.01736450	0.19833205	0.66187917
8	−89.42102623	0.02021484	0.21561932	0.68272648
16	−87.40464020	0.02316284	0.23232431	0.70110354
24	−88.17846870	0.01965942	0.22132125	0.69493783
32	−90.52461393	0.02318115	0.23939702	0.71174777
40	−90.07130432	0.02406006	0.24375106	0.71589811
48	−90.97514629	0.02312622	0.23958630	0.71239267
56	−88.51494217	0.02221069	0.23021766	0.70169053
60	−147.38232040	0.00330200	0.11666346	0.58657999
		$N=256$		
1	−109.40897179	0.00632324	0.10524188	0.49282388
2	−103.89405537	0.01022339	0.14791087	0.58696792
3	−93.76251125	0.01429443	0.17487576	0.62793383
5	−91.26832485	0.01748047	0.19825172	0.66139144
9	−88.72007465	0.02074585	0.21817298	0.68543746
16	−88.28249550	0.02236328	0.22890808	0.69809890
32	−91.24586964	0.02164307	0.23139950	0.70409545
48	−91.01594257	0.02251587	0.23625686	0.70892584
64	−90.69410038	0.02330322	0.24007374	0.71251523
80	−89.23360157	0.02214355	0.23625978	0.70987380
96	−89.70531368	0.02456665	0.24640362	0.71851692
112	−90.55468941	0.02351685	0.24150463	0.71425258
120	−89.99984550	0.02116089	0.22640173	0.69891484
121	−101.17623997	0.01504517	0.19942922	0.67579737
122	−100.22463131	0.00847168	0.16540943	0.64397531
123	−100.61959362	0.00181885	0.12179494	0.59890476
124	−162.43343735	0.00360641	0.11918625	0.58952593

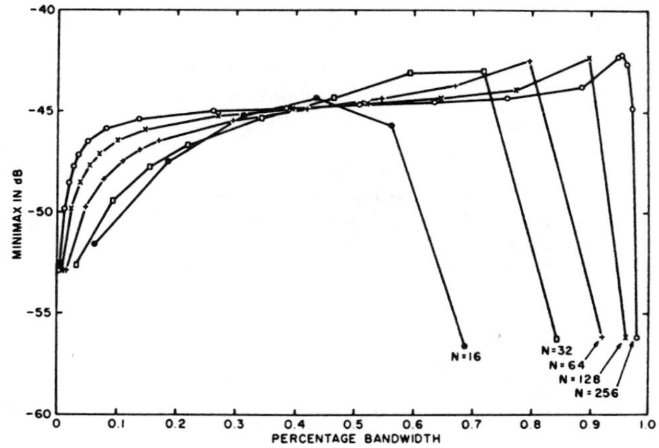

Fig. 15. The minimax as a function of percentage bandwidth for type-2 low-pass filters with one transition coefficient.

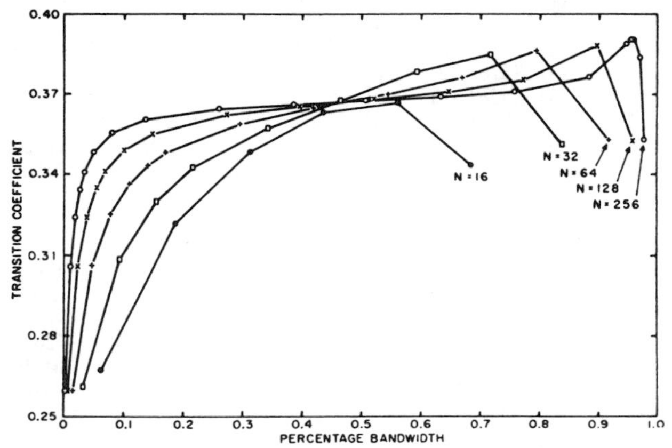

Fig. 16. The value of the transition coefficient as a function of percentage bandwidth for type-2 low-pass filters with one transition coefficient.

Fig. 17. The minimax as a function of percentage bandwidth for type-2 low-pass filters with two transition coefficients.

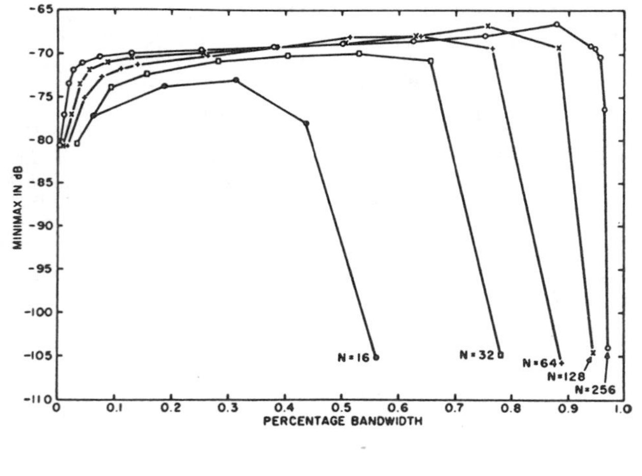

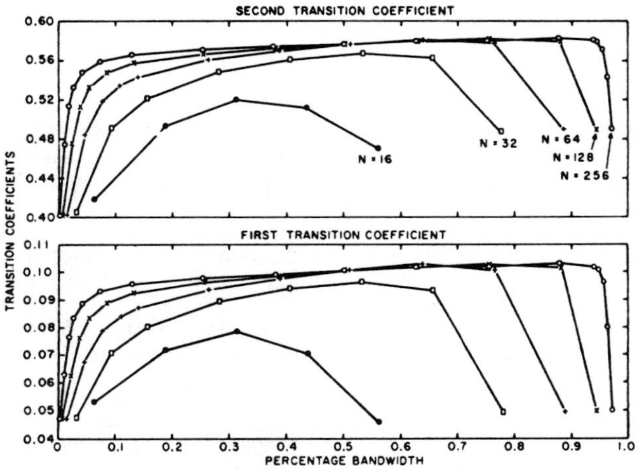

Fig. 18. The values of the transition coefficients as a function of percentage bandwidth for type-2 low-pass filters with two transition coefficients.

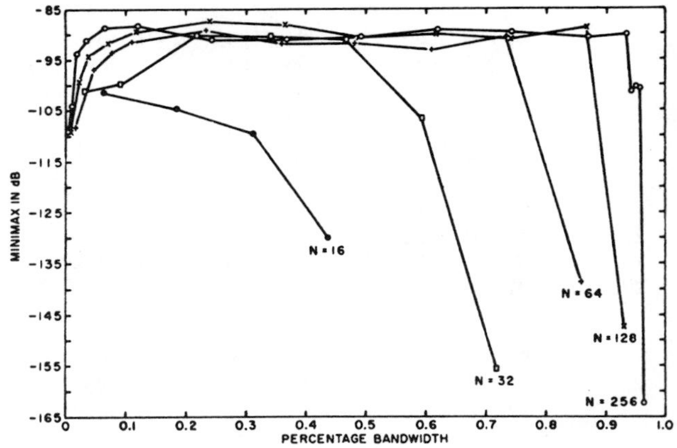

Fig. 19. The minimax as a function of percentage bandwidth for type-2 low-pass filters with three transition coefficients.

Fig. 20. The values of the transition coefficients as a function of percentage bandwidth for type-2 low-pass filters with three transition coefficients.

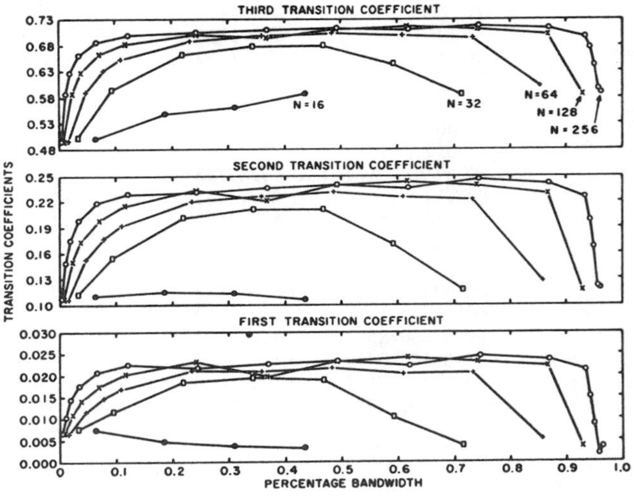

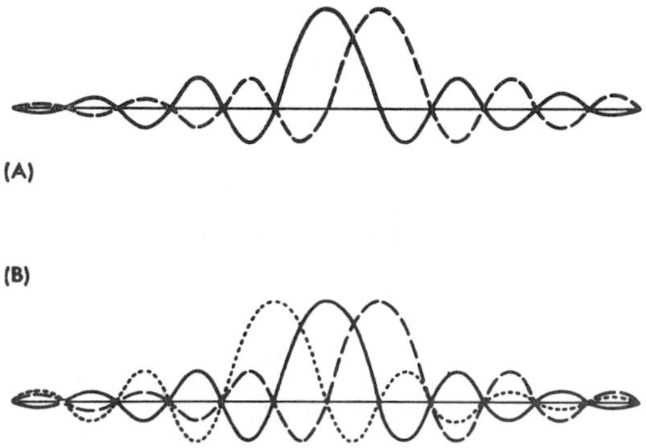

Fig. 21. The summation of an even number (2) and an odd number (3) of sin (x)/x curves.

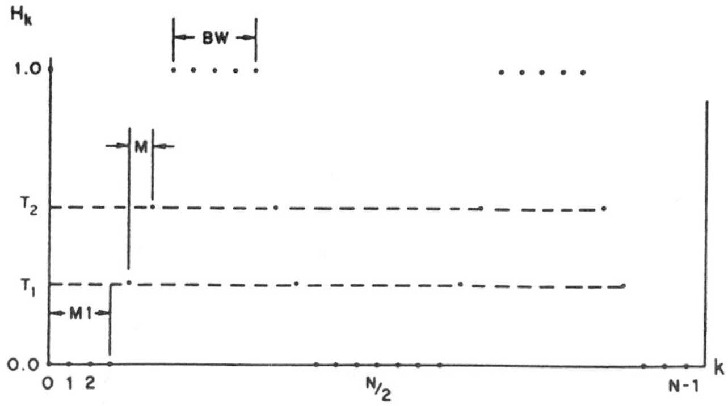

Fig. 22. Typical specifications for a type-1 bandpass filter.

type-1 filters. This can best be explained by referring to Fig. 21. In Fig. 21(A) the ripple from two (sin ω/ω) functions is shown. This case corresponds to a type-2 filter with no transition samples. The ripple peaks from each of the functions tend to cancel uniformly. In Fig. 21(B) the ripple from three (sin ω/ω) functions is shown. This case corresponds to a type-1 filter where the odd term comes from the unpaired frequency sample at zero frequency. The sidelobes from the additional (sin ω/ω) function are seen to add uniformly to all the ripples from Case A. Thus before trying to cancel the ripple with the transition coefficients, the ripple of Fig. 21(A) is significantly less than the ripple of (B). Experimentally it turns out that ripple cancellation for the data of Fig. 21(A) is also much better than for the data of (B). The reason all type-2 filters are not better than all type-1 filters is that as the number of elementary (sin ω/ω) functions increase, the difference in ripple heights between the sum of an even number and the sum of an odd number of such functions becomes smaller and smaller and is negligible for larger bandwidths.

Bandpass Filters

The nomenclature for defining a bandpass filter in terms of its frequency samples is given in Fig. 22; in addition to the parameters N, BW, and M, there is also the center frequency of the filter. We have defined the parameter $M1$ as the number of zero-valued samples preceding the first transition sample. Furthermore, for all cases considered, the bandpass filter samples were considered to be symmetrical about the center frequency. This arbitrary constraint is desirable for computational purposes since it reduces the number of variables by one half. In general, nonsymmetric transition samples lead to a somewhat lower minimax sidelobe, but this advantage seems canceled out by the increased computational cost.

We have approached the design problem in two ways. First, given a version of the optimization program, one can choose the parameters M, N, BW, and $M1$ and run the program to give any desired optimum bandpass filter. We have tabulated the results of a few runs for various values of N, $M1$, and BW, for one, two, and three symmetric transition coefficients. These data are shown in Tables XI through XIII. The most striking observation from these tables is the difference in minimax between odd and even values of bandwidth for small values of bandwidth. This effect is similar to the one discussed earlier for low-pass filters, and it is worthwhile for the designer to keep it in mind.

The second approach to the design of bandpass filters is to define "suboptimum" bandpass filters, which are derived very simply from the low-pass prototype by appropriately rotating the low-pass frequency samples (including the optimized transition coefficients) to the desired center frequency. An example is given in Fig. 23; the sampled passbands of the derived filter are identical with those of the low-pass, but at different locations. The resulting interpolated bandpass response can be obtained by adding the interpolated low-pass response which has been rotated counterclockwise to the same response rotated clockwise. Therefore, it is clear that the suboptimum filter minimax can never be more than 6 dB worse than the low-pass prototype. However, a truly optimum bandpass filter, as designed by our first approach, may be better than this low-pass prototype; therefore, there is no guarantee that suboptimum bandpass filters are within 6 dB of the optimum. Our experimental results show that a 3-dB loss of suboptimum (relative to optimum) is the usual case.

By allowing rotations of an integer $+\frac{1}{2}$ number of samples, as well as integer rotations, one can design either type-1 or type-2 bandpass filters from either type-1 or type-2 low-pass prototypes. It can be shown that in many cases one of the possible frequency transformations is superior to other transformations. A schematized example is shown in Fig. 24. Fig. 24(A) shows a type-2 low-pass filter with a double frequency ripple peak pass near the band edge. (This situation is typical of many type-2 low-pass filters.) The result of a frequency transformation of an integer number of rotations is shown in Fig. 24(B). The sidelobes add almost everywhere in the out-of-band region. The resulting design is a type-2 bandpass filter. The result of a frequency transformation of an integer

TABLE XI

Bandpass Filter Designs, One Transition Coefficient (Type-1 Data, *N* Even)

BW	*M*1	Minimax	T_1
		$N=16$	
3	2	−34.175276	0.45593262
		$N=32$	
5	2	−35.767563	0.40270386
5	3	−35.668758	0.39149780
5	4	−36.150829	0.39191895
5	5	−36.577216	0.39454346
6	2	−47.610400	0.30016480
6	3	−49.023605	0.30545654
6	4	−50.470645	0.30634766
6	5	−50.470645	0.30634766
7	2	−35.624309	0.40203247
7	3	−35.433974	0.38943481
7	4	−35.872498	0.38916626
7	5	−35.433974	0.38943481
		$N=128$	
13	10	−41.259604	0.38093262
18	20	−46.386827	0.34979248
19	20	−42.807318	0.37623901
20	20	−46.300830	0.35097046
30	8	−45.528528	0.35111694
35	2	−40.743021	0.35415039
35	8	−42.548930	0.36517944
35	16	−43.227605	0.36619873
35	20	−42.548930	0.36517944

TABLE XII

Bandpass Filter Designs, Two Transition Coefficients (Type-1 Data, *N* Even)

BW	*M*1	Minimax	T_1	T_2
		$N=16$		
1	2	−74.261646	0.09519653	0.59498585
		$N=32$		
3	2	−60.824677	0.11812134	0.61574359
3	3	−61.426695	0.11731567	0.61224050
3	4	−62.291398	0.12105713	0.61825728
3	5	−62.859379	0.12384644	0.62291631
4	2	−80.477118	0.05566406	0.45630774
4	3	−76.145997	0.06130981	0.46821324
4	4	−79.328351	0.05487671	0.45372102
4	5	−79.328351	0.05487671	0.45372102
5	2	−59.856571	0.12357788	0.62173523
5	3	−61.337764	0.12523804	0.62286495
5	4	−62.965949	0.12827148	0.62855407
5	5	−61.337764	0.12523804	0.62286495
		$N=128$		
11	10	−62.184930	0.11203003	0.59676391
16	20	−70.834468	0.09031982	0.55215414
17	20	−63.214016	0.10917969	0.59054608
18	20	−70.729173	0.09111938	0.55425376
28	8	−69.713386	0.09022827	0.55326966
33	2	−62.526547	0.10723877	0.58803040
33	8	−61.611602	0.11864624	0.60405885
33	16	−62.165339	0.11191406	0.59243833
33	20	−61.611602	0.11864624	0.60405885

TABLE XIII

Bandpass Filter Designs, Three Transition Coefficients (Type-1 Data, *N* Even)

BW	*M*1	Minimax	T_1	T_2	T_3
			$N=32$		
1	2	−92.550184	0.01790771	0.20360144	0.68572443
1	3	−90.696831	0.01699829	0.19934001	0.68240900
1	4	−90.352082	0.01849976	0.20508501	0.68649875
1	5	−96.630682	0.01597290	0.19530278	0.67931499
2	2	−113.580111	0.00490723	0.09406196	0.47584767
2	3	−110.531433	0.00568848	0.10004413	0.48475702
2	4	−113.033724	0.00422363	0.08800332	0.46619571
2	5	−113.033724	0.00422363	0.08800332	0.46619571
3	2	−89.015630	0.02207642	0.23341023	0.71117572
3	3	−85.383010	0.02587891	0.24743934	0.72135515
3	4	−95.484849	0.01950073	0.22380012	0.70428223
3	5	−85.383010	0.02587891	0.24743934	0.72135515
			$N=128$		
9	10	−86.703837	0.02807617	0.26306832	0.73460394
14	20	−90.302061	0.01867676	0.20640635	0.67233389
15	20	−84.654966	0.03029785	0.27145008	0.74047619
16	20	−90.789648	0.01875610	0.20830675	0.67549529
26	8	−91.905838	0.01835937	0.21150551	0.68240265
31	2	−83.873835	0.02974243	0.27190412	0.74186481
31	8	−84.547179	0.02946167	0.26787226	0.73722876
31	16	−85.064596	0.03010254	0.27143276	0.74060358
31	20	−84.547179	0.02946167	0.26787226	0.73722876

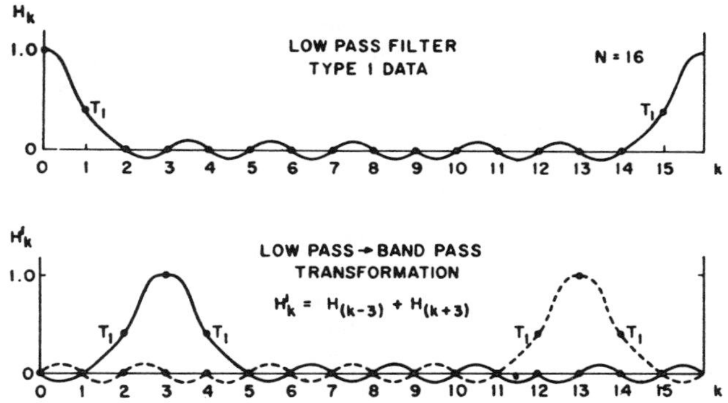

Fig. 23. An illustration of the low-pass to bandpass transformation.

Fig. 24. Transformation of a type-2 (*N* even) low-pass filter (A) to both type-2 (B) and type-1 (C) bandpass filters. (B) Integer number of rotations. (C) Integer + 1/2 number of rotations.

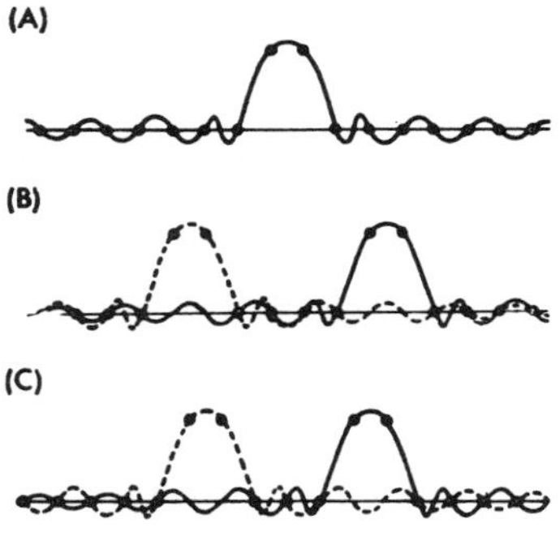

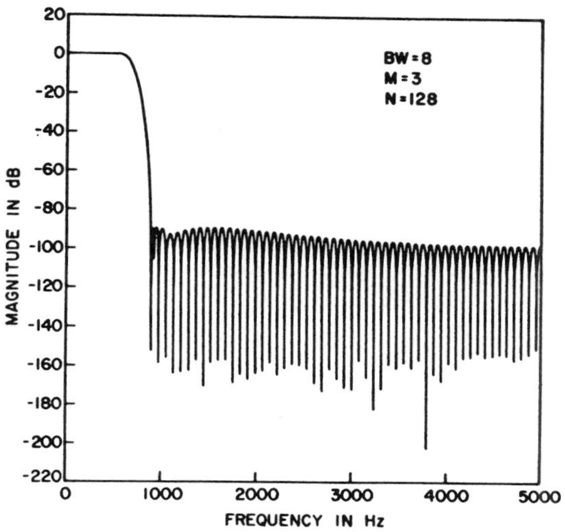

Fig. 25. The frequency response for a type-2 low-pass filter.

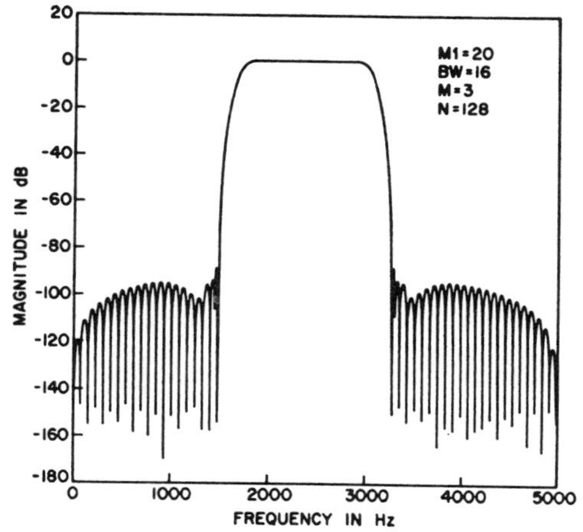

Fig. 27. The frequency response for a type-1 bandpass filter obtained from a transformation of the type-2 low-pass filter of Fig. 25 (integer+1/2 number of rotations).

Fig. 26. The frequency response for a type-2 bandpass filter obtained from a transformation of the type-2 low-pass filter of Fig. 25 (integer number of rotations).

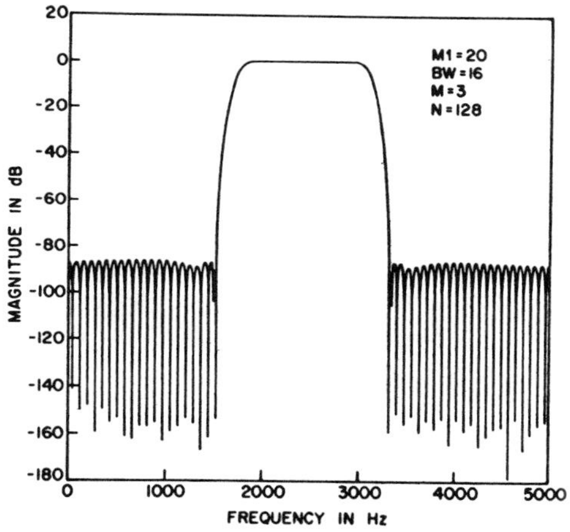

Fig. 28. The frequency response for an optimum type-1 bandpass filter.

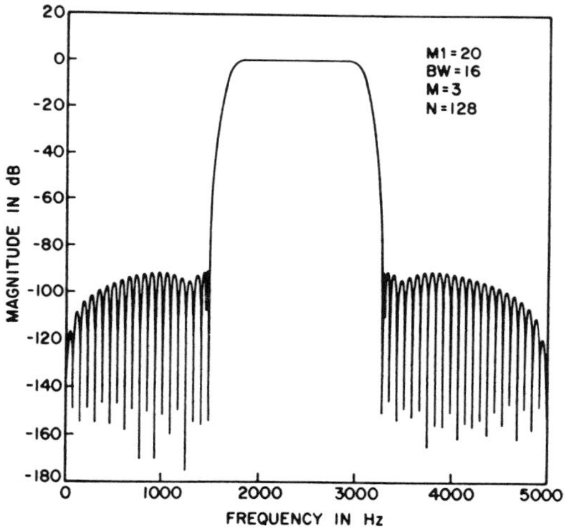

$+\frac{1}{2}$ number of rotations is shown in Fig. 24(C). The sidelobes cancel almost everywhere. The resulting design is a type-1 bandpass filter, the characteristics of which are superior to the filter in part (B).

A practical demonstration of these ideas is shown in Figs. 25 through 27. Fig. 25 shows an optimum type-2 low-pass design. The result of an integer number of rotations is shown in Fig. 26; the result of an integer $+\frac{1}{2}$ number of rotations is shown in Fig. 27. The minimax of the low-pass design is -89.4 dB, and the ripple envelope falls to -98 dB at high frequency. The peak ripple of Fig. 26 is about -86 dB, and the ripple envelope falls to -87 dB at high frequency. The peak ripple of Fig. 27 is -89.2 dB, and the ripple envelope falls rapidly at high frequency to about -119 dB. Thus the second frequency transformation is far superior to the first in this case.

For comparison purposes Fig. 28 shows the optimum type-1 bandpass filter as designed by our first approach. The minimax is -90.8 dB, and the ripple envelope drops off rapidly at high frequencies to about -120 dB. It is clear that the suboptimum filter of Fig. 27 is quite similar to the optimum design.

Comparison Between Window and Sampling Design

Direct comparison between the classical window design technique and the frequency-sampling technique described here is difficult; however, we have enough numerical design information to present some comparisons for the low-pass filter case. In addition, since Kaiser [2] in his exposition of the window method surveys the work of himself and others on wide-band differentiators, we thought it useful to design a few differentiators, using the sampling method for further comparisons.

Low-Pass Design Comparisons

The window function introduced by Kaiser has properties very close to those of the prolate spheroidal window [14] and is thus quite close to optimum, given the constraints of a window function design. Kaiser has given an approximate formula for the number of terms N required for a 0.01 percent of peak overshoot in the response. (For the design of a low-pass filter this corresponds to a peak ripple of -80 dB.) The number of terms required is

$$N \approx \frac{11.5}{\text{percentage transition band—windowing}} = \frac{11.5}{(\omega_2 - \omega_1)/(\omega_s/2)}. \tag{20}$$

For the sampling technique the percentage transition band, assuming three transition points, is

$$\text{percentage transition band—sampling} = \frac{4}{N/2} = \frac{8}{N}. \tag{21}$$

All of the low-pass designs of the sampling method have their peak ripple lower than -85 dB, hence somewhat better than the design constraint of (20). Yet (20) implies that for a given N, to achieve -80 dB peak ripple requires a percentage transition band of about $11.5/N$ for Kaiser's window, or about 50 percent bigger than that required by the sampling technique. It should be pointed out, however, that, according to Kaiser, the in-band ripple characteristics of filters designed using the Kaiser windows are equally as good as the out-of-band characteristics. No such claim can be made for the sampling technique because no constraint was placed on the in-band ripple in the design. However, we have found that the largest value of in-band ripple for any of the filters we have designed was less than 0.15 dB.

Helms recently proposed another window possessing certain desirable properties—the Dolph–Chebyshev window. For this window the number of terms N needed to achieve a peak ripple of -80 dB is

$$N \approx \frac{13.0}{(\omega_2 - \omega_1)/(\omega_s/2)} \tag{22}$$

where $(\omega_2-\omega_1)/(\omega_s/2)$ is the percentage transition band of the filter. Equation (22) shows that the window requires slightly more terms (larger N) than the equivalent Kaiser window and again about 50 percent more terms than the sampling method to achieve this design constraint.

Wide-Band Differentiators

As mentioned earlier, the sampling technique is amenable to filter designs other than standard low-pass or bandpass filters. To illustrate how to apply this procedure to a more general frequency-response characteristic, various wide-band differentiators were designed.

The basic design used data for type-1 filters. Since the ideal frequency response for a differentiator has characteristic response

$$H(\omega) = j\omega, \tag{23}$$

the frequency samples H_k were set to values

$$H_k = H(e^{j2\pi k/N}) = \begin{cases} j\dfrac{k}{N} & k = 0, 1, \cdots, L \\ -\dfrac{j(N-k)}{N} & \\ \quad k = N - L, \cdots, N-1 & \\ \text{optimally chosen, all other } k. & \end{cases} \tag{24}$$

A single value of 19 was used for N in order to compare the resulting differentiators with those described by Kaiser [1].

The design criterion used was one which sought to minimize either the maximum absolute deviation or the maximum absolute relative deviation between the interpolated frequency response and the ideal differentiator frequency response over some specified range. For the case studied ($N=19$) there were seven fixed values of H_k and three variable samples. Various normalized in-band frequency ranges were used for the minimization. These frequency ranges included:

1) 0 to 0.737 full band
2) 0 to 0.789 full band
3) 0 to 0.842 full band.

The resulting differentiators are tabulated with respect to the maximum absolute error and transition coefficients in Table XIV. A typical interpolated frequency response and the absolute error for a minimum absolute error differentiator in the range 0 to 0.737 full band are shown in Fig. 29. The peak error in this range is 0.00019 and occurs at a normalized frequency near the edge of the differentiator band. However, as seen in Fig. 29, the peak error remains large even for low frequencies. Fig. 30 shows the frequency response and absolute error for a minimum relative error differentiator. The peak error here is 0.0003; however the error is much smaller at low frequencies (on the order of 10^{-5} to 10^{-4}) and remains small for most of the frequency range.

Kaiser [1] has compared six techniques for designing nonrecursive wide-band differentiators. The best result among those presented uses a Kaiser window ($\omega_a\tau=6.0$) with differentiation bandwidth of about 0.8 full band and

TABLE XIV

Differentiator Design (Type-1 Data, $N=19$)

Percent Band-width	Peak Error	T_1	T_2	T_3
	Minimized Absolute In-Band Error			
0.737	0.0001891	0.37163696	0.76372207	0.73665305
0.789	0.0010745	0.42243652	0.80468043	0.73684211
0.842	0.0051854	0.48053589	0.83691982	0.73684211
	Minimized Relative In-Band Error			
0.737	0.0003032	0.36164551	0.75508157	0.73642816
0.789	0.0017759	0.39659424	0.78972235	0.73752642
0.842	0.0136256	0.43627930	0.82847971	0.73684211

Fig. 29. The frequency response (A) and absolute error curve (B) for a wide-band differentiator whose transition coefficients were chosen so as to minimize the maximum absolute error.

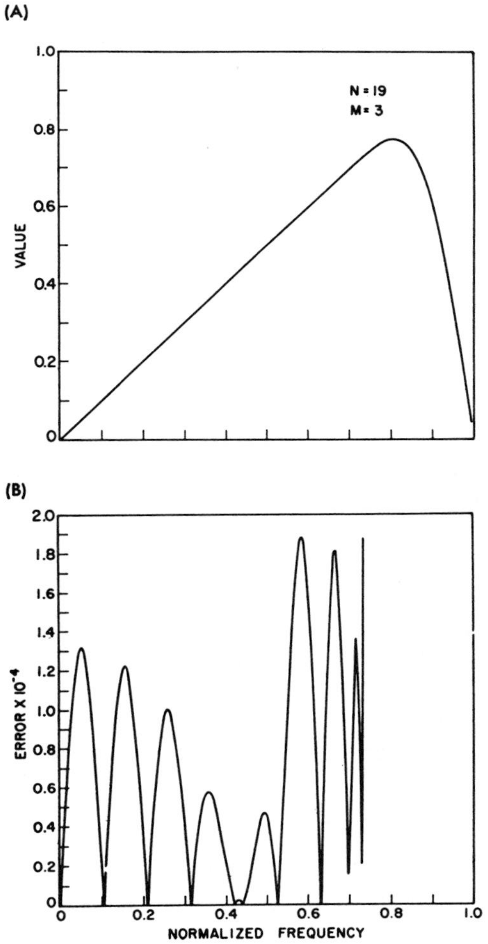

Fig. 30. The frequency response (A) and relative error curve (B) for a wide-band differentiator whose transition coefficients were chosen so as to minimize the maximum relative error.

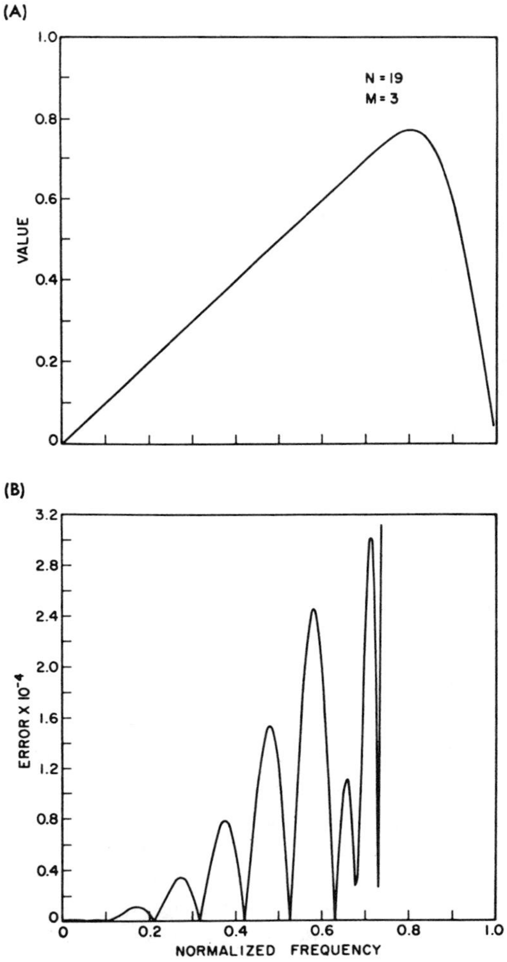

peak error of 0.0016. Both the peak error and the shape of the error curve are similar to the minimum relative error design presented in Table XIV for a bandwidth of 0.789 full band. Hence in this case these very different techniques yield filter designs which are quite similar.

Experimental Results Obtained When Finite Register Length Is Taken into Account

In performing the search for the optimum filter designs, a 60-bit word length computer (CDC-6600) was used and the results checked with a 36-bit word length machine (GE-635). We can therefore assume no significant truncation errors occurred in this computation. However, the synthesis of a given filter could conceivably be performed on an 18-, 16-, or even 12-bit machine, or perhaps with special purpose hardware where the shortest possible word length is desirable. Much work has been done recently on the subject of the effects of finite register length. This work can roughly be divided into two parts:

1) truncation of the parameters, which changes the filter shape;
2) truncation of the variables, which introduces noise into the output.

In an earlier section we saw that there are three standard nonrecursive filter realizations: direct convolution, frequency sampling, and fast convolution. Weinstein [15] has treated the latter two realizations for case 2, both theoretically and experimentally. Noise in the direct-convolution realization is easily computed by assuming that each multiplication introduces an independent noise of variance $E_0^2/12$, where E_0 is a single quantization level. The total noise variance is $E_0^2 \cdot N/12$ where N is the number of multiplications in the realization, i.e., the length of the filter impulse response.

For parameter truncation simple models are not readily available so that theoretical prediction cannot safely be made. Therefore we performed measurements for the standard realizations.

1) Direct Convolution: The impulse response of several of the type-1 designs of low-pass filters was accurately computed and the coefficients were then truncated. Values for N of 16 and 32 were used since a direct convolution realization would not generally be used for larger values of N. The results of truncation are shown in Table XV. The maintenance of at least −80 dB rejection required 17 bits, and the maintenance of −73 dB rejection required 14 bits for three transition samples.

2) Frequency Sampling: The frequency samples for several type-1 low-pass filters were truncated. Since most of the frequency samples for the low-pass case were either 0 or 1.0, only three coefficients were actually affected by the truncation. The results of truncation are presented in Table XVI. Truncation to 17 bits did not seriously affect the peak ripple. The maintenance of at least −75 dB rejection required only 11 bits for the coefficients. It should be noted that the coefficients of the resonators in the frequency sampling realization (Fig. 6) were not truncated. Hence the results here are an overbound on the actual results of coefficient truncation.

3) Fast Convolution: The effects of truncation are straightforward. Each of the interpolated frequency response coefficients are truncated; hence coefficients falling below the quantization level are truncated to have 0 value.

TABLE XV

Truncation of Frequency Samples of Type-1 Low-Pass Filters, Three Transition Coefficients

Number of Bits	Minimax	Number of Bits	Minimax
$N=16$, $BW=1$		$N=64$, $BW=28$	
36	−96.63	36	−115.24
17	−95.05	17	−112.80
14	−88.60	14	−111.43
11	−92.74	11	−89.84
8	−75.57	8	−70.75
5	−38.86	5	−62.35
$N=16$, $BW=4$		$N=128$, $BW=8$	
36	−114.12	36	−87.41
17	−113.53	17	−87.35
14	−113.33	14	−83.95
11	−76.51	11	−75.71
8	−59.29	8	−72.10
5	−41.81	5	−62.66
$N=32$, $BW=2$		$N=128$, $BW=60$	
36	−89.37	36	−114.44
17	−88.51	17	−110.66
14	−84.75	14	−99.74
11	−88.23	11	−89.86
8	−75.99	8	−70.75
5	−59.03	5	−62.35
$N=32$, $BW=12$		$N=256$, $BW=8$	
36	−112.50	36	−89.21
17	−111.57	17	−88.41
14	−109.60	14	−87.23
11	−79.18	11	−75.80
8	−62.24	8	−72.10
5	−62.34	5	−62.66
$N=64$, $BW=4$		$N=256$, $BW=124$	
36	−87.48	36	−112.80
17	−87.39	17	−109.22
14	−86.45	14	−100.43
11	−81.26	11	−82.60
8	−59.53	8	−64.07
5	−39.72	5	−62.35

Nonuniform Frequency Samples

In this section we will show that a finite-duration impulse-response filter could be designed from frequency samples placed *anywhere* in the z-plane. Whereas in the

TABLE XVI

Truncation of Impulse-Response Coefficients of Type-1 Low-Pass Filters, Three Transition Coefficients

Number of Bits	Minimax	Number of Bits	Minimax
$N=16$, $BW=1$		$N=32$, $BW=4$	
36	−96.63	36	−86.97
17	−96.33	17	−85.74
14	−83.67	14	−75.23
11	−67.34	11	−58.75
8	−51.14	8	−39.76
5	−33.18	5	−31.44
$N=16$, $BW=2$		$N=32$, $BW=6$	
36	−95.47	36	−86.69
17	−92.73	17	−82.39
14	−78.52	14	−73.92
11	−67.26	11	−55.41
8	−46.06	8	−36.03
5	−27.32	5	−24.76
$N=16$, $BW=3$		$N=32$, $BW=8$	
36	−98.70	36	−88.41
17	−94.57	17	−83.14
14	−73.63	14	−78.70
11	−55.53	11	−56.21
8	−38.73	8	−38.11
5	−39.71	5	−27.86
$N=16$, $BW=4$		$N=32$, $BW=10$	
36	−114.12	36	−92.85
17	−95.15	17	−87.03
14	−83.53	14	−74.11
11	−63.30	11	−53.65
8	−42.20	8	−36.31
5	−29.31	5	−25.38
$N=32$, $BW=2$		$N=32$, $BW=12$	
36	−89.37	36	−112.50
17	−88.26	17	−89.05
14	−82.35	14	−68.20
11	−62.82	11	−61.38
8	−50.18	8	−44.26
5	−30.21	5	−26.32

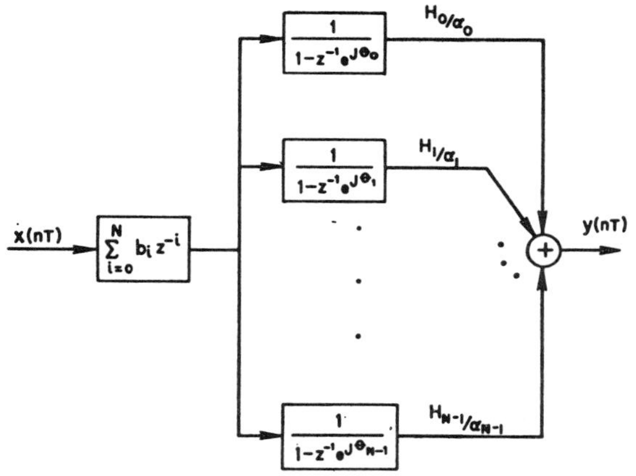

Fig. 31. A method of realization of the nonuniform sampled nonrecursive digital filter.

previous sections we have restricted ourselves to the case of uniformly spaced samples around the unit circle, in this section we will discuss an extension of the techniques to nonuniformly spaced samples around the unit circle.

Let $h(n)$, $n=0, 1, \cdots, N-1$, be the impulse response of a nonrecursive filter with z-transform $H(z)$. It can be shown that N independent values of $H(z)$ can be specified for this filter by writing $H(z)$ in the form [16]

$$H(z) = \sum_{k=0}^{N-1} \frac{H_k}{\alpha_k} \frac{\prod_{i=0}^{N-1} (1 - z^{-1}a_i)}{(1 - z^{-1}a_k)} \tag{25}$$

where

$$\alpha_k = \prod_{i=0,\, i\neq k}^{N-1} (1 - a_i/a_k) \tag{26}$$

and $\{a_k\}$ are the z-plane positions at which $H(a_k)=H_k$. $H(z)$ in (25) can be shown to be an $(N-1)$st order polynomial in z^{-1}. Thus the design of a nonrecursive filter can be thought of in terms of deriving suitable values for $\{a_k\}$ and $\{H_k\}$ of (25).

In this section we will consider the nonuniform case where

$$a_k = e^{j2\pi\theta_k/N}, \tag{27}$$

i.e., nonuniformly spaced samples around the unit circle. For this set of samples (25) can be manipulated into the form

$$H(z) = \sum_{k=0}^{N-1} \frac{H_k \sum_{i=0}^{N-1} b_i z^{-i}}{\alpha_k (1 - z^{-1} e^{j2\pi\theta_k/N})} \tag{28}$$

where

$$\alpha_k = \prod_{i=0,\, i\neq k}^{N-1} (i - e^{j2\pi(\theta_i - \theta_k)N}) \tag{29}$$

and the internal summation in (28) comes from expanding the product in the numerator of (25). The realization of (28) is shown in Fig. 31. The internal summation is realized as a nonrecursive filter, whose output is fed into N parallel channels, each consisting of a complex resonator followed by a complex multiplication. The outputs of the parallel channels are summed to give the filter output.

To obtain the interpolated frequency response of networks of the form of (28), the network realization of Fig. 31 is first excited by an impulse to give the impulse response; zero-valued samples are added to the impulse response; and the entire array is transformed using the FFT. Since (28) is still linear in the coefficients, the H_k, the techniques for finding optimum values of transitions are still valid.

Two sets of nonuniform data were investigated. These data are shown in Fig. 32. The first set, shown in Fig. 32(A), consisted of 16 uniform samples with an extra sample placed between the third and fourth uniform

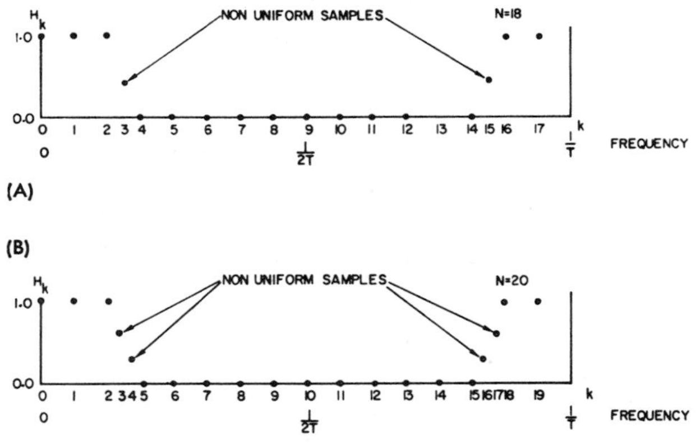

Fig. 32. Two cases of nonuniform frequency samples investigated in this paper.

samples. The design criterion was to choose an optimum position for the nonuniform sample to minimize out-of-band ripple. The optimum value turned out to be 0.468 yielding a peak out-of-band ripple of −20.3 dB, a peak in-band ripple of 1.2 dB, and a flat ripple envelope. For comparison purposes, the case of 16 uniformly spaced samples with no transitions was examined. Here the peak out-of-band ripple was −15.7 dB, the peak in-band ripple was 0.8 dB; and the ripple envelope fell to −24 dB at high frequencies.

Since the peak ripple was reduced by about 6 dB from the uniform case, a second nonuniform case, Fig. 32(B), was studied. Two additional samples were placed between the third and fourth uniform samples. The design program chose optimum values for these transitions to minimize the peak ripple. Here the results were discouraging as the peak ripple was increased to −11.3 dB.

The results obtained with nonuniformly spaced samples have not been entirely encouraging. Further work must be done before any conclusions can be arrived at as to the advantages over uniform sampling.

Conclusion

This paper has presented a technique for designing many types of finite-duration impulse-response digital filters from considerations strictly in the digital frequency domain. The ideal frequency response of the filter is approximated by placing appropriate frequency samples in the z-plane and then choosing the remaining frequency samples to satisfy an optimization criterion. This technique has been applied successfully to the design of low-pass and bandpass filters, as well as wide-band differentiators. The extension of this procedure to standard filters, such as bandstop and high-pass filters, as well as Hilbert transform filters, notch filters, double differentiators, and many others is straightforward.

The design program is sufficiently simple to implement so that it can be programmed to meet the requirements of the individual user. However, should the user merely desire a standard filter with good out-of-band characteristics, he can use the data included in the tables of this paper and proceed from there. Should the user desire a value of bandwidth which is not in the tables, a simple technique would be to interpolate linearly between the nearest values in the table. This will generally yield a suboptimum filter which is almost as good as the optimum.

The design of bandpass, bandstop, and high-pass filters can be treated as a separate design problem, using the frequency sampling technique described; or else simple frequency transformations of low-pass filters can be used to derive suboptimum designs. In many cases these suboptimum designs are nearly optimum.

The frequency sampling technique has been shown to be competitive with the standard window technique in that the number of terms needed to achieve a desired peak ripple in the stopband using this technique is about 50 percent less than the number of terms using the optimum windows described by Kaiser and Helms.

The extension of the frequency sampling technique to include nonuniform sampling points has been discussed briefly. More work must be done before the limitations and advantages of nonuniform samples are fully understood and appreciated.

Acknowledgment

The authors would like to express their appreciation for helpful technical discussions of the material in this paper with Dr. K. Jordan of M.I.T. Lincoln Lab. and Dr. R. Schafer of Bell Telephone Labs.

We would also like to thank Miss June Ley for valuable clerical assistance in the preparation of this manuscript.

References

[1] J. F. Kaiser, "Digital filters," ch. 7 in *System Analysis by Digital Computers*, F. F. Kuo and J. F. Kaiser, Eds. New York: Wiley, 1966.

[2] B. Gold and C. M. Rader, *Digital Processing of Signals*. New York: McGraw-Hill, 1969, ch. 3.

[3] H. D. Helms, "Nonrecursive digital filters: Design methods for achieving specifications on frequency response," *IEEE Trans. Audio and Electroacoustics*, vol. AU-16, pp. 336–342, September 1968.

[4] M. A. Martin, "Digital filters for data processing," Missile and Space Div., General Electric Co., Tech. Information Series Rept. 62-SD484, 1962.

[5] B. Gold and K. L. Jordan, Jr., "A direct search procedure for designing finite duration impulse response filters," *IEEE Trans. Audio and Electroacoustics*, vol. AU-17, pp. 33–36, March 1969.

[6] T. G. Stockham, "High-speed convolution and correlation," *1966 Spring Joint Computer Conf., AFIPS Proc.*, vol. 28. Washington, D.C.: Spartan, 1965, pp. 229–233.

[7] H. D. Helms, "Fast Fourier transform method of computing difference equations and simulating filters," *IEEE Trans. Audio and Electroacoustics*, vol. AU-15, pp. 85–90, June 1967.

[8] C. M. Rader and B. Gold, "Digital filter design techniques in the frequency domain," *Proc. IEEE*, vol. 55, pp. 149–171, February 1967.

[9] J. W. Cooley and J. W. Turkey, "An algorithm for machine computation of complex Fourier series," *Math. Computation*, vol. 19, pp. 297–301, April 1965.

[10] L. R. Rabiner, R. W. Schafer, and C. M. Rader, "The chirp z-transform algorithm and its application," *Bell Sys. Tech. J.*, vol. 48, pp. 1249–1292, May 1969.
[11] G. Hadley, *Linear Programming*. Reading, Mass.: Addison-Wesley, 1963, ch. 2.
[12] R. G. Gallager, *Information Theory and Reliable Communication*. New York: Wiley, 1968, ch. 4.
[13] A. Papoulis, "On the approximation problem in filter design," *IRE Conv. Rec.*, pt. 2, pp. 175–185, 1957.
[14] D. Slepian and H. O. Pollak, "Prolate spheroidal wave functions, Fourier analysis and uncertainty—I and II," *Bell Sys. Tech. J.*, vol. 40, pp. 43–84, 1961.
[15] C. J. Weinstein, "Quantization effects in digital filters," Ph.D. dissertation, Dept. of Elec. Engrg., M.I.T., Cambridge, Mass., July 1969.
[16] A. Cauchy, "Analyse mathématique-memoire sur diverses formules d'analyse," in *Oeuvres Complètes*, ser. 1, vol. 6. Paris, France: Gauthier-Villars, 1888, pp. 63–78.

DESIGN OF NONRECURSIVE DIGITAL FILTERS WITH LINEAR PHASE

Indexing term: Digital filters

A new class of selective nonrecursive digital filters with independently prescribed equiripple passband and stopband attenuation and linear phase is obtained by numerical solution of a set of nonlinear equations. Some examples are given, and a comparison is made of the new solutions and those previously known.

Nonrecursive digital filters have been used as systems with special frequency responses not obtainable with other devices. The design of differentiating systems has been frequently described, and in Reference 2 the design of filters suitable for the Hilbert transformation has been described. A catalogue of filters of this type has been published.[3]

This letter deals with the design of nonrecursive digital filters having equal-ripple attenuation in the passband, as well as in the stopband, and linear phase. Filters of this type are described by a transfer function

$$H(z) = \frac{1}{z^{2n}} \frac{1}{2} \sum_{\mu=0}^{n} d_\mu (z^{n+\mu} + z^{n-\mu}) = \frac{1}{z^{2n}} H_0(z)$$

$$= \frac{1}{z^{2n}} \frac{d_n}{2} \prod_{\mu=1}^{n} (z - z_{0\mu})(z - z_{0\mu}^{-1}) \quad \text{with} \quad |z_{0\mu}| \leqslant 1$$

With $z = \exp(j\omega T) = \exp(j2\pi\Omega)$, we obtain the frequency response

$$H(\Omega) = \frac{1}{\exp(jn2\pi\Omega)} \sum_{\mu=0}^{n} d_\mu \cos \mu 2\pi\Omega$$

$$= \frac{1}{\exp(jn2\pi\Omega)} H_0(\Omega)$$

Here $H_0(z)$ is a mirror-image polynomial (see, for example,

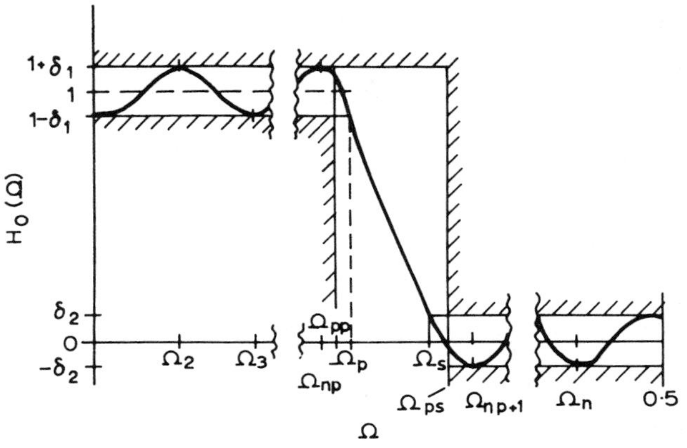

Fig. 1 *Tolerance scheme and wanted frequency response $H_0(\Omega)$*

Reference 4), having zeros which lie on the unit circle or reciprocal to it. The coefficients d_μ have to be found in such a way that $H_0(\Omega)$ satisfies the tolerance scheme indicated in Fig. 1 for the lowpass case. This tolerance scheme can be described by four parameters:

δ_1 = tolerated deviation of 1 or $a_{max} = 20 \log (1+\delta_1)$
= tolerated attenuation in passband

δ_2 = tolerated deviation of 0 or $a_{min} = -20 \log \delta_2$
= prescribed minimum attenuation in stopband

Ω_{pp} = prescribed cutoff frequency of passband

Ω_{ps} = prescribed cutoff frequency of stopband.

Problems of this type have been treated using Fourier-series expansions of the wanted frequency response and appropriate window functions.[1] Another method has recently been published.[5] Here equally spaced points Ω_ν are prescribed, where $H_0(\Omega_\nu) = 1$ in the passband or $H_0(\Omega_\nu) = 0$ in the stopband, leaving only three or four values $H_0(\Omega_\nu)$ in the transition range as free parameters to control the frequency response in the passband or stopband. The method leads to good results for large n.

The solution presented here uses all the free parameters of the system and gives independent control over δ_1 and δ_2.

Disregarding the cutoff frequencies to begin with, we write down a set of $2n$ equations for the unknown parameters $d_0, \ldots, d_n$ and the unknown frequencies $\Omega_2, \ldots, \Omega_n$, where $H_0(\Omega)$ has extremums (see Fig. 1)

$$H_0(\Omega_1 = 0) = 1(\mp)\delta_1$$

$$H_0(\Omega_2) = 1(\pm)\delta_1 \qquad H_0'(\Omega_2) = 0$$

$$\vdots \qquad \vdots$$

$$H_0(\Omega_{n_p}) = 1+\delta_1 \qquad H_0'(\Omega_{n_p}) = 0$$

$$H_0(\Omega_{n_p+1}) = -\delta_2 \qquad H_0'(\Omega_{n_p+1}) = 0$$

$$H_0(\Omega_{n_p+2}) = +\delta_2 \qquad H_0'(\Omega_{n_p+2}) = 0$$

$$\vdots \qquad \vdots$$

$$H_0(\Omega_n) = (\mp)\delta_2 \qquad H_0'(\Omega_n) = 0$$

$$H_0(\Omega_{n+1} = 0{\cdot}5) = (\pm)\delta_2$$

Here n_p is the number of extremums in the passband $0 \leqslant \Omega < \Omega_p$ which will be picked arbitrarily. If n_s is the number of extremums in the stopband $\Omega_s < \Omega \leqslant 0{\cdot}5$, then $n_p + n_s = n+1$.

This set of $2n$ equations can be solved numerically using an iterative procedure. We obtain a solution having the required δ_1 and δ_2 and certain cutoff frequencies Ω_p and Ω_s, which, in general, are unequal to the prescribed Ω_{pp} and Ω_{ps}. Fig. 2 shows two examples for $2n = 20$, $\delta_1 = 0{\cdot}05$,

Reprinted with permission from *Electron. Lett.*, vol. 6, May 28, 1970.

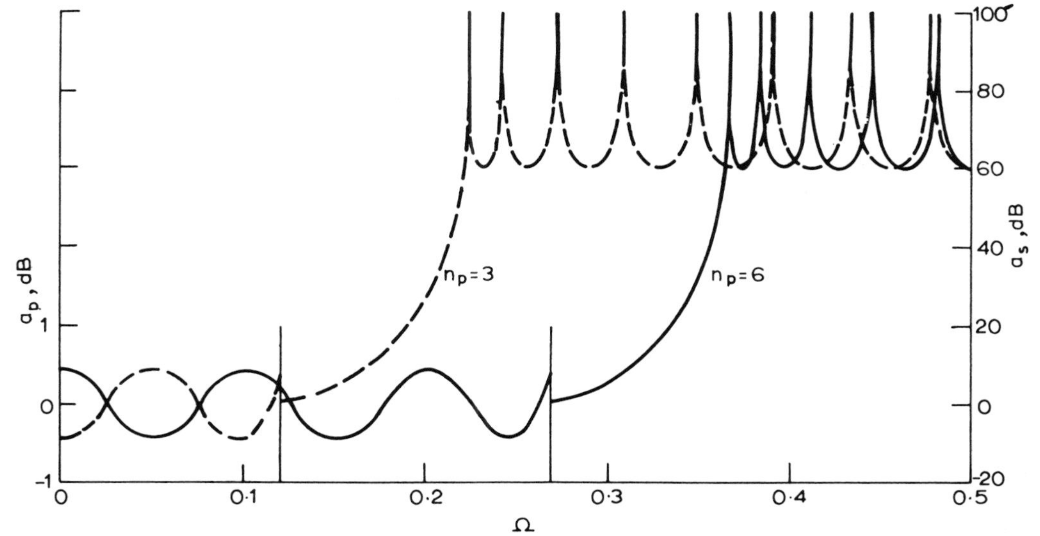

Fig. 2 *Attenuation of filters with* $2n = 20$, $\delta_1 = 0{\cdot}05$, $\delta_2 = 0{\cdot}001$

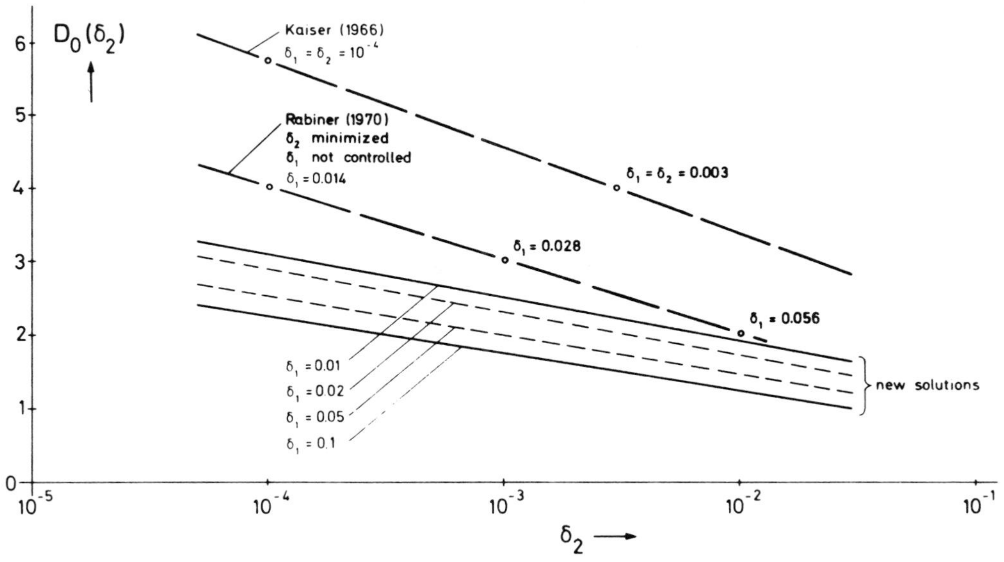

Fig 3 Figure of merit D_0 as a function of tolerated deviation δ_2 in stopband

$\delta_2 = 0{\cdot}001$ and $n_p = 3$ or 6, respectively. Obviously, for fixed δ_1 and δ_2 and fixed degree $2n$, only fixed pairs of cutoff frequencies Ω_p and Ω_s can be obtained choosing different values n_p. Since only n different values for n_p can be chosen, there exist only n different nonrecursive filters for fixed n, δ_1 and δ_2. A lowpass–lowpass transformation usable in the design of recursive filters to obtain any cutoff frequency is not applicable here, since it transforms the nonrecursive filter into a recursive one with nonlinear phase. However, the transformation into highpass, bandpass or bandstop with fixed cutoff frequencies is easily effected.

Using this method, roughly 400 filters have been calculated with $2n$ between 6 and 42, $\delta_1 = 10^{-2} \ldots 10^{-1}$, $\delta_2 = 5 \times 10^{-5} \ldots 10^{-1}$ and different values of n_p, and the parameters of these filters are available.

As in References 1 and 5, a figure of merit can be defined by

$$D = 2n(\Omega_s - \Omega_p) = D(n, n_p, \delta_1, \delta_2)$$

The smaller is D, the better is the filter.

As indicated, D depends in general on n, n_p and the tolerated deviation in the passband and stopband. An examination of the results shows this expression to become independent of n and nearly independent of n_p for $2n > 30$. We define

$$D_0 = 2n(\Omega_s - \Omega_p) = D_0(\delta_1, \delta_2)$$

This limit, achieved for the filters calculated so far, has been drawn in Fig. 3 as a function of δ with δ_1 as parameter. The corresponding values published in References 1 and 5 are indicated. A comparison with a recursive filter may be of interest. A frequency response, similar to the two examples of Fig. 2, which requires $2n = 20$ unit delays in the nonrecursive case takes only five delays in the recursive case if an elliptic approximation is chosen. The number of multiplications for each output value is 10 for the nonrecursive structure and seven for the recursive one. This comparison indicates the price to be paid for the linear phase we obtain with the nonrecursive filters treated here.

The author is very grateful to Prof. H. W. Schuessler for his suggestions and encouragement.

O. HERRMANN

27th April 1970

Institut für Nachrichtentechnik
Universität Erlangen-Nürnberg
852 Erlangen, W. Germany

References

1 KAISER, J. F., and KUO, F. F.: 'System analysis by digital computer' (Wiley, 1966), chap. 7

2 HERRMANN, O.: 'Transversalfilter zur Hilbert-Transformation', *Arch. Elekt. Ubertragung*, 1969, **23**, pp. 581–587

3 HERRMANN, O. *in* SCHUESSLER, W. (Ed.): 'Ausgewählte Arbeiten über Nachrichtensysteme. no. 11' (Institut für Nachrichtentechnik der Universität Erlangen–Nürnberg)

4 GOLD, B., and RADER, C. M.: 'Digital processing of signals' (McGraw-Hill, 1969), pp. 93–97

5 RABINER, R. L.: 'The approximation problem for nonrecursive digital filters'. Presented at IEEE workshop on digital filtering, Jan. 1970, Harriman, NY, USA

DESIGN OF NONRECURSIVE DIGITAL FILTERS WITH MINIMUM PHASE

Indexing term: Digital filters

A method is described of transforming nonrecursive filters with equal-ripple attenuation in the passband, stopband and linear phase into those with minimum phase and half the degree, but again with equal-ripple attenuation in the passband and stopband.

In another letter the design of nonrecursive digital filters with linear phase and equal-ripple attenuation in the passband and the stopband has been considered.[1] This letter presents a method of transforming these filters to those with the same type of attenuation behaviour, but with minimum phase and half the degree. The filter with linear phase can, to begin with, be described by

$$H(z) = \frac{1}{z^{2n}}\frac{1}{2}\sum_{\mu=0}^{n} d_\mu(z^{n+\mu}+z^{n-\mu}) = \frac{1}{z^{2n}}H_0(z)$$

With $z = \exp(j2\pi\Omega)$, we obtain

$$H(\Omega) = \frac{1}{\exp(jn2\pi\Omega)}\sum_{\mu=0}^{n} d_\mu \cos \mu 2\pi\Omega$$

$$= \frac{1}{\exp(jn2\pi\Omega)} H_0(\Omega)$$

Let the coefficients d_μ be chosen in such a way that $H_0(\Omega)$ has the equal-ripple behaviour shown in Fig. 1 with tolerated deviations δ_1 in the passband and δ_2 in the stopband. The Figure indicates additionally the poles and zeros of $H(z)$ in the z plane.

We now define a transfer function $H_1(z)$ by

$$H_1(z) = H(z)+\delta_2\frac{1}{z^n}$$

which has a frequency response

$$|H_1(\Omega)| = H_0(\Omega)+\delta_2$$

The frequency response and the pole-zero pattern are shown in Fig. 2. As indicated, the filter has zeros of second order

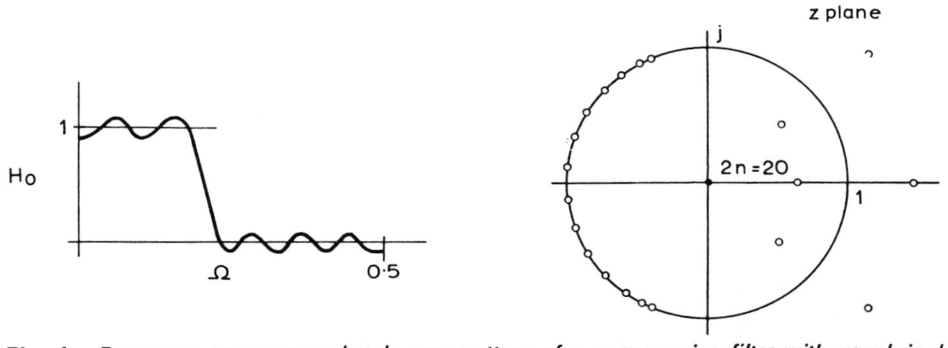

Fig. 1 *Frequency response and pole–zero pattern of a nonrecursive filter with equal-ripple attenuation and linear phase*

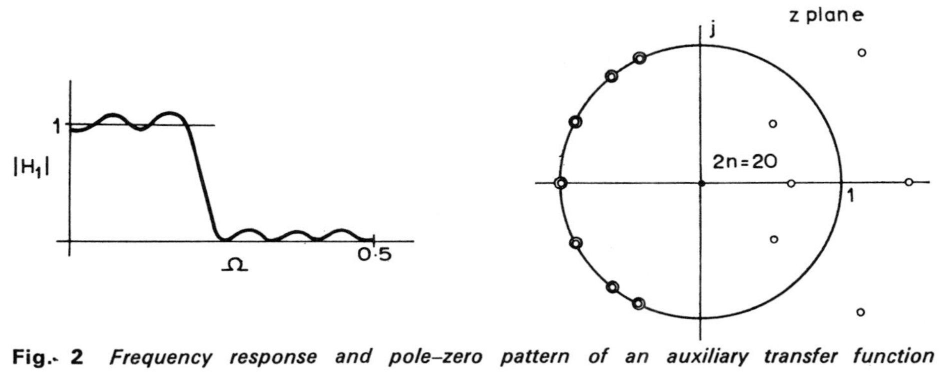

Fig. 2 *Frequency response and pole–zero pattern of an auxiliary transfer function $H_1(z)$*

Reprinted with permission from *Electron. Lett.*, vol. 6, May 28, 1970.

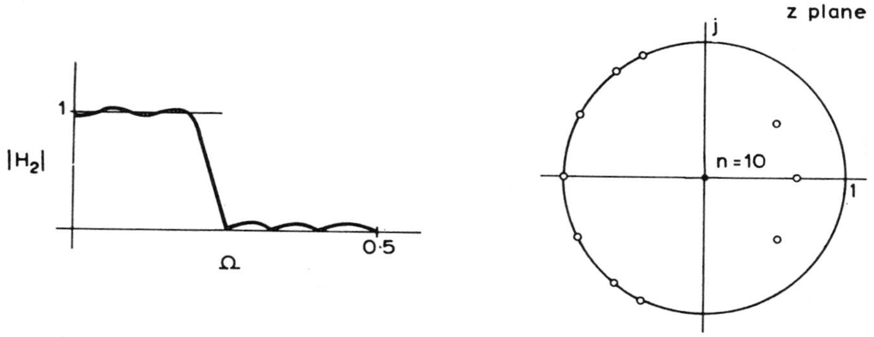

Fig. 3 *Frequency response and pole–zero pattern of a nonrecursive filter with equal-ripple attenuation and minimum phase*

on the unit circle. Expressed with its poles and zeros, the transfer function $H_1(z)$ becomes

$$H_1(z) = \frac{d_n}{2z^{2n}} \prod_{\mu=1}^{n} (z-z_{0\mu})(z-z_{0\mu}^{-1}) \quad \text{with} \quad |z_{0\mu}| \leqslant 1$$

which can be written as

$$H_1(z) = \frac{(-1)^n d_n}{2z^n \prod_{\mu=1}^{n} z_{0\mu}} \prod_{\mu=1}^{n} (z-z_{0\mu})(z^{-1}-z_{0\mu})$$

For $z = \exp(j2\pi\Omega)$, the magnitude of H_1 turns out to be a square:

$$|H_1(\Omega)| = \frac{|d_n|}{2\left|\prod_{\mu=1}^{n} z_{0\mu}\right|} \prod_{\mu=1}^{n} |\{\exp(j2\pi\Omega)-z_{0\mu}\}|^2$$

We now define a transfer function $H_2(z)$ with frequency response

$$|H_2(\Omega)| = \sqrt{|H_1(\Omega)|}$$

To obtain a transfer function with minimum phase we choose those zeros of $H_1(z)$ which are inside the unit circle and a simple zero at those points of the unit circle where $H_1(z)$ has a pair of zeros. In addition, we pick half the poles located at the origin (see Fig. 3). We obtain

$$H_2(z) = \sqrt{\left(\frac{d_n}{2\prod_{\mu=1}^{n} z_{0\mu}}\right)} \prod_{\mu=1}^{n} (z-z_{0\mu}) \frac{1}{z^n}$$

Since $|H_1(\Omega)|$ approximates to $1+\delta_2$ in the passband, we have to divide by $\sqrt{(1+\delta_2)}$ to obtain the wanted approximation of 1. Finally, we obtain

$$H_3(z) = \sqrt{\left\{\frac{d_n}{(1+\delta_2)\, 2\prod_{\mu=1}^{n} z_{0\mu}}\right\}} \prod_{\mu=1}^{n} (z-z_{0\mu}) \frac{1}{z^n}$$

whose frequency response approximates to 1 in the passband with a deviation

$$\delta_1' = \sqrt{\left(1+\frac{\delta_1}{1+\delta_2}\right)} - 1$$

and to zero in the stopband with a deviation

$$\delta_2' = \sqrt{\left(\frac{2\delta_2}{1+\delta_2}\right)}$$

O. HERRMANN
W. SCHUESSLER

Institut für Nachrichtentechnik
Universität Erlangen-Nürnberg
852 Erlangen, W. Germany

27th April 1970

Reference

1 HERRMANN, O.: 'Design of nonrecursive digital filters with linear phase', see p. 328

A NEW TECHNIQUE FOR THE DESIGN OF NON-RECURSIVE DIGITAL FILTERS

E. Hofstetter *
Lincoln Laboratory, Massachusetts Institute of Technology
Lexington, Massachusetts 02173

A. Oppenheim +
J. Siegel
Massachusetts Institute of Technology
Research Laboratory of Electronics
Cambridge, Massachusetts

Introduction

Over the past several years, non-recursive digital filters, that is digital filters for which the unit sample response is of finite duration, have become increasingly important. The disclosure of the Fast Fourier Transform algorithm and its application to digital filtering provided an efficient means for implementing high order digital filters. The use of this technique, however, requires that the filter be non-recursive. Consequently, there has been a need for designing high order non-recursive digital filters with good frequency characteristics. As an additional consideration, non-recursive filters can be designed to have exactly linear phase. This is a consequence of the fact that if the unit sample response is an even function then, since it is of finite duration, it can always be cascaded with a finite delay to correspond to a realizable system. Furthermore, since non-recursive filters have no poles in their transfer function, a problem of instability due to coefficient truncation cannot arise as it can for recursive filters.

In addition to their application to digital filtering, results with regard to non-recursive filter design are useful in a number of related problems. The design of phased array antenna patterns has for a long time been recognized as identical to the problem of designing non-recursive filters. Similarly, the design of finite duration time windows for short-time spectral analysis and spectral estimation can be regarded as being similar to that of designing a finite duration impulse response for non-recursive digital filters.

Historically the most widely used technique for designing non-recursive filters begins with an ideal frequency response or the frequency response corresponding to a recursive filter which is to be approximated with a non-recursive design. The non-recursive design is obtained by truncating or modifying the corresponding unit sample response with an appropriate window. The window is chosen in such a way that its transform has a narrow main lobe with small side lobes. A number of windows have been effective in this way including the Kaiser window and the Dolph-Chebyschev window as well as a variety of others.[1]

Another approach to the design of non-recursive filters, which is distinct from the technique utilizing a window involves an initial specification of frequency samples followed by an adjustment of the frequency samples on the basis of some algorithm. An example of such a procedure is the one introduced by Gold and Jordan[2] and developed by Rabiner, Gold and McGonegal[3]. In this technique equally spaced samples of the desired frequency response are specified with the total number of samples equal to the desired duration for the unit sample response. These frequency samples are considered to be the discrete Fourier transform of the filter unit sample response so that the continuous frequency response is an interpolation between these frequency samples. In general, sampling of the desired frequency response will not provide good frequency characteristics in between the samples. According to the procedure by Gold and Jordan, a number of the frequency samples are then varied in amplitude to improve the filter characteristics by minimizing the maximum deviation between it and the desired frequency response over some frequency range of interest. For example in designing a low pass filter by this procedure, the initial frequency samples may be chosen to have unity magnitude in the pass band and to be zero in the stop band. Because of the sharp transition band implied by these frequency samples, however, the interpolated frequency response will have considerable ripple in both the pass band and stop band near the pass band and stop band edges. To reduce this ripple a number of frequency samples at the edge of the pass band or stop band can be varied in magnitude to reduce the ripple by broadening the transition band. Such a procedure is implemented on a computer using a linear search for the optimum values of the frequency samples being varied. For practical reasons the number of frequency samples which can be varied simultaneously must be small. Because all of the frequency samples cannot be varied, this procedure can only converge to the optimum non-recursive filter if the frequency samples which are not varied are already optimum, which will generally not be the case.

A third approach to non-recursive filter design corresponds to a non-linear optimization of the filter characteristics. This technique, proposed by Herrmann and Schuessler,[4,5] is directed toward the design of filters with equiripple pass band and stop band characteristics. The design is carried out by constraining the equiripple frequency characteristic by means of a set of non-linear equations. The equations are those that the frequency response must satisfy in order to have its extrema achieve the maximum allowable ripple values in both pass band and stop band. The unknowns in these equations are the filter coefficients and the

*Sponsored by the Department of the Air Force.
+Supported in part by the U.S. Air Force (office of Aerospace Research under contract F19628-69-C-0044)

Reprinted with permission from *Proc. 5th Annu. Princeton Conf. Inform. Sci. Systems*, pp. 64–72, 1971.

frequencies at which the extrema occur. This approach thus far has been restricted in the order of the filters which can be designed because available algorithms exhibit computation difficulties when the number of unknowns is large.

In this paper a new algorithm for the design of equiripple non-recursive digital filters is discussed. The discussion is phrased primarily in terms of the design of non-recursive lowpass filters, although as will be indicated in a later section, the ideas are easily extended to the design of frequency characteristics that are piecewise constant in a number of frequency bands, and to some non-piecewise constant frequency characteristics.

Equiripple Non-recursive Digital Filters

In the design of a digital lowpass filter the ideal frequency characteristic to be approximated is unity in the pass band and zero in the stop band. Thus if $H(e^{j2\pi F})$ represents the z-transform of the non-recursive filter on the unit circle, then the filter design is directed toward choosing $H(e^{j2\pi F})$ to approximate an ideal lowpass characteristic. Since the ideal frequency characteristic cannot be obtained exactly, some deviation from unity in the pass band and from zero in the stop band must be allowed. Furthermore a transition band between the pass band and the stop band must be included. In the case of recursive filters, it is well known that for a given filter order and for given allowable tolerances in the pass band and the stop band the narrowest transition band results when the filter is equiripple in both pass band and stop band, and has the maximum number of ripples consistent with the filter order. In a similar manner it can be shown that the non-recursive filter with an equiripple frequency characteristic in the pass band and the stop band will have the narrowest transition band.* Stated more precisely, following Herrmann and Schuessler, the approximation will be specified in terms of a maximum allowable deviation δ_p in the pass band and a maximum allowable deviation δ_s in the stop band, and the number of ripples (extrema excluding the endpoints) in the pass band and in the stop band designated by n_p and n_s respectively. The minimum duration of the filter's impulse response, which can result in the designated number of pass band and stop band ripples is $2n + 1$, where $n = n_p + n_s + 1$. Since interest lies in designing a filter of minimum order consistent with a given set of specifications, n will always be chosen to satisfy, $n = n_p + n_s + 1$. The pass band edge F_p will be defined as the frequency for which the frequency characteristic first leaves the tolerance band associated with the pass band, and the stop band edge F_s will be defined as the frequency for which the frequency characteristic enters the tolerance band associated with the stop band and remains there. With the problem stated in these terms, different coefficients satisfying the same specifications are compared on the basis of the width of the transition band measured as $|F_s - F_p|$. The optimum filter meeting the specifications will be considered to be the one for which the width of the transition band is minimized. It can be shown that among the set of filters satisfying such a set of specifications, the narrowest transition band occurs when the filter characteristic is equiripple and achieves the tolerance limits in both the pass band and stop band. Furthermore, it can be shown that for a filter having a frequency response that is real, a set of filter coefficients resulting in this equiripple property exists and is unique.

*These results along with other related results to be mentioned shortly are being prepared for publication.

When the filter is restricted to have a real frequency characteristic, the approximation problem corresponds to approximation by a trigonometric polynomial. i.e. $G(F) = H(e^{j2\pi F})$ will be of the form

$$G(F) = \sum_{k=0}^{n} c_k \cos 2\pi kF \quad (1)$$

which can also be written in the form

$$G(F) = \sum_{k=0}^{n} d_k (\cos 2\pi F)^k \quad (2)$$

$$= \sum_{k=0}^{n} d_k x^k$$

where $x = \cos 2\pi F$

Thus, a trigonometric polynomial approximation to the desired frequency characteristic is being sought. It is important to keep in mind, however, that the determination of an equiripple non-recursive filter does not correspond to a Chebychev approximation to the ideal filter characteristics since the equiripple non-recursive filter will in general have different ripple heights in the pass band and in the stop band and will not have any ripples in the transition band.

As pointed out by Herrmann and Schuessler,[6] a design procedure for zero phase non-recursive equiripple filters can also be applied to the design of non-zero phase filters. For this case, a trigonometric polynomial of tne form of eq. (2) is used to represent the squared magnitude function of the filter, and the coefficients c_k are chosen so that G(F) has ripples between $(1 \pm \delta_p)^2$ in the pass band and between 0 and δ_s^2 in the stop band. Since the resulting G(F) must be non-negative, it can only be zero with zero slope. Consequently the zeros of the z-transform corresponding to G(F) will always occur in reciprocal pairs and thus can be factored in a number of ways to obtain a frequency response having the desired squared magnitude function G(F) on the unit circle. Clearly, the duration of the unit sample response of such a filter is $n + 1$, where n is the order of the underlying trigonometric polynomial, rather than $2n + 1$ which was the case when designing a filter having a real frequency response. Finally, it should be noted that as a practical matter, the procedure just outlined is limited to the design of relatively low order filters because of the difficulty of finding the zeros of a high order polynomial.

In a variety of filter design problems, the frequency response to be approximated is purely imaginary as is the case when the desired filter is a differentiator. When the frequency response of a non-recursive filter of duration $2n + 1$ is purely imaginary, it can be written as a trigonometric sine polynomial,

$$G(F) = j \sum_{k=1}^{n} c_k \sin 2\pi kF \tag{3}$$

It is well known that a trigonometric sine polynomial can be expressed as an ordinary polynomial in the variable $x = \cos 2\pi F$ multiplied by the factor $\sqrt{1-x^2}$. Therefore, eq. (3) can be written in the form,

$$G(F) = j \sin 2\pi F \sum_{k=0}^{n-1} d_k (\cos 2\pi F)^k \tag{4}$$

An ideal lowpass differentiator has a frequency response given by,

$$G(F) = \begin{cases} j2\pi F, & F \text{ in pass band} \\ 0\ , & F \text{ in stop band} \end{cases} \tag{5}$$

Comparison of equations (4) and (5) now shows that the present approximation problem consists of finding a polynomial in $x = \cos 2\pi F$ that is a "good" approximation to zero in the stop band and, at the same time, a good approximation to $2\pi F/\sin 2\pi F$ in the pass band. A good approximation will mean that the frequency response of the filter remains within $\pm\delta_s$ in the stop band and within $2\pi F\ (1 \pm \delta_p)$ in the pass band. Note that this last requirement controls the percentage error in the pass band rather than the absolute error. In addition the number of allowable ripples in the two bands will be specified to be n_p and n_s. A stop band ripple is still defined to be a non-end-point extremum of G(F) but a pass band ripple is now taken to mean a non-end-point extremum of the percentage error, $[2\pi F - G(F)]/2\pi F$. In other words, stop band ripples occur at non-end-point frequencies at which the slope of G(F) is equal to $2\pi\ (1 \pm \delta_p)$. It should be noted that the percentage error $[2\pi F - G(F)]/2\pi F$ automatically has an extremum at $F=0$ so that the total allowable number of pass band ripples including the endpoint is $n_p + 1$. On the other hand, G(F) does not have an extremum at $F = 1/2$ so that the total allowable number of stop band ripples is still n_s.

It is conjectured that, just as in the case of a lowpass filter, among all filters meeting the above ripple and error specifications, the one having the smallest transition band is the equiripple filter; i.e., that filter whose stop band ripples alternatingly go through the points $\pm\delta_s$ a total of n_s times and whose percentage error, $[2\pi F - G(F)]/2\pi F$, has extrema that alternatingly go through the points $\pm\delta_p$, $(n_p + 1)$ times in the pass band.

Obvious modifications of the above argument yield corresponding results for band pass, high pass or other frequency selective differentiators.

The Design Algorithm

The design technique used by Herrmann and Schuessler can be applied to all of the filter design problems just discussed. The chief drawback to this method is that, until recently, computational difficulties limited the order of the filter that could be designed to about 65. The algorithm that is to be presented here has no presently known limitation on order and filters having orders in the hundreds already have been designed with it.

The proposed design algorithm is basically an iterative technique for producing a polynomial that has extrema of preassigned value. It begins by making an initial estimate of the frequencies at which these extrema will occur and then uses the Lagrange interpolation formula to obtain a polynomial that alternatingly goes through the maximum allowable ripple values at these frequencies. The details of this initial guess do not seem to affect the ultimate convergence of the algorithm but the number of iterations required to achieve the desired degree of convergence is somewhat dependent on how close the initial frequencies are to the final answer.

For the case of a lowpass filter design, this initial set of frequencies along with the associated Lagrange interplation polynomial are sketched in Fig. (1). Note that the polynomial associated with the initial guess does not have extrema that achieve the maximum allowable ripples but, rather, extrema that exceed these values. The next stage of the algorithm is to locate the frequencies at which the extrema of the first Lagrange interpolation polynomial occur. These frequenices are taken to be a second, hopefully improved, guess as to the frequencies at which the extrema of the filter response will achieve the desired ripple values. This second set of frequencies is indicated in Fig. (1). The algorithm now "closes the loop" by using these new frequencies to construct a Lagrange interpolation polynomial that achieves the desired ripple values at these frequencies. The extrema of this new polynomial are then located and used to start the next cycle of the algorithm. The algorithm is reminiscent of, but different from the Remes exchange algorithm used in the theory of Chebychev approximation.

The formal mathematical description of the design algorithm can be written as follows. Denote the ordered set of frequencies obtained at the ith iteration of the algorithm by $F_k^{(i)}$, $i=0, 1, \ldots;\ k = 1, \ldots, n+1$. Since the endpoints of the frequency interval are always extrema of the frequency response G(F), it is always true that $f_1^{(i)} = 0$ and $F_{n+1}^{(i)} = 0.5$. The frequency response of the filter obtained at the ith stage can be written in the form,

$$G^{(i)}(F) = \frac{\displaystyle\sum_{k=1}^{n+1} \frac{A_k^{(i)}}{(x - x_k^{(i)})} y_k}{\displaystyle\sum_{k=1}^{n+1} \frac{A_k^{(i)}}{(x - x_k^{(i)})}} \tag{6}$$

where

$$x = \cos 2\pi F,\quad x_k^{(i)} = \cos 2\pi F_k^{(i)} \tag{7}$$

and the coefficient $A_k^{(i)}$ are given by,

$$A_k^{(i)} = \left[\prod_{\substack{q=1 \\ q\neq k}}^{n+1} (x_k^{(i)} - x_q^{(i)}) \right]^{-1} \tag{8}$$

The y_k's are given by

$$y_1 + 1 \pm \delta_p,\ y_2 = 1 \mp \delta_p, \ldots, y_{n_p+1} = 1 + \delta_p$$

$$y_{n_p+2} = -\delta_s,\ y_{n_p+3} = \delta_s, \ldots y_n = \pm\delta_s,\ y_{n+1} = \mp\delta_s \tag{9}$$

Equation (6) for the frequency response of the ith filter is essentially the Lagrange interpolation formula in what is called barycentric form. The reason for using this form rather than the more familiar form is that the barycentric form requires a factor of n fewer multiplications to evaluate at a given value of F than straightforward Lagrange interpolation formula.

The frequencies $F_k^{(i+1)}$ associated with the next stage of the algorithm are obtained by locating the n+1 extrema (endpoints included) of $G^{(i)}(F)$. This is accomplished by evaluating $G^{(i)}(F)$ at a grid of points $F_q = q/2N$, q=0,...N for a suitably large value of N and calling the endpoints F=0, F=0.5 and those frequencies for which the difference product $[G^{(i)}(F_q) - G^{(i)}(F_{q-1})][G^{(i)}(F_{q+1}) - G^{(i)}(F_q)]$, q=2,...N-2, is negative the extrema of $G^{(i)}(F)$. This simple procedure seems to yield satisfactory results as long as $N \cong 20$ n. The frequencies obtained by this process are denoted $F_k^{(i+1)}$, k=1,... n+1, and the $(i+1)^{st}$ frequency response is obtained by using $F_k^{(i+1)}$ in equation (6) instead of $F_k^{(i)}$.

An important detail concerning the process of finding the extrema $G^{(i)}(F)$ must be mentioned. It is a peculiarity of the barycentric form of the Lagrange interpolation formula that it assumes the indeterminate form ∞/∞ at the constraint points $F=F_k^{(i)}$. This computational difficulty can be avoided by the simple device of adding a small increment such as .01/2N to the frequencies for which the difference product is negative and calling these slightly shifted frequencies the new extrema $F_k^{(i+1)}$. The constraint points $F_k^{(i+1)}$ are now never equal to any of the frequencies F_q, q=2,...,N-2 at which G(F) must be evaluated and the indeterminate case of the Lagrange formula does not arise.

It should be obvious that if the procedure just described converges, it must converge to the desired equiripple filter. No mathematical proof of convergence is available at this time; however, a large amount of computational experience with the algorithm has convinced the authors that, in a practical sense, the algorithm converges rapidly to the desired solution. The rule used by the authors for terminating the iteration process is to call a halt when all of the extremum frequencies $F_k^{(i)}$, k=1,... n+1 are equal to their counterparts $F_k^{(i+1)}$ from the previous iteration.

The above description of the algorithm has been couched in terms of a lowpass filter design; however, the minor modifications necessary to design any frequency selective filter (band pass, band elimination, etc. should be obvious). The algorithm has been programmed for the Hewlett-Packard 9100A desk calculator and the IBM 360 computer. An example of a 41st order band pass filter designed on the Hewlett-Packard machine is shown in Fig. 2. The additional subscripts "u" and "ℓ" refer to the upper and lower stop band parameter respectively. Fig. 3 shows an example of a 251st order lowpass filter designed on the IBM 360. The later design required 12 iterations starting from an essentially uniform set of initial frequencies. It is possible to reduce the required number of iterations somewhat by means of a cleverly chosen set of initial frequencies but the savings obtained are not great. Another band pass filter design is shown in Fig. 4 This example was chosen to illustrate the fact that the various bandwidths and rippler tolerances can be controlled independently in this design technique.

Only slight modifications of the algorithm are necessary to make it capable of designing filters whose frequency response is purely imaginary. For the case of a lowpass differentiator, for example, the frequency response at the ith stage $G^{(i)}(F)$ is given by the right-hand side of equation (6) multiplied by the factor $\sqrt{1-x^2}$. The coefficients A_k are still given by equation (8) but with n+1 replaced by n and the frequency $F_1^{(i)}$ is still equal to 0 but the y_k's must be chosen according to the following scheme,

$$y_1 = (1 \pm \delta_p),\ y_2 = (1 \mp \delta_p)\ \frac{2\pi F_2^{(i)}}{\sin 2\pi F_2^{(i)}}$$

$$\ldots y_{n_p+1} = (1 + \delta_p)\ \frac{2\pi F_{n_p+1}^{(i)}}{\sin 2\pi F_{n_p+1}^{(i)}},$$

$$y_{n_p+2} = -\delta_s,\ y_{n_p+3} = \delta_s, \ldots y_n = \pm\delta_s \qquad (10)$$

The frequencies for the next stage of the algorithm $F_k^{(i+1)}$ are now obtained by evaluating G(F) on the same grid of points used earlier and using the difference-product scheme to locate the extrema of G(F) in the stop-band. This results in a set of n_s new frequencies. The difference-product scheme described above is used to locate the non-end-point extrema of the percentage error$[2\pi F - G(F)]/2\pi F$, in the pass band and results in an additional set of n_p new frequencies. These new frequencies plus the point F=0 are arranged in ascending order and labeled $0=F_1^{(i+1)}, < F_2^{(i+1)} < \ldots < F_{n+1}^{(i+1)}$. The algorithm is now repeated with $F_k^{(i)}$ replaced by $F_k^{(i+1)}$ and the process continued until the desired degree of convergence is obtained. The further modifications of the algorithm needed for the design of other types of frequency selective differentiators are obvious and will not be discussed here.

A low pass differentiator designed by the above technique is shown in Fig. 5 and a band pass differentiator in Fig. 6. In both examples, the linearly increasing error magnitude in the pass band is indicative of a constant percentage error magnitude.

The design process for each of the cases discussed above results in a set of coefficients A_k and a set of frequencies F_k. These quantities serve to completely define the frequency response G(F) of the desired filter via the Lagrange interpolation formula. The actual filter coefficients c_k (which are very simply related to the filter's unit sample response) in equation (1) or equation (3) can be obtained from these quantities by a direct, but very lengthy algebraic process. A much more efficient method for calculating the c_k's utilizes the fact that the unit sample response of a non-recursive filter of duration M is the M-point discrete Fourier transform (DFT) of the sequence G(k/M), =0, 1... M-1 where $G(F) = H(e^{j2\pi F})$ is the unit circle frequency response of the filter. If M is a power of two, then the fast Fourier transform (FFT) algorithm is an efficient means of evaluating the DFT. The FFT algorithm also can be used to compute the unit sample response when

M is not a power of two simply by choosing M_o to be the smallest power of two greater than M and then using the FFT algorithm to computer M_o -point DFT of the sequence G(k/M), k=0,... M_o -1. The resulting M_o numbers will be zero except for a subset of M numbers which represent the desired unit sample response.

References

1. J.F. Kaiser, "Digital Filters" in System Analysis by Digital Computer, F.F. Kuo and J.F. Kaiser, J. Wiley and Sons, N.Y. 1966.

2. B. Gold and K. L. Jordan, "A Direct Search Procedure for Designing Finite Duration Impulse Response Filter," IEEE Trans. on Audio and Electroacoustics, Vol. AU-17, No. 1, March 1969.

3. L. R. Rabiner, B. Gold and C.A. McGonegal, "An Approach to the Approximation Problem for Nonrecursive Digital Filters," IEEE Trans. on Audio and Electroacoustics, Vol. AU-18, No. 2, June 1970.

4. O. Herrmann and H. W. Schuessler, "On the Design of Selective Nonrecursive Digital Filters," IEEE Arden House Workshop, Jan. 1970, Harriman, N.Y.

5. O. Herrmann, "Design of Nonrecursive Digital Filters with Linear Phase," Electronics Letters, Vol. 6, No. 11, 28 May 1970.

6. O. Herrmann and H. W. Schuessler "Design of Nonrecursive Digital Filters with Minimum Phase", Electronics Letters, Vol. 6, No. 11 28 May 1970.

7. E. Hofstetter, "A New Technique for the Design of Non-recursive Digital Filters", Lincoln Laboratory Technical Note 1970-42, 15 December 1970.

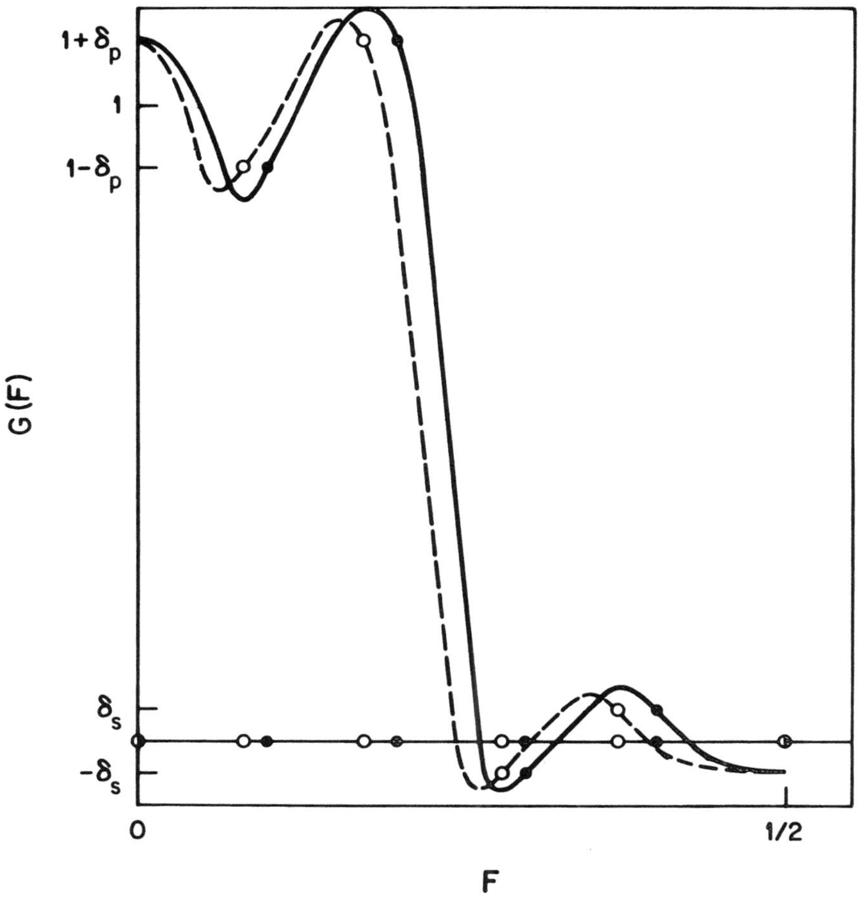

Fig. 1 The design algorithm for $n_p = n_s = 2$.

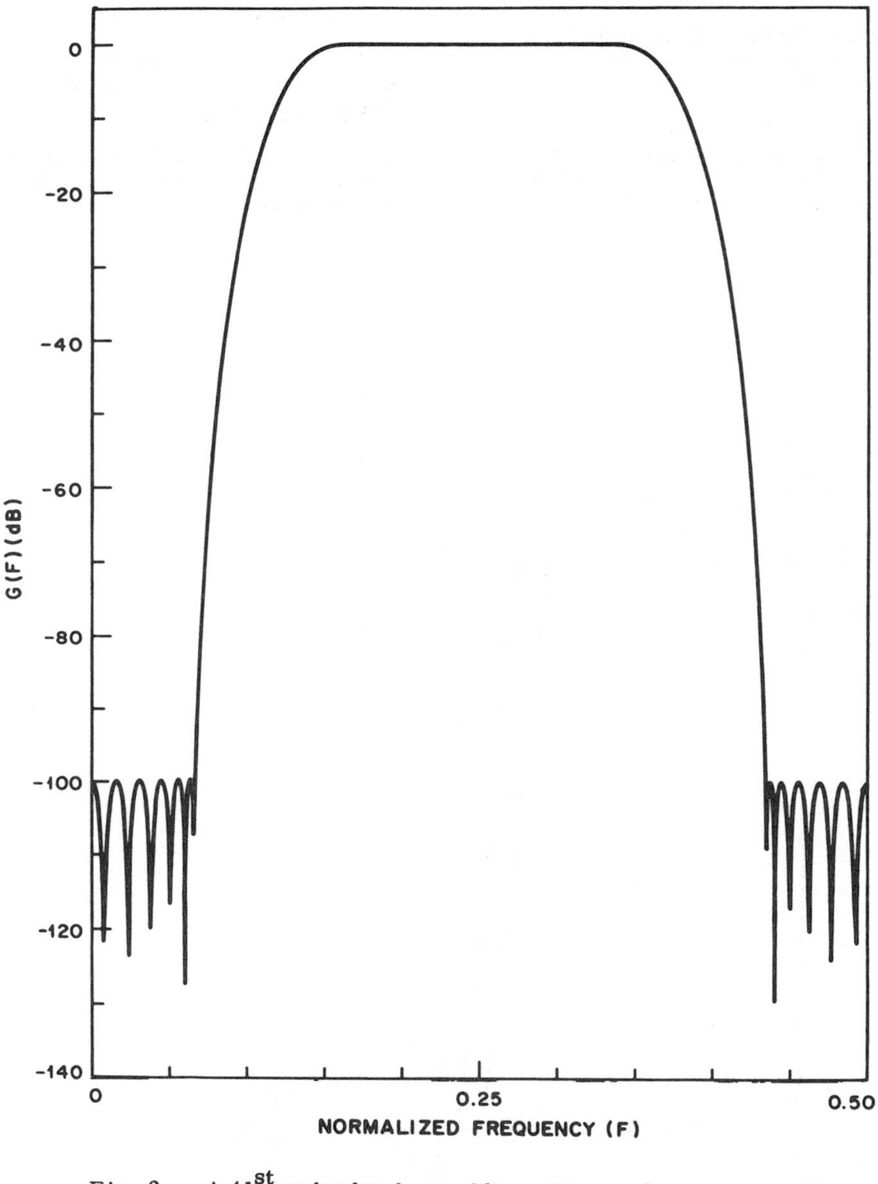

Fig. 2 A 41st order band pass filter with $n_p = 9$, $n_{s\ell} = n_{su} = 5$, $\delta_p = 0.005$, $\delta_{s\ell} = \delta_{su} = 10^{-5}$.

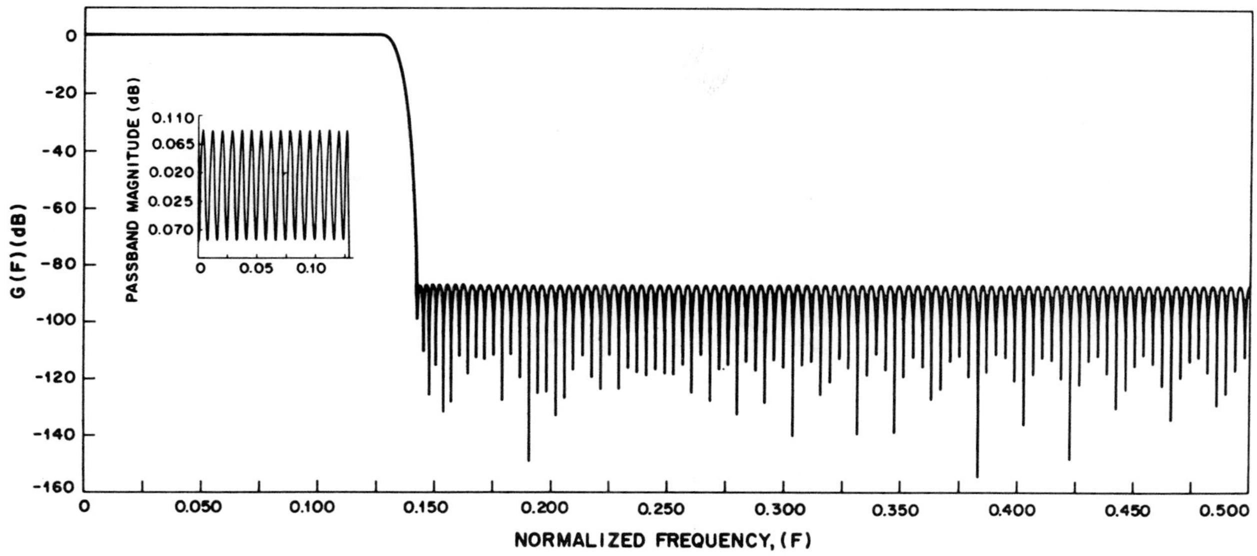

Fig. 3 A 251st order lowpass filter with $n_p = 31$ $n_s = 93$, $\delta_p = 0.01$, $\delta_s = 4 \times 10^{-5}$.

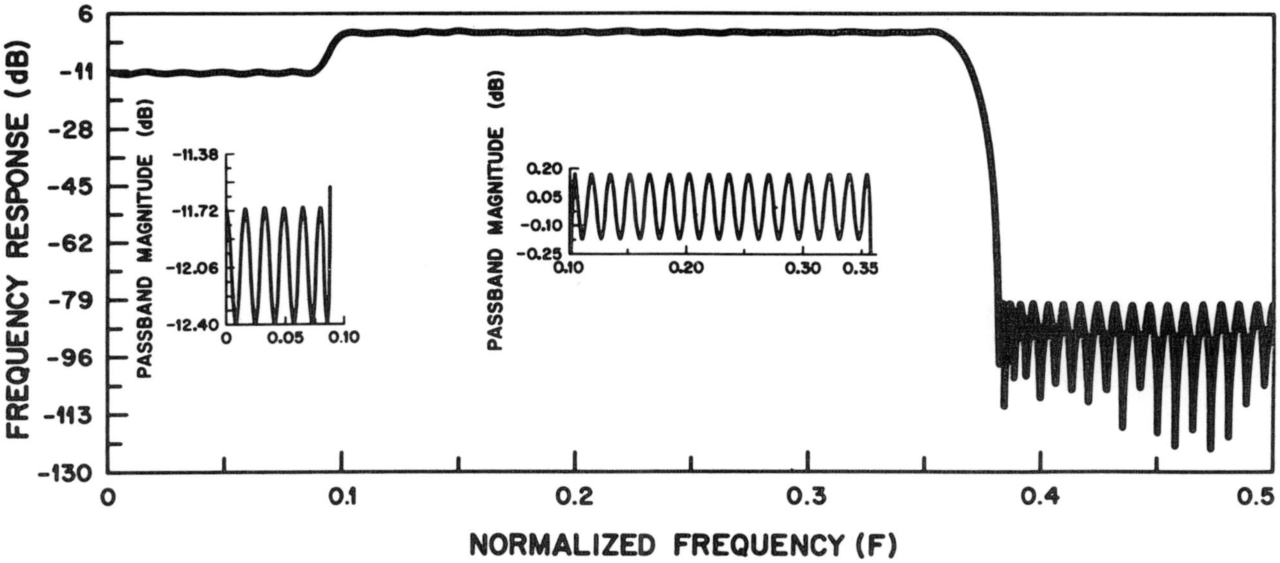

Fig. 4 A 121st order band pass filter with $n_{p\ell} = 11$, $n_{pu} = 31$, $n_s = 17$ $\delta_{p\ell} = 0.01$, $\delta_{pu} = 0.02$, $\delta_s = 0.001$

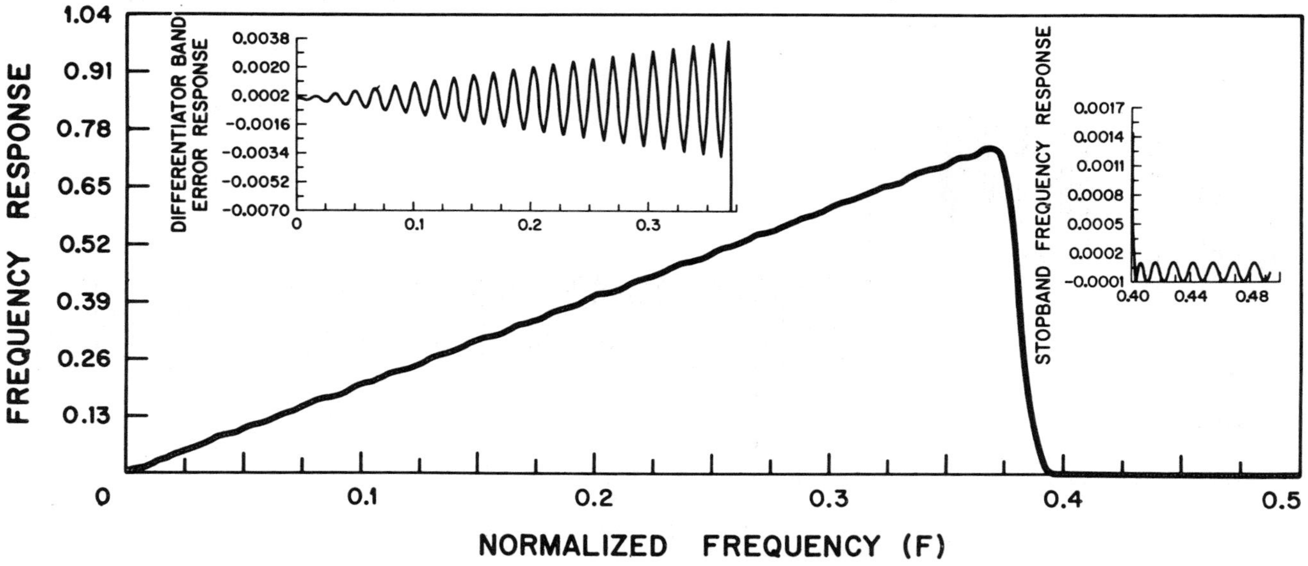

Fig. 5 A lowpass differentiator with $n_p = 44$, $n_s = 15$, $\delta_p = 0.005$, $\delta_s = 0.0001$

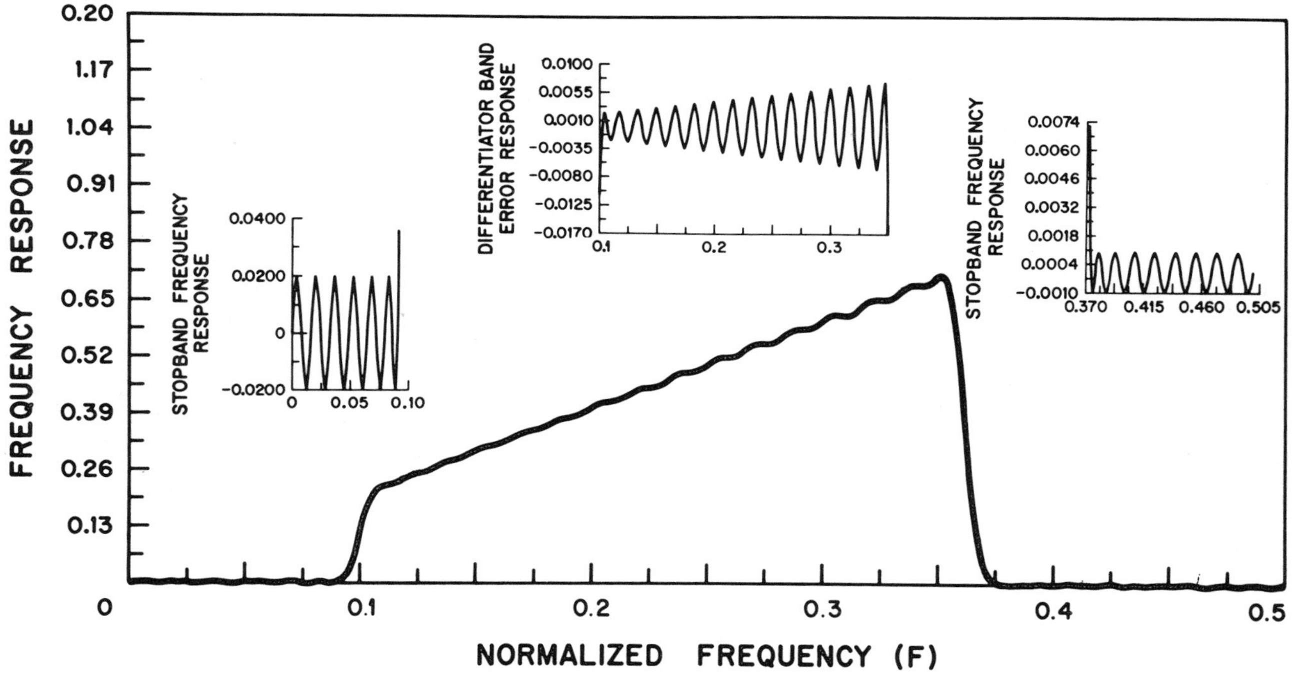

Fig. 6 A band pass differentiator with $n_{s\ell} = 12$, $n_p = 31$, $n_{su} = 17$, $\delta_{s\ell} = 0.002$, $\delta_p = 0.01$, $\delta_{su} = 0.001$.

ON OPTIMUM NONRECURSIVE DIGITAL FILTERS

E. M. HOFSTETTER*
Lincoln Laboratory, Massachusetts Institute of Technology
Lexington, Massachusetts 02173

A. V. OPPENHEIM+
J. SIEGEL
Massachusetts Institute of Technology
Research Laboratory of Electronics
Cambridge, Massachusetts

ABSTRACT

Several optimality criteria for the design of nonrecursive digital filters are discussed. Attention is then focussed on the criterion of minimum transition bandwidth subject to constraints on the maximum allowable errors in the pass and stop band and the number of allowable pass and stop band ripples. It is conjectured that the optimum filter in this class is equiripple in both pass and stop bands and the available mathematical evidence for this conjecture is sketched.

INTRODUCTION

With the exception of tapped delay line filters, which can be implemented exactly as nonrecursive digital filters, analog filters with finite duration impulse response are of minor importance. In contrast, digital filters with a finite duration impulse response, i.e. nonrecursive digital filters have received an increasing amount of attention over the last several years. With this increased emphasis, attention has focussed on the approximation problem for this class of filters.

As will be described, the squared magnitude of the frequency response of a nonrecursive filter can be represented as a trigonometric polynomial and consequently the design of a nonrecursive filter to approximate a desired characteristic corresponds to approximating the specified characteristic with a polynomial. In this paper the design of optimum nonrecursive lowpass digital filters will be discussed. There are of course a number of ways in which optimality can be defined for lowpass filter design. Some choices coincide with optimality criteria which classically have been applied to the polynomial approximation problem in which case such theory can be applied to the design of nonrecursive digital filters. This paper considers primarily optimal approximation for a nonrecursive lowpass digital filter for which classical results in approximation theory do not seem to be available. Some conjectures regarding the nature of the optimum approximation will be discussed.

THE APPROXIMATION PROBLEM FOR NONRECURSIVE LOWPASS FILTERS

The approximation problem as it is generally considered for nonrecursive digital filters consists of approximating the ideal filter characteristic with a trigonometric polynomial. If we consider the filter to have zero phase, so that the imaginary part of the frequency response is zero, then the unit sample response is even about $k=0$. Such a filter is easily converted to a causal filter with exactly linear phase by cascading with a suitable delay. Alternatively the design of a nonrecursive zero phase

*Sponsored by the Department of the Air Force.

+This work was supported in part by the Joint Services Electronics Program (Contract DAAB07-71-C-0300.)

Reprinted from *Proc. 9th Annu. Allerton Conf. Circuit System Theory*, pp. 789–798, 1971.

filter can be converted to the design of a minimum phase filter.[1] Thus when we consider the approximation with a zero phase characteristic we may either utilize this design for a linear phase filter or a minimum phase filter. For convenience, however, the discussion will be phrased in terms corresponding to the design of a linear phase filter.

Since the unit sample response h(k) is even, i.e. h(k) = h(-k) the frequency response can be expressed as a cosine series of the form

$$G(F) = H(e^{j2\pi F}) = h(0) + 2 \sum_{k=1}^{n} h(k) \cos k\, 2\pi F \tag{1}$$

where the total duration of the unit sample response is 2n + 1. Equivalently Eq. (1) can be expressed in terms of trigonometric polynomial of the form

$$G(F) = P(\cos 2\pi F)$$

$$\text{where } P(x) = \sum_{k=0}^{n} c_k x^k \tag{2}$$

Because of the form of eq. (1) or (2) the frequency response is symmetric about F=1/2 and consequently we need only consider a specification of the filter for F in the interval 0 to 1/2. Then according to eq. (2), we can equivalently approximate the desired characteristic with a polynomial P(x). By mapping the interval $-1 \le x \le 1$ to the interval $0 \le F \le 1/2$ according to $x = \cos 2\pi F$, the polynomial approximation is mapped to the filter frequency response. Figure 1 illustrates an acceptable lowpass frequency characteristic versus F and Figure 2 illustrates the same filter as a function of the variable x. It should be noted that in figure 1 the frequency response has zero slope at F=0 and F=1/2 whereas in Fig. 2 it does not. From the form of eq. (2) it should be clear that whether or not the polynomial P(x) has zero slope at x= -1 and x=1, the trigonometric polynomial P (cos 2πF) always has zero slope at F=0 and F=1/2.

The parameters typically of interest in characterizing a lowpass filter are the filter order, the passband and stopband edges F_p and F_s, the transition bandwidth $|F_p - F_s|$, the passband ripple δ_p and the stopband ripple δ_s. Clearly it would be desirable to design the filter with the smallest order, transition bandwidth and passband and stopband ripple. Since there are tradeoffs between these parameters, a particular optimization strategy can only be directed toward one parameter. One approach to the optimization is based on the use of linear programming as pursued by Helms[2] and Rabiner[3]. In that approach the filter order, and the parameters F_p, F_s and δ_p are specified and linear programming is used to minimize the stopband ripple. In an approach pursued by Parks[4], the filter order, the passband and stopband edges, and the ratio of δ_p to δ_s is fixed. An exchange algorithm is then applied to minimizing δ_p and consequently also δ_s subject to the constraint that the ratio of passband to stopband ripple remain constant. A third approach is directed toward a minimization of the transition bandwidth. This approach is suggested in part by the importance of transition bandwidth in filter design but primarily by the availability of design algorithms[5,6,7] that lead to filters that appear to be optimum according to this criterion. In particular, there are several related design techniques for which the design specifications are the number of passband ripples n_p, the number of stopband ripples n_s and the size of the passband and stopband ripples δ_p, and δ_s respectively. These design algorithms result in filters for which the characteristics in the passband are equiripple and the characteristics in the stopband are equiripple. The remainder of this paper is directed toward a discussion of the relationship between the optimality criterion of minimizing the transition bandwidth and these equiripple designs.

MINIMUM TRANSITION BANDWIDTH FILTERS

The first class of filters to be considered consists of those for which the polynomial P(x) has exactly n_p ripples in the passband $x_p \le x \le 1$ and exactly n_s ripples in the stopband $-1 \le x \le x_s$ and an order $n = n_p + n_s + 1$. A ripple is defined to be a zero of P'(x) counting multiplicity. Furthermore, we require that within the stop band $|P(x)| \le \delta_s$ and within the passband $|P(1) - 1| \le \delta_p$. Filters within this class are to be compared with regard to their transition bandwidth $|F_p - F_s|$ where $x_p = \cos 2\pi/F$, and $x_s = \cos 2\pi F_s$. It is not hard to show that there is at least one polynomial $P^*(x)$ in the class for which the transition bandwidth is smallest, i.e. for which $|F_p - F_s|$ is equal to the greatest lower bound of all the transition bandwidths associated with filters in the class. It is conjectured that $P^*(x)$ has the following characteristics:

<u>Conjecture:</u>

P^* alternatingly achieves the values $1 \pm \delta_p$ exactly $n_p + 2$ times in the passband and alternatingly achieves the values $\pm \delta_s$ exactly $n_s + 2$ times in the stopband. This type of behavior (which we will refer to as maximally equiripple behavior) is illustrated in Fig. 3 for the case $n_s = 2$, $n_p = 3$.

<u>Discussion:</u> We have not succeeded in proving this conjecture in the full generality stated above; however, a proof can be given if the conjecture is weakened by making the additional assumption that all the ripples of $P^*(x)$ are <u>simple</u> (of multiplicity one). This in fact is the class of filters generated by the algorithms in references (5, 6, 7). The nature of the argument will be sketched but a formal proof is omitted.

First we note that all the ripples of $P^*(x)$ must lie in the interior of the passband and stopband and that $P^*(-1) = \pm \delta_s$ and $P^*(1) = 1 \pm \delta_p$. If this were not the case, then $P^*(x)$ could be subjected to a linear compression in x of the form $x' = a x + b$, $0 < a < 1$ resulting in a new polynomial $P^*(ax+b)$ that meets the above requirements and has a smaller transition bandwidth than $P^*(x)$. Since $P^*(x)$ has the minimum possible transition bandwidth, this is impossible and the above statement must be true.

Now let us denote the locations of the passband ripples by x_{pi}, $i = 1, \ldots n_p$ and the locations of the stopband ripples by x_{si}, $i = 1, \ldots n_s$. If $P^*(x_{pi}) = 1 \pm \delta_p$, $i=1, \ldots n_p$ and $P^*(x_{si}) = \pm\delta_s$, $i = 1, \ldots n_s$, then the conjecture is true because it is easy to verify that $P^*(x)$ must exhibit the desired alternating behavior at the $n_p + 2$ passband points x_p, $x_{p'}, \ldots n_{pn_p}$, 1 and the $n_s + 2$ stopband points -1, $x_{s1}, \ldots x_{sn_s} x_s$. Therefore, in order for the conjecture to be false, there must exist at least one ripple x_{si} or x_{pi} for which $P^*(x_{si}) \neq \pm \delta_s$ or $P^*(x_{pi}) \neq 1 \pm \delta_p$. Assume that this is the case as is illustrated in Fig. 4. It will now be argued that $P^*(x)$ can be modified in such a way as to reduce its transition bandwidth which contradicts the optimality of $P^*(x)$. This contradiction establishes the truth of the modified conjecture.

This modification procedure is illustrated in Fig. 4 in which a stopband ripple has been assumed not to achieve the tolerance limit $\pm\delta_s$. It is now possible to draw a line such as aa' that lies in the tolerance band but does not intersect this ripple. The polynomial q(x) is now constructed as shown in Fig. 4 and the modified polynomial $P^*(x) + \lambda q(x)$, $\lambda > 0$ is formed. Note that the order of q(x) will not be greater than n so that the modified polynomial $P^*(x) + \lambda q(x)$ has the correct order. Furthermore, since the ripples of q(x) are out of phase with all the ripples of $P^*(x)$ except the one that did not intersect the line aa', it is obvious that, as long as λ is made sufficiently small, $P^*(x) + \lambda q(x)$ will meet the tolerance specifications in $(-1, x_s)$ and $(x_p, 1)$. In addition it is easy to verify that choosing λ small enough leads to the conclusion that $P^*(x) + \lambda q(x)$ like $P^*(x)$ itself, has n_s simple ripples in the stopband and n_p simple ripples in the passband. But, since $q(x_s) < 0$ and $q(x_p) > 0$, the stopband edge of $P*(x) + \lambda\, q(x)$ is greater than x_s and the passband edge of $P*(x) + \lambda\, q(x)$ less than x_p. This means that the transition bandwidth of $P*(x) + \lambda\, q(x)$ is smaller than that of $P*(x)$ and this contradiction establishes the weakened form of the conjecture.

The argument sketched above has not been used to prove the conjecture in its full generality because, in this more general case, no way has been found to show that the modified polynomial $P*(x) + \lambda\, q(x)$ has exactly n_p ripples in the passband and n_s ripples in the stopband.

In the class of filters discussed above the filter characteristic was constrained to have the maximum allowable number of ripples for real x between -1 and 1. We next consider the class of polynomials defined exactly as above with the exception that now the polynomials are only constrained to have no more than n_s ripples in the stopband and no more than n_p ripples in the passband. At first glance, it might appear that, even in this modified class of polynomials, the polynomial with the minimum transition bandwidth is again a maximally equiripple polynomial. This is not true however as demonstrated in the examples prescribed in Figs. 5 and 6. Fig. 5 depicts a maximally equiripple filter response. Fig. 6 shows a filter response with one ripple missing in the stopband. The transition bandwidth of the first filter is $|F_p - F_s| = 0.13$ and that of the second filter $|F_p - F_s| = 0.115$. The latter transition bandwidth is smaller and so we must conclude that the optimum filter in this new class of filters does not necessarily have the maximum allowable number of ripples.

In the example shown in Fig. 6 the ripple that is mssing from the stopband is located just to the right of the passband at x = 1. This ripple is not obvious on a plot of G(F) as a function of F because it is obscured by the cos $2\pi F$ term in the frequency response; however, in the polynomial domain, such a ripple would appear as sketched in Fig. 7. The decision to place this ripple just outside of the passband was arrived at after a numerical search through a set of polynomials that had exactly n_p ripples in the passband and exactly $n_s - 1$ ripples in the stopband so that there is one "invisible" ripple of unconstrained amplitude lying outside the interval (-1, 1). A typical such polynomial is sketched in Fig. 8. The narrowest transition band was found to occur when the "invisible" ripple was brought right up to the passband edge.

Numerical experience with many examples has led the authors to speculate that if the maximally equiripple filter having exactly n_p ripples in the passband and exactly n_s ripples in the stopband is not optimum, then the greatest lower bound of all transition bandwidths in the class can not be achieved by a filter in the class. It appears however that this greatest lower bound can be approached by filters that have one or more invisible ripples just outside the one band edge and that same number of ripples missing from the opposite band.

To partially illustrate a mechanism for this behavior consider Fig. 9 which shows transition bandwidth vs. n_p for fixed n = 10 with $n = n_p + n_s + 1$ so that no ripples are missing. The passband ripple δ_p is 0.01 and the stopband ripple δ_s is 0.0001. We note that as a ripple is exchanged from the stopband to the passband the width of the transition band decreases. Let us consider a specific example with $n_p = 6$ and $n_s = 3$. For this case the width of the transition band is 0.139. If $n_p = 7$ and $n_s = 2$ then the transition bandwidth has been reduced to 0.130 but the number of passband ripples has been increased by one. To remove this ripple the polynomial can be linearly scaled. As the amount of scaling increases the transition width increases. To remove the ripple from the passband, then with the minimum amount of scaling, this ripple should be scaled to the edge of the passband. In certain cases the numbers are such that the decrease in transition bandwidth produced by the removal of a stopband ripple is greater than the increase produced by the subsequent scaling thus resulting in a filter that is better than the maximally equiripple filter having the maximum allowable number of ripples.

REFERENCES

1. O. Herrmann, W. Schuessler, "Design of Nonrecursive Digital Filters with Minimum Phase", Electronics Letters, Vol. 6 No. 11 (May 1970).

2. H. D. Helms, "Digital Filters with Equiripple or Minimax Responses", IEEE Trans. Audio and Electroacoustics, AU-19, No. 1 (March 1971).

3. L. R. Rabiner, J. V. Hu, "Proc. 1971 Symposium on Digital Filters", Imperial College, England (August 1971).

4. T. W. Parks, Rice University, Houston, Texas, Private Communication.

5. O. Herrmann, "Design of Nonrecursive Digital Filters with Linear Phase", Electronics Letters, Vol. 6 No. 11 (May 1970).

6. E. M. Hofstetter, "A New Technique for the Design of Non-Recursive Digital Filters", MIT Lincoln Laboratory Technical Note No. 1970-42 (December 1970).

7. E. M. Hofstetter, A. V. Oppenheim, J. Siegel, "A New Technique for the Design of Nonrecursive Digital Filters", Proceedings, Fifth Annual Princeton Conference on Information Sciences and Systems (March 1971).

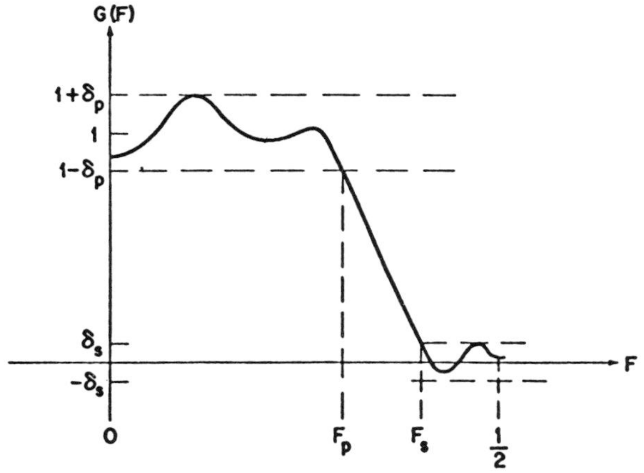

Fig. 1. ACCEPTABLE FILTER RESPONSE

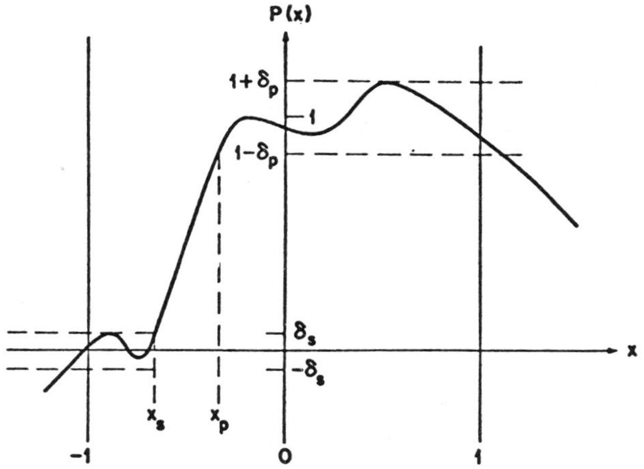

Fig. 2. ACCEPTABLE POLYNOMIAL

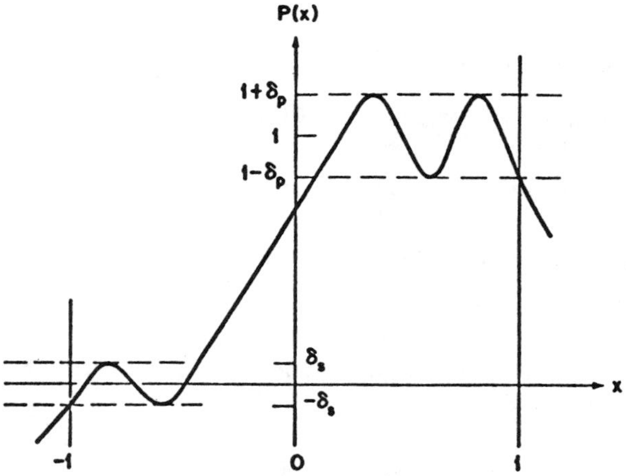

Fig. 3. MAXIMALLY EQUIRIPPLE FILTER

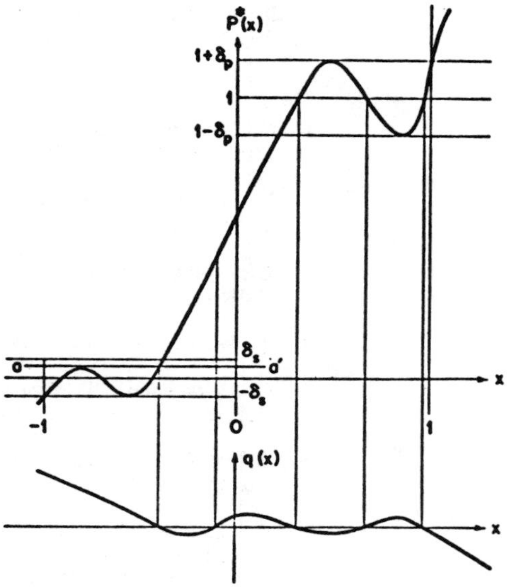

Fig. 4. ILLUSTRATING THE CONSTRUCTION OF q(x)

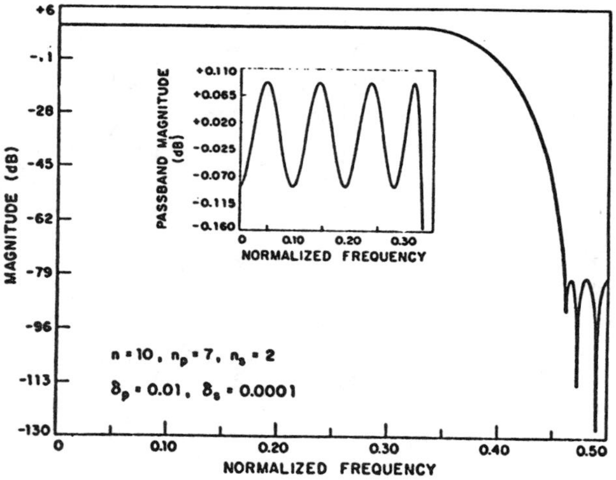

Fig. 5. FILTER WITH ALL ALLOWABLE RIPPLES

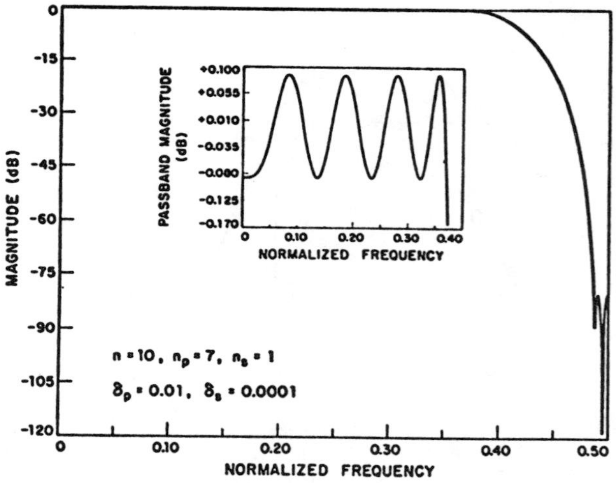

Fig. 6. FILTER WITH ONE RIPPLE MISSING

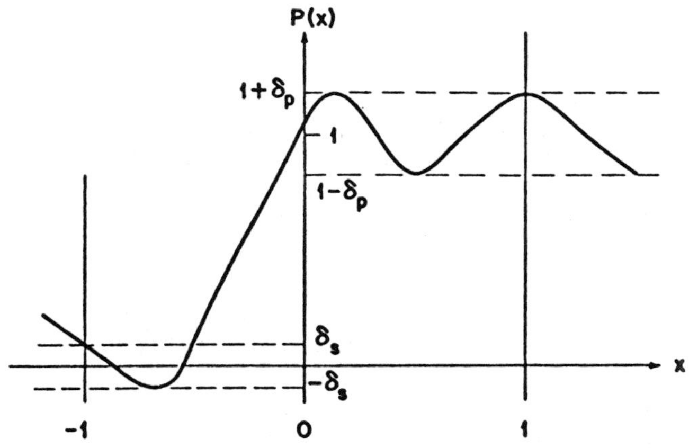

Fig. 7. POLYNOMIAL WITH ENDPOINT RIPPLE

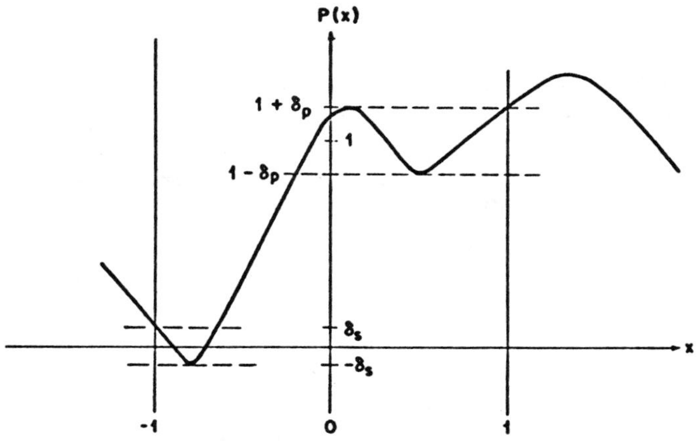

Fig. 8. POLYNOMINAL WITH "INVISIBLE" RIPPLE

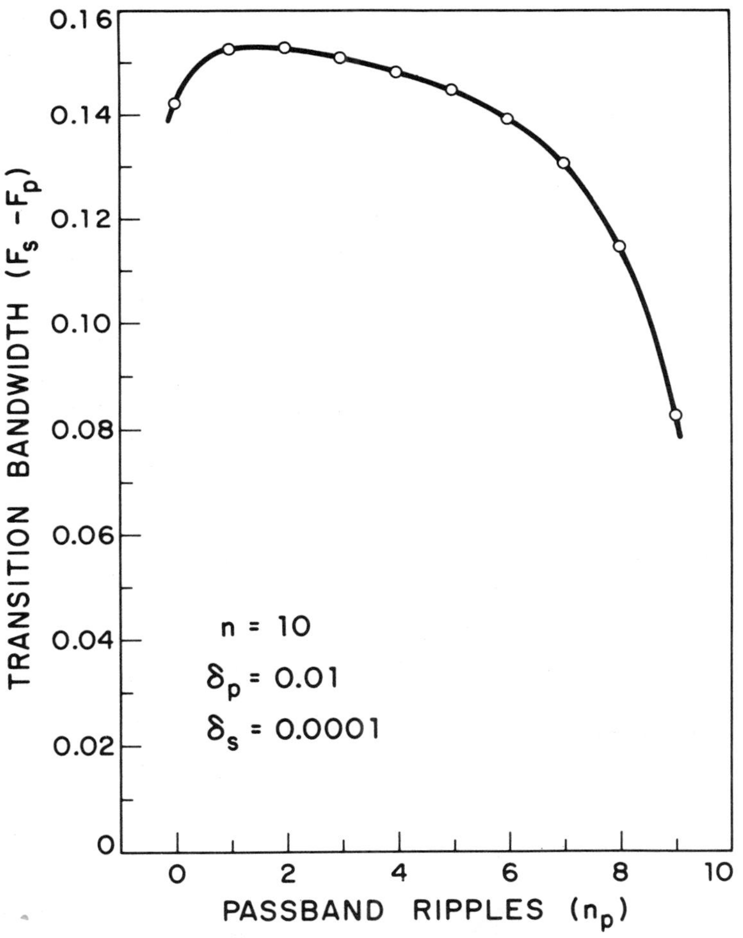

Fig. 9. TRANSITION BANDWIDTH vs n_p

On the Approximation Problem in Nonrecursive Digital Filter Design

Abstract—A new class of selective nonrecursive digital filters with a maximally flat frequency response in the passband and stopband is introduced. The proposed design method is based on a special solution of the general Hermite polynomial interpolation and allows computation of the parameters of these filters in closed form. Therefore it yields some advantage over numerical iterative methods. Design examples are given and an extension to the design of unsymmetrical bandpass systems is made.

In a recent paper, the design of selective nonrecursive digital filters with linear phase has been considered, having equal ripple behavior in the passband as well as in the stopband [1]. These filters are optimal since they satisfy a given tolerance scheme with the lowest possible degree. However, the parameters of these filters can be found by a numerical procedure only. Therefore a solution of a related problem will be considered here, where a maximally flat frequency response is of interest. In this case a solution in closed form can be found.

A nonrecursive digital filter with linear phase is described by a transfer function

$$H(z) = \frac{1}{z^{2n}} \sum_{\mu=0}^{n} \frac{1}{2} d_\mu (z^{n+\mu} + z^{n-\mu}) \tag{1}$$

with a mirror-image polynomial in the numerator. The frequency response of this system is given by

$$H(z = \exp(j\omega T)) = \frac{\sum_{\mu=0}^{n} d_\mu \cos \mu\omega T}{\exp(jn\omega T)} = \frac{H_0(\omega)}{\exp(jn\omega T)}. \tag{2}$$

Restricting ourself to the low-pass case, we are looking for a set of coefficients d_μ such that

$$\begin{aligned} &H_0(\omega)\Big|_{\omega=0} = 1 \\ &\frac{d^\nu}{d\omega^\nu} H_0(\omega)\Big|_{\omega=0} = 0, \qquad \nu = 1, 2, \cdots, 2(n-k)+1 \\ &\frac{d^\mu}{d\omega^\mu} H_0(\omega)\Big|_{\omega=\pm\pi/T} = 0, \qquad \mu = 0, 1, \cdots, 2k-1. \end{aligned} \tag{3}$$

Here k is an integer constant to be chosen arbitrarily within the limits $1 \leq k \leq n$ to obtain the required order of "flatness" at the points $\omega = 0$ and $\omega = \pm\pi/T$. With

$$\cos \omega T = 1 - 2x, \qquad x = \tfrac{1}{2}(1 - \cos \omega T) \tag{4}$$

the function $H_0(\omega)$ is transformed into a simple polynomial of degree n,

$$P_{n,k}(x) = \sum_{\nu=0}^{n} a_\nu x^\nu \tag{5}$$

Manuscript received July 6, 1970; revised November 24, 1970.

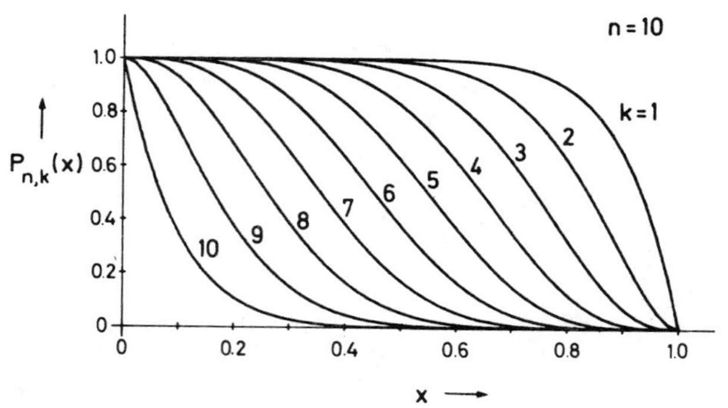

Fig. 1. Maximally flat polynomial approximation to ideal low-pass with n constant.

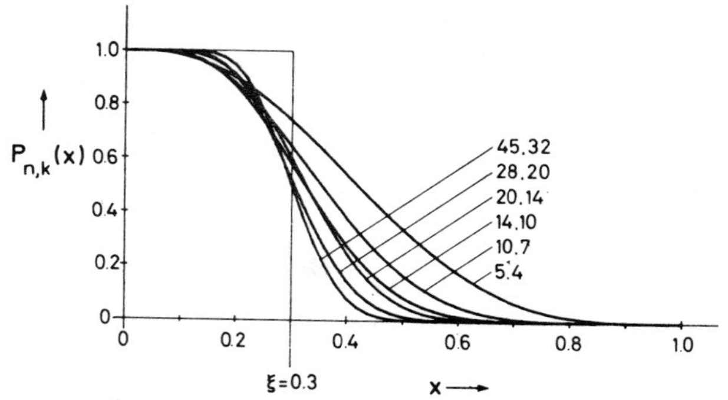

Fig. 2. Maximally flat polynomial approximation to ideal low-pass with ξ constant.

with an approximation interval $0 \leq x \leq 1$ and the properties

$$\begin{aligned} &P_{n,k}(x) \qquad \text{having zeros of order } k \text{ at } x = 1 \\ &P_{n,k}(x) - 1 \quad \text{having zeros of order } n - k + 1 \text{ at } x = 0. \end{aligned} \tag{6}$$

This is a special case of the Hermite interpolation problem [2] and can be solved by using the Newton interpolation formula. But there exists an explicit solution of this problem which is given by the expression

$$\begin{aligned} P_{n,k}(x) &= (1-x)^k \frac{1}{(k-1)!} \frac{d^{k-1}}{dx^{k-1}} \sum_{\nu=0}^{n-1} x^\nu \\ &= (1-x)^k \sum_{\nu=0}^{n-k} \binom{k+\nu-1}{\nu} x^\nu. \end{aligned} \tag{7}$$

It is easy to show that this equation satisfies (6) with $k = 1$, since

$$P_{n,1}(x) = (1-x)(1 + x + \cdots + x^{n-1}) = 1 - x^n \tag{8}$$

obviously has a zero of order 1 at $x = 1$ and

$$P_{n,1}(x) - 1 = -x^n$$

has zeros of order n at $x = 0$. The complete proof follows immediately by induction from k to $k + 1$.

Fig. 1 shows the polynomial $P_{n,k}$ for $n = 10$ with k as a parameter. There exist exactly n different approximations for a given degree n with different

Reprinted from *IEEE Trans. Circuit Theory*, vol. CT-18, pp. 411–413, May 1971.

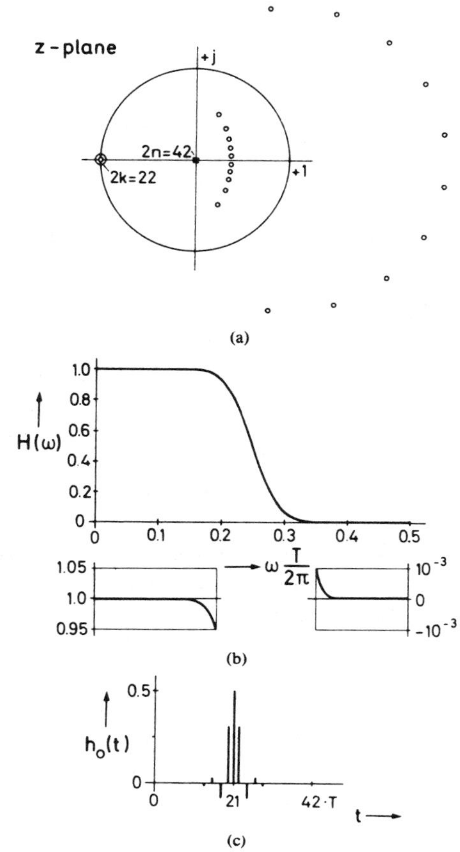

Fig. 3. Nonrecursive digital filter with maximally flat approximation to ideal low-pass ($n=21$, $k=11$). (a) Pole-zero pattern. (b) Frequency response. (c) Impulse response.

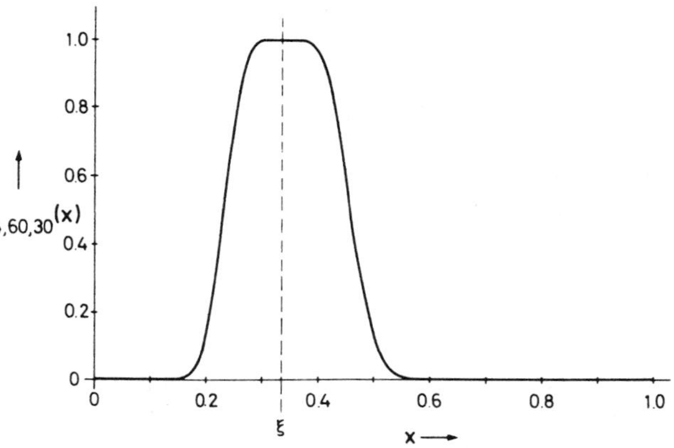

Fig. 4.

"flatness" at $x=0$ and $x=1$. Hence it is necessary to select the parameter k carefully so that the transition zone is as near as possible to the required cutoff point. A more thorough investigation leads to the result, that to satisfy a certain cutoff point $x=\xi$, the parameter k has to be chosen as

$$k = n - [n\xi + 0.5]. \tag{9}$$

Fig. 2 shows $P_{n,k}(x)$ for $\xi=0.3$ and increasing n, where k has been chosen according to (9).

If a suitable polynomial $P_{n,k}(x)$ has been found, the transformation (4) leads to the frequency response $H_0'(\omega)$ and the transfer function $H(z)$. The relation between the coefficients d_μ in (2) and the coefficients a_ν in (5) is given by

$$d_0 = \sum_{k=0}^{[N/2]}\left[2^{-2k}\binom{2k}{k}\sum_{i=2k}^{N} 2^{-i}\binom{i}{2k}a_i\right]$$

$$d_\mu = 2\sum_{k=0}^{[(N-\mu)/2]}\left[2^{-(2k+\mu)}\binom{2k+\mu}{k}\sum_{i=2k+\mu}^{N} 2^{-i}\binom{i}{2k+\mu}a_i\right].$$

$$\mu = 1, 2, \cdots, N \tag{10}$$

but with respect to numerical accuracy it seems to be more advantageous to compute first the complex roots $p_{0\mu}$ of $P_{n,k}(x)$ and transform them by the relation

$$z_{0\nu,\nu+1} = 1 - 2p_{0\mu} \pm \sqrt{(1-2p_{0\mu})^2 - 1} \tag{11}$$

into the z plane. Only in the special case $k=1$ we get a closed-form expression for the roots of $P_{n,k}(x)$ which follows immediately from (8):

$$p_{0\mu} = \sqrt[n]{1} \quad \text{with } k = 1.$$

Fig. 3 shows the pole-zero pattern in the complex z plane, the frequency response, and the impulse response for an example with $n=21$, $k=11$. It is to be seen that the coefficients of the impulse response tend rapidly to zero. The variation between the greatest and the lowest term in the impulse response is about 10^{-7}. Hence a realization of nonrecursive filters of this type in cascade form seems to be much more appropriate than in direct form. Naturally the filter degree required is considerably higher than in the equal-ripple case. In a couple of examples it turned out to be 3–4 times larger than in the case considered in [1].

The low-pass case considered so far can be transformed into the high-pass case using the transformation $z := -z$, and into the bandpass region by the substitution $z := -z^2$. In the case of the bandpass transformation we get a midband frequency at $\omega=\pi/2T$ and a frequency response symmetric about this point. For other midband frequencies and asymmetric frequency response, (7) can be extended to

$$P_{n,k,m} = x^m(1-x)^k R_{n-k-m}(x) \tag{12}$$

a polynomial with zeros of order m at $x=0$ and k at $x=1$, whose first $(n-k-m)$ derivatives are zero at a certain point $x=\xi$. But in this case, the explicit expression for the parameters a_ν turned out to be very complicated. Therefore it was preferred to calculate them directly from the Newton interpolation formula. Fig. 4 shows an example.

A more comprehensive paper on the subject of nonrecursive filters with maximally flat frequency response giving other suitable representations of the solution will be published elsewhere.

Acknowledgment

The author wishes to thank Prof. Meinardus and Prof. Schuessler for very helpful discussions on the subject.

O. Herrmann
Institut f. Nachrichtentechnik
Universität Erlangen–Nürnberg
852 Erlangen, Germany

References

[1] O. Herrmann, "Design of nonrecursive digital filters with linear phase," *Electron. Lett.*, vol. 6, May 1970, pp. 328–329.
[2] P. J. Davis, *Interpolation and Approximation*. New York: Blaisdell, 1965, ch. 3, sec. 3.5.

The Design of Wide-Band Recursive and Nonrecursive Digital Differentiators

LAWRENCE R. RABINER, Member, IEEE
Bell Telephone Laboratories, Inc.
Murray Hill, N. J. 07974

KENNETH STEIGLITZ, Member, IEEE
Department of Electrical Engineering
Princeton University
Princeton, N. J. 08540

Abstract

Designs for recursive and nonrecursive wide-band differentiators are presented. The coefficients for the recursive differentiators were optimally chosen to minimize a square-error criterion based on the magnitude of the frequency response. The coefficients for the nonrecursive differentiators were chosen using a frequency sampling technique. One or more of the coefficients were optimally selected to minimize the peak absolute error between the obtained frequency response and the response of an ideal differentiator. The frequency response characteristics of the recursive differentiators had small magnitude errors but significant phase errors. The nonrecursive differentiators required on the order of 16 to 32 terms for the magnitude error of the frequency response to be as small as the magnitude errors for the recursive differentiators; however, there were *no* phase errors for the nonrecursive case. The delay of the recursive differentiators was small compared to the delay of the nonrecursive differentiators.

Manuscript received January 19, 1970.

The work of K. Steiglitz was supported by the U. S. Army Research Office, Durham, N. C., under Contract DA-HCO4-69-C-0012.

Introduction

A differentiator forms an integral part of many physical systems. Therefore, the design of adequate wide-band differentiators has always been of considerable interest. With the increased trend towards digital simulation of systems, optimal techniques for designing wide-band digital differentiators are being more widely investigated. Kaiser [1] has presented a review of several techniques for designing both nonrecursive and recursive differentiators.

Recent work by Steiglitz [2] has concentrated on the optimal design of recursive digital filters (i.e., filters synthesized with both poles and zeros) with the aid of a large digital computer. The computer optimally chose z-plane positions of poles and zeros to minimize a square-error criterion based on the magnitude of the frequency response. The design problem for nonrecursive digital filters (i.e., filters synthesized with only zeros) has recently been considered by Gold and Jordan [3], and Rabiner, Gold, and McGonegal [4]. They used a digital computer to determine optimal values of a few samples of the discrete Fourier transform of the finite impulse response in order to minimize peak magnitude deviation from the prescribed frequency response.

The work done by Steiglitz indicated that by designing ideal differentiators and allowing a noninteger number of samples of delay, differentiators could be designed with usable bandwidths up to 100 percent full band. In this paper we present specific designs for several recursive and nonrecursive differentiators using the optimal design techniques of the earlier work. These designs are evaluated and compared with respect to approximation errors and realizations, whenever possible.

Theory

The ideal frequency response characteristics of a digital differentiator are shown in Fig. 1. The first two curves in the top line show the magnitude and phase of the frequency response, and the third curve shows the resulting imaginary part of the frequency response (the real part is identically zero in this case). The magnitude response increases linearly up to a normalized frequency of 1.0 (the Nyquist frequency[1]) and then decreases linearly back to 0.0 at the sampling frequency. The magnitude response is periodic in frequency, as shown, because of the discreteness property. The phase is $\pi/2$ radians for frequencies up to the Nyquist frequency and $-\pi/2$ radians from the Nyquist frequency to the sampling frequency, and is also periodic. The resulting imaginary part of the frequency response increases linearly to 1.0 at the Nyquist frequency, jumps discontinuously to -1.0, and then increases linearly to 0.0 at the sampling frequency.

[1] The Nyquist frequency is the highest frequency allowable at the input to the digital system. Thus the input is sampled at twice the Nyquist frequency.

Reprinted from *IEEE Trans. Audio Electroacoust.*, vol. AU-18, pp. 204–209, June 1970.

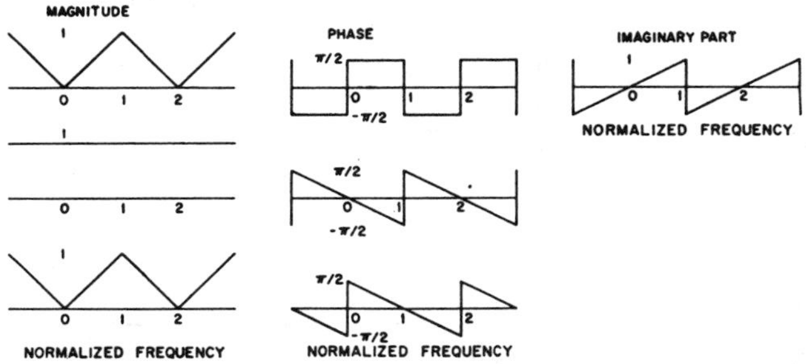

Fig. 1. The frequency response curves for the ideal differentiator. In the first line are shown the magnitude curve, phase curve, and imaginary part of the frequency response. The second line shows the magnitude and phase curves for an ideal one-half sample delay network. The third line shows the magnitude and phase curves for an ideal differentiator with half a sample of delay.

It is the discontinuity in the imaginary part of the response at the Nyquist frequency which makes it difficult to design very wide-band differentiators. It is impossible to obtain a discontinuity in the imaginary part, like the one shown at the top of Fig. 1, in a zero bandwidth region. Therefore a certain amount of the frequency band is designated to be a transition region. Typical designs which result from this method are shown in [1] and [4]. The widest bandwidth differentiators obtained (with reasonable approximation error) are about 95 percent.

One way around the problem of a discontinuity at the Nyquist frequency is to add a delay of a half sample to the differentiator, i.e., to consider the design of an ideal differentiator with half a sample delay. The magnitude response and phase response curves of an ideal half-sample delay network are shown in the middle of Fig. 1. The magnitude is 1 for all frequencies; the phase response is linear with a discontinuity of π radians at the Nyquist frequency. The overall differentiator frequency response with half a sample delay is shown at the bottom of Fig. 1. The magnitude response curve is identical to the original differentiator magnitude response curve, but the phase response curve is now linear with a discontinuity of π radians at 0 frequency. At first thought it would seem that all that this procedure has accomplished is a shifting of a discontinuity from the Nyquist frequency to 0 frequency. However, if one takes into consideration that the magnitude at 0 frequency is exactly 0, it is seen that there must be a zero of the differentiator on the unit circle at zero frequency. A zero on the unit circle will automatically give a phase discontinuity of π radians. Thus the shifting of the discontinuity to 0 frequency, along with the zero of the magnitude response at this frequency, has alleviated the approximation difficulties.

Using this result, both recursive and nonrecursive approximations to the half-sample delay differentiator can be designed with bandwidths up to 100 percent full band. Some specific designs are presented in the following sections.

Recursive Designs

The canonic form used to describe the transfer function of a recursive differentiator is seen in (1):

$$H(z) = A \prod_{i=1}^{K} \frac{(1 - z^{-1}a_{i1})(1 - z^{-1}a_{i2})}{(1 - z^{-1}b_{i1})(z - z^{-1}b_{i2})} \cdot \qquad (1)$$

The transfer function of (1) describes a cascade of K sections each containing two zeros (at $z = a_{i1}, a_{i2}$) and two poles (at $z = b_{i1}, b_{i2}$). The poles and zeros are chosen optimally by computer to minimize a square-error criterion based on the magnitude of the frequency response. The sum of the squares of the magnitude error at 21 equally spaced frequencies from 0 to the Nyquist frequency was minimized with respect to positions of the fixed number of poles and zeros used in the approximation. In Figs. 2 through 4 are shown the error curves (both magnitude and phase error) for one-section ($K = 1$), two-section ($K = 2$), and three-section ($K = 3$) differentiators. It should be noted that these designs are for full-band differentiators. The actual values of the poles and zeros, as well as the constant multiplier A, are given in Table I.

Fig. 2 shows that the peak magnitude error for the one-section design is about 1.1×10^{-2} and occurs near the Nyquist frequency. The peak phase error for this design is about 10.5 degrees and occurs at a normalized frequency of about 0.6. For the two-section design of Fig. 3, the peak magnitude error is about 6.3×10^{-3} occurring near the Nyquist frequency, and the peak phase error is again 10.5 degrees. It should be noted that the magnitude error is under 1.0×10^{-3} for about 95 percent of the bandwidth.

TABLE I

Poles and Zeros of Recursive Differentiators

One-Section		
Zeros:	1.00000000,	−0.67082621
Poles:	−0.14240300,	−0.71698670
A:	0.36637364	
Two-Section		
Zeros:	0.99999949,	−0.86810806
	0.32672838,	−0.44183252
Poles:	−0.10779165,	−0.87602073
	0.33494085,	−0.51312758
A:	0.36804011	
Three-Section		
Zeros:	0.99999956,	−0.87737870
	0.36692749,	−0.49648721
	$-0.15993072\times10^{-2}\pm j0.72664683\times10^{-1}$	
Poles:	−0.11127243,	−0.88411119
	0.35896158,	−0.55390515
	0.63811464×10^{-1}	-0.63137788×10^{-1}
A:	0.36789870	

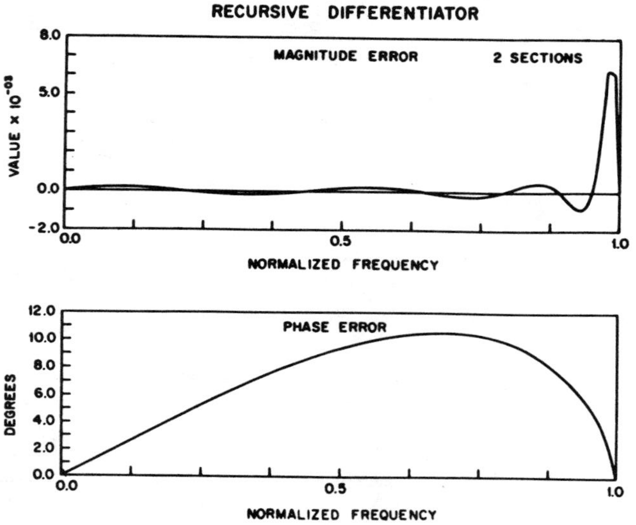

Fig. 3. The magnitude and phase error curves for the optimum two-section recursive approximation to the ideal differentiator.

Fig. 2. The magnitude and phase error curves for the optimum one-section recursive approximation to the ideal differentiator.

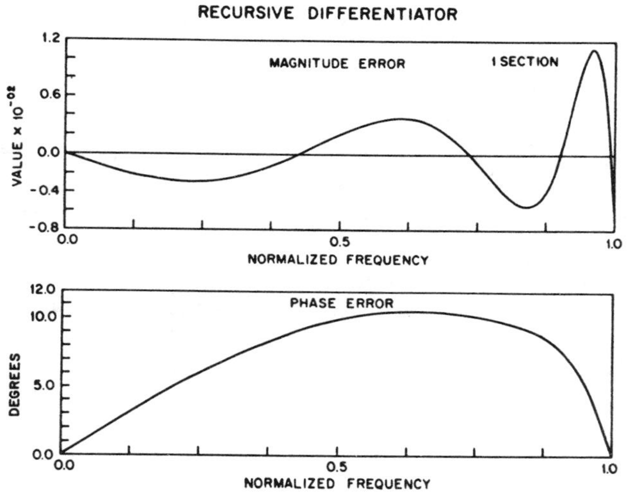

Fig. 4. The magnitude and phase error curves for the optimum three-section recursive approximation to the ideal differentiator.

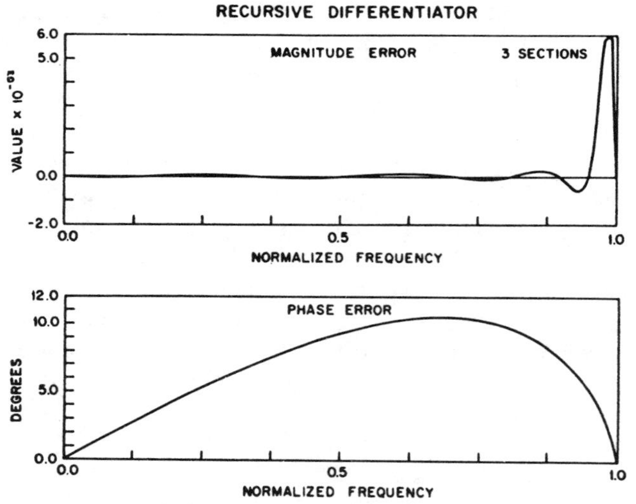

The errors of the three-section differentiator of Fig. 4 are quite similar to the two-section differentiator. The most notable improvement is a halving of the magnitude error in the low-frequency region. It seems clear that the approximation error reduction in going from two to three sections is small compared to the error reduction in going from one to two sections. Thus further increases in the number of sections would probably not change the results dramatically. Furthermore, since the design technique approximated only the magnitude characteristics of the differentiator, the phase error did not change appreciably as the number of sections increased, as can easily be seen from Figs. 2 through 4.

Nonrecursive Designs

An optimal design method for nonrecursive filters was recently presented by Gold and Jordan, and developed by Rabiner, Gold, and McGonegal. In this technique, linearly spaced samples of the frequency response of the desired filter were specified and the continuous frequency response was determined using the discrete Fourier transform. The interpolation formula obtained was

$$H(e^{j\omega T}) = \frac{\exp\left[-\frac{j\omega NT}{2}\left(1-\frac{1}{N}\right)\right]}{N} \cdot \sum_{k=0}^{N-1} \frac{H_k e^{-j\pi k/N} \sin\left(\frac{\omega NT}{2}\right)}{\sin\left(\frac{\omega T}{2}-\frac{\pi k}{N}\right)} \tag{2}$$

where

$$H_k = H(e^{j\omega T})\big|_{\omega=2\pi k/NT} \qquad k = 0, 1, \cdots, N-1; \tag{3}$$

i.e., $\{H_k\}$ are values of the continuous frequency response at equally spaced points around the unit circle; T is the sampling period; and N is the duration of the impulse response in samples. By making the substitution

$$H_k = jG_k e^{j\pi k/N}, \tag{4}$$

each of the terms inside the summation of (2) becomes imaginary, and thus the entire sum is imaginary. For N even, the complex factor outside the summation in (2) represents a pure delay of an integer ($e^{-j\omega T(N/2)}$) plus one-half ($e^{-j\omega T(1/2)}$) number of samples. Thus (2) suggests that a differentiator with exactly half a sample delay can be designed nonrecursively by setting

$$G_k = \begin{cases} k/(N/2), & k = 0, 1, \cdots, N/2 \\ (N-k)/(N/2), & k = \frac{N}{2}+1, \cdots, N-1 \end{cases} \tag{5}$$

and applying the substitution of (4) into (2). Equation (2) shows that in the resultant interpolated frequency response, the magnitude response approximates the differentiator magnitude response (with no approximation error at the frequencies where the exact values were specified), and the phase response is *exactly* the phase response shown at the bottom of Fig. 1, i.e., half a sample delay curve $\pm\pi$ radians. [It should be noted that the H_k are complex, as seen from (4).] By varying some of the G_k, or equivalently the H_k, the magnitude approximation error can be reduced without affecting the phase curve at all.

Examples of full-band nonrecursive differentiators for values of N from 16 to 256 in powers of 2 are given in Figs. 5 through 9, and the relevant data are tabulated in Table II. For each of these cases a single frequency sample (one of the G_k) was optimally chosen to minimize peak error. Table II lists the value of this variable frequency sample ($G_{N/2}$) as well as the peak magnitude error for the various values of N used. In each of Figs. 5 through 9 are shown the impulse response at the top of the figure, the magnitude response curve in the middle, and the magnitude error curve at the bottom. Since the phase curve is always identical to the phase curve at the bottom of Fig. 1, there was no need to plot it again.

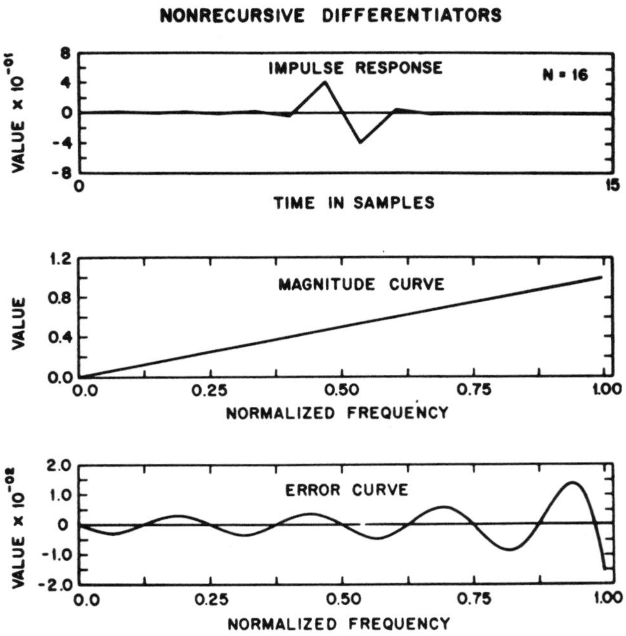

Fig. 5. The impulse response, magnitude curve, and magnitude error curve for the optimum nonrecursive approximation to the ideal differentiator with $N=16$.

Fig. 6. The impulse response, magnitude curve, and magnitude error curve for the optimum nonrecursive approximation to the ideal differentiator with $N=32$.

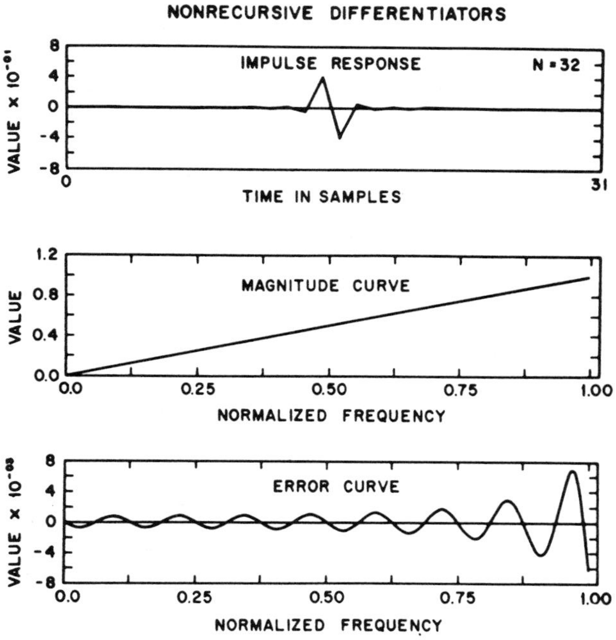

As seen in Fig. 5, the peak magnitude error for $N=16$ is about 1.4×10^{-2} and occurs at about 95 percent full band. For $N=32$ the peak magnitude error is 6.9×10^{-3} again occurring near the Nyquist frequency. Peak magnitude

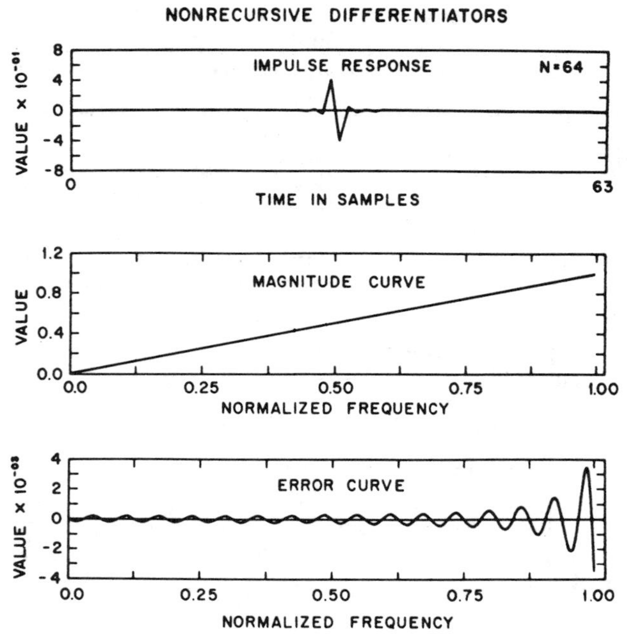

Fig. 7. The impulse response, magnitude curve, and magnitude error curve for the optimum nonrecursive approximation to the ideal differentiator with $N=64$.

Fig. 8. The impulse response, magnitude curve, and magnitude error curve for the optimum nonrecursive approximation to the ideal differentiator with $N=128$.

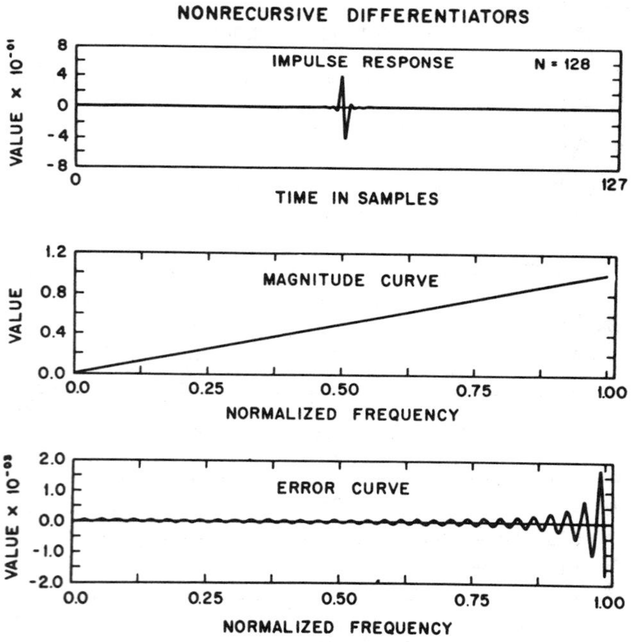

errors for $N=64$, 128, and 256 are 3.5×10^{-3}, 1.7×10^{-3}, and $9.\times10^{-4}$, respectively. Thus a doubling of N tends to halve the magnitude error. Therefore at the expense of increased N it would seem that the error can be made arbitrarily small.

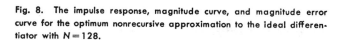

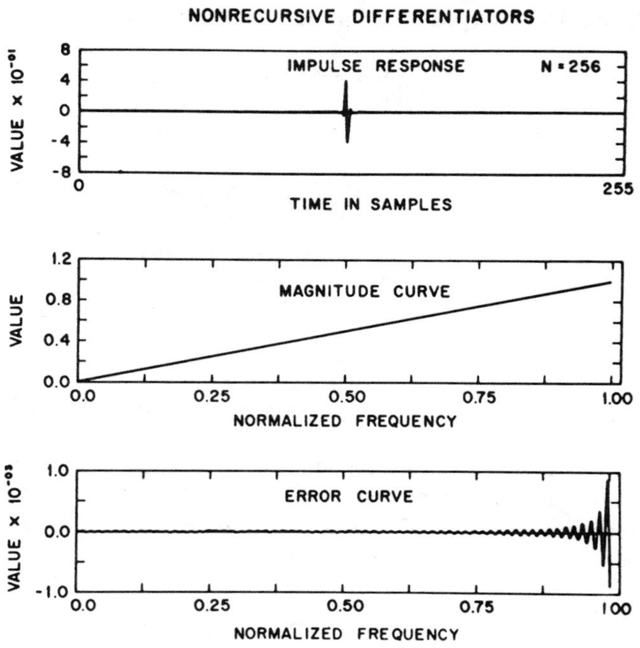

Fig. 9. The impulse response, magnitude curve, and magnitude error curve for the optimum nonrecursive approximation to the ideal differentiator with $N=256$.

TABLE II

Coefficients for Full-Band Nonrecursive Differentiators

N	$G_{N/2}$	Peak Magnitude Error
16	0.98609619	0.013909
32	0.99306030	0.006944
64	0.99653320	0.003472
128	0.99826661	0.001735
256	0.99913330	0.000868

Additional Nonrecursive Designs

It was of interest to consider designs of nonrecursive differentiators whose bandwidth was less than 100 percent full band for small values of $N (N=16, 32)$. For these cases, the three largest frequency samples ($G_{N/2}$, $G_{(N/2)-1}$, $G_{(N/2)-2}$) were optimally varied to minimize the peak error over the band of interest. The resulting designs are given in Table III. In this table the three coefficients are given, along with the peak error for several values of percentage bandwidth.

For $N=16$ the results in Table III show that the peak magnitude error drops very fast as percentage bandwidth decreases, going from 2.7×10^{-3} for 95 percent bandwidth, to 0.7×10^{-5} for 80 percent bandwidth. For $N=32$ the peak magnitude error drops even faster going from 3.8×10^{-4} for 95 percent bandwidth to 8×10^{-7} for 80 percent bandwidth. Thus merely by restricting the bandwidth of differentiation to a reasonable value (less than

TABLE III

Coefficients for Wide-Band Nonrecursive Differentiators

Percent Band-width	Peak Magnitude Error	$G_{N/2}$	$G_{(N/2)-1}$	$G_{(N/2)-2}$
		$N=16$		
95	0.00269	0.96256714	0.87614811	0.75000559
90	0.00072	0.94945069	0.87565707	0.74964216
85	0.00022	0.93826903	0.87324582	0.74978209
80	0.00007	0.92890015	0.86994255	0.75000000
		$N=32$		
95	0.00038	0.97510987	0.93785916	0.87484839
90	0.00003	0.96475830	0.93508185	0.87500000
85	0.000002	0.95614625	0.93098622	0.87483514
80	0.0000008	0.95259399	0.92893748	0.87453343

100 percent), the peak magnitude error can be made quite small even for small values of N.

Comparison of these data with earlier results [1], [4] on nonrecursive differentiator design shows considerable improvements. For a given N and a given bandwidth, the peak magnitude error of designs given in this paper is considerably less than the peak magnitude error resulting from earlier designs.

Comparison Between Recursive and Nonrecursive Designs

It is very difficult to compare the different designs for differentiators (or any other filter for that matter) because there are many issues which must be considered. For example, the recursive designs are easily realized with only a few multiplications per output sample; whereas for a large value of N a large number of multiplications per sample are required in the nonrecursive case (for realization by direct convolution). However, for realization by fast convolution, using the fast Fourier transform, the processing time is only weakly sensitive to the value of N used.

In terms of the accuracy of approximation, the one-section recursive differentiator has somewhat smaller peak magnitude error than the $N=16$ full-band nonrecursive case. However, there is a large phase error in the recursive design and none in the nonrecursive one. The two- and three-section recursive designs have peak magnitude errors comparable to the $N=32$ nonrecursive design.

Finally, recursive differentiators can be designed to have a small delay, whereas the delay for nonrecursive differentiators is generally about $N/2$ samples.

Conclusions

Designs for full-band recursive and nonrecursive differentiators have been presented and discussed. Additional designs for small values of N, where the differentiation bandwidth is less than 100 percent, have been presented for the nonrecursive case.

References

[1] J. F. Kaiser, "Digital filters," ch. 7 in *System Analysis by Digital Computer*, F. F. Kuo and J. F. Kaiser, Eds. New York: Wiley, 1966.

[2] K. Steiglitz, "Computer aided design of recursive digital filters," this issue, pp. 123–129.

[3] B. Gold and K. L. Jordan, Jr., "A direct search procedure for designing finite duration impulse response filters," *IEEE Trans. Audio and Electroacoustics*, vol. AU-17, pp. 33–36, March 1969.

[4] L. Rabiner, B. Gold, and C. McGonegal, "An approach to the approximation problem for nonrecursive digital filters," this issue, pp. 83–106.

An Approach to the Implementation of Digital Filters

LELAND B. JACKSON, Member, IEEE
JAMES F. KAISER, Associate Member, IEEE
HENRY S. McDONALD, Member, IEEE
Bell Telephone Laboratories, Inc.
Murray Hill, N. J.

Abstract

An approach to the implementation of digital filters is presented that employs a small set of relatively simple digital circuits in a highly regular and modular configuration, well suited to LSI construction. Using parallel processing and serial, two's-complement arithmetic, the required arithmetic circuits (adders and multipliers) are quite simple, as are the remaining circuits, which consist of shift registers for delay and small read-only memories for coefficient storage. The arithmetic circuits are readily multiplexed to process multiple data inputs or to effect multiple, but different, filters (or both), thus providing for efficient hardware utilization. Up to 100 filter sections can be multiplexed in audio-frequency applications using presently available digital circuits in the medium-speed range. The filters are also easily modified to realize a wide range of filter forms, transfer functions, multiplexing schemes, and round-off noise levels by changing only the contents of the read-only memory and/or the timing signals and the length of the shift-register delays. A simple analog-to-digital converter, which uses delta modulation as an intermediate encoding process is also presented for audio-frequency applications.

Manuscript received May 6, 1968.

Introduction

The basic theory underlying the analysis and design of digital filters is well advanced (although by no means complete) and quite a few summaries of theoretical results are now available in the engineering literature [1]–[4]. However, the impact of digital filtering theory has not yet been felt by most of the engineers and technicians who design and use the wide variety of filters presently constructed from RLC or crystal circuits. This has been due, in part, to a general unawareness of the possibilities of digital filtering and also, until recently, to the prohibitive complexity and cost of constructing most digital filters. Hence, digital filter implementation has been confined primarily to computer programs for simulation work or for processing relatively small amounts of data, usually not in real time. However, the rapid development of the integrated-circuit technology and especially the potential for large-scale integration (LSI) of digital circuits now promise to reverse this situation in many instances and to make many digital filters more attractive than their analog counterparts, from the standpoints of cost, size, and reliability.

In this paper, we present an approach to the physical implementation of digital filters which has the following features.

1) The filters are constructed from a small set of relatively simple digital circuits, primarily shift registers and adders.
2) The configuration of the digital circuits is highly modular in form and thus well suited to LSI construction.
3) The configuration of the digital circuits has the flexibility to realize a wide range of filter forms, coefficient accuraices, and round-off noise levels (i.e., data accuracies).
4) The digital filter may be easily multiplexed to process multiple data inputs or to effect multiple, but different, filters with the same digital circuits, thus providing for efficient hardware utilization.

After a brief review of general digital filter forms, the advantages of serial, two's-complement, binary arithmetic in the implementation of digital filters are discussed. The required arithmetic circuits (adder/subtractor, complementer, and multiplier) are then described, followed by the techniques for multiplexing. Finally, several examples are presented of multiplexed digital filters that have been constructed and tested. A description of a simple analog-to-digital converter for relatively low-frequency applications is also included.

Canonical Forms

The transfer characteristics of a digital filter are commonly described in terms of its z-domain transfer function [1],

Reprinted from *IEEE Trans. Audio Electroacoust.*, vol. AU-16, pp. 413–421, Sept. 1968.

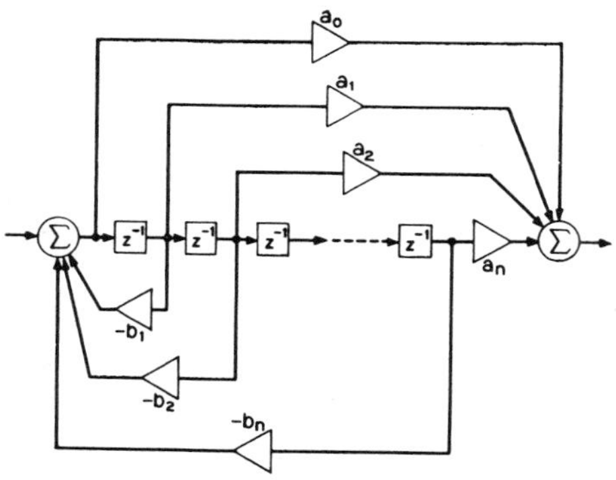

Fig. 1. The direct form for a digital filter.

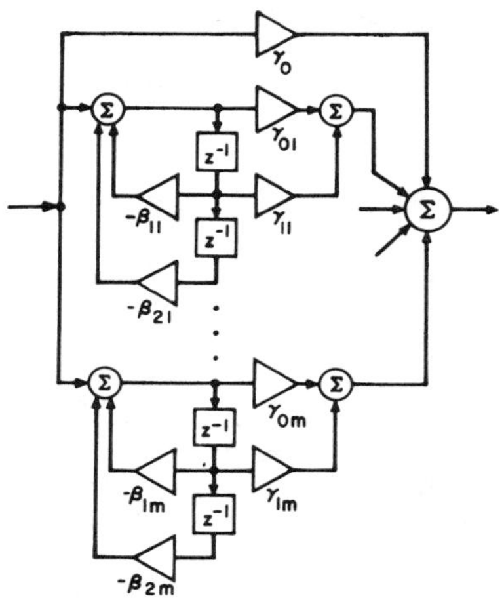

Fig. 3. The parallel form for a digital filter

Fig. 2. The cascade form for a digital filter.

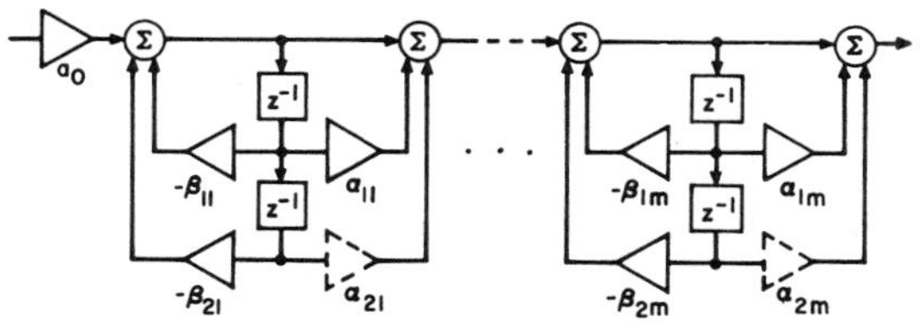

Fig. 4. (A) Second-order section for digital all-pass filter in cascade form. (B) Alternate configuration for digital all-pass filter in cascade form.

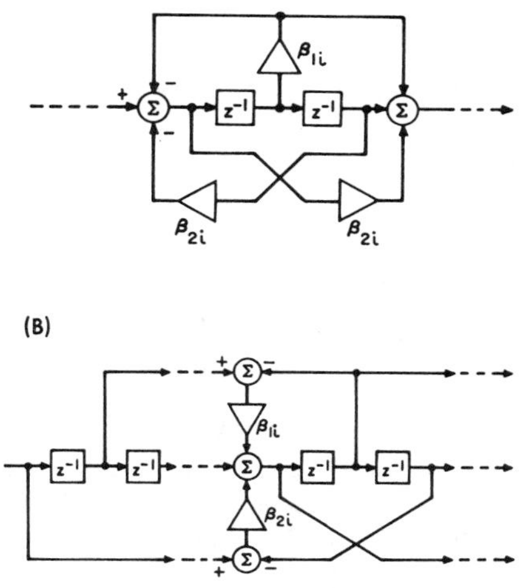

$$H(z) = \frac{\sum_{i=0}^{n} a_i z^{-i}}{1 + \sum_{i=1}^{n} b_i z^{-i}}, \tag{1}$$

where z^{-1} is the unit delay operator. There are a multitude of equivalent digital circuit forms in which (1) may be realized, but three canonical forms, or variations thereof, are most often employed. These forms are canonical in the sense that a minimum number of adders, multipliers, and delays are required to realize (1) in the general case. The first of these forms, shown in Fig. 1, is a direct realization of (1) and as such is called the *direct form*. It has been pointed out by Kaiser [5] that use of the direct form is usually to be avoided because the accuracy requirements on the coefficients $\{a_i\}$ and $\{b_i\}$ are often severe. Therefore, although the implementation techniques presented here are applicable to any filter form, we will not specifically consider the direct form.

The second canonical form corresponds to a factorization of the numerator and denominator polynomials of (1) to produce an $H(z)$ of the form

$$H(z) = a_0 \prod_{i=1}^{m} \frac{\alpha_{2i} z^{-2} + \alpha_{1i} z^{-1} + 1}{\beta_{2i} z^{-2} + \beta_{1i} z^{-1} + 1}, \tag{2}$$

where m is the integer part of $(n+1)/2$. This is the *cascade form* for a digital filter, depicted in Fig. 2. Second-order factors (with real coefficients) have been chosen for (2) rather than a mixed set of first- and second-order factors for real and complex roots, respectively, to simplify the implementation of the cascade form, especially when multiplexing is employed. If n is odd, then the coefficients α_{2i} and β_{2i} will equal zero for some i. The α_{2i} multipliers are shown in dotted lines in Fig. 2, because for the very common case of zeros on the unit circle in the z-plane (corresponding to zeros of transmission in the frequency response of the filter) the associated α_{2i} coefficients are unity. Thus, for these α_{2i} coefficients, no multiplications are actually required.

The third canonical form is the *parallel form*, shown in Fig. 3, which results from a partial faction expansion of

(1) to produce

$$H(z) = \gamma_0 + \sum_{i=1}^{m} \frac{\gamma_{1i}z^{-1} + \gamma_{0i}}{\beta_{2i}z^{-2} + \beta_{1i}z^{-1} + 1}, \quad (3)$$

where $\gamma_0 = a_n/b_n$ and we have again chosen to use all second-order (denominator) factors. Note that all three canonical forms are entirely equivalent with regard to the amount of storage required (n unit delays) and the number of arithmetic operations required ($2n+1$ multiplications and $2n$ additions per sampling period). As previously noted, however, the cascade form requires significantly fewer multiplications for zeros on the unit circle and is thus especially appropriate for filters of the band-pass and the band-stop variety (including low-pass and high-pass filters).

Another interesting filter form may be derived for the special case of an all-pass filter (APF), i.e., a filter or "equalizer" with unity gain at all frequencies. The transfer function for a discrete APF has the general form [6]

$$H_A(z) = \frac{\sum_{i=0}^{n} b_{n-i}z^{-i}}{\sum_{i=0}^{n} b_iz^{-i}}. \quad (4)$$

Thus, with $a_i = b_{n-i}$ and $b_0 = 1$, the direct form can be used to implement (4). However, to reduce the accuracy requirements on the filter coefficients, a modified cascade form can be derived for the APF, corresponding to an $H_A(z)$ of the form

$$H_A(z) = \prod_{i=1}^{m} \frac{z^{-2} + \beta_{1i}z^{-1} + \beta_{2i}}{\beta_{2i}z^{-2} + \beta_{1i}z^{-1} + 1}. \quad (5)$$

Second-order sections for the cascade form of the APF are shown in Fig. 4. Fig. 4(A) is a straightforward modification of the standard cascade form in Fig. 2. Note that because the β_{1i} multiplier may be shared by both the feed-forward and feedback paths, only three multiplications are required per second-order section rather than four. The number of multiplications may be further reduced by using the form of Fig. 4(B), which requires only two multiplications per second-order section. But now, two additional delays are required preceding the first second-order section to supply appropriately delayed inputs to the first section. Therefore, the cascade form of Fig. 4(B) requires a total of n multiplications and $n+2$ delays for an nth-order APF.

Serial Arithmetic

Using any of the canonical forms described in the preceding section, all of the coefficient multiplications and many of the additions during a given Nyquist interval may be performed simultaneously. Therefore, a high degree of parallel processing is possible in the implementation of a digital filter and this may be achieved by providing multiple adders and multipliers with appropriate interconnections. Economy is then realized by using serial arithmetic, and by sharing the adders and multipliers (using the multiplexed circuit configurations to be described) insofar as circuit speed will allow.

In addition to a significant simplification of the hardware, serial arithmetic provides for an increased modularity and flexibility in the digital circuit configurations. Also, the processing rate is limited only by the speed of the basic digital circuits and not by carry-propagation times in the adders and multipliers. Finally, with serial arithmetic, sample delays are realized simply as single-input single-output shift registers.

The two's-complement representation [7] of binary numbers is most appropriate for digital filter implementation using serial arithmetic because additions may proceed (starting with the least significant bits) with no advance knowledge of the signs or relative magnitudes of the numbers being added (and with no later corrections of the obtained sums as with one's-complement). We will assume a two's-complement representation of the form

$$\delta_0 \cdot \delta_1\delta_2 \cdots \delta_{N-1}, \quad (6)$$

which represents a number (δ) having a value of

$$\delta = -\delta_0 + \sum_{i=1}^{N-1} \delta_i 2^{-i}, \quad (7)$$

where each δ_i is either 0 or 1. Thus, the data is assumed to lie in the interval

$$-1 \leq \delta < 1, \quad (8)$$

with the sign of the number δ being given by the last bit (in time) δ_0.

An extremely useful property of two's-complement representation is that in the addition of more than two numbers, if the magnitude of the correct total sum is small enough to allow its representation by the N available bits, then the correct total sum will be obtained regardless of the order in which the numbers are added, even if an overflow occurs in some of the partial sums. This property is illustrated in Fig. 5, which depicts numbers in two's-complement representation as being arrayed in a circle, with positive full scale $(1-2^{-N+1})$ and negative full scale (-1) being adjacent. The addition of positive addends produces counterclockwise rotation about the circle, whereas negative addends produce clockwise rotation. Thus, if the correct total sum satisfies (8), no information is lost with positive or negative overflows and the correct total sum will be obtained.

This overflow property is important for digital filter implementation because the summation points in the filters often contain more than two inputs (see Fig. 3);

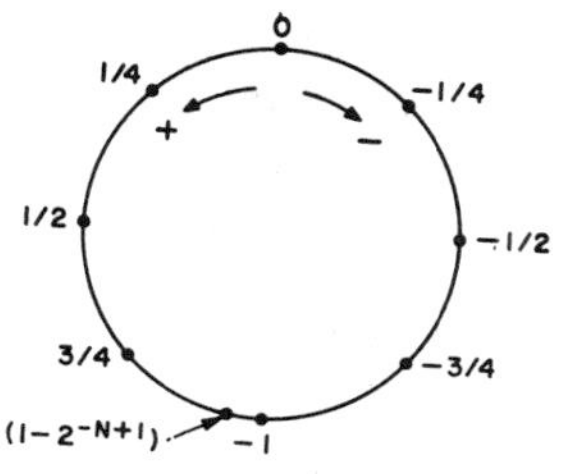

Fig. 5. Illustration of overflow property in two's-complement binary representation.

Fig. 6. Serial two's-complement adder.

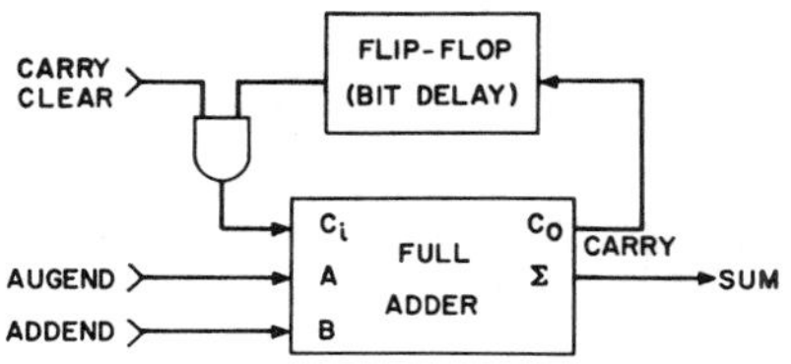

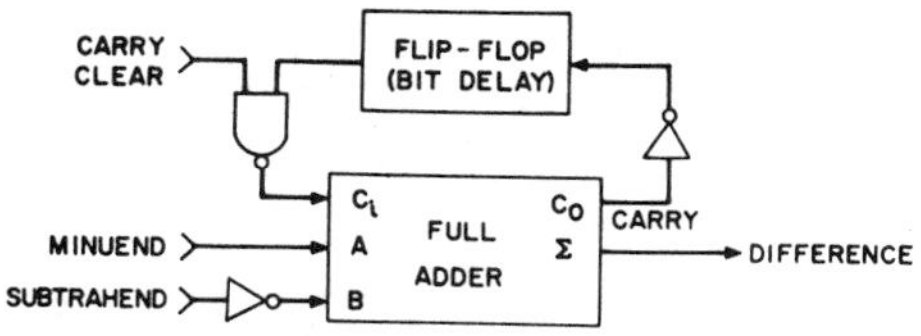

Fig. 7. Serial two's-complement subtractor.

Fig. 8. Serial two's complementer.

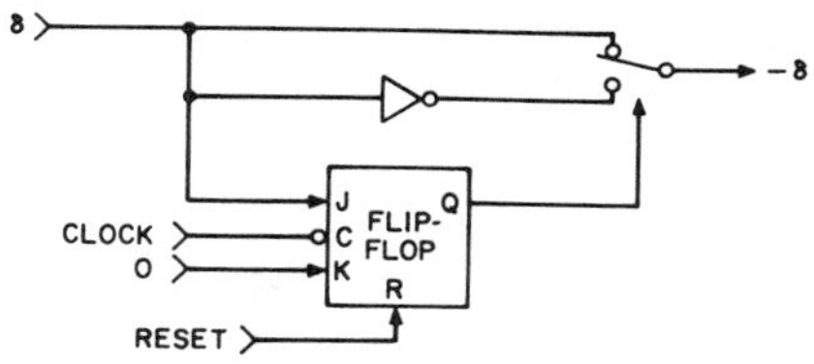

although it may be possible to argue that because of gain considerations the output of the summation point cannot overflow, there is no assurance that an overflow will not occur in the process of performing the summation. Note that this property also applies when one of the inputs to the summation has itself overflowed as a result of a multiplication by a coefficient of magnitude greater than one.

Arithmetic Unit

The three basic operations to be realized in the implementation of a digital filter are delay, addition (or subtraction), and multiplication. As previously mentioned, serial delays (z^{-1}) are realized simply as single-input single-output shift registers. Realizations for a serial adder (subtractor) and a serial multiplier are described in this section. The adders and multipliers, including their interconnections, will be said to comprise the arithmetic unit of the digital filter.

A serial adder for two's-complement addition is extremely simple to construct [7]. As shown in Fig. 6, it consists of a full binary adder with a single-bit delay (flip-flop) to transfer the carry output back to the carry input. A gate is also required in the carry feedback path to clear the carry to zero at the beginning of each sample. Accordingly, the carry-clear input is a timing signal, which is zero during the least significant bit of each sample and is one otherwise.

A serial two's-complement subtractor is implemented by first complementing (negating) the subtrahend input and then adding the complemented subtrahend to the minuend. To complement a number in two's-complement representation, each bit of the representation is inverted and a one is then added to the least significant bit of the inverted representation (i.e., 2^{-N+1} is added to the inverted number). The corresponding serial subtractor circuit is shown in Fig. 7. The subtrahend is inverted and a one is added to the least significant bit by clearing the initial carry bit to one, rather than to zero as in the adder. This is accomplished by means of two inverters in the carry feedback path, as shown.

A separate two's complementer (apart from a subtractor) may also be constructed; such circuits are required in the multiplier to be described. This operation is implemented with a simple sequential circuit which, for each sample, passes unchanged all initial (least significant) bits up to and including the first "1" and then inverts all succeeding bits. A corresponding circuit is depicted in Fig. 8.

A serial multiplier may be realized in a variety of configurations, but a special restriction imposed by this implementation approach makes one configuration most appropriate. This restriction is that no more than N bits per sample may be generated at any point in the digital network because successive samples are immediately adjacent in time and there are no "time slots" available for more than N bits per sample. Hence, the full $(N+K)$-bit product of the multiplication of an N-bit sample by a K-bit (fractional) coefficient may not be accumulated before rounding is performed. However, using the multiplication scheme described below, it is possible to obtain the same rounded N-bit product without ever generating more than N bits per sample. Rounding is usually preferable, rather than truncation, to limit the introduction of extraneous low-frequency components (dc drift) into the filter.

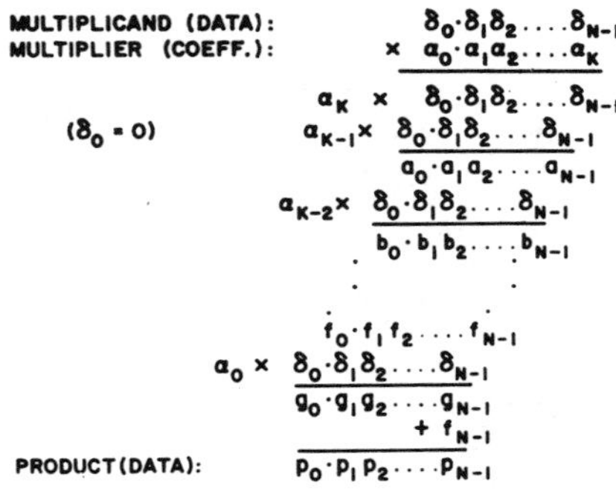

Fig. 9. Serial multiplication using no more than N bits per data word.

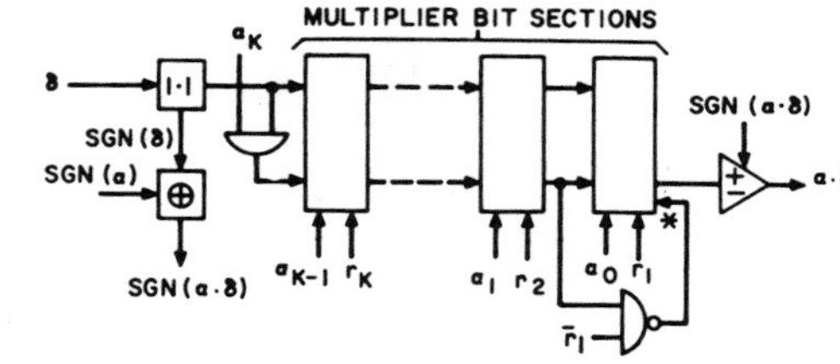

Fig. 10. Serial multiplier, showing modularity.

Fig. 11. A multiplier bit section.

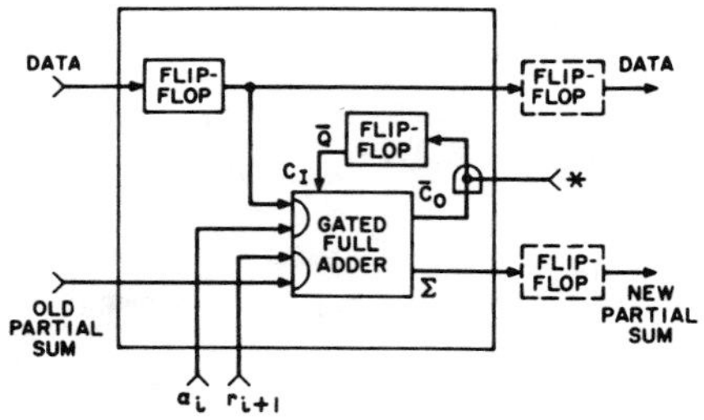

The serial multiplication scheme is depicted in Fig. 9. To simplify the required hardware, both the multiplicand (data) and the multiplier (coefficient) are constrained to be positive, with appropriate sign changes being made before and after the multiplication. Thus $\delta_0=0$ in Fig. 9 and the sign of the multiplicand (δ) is stored separately as SGN δ. For convenience, the multiplier (α) will be assumed to lie in the interval.

$$-2 < \alpha < 2. \tag{9}$$

Although (9) is not necessarily applicable in the general case, it does hold for the denominator coefficients of the cascade and parallel forms, and usually for the numerator coefficients of the cascade form as well. The magnitude of the multiplier is thus represented in Fig. 9 as

$$\alpha_0 \cdot \alpha_1 \alpha_2 \cdots \alpha_K, \tag{10}$$

which represents a value of

$$|\alpha| = \sum_{i=0}^{K} \alpha_i 2^{-i}. \tag{11}$$

The restriction in (9) and the resulting representation in (10) and (11) are in no way essential to the serial multipliers to be described, but are meant only to be representative of the multiplication scheme. The sign of the multiplier is also stored separately as SGN α.

The multiplication scheme in Fig. 9 proceeds as follows. The multiplicand is successively shifted (delayed) and multiplied (gated) by the appropriate bit (α_i) of the multiplier. These delayed gated instances of the multiplicand are then added in the order indicated. After each addition (including the "addition" of $\alpha_K\delta$ to 0), the least significant bit of the resulting partial sum (i.e., $\delta_{N-1}, a_{N-1}, b_{N-1}, \cdots, f_{N-1}$) is truncated to prevent the succeeding partial sum from exceeding N bits in length. Note that these bits may be truncated because the full unrounded product would be

$$g_0 \cdot g_1 g_2 \cdots g_{N-1} f_{N-1} \cdots b_{N-1} a_{N-1} \delta_{N-1} \tag{12}$$

and to round (12) to N bits, only the value of the bit f_{N-1} is required. Thus, before truncating f_{N-1}, its value is stored elsewhere to be added in the final step to g, as shown in Fig. 9, to obtain the rounded product (p).

The serial multiplier corresponding to the scheme described above is shown in Fig. 10. The absolute value of each incoming datum (δ) is taken and its sign (SGN δ) is added modulo-2 to the coefficient sign (SGN α) to determine the product sign (SGN $\alpha \cdot \delta$). The (positive) multiplicand is then successively delayed and gated by the appropriate multiplier bits (α_i) and the partial sums are accumulated in the multiplier bit sections. A single multiplier bit section is shown in Fig. 11. The least significant bit of each partial sum is truncated (gated to zero) by the appropriate timing signal r_{i+1}. Rounding is accomplished by adding in the last truncated bit (f_{N-1}) via the * input to the last bit section. Finally, the sign of the product is inserted using a two's-complementer such as that in Fig. 8. At high data rates, it may be necessary to insert extra flip-flops between some or all of the multiplier bit sections, as shown in dotted lines in Fig. 11, to keep the propagation delay through the adder circuits from becoming excessive.

Several observations concerning the serial multiplier should be made at this point. First, there is a delay of K bits in going through the multiplier and this delay must be deducted from a delay (z^{-1}) that precedes the multiplier. (If the extra flip-flops in Fig. 11 are required, then the multiplier will yield a delay of up to $2K$ bits.) In addition, the absolute value operation at the first of the multiplier requires a delay of N bits (to determine the sign of each incoming datum) and this must be deducted from a preceding delay as well. Thus, to use this serial multiplier, the z^{-1} delays of the digital filter must be at least $N+K$ (or up to $N+2K$) bits in length. This in turn implies, as we shall see in the next section, that some form of multiplexing is required if the multipliers are to be implemented in this manner.

Another observation is that the adders in the multiplier bit sections do not require carry-clear inputs because only positive numbers are being added. However, output product overflows (in the sense of Fig. 5) are possible with coefficients (α) of magnitude greater than one. It may thus be necessary to restrict the amplitude of the data into certain multipliers to prevent output overflows; while in certain other multipliers, these overflows may be perfectly allowable as discussed in the preceding section. In general, however, the inputs to a summation must be scaled so that an overflow will not occur in the *final* output of the summation. Such overflows represent a severe nonlinearity in the system, and very undesirable effects can result in the output of the filter.

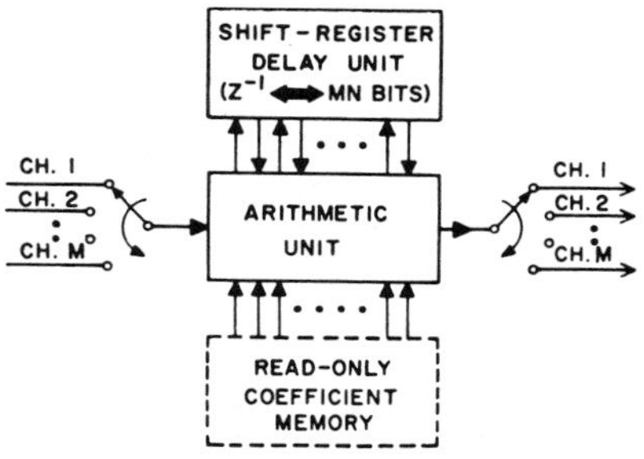

Fig. 12. Type-1 multiplexing for M input channels.

Multiplexing

Having realized the three basic digital filter components (delays, adders, and multipliers), the filter itself may be implemented by simply interconnecting these components in a configuration corresponding to one of the digital forms, canonical or otherwise, for the filter. However, if the input rate bit (sampling rate times bits per sample) is significantly below the capability of the digital circuits, the digital filter can be multiplexed to utilize the circuits more efficiently. The various multiplexing schemes are of two main types: 1) the multiplexed filter may operate upon a number of input signals simultaneously or 2) the multiplexed filter may effect a number ot (different) filters for a single input signal. A combination of these two types is also possible.

To multiplex the filter to process M simultaneous inputs (type 1), the input samples from the M sources are interleaved sample by sample and fed (serially) into the filter. The bit rate in the filter is thus increased by a factor of M. The shift-register delays must also be increased by a factor of M to a length of MN bits. Otherwise, the filter is identical in its construction to the single-input case. In particular, the arithmetic unit containing the adders and multipliers is the same; it just operates M times faster. The output samples emerge in the same interleaved order as the input and are thus easily separated. Type-1 multiplexing is depicted in Fig. 12.

If the M channels in Fig. 12 are to be filtered differently or if type-2 multiplexing is also employed, the filter coefficients are stored in a separate read-only coefficient memory and are read-out as required by the multiplexed filter. A diode matrix provides a very fast and inexpensive form of read-only memory (ROM) for this purpose. If, however, all M channels are to be filtered identically and no type-2 multiplexing is employed, the coefficients may be wired into the multipliers and no ROM is then required. This is indicated by the dotted lines enclosing the ROM in Fig. 12. In this case, adders must be included only in those multiplier bit sections of the arithmetic unit for which the corresponding multiplier bits (α_i) equal one.

In many cases, a number of different, but similar, filters or subfilters are required for the same input signal. For example, all of the second-order subfilters comprising the cascade or parallel forms are similar in form, differing only in the values for the multiplying coefficients (see Figs. 2 and 3). Type-2 multiplexing refers to the implementation of these different (sub)filters with a signal multiplexed filter. An example of a multiplexed second-order filter is shown in Fig. 13. As with type-1 multiplexing, the combining of M separate filters into one multiplexed filter requires that the bit rate in the filter be increased by a factor of M and that the shift-register delays (z^{-1}) also be increased by a factor of M to MN bits in length. The coefficients are supplied from the read-only coefficient memory, which cycles through M values for each coefficient during every Nyquist interval. Data are routed in, around, and out of the filter by external routing switches, which are also controlled from the ROM.

As an example of type-2 multiplexing, consider the implementation of a 12th-order filter in cascade form, using the multiplexed second-order filter in Fig. 13. Here $M=6$, so the bit rate in the filter must be (at least) $6N$ bits per Nyquist interval. During the first N bits of each Nyquist interval, the input sample is introduced into and is processed by the arithmetic unit with the multiplying coefficients (α_1, α_2, β_1, β_2) of the first subfilter in the cascade form. The resulting output is delayed by N bits ($z^{-1/M}$) and fed back via the input routing switch to become the input to the filter during the second N-bit portion of the Nyquist interval. This feedback process is repeated four

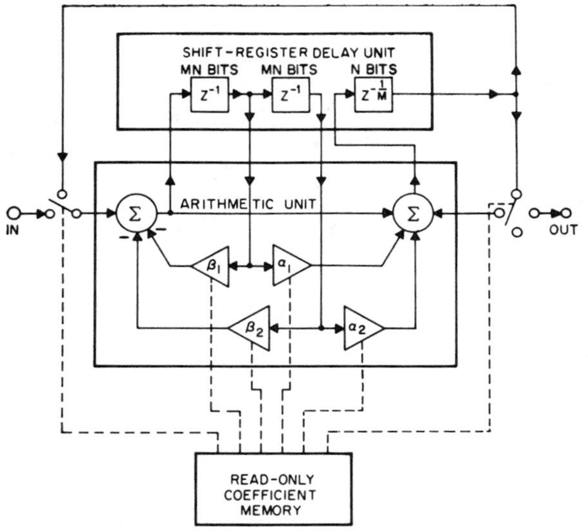

Fig. 13. General second-order filter for type-1 and type-2 multiplexing.

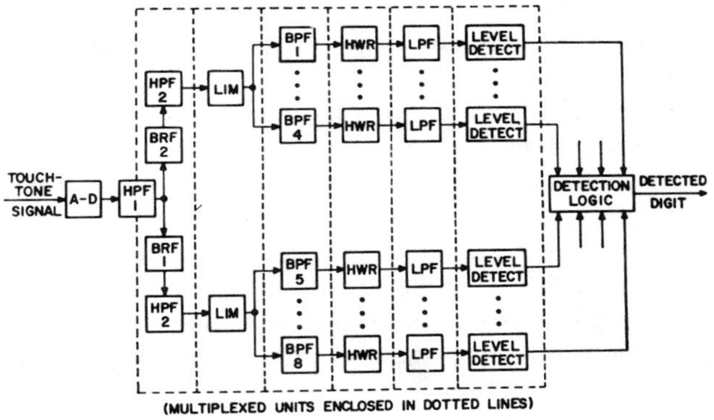

Fig. 14. Digital touch-tone receiver, showing multiplexed filters and nonlinear units.

more times, with the filter coefficients from the ROM being changed each time to correspond to the appropriate subfilter in the cascade form. The sixth (last) filter output during each Nyquist interval is the desired 12th-order filter output. The parallel form, or a combination of cascade and parallel filters, may be realized using the filter in Fig. 13 by simply changing the bits in the ROM which control the switching sequences of the input and output routing switches.

Sample System

As an example of this approach to the implementation of digital filters, we will take an experimental, all-digital touch-tone receiver (TTR) which has been designed and constructed at Bell Telephone Laboratories, Inc. The digital TTR is depicted in block-diagram form in Fig. 14 This is a straightforward digital version of the standard analog TTR described elsewhere [8]. Without going into the detailed operation of the system, we simply note that the combined high-pass filters (HPF's) are third order, the band-rejection filters (BRF's) are each sixth order, the bandpass filters (BPF's) are each second order, and the low-pass filters (LPF's) are each first order. The other signal-processing units required are the limiters (LIM's), half-wave rectifiers (HWR's), and level detectors. These nonlinear operations are, of course, easily implemented in digital form.

A multiplexing factor of $M=8$ is employed in the experimental TTR to combine all of the units enclosed in dotted lines into single multiplexed units. In particular, all of the HPF's and BRF's are multiplexed into one second-order filter (combined type-1 and type-2 multiplexing), the eight BPF's are multiplexed into another second-order filter (type-1 multiplexing with ROM coefficients), and the eight LPF's are multiplexed into one first-order section (type-1 multiplexing with wired-in coefficients). The nonlinear units are readily multiplexed as well and operate directly upon the interleaved output samples from the filters.

Some of the parameters of the experimental TTR design are as follows: the sampling rate is 10 K samples/second with an initial quantization (A/D conversion) of 7 bits/sample; the data word length (N) within the filter is 10 bits/sample; the filter coefficients have 6-bit fractional parts (K); and, as previously stated, the multiplexing factor (M) is eight. Thus, the bit rate within the filter (sampling rate × bits/sample × M) is 800 K bits/second. The number of bits required to represent the data and the coefficients of the TTR were determined through computer simulation of the system. The hardware required to implement this design consists primarily of about 40 serial adders and 400 bits of shift-register storage.

Analog-to-Digital Converter

In most applications of digital filters, the initial input signal is in analog form and must be converted to digital form for processing. It may or may not be necessary to reconvert the digital output signal to analog form, depending upon the application. Digital-to-analog (D/A) conversion is a relatively straightforward and inexpensive process, but the initial analog-to-digital (A/D) conversion is often quite a different situation. For audio-frequency applications, however, a simple and very accurate A/D converter may be implemented using delta modulation (Δ-mod) as an intermediate encoding process. This A/D converter will now be described.

The A/D converter is depicted in Fig. 15, with a Δ-mod encoder being shown in Fig. 16. The Δ-mod encoder produces a series of bivalued pulses (0's and 1's) which, when integrated, constitute an approximation to the input analog signal. The number of 1's (or 0's) occurring during each (eventual) sampling interval is accumulated in the counter as a measure of the change in signal amplitude during that interval. At the end of the interval, this number is transferred to the storage register and the counter is then reset to its initial value to begin counting during the next interval. The appropriate initial value for the counter is minus one-half the number of Δ-mod pulses per sampling interval.

The number stored in the storage register for each sampling interval is the difference between the desired sample value and the preceding sample value. Thus, if these difference samples are accumulated in a simple first-order accumulator, as shown, the full digital sample values result. A small "leak" is introduced into the accumulator by making the feedback gain slightly less than one ($1-2^{-K}$) to keep the dc gain of the accumulator from being infinite. This prevents a small dc bias in the Δ-mod output from generating an unbounded accumulator output. The accumulator leak should be matched by a similar leak in the integrator of the Δ-mod encoder.

Since the accumulator is itself a first-order digital filter, it can be implemented and multiplexed using the same circuits and techniques as previously described. The multiplexing may be either with other A/D conversion channels or with other filters, or both. Note, however, that if the digital filter following the A/D converter has, or can have, a zero at $z=1$ (corresponding to zero of transmission at dc), this zero would cancel the pole supplied by the accumulator (with no leak). Therefore, in this case, the accumulator may be eliminated from the A/D converter (along with the zero at $z=1$ from the following filter), making the A/D conversion even simpler.

The accuracy of A/D conversion implemented in this manner is a function of the following factors: 1) the ratio of the Δ-mod rate to the sampling rate, 2) the sensitivity of the input comparison amplifier in the Δ-mod encoder, and 3) the match between the accumulator leak (if an accumulator is required) and the integrator leak in the Δ-mod encoder. There is also a maximum-slope limitation with delta modulation and an accompanying slope overload noise results if this slope limitation is exceeded [9]. Assuming that the error resulting from 2) and 3) and from slope overload is negligible, a useful rule of thumb for the A/D conversion accuracy is that the number of quantization levels effected equals approximately the

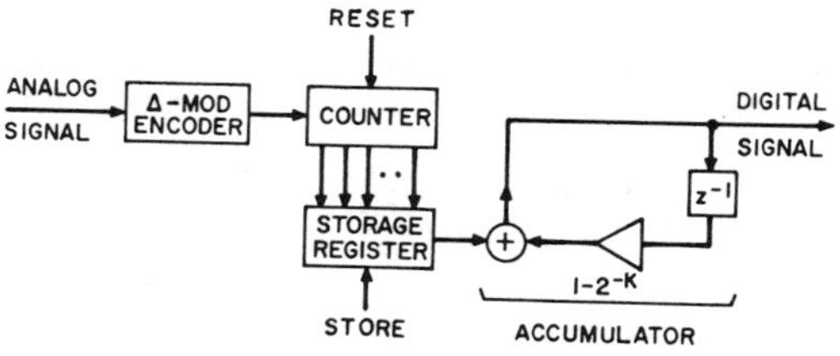

Fig. 15. Simple A/D converter using delta modulation.

Fig. 16. Delta-modulation encoder.

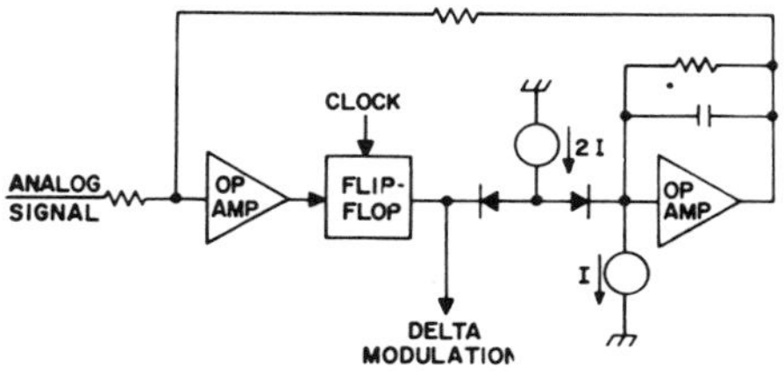

ratio of the Δ-mod rate to the sampling rate. Thus, for example, to effect 10-bit A/D conversion using this scheme, the Δ-mod rate must be approximately 1000 times the sampling rate.

Conclusions

An approach to the implementation of digital filters has been described that employs a small set of relatively simple digital circuits in a highly regular and modular configuration, well suited to LSI construction. By using parallel processing and serial, two's-complement arithmetic, the required arithmetic circuits (adders and multipliers) are greatly simplified, and the processing rate is limited only by the speed of the basic digital circuits and not by carry-propagation times. The resulting filters are readily multiplexed to process multiple data inputs or to effect multiple, but different, filters (or both) using the same arithmetic circuits, thus providing for efficient hardware utilization. A multiplexing factor of 100 or so is possible in audio-frequency applications, using presently available digital logic in the medium-speed range. The filters are also easily modified to realize a wide range of filter forms, transfer functions, multiplexing schemes, and round-off noise levels (i.e., data accuracies) by changing only the contents of the read-only coefficient memory and/or the timing signals and the length of the shift-register delays. For audio-frequency applications, a simple A/D converter may be implemented using delta modulation as an intermediate encoding process.

REFERENCES

[1] J. F. Kaiser, "Digital filters," in *System Analysis by Digital Computer*, J. F. Kaiser and F. F. Kuo, Eds. New York: Wiley, 1966, pp. 218–85.
[2] C. M. Rader and B. Gold, "Digital filter design techniques in the frequency domain," *Proc. IEEE*, vol. 55, pp. 149–171, February 1967.
[3] R. M. Golden, "Digital filter synthesis by sampled-data transformation," this issue, pp. 321–329.
[4] B. Gold and C. M. Rader, *Digital Processing of Signals.* New York: McGraw-Hill, 1969.
[5] J. F. Kaiser, "Some practical considerations in the realization of linear digital filters," *1965 Proc. 3rd Allerton Conf. on Circuit and System Theory*, pp. 621–633.
[6] R. B. Blackman, unpublished memorandum.
[7] Y. Chu, *Digital Computer Design Fundamentals.* New York: McGraw-Hill, 1962.
[8] R. N. Battista, C. G. Morrison, and D. H. Nash: "Signaling system and receiver for touch-tone calling," *IEEE Trans. Communications and Electronics*, vol. 82, pp. 9–17, March 1963.
[9] E. N. Protonotarios, "Slope overload noise in differential pulse code modulation systems," *Bell Sys. Tech. J.*, vol. 46, pp. 2119–2161, November 1967.

Variable Digital Filters

Based on the general rational transformation described in an earlier paper and the frequency transformations for digital filters a realization scheme is derived for variable digital filters satisfying a tolerance scheme with constant attenuation in the passband and in the stopband, that renders possible a variation of the cutoff-frequencies all over the base band $|f| < f_s/2$. Measured frequency responses of a built model are shown; some problems in digital implementations are discussed.

Veränderbare digitale Filter

Ein Blockschaltbild für variable digitale Filter mit einem Toleranzschema jeweils konstanter Dämpfung im Sperr- und Durchlaßbereich wird entwickelt, das es ermöglicht, die Grenzfrequenzen derartiger Filter im gesamten Basisband $|f| < f_s/2$ zu variieren. Das Verfahren ist ein Spezialfall der früher beschriebenen allgemeinen rationalen Transformationen und macht Gebrauch von den Frequenztransformationen für digitale Filter. Es werden Meßergebnisse von einem realisierten Modellfilter gebracht; einige Probleme bei der digitalen Realisierung werden erörtert.

In some previous papers [2], [3] A. G. Constantinides outlined the principles of the frequency transformation of digital filters. It has been shown there, that by replacing the variable z^{-1} of a given lowpass transfer function by an appropriate specification of the unit function

$$G(z^{-1}) = \prod_{\nu=1}^{n} \frac{z^{-1} - \alpha_\nu}{1 - \alpha_\nu z^{-1}}, \quad |\alpha_\nu| < 1 \tag{1}$$

one obtains a new stable and realizable transfer function with the same passband and stopband attenuation, but at different frequencies. The relations between the cutoff frequency of the lowpass to be transformed, the coefficients a_ν, and the obtained cutoff frequency (frequencies for bandpass and bandstop filters, respectively) are given in these references.

In the most interesting cases: lowpass-to-lowpass, lowpass-to-highpass, lowpass-to-bandpass, and lowpass-to-bandstop transformation, the unit functions to be applied for transformation turn out to be of the type

$$F(z^{-1}) = \pm \sum_{\nu=0}^{n} b_{n-\nu} z^{-\nu} \Big/ \sum_{\nu=0}^{n} b_\nu z^{-\nu} \tag{2}$$

where all b_ν are real, and the roots of the denominator lie outside of the unit circle in the z^{-1}-plane.

Eq. (2) is the transfer function of a digital allpass; hence, the method described so far is not only suitable to generate other transfer functions from a given one, but it offers also a scheme for realization according to the following rule: A digital lowpass filter will be transformed into another lowpass or a highpass, bandpass or bandstop filter by replacing each delay-element by the same appropriate digital allpass (Fig. 1). This is a special case of the general rational transformation described in [1].

lowpass

$$A_{LP}(z) = \frac{b_2 z^2 + b_1 z + b_0}{z^2 + c_1 z + c_0}$$

$$H(z) = \frac{\alpha z + 1}{z + \alpha} \quad \text{lowpass}$$

$$H(z) = -\frac{\alpha z + 1}{z + \alpha} \quad \text{highpass}$$

$$H(z) = -\frac{\alpha_2 z^2 + \alpha_1 z + 1}{z^2 + \alpha_1 z + \alpha_2} \quad \text{bandpass}$$

$$H(z) = \frac{\alpha_2 z^2 + \alpha_1 z + 1}{z^2 + \alpha_1 z + \alpha_2} \quad \text{bandstop}$$

Fig. 1. Frequency transformation of a digital lowpass.

The advantage in this scheme is that by varying the coefficients in all allpasses simultaneously the frequency response of the filter can be varied in such a way that the cutoff frequency can be put to any frequency within the baseband, whereas the general behavior of the frequency response, e.g. the number of ripples in the passband and the number of poles in the stopband, is fixed. As a specification of eq. (2) the allpass transfer function for the lowpass-to-bandpass or lowpass-to-bandstop transformation can be written as

$$H(z) = \pm \frac{\alpha z^2 + \beta(1-\alpha) z + 1}{z^2 + \beta(1-\alpha) z + \alpha}, \quad |\alpha| < 1, \quad |\beta| < 1.$$

It can be shown, that the midband frequency is controlled by β only, whereas α affects the bandwidth of the bandpass (or bandstop) filter. Hence we have a realization scheme for bandpass or bandstop filters with independently variable midband frequency and bandwidth, if we assume α and β to be variable.

A model for variable bandwidth only was built with sample-and-hold devices [4], [5]. The characteristics are:

Reprinted with permission from *Arch. Elek. Übertragung*, vol. 24, pp. 524–525, 1970.

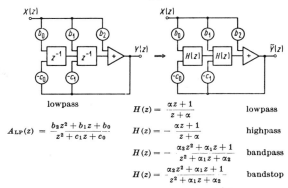

Basic lowpass filter: C 04 25 35 a [6]

$$a_{p\,max} = 0.28\ \text{dB},$$
$$a_{st\,min} = 40.34\ \text{dB}.$$

Allpass transfer function:

$$H(z) = \frac{\alpha z^2 - 1}{z^2 - \alpha} \quad (\text{i.e.} \quad \beta = 0)\,.$$

From $H(\pm j) = 1$ it is to be seen that the bandpass midband frequency is independent of α and fixed to $f_c = f_s/4$.

The lowpass found in [6] was transformed into the z-plane by bilinear transformation in such a way that the passband cutoff frequency was set to $f_p = 1.5$ kHz while the sampling frequency was $f_s = 24$ kHz.

The variation range of α is $-0.52 \leqq \alpha \leqq 0.52$, yielding a bandwidth variation of $0.44\ \text{kHz} \leqq B \leqq 4\ \text{kHz}$. The measured frequency responses for $\alpha = \alpha_{min}$ and $\alpha = \alpha_{max}$ are shown in Fig. 2.

In general the method described so far cannot be implemented digitally since the allpass (eq. (2)) has a straight feedthrough for $b_n \neq 0$. By implementing the allpass in a recursive lowpass filter one obtains a network with algebraic feedback loops, which are not digitally realizable. This difficulty can be avoided in two cases:

1. Only allpasses with $b_n = 0$ are used. This means that the lowpass-to-lowpass and the lowpass-to-highpass transformations are impossible. For the lowpass-to-bandpass and the lowpass-to-bandstop transformations only the variation of the center frequency is possible, but the bandwidth of the filter is fixed. A variable lowpass can be realized with such a bandstop filter with bandwidth $B = f_s/4$, but the passband variation range of the lowpass is then only $f_s/4$ instead of $f_s/2$ in the general form.
2. Using a nonrecursive structure for the basic filter, the algebraic feedback loop will be avoided with any allpass. A possible linear phase property of these filters [7] is lost, since the resulting transformed filter is a recursive one. But the variation of the cutoff-frequencies can be accomplished as described above.

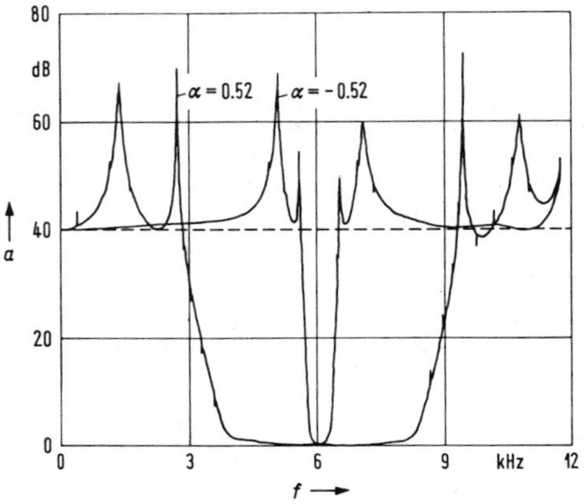

Fig. 2. Frequency responses of the variable bandpass.

(Received August 27, 1970.)

Prof. Dr. W. Schüssler,
Dipl.-Ing. W. Winkelnkemper
Institut für Nachrichtentechnik
der Universität
D-852 Erlangen
Egerlandstrasse 5

References

[1] Schüssler, W., Zur allgemeinen Theorie der Verzweigungsnetzwerke. AEÜ **22** [1968], 361–369.
[2] Constantinides, A. G., Frequency transformations for digital filters. Electron. Letters **3** [1967], 487–489.
[3] Constantinides, A. G., Frequency transformations for digital filters. Electron. Letters **4** [1968], 115–116.
[4] Galpin, R. K. P., Variable electronic allpass delay network. Electron. Letters **4** [1968], 137–139.
[5] Kuntz, W., Analoge Realisierung digitaler Filter. Ausgewählte Arbeiten über Nachrichtensysteme, No. 10, 1969, pp. 47–74.
[6] Christian, E. and Eisenstein, E., Filter design tables and graphs. J. Wiley & Sons, New York 1966.
[7] Herrmann, O., Design of nonrecursive digital filters with linear phase. Electron. Letters **6** [1970], 328–329.

Part 2
Fast Fourier Transforms

An Algorithm for the Machine Calculation of Complex Fourier Series

By James W. Cooley and John W. Tukey

An efficient method for the calculation of the interactions of a 2^m factorial experiment was introduced by Yates and is widely known by his name. The generalization to 3^m was given by Box et al. [1]. Good [2] generalized these methods and gave elegant algorithms for which one class of applications is the calculation of Fourier series. In their full generality, Good's methods are applicable to certain problems in which one must multiply an N-vector by an $N \times N$ matrix which can be factored into m sparse matrices, where m is proportional to $\log N$. This results in a procedure requiring a number of operations proportional to $N \log N$ rather than N^2. These methods are applied here to the calculation of complex Fourier series. They are useful in situations where the number of data points is, or can be chosen to be, a highly composite number. The algorithm is here derived and presented in a rather different form. Attention is given to the choice of N. It is also shown how special advantage can be obtained in the use of a binary computer with $N = 2^m$ and how the entire calculation can be performed within the array of N data storage locations used for the given Fourier coefficients.

Consider the problem of calculating the complex Fourier series

$$(1) \qquad X(j) = \sum_{k=0}^{N-1} A(k) \cdot W^{jk}, \qquad j = 0, 1, \cdots, N-1,$$

where the given Fourier coefficients $A(k)$ are complex and W is the principal Nth root of unity,

$$(2) \qquad W = e^{2\pi i/N}.$$

A straightforward calculation using (1) would require N^2 operations where "operation" means, as it will throughout this note, a complex multiplication followed by a complex addition.

The algorithm described here iterates on the array of given complex Fourier amplitudes and yields the result in less than $2N \log_2 N$ operations without requiring more data storage than is required for the given array A. To derive the algorithm, suppose N is a composite, i.e., $N = r_1 \cdot r_2$. Then let the indices in (1) be expressed

$$(3) \qquad \begin{aligned} j &= j_1 r_1 + j_0, \qquad j_0 = 0, 1, \cdots, r_1 - 1, \qquad j_1 = 0, 1, \cdots, r_2 - 1, \\ k &= k_1 r_2 + k_0, \qquad k_0 = 0, 1, \cdots, r_2 - 1, \qquad k_1 = 0, 1, \cdots, r_1 - 1. \end{aligned}$$

Then, one can write

$$(4) \qquad X(j_1, j_0) = \sum_{k_0} \sum_{k_1} A(k_1, k_0) \cdot W^{jk_1 r_2} W^{jk_0}.$$

Received August 17, 1964. Research in part at Princeton University under the sponsorship of the Army Research Office (Durham). The authors wish to thank Richard Garwin for his essential role in communication and encouragement.

Since

$$W^{jk_1r_2} = W^{j_0k_1r_2}, \tag{5}$$

the inner sum, over k_1, depends only on j_0 and k_0 and can be defined as a new array,

$$A_1(j_0, k_0) = \sum_{k_1} A(k_1, k_0)\cdot W^{j_0k_1r_2}. \tag{6}$$

The result can then be written

$$X(j_1, j_0) = \sum_{k_0} A_1(j_0, k_0)\cdot W^{(j_1r_1+j_0)k_0}. \tag{7}$$

There are N elements in the array A_1, each requiring r_1 operations, giving a total of Nr_1 operations to obtain A_1. Similarly, it takes Nr_2 operations to calculate X from A_1. Therefore, this two-step algorithm, given by (6) and (7), requires a total of

$$T = N(r_1 + r_2) \tag{8}$$

operations.

It is easy to see how successive applications of the above procedure, starting with its application to (6), give an m-step algorithm requiring

$$T = N(r_1 + r_2 + \cdots + r_m) \tag{9}$$

operations, where

$$N = r_1\cdot r_2 \cdots r_m. \tag{10}$$

If $r_j = s_jt_j$ with $s_j, t_j > 1$, then $s_j + t_j < r_j$ unless $s_j = t_j = 2$, when $s_j + t_j = r_j$. In general, then, using as many factors as possible provides a minimum to (9), but factors of 2 can be combined in pairs without loss. If we are able to choose N to be highly composite, we may make very real gains. If all r_j are equal to r, then, from (10) we have

$$m = \log_r N \tag{11}$$

and the total number of operations is

$$T(r) = rN \log_r N. \tag{12}$$

If $N = r^m s^n t^p \cdots$, then we find that

$$\begin{aligned} \frac{T}{N} &= m\cdot r + n\cdot s + p\cdot t + \cdots, \\ \log_2 N &= m\cdot\log_2 r + n\cdot\log_2 s + p\cdot\log_2 t + \cdots, \end{aligned} \tag{13}$$

so that

$$\frac{T}{N \log_2 N}$$

is a weighted mean of the quantities

$$\frac{r}{\log_2 r}, \quad \frac{s}{\log_2 s}, \quad \frac{t}{\log_2 t}, \cdots,$$

whose values run as follows

r	$\frac{r}{\log_2 r}$
2	2.00
3	1.88
4	2.00
5	2.15
6	2.31
7	2.49
8	2.67
9	2.82
10	3.01.

The use of $r_j = 3$ is formally most efficient, but the gain is only about 6% over the use of 2 or 4, which have other advantages. If necessary, the use of r_j up to 10 can increase the number of computations by no more than 50%. Accordingly, we can find "highly composite" values of N within a few percent of any given large number.

Whenever possible, the use of $N = r^m$ with $r = 2$ or 4 offers important advantages for computers with binary arithmetic, both in addressing and in multiplication economy.

The algorithm with $r = 2$ is derived by expressing the indices in the form

$$(14)\qquad \begin{aligned} j &= j_{m-1}\cdot 2^{m-1} + \cdots + j_1\cdot 2 + j_0 , \\ k &= k_{m-1}\cdot 2^{m-1} + \cdots + k_1\cdot 2 + k_0 , \end{aligned}$$

where j_v and k_v are equal to 0 or 1 and are the contents of the respective bit positions in the binary representation of j and k. All arrays will now be written as functions of the bits of their indices. With this convention (1) is written

$$(15)\qquad X(j_{m-1}, \cdots, j_0) = \sum_{k_0}\sum_{k_1}\cdots\sum_{k_{m-1}} A(k_{m-1}, \cdots, k_0)\cdot W^{jk_{m-1}\cdot 2^{m-1}+\cdots+jk_0},$$

where the sums are over $k_v = 0, 1$. Since

$$(16)\qquad W^{jk_{m-1}\cdot 2^{m-1}} = W^{j_0k_{m-1}\cdot 2^{m-1}},$$

the innermost sum of (15), over k_{m-1}, depends only on $j_0, k_{m-2}, \cdots, k_0$ and can be written

$$(17)\qquad A_1(j_0, k_{m-2}, \cdots, k_0) = \sum_{k_{m-1}} A(k_{m-1}, \cdots, k_0)\cdot W^{j_0k_{m-1}\cdot 2^{m-1}}.$$

Proceeding to the next innermost sum, over k_{m-2}, and so on, and using

$$(18)\qquad W^{j\cdot k_{m-l}\cdot 2^{m-l}} = W^{(j_{l-1}\cdot 2^{l-1}+\cdots+j_0)k_{m-l}\cdot 2^{m-l}},$$

one obtains successive arrays,

$$(19)\qquad \begin{aligned} &A_l(j_0, \cdots, j_{l-1}, k_{m-l-1}, \cdots, k_0) \\ &\quad = \sum_{k_{m-l}} A_{l-1}(j_0, \cdots, j_{l-2}, k_{m-l}, \cdots, k_0)\cdot W^{(j_{l-1}\cdot 2^{l-1}+\cdots+j_0)\cdot k_{m-l}\cdot 2^{m-l}} \end{aligned}$$

for $l = 1, 2, \cdots, m$.

Writing out the sum this appears as

$$
\begin{aligned}
(20) \qquad & A_l(j_0, \cdots, j_{l-1}, k_{m-l-1}, \cdots, k_0) \\
&= A_{l-1}(j_0, \cdots, j_{l-2}, 0, k_{m-l-1}, \cdots, k_0) \\
&\quad + (-1)^{j_{l-1}} i^{j_{l-2}} A_{l-1}(j_0, \cdots, j_{l-2}, 1, k_{m-l-1}, \cdots, k_0) \\
&\quad \cdot W^{(j_{l-3}\cdot 2^{l-3} + \cdots + j_0)\cdot 2^{m-l}}, \qquad j_{l-1} = 0, 1.
\end{aligned}
$$

According to the indexing convention, this is stored in a location whose index is

$$(21) \qquad j_0 \cdot 2^{m-1} + \cdots + j_{l-1} \cdot 2^{m-l} + k_{m-l-1} \cdot 2^{m-l-1} + \cdots + k_0 .$$

It can be seen in (20) that only the two storage locations with indices having 0 and 1 in the 2^{m-l} bit position are involved in the computation. Parallel computation is permitted since the operation described by (20) can be carried out with all values of $j_0, \cdots, j_{l-2}$, and $k_0, \cdots, k_{m-l-1}$ simultaneously. In some applications* it is convenient to use (20) to express A_l in terms of A_{l-2}, giving what is equivalent to an algorithm with $r = 4$.

The last array calculated gives the desired Fourier sums,

$$(22) \qquad X(j_{m-1}, \cdots, j_0) = A_m(j_0, \cdots, j_{m-1})$$

in such an order that the index of an X must have its binary bits put in reverse order to yield its index in the array A_m.

In some applications, where Fourier sums are to be evaluated twice, the above procedure could be programmed so that no bit-inversion is necessary. For example, consider the solution of the difference equation,

$$(23) \qquad aX(j + 1) + bX(j) + cX(j - 1) = F(j).$$

The present method could be first applied to calculate the Fourier amplitudes of $F(j)$ from the formula

$$(24) \qquad B(k) = \frac{1}{N} \sum_j F(j) W^{-jk}.$$

The Fourier amplitudes of the solution are, then,

$$(25) \qquad A(k) = \frac{B(k)}{aW^k + b + cW^{-k}}.$$

The $B(k)$ and $A(k)$ arrays are in bit-inverted order, but with an obvious modification of (20), $A(k)$ can be used to yield the solution with correct indexing.

A computer program for the IBM 7094 has been written which calculates three-dimensional Fourier sums by the above method. The computing time taken for computing three-dimensional $2^a \times 2^b \times 2^c$ arrays of data points was as follows:

* A multiple-processing circuit using this algorithm was designed by R. E. Miller and S. Winograd of the IBM Watson Research Center. In this case $r = 4$ was found to be most practical.

a	b	c	No. Pts.	Time (minutes)
4	4	3	2^{11}	.02
11	0	0	2^{11}	.02
4	4	4	2^{12}	.04
12	0	0	2^{12}	.07
5	4	4	2^{13}	.10
5	5	3	2^{13}	.12
13	0	0	2^{13}	.13

IBM Watson Research Center
Yorktown Heights, New York

Bell Telephone Laboratories,
Murray Hill, New Jersey

Princeton University
Princeton, New Jersey

1. G. E. P. Box, L. R. Connor, W. R. Cousins, O. L. Davies (Ed.), F. R. Hirnsworth & G. P. Silitto, *The Design and Analysis of Industrial Experiments*, Oliver & Boyd, Edinburgh, 1954.

2. I. J. Good, "The interaction algorithm and practical Fourier series," *J. Roy. Statist. Soc. Ser. B.*, v. 20, 1958, p. 361–372; Addendum, v. 22, 1960, p. 372–375. MR **21** #1674; MR **23** #A4231.

A guided tour of the fast Fourier transform

The fast Fourier transform algorithm can reduce the time involved in finding a discrete Fourier transform from several minutes to less than a second, and also can lower the cost from several dollars to several cents

G. D. Bergland Bell Telephone Laboratories, Inc.

For some time the Fourier transform has served as a bridge between the time domain and the frequency domain. It is now possible to go back and forth between waveform and spectrum with enough speed and economy to create a whole new range of applications for this classic mathematical device. This article is intended as a primer on the fast Fourier transform, which has revolutionized the digital processing of waveforms. The reader's attention is especially directed to the IEEE Transactions on Audio and Electroacoustics for June 1969, a special issue devoted to the fast Fourier transform.

This article is written as an introduction to the fast Fourier transform. The need for an FFT primer is apparent from the barrage of questions asked by each new person entering the field. Eventually, most of these questions are answered when the person gains an understanding of some relatively simple concept that is taken for granted by all but the uninitiated. Here the basic concepts will be introduced by the use of specific examples. The discussion is centered around these questions:

1. What is the fast Fourier transform?
2. What can it do?
3. What are the pitfalls in using it?
4. How has it been implemented?

Representative references are cited for each topic covered so that the reader can conveniently interrupt this fast guided tour for a more detailed study.

What is the fast Fourier transform?

The Fourier transform has long been used for characterizing linear systems and for identifying the frequency components making up a continuous waveform. However, when the waveform is sampled, or the system is to be analyzed on a digital computer, it is the finite, discrete version of the Fourier transform (DFT) that must be understood and used. Although most of the properties of the continuous Fourier transform (CFT) are retained, several differences result from the constraint that the DFT must operate on sampled waveforms defined over finite intervals.

The fast Fourier transform (FFT) is simply an efficient method for computing the DFT. The FFT can be used in place of the continuous Fourier transform only to the extent that the DFT could before, but with a substantial reduction in computer time. Since most of the problems associated with the use of the fast Fourier transform actually stem from an incomplete or incorrect understanding of the DFT, a brief review of the DFT will first be given. The degree to which the DFT approximates the continuous Fourier transform will be discussed in more detail in the section on "pitfalls."

The discrete Fourier transform. The Fourier transform pair for continuous signals can be written in the form

$$X(f) = \int_{-\infty}^{\infty} x(t)e^{-i2\pi ft}\,dt \tag{1}$$

$$x(t) = \int_{-\infty}^{\infty} X(f)e^{i2\pi ft}\,df \tag{2}$$

for $-\infty < f < \infty$, $-\infty < t < \infty$, and $i = \sqrt{-1}$. The uppercase $X(f)$ represents the frequency-domain function; the lowercase $x(t)$ is the time-domain function.

The analogous discrete Fourier transform pair that applies to sampled versions of these functions can be written in the form

$$X(j) = \frac{1}{N}\sum_{k=0}^{N-1} x(k)e^{-i2\pi jk/N} \tag{3}$$

$$x(k) = \sum_{j=0}^{N-1} X(j)e^{i2\pi jk/N} \tag{4}$$

for $j = 0, 1, \cdots, N-1$; $k = 0, 1, \cdots, N-1$. Both

Reprinted from *IEEE Spectrum*, vol. 6, pp. 41–52, July 1969.

$X(j)$ and $x(k)$ are, in general, complex series. A derivation of the discrete Fourier transform from the continuous Fourier transform can be found in Refs. 12 and 23.

When the expression $e^{2\pi i/N}$ is replaced by the term W_N, the DFT transform pair takes the form

$$X(j) = \frac{1}{N}\sum_{k=0}^{N-1} x(k) W_N{}^{-jk} \tag{5}$$

$$x(k) = \sum_{j=0}^{N-1} X(j) W_N{}^{+jk} \tag{6}$$

FIGURE 1. A real signal and its complex discrete Fourier transform displayed in the FFT algorithm format.

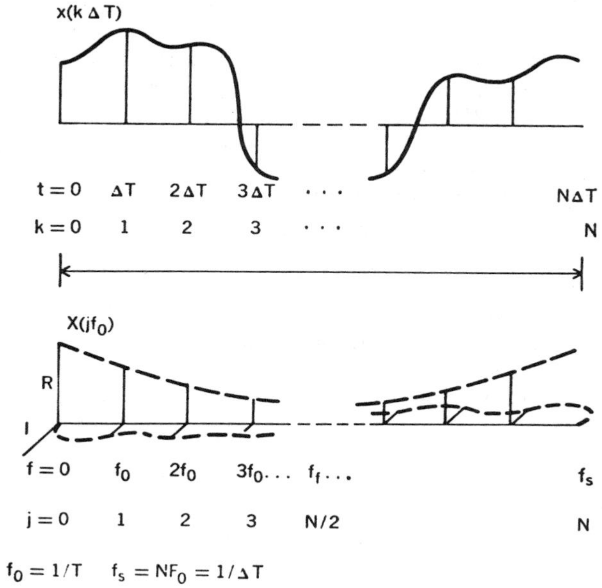

An example of a real-valued time series and its associated DFT is shown in Fig. 1. The time series $x(k\Delta T)$ is assumed to be periodic in the time domain of period T seconds, and the set of Fourier coefficients $X(jf_0)$ is assumed to be periodic over the sample frequency f_s. Only one complete period of each function is shown.

The fundamental frequency f_0 and the sample period ΔT do not appear explicitly in Eqs. (5) and (6), but each j should still be interpreted as a harmonic number and each k still refers to a sample period number. That is, the true frequency is the product of j and f_0 and the true time is the product of k and ΔT.

When the $x(k)$ series is real, the real part of $X(j)$ is symmetric about the folding frequency f_f (where $f_f = f_s/2$) and the imaginary part is antisymmetric. Since $X(j)$ has been interpreted as being periodic, these symmetries are equivalent to saying that the real part of $X(j)$ is an even function, and that the imaginary part of $X(j)$ is an odd function. This also means that the Fourier coefficients between $N/2$ and $N - 1$ can be viewed as the "negative frequency" harmonics between $-N/2$ and -1. Likewise, the last half of the time series can be interpreted as negative time (that is, as occurring before $t = 0$).

Derivation of the Cooley–Tukey FFT algorithm. A derivation of the Cooley–Tukey FFT algorithm[8] for evaluating Eq. (6) is given in this section for the example of $N = 8$. This derivation is also appropriate to the forward transform, since Eq. (5) can be rewritten in the form[9]

$$X(j) = \frac{1}{N}\left[\sum_{k=0}^{N-1} x(k)^* W_N{}^{jk}\right]^* \tag{7}$$

where the asterisk refers to the complex conjugate operation. Alternatively, the FFT algorithm used for comput-

FIGURE 2. A flow diagram of the Cooley–Tukey FFT algorithm for performing an eight-point transform.

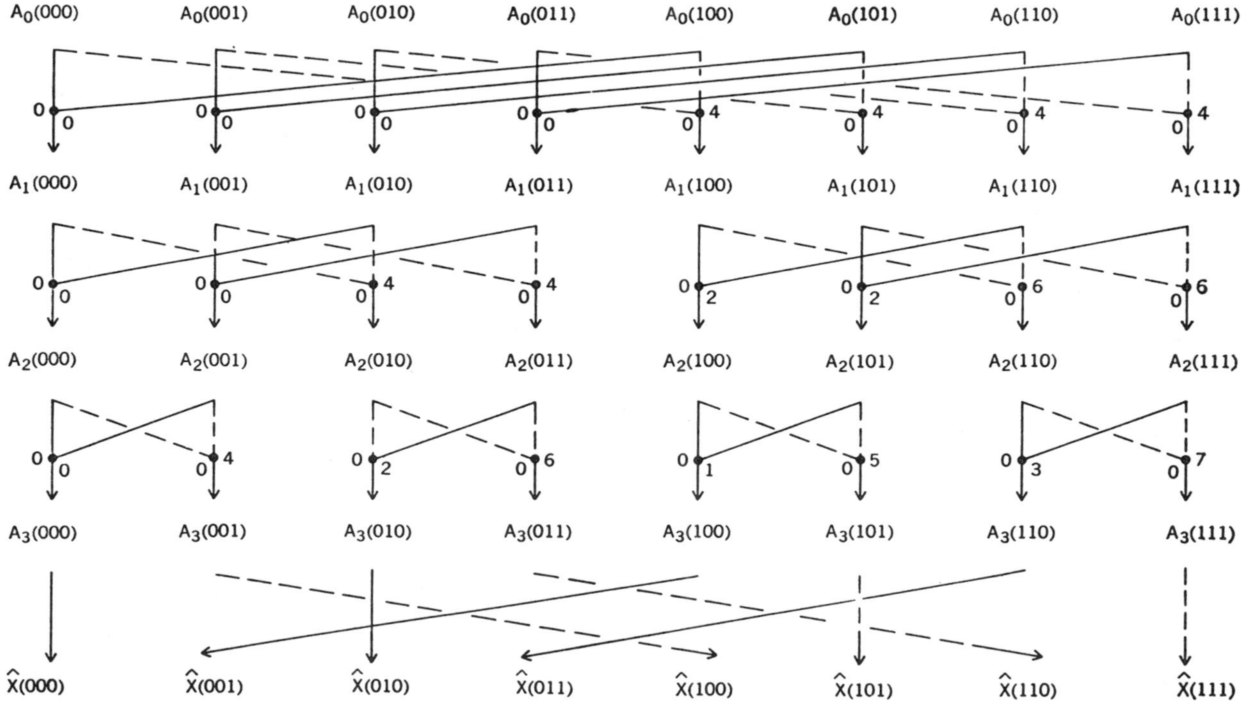

ing Eq. (6) can be altered by redefining W_N to be exp $(-2\pi i/N)$ and dividing each result by N.

Using Cooley's notation,[8] the FFT algorithm involves evaluating the expression

$$\hat{X}(j) = \sum_{k=0}^{N-1} A(k) W^{jk} \tag{8}$$

where $j = 0, 1, \ldots, N-1$, and $W = \exp(2\pi i/N)$. Note that $\hat{X}$ and A can be interpreted as X^* and x^*/N, respectively, if the forward transform is being computed, and can be interpreted as x and X, respectively, if the inverse transform is being computed.

When N is equal to 8, it is convenient to represent both j and k as binary numbers; that is, for

$$j = 0, 1, \cdots, 7 \qquad k = 0, 1, \cdots, 7$$

we can write

$$j = j_2 4 + j_1 2 + j_0 \qquad k = k_2 4 + k_1 2 + k_0 \tag{9}$$

where j_0, j_1, j_2, k_0, k_1, and k_2 can take on values of 0 and 1 only. Using this representation of j and k, Eq. (8) becomes

$$\hat{X}(j_2,j_1,j_0) = \sum_{k_0=0}^{1} \sum_{k_1=0}^{1} \sum_{k_2=0}^{1} A(k_2,k_1,k_0) W^{(j_2 4+j_1 2+j_0)(k_2 4+k_1 2+k_0)} \tag{10}$$

Noting that $W^{m+n} = W^m \cdot W^n$, we have

$$W^{(j_2 4+j_1 2+j_0)(k_2 4+k_1 2+k_0)} = W^{(j_2 4+j_1 2+j_0)k_2 4} W^{(j_2 4+j_1 2+j_0)k_1 2} W^{(j_2 4+j_1 2+j_0)k_0} \tag{11}$$

If we look at these terms individually, it is clear that they can be written in the form

$$W^{(j_2 4+j_1 2+j_0)k_2 4} = [W^{8(j_2 2+j_1)k_2}] W^{j_0 k_2 4} \tag{12}$$

$$W^{(j_2 4+j_1 2+j_0)k_1 2} = [W^{8 j_2 k_1}] W^{(j_1 2+j_0)k_1 2} \tag{13}$$

$$W^{(j_2 4+j_1 2+j_0)k_0} = W^{(j_2 4+j_1 2+j_0)k_0} \tag{14}$$

Note, however, that

$$W^8 = [e^{2\pi i/8}]^8 = e^{2\pi i} = 1 \tag{15}$$

Thus, the bracketed portions of Eqs. (12) and (13) can be replaced by a one. This means that Eq. (10) can be written in the form

$$\hat{X}(j_2,j_1,j_0) = \underbrace{\sum_{k_0=0}^{1} \underbrace{\sum_{k_1=0}^{1} \underbrace{\sum_{k_2=0}^{1} A(k_2,k_1,k_0) W^{j_0 k_2 4}}_{A_1(j_0,k_1,k_0)} W^{(j_1 2+j_0)k_1 2}}_{A_2(j_0,j_1,k_0)} W^{(j_2 4+j_1 2+j_0)k_0}}_{A_3(j_0,j_1,j_2)} \tag{16}$$

In this form it is convenient to perform each of the summations separately and to label the intermediate results. Note that each set consists of only eight terms and that only the latest set needs to be saved. Thus the equations can be rewritten in the form

$$A_1(j_0,k_1,k_0) = \sum_{k_2=0}^{1} A(k_2,k_1,k_0) W^{j_0 k_2 4} \tag{17}$$

$$A_2(j_0,j_1,k_0) = \sum_{k_1=0}^{1} A_1(j_0,k_1,k_0) W^{(j_1 2+j_0)k_1 2} \tag{18}$$

$$A_3(j_0,j_1,j_2) = \sum_{k_0=0}^{1} A_2(j_0,j_1,k_0) W^{(j_2 4+j_1 2+j_0)k_0} \tag{19}$$

$$\hat{X}(j_2,j_1,j_0) = A_3(j_0,j_1,j_2) \tag{20}$$

The terms contributing to each sum are shown in Fig. 2. Each small number refers to a power of W applied along the adjacent path. The last operation shown in Fig. 2 is the reordering. This is due to the bit reversal in the arguments of Eq. (20).

This set of recursive equations represents the original Cooley–Tukey formulation of the fast Fourier transform algorithm for $N = 8$. Although a direct evaluation of Eq. (8) for $N = 8$ would require nearly 64 complex multiply-and-add operations, the FFT equations show 48 operations. By noting that the first multiplication in each summation is actually a multiplication by $+1$, this number becomes only 24. By further noting that $W^0 = -W^4$, $W^1 = -W^5$, etc., the number of multiplications can be reduced to 12. These reductions carry on to the more general case of $N = 2^m$, reducing the computation from nearly N^2 operations to $(N/2) \log_2 N$ complex multiplications, $(N/2) \log_2 N$ complex additions, and $(N/2) \log_2 N$ subtractions. For $N = 1024$, this represents a computational reduction of more than 200 to 1. This difference is represented graphically in Fig. 3.

What can it do?

The operations usually associated with the FFT are: (1) computing a spectrogram (a display of the short-term power spectrum as a function of time); (2) the convolution of two time series to perform digital filtering; and (3) the correlation of two time series. Although all of these operations can be performed without the FFT, its computational savings have significantly increased the interest in performing these operations digitally.

Spectrograms. The diagram in Fig. 4 represents a method of obtaining estimates of the power spectrum of a time signal through the use of the fast Fourier transform.

FIGURE 3. The number of operations required for computing the discrete Fourier transform using the FFT algorithm compared with the number of operations required for direct calculation of the discrete Fourier transform.

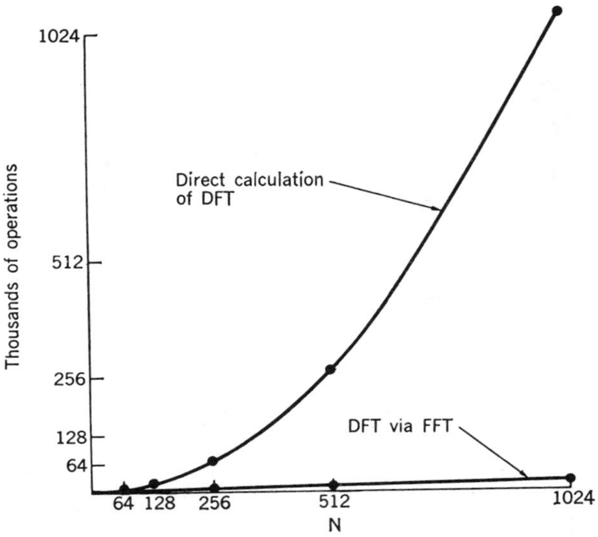

In this case the square of the magnitude of the set of complex Fourier coefficients (that is, the periodogram) is used to estimate the power spectrum of the original signal.

A snapshot of the spectrum of the signal can always be computed from the last T seconds of data. By taking a series of these snapshots, estimates of the power spectrum can be displayed as a function of time as shown in Fig. 5. When the audio range is displayed, this is usually called a sound spectrogram or sonogram.

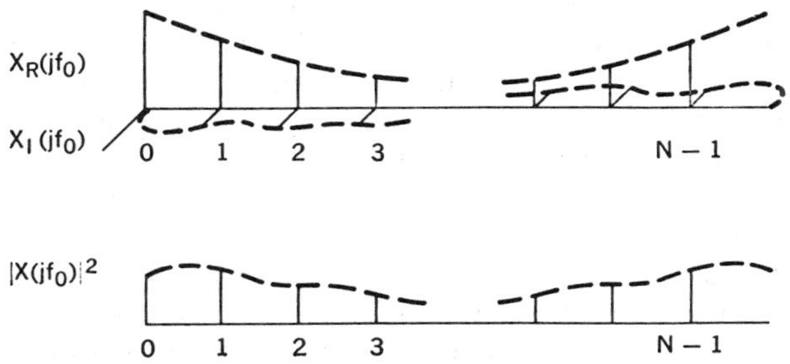

FIGURE 4. The power spectrum of a real function computed by taking the sum of the squares of the real and imaginary components of the discrete Fourier transform Fourier coefficients.

FIGURE 5. A spectrogram made by using the fast Fourier transform algorithm.

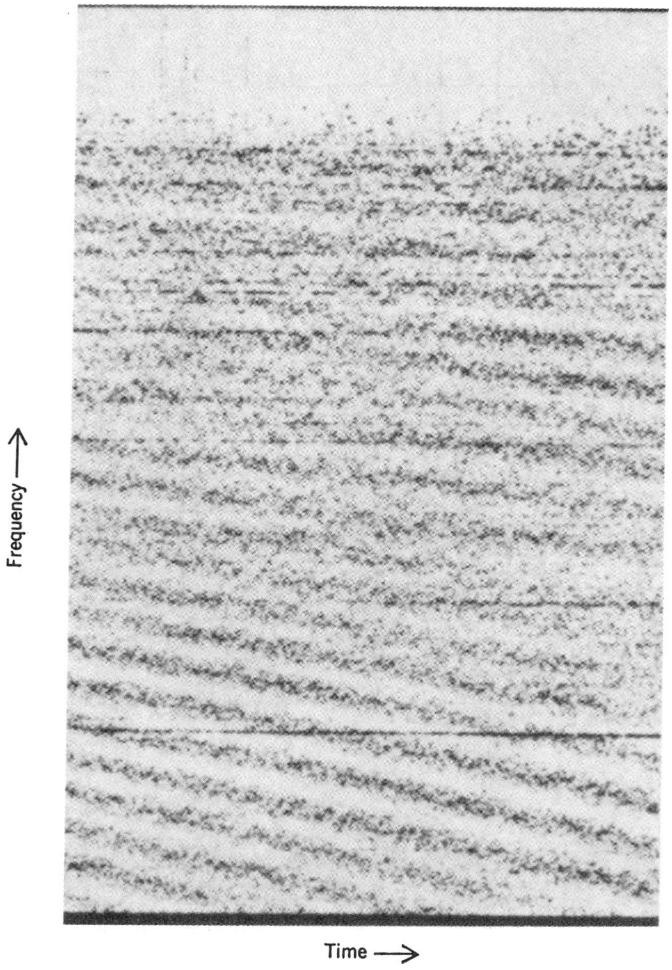

In some cases it is of interest to go one step further. When the spectrum of a signal contains a periodic component, this spectrum can be compressed by taking the logarithm, and then the fast Fourier transform can be taken again. The result is called a cepstrum (pronounced "kepstrum").[13, 19] An example of using the cepstrum to determine the pitch period of a speaker is described in Ref. 14. For a more complete discussion of short-term spectrum and cepstrum analysis see Refs. 10–21.

Digital filtering. In a linear system, one is frequently confronted with the problem of either (1) determining the output, given the input and the impulse response, or (2) finding the impulse response, given the input and the output. Both of these problems can be approached rather easily in the frequency domain.

In Fig. 6, the output $c(t)$ is formed by convolving the input $g(t)$ with the impulse response of the system $h(\tau)$. For sampled functions this convolution takes the form

$$c(k) = \frac{1}{N}\sum_{\tau=0}^{N-1} g(\tau)h(k-\tau) \tag{21}$$

This equation represents the linear system in Fig. 6, as long as $h(\tau)$ is assumed to be zero for $\tau < 0$. If both $g(k)$ and $h(k)$ consist of N consecutive nonzero samples, the series $c(k)$ can consist of $2N - 1$ nonzero terms.

Since the FFT algorithm gives us a fast way of getting to the frequency domain, it is interesting to consider

$$C(j) = G(j) \cdot H(j) \tag{22}$$

where $C(j)$ is the DFT of $c(k)$, $G(j)$ is the DFT of $g(k)$, and $H(j)$ is the DFT of $h(k)$. By Eq. (6), this is equivalent to

$$c(k) = \sum_{j=0}^{N-1} [G(j) \cdot H(j)] W^{jk} \tag{23}$$

By Eq. (5), this can be written as

$$c(k) = \sum_{j=0}^{N-1} \left(\frac{1}{N}\sum_{\tau=0}^{N-1} g(\tau)W^{-j\tau}\right)\left(\frac{1}{N}\sum_{\hat{\tau}=0}^{N-1} h(\hat{\tau})W^{-j\hat{\tau}}\right)W^{jk} \tag{24}$$

Since all of the sums are finite, this can be rewritten as

$$c(k) = \frac{1}{N}\sum_{\tau=0}^{N-1}\sum_{\hat{\tau}=0}^{N-1} g(\tau)h(\hat{\tau})\left[\frac{1}{N}\sum_{j=0}^{N-1} W^{+j(k-\tau-\hat{\tau})}\right] \tag{25}$$

FIGURE 6. The response of a linear system to a driving function g(k) expressed as the convolution of the input signal with the impulse response of the system.

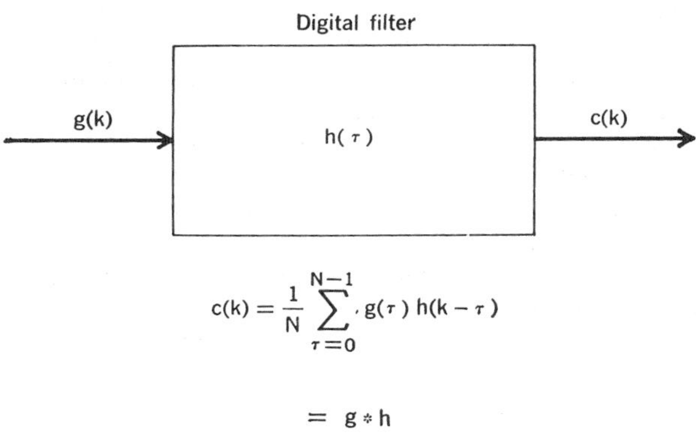

This can be simplified through the use of the orthogonality relationship

$$\sum_{j=0}^{N-1} W_N^{nj} W_N^{-mj} = N \qquad \text{if } n = m \bmod N$$
$$= 0 \qquad \text{otherwise} \qquad (26)$$

Thus Eq. (25) is equal to zero unless $\hat{\tau} = k - \tau$, for which we have

$$c(k) = \frac{1}{N} \sum_{\tau=0}^{N-1} g(\tau)h(k - \tau) \qquad (27)$$

This equation is identical to the desired Eq. (21) but the requirement that $h(\tau)$ be zero for $\tau < 0$ is not met. In representing $g(k)$ and $h(k)$ by their Fourier coefficients the assumption is made that they are periodic functions. A method of sidestepping this problem is described later.

Correlation. By considering the equation

$$C(j) = G(j) \cdot H^*(j) \qquad (28)$$

and following a development similar to that of the preceding section, we obtain the result

$$c(k) = \frac{1}{N} \sum_{\tau=0}^{N-1} g(\tau)h(\tau - k) \qquad (29)$$

This is the form of the cross-correlation function of $g(k)$ and $h(k)$. When $h(k) = g(k)$ we obtain the autocorrelation function. The problem again is that both $g(k)$ and $h(k)$ were assumed to be periodic in finding their Fourier coefficients. This problem is discussed further in the next section.

From Eqs. (29) and (27), it is clear that convolution is simply the process of correlating one time series with another time series that has been reversed in time.

What are the pitfalls?

The three problems most often encountered in using the discrete Fourier transform appear to be aliasing, leakage, and the picket-fence effect. Also of interest are the problems associated with the blind use of Eqs. (22) and (28) to perform convolutions and correlations. An ever-present problem, which is not discussed here, concerns finding the statistical reliability of an individual power spectral estimate when the signal being analyzed is noise-like. A discussion of this last problem can be found in Refs. 11, 16, 17, 19, and 21.

Relating the DFT to the CFT. Most of the time, engineers are interested in the discrete Fourier transform only because it approximates the continuous Fourier transform. Most of the problems in using the DFT are caused by a misunderstanding of what this approximation involves.

In Fig. 7, the results of the DFT are treated as a corrupted estimate of the CFT. The example considers a cosine-wave input, but the results can be extended to any function expressed as a sum of sine and cosine waves.

Line (a) of Fig. 7 represents the input signal $s(t)$ and its continuous Fourier transform $S(f)$. Since $s(t)$ is shown to be a real cosine waveform, the continuous Fourier transform consists of two impulse functions that are symmetric about zero frequency.

The finite portion of $s(t)$ to be analyzed is viewed through the unity amplitude data window $w(t)$. This rectangular data window has a continuous Fourier transform, which is in the form of a $(\sin x)/x$ function. [When this function takes the form $(\sin \pi x)/\pi x$, it is referred to as the sinc x function.[2]] The portion of $s(t)$ that will be analyzed is represented as the product of $s(t)$ and $w(t)$ in line (c) of Fig. 7. The corresponding convolution in the frequency domain results in a blurring of $S(f)$ into two $(\sin x)/x$-shaped pulses. Thus our estimate of $S(f)$ is already corrupted considerably.

The sampling of $s(t)$ is performed by multiplying by $c(t)$. (In Ref. 12 this infinite train of impulses is called a Dirac comb.) The resulting frequency-domain function is shown in line (e).

The continuous frequency-domain function shown in line (e) can also be made discrete if the time function is treated as one period of a periodic function. This assumption forces both the time-domain and frequency-domain functions to be infinite in extent, periodic, and discrete, as shown in line (f).

The discrete Fourier transform is simply a reversible mapping of N terms of $\hat{s}(k)$ into N terms of $\hat{S}(j)$. In this example, the N terms of $\hat{s}(k)$ and $\hat{S}(j)$ approximate $s(t)$ and $S(f)$ extremely well. This is an unusual case, however, in that the frequency-domain function of line (e) of Fig. 7 was sampled at exactly the peaks and zeros. The problems

FIGURE 7. The Fourier coefficients of the discrete Fourier transform viewed as a corrupted estimate of the continuous Fourier transform.

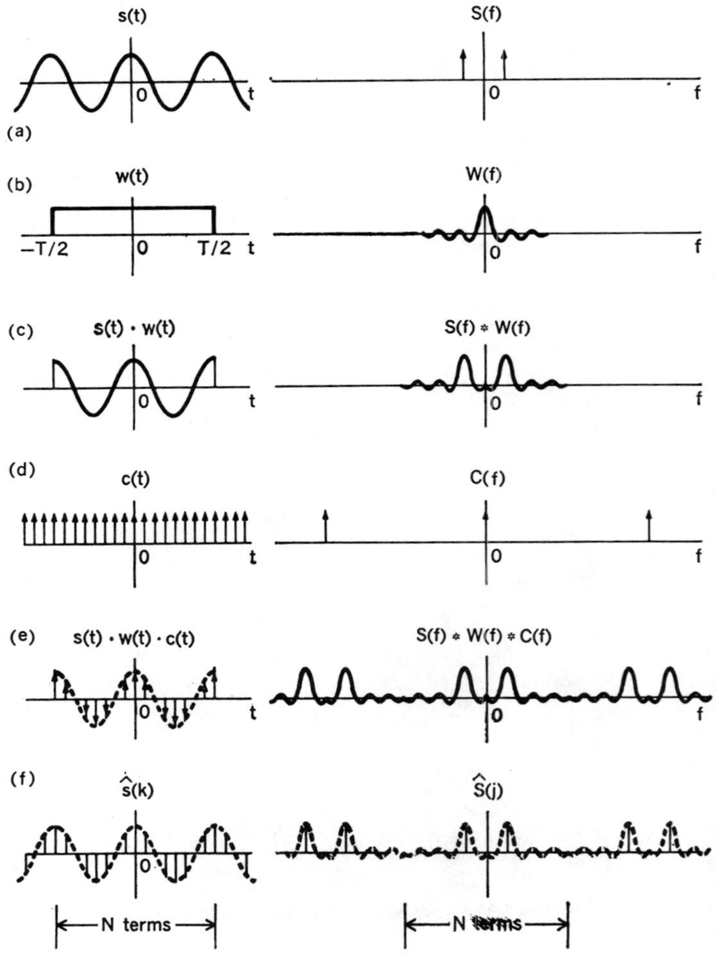

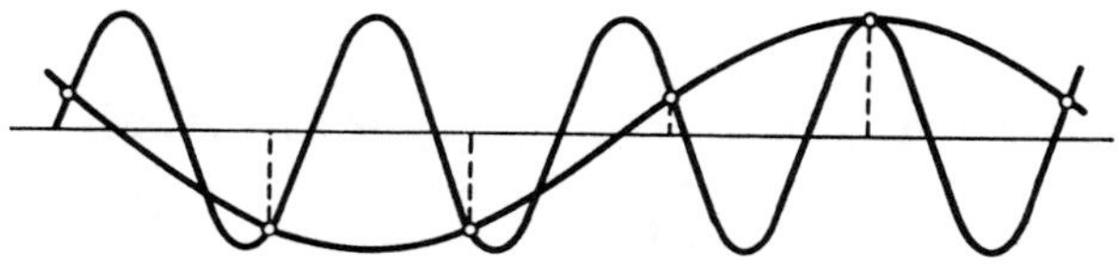

FIGURE 8. An example of a high frequency "impersonating" a low frequency.

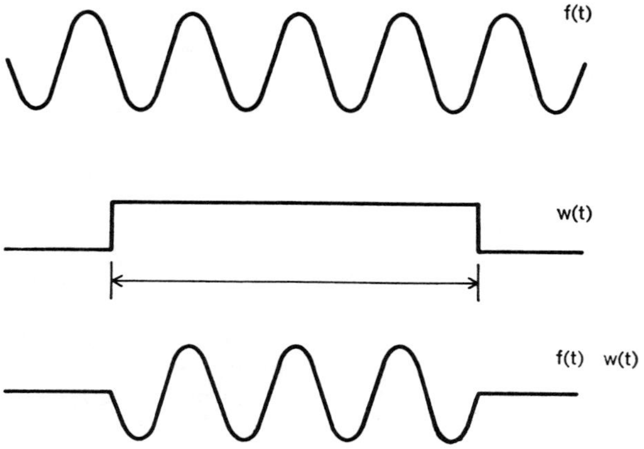

FIGURE 9. The rectangular data window implied when a finite record of data is analyzed.

FIGURE 10. The leakage of energy from one discrete frequency into adjacent frequencies resulting from the analysis of a finite record.

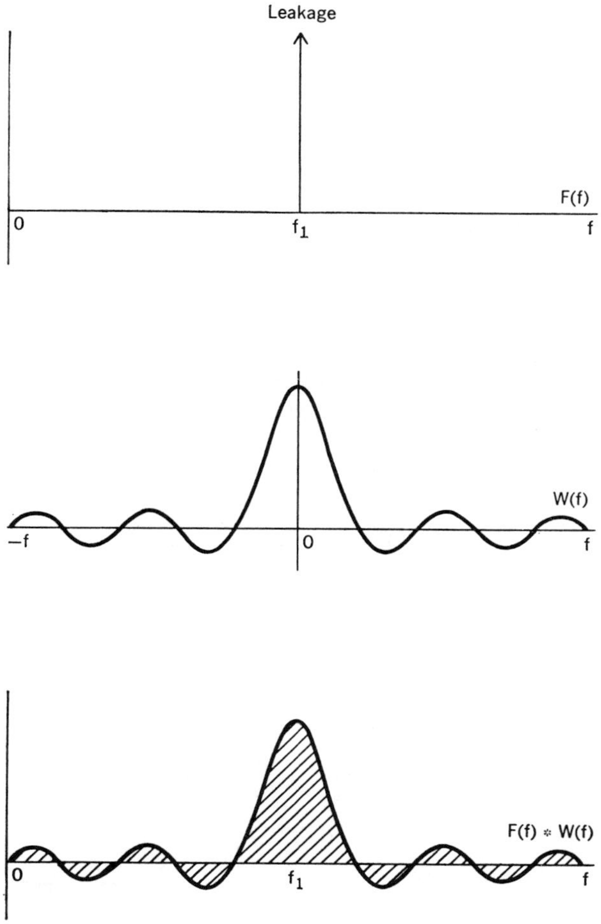

of leakage, aliasing, and picket-fence effects are associated with variations from this ideal condition.

Since both $\hat{s}(k)$ and $\hat{S}(j)$ are periodic, with a period N, any set of N adjacent terms can be used to characterize either set. The magnitude of $\hat{S}(j)$ is the same for samples of $\hat{s}(k)$ that are symmetric about the origin as it is for the DFT of a set of samples starting at the origin. The change in the phase of $\hat{S}(j)$ for different time origins can be determined by application of the DFT shifting theorem.[9]

Aliasing. The term "aliasing" refers to the fact that high-frequency components of a time function can impersonate low frequencies if the sampling rate is too low. This is demonstrated in Fig. 8 by showing a relatively high frequency and a relatively low frequency that share identical sample points. This uncertainty can be removed by demanding that the sampling rate be high enough for the highest frequency present to be sampled at least twice during each cycle.

In the diagram of Fig. 1, the sampling frequency f_s is shown to be $1/\Delta T$ samples per second. The folding frequency (or Nyquist frequency) f_f is shown to be equal to $f_s/2$. In this example, an input signal $x(t)$ will be represented correctly if its highest frequency component is less than the folding frequency. A frequency component 5 Hz higher than the folding frequency will impersonate a frequency 5 Hz lower than the folding frequency. Thus any components in $x(t)$ that are higher than the folding frequency are aliased (or folded) into the frequency interval below the folding frequency. In Fig. 7, aliasing occurs when $S(f)$ extends over a wider range of frequency than one period of $\hat{S}(j)$.

The cure to this problem involves sampling the signal at a rate at least twice as high as the highest frequency present. If the signal is known to be restricted to a certain band, the sampling rate can be picked accordingly. If the signal has been passed through a low-pass filter, a sampling rate can be chosen so that the components above the Nyquist frequency are negligible. Aliasing is discussed in more detail in Refs. 1–3, 10, and 12.

Leakage. The problem of leakage is inherent in the Fourier analysis of any finite record of data. The record has been formed by looking at the actual signal for a period of T seconds and by neglecting everything that happened before or after this period. As shown in Figs. 7 and 9, this is equivalent to multiplying the signal by a rectangular data window.

Had the continuous Fourier transform of the pure cosine wave in Fig. 9 been found, its contribution would have been limited to only one point on the frequency axis. (This is represented by the impulse at frequency f_1 in Fig. 10.) The multiplication by the data window in the time domain, however, is equivalent to performing a convolution in the frequency domain. Thus this impulse function is convolved with the Fourier transform of the square data window, resulting in a function with an amplitude of the $(\sin x)/x$ form centered about f_1.

This function is not localized on the frequency axis and in fact has a series of spurious peaks called sidelobes. The objective is usually to localize the contribution of a given frequency by reducing the amount of "leakage" through these sidelobes.

The usual approach consists of applying a data window to the time series, which has lower sidelobes in the frequency domain than the rectangular data window. This is analogous to weighing the outputs of a linear antenna array to reduce the sidelobe levels of the antenna pattern.

A host of different data windows have been discussed in the literature; see Refs. 11, 12, 19, and 21. An example of Tukey's "interim" data window is shown in Fig. 11. In this window, a raised cosine wave is applied to the first and last 10 percent of the data and a weight of unity is applied in between. This window was suggested with reservations in Refs. 11 and 19. Since only 20 percent of the terms in the series are given a weight other than unity, the computation required to apply this window in the time domain is relatively small. Another window that can be applied conveniently is the Hanning window, described by Blackman and Tukey.[12] This window is a cosine bell on a pedestal, but it can be applied by convolving the DFT coefficients with the weights $-\frac{1}{4}$, $\frac{1}{2}$, and $-\frac{1}{4}$.[11]

When the computation required is not the overriding consideration, one can find a number of windows that give rise to more rapidly decreasing sidelobes than the cosine tapers described previously. An application of the Parzen window (which has a shape of $1 - |t|$, for $-1 \leq t \leq 1$) is described in Ref. 21, and an application of the Dolph–Chebyshev function is described in Ref. 31. The spectral window arising from an application of the Dolph–Chebyshev function has a principal lobe width that is as small as possible for a given sidelobe ratio and a given number of terms.[31]

Whatever the form of the data window, it should be applied only to the actual data and not to any artificial data generated by filling out a record with zeros.

Picket-fence effect. In Fig. 12, an analogy is drawn between the output of the FFT algorithm and a bank of bandpass filters. Although each Fourier coefficient would ideally act as a filter having a rectangular response in the frequency domain, the amplitude response is in fact of the form shown in Fig. 10 because of the multiplication of the input time series by the T-second data window. In Fig. 12, the main lobes of the resulting set of spectral windows have been plotted to represent the output of the FFT. The sidelobes are not shown here. The width of each main lobe is inversely proportional to the original record length T.

At the frequencies computed, these main lobes appear to be N independent filters; that is, a unity-amplitude complex exponential (of the form $e^{i\omega t}$) with a frequency that is an integral multiple of $1/T$, would result in a response of unity at the appropriate harmonic frequency and zero at all of the other harmonics.

The picket-fence effect becomes evident when the signal being analyzed is not one of these discrete orthogonal frequencies. A signal between the third and fourth harmonics, for example, would be seen by both the third- and fourth-harmonic spectral windows but at a value lower than one. In the worst case (exactly halfway between the computed harmonics) the amplitude of the signal is reduced to 0.637 in both of the spectral windows. When this result is squared, the apparent peak power of the signal is only 0.406. Thus the power spectrum seen by this "set of filters" has a ripple that varies by a factor of 2.5 to 1. One seems to be viewing the true spectrum through a picket fence.

A cure to this problem involves performing complex interpolation on the complex Fourier coefficients. This can be accomplished by the use of an interpolation function[29] or through modification of the DFT. The latter approach will be described here.

By extending the record analyzed with a set of samples that are identically zero, one can make the redundant FFT algorithm compute a set of Fourier coefficients with terms lying between the original harmonics. Since the width of the spectral window associated with each coefficient is related solely to the reciprocal of the true record length (T), the width of these new spectral windows remains unchanged. As shown in Fig. 13, this means that the spectral windows associated with this new set of Fourier coefficients overlap considerably.

If the original time series is represented by $g(k)$ for $k = 0, 1, \cdots, N-1$, then the series analyzed in this example can be represented by $\hat{g}(k)$, where

$$\hat{g}(k) = g(k) \qquad \text{for } 0 \leq k < N$$
$$\hat{g}(k) = 0 \qquad \text{for } N \leq k < 2N \qquad (30)$$

The additional Fourier coefficients that result are interleaved with the original set.

As shown in Fig. 13, the ripple in the power spectrum has been reduced from approximately 60 percent to 20 percent. The ripple can be made larger or smaller than 20 percent by the use of less or more than N additional zeros.

In practice the picket-fence problem is not as great as this discussion implies. In many cases the signal being processed will not be a pure sinusoid but will be broad enough to fill several of the original spectral windows. Moreover, the use of any data window other than the rectangular (or boxcar) data window discussed here,

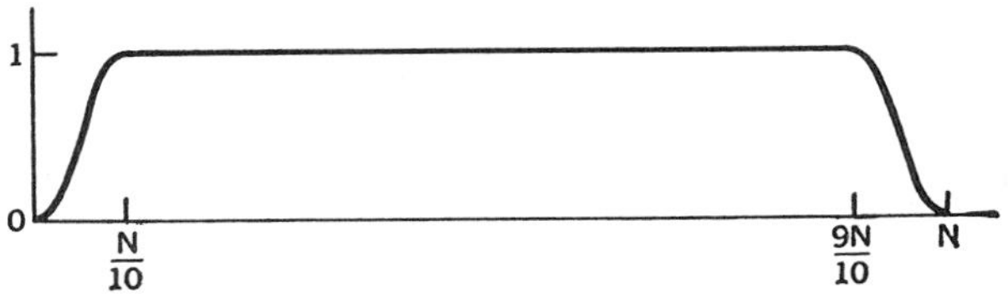

FIGURE 11. An extended cosine-bell data window.

FIGURE 12. The response of the discrete Fourier transform Fourier coefficients viewed as a set of bandpass filters (picket-fence effect).

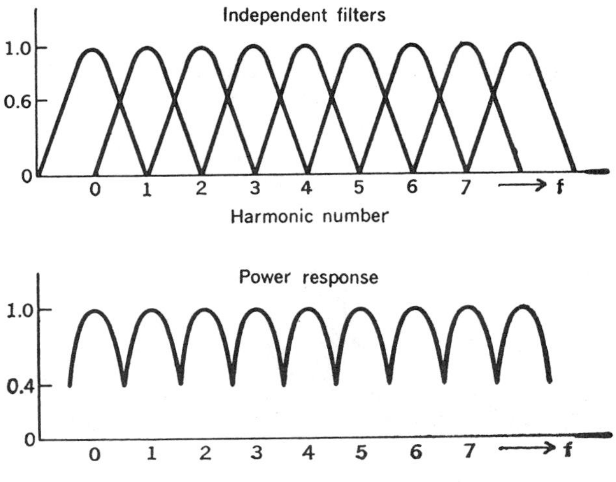

usually tends to reduce the picket-fence effect by widening the main lobe of each spectral window.

Convolution and correlation. The blind use of Eqs. (22) and (28) to perform correlation and convolution is often inconvenient and usually incorrect. An example of incorrect usage is discussed first.

Given a function $g(t)$, one is frequently interested in convolving this function with another function $h(t)$. As shown in Fig. 14, this involves reversing (or flipping) $h(t)$, and sliding it by $g(t)$. The convolution of these functions is formed by computing and integrating the product $g(\tau)\cdot h(t-\tau)$ as a function of the relative displacement t. As indicated by Fig. 14, both functions are considered to be identically zero outside of their domain of definition. Where $g(t)$ and $h(t)$ are finite and sampled, this corresponds to the sliding, multiplying, and summing operations of Eq. (21).

When convolutions are computed with the aid of Eq. (22) and the fast Fourier transform, both of the functions are treated as being periodic. The corresponding forms of $g(\tau)$ and $h(t-\tau)$ are shown in Fig. 15. The result is a cyclical convolution, which is entirely different from the noncyclical convolution described previously.

Fortunately, this problem is easily sidestepped by simply defining and convolving $\hat{g}(\tau)$ and $\hat{h}(t-\tau)$ as shown in Fig. 16; see Refs. 7, 10, 23, 24, 28, and 31. In this case

$$\hat{g}(k) = g(k) \qquad 0 \le k < N$$
$$\hat{g}(k) = 0 \qquad N \le k < 2N \qquad (31)$$

Thus the appropriate procedure is to

(1) Form $\hat{g}(k)$ and $\hat{h}(k)$.
(2) Find $\hat{G}(j)$ and $\hat{H}(j)$ via the FFT.
(3) Compute $\hat{C}(j) = \hat{G}(j)\cdot\hat{H}(j)$.
(4) Find $\hat{c}(k) = F^{-1}[\hat{C}(j)]$ via the FFT.

The series $\hat{c}(k)$ represents the $2N$-term series resulting from the convolution of the two N-term series $g(k)$ and $h(k)$. This technique is expressed in general terms in the section on "select-saving" in Ref. 24, and it is appropriate to correlation as well as convolution.

When one of the series convolved or correlated is much longer than the other, transforming the entire record of both functions is inconvenient and unnecessary.

If $h(k)$ is an N-term series and $g(k)$ is much longer, the procedure shown in Fig. 17 can be used. The $h(k)$ series can be thought of as the impulse response of a filter that is acting on the series $g(k)$. As shown in Fig. 14, the "present" output of the digital filter is merely a weighted sum of the last N samples it has seen. In the example of Fig. 17, this means that the $2N$-term series $h(k)$ convolved with one of the $2N$-term segments of $g(k)$ would result in N lags (or displacements), for which the correct N-term history is not available, and N lags for which the correct N-term history is available. Thus the first N lags computed are meaningless and the last N lags are correct.

In Fig. 17, overlapping sets of $2N$-term segments of $g(k)$ are shown. When convolved with $h(k)$, each of these segments contributes N correct terms to the final convolution. When all of these sets of N terms are pieced together, the result is the desired convolution of $g(k)$ and $h(k)$. Thus the length of the required FFTs is determined by the length of the short series rather than the long one. The length of the long series doesn't need to be specified.

Having exactly half of the values of $\hat{h}(\tau)$ be zero is a specific example of the "select-saving" method[24] and is made a requirement only to limit the discussion.

The segmenting technique used in convolving a short series with a long series can also be applied to the problem

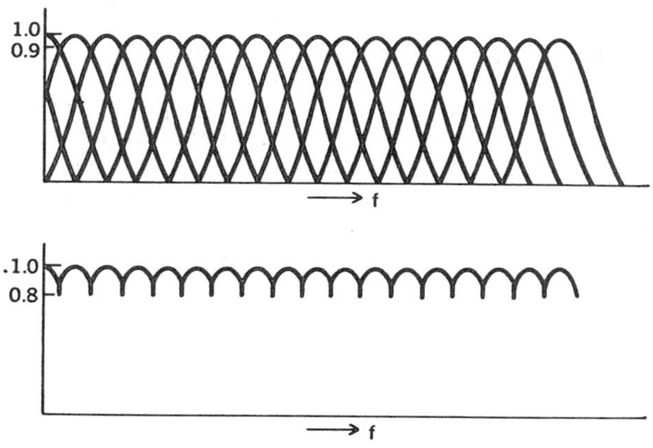

FIGURE 13. The reduction of the picket-fence effect, brought about by computing redundant overlapping sets of Fourier coefficients.

FIGURE 14. The procedure for performing noncyclical convolution on two finite signals.

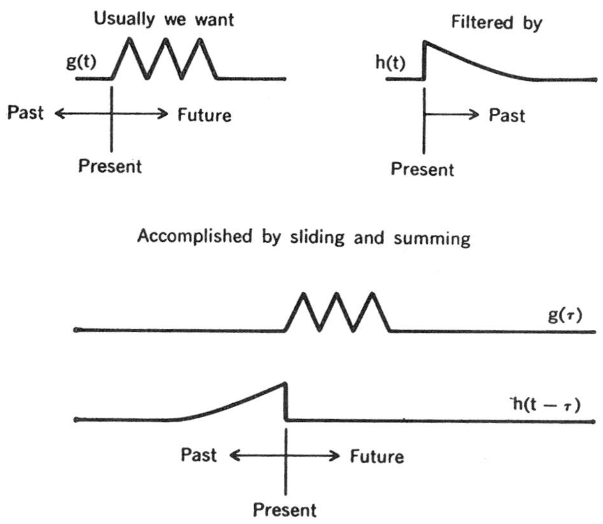

FIGURE 15. The cyclical convolution of two finite signals analogous to that performed by the FFT algorithm.

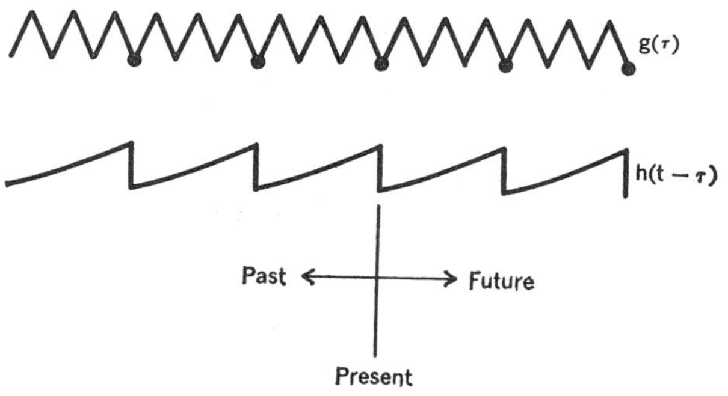

of computing N lags of the autocorrelation function of an M-term series when $N \ll M$. Figure 18 shows a function $g(t)$, which can be sampled to form an M-term time series. To compute N lags of the autocorrelation function of $g(t)$, this series is broken into overlapping segments of at least $2N$ terms. For each segment, the functions $\hat{g}(t)$ and $\hat{\hat{g}}(t)$ can be formed and sampled such that

$$\hat{g}(k) = g(k) \qquad 0 \leq k < 2N \tag{32}$$

$$\begin{aligned} \hat{\hat{g}}(k) &= g(k) \qquad 0 \leq k < N \\ &= 0 \qquad N \leq k < 2N \end{aligned} \tag{33}$$

An appropriate procedure is to

(1) Form $\hat{G}(j)$ and $\hat{\hat{G}}(j)$ via the FFT.

(2) Compute $\hat{C}(j) = \hat{G}(j) \cdot \hat{\hat{G}}^*(j)$.

(3) Find $\hat{c}(k) = F^{-1}[\hat{C}(j)]$ via the FFT.

The first half of the $c(k)$ series represents the contribution of the first half of the $2N$-term segment to the first N lags of the autocorrelation function of $g(k)$. When the contributions from all of the segments are added together, the result is the first N lags of the autocorrelation function of all M terms of $g(k)$. This technique, called "overlap-adding," is described in detail by Helms.[24]

The choice of $2N$ zeros is made to limit the present discussion and can be changed. The optimum choice with respect to computational effort is discussed in Refs. 22 and 24. It is also a simple matter to compute any set of N lags and to apply this same technique to computing cross-correlation functions.

Frequency-domain filter design. Given the capability for convenient and efficient digital filtering, one possesses considerable degree of freedom in specifying the form of the filter. It becomes very tempting to make use of the ideal low-pass, high-pass, and bandpass filters that are so difficult to come by in the real world. This should not be done, however, without exercising great caution.

When a digital filter is specified in the frequency domain, this is equivalent to multiplying the Fourier coefficients by a window. This multiplication in the frequency domain is equivalent to performing a convolution in the time domain. Thus, the constraints discussed previously still apply but are not as readily apparent. For this discussion, the constraint is that at least half of the impulse response implied by the frequency domain filter be identically zero.

When an ideal filter, such as $H(f)$ in Fig. 19, is specified in the frequency domain, this implies a $(\sin x)/x$ impulse response in the time domain. In this example of an ideal low-pass filter, $h(t)$ does not go to zero, meaning that the tails of the $(\sin x)/x$ impulse response fold back into the T-second region. The result is aliasing in the time domain caused by undersampling the filter function in the frequency domain. This is directly analogous to the aliasing in the frequency domain caused by undersampling in the time domain.

The cure to this problem is always to choose a filter with an impulse response that dies out or can be truncated so that at least half of its terms are essentially zero. Two procedures based on truncating the impulse response are described in detail by Helms[30,31]; see also Refs. 32 and 33. In specifying a filter in the frequency domain, remember that a filter with a real impulse response implies a set of Fourier coefficients whose real part is an even function and whose imaginary part is an odd function. In the fast Fourier transform format, this implies that the real part be symmetric about the folding frequency [the (N/2)th harmonic] and the imaginary part be

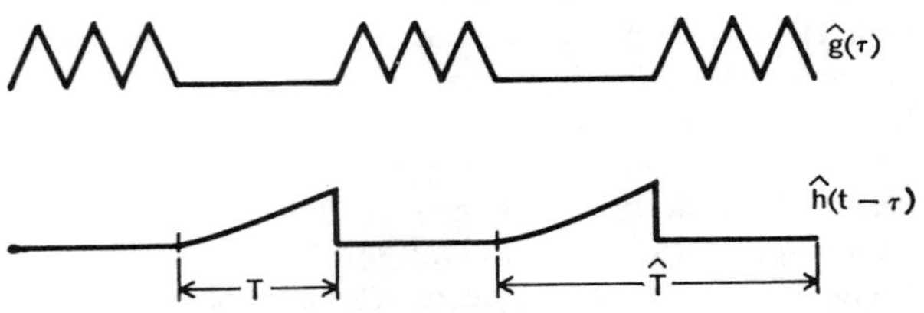

FIGURE 16. Noncyclical convolution of two finite signals analogous to that performed by the FFT algorithm.

FIGURE 17. A method for convolving a finite impulse response with an infinite time function by performing a series of fast Fourier transforms.

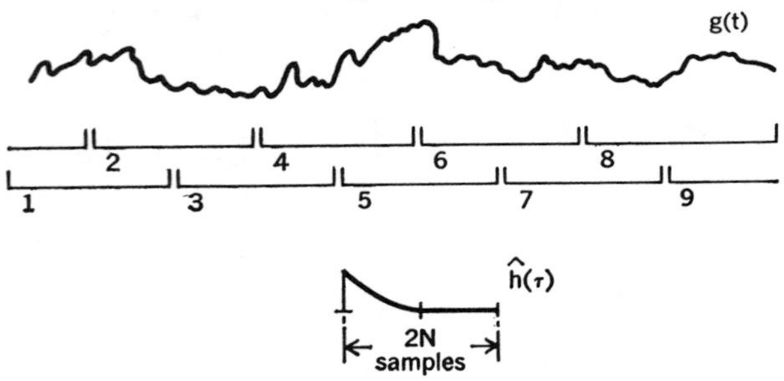

FIGURE 18. A method of using the fast Fourier transform algorithm to compute N lags of the autocorrelation function of an M-term series.

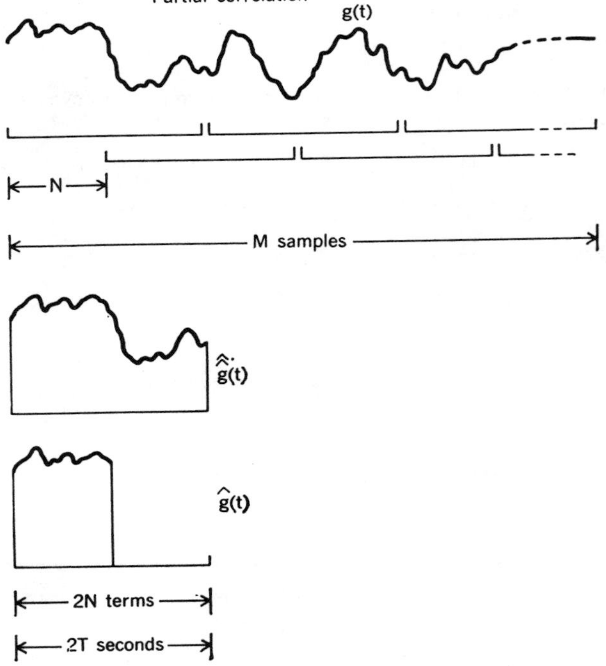

antisymmetric, as shown in the example of Fig. 1.

Applications. An example of a radar signal processing system is shown in the top half of Fig. 20. A chirped radar pulse, $s(t)$, is shown entering a matched dechirping filter followed by a Taylor weighting network.[62] In practice the input signal is corrupted by noise, the atmosphere, the equipment, etc., and the problem is to determine the effect this has on system performance.[61] In Fig. 20 this degradation is expressed in the form of a weighting function applied to $s(t)$ before it enters the matched filter. In this example the weighting is of the form $1 - x(t)$.

For our discussion, the matched filter can be viewed as finding the autocorrelation function of the chirped signal, and the Taylor weighting can be viewed as applying a weighting function in the frequency domain. The autocorrelation can be thought of as convolving the waveform $s(t)$ with itself reversed in time—that is, with $f(t)$. The multiplication by the Taylor weighting coefficients in the frequency domain corresponds to convolving the result of the first convolution with the impulse response of the Taylor weighting network $w(t)$. Thus the response of the system, $r(t)$, is actually the result of performing two convolutions. If $s(t)$, $f(t)$, and $w(t)$ can all be represented by N-term series, they each should be augmented with $3N$ zeros to assure that the final convolution is noncyclical.

The application of the fast Fourier transform to radar and sonar ranging systems can be viewed in terms of performing a correlation. Since correlation can be viewed as one function searching to find itself in another, it is clear how echo-ranging systems using a chirped or even a random-noise signal can operate. Since these correlations can now be done by means of the FFT, many of these operations can be performed digitally in real time.

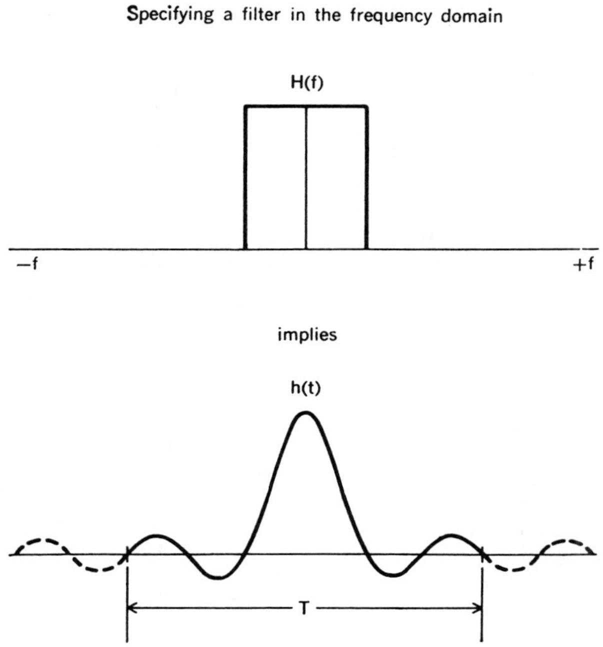

FIGURE 19. The (sin x)/x impulse response of a filter implied when a rectangular frequency response is specified.

FIGURE 20. Block diagrams illustrating the use of the FFT for simulating a radar signal processing system.

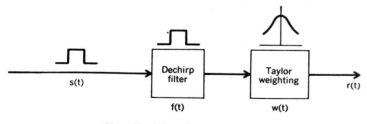

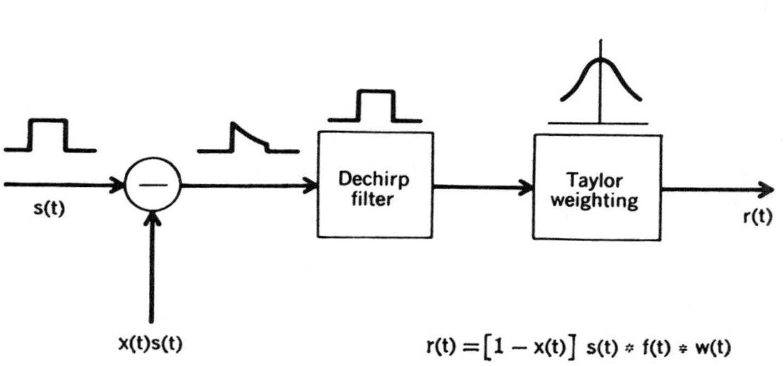

How has it been implemented?

Software. The number of variations of the FFT algorithm appears to be directly proportional to the number of people using it.[34–53] Most of these algorithms are based on either the Cooley–Tukey or the Sande–Tukey algorithm,[11] but are formulated to exploit properties of the series being analyzed or properties of the computer being used.

Equations (17) through (20) represent the Cooley–Tukey formulation of the FFT algorithm. By separating the components of j instead of the components of k in Eq. (11), the Sande–Tukey equations would have resulted. In either case, the recursive equations can be written in the form of a two-point transform followed by a rereferencing (or twiddling[9]) operation. The resulting counterparts of Eqs. (17)–(20) are as follows:

$$\hat{A}_1(j_0,k_1,k_0) = \left\{\sum_{k_2=0}^{1} A(k_2,k_1,k_0) W_2^{j_0k_2}\right\} W_4^{j_0k_1} \tag{34}$$

$$\hat{A}_2(j_0,j_1,k_0) = \left\{\sum_{k_1=0}^{1} \hat{A}_1(j_0,k_1,k_0) W_2^{j_1k_1}\right\} W_8^{(j_12+j_0)k_0} \tag{35}$$

$$\hat{A}_3(j_0,j_1,j_2) = \left\{\sum_{k_0=0}^{1} \hat{A}_2(j_0,j_1,k_0) W_2^{j_2k_0}\right\} \tag{36}$$

$$\hat{X}(j_2,j_1,j_0) = \hat{A}_3(j_0,j_1,j_2) \tag{37}$$

The extension of the FFT algorithm from radix-two algorithms to arbitrary-radix algorithms is accomplished by representing the j and k variables of Eq. (8) in a mixed radix number system.[34,51] For the example of $N = r_1r_2r_3$, they take the form

$$j = j_2(r_1r_2) + j_1(r_1) + j_0$$
$$k = k_2(r_2r_3) + k_1(r_3) + k_0 \tag{38}$$

where j_0, $k_2 = 0, 1, \cdots, r_1 - 1$; j_1, $k_1 = 0, 1, \cdots, r_2 - 1$; and j_2, $k_0 = 0, 1, \cdots, r_3 - 1$. The resulting recursive equations can functionally be separated into r_1 point transforms, r_2 point transforms, r_3 point transforms, and rereferencing operations.

Since four-point transforms can be performed with only additions and subtractions, an algorithm with a large number of factors that are four requires less computation than a radix-two algorithm.[44] In much the same manner, a further reduction can be made by forcing the algorithm into a form where a large number of eight-point transforms can be done very efficiently.[35] The number of multiplications required by the resulting algorithms is reduced by 30 percent and 40 percent, respectively, compared with radix-two algorithms.

Several variants of the FFT algorithm have been motivated by characteristics of the series being transformed. When the series is real, the expected two-to-one reduction in computation and storage can be obtained by either putting the N-term real record in the form of an artificial $N/2$-term complex record,[7] or by restructuring the FFT algorithm.[36]

An algorithm that allows the value of N to take on any value (including a prime number) has recently been proposed.[25, 38] For any value of N, the discrete Fourier transform can be written

$$X(j) = \frac{1}{N}\sum_{k=0}^{N-1} x(k)W^{-jk} \tag{39}$$

$$X(j) = \frac{1}{N}\sum_{k=0}^{N-1} x(k)W^{-jk+[(k^2-k^2+j^2-j^2)/2]} \tag{40}$$

$$X(j) = \frac{W^{-j^2/2}}{N}\left\{\sum_{k=0}^{N-1} [W^{-k^2/2}x(k)]W^{(j-k)^2/2}\right\} \tag{41}$$

Note that Eq. (41) is in the form of a convolution; i.e.,

$$X(j) = \frac{W^{-j^2/2}}{N}\sum_{k=0}^{N-1} g(k)h(j-k) \tag{42}$$

where $g(k) = W^{-k^2/2}x(k)$, and $h(j-k) = W^{(j-k)^2/2}$. In performing this convolution, both of the series can be augmented with zeros until they contain a highly factorable number of terms $\hat{N}$. These extended series can be transformed by a conventional $\hat{N}$-term FFT algorithm to obtain the noncyclical convolution required by Eq. (42). The Fourier coefficients corresponding to the original N-term DFT of Eq. (39) are obtained by multiplying the results of this convolution by the unit amplitude complex exponential $W^{-j^2/2}$, where $W = \exp(2\pi i/N)$.

This algorithm should be very useful when employed with a hardware FFT processor that otherwise would be restricted to values of N that are powers of two.

A variety of different FFT programs for performing one-dimensional and multidimensional fast Fourier transforms have been made available by Cooley,[41, 42] Sande,[48] Singleton,[49-52] and Brenner,[39] who have programmed most of the options described here plus several others as well.

Hardware. The advent of the FFT algorithm reduced the time required for performing a 2^{10}-point discrete Fourier transform from several minutes to less than a second. The advent of special-purpose digital hardware further reduced this time to tens of milliseconds.[54-60]

The cost of finding a 2^{10}-point transform used to be measured in dollars. The FFT algorithm reduced this cost to a few cents. Special-purpose hardware has reduced it further to hundredths of a cent.

Thus in the past five years, both the cost and execution time of computing a discrete Fourier transform have been reduced by nearly four orders of magnitude. Therefore, we can now build a relatively inexpensive special-purpose processor that is able to meet the real-time constraints of a wide variety of signal-processing problems.

A survey of FFT processors and their characteristics is reported in Ref. 55. Most of these processors can be classified in one of the four families of FFT processor machine organizations discussed in Ref. 56. These four families represent varying degrees of parallelism, performance, and cost.

Most of the hardware implementations built to date have relied on either a radix-two or a radix-four algorithm. The regularity of these algorithms and the resulting simplification in control have tended to discourage the use of the more general arbitrary-radix algorithms. In addition, most users of real-time FFT hardware seem to have adapted quickly to thinking in powers of two.

The author would like to thank D. E. Wilson for the loan of the title; R. A. Kaenel and W. W. Lang for their encouragement; and B. P. Bogert, W. T. Hartwell, H. D. Helms, C. M. Rader, and P. T. Rux for suggesting several improvements, which were incorporated in this article.

REFERENCES

Introduction to Fourier analysis

1. Arsac, J., *Fourier Transforms.* Englewood Cliffs, N.J.: Prentice-Hall, 1966.
2. Bracewell, R., *The Fourier Transform and Its Applications.* New York: McGraw-Hill, 1965.
3. Papoulis, A., *The Fourier Integral and Its Applications.* New York: McGraw-Hill, 1962.

Historical development of the fast Fourier transform

4. Cooley, J. W., Lewis, P. A. W., and Welch, P. D., "Historical notes on the fast Fourier transform," *IEEE Trans. Audio and Electroacoustics*, vol. AU-15, pp. 76–79, June 1967.

Introduction to the fast Fourier transform

5. Brigham, E. O., and Morrow, R. E., "The fast Fourier transform," *IEEE Spectrum*, vol. 4, pp. 63–70, Dec. 1967.
6. Cochran, W. T., *et al.*, "What is the fast Fourier transform?" *IEEE Trans. Audio and Electroacoustics*, vol. AU-15, pp. 45–55, June 1967.
7. Cooley, J. W., Lewis, P. A. W., and Welch, P. D., "The fast Fourier transform algorithm and its applications," IBM Research Paper RC-1743, Feb. 1967.
8. Cooley, J. W., and Tukey, J. W., "An algorithm for the machine calculation of complex Fourier series," *Math. Comput.*, vol. 19, pp. 297–301, Apr. 1965.
9. Gentleman, W. M., and Sande, G., "Fast Fourier transforms—for fun and profit," *1966 Fall Joint Computer Conf., AFIPS Proc.*, vol. 29. Washington, D.C.: Spartan Books.
10. Gold, B., and Rader, C. M., *Digital Processing of Signals.* New York: McGraw-Hill, 1969.

Spectrum and cepstrum analysis

11. Bingham, C., Godfrey, M. D., and Tukey, J. W., "Modern techniques of power spectrum estimation," *IEEE Trans. Audio and Electroacoustics*, vol. AU-15, pp. 56–66, June 1967.
12. Blackman, R. B., and Tukey, J. W., *The Measurement of Power Spectra.* New York: Dover, 1958.
13. Bogert, B. P., Healy, M. J. and Tukey, J. W., "The frequency analysis of time series for echoes: cepstrum, pseudoautocovariance, cross-cepstrum and saphe-cracking," *Time Series Analysis*, Murray Rosenblatt, ed. New York: Wiley, 1963, pp. 201–243.
14. Noll, A. Michael, "Short-time spectrum and 'cepstrum' techniques for vocal-pitch detection," *J. Acoust. Soc. Am.*, vol. 36, pp. 296–302, 1964.
15. Oppenheim, A. V., Schafer, R. W., and Stockham, T. G., Jr., "Nonlinear filtering of multiplied and convolved signals," *Proc. IEEE*, vol. 56, pp. 1264–1291, Aug. 1968. (Reprinted in *IEEE Trans. Audio and Electroacoustics*, vol. AU-16, pp. 437–465, Sept. 1968.)
16. Parzen, E., "Statistical spectral analysis (single channel case) in 1968," Tech. Rep. 11, ONR Contract Nonr-225(80)(NR-042-234), Stanford University, Dept. of Statistics, Stanford, Calif., June 10, 1968.
17. Richards, P. I., "Computing reliable power spectra," *IEEE Spectrum*, vol. 4, pp. 83–90, Jan. 1967.

18. Tukey, J. W., "An introduction to the measurement of spectra," in *Probability and Statistics*, Ulf Grenander, ed. New York: Wiley, 1959, pp. 300–330.

19. Tukey, J. W., "An introduction to the calculations of numerical spectrum analysis," in *Spectral Analysis of Time Series*, Bernard Harris, ed. New York: Wiley, 1967, pp. 25–46.

20. Welch, P. D., "A direct digital method of power spectrum estimation," *IBM J. Res. Develop.*, vol. 5, pp. 141–156, 1961.

21. Welch, P. D., "The use of the fast Fourier transform for the estimation of power spectra: a method based on time averaging over short, modified periodograms," *IEEE Trans. Audio and Electroacoustics*, vol. AU-15, pp. 70–73, June 1967.

Use of the fast Fourier transform in convolution, correlation, digital filtering, etc.

22. Cooley, J. W., "Applications of the fast Fourier transform method," *Proc. IBM Scientific Computing Symp. on Digital Simulation of Continuous Systems*, Thomas J. Watson Research Center, Yorktown Heights, N.Y., 1966.

23. Cooley, J. W., Lewis, P. A. W., and Welch, P. D., "Application of the fast Fourier transform to computation of Fourier integrals, Fourier series, and convolution integrals," *IEEE Trans. Audio and Electroacoustics*, vol. AU-15, pp. 79–84, June 1967.

24. Helms, H. D., "Fast Fourier transform method of computing difference equations and simulating filters," *IEEE Trans. Audio and Electroacoustics*, vol. AU-15, pp. 85–90, June 1967.

25. Rabiner, L. R., Schafer, R. W., and Rader, C. M., "The chirp *Z*-transform algorithm and its applications," *Bell System Tech. J.*, vol. 48, pp. 1249–1292, May–June 1969.

26. Sande, G., "On an alternative method of calculating covariance functions," unpublished technical note, Princeton University, Princeton, N.J., 1965.

27. Singleton, R. C., "Algorithm 345, an Algol convolution procedure based on the fast Fourier transform," *Commun. Assoc. Comput. Mach.*, vol. 12, Mar. 1969.

28. Stockham, T. G., "High-speed convolution and correlation," *1966 Spring Joint Computer Conf., AFIPS Proc.*, vol. 28. Washington, D.C.: Spartan Books, pp. 229–233.

The picket-fence effect

29. Hartwell, W. T., "An alternate approach to the use of discrete Fourier transforms," to be published.

Frequency-domain filter design

30. Helms, H. D., "Fast Fourier transform method for computing difference equations and simulating filters," *IEEE Trans. Audio and Electroacoustics*, vol. AU-15, pp. 85–90, June 1967.

31. Helms, H. D., "Nonrecursive digital filters: design methods for achieving specifications on frequency response," *IEEE Trans. Audio and Electroacoustics*, vol. AU-16, pp. 336–342, Sept. 1968.

32. Kaiser, J. F., "Digital filters," in *System Analysis by Digital Computer*, F. F. Kuo and J. F. Kaiser, eds. New York: Wiley, 1966.

33. Otnes, R. K., "An elementary design procedure for digital filters," *IEEE Trans. Audio and Electroacoustics*, vol. AU-16, pp. 336–342, Sept. 1968.

Algorithms and software

34. Bergland, G. D., "The fast Fourier transform recursive equations for arbitrary length records," *Math. Comput.*, vol. 21, pp. 236–238, Apr. 1967.

35. Bergland, G. D., "A fast Fourier transform algorithm using base 8 iterations," *Math. Comput.*, vol. 22, pp. 275–279, Apr. 1968.

36. Bergland, G. D., "A fast Fourier transform algorithm for real-valued series," *Commun. Assoc. Comput. Mach.*, vol. 11, pp. 703–710, Oct. 1968.

37. Bergland, G. D., and Wilson, D. E., "An FFT algorithm for a global, highly parallel processor," *IEEE Trans. Audio and Electroacoustics*, vol. AU-17, June 1969.

38. Bluestein, L. I., "A linear filtering approach to the computation of the discrete Fourier transform," *1968 NEREM Record*, pp. 218–219.

39. Brenner, N. M., "Three Fortran programs that perform the Cooley–Tukey Fourier transform," Tech. Note 1967-2, Lincoln Laboratory, M.I.T., Lexington, Mass., July 1967.

40. Cooley, J. W., and Tukey, J. W., "An algorithm for the machine calculation of complex Fourier series," *Math. Comput.*, vol. 19, pp. 297–301, Apr. 1965.

41. Cooley, J. W., "Harmonic analysis complex Fourier series," SHARE Doc. 3425, Feb. 7, 1966.

42. Cooley, J. W., "Complex finite Fourier transform subroutine," SHARE Doc. 3465, Sept. 8, 1966.

43. Danielson, G. C., and Lanczos, C., "Some improvements in practical Fourier analysis and their application to X-ray scattering from liquids," *J. Franklin Inst.*, vol. 233, pp. 365–380, 435–452.

44. Gentleman, W. M., and Sande, G., "Fast Fourier transforms—for fun and profit," *1966 Fall Joint Computer Conf., AFIPS Proc.*, vol. 29. Washington, D.C.: Spartan Books.

45. Good, I. J., "The interaction algorithm and practical Fourier series," *J. Roy. Statist. Soc.*, vol. 20, series B, pp. 361–372, 1958; addendum, vol. 22, pp. 372–375, 1960.

46. Pease, M. C., "An adaption of the fast Fourier transform for parallel processing," *J. Assoc. Comput. Mach.*, vol. 15, pp. 252–264, Apr. 1968.

47. Rader, C. M., "Discrete Fourier transforms when the number of data samples is prime," *Proc. IEEE*, vol. 56, pp. 1107–1108, June 1968.

48. Sande, G., "Arbitrary radix one-dimensional fast Fourier transform subroutines," University of Chicago, Ill., 1968.

49. Singleton, R. C., "On computing the fast Fourier transform," *Commun. Assoc. Comput. Mach.*, vol. 10, pp. 647–654, Oct. 1967.

50. Singleton, R. C., "Algorithm 338, Algol procedures for the fast Fourier transform," *Commun. Assoc. Comput. Mach.*, vol. 11, pp. 647–654, Nov. 1968.

51. Singleton, R. C., "Algorithm 339, an Algol procedure for the fast Fourier transform with arbitrary factors," *Commun. Assoc. Comput. Mach.*, vol. 11, pp. 776–779, Nov. 1968.

52. Singleton, R. C., "Algorithm 345, an Algol convolution procedure based on the fast Fourier transform," *Commun. Assoc. Comput. Mach.*, vol. 12, Mar. 1969.

53. Yavne, R., "An economical method for calculating the discrete Fourier transform," *1968 Fall Joint Computer Conf., IFIPS Proc.*, vol. 33. Washington, D.C.: Spartan Books, pp. 115–125.

Fast Fourier transform hardware

54. Bergland, G. D., and Hale, H. W., "Digital real-time spectral analysis," *IEEE Trans. Electronic Computers*, vol. EC-16, pp. 180–185, Apr. 1967.

55. Bergland, G. D., "Fast Fourier transform hardware implementations—a survey," *IEEE Trans. Audio and Electroacoustics*, vol. AU-17, June 1969.

56. Bergland, G. D., "Fast Fourier transform hardware implementations—an overview,"' *IEEE Trans. Audio and Electroacoustics*, vol. AU-17, June 1969.

57. McCullough, R. B., "A real-time digital spectrum analyzer," Stanford Electronics Laboratories Sci. Rept. 23, Stanford University, Calif., Nov. 1967.

58. Pease, M. C., III, and Goldberg, J., "Feasibility study of a special-purpose digital computer for on-line Fourier analysis," Order No. 989, Advanced Research Projects Agency, Washington, D.C., May 1967.

59. Shively, R. R., "A digital processor to generate spectra in real time," *IEEE Trans. Computers*, vol. C-17, pp. 485–491, May 1968.

60. Smith, R. A., "A fast Fourier transform processor," Bell Telephone Laboratories, Inc., Whippany, N.J., 1967.

Other

61. Gilbert, S. M., Private communication, Bell Telephone Laboratories, Inc., Whippany, N.J.

62. Klauder, J. R., Price, A. C., Darlington, S., and Albersheim, W. J., "The theory and design of chirp radars," *Bell System Tech. J.*, vol. 39, pp. 745–808, July 1960.

What is the Fast Fourier Transform?

G-AE Subcommittee on Measurement Concepts

WILLIAM T. COCHRAN
JAMES W. COOLEY
DAVID L. FAVIN, MEMBER, IEEE
HOWARD D. HELMS, MEMBER, IEEE
REGINALD A. KAENEL, SENIOR MEMBER, IEEE
WILLIAM W. LANG, SENIOR MEMBER, IEEE
GEORGE C. MALING, JR., ASSOCIATE MEMBER, IEEE
DAVID E. NELSON, MEMBER, IEEE,
CHARLES M. RADER, MEMBER, IEEE
PETER D. WELCH

***Abstract*—The fast Fourier transform is a computational tool which facilitates signal analysis such as power spectrum analysis and filter simulation by means of digital computers. It is a method for efficiently computing the discrete Fourier transform of a series of data samples (referred to as a time series). In this paper, the discrete Fourier transform of a time series is defined, some of its properties are discussed, the associated fast method (fast Fourier transform) for computing this transform is derived, and some of the computational aspects of the method are presented. Examples are included to demonstrate the concepts involved.**

Introduction

AN ALGORITHM for the computation of Fourier coefficients which requires much less computational effort than was required in the past was reported by Cooley and Tukey [1] in 1965. This method is now widely known as the "fast Fourier transform," and has produced major changes in computational techniques used in digital spectral analysis, filter simulation, and related fields. The technique has a long and interesting history that has been summarized by Cooley, Lewis, and Welch in this issue [2].

The fast Fourier transform (FFT) is a method for efficiently computing the discrete Fourier transform (DFT) of a time series (discrete data samples). The efficiency of this method is such that solutions to many problems can now be obtained substantially more economically than in the past. This is the reason for the very great current interest in this technique.

The discrete Fourier transform (DFT) is a transform in its own right such as the Fourier integral transform or the Fourier series transform. It is a powerful reversible mapping operation for time series. As the name implies, it has mathematical properties that are entirely analogous to those of the Fourier integral transform. In particular, it defines a spectrum of a time series; multiplication of the transform of two time series corresponds to convolving the time series.

If digital analysis techniques are to be used for analyzing a continuous waveform then it is necessary that the data be sampled (usually at equally spaced intervals of time) in order to produce a time series of discrete samples which can be fed into a digital computer. As is well known [6], such a time series completely represents the continuous waveform, provided this waveform is frequency band-limited and the samples are taken at a rate that is at least twice the highest frequency present in the waveform. When these samples are equally spaced they are known as Nyquist samples. It will be shown that the DFT of such a time series is closely related to the Fourier transform of the continuous waveform from which samples have been taken to form the time series. This makes the DFT particularly useful for power spectrum analysis and filter simulation on digital computers.

The fast Fourier transform (FFT), then, is a highly efficient procedure for computing the DFT of a time series. It takes advantage of the fact that the calculation of the coefficients of the DFT can be carried out iteratively, which results in a considerable savings of computation time. This manipulation is not intuitively obvious, perhaps explaining why this approach was overlooked for such a long time. Specifically, if the time series consists of $N=2^n$ samples, then about $2nN=2N\cdot\log_2 N$ arithmetic operations will be shown to be required to evaluate all N associated DFT coefficients. In comparison with the number of operations required for the calculation of the DFT coefficients with straightforward procedures (N^2), this number is so small when N is large as to completely change the computationally economical approach to various problems. For example, it has been reported that for $N=8192$ samples, the computations require about five seconds

Manuscript received March 10, 1967.
W. T. Cochran, D. L. Favin, and R. A. Kaenel are with Bell Telephone Laboratories, Inc., Murray Hill, N. J.
H. D. Helms is with Bell Telephone Laboratories, Inc., Whippany, N. J.
J. W. Cooley and P. D. Welch are with the IBM Research Center, Yorktown Heights, N. Y.
W. W. Lang and G. C. Maling are with the IBM Corporation, Poughkeepsie, N. Y.
C. M. Rader is with Lincoln Laboratory, Massachusetts Institute of Technology, Lexington, Mass. (Operated with support from the U. S. Air Force.)
D. E. Nelson is with the Electronics Division of the General Dynamics Corporation, Rochester, N. Y.

Reprinted from *IEEE Trans. Audio Electroacoust.*, vol. AU-15, pp. 45–55, June 1967.

for the evaluation of all 8192 DFT coefficients on an IBM 7094 computer. Conventional procedures take on the order of half an hour.

The known applications where a substantial reduction in computation time has been achieved include: 1) computation of the power spectra and autocorrelation functions of sampled data [4]; 2) simulation of filters [5]; 3) pattern recognition by using a two-dimensional form of the DFT; 4) computation of bispectra, cross-covariance functions, cepstra and related functions; and 5) decomposing of convolved functions.

The Discrete Fourier Transform (DFT)

Definition of the DFT and its Inverse

Since the FFT is an efficient method for computing the DFT it is appropriate to begin by discussing the DFT and some of the properties that make it so useful a transformation. The DFT is defined by[1]

$$A_r = \sum_{k=0}^{N-1} X_k \exp(-2\pi jrk/N) \qquad r = 0, \cdots, N-1 \quad (1)$$

where A_r is the rth coefficient of the DFT and X_k denotes the kth sample of the time series which consists of N samples and $j=\sqrt{-1}$. The X_k's can be complex numbers and the A_r's are almost always complex. For notational convenience (1) is often written as

$$A_r = \sum_{k=0}^{N-1} (X_k) W^{rk} \qquad r = 0, \cdots, N-1 \quad (2)$$

where

$$W = \exp(-2\pi j/N). \quad (3)$$

Since the X_k's are often values of a function at discrete time points, the index r is sometimes called the "frequency" of the DFT. The DFT has also been called the "discrete Fourier transform" or the "discrete time, finite range Fourier transform."

There exists the usual inverse of the DFT and, because the form is very similar to that of the DFT, the FFT may be used to compute it.

The inverse of (2) is

$$X_l = (1/N) \sum_{r=0}^{N-1} A_r W^{-rl} \qquad l = 0, 1, \cdots, N-1. \quad (4)$$

This relationship is called the inverse discrete Fourier transform (IDFT). It is easy to show that this inversion is valid by inserting (2) into (4)

$$X_l = \sum_{r=0}^{N-1} \sum_{k=0}^{N-1} (X_k/N) W^{r(k-l)}. \quad (5)$$

Interchanging in (5) the order of summing over the indices r and k, and using the orthogonality relation

$$\sum_{r=0}^{N-1} \exp(2\pi j(n-m)r/N) = N, \text{ if } n \equiv m \bmod N$$
$$= 0, \text{ otherwise} \quad (6)$$

establishes that the right side of (5) is in fact equal to X_k.

It is useful to extend the range of definition of A_r to all integers (positive and negative). Within this definition it follows that

$$A_r = A_{N+r} = A_{2N+r} = \cdots \quad (7)$$

Similarly,

$$X_l = X_{N+l} = X_{2N+l} = \cdots. \quad (8)$$

Relationships between the DFT and the Fourier Transform of a Continuous Waveform

An important property that makes the DFT so eminently useful is the relationship between the DFT of a sequence of Nyquist samples and the Fourier transform of a continuous waveform, that is represented by the Nyquist samples. To recognize this relationship, consider a frequency band-limited waveform $g(t)$ whose Nyquist samples, X_k, vanish outside the time interval $0 \leq t \leq NT$

$$g(t) = \sum_{k=0}^{N-1} \frac{\sin(\pi(t-kT)/T)}{(\pi(t-kT)/T)} \cdot X_k \quad (9)$$

where T is the time spacing between the samples. A periodic repetition of $g(t)$ can be constructed that has identically the same Nyquist samples in the time interval $0 \leq t \leq NT$

$$g_p(t) = \sum_l \sum_{k=0}^{N-1} X_k \cdot \frac{\sin(\pi(t-kT-lNT)/T)}{(\pi(t-kT-lNT)/T)} \cdot \quad (10)$$

Let the Fourier transform of $g(t)$ be $G(f)$. As is well known [6], this transform is exactly specified at discrete frequencies by the complex Fourier series coefficients of $g_p(t)$. From this it follows:

$$\frac{G(n/NT)}{NT} = D_n$$
$$= (1/NT) \int_0^{NT} g_p(t) \cdot \exp(-2\pi jnt/NT) \cdot dt$$
$$= (1/NT) \sum_{k=0}^{N-1} X_k \cdot \exp(-2\pi jnkT/NT) \quad (11)$$

where $|n| \leq N/2$ due to the spectral bandwidth limitation implicitly assumed by the sampling theorem underlying the validity of Nyquist samples.

Comparing (11) and (1) it is seen that they are exactly the same except for a factor of NT and (r, n) are both unbounded. That is,

$$N \cdot A_r = D_n \text{ for } r = n \text{ and } T = 1 \text{ second.} \quad (12)$$

The bounds specified for r and n require a correspondence which depends on (7)

[1] The definition of the DFT is not uniform in the literature. Some authors use A_r/N as the DFT coefficients, others use $A_r/\sqrt{N}$, still others use a positive exponent.

$$\frac{G(n/NT)}{NT} = D_n = N \cdot A_r$$

where

$$n = r \qquad \text{for } n = 0, 1, \cdots, q < N/2,$$

and

$$n = N - r \quad \text{for } n = -1, -2, \cdots, -q > -N/2 \qquad (13)$$

and

$$\frac{G(n/NT)}{NT} = D_n = N \cdot A_r/2 \qquad \text{for } n = N/2. \qquad (14)$$

Equations (13) and (14) give a direct relationship between the DFT coefficients and the Fourier transform at discrete frequencies for the waveform stipulated by (9). A one-to-one correspondence could have been obtained if the running variable r had been bounded by $\pm N/2$. This, however, would have required distinguishing between even and odd values of N, a distinction avoided by keeping r positive.

A waveform of the type considered by (9) is shown in Fig. 1(e). It is usually obtained as an approximation of a frequency band-limited source waveform [such as the one sketched in Fig. 1(a)] by truncating the Nyquist sample series of this waveform, and reconstructing the continuous waveform corresponding to the truncated Nyquist sample series [Fig. 1(b), (d), and (e)]. Notwithstanding the identity of the Nyquist samples of this reconstructed waveform and the frequency band-limited source waveform, these waveforms differ in the truncation interval [Fig. 1(c) and (e)]. The difference is usually referred to as aliasing distortion; the mechanics of this distortion is most apparent in the frequency domain [Fig. 1(c)–(e)]. It can be made negligibly small by choosing a sufficiently large product of the frequency bandwidth of the source waveform and the duration of the truncation interval [6] (e.g., N is greater than ten).

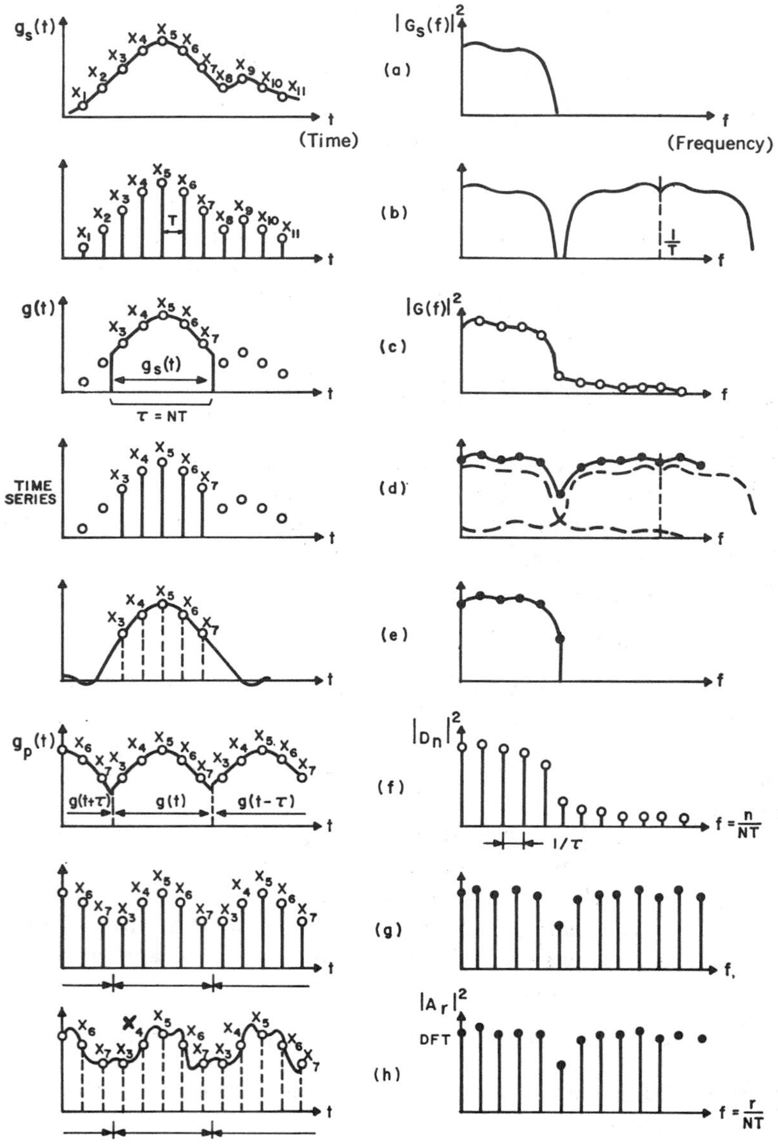

(a) Frequency-band-limited source waveform.

(b) Nyquist samples of the frequency band-limited source waveform.

(c) Truncated source waveform.

(d) Truncated series of Nyquist samples of the source waveform.

(e) Frequency-band-limited waveform whose Nyquist samples are identical to the truncated series of Nyquist samples of the source waveform.

(f) Periodic continuation of the truncated source waveform.

(g) Periodic continuation of the truncated series of Nyquist samples of the source waveform.

(h) DFT coefficients interpreted as Fourier series coefficients producing complex waveform.

Fig. 1. Related waveforms and their corresponding spectra as defined by the Fourier transforms (integral transforms for energy-limited waveforms; series transform for periodic waveforms).

These aliasing distortions are carried over directly to the discrete spectra of the periodically repeated waveforms [Fig. 1(f) and (g)], and appear correspondingly in the DFT of the truncated series of Nyquist samples [Fig. 1(h)]. It may be of interest to observe that the waveform corresponding to the DFT coefficients interpreted as Fourier series coefficients is complex [Fig. 1(h)].

Some Useful Properties of the DFT

Another property that makes the DFT eminently useful is the convolution relationship. That is, the IDFT of the product of two DFTs is the periodic mean convolution of the two time series of the DFTs. This relationship proves very useful when computing the filter output as a result of an input waveform; it becomes especially effective when computed by the FFT. A derivation of this property is given in Appendix A.

Other properties of the DFT are in agreement with the corresponding properties of the Fourier integral transform, perhaps with slight modifications. For example, the DFT of a time series circularly shifted by h is the DFT of the time series multiplied by W^{-rh}. Furthermore, the DFT of the sum of two functions is the sum of the DFT of the two functions. These properties are readily derived using the definition of the DFT. These and other properties have been compiled by Gentleman and Sande [7].

The Fast Fourier Transform

General Description of the FFT

As mentioned in the Introduction, the FFT is an algorithm that makes possible the computation of the DFT of a time series more rapidly than do other algorithms available. The possibility of computing the DFT by such a fast algorithm makes the DFT technique important. A comparison of the computational savings that may be achieved through use of the FFT is summarized in Table I for various computations that are frequently performed. It is important to add that the computational efforts listed represent comparable upper bounds; the actual efforts depend on the number N and the programming ingenuity applied [7].

It may be useful to point out that the FFT not only reduces the computation time; it also substantially reduces round-off errors associated with these computations. In fact, both computation time and round-off error essentially are reduced by a factor of $(\log_2 N)/N$ where N is the number of data samples in the time series. For example, if $N=1024=2^{10}$, then $N\cdot\log_2 N=10\ 240$ [7], [9]. Conventional methods for computing (1) for $N=1024$ would require an effort proportional to $N^2=1\ 048\ 576$, more than 50 times that required with the FFT.

The FFT is a clever computational technique of sequentially combining progressively larger weighted sums of data samples so as to produce the DFT coefficients as defined by (2). The technique can be interpreted in terms of combining the DFTs of the individual data samples such that the occurrence times of these samples are taken into account sequentially and applied to the DFTs of progressively larger mutually exclusive subgroups of data samples, which are combined to ultimately produce the DFT of the complete series of data samples. The explanation of the FFT algorithm adopted in this paper is believed to be particularly descriptive for programming purposes.

TABLE I

Comparison of the Number of Multiplications Required Using "Direct" and FFT Methods

Operation	Formula	Approximate Number of Multiplications (upper comparable bounds)	
		Direct	FFT
Discrete Fourier Transform (DFT)	$\sum_{k=0}^{N-1} X_k e^{-2\pi jrk/N}$ $r=1,2,\cdots,N-1$	N^2	$2N\log_2 N$
Filtering (Convolution)	$\sum_{k=0}^{N-1} X_k Y_{u-k}$ $u=0,1,\cdots,N-1$	N^2	$3N\log_2 N$
Autocorrelation Functions	$\sum_{k=0}^{N-1-r} X_k X_{r+k}$ $r=0,1,\cdots,N-1$	$\frac{N}{4}\left(\frac{N}{2}+3\right)$	$3N\log_2 N$
Two-Dimensional Fourier Transform (Pattern Analysis)	$\sum_{k=0}^{N-1}\sum_{l=0}^{N-1} X_{k,l}e^{-2\pi j(kq+rl/N)}$ $r,q=0,1,\cdots,N-1$	N^4	$4N^2\log_2 N$
Two-Dimensional Filtering	$\sum_{k=0}^{N-1}\sum_{l=0}^{N-1} X_{k,l}Y_{q-k,r-l}$ $q,r=1,2,\cdots,N-1$	N^4	$3N^2\log_2 N$

Conventional Forms of the FFT

Decimation in Time: The DFT [as per (2)] and its inverse [see (4)] are of the same form so that a procedure, machine, or sub-routine capable of computing one can be used for computing the other by simply exchanging the roles of X_k and A_r, and making appropriate scale-factor and sign changes. The two basic forms of the FFT, each with its several modifications, are therefore equivalent. However, it is worth distinguishing between them and discussing them separately. Let us first consider the form used by Cooley and Tukey [1] which shall be called *decimation in time.* Reversing the roles of A_r and X_k gives the form called *decimation in frequency*, which will be considered afterwards.

Suppose a time series having N samples [such as X_k shown in Fig. 2(a)] is divided into two functions, Y_k and Z_k, each of which has only half as many points ($N/2$). The function Y_k is composed of the even-numbered points ($X_0, X_2, X_4 \cdots$), and Z_k is composed of the odd numbered points ($X_1, X_3, X_5 \cdots$). These functions are shown in Fig. 2(b) and (c), and we may write them formally as

$$\left.\begin{aligned} Y_k &= X_{2k} \\ Z_k &= X_{2k+1} \end{aligned}\right\} \quad k = 0, 1, 2, \cdots, \frac{N}{2} - 1. \tag{15}$$

Since Y_k and Z_k are sequences of $N/2$ points each, they have discrete Fourier transforms defined by

$$\left.\begin{aligned} B_r &= \sum_{k=0}^{(N/2)-1} Y_k \exp(-4\pi jrk/N) \\ C_r &= \sum_{k=0}^{(N/2)-1} Z_k \exp(-4\pi jrk/N) \end{aligned}\right\} \quad r = 0, 1, 2, \cdots, \frac{N}{2} - 1. \tag{16}$$

The discrete Fourier transform that we want is A_r, which we can write in terms of the odd- and even-numbered points

$$A_r = \sum_{k=0}^{(N/2)-1} \left\{ Y_k \exp(-4\pi jrk/N) + Z_k \exp\left(-\frac{2\pi jr}{N}[2k+1]\right)\right\} \qquad r = 0, 1, 2, \cdots, N-1 \tag{17}$$

or

$$A_r = \sum_{k=0}^{(N/2)-1} Y_k \exp(-4\pi jrk/N) + \exp(-2\pi jr/N) \sum_{k=0}^{(N/2)-1} Z_k \exp(-4\pi jrk/N) \tag{18}$$

which, using (16), may be written in the following form:

$$A_r = B_r + \exp(-2\pi jr/N) C_r \quad 0 \le r < N/2. \tag{19}$$

For values of r greater than $N/2$, the discrete Fourier transforms B_r and C_r repeat periodically the values taken on when $r < N/2$. Therefore, substituting $r + N/2$ for r in (19), we obtain

$$A_{r+N/2} = B_r + \exp\left(-2\pi j\left[r + \frac{N}{2}\right]/N\right) C_r \qquad 0 \le r < N/2$$

$$= B_r - \exp(-2\pi jr/N) C_r \qquad 0 \le r < N/2. \tag{20}$$

By using (3), (19) and (20) may be written as

$$A_r = B_r + W^r C_r \qquad 0 \le r < N/2 \tag{21}$$

$$A_{r+N/2} = B_r - W^r C_r \qquad 0 \le r < N/2. \tag{22}$$

From (21) and (22), the first $N/2$ and last $N/2$ points of the discrete Fourier transform of X_k (a sequence having N samples) can be simply obtained from the DFT of Y_k and Z_k, both sequences of $N/2$ samples.

Assuming that we have a method which computes discrete Fourier transforms in a time proportional to the square of the number of samples, we can use this algorithm to compute the transforms of Y_k and Z_k, requiring a time proportional to $2(N/2)^2$, and use (21) and (22) to find A_r with additional N operations. This is illustrated in the signal flow graph of Fig. 3. The points on the left are the values of X_k (i.e., Y_k and Z_k), and the points on the right are the points of the discrete Fourier transform, A_r. For simplicity, Fig. 3 is drawn for the case where X_k is an eight-point function, and advantage is taken of the fact that $W^n = -W^{n-N/2}$, as per (3).

However, since Y_k and Z_k are to be transformed, and since we have shown that the computation of the DFT of N samples can be reduced to computing the DFTs of two sequences of $N/2$ samples each, the computation of B_k (or C_k) can be reduced to the computation of sequences of $N/4$ samples. These reductions can be carried out as long as each function has a number of samples that is divisible by 2. Thus, if $N = 2^n$ we can make n such reductions, applying (15), (21), and (22) first for N, then for $N/2, \cdots$, and finally for a two-point function. The discrete Fourier transform of a one-point function is, of course, the sample itself. The successive reduction of an eight-point discrete Fourier transform, begun in Fig. 3, is continued in Figs. 4 and 5. In Fig. 5 the operation has been completely reduced to complex multiplications and additions. From the signal flow graph there are 8 by 3 terminal nodes and 2 by 8 by 3 arrows, corresponding to 24 additions and 48 multiplications. Half of the multiplications can be omitted since the transmission indicated by the arrow is unity. Half of the remaining multiplications are also easily eliminated, as we shall see below. Thus, in general, $N \cdot \log_2 N$ complex additions and, at most, $\frac{1}{2}N \cdot \log_2 N$ complex multi-

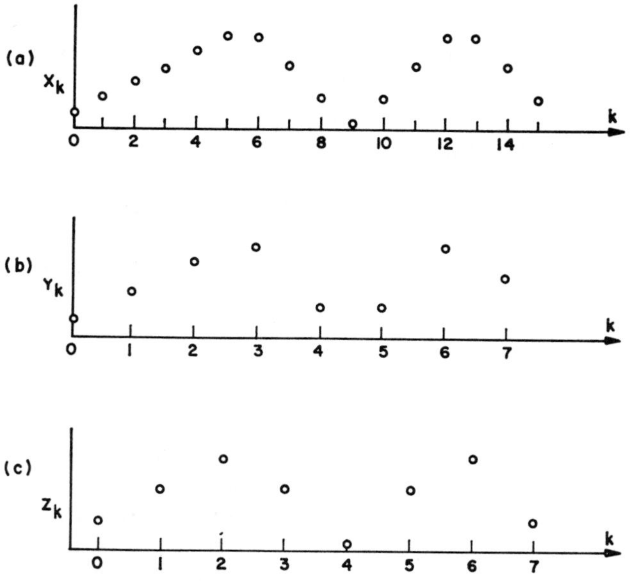

Fig. 2. Decomposition of a time series into two part-time series, each of which consists of half the samples.

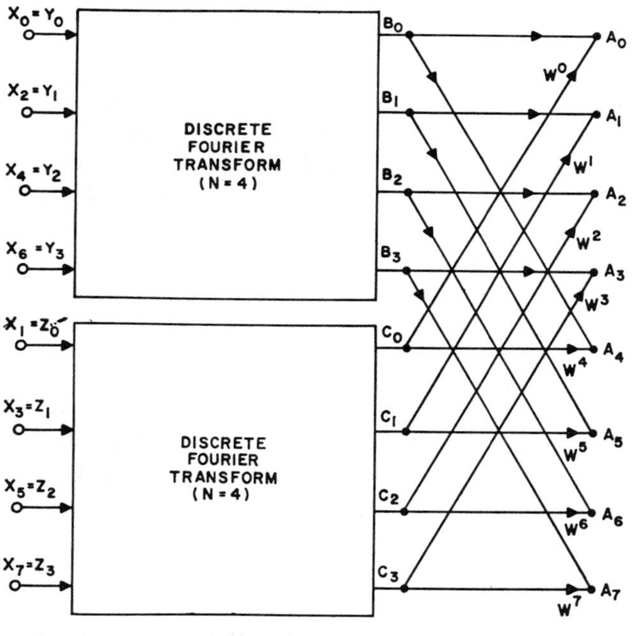

Fig. 3. Signal flow graph illustrating the reduction of endpoint DFT to two DFTs of $N/2$ points each, using decimation in time. The signal flow graph may be unfamiliar to some readers. Basically it is composed of dots (or nodes) and arrows (transmissions). Each node represents a variable, and the arrows terminating at that node originate at the nodes whose variables contribute to the value of the variable at that node. The contributions are additive, and the weight of each contribution, if other than unity, is indicated by the constant written close to the arrowhead of the transmission. Thus, in this example, the quantity A_7 at the bottom right node is equal to $B_3+W_7\times C_3$. Operations other than addition and constant multiplication must be clearly indicated by symbols other than · or ⟶.

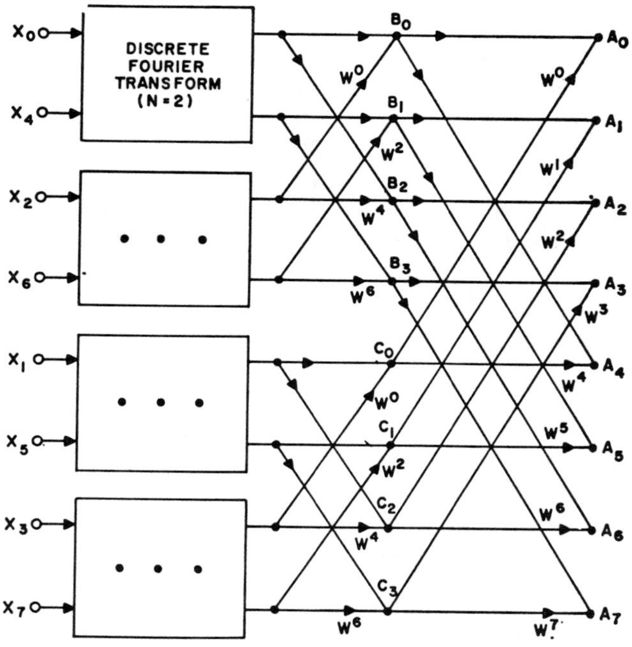

Fig. 4. Signal flow graph illustrating further reduction of the DFT computation suggested by Fig. 3.

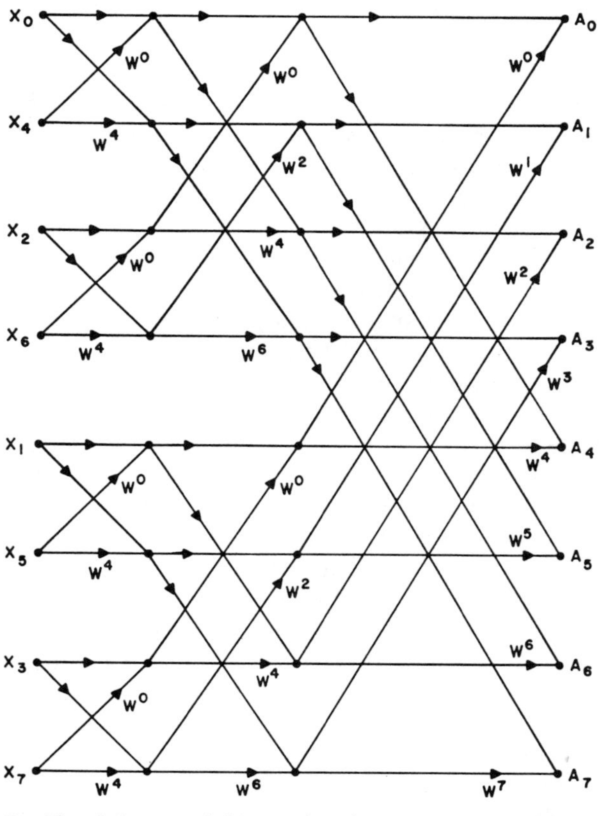

Fig. 5. Signal flow graph illustrating the computation of the DFT when the operations involved are completely reduced to multiplications and additions

plications are required for computation of the discrete Fourier transform of an N point sequence, where N is a power of 2.

When N is not a power of 2, but has a factor p, the development of equations analogous to (15) through (22) is possible by forming p different sequences, $Y_k^{(i)} = X_{pk+i}$, each having N/p samples. Each of these sequences has a DFT $B_r^{(i)}$, and the DFT of the sequence X_k can be computed from the p simpler DFTs with pN complex multiplications and additions. That is,

$$A_{r+m(N/p)} = \sum_{i=0}^{p-1} B_r^{(i)} W^{i[r+m(N/p)]}$$

$$m = 0, 1, 2, \cdots, p-1$$

$$r = 0, 1, 2, \cdots, \frac{N}{p} - 1. \quad (23)$$

The computation of the DFTs can be further simplified if N has additional prime factors.

Further information about the fast Fourier transform can be extracted from Fig. 5. For example, if the input sequence X_k is stored in computer memory in the order

$$X_0, X_4, X_2, X_6, X_1, X_5, X_3, X_7, \quad (24)$$

as in Fig. 5, the computation of the discrete Fourier transform may be done "in place," that is, by writing all intermediate results over the original data sequence, and writing the final answer over the intermediate results. Thus, no storage is needed beyond that required for the original N complex numbers. To see this, suppose that each node corresponds to two memory registers (the quantities to be stored are complex). The eight nodes farthest to the left in Fig. 5 then represent the registers containing the shuffled order input data. The first step in the computation is to compute the contents of the registers represented by the eight nodes just to the right of the input nodes. But each pair of input nodes affects only the corresponding pair of nodes immediately to the right, and if the computation deals with two nodes at a time, the newly computed quantities may be written into the registers from which the input values were taken, since the input values are no longer needed for further computation. The second step, computation of the quantities associated with the next vertical array of nodes to the right, also involves pairs of nodes although these pairs are now two locations apart instead of one. This fact does not change the property of "in place" computation, since each pair of nodes affects only the pair of nodes immediately to the right. After a new pair of results is computed, it may be stored in the registers which held the old results that are no longer needed. In the computation for the final array of nodes, corresponding to the values of the DFT, the computation involves pairs of nodes separated by four locations, but the "in place" property still holds.

For this version of the algorithm, the initial shuffling of the data sequence, X_k, was necessary for the "in place" computation. This shuffling is due to the repeated movement of odd-numbered members of a sequence to the end of the sequence during each stage of the reduction, as shown in Figs. 3, 4, and 5. This shuffling has been called *bit reversal*[2] because the samples are stored in bit-reversed order; i.e., $X_4 = X_{(100)_2}$ is stored in position $(011)_2 = 3$, etc. Note that the initial data shuffling can also be done "in place."

Variations of Decimation in Time: If one so desires, the signal flow graph shown in Fig. 5 can be manipulated to yield different forms of the *decimation in time* version of the algorithm. If one imagines that in Fig. 5 all the nodes on the same horizontal level as A_1 are interchanged with all the nodes on the same horizontal level as A_4, and all the nodes on the level of A_3 are interchanged with the nodes on the level of A_6, *with the arrows carried along with the nodes*, then one obtains a flow graph like that of Fig. 6.

For this rearrangement one need not shuffle the original data into the bit-reversed order, but the resulting spectrum needs to be *unshuffled*. An additional disadvantage might be that the powers of W needed in the computation are in bit-reversed order. Cooley's original description of the algorithm [1] corresponds to the flow graph of Fig. 6.

A somewhat more complicated rearrangement of Fig. 5 yields the signal flow graph of Fig. 7. For this case both the input data and the resulting spectrum are in "natural" order, and the coefficients in the computation are also used in a natural order. However, the computation may no longer be done "in place." Therefore, at least one other array of registers must be provided. This signal flow graph, and a procedure corresponding to it, are due to Stockham [8].

Decimation in Frequency: Let us now consider a second, quite distinct, form of the fast Fourier transform algorithm, *decimation in frequency*. This form was found independently by Sande [7], and Cooley and Stockham [8]. Let the time series X_k have a DFT A_r. The series and the DFT both contain N terms. As before, we divide X_k into two sequences having $N/2$ points each. However, the first sequence, Y_k, is now composed of the first $N/2$ points in X_k, and the second, Z_k, is composed of the last $N/2$ points in X_k. Formally, then

$$Y_k = X_k$$

$$k = 0, 1, 2, \cdots, \frac{N}{2} - 1. \quad (25)$$

$$Z_k = X_{k+N/2}$$

[2] This is a special case of digit reversal where the radix of the address is 2; more general digit reversals are available for transforms with other radices.

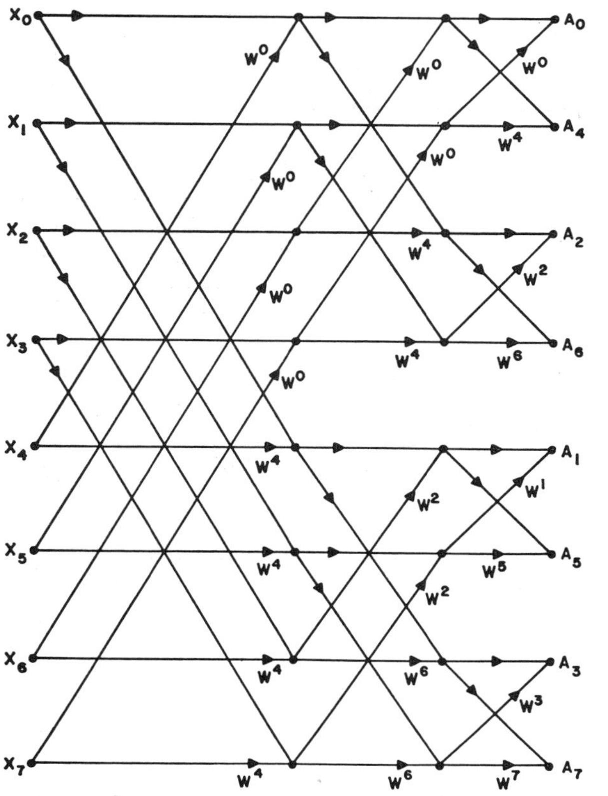

Fig. 6. Rearrangement of the flow graph of Fig. 5 illustrating the DFT computation from naturally ordered time samples.

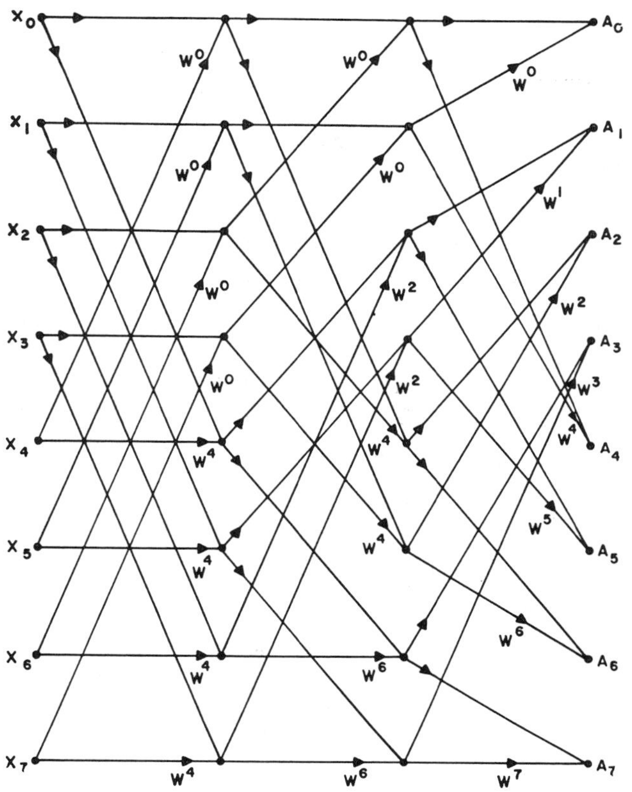

Fig. 7. Rearrangement of the flow graph of Fig. 5 illustrating the DFT computation without bit reversal.

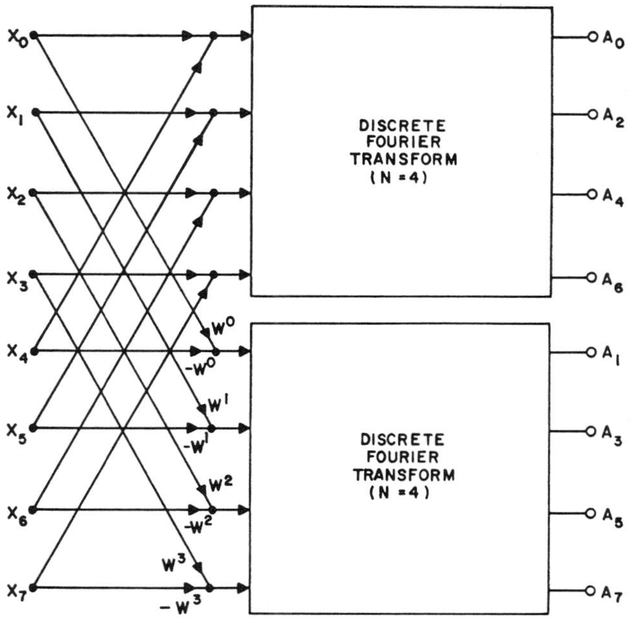

Fig. 8. Signal flow graph illustrating the reduction of endpoint DFT to two DFTs of $N/2$ points each, using decimation in frequency.

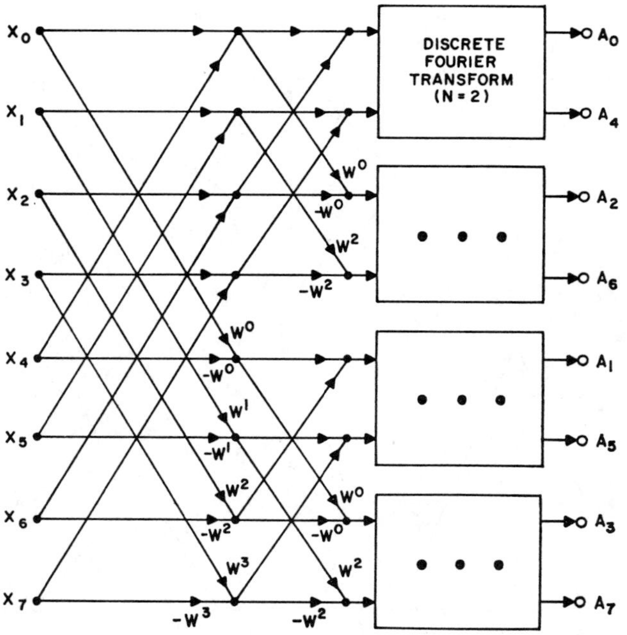

Fig. 9. Signal flow graph illustrating further reduction of the DFT computation suggested by Fig. 8.

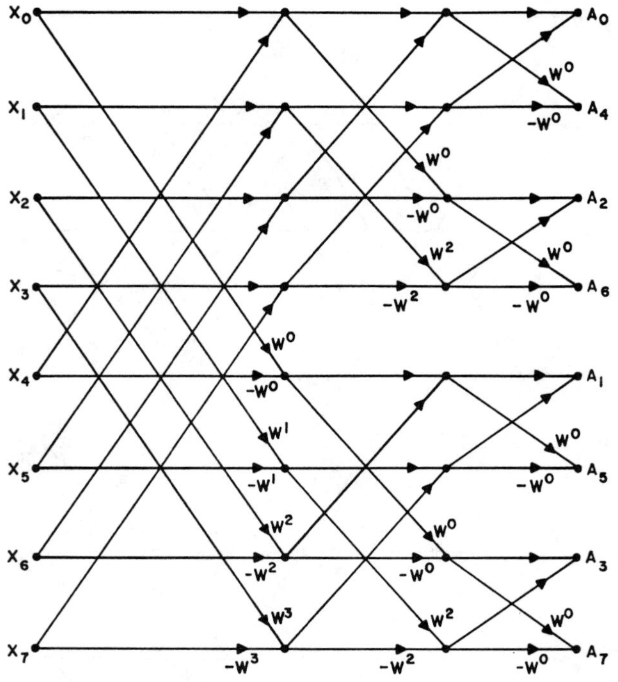

Fig. 10. Signal flow graph illustrating the computation of the DFT when the operations involved are completely reduced to multiplications and additions.

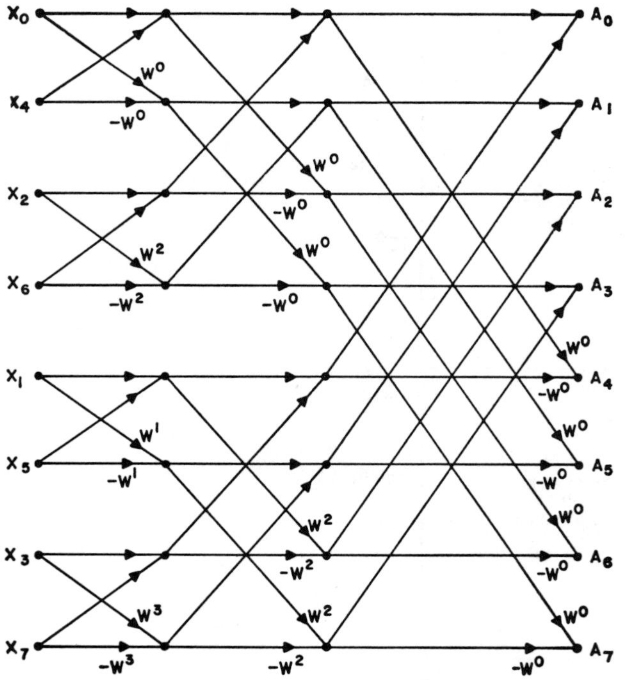

Fig. 11. Rearrangement of the flow graph of Fig. 10 illustrating the computation of the DFT to yield naturally ordered DFT coefficients.

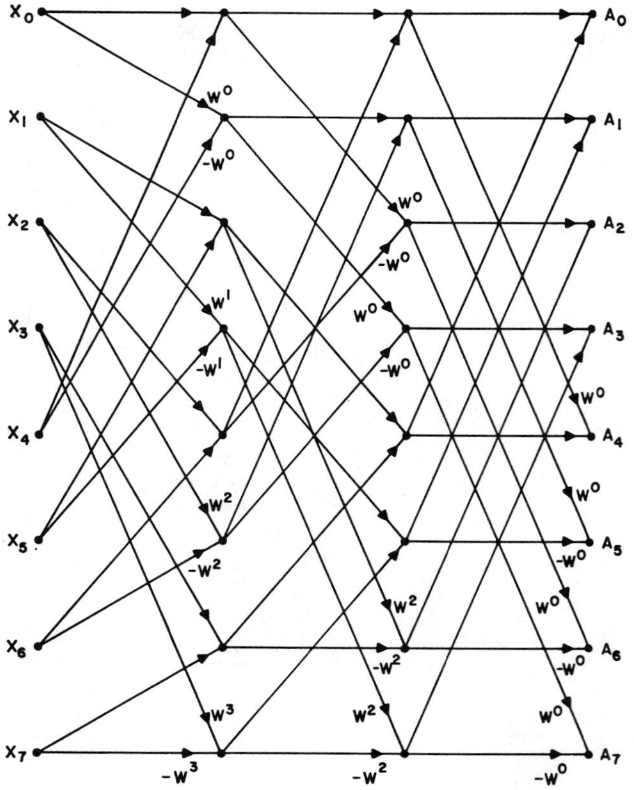

Fig. 12. Rearrangement of the flow graph of Fig. 10 illustrating the DFT computation without bit reversal.

The N point DFT of X_k may now be written in terms of Y_k and Z_k

$$A_r = \sum_{k=0}^{(N/2)-1} \left\{ Y_k \exp(-2\pi jrk/N) + Z_k \exp\left(-2\pi jr\left[k + \frac{N}{2}\right]/N\right)\right\} \tag{26}$$

$$A_r = \sum_{k=0}^{(N/2)-1} \{ Y_k + [\exp(-\pi jr)] Z_k \} \exp(-2\pi jrk/N). \tag{27}$$

Let us consider separately the even-numbered and odd-numbered points of the transform. Let the even-numbered points be R_r and the odd-numbered points be S_r, where

$$\begin{aligned} R_r &= A_{2r} \\ & \qquad 0 \le r < N/2. \\ S_r &= A_{2r+1} \end{aligned} \tag{28}$$

It is this step that may be called *decimation in frequency*. Note that for computing the even-numbered spectrum points, (27) becomes simply

$$R_r = A_{2r} = \sum_{k=0}^{(N/2)-1} \{ Y_k + Z_k \} e^{(-2\pi jrk)/(N/2)} \tag{29}$$

which we recognize as the $N/2$ point DFT of the function $(Y_k + Z_k)$, the sum of the first $N/2$ and the last $N/2$ time samples. Similarly, for the odd-numbered spectrum points, (27) becomes

$$\begin{aligned} S_r = A_{2r+1} &= \sum_{k=0}^{(N/2)-1} \{ Y_k + Z_k \exp(-\pi j[2r+1]) \} \\ & \qquad \cdot \exp(-2\pi j[2r+1]k/N) \\ &= \sum_{k=0}^{(N/2)-1} \{ Y_k - Z_k \} e^{(-2\pi jk)/N} e^{[(-2\pi jrk)/(N/2)]} \end{aligned} \tag{30}$$

which we recognize as the $N/2$ point DFT of the function $(Y_k - Z_k) \exp(-2\pi jk/N)$.

It can be concluded from (29) and (30) that the DFT of an N-sample sequence, X_k, may be determined as follows. For even-numbered transform points, it may be computed as an $N/2$ point DFT of a simple combination of the first $N/2$ and last $N/2$ samples of X_k. For odd-numbered transform points, it may be computed as another $N/2$ point DFT of a different simple combination of the first and last $N/2$ samples of X_k. This is illustrated in the signal flow graph of Fig. 8 for an eight-point function. W has been defined in (3).

As was the case with decimation in time, we can replace each of the DFTs indicated in Fig. 8 by two 2-point DFTs, and each of the 2-point DFTs by two 1-point transforms, these last being equivalency operations. These steps are indicated in Figs. 9 and 10.

Examination of Fig. 10 gives us much information about the method of decimation in frequency, and allows us to compare it with decimation in time. Both methods require $N/2 \cdot \log N$ complex additions, complex subtractions, and complex multiplications. Both computations can be done in place. If the coefficients in the computation are to be used in a "natural" rather than "bit-reversed" order, as in Figs. 5 and 10, then the decimation in frequency method works on time samples in unshuffled order and yields frequency samples in shuffled (bit-reversed) order. Recall that Fig. 5 yielded the opposite result.

We are also able to rearrange the nodes in Fig. 10 to obtain the signal flow graph, Fig. 11, which works on shuffled time samples and yields naturally ordered frequency samples, but the coefficients are needed by the computation in bit-reversed order. The geometry of this signal flow graph is identical to the geometry of Fig. 5, just as the geometry of Fig. 10 is identical to the geometry of Fig. 6. The differences lie in the transmissions.

A somewhat more complicated rearrangement of Fig. 10 (shown in Fig. 12) yields a signal flow graph that takes unshuffled samples of the time series and produces a set of Fourier coefficients that are ***not*** in bit-reversed order. The computation cannot, however, be done "in place," and at least one other array of registers must be provided. The method is similar to that shown in Fig. 7 for decimation in time. The forms of Figs. 5, 6, 7, 10, 11, and 12 constitute a set of what we might call canonic forms of the fast Fourier transform. We may choose among these forms to find an algorithm with the properties of "in place" computation, normally ordered input, normally ordered output, or normally ordered coefficients, but not all four at once. To achieve "in place" computation, we must deal with bit reversal, and to eliminate bit reversal we must give up "in place" computation. The two methods most effective when using homogeneous storage facilities are those providing in right order the sine and cosine coefficients needed in the computation. The other methods seem less desirable since they require wasteful tables. Still, all six methods have about equal usefulness, and the method used best will depend on the problem at hand. For example, the method shown in Fig. 10 may be used to transform from the time to the frequency domain, and the method shown in Fig. 4 may be used for the inverse transform. Any of the methods described above may be used for the inverse discrete Fourier transform if the coefficients are replaced by their complex conjugates, and if the result of the computation is multiplied by $1/N$.

The six forms mentioned are, in a sense, canonic, but one could also employ a combination of decimation in time and decimation in frequency at different stages in the reduction process, yielding a hybrid signal flow graph.

A Useful Computational Variation: It may be worth pointing out here how some programming simplicity is realized when the factors p and $q=N/p$ are relatively prime. As described by Cooley, Lewis, and Welch, [2], the "twiddle factor" W^{ir} of (23) can be eliminated by choosing subsequences of the X_k's that are different than those used before. The DFT computations are then conveniently performed in two stages.

1) Compute the q-point transforms

$$B_r^{(i)} = \sum_{k=0}^{q-1} Y_k^{(i)} \cdot W^{pkr} \qquad \begin{array}{l} i = 0, 1, \cdots, p-1 \\ r = 0, 1, \cdots, q-1 \end{array} \tag{31}$$

of each of the p sequences

$$Y_k^{(i)} = X_{pk+qi} \qquad \begin{array}{l} i = 0, 1, \cdots, p-1 \\ k = 0, 1, \cdots, q-1. \end{array} \tag{32}$$

2) Compute, then, the p-point transforms

$$A_s = \sum_{i=0}^{p-1} B_r^{(i)} \cdot W^{qim} \tag{33}$$

of the q sequences $B_r^{(i)}$, where

$$s = r \cdot p(p)_q^{-1} + m \cdot q(q)_p^{-1} \qquad (\text{mod } N, 0 \leq s < N) \tag{34}$$

and the notation $(p)_q^{-1}$ is meant to represent the reciprocal of p, mod q, i.e., the solution of $p(p)_q^{-1} > 1$ (mod q).

Conclusion

The integral transform method has been one of the foundations of analysis for many years because of the ease with which the transformed expressions may be manipulated, particularly in such diverse areas as acoustic wave propagation. speech transmission, linear network theory, transport phenomena, optics, and electromagnetic theory. Many problems which are particularly amenable to solution by integral transform methods have not been attacked by this method in the past because of the high cost of obtaining numerical results this way.

The fast Fourier transform has certainly modified the economics of solution by transform methods. Some new applications are presented in this special issue, and further interesting and profitable applications probably will be found during the next few years.

Appendix

As is well known, if the filter impulse response is frequency-band-limited to $1/2T$ Hz and is given by its Nyquist samples Y_h spaced T second apart, and furthermore, if the input waveform is also frequency-band-limited to $1/2T$ Hz and given by its Nyquist samples X_k spaced T second apart, then the filter output waveform is also frequency-band-limited to $1/2T$ Hz and completely specified by its Nyquist samples Z_s spaced T second apart

$$Z_s = \sum_{k=0}^{s} X_k \cdot Y_{s-k} = \sum_{l=0}^{s} X_{s-l} \cdot Y_l. \tag{35}$$

The convolution relationship facilitates computation of this equation.

To prove the convolution relationship, let the DFT of the X_k's be A_r and correspondingly the DFT of the Y_h's be B_r. The IDFT of the product of A_r B_r then becomes [see (4)]

$$\left(\frac{1}{N^2}\right) \sum_{r=0}^{N-1} A_r B_r W^{-rs}$$

$$= \frac{1}{N^2} \sum_{r=0}^{N-1} \sum_{k=0}^{N-1} \sum_{l=0}^{N-1} X_k Y_l W^{r(k+l-s)}$$

$$= \frac{1}{N} \sum_{k=0}^{N-1} \sum_{l=0}^{N-1} X_k Y_l \sum_{r=0}^{N-1} \frac{W^{r(k+l-s)}}{N}$$

$$= \frac{1}{N} \sum_{k=0}^{s} X_k Y_{s-k} + \frac{1}{N} \sum_{k=s+1}^{N-1} X_k Y_{N+s-k}$$

$$= \left(\frac{1}{N}\right) \cdot Z_s + \text{perturbation term.} \tag{36}$$

If the first $N/2$ samples of each of the two time series (X_k) and (Y_h) are assumed to be identically zero, then the perturbation term of (36) is zero so that the IDFT of the product of the two DFTs multiplied by N is equal to the convolution product Z_s of (35). Since it is always possible to select the time series to be convolved such that half of the samples are zero, the convolution relationship for the DFT can be used to compute the convolution product [see (35)] of two time series.

It is useful to point out that if $A_r = B_r$, a periodic autocorrelation function emerges.

References

[1] J. W. Cooley and J. W. Tukey, "An algorithm for the machine calculation of complex Fourier series," *Math. of Comput.*, vol. 19, pp. 297–301, April 1965.

[2] J. W. Cooley, P. A. W. Lewis, and P. D. Welch, "Historical notes on the fast Fourier transform" this issue, p. 76–79.

[3] R. B. Blackman and J. W. Tukey, *The Measurement of Power Spectra*. New York: Dover, 1959.

[4] C. Bingham, M. D. Godfrey, and J. W. Tukey, "Modern techniques of power spectrum estimation," this issue, p. 56–66.

[5] T. G. Stockham, "High speed convolution and correlation," *1966 Spring Joint Computer Conf., AFIPS Proc.*, vol. 28. Washington, D. C.: Spartan, 1966, pp. 229–233.

[6] W. T. Cochran, J. J. Downing, D. L. Davin, H. D. Helms, R. A. Kaenel, W. W. Lang, and D. E. Nelson, "Burst measurements in the frequency domain," *Proc. IEEE*, vol. 54, pp. 830–841, June 1966.

[7] W. M. Gentleman and G. Sande, "Fast Fourier transforms—for fun and profit," *1966 Fall Joint Computer Conf. AFIPS Proc.*, vol. 29. Washington, D. C.: Spartan, 1966, pp. 563–578.

[8] Private communication.

[9] J. W. Cooley, "Applications of the fast Fourier transform method," *Proc. of the IBM Scientific Computing Symp.*, June 1966.

The Finite Fourier Transform

J. W. COOLEY
P. A. W. LEWIS
P. D. WELCH, Member, IEEE
IBM Watson Research Center
Yorktown Heights, N. Y. 10598

Abstract

The finite Fourier transform of a finite sequence is defined and its elementary properties are developed. The convolution and term-by-term product operations are defined and their equivalent operations in transform space are given. A discussion of the transforms of stretched and sampled functions leads to a sampling theorem for finite sequences. Finally, these results are used to give a simple derivation of the fast Fourier transform algorithm.

I. Introduction

In mathematical analysis and in scientific theory it is natural, or perhaps conventional, to think in terms of functions defined on a variable taking values in a continuum. The continuum is usually infinite or doubly infinite in extent and the variable is very often time. However, in numerical mathematics or data analysis we usually deal with ordered sets of numbers—functions of a discrete variable—known as sequences. Moreover, these sets or sequences are finite in extent. The discreteness and finiteness are imposed by practical computing considerations, and by practical limitations on the duration and frequency of experimental observation.

Manuscript received February 21, 1969.

Given that we have to operate on such sequences, say $X(j)$ for $j=0, 1, \cdots, N-1$, it is natural to develop a spectral theory for them, i.e., to define a particular orthogonal transformation which takes the sequence $X(j)$ into another sequence, say $A(n)$, of the same length as $X(j)$ and which describes the frequency structure of $X(j)$. This transformation, called the (discrete) finite Fourier transform, is defined in Section II of this paper.

There are very few specific and extensive expositions of the theory of the finite Fourier transform in the literature and in particular in books on Fourier analysis. This is probably because the theory is relatively simple, and because it can be subsumed under the general theory of Fourier analysis on locally compact Abelian groups [7]. In fact, it will be seen below that both computationally and mathematically the defining relationship of the finite Fourier transform automatically forces the assumption that the sequence $X(j)$ repeats itself periodically on all the integers outside of the range $j=0, 1, \cdots, N-1$. Any other assumption, say that $X(j)$ has the value zero for values of j outside the range $0, 1, \cdots, N-1$, produces a different transform theory, i.e., the theory of Fourier series. The periodic repetition of $X(j)$ is equivalent to considering the range of j to be circularly defined, or the value of j to be considered modulo N. This is such a simple group, and the Fourier theory on it is so straightforward compared to the general theory, that it seems well worth presenting a separate development for the finite Fourier transform.

Schoenberg [8] has given the essentials of the theory of the finite Fourier transform, and has considered the geometric interpretation of the transform and the application of the theory to the solution of several geometric problems. A number of references to earlier applications are given.

The present paper is of a tutorial nature, considering the theory of the finite Fourier transform and the operations that can be performed on it in more detail than is given in Schoenberg [8]. It also presents parts of the theory that relate to applications of the finite Fourier transform in signal processing and computing.

The finite Fourier transform is defined in Section II and elementary properties such as linearity are given in Section III. In Section IV we examine the important concepts

Reprinted from *IEEE Trans. Audio Electroacoust.*, vol. AU-17, pp. 77–85, June 1969.

of convolution and term-by-term product of two sequences and give the equivalent operations in transform space. The section ends with a derivation of Parseval's theorem. Section V considers the transforms of stretched and sampled functions and derives a sampling theorem for finite sequences. In Section VI the previous development is used to give a very simple and illuminating derivation of the fast Fourier transform algorithm. It is, of course, this algorithm which has generated much of the current interest in finite Fourier transforms.

Finally, we note that the presentation of the finite Fourier transform given in this paper can be considered an ideal way of introducing the basic ideas of Fourier analysis to a student or even a scientist who does not yet have (or may never have) the mathematical sophistication to appreciate Fourier theory in its full generality. Thus, all the proofs in this paper are straightforward algebraic manipulations of complex numbers and require no inordinate mathematical sophistication, e.g., the theory of infinite series, or the theory of integration and measure.

II. Definition of the Finite Fourier Transform

Let $X(j), j=0, 1, \cdots, N-1$, be a sequence of N finite valued *complex* numbers. The finite Fourier transform of $X(j)$ is defined as

$$A(n) = \frac{1}{N}\sum_{j=0}^{N-1} X(j)e^{-2\pi inj/N},$$

where $i=(-1)^{1/2}$. From here on in this paper we will let $W_N = e^{2\pi i/N}$. Hence,

$$A(n) = \frac{1}{N}\sum_{j=0}^{N-1} X(j)W_N^{-nj}. \qquad (1)$$

Similarly, $X(j)$ can be expressed as the inverse finite Fourier transform of $A(n)$; i.e.,

$$X(j) = \sum_{n=0}^{N-1} A(n)W_N^{nj} \qquad (j = 0, i, \cdots, N-1). \qquad (2)$$

That (1) and (2) are a transform pair, i.e., that substituting $A(n)$ from (1) into (2) gives back $X(j)$, comes from the following orthogonality relationships of the exponential function W_N^{nj}:

$$\sum_{j=0}^{N-1} W_N^{nj}W_N^{-mj} = \begin{cases} N & \text{if } n \equiv m \pmod N \\ 0 & \text{otherwise.} \end{cases} \qquad (3)$$

In other words, we are saying that the representation of $X(j)$ as finite polynomials in the roots of unity W_N^n is unique (Schoenberg, [8]).

The exponential function W_N^{nj}, as a function of n and j, is periodic of period N; i.e.,

$$W_N^{nj} = W_N^{(n+N)j} = W_N^{n(j+N)}$$

A direct consequence of this is that the sequences $A(n)$ and $X(j)$, as defined by their transforms, (1) and (2), are periodic of period N. From here on we will consider $A(n)$ and $X(j)$ to be defined by (1) and (2) for all integers. Thus, we have $X(j)$ $(j=0, \pm 1, \pm 2, \cdots)$, $A(n)$ $(n=0, \pm 1, \pm 2, \cdots)$, and

$$\begin{aligned} X(j) &= X(kN + j) \qquad (k = 0, \pm 1, \pm 2, \cdots), \\ A(n) &= A(kN + n) \qquad (k = 0, \pm 1, \pm 2, \cdots), \end{aligned} \qquad (4)$$

Another viewpoint is to consider $X(i)$ and $A(n)$ to be defined circularly on integer points of a circle with diameter N. However, the periodic extension (4) of the finite sequences to infinite sequences is more convenient. It simplifies the proofs of the properties of the finite Fourier transform and makes the results easier to visualize. The finite sequences are always recovered by considering the values of the finite sequences at the N points $0, 1, \cdots, N-1$.

Following directly from (4) as special cases are two relationships we will use extensively, namely,

$$\begin{aligned} X(-j) &= X(N - j), \\ A(-n) &= A(N - n). \end{aligned} \qquad (5)$$

The following convention will be useful in the remainder of the paper. By a double-headed arrow connecting two sequences, i.e., by $X(j) \leftrightarrow A(n)$, we indicate that the two sequences are a finite Fourier pair.

III. Elementary Properties of the Finite Fourier Transform

A fundamental property of the finite Fourier transformation is that it is a linear operation. This is expressed by Theorem 1.

Theorem 1

The finite Fourier transform is linear; i.e., if

$$X_1(j) \leftrightarrow A_1(n)$$

and

$$X_2(j) \leftrightarrow A_2(n),$$

then for any complex constants a and b,

$$aX_1(j) + bX_2(j) \leftrightarrow aA_1(n) + bA_2(n).$$

The proof follows directly from the defining equations (1) and (2). Q.E.D.

We now establish some additional results concerning these finite Fourier transform pairs. Denote by $\bar{Y}(j)$ the complex conjugate of $Y(j)$.

Theorem 2

If

$$X(j) \leftrightarrow A(n),$$

then

$$X(-j) \leftrightarrow A(-n).$$

Proof: From the definition (2),

$$X(j) = \sum_{n=0}^{N-1} A(n) W_N^{nj}.$$

Hence

$$X(-j) = \sum_{n=0}^{N-1} A(n) W_N^{-nj}.$$

Now let $m = -n$. Then

$$X(-j) = \sum_{m=0}^{-(N-1)} A(-m) W_N^{mj}.$$

However, because of the periodic nature of the functions $A(-m)$ and W_N^{mj}, the sum over the period $(-N+1, 0)$ is the same as the sum over the period $(0, N-1)$ and we have

$$X(-j) = \sum_{m-0}^{N-1} A(-m) W_N^{mj}. \qquad \text{Q.E.D.}$$

Since the periodicity of $X(j)$ gives the result (5) that $X(-j) = X(N-j)$, Theorem 2 implies that the finite Fourier operation is not time (j) reversible. This property of the transform will be in evidence in Section IV when we consider the convolution and lagged-product operations.

A sequence $Y(j)$ is said to be *even* if $Y(j) = Y(-j)$. A sequence $Y(j)$ is said to be *odd* if $Y(j) = -Y(-j)$. These definitions form the basis for Corollary 1.

Corollary 1

$X(j)$ is even if and only if $A(n)$ is even. $X(j)$ is odd if and only if $A(n)$ is odd.

Proof: The corollary follows directly from Theorem 2. Q.E.D.

The sequences $X(j)$ and $A(n)$ are periodic of period N. Now if $Y(j)$ is even and periodic of period N, then $Y(j) = Y(-j) = Y(N-j)$. If $Y(j)$ is odd and periodic of period N, then $Y(j) = -Y(-j) = -Y(N-j)$. Hence, over the interval $[0, N-1]$ the evenness and oddness of sequences such as $X(n)$ and $A(n)$ show up in relationships or symmetries between the values of the sequences at j and $N-j$.

We now consider the relationship between the finite Fourier transform of a sequence $X(j)$ and the finite Fourier transform of its (term-by-term) complex conjugate.

Theorem 3

If

$$X(j) \leftrightarrow A(n),$$

then

$$\tilde{X}(j) \leftrightarrow \tilde{A}(-n)$$

and

$$\tilde{X}(-j) \leftrightarrow \tilde{A}(n).$$

Proof: From the definition (2),

$$X(j) = \sum_{n=0}^{N-1} A(n) W_N^{nj}.$$

Hence

$$\tilde{X}(j) = \sum_{n=0}^{N-1} \tilde{A}(n) W_N^{-nj}.$$

Now let $m = -n$. Then

$$\tilde{X}(j) = \sum_{m=0}^{-(N-1)} \tilde{A}(-m) W_N^{mj} = \sum_{m=0}^{N-1} \tilde{A}(-m) W_N^{mj}.$$

This establishes the first half of the theorem. The second half follows from Theorem 2. Q.E.D.

Corollary 2

1) $X(j)$ is real if and only if $A(n) = \tilde{A}(-n) = \tilde{A}(N-n)$.
2) $A(n)$ is real if and only if $X(j) = \tilde{X}(-j) = \tilde{X}(N-j)$.
3) $X(j)$ is pure imaginary if and only if $A(n) = -\tilde{A}(-n) = -\tilde{A}(N-m)$.
4) $A(n)$ is pure imaginary if and only if $X(j) = -\tilde{X}(-j) = -\tilde{X}(m-j)$.

Proof of 1): Suppose $X(j)$ is real; then $X(j) = \tilde{X}(j)$ and, from Theorem 3, $A(n) = \tilde{A}(-n)$. Suppose $A(n) = \tilde{A}(-n)$; then, from Theorem 3, $X(j) = \tilde{X}(j)$, and hence $X(j)$ is real.

Proof of 3): Suppose $X(j)$ is pure imaginary; then $X(j) = -\tilde{X}(j)$ and, from Theorem 3, $A(n) = -\tilde{A}(-n)$. Suppose $A(n) = -\tilde{A}(-n)$; then, from Theorem 3, $\tilde{X}(j) = -\tilde{X}(j)$, and hence $X(j)$ is pure imaginary.

The proofs of 2) and 4) are similar. Q.E.D.

Corollary 3

1) $X(j)$ is real and even if and only if $A(n)$ is real and even.
2) $X(j)$ is real and odd if and only if $A(n)$ is pure imaginary and odd.
3) $X(j)$ is pure imaginary and even if and only if $A(n)$ is pure imaginary and even.
4) $X(j)$ is pure imaginary and odd if and only if $A(n)$ is real and odd.

Proof of 1): Suppose $X(j)$ is real and even; then, from Corollary 1, $A(n)$ is even, and, from Corollary 2, $A(n) = \tilde{A}(-n)$. Hence, $A(n) = A(-n)$ and $A(n) = \tilde{A}(-n)$ or $A(n) = \tilde{A}(n)$ and $A(n)$ is real. The converse is established in a parallel fashion.

2), 3), and 4) are proved similarly. Q.E.D.

Using the linearity of the transform and Corollary 2 it is easy to show how one can obtain the transform of two real functions simultaneously. Let $X_1(j)$ and $X_2(j)$ be real and let $X_1(j) \leftrightarrow A_1(n)$ and $X_2(j) \leftrightarrow A_2(n)$. If we form the function $X(j) = X_1(j) + iX_2(j)$ and let $A(n)$ be its transform, then from the linearity property,

$$A(n) = A_1(n) + iA_2(n).$$

Clearly we have

$$A(N-n) = A_1(N-n) + iA_2(N-n)$$

and

$$\tilde{A}(N-n) = \tilde{A}_1(N-n) + \tilde{i}\tilde{A}_2(N-n)$$
$$= \tilde{A}_1(N-n) - i[\tilde{A}_2(N-n)].$$

However, since $X_1(j)$ and $X_2(j)$ are real we have, by 1) of Corollary 2, that $\tilde{A}_1(N-n) = A_1(n)$ and $\tilde{A}_2(N-n) = A_2(n)$, so that the expression for $\tilde{A}(N-n)$ becomes

$$\tilde{A}(N-n) = A_1(n) - iA_2(n).$$

Combining this with the expression for $A(n)$ above we obtain $A_1(n)$ and $A_2(n)$ as

$$A_1(n) = \frac{A(n) + \tilde{A}(N-n)}{2} \tag{6}$$

$$A_2(n) = \frac{A(n) - \tilde{A}(N-n)}{2i}. \tag{7}$$

The following theorem describes the behavior of the Fourier pair when one of the sequences is shifted along its j (time) axis or n (frequency) axis.

Theorem 4

If

$$X(j) \leftrightarrow A(n),$$

then

$$X(j-k) \leftrightarrow W_N^{-nk} A(n)$$

and

$$W_N^{mj} X(j) \leftrightarrow A(n-m).$$

Proof:

$$X(j-k) = \sum_{n=0}^{N-1} A(n) W_N^{n(j-k)}$$
$$= \sum_{n=0}^{N-1} \{A(n) W_N^{-nk}\} W_N^{nj}.$$

Therefore, $X(j-k) \leftrightarrow W_N^{-nk} A(n)$ by the definition (2). The other half of the theorem follows similarly. Note that shifting the sequence $X(j)$ along its time axis changes the phase but not the amplitudes of the components of the sequence $A(n)$. Q.E.D.

We define the sequence $\delta(j)$ by

$$\begin{aligned} \delta(j) &= 1 \quad \text{if } j \equiv 0 \pmod{N} \\ \delta(j) &= 0 \quad \text{otherwise.} \end{aligned} \tag{8}$$

Then we have Theorem 5.

Theorem 5

$$\delta(j) \leftrightarrow 1/N$$
$$1 \leftrightarrow \delta(n).$$

Proof: The proof follows directly from the definitions. Q.E.D.

This theorem is very useful. For example, suppose we have the transform of $X(j)$ and wish the transform of $X(j) - a$; that is, we wish to move $X(j)$ up or down. Then, if $X(j) \leftrightarrow A(n)$, $X(j) - a \leftrightarrow A(n) - a\delta(n)$. Now $A(n) - a\delta(n)$ is just $A(n)$ with $A(0)$ replaced by $A(0) - a$. Hence, subtracting a constant from all the values of $X(j)$ is equivalent, in the frequency domain, to subtracting that same constant only from $A(0)$.

Theorem 6

If

$$X(j) \leftrightarrow A(n),$$

then

$$X(0) = \sum_{n=0}^{N-1} A(n) \tag{9}$$

and

$$A(0) = \frac{1}{N} \sum_{j=0}^{N-1} X(j). \tag{10}$$

Proof: The proof follows directly from the definitions (1) and (2). Q.E.D.

This result is useful in statistical applications. If $X(j)$ is a set of observations of a random variable, then $A(0)$ is the sample mean.

IV. Convolutions and Term-by-Term Products of Sequences

Next we will treat the subject of the transform of the term-by-term product of two sequences. Let $X_1(j)$ and $X_2(j)$ be two sequences with finite Fourier transforms $A_1(n)$ and $A_2(n)$. The term-by-term product of the sequences $A_1(n)$ and $A_2(n)$ is the sequence whose nth term is $A_1(n)A_2(n)$. The inverse transform (2) of this product sequence turns out to be the *convolution* of the sequences $X_1(j)$ and $X_2(j)$ as in the usual Fourier analysis. Similarly, the transform of the product sequence $X_1(j)X_2(j)$ is the convolution of $A_1(n)$ and $A_2(n)$. This is one of the most important properties of the finite Fourier transform. To be explicit, we have the following theorem.

Theorem 7

If

$$X_1(j) \leftrightarrow A_1(n)$$

and

$$X_2(j) \leftrightarrow A_2(n),$$

then

$$\frac{1}{N}\sum_{k=0}^{N-1} X_1(k)X_2(j-k) = \frac{1}{N}\sum_{k=0}^{N-1} X_1(j-k)X_2(k) \leftrightarrow A_1(n)A_2(n) \tag{11}$$

and

$$\begin{aligned} X_1(j)X_2(j) &\leftrightarrow \sum_{m=0}^{N-1} A_1(m)A_2(n-m) \\ &= \sum_{m=0}^{N-1} A_1(n-m)A_2(m). \end{aligned} \tag{12}$$

Proof: By direct substitution from (2),

$$\frac{1}{N}\sum_{k=0}^{N-1} X_1(k)X_2(j-k)$$

$$= \frac{1}{N}\sum_{k=0}^{N-1}\left\{\sum_{n=0}^{N-1} A_1(n)W_N^{nk}\right\}\left\{\sum_{m=0}^{N-1} A_2(m)W_N^{m(j-k)}\right\}$$

$$= \frac{1}{N}\sum_{n=0}^{N-1}\sum_{m=0}^{N-1} A_1(n)A_2(m)W_N^{mj}\left\{\sum_{k=0}^{N-1} W_N^{k(n-m)}\right\}.$$

Now, using the orthogonality relationships (3), the sum in the brackets is nonzero and equal to N if and only if $n=m$. Thus

$$\frac{1}{N}\sum_{k=0}^{N-1} X_1(k)X_2(j-k) = \sum_{n=0}^{N-1} A_1(n)A_2(n)W_N^{nj}.$$

The other half of the theorem is proved similarly.

Q.E.D.

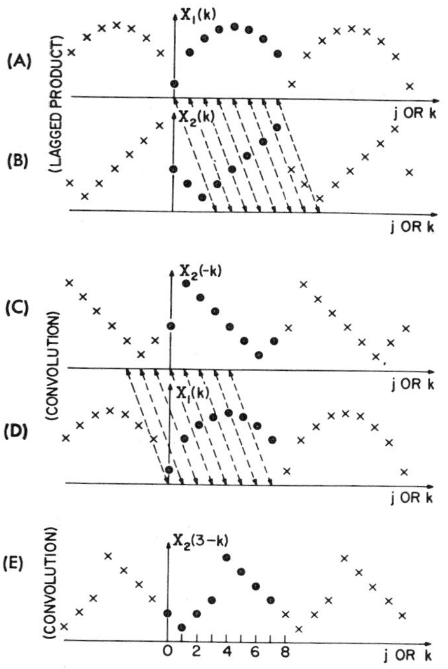

Fig. 1. The lagged-product and convolution operations on two sequences. The fourth term ($j=3$) in the lagged-product sequence is the sum of the products of the terms in $X_1(k)$ and $X_2(k)$ connected by the arrows between parts (A) and (B). The fourth term in the convolution sequence is the sum of the products of the terms in $X_1(k)$ and $X_2(k)$ connected by the arrows between parts (C) and (D). The shifted sequence $X_2(3-k)$ is shown in part (E). The convolution is the sum of products of matching terms in parts (D) and (E).

Thus, in the theory of the (discrete) finite Fourier transformation, we have convolution corresponding to multiplication. The equalities on the left side of (11) and on the right side of (12) indicate that convolution is a commutative operation. This can easily be proved directly, but once the theorem is proved it follows immediately from the commutativity of the multiplication operation. Note also that the convolution defined in (11) and (12) is cyclic; i.e., when one sequence moves over the end of the other, it does not encounter zeros, but rather the periodic extension of the sequence. This is consistent with the periodic extension property of the finite Fourier transform which was noted in Section II.

The detailed makeup of the convolution sequence is worth examining. For instance, its first term ($j=0$) is the sum of the term-by-term products of the members of the sequence $X_1(k)$ and the members of the sequence $X_2(-k)$. Its second term ($j=1$) is the sum of the term-by-term products of the members of the sequence $X_1(k)$ and members of the sequence $X_2(-k)$ shifted forward (cyclically) in time by one step or shifted backward cyclically $N-1$ steps. If viewed in terms of a sequence defined only for the values $0, 1, \cdots, N-1$, the convolution is said to "wrap around." For $N=8$ and $j=3$. this "wrap around" is illustrated in parts (C) and (D) of Fig. 1. Thus the fourth term ($j=3$) in the convolution is $X_1(0)\cdot X_2(3) + X_1(1)\cdot X_2(2) + X_1(2)\cdot X_2(1) + \cdots + X_1(N-1)\cdot X_2(4-N)$. This is the sum of the products of the sequence values connected by the arrows. The commutativity of the convolution operation is obvious since shifting $X_2(-j)$ forward cyclically in time by 1 step is equivalent to shifting $X_1(j)$ back (cyclically) in time by $N-1$ steps in the operation.

Now consider the operation of forming the *lagged product* of two sequences. To get the lagged product of two sequences, we multiply the transform of the one by the transform of the other with a negative argument. This is made explicit in the following corollary.

Corollary 4

If

$$X_1(j) \leftrightarrow A_1(n)$$

and

$$X_2(j) \leftrightarrow A_2(n),$$

then

$$\frac{1}{N}\sum_{k=0}^{N-1} X_1(k+j)X_2(k) = \frac{1}{N}\sum_{k=0}^{N-1} X_1(k)X_2(k-j) \leftrightarrow A_1(n)A_2(-n), \tag{13}$$

$$\frac{1}{N}\sum_{k=0}^{N-1} X_1(k)X_2(k+j) = \frac{1}{N}\sum_{k=0}^{N-1} X_1(k-j)X_2(k) \leftrightarrow A_1(-n)A_2(n), \tag{14}$$

$$X_1(j)X_2(-j) \leftrightarrow \sum_{m=0}^{N-1} A_1(m+n)A_2(m) = \sum_{m=0}^{N-1} A_1(m)A_2(m-n), \tag{15}$$

$$X_1(-j)X_2(j) \leftrightarrow \sum_{m=0}^{N-1} A_1(m)A_2(m+n) = \sum_{m=0}^{N-1} A_1(m-n)A_2(m). \tag{16}$$

Proof: The corollary follows directly from Theorems 7 and 2, since the lagged product of $X_1(j)$ and $X_2(j)$ is just the convolution of $X_1(j)$, with transform $A_1(n)$, and the reverse of $X_2(j)$, namely $X_2(N-j)$, with transform $A_2(-n)$.
Q.E.D.

While the proof of the corollary makes the relationship between the lagged product and convolution operations evident, there are essential differences. In particular, the lagged product operation is *not* commutative.

The convolution and lagged product operations are illustrated in Fig. 1. The sequence $X_1(k)$ is shown in part (A), with the periodic repetition outside the range k (or j) $=0, 1, \cdots, N-1$. In part (B) we show $X_2(k)$; the fourth term ($j=3$) in the lagged product sequence (14) is proportional to the sum of the products of the terms connected by arrows. Alternatively, we could shift $X_2(k)$ back three points and form the sum of the products of terms at matching time points, i.e., $X_1(0)X_2(0+3)+X_1(1)X_2(1+3) + \cdots + X_1(N-1)X_2(N-1+3)$. The time reversed sequence $X_2(-k)=X_2(N-k)$ is shown in part (C) and $X_1(k)$ is repeated under it in part (D). The fourth term ($j=3$) in the convolution is then the lagged product of the sum of the products of the terms connected by arrows. Again we could shift $X_2(-k)$ forward in time by three points to get $X_2(-(k-3))=X_2(3-k)$, as shown in part (E), and form the sum of the products of terms at matching time points, i.e., $X_1(0)X_2(3-0)+X_1(1)X_2(3-1) + \cdots + X_1(N-1)\cdot X_2(2-N)$. Note the wrap-around effect; the convolution and lagged product operations we have defined are cyclic operations.

With regard to applying Corollary 4 it should be kept in mind that $A(-n)=A(N-n)$, giving, for (13),

$$\frac{1}{N}\sum_{k=0}^{N-1} X_1(k+j)X_2(k) \leftrightarrow A_1(n)A_2(N-n).$$

Also, if $X(j)$ is real, then $A(-n)=A(N-n)=\tilde{A}(n)$ and we have, for (13),

$$\frac{1}{N}\sum_{k=0}^{N-1} X_1(k+j)X_2(k) \leftrightarrow A_1(n)\tilde{A}_2(n).$$

Among other things, the last expression and Corollary 3 show that the lagged product (13) and convolution are the same if $X_2(j)$ is not only real but also even. The lagged product (14) is equal to convolution if $X_1(j)$ is not only real but also even. If both series are real and even, all three series are equal.

The very important Parseval's theorem can be obtained from the preceding results.

Corollary 5 (Parseval's Theorem)

If

$$X(j) \leftrightarrow A(n),$$

then

$$\frac{1}{N}\sum_{j=0}^{N-1} |X(j)|^2 = \sum_{n=0}^{N-1} |A(n)|^2. \tag{17}$$

Proof: Let

$$X_1(j) = X(j)$$

and

$$X_2(j) = \tilde{X}(j);$$

then, from (13) of Corollary 4 and the first part of Theorem 3, we have

$$\frac{1}{N}\sum_{k=0}^{N-1} X(k+j)\tilde{X}(k) \leftrightarrow A(n)\tilde{A}(n).$$

The corollary now follows directly from Theorem 6.
Q.E.D.

V. The Transforms of Stretched and Sampled Functions

Given a sequence $X(j): j=0, 1, \cdots, N-1$, we define the sequence $\text{Stretch}_K\{j: X\}$ for any positive integer K to be $X(j)$ stretched by a factor K with zeros filling in the gaps; i.e.,

$$\text{Stretch}_K\{j: X\} = X(j/K) \quad \text{for } j = lK, \quad l = 0, \cdots, N-1; \quad = 0 \quad \text{otherwise}. \tag{18}$$

For $N=8$ and $K=2$, the definition is illustrated in Fig. 2. Note that $\text{Stretch}_K\{j: K\}$ is a sequence of length NK.

Theorem 8

If

$$X(j) \leftrightarrow A(n), \quad (n = 0, 1, \cdots, N-1; \; j = 0, 1, \cdots, N-1),$$

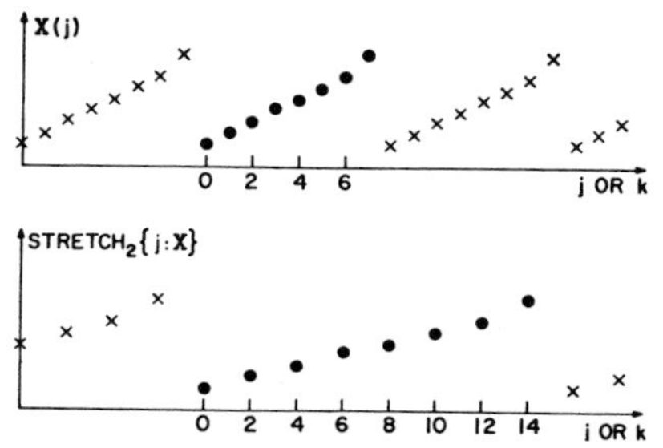

Fig. 2. The stretch function illustrated for $K=2$ on a sequence $X(j)$ of length $N=8$.

then

$$\text{Stretch}_K\{j:X\} \leftrightarrow A(n)/K \qquad (n = 0, 1, \cdots, NK-1)$$
$$X(j) \leftrightarrow \text{Stretch}_K\{n:A\} \qquad (j = 0, 1, \cdots, NK-1).$$

Proof:

$$\frac{1}{KN}\sum_{j=0}^{KN-1} \text{Stretch}_K\{j:X\} W_{KN}^{-nj}$$
$$= \frac{1}{KN}\sum_{j'=0}^{N-1} X(j') W_{KN}^{-Kj'n}$$

where $j'=j/K$. Since $W_{KN}^{K}=W_N$, this becomes

$$\frac{1}{K}\frac{1}{N}\sum_{j'=0}^{N-1} X(j') W_N^{-nj'} = A(n)/K.$$

The other half of the theorem is proved similarly.

Q.E.D.

Thus, any sequence which is nonzero only at multiples of an integer K has a transform which is periodic with a period $1/K$ times the length of the sequence of nonzero elements. Conversely, any sequence with a period N/K has a transform which has only N/K equally spaced nonzero values.

Given a sequence $X(j)$, $(j=0, 1, \cdots, N-1)$, and a positive integer K which divides N, we define the sequence

$$\text{Sample}_K\{j:X\} \qquad \left(j = 0, 1, \cdots, \frac{N}{K} - 1\right) \tag{19}$$

by

$$\text{Sample}_K\{j:X\} = X(jK).$$

This sequence, $\text{Sample}_K\{j:K\}$, of period N/K, is initially defined for $j=0, 1, \cdots, (N/K)-1$, but extended to all integers in a periodic fashion.

Theorem 9 (Sampling Theorem)

If

$$X(j) \leftrightarrow A(n),$$

then

$$\text{Sample}_K\{j:X\} \leftrightarrow \sum_{m=0}^{K-1} A\left(n + \frac{mN}{K}\right)$$

and

$$\frac{1}{K}\sum_{k=0}^{K-1} X\left(j + \frac{kN}{K}\right) \leftrightarrow \text{Sample}_K\{n:A\}$$

where the sequences have period N/K.

Further, if $A(n)=0$ for $N/K \le n \le N-1$, then $\text{Sample}_K\{j:X\}$ completely determines $X(j)$ through the relationship

$$X(j) = \sum_{l=0}^{(N/K)-1} \text{Sample}_K\{l:X\}\, G(j - lK)$$

where

$$G(j) = 1 \qquad \text{if } j \equiv 0 \pmod N$$

and

$$G(j) = \frac{K}{N}\frac{1 - W_K^{j}}{1 - W_N^{j}} \qquad \text{if } j \not\equiv 0 \pmod N.$$

Proof:

$$\text{Sample}_K\{j:X\} = X(jK) = \sum_{n=0}^{N-1} A(n) W_N^{jnK}$$
$$= \sum_{n=0}^{N-1} A(n) W_{N/K}^{nj}.$$

However, $W_{N/K}^{nj}$ is periodic of period N/K; hence, letting $n=(N/K)\,m+n'$,

$$\text{Sample}_k\{j:X\} = \sum_{n'=0}^{(N/K)-1}\left\{\sum_{m=0}^{K-1} A\left(n' + \frac{mN}{K}\right)\right\} W_{N/K}^{n'j}.$$

The other half of the first part of the theorem is established similarly.

Q.E.D.

To establish the second part of the theorem, suppose $A(n)=0$ for $N/K \le n \le N-1$; then

$$X(j) = \sum_{n=0}^{(N/K)-1} A(n) W_N^{nj}. \tag{20}$$

However,

$$A(n) = \sum_{m=0}^{K-1} A\left(n + \frac{mN}{K}\right),$$

so that from the first part of the theorem, $A(n)$ and $\text{Sample}_K\{j:X\}$ are a finite Fourier transform pair. By (1), then,

$$A(n) = \frac{K}{N}\sum_{l=0}^{(N/K)-1} \text{Sample}_K\{l:X\} W_{N/K}^{-nl} \qquad (n = 0, 1, \cdots, N/K - 1). \tag{21}$$

Hence by substitution of (21) in (20),

$$X(j) = \sum_{n=0}^{(N/K-1)} \frac{K}{N} \sum_{l=0}^{(N/K)-1} \text{Sample}_K \{l: X\} W_{N/K}^{-nl} W_N^{nj}$$

$$= \sum_{l=0}^{(N/K)-1} \text{Sample}_K \{l: X\} \left[\frac{K}{N} \sum_{n=0}^{(N/K)-1} W_{N/K}^{-nl} W_N^{nj} \right].$$

Now

$$\frac{K}{N} \sum_{n=0}^{(N/K)-1} W_{N/K}^{-nl} W_N^{nj} = \frac{K}{N} \sum_{n=0}^{(N/K)-1} W_N^{n(j-lK)}$$

$$= \frac{K}{N} \frac{1 - W_N^{N(j-lK)/K}}{1 - W_N^{j-lK}}$$

$$= \frac{K}{N} \frac{1 - W_K^{j-lK}}{1 - W_N^{j-lK}}$$

$$= G(j - lK)$$

where

$$G(j) = \frac{K}{N} \sum_{n=0}^{(N/K)-1} e^{2\pi i n j/N} = \frac{K}{N} \sum_{n=0}^{(N/K)-1} W_N^{nj}.$$

The sequence $G(j)$ can be easily transformed to the form given in the statement of the theorem. Q.E.D.

Note that the first part of Theorem 9 says that we can obtain the transform of a sampling of a sequence $X(j)$ at a sampling period which divides the period of $X(j)$ by simply overlapping and summing the transform of $X(j)$. This overlapped and summed version of $A(n)$ is called an "aliased" version of $A(n)$. One example of an application of the second part of Theorem 9 is in the situation where one would be satisfied with lower resolution in the frequency domain than that given by a full N-point Fourier transformation. In this case an N/K-point Fourier transform of the "aliased" version of $X(j)$ gives values of $A(n)$ at every Kth point, with a corresponding saving in computation time.

Now let $Z(j)$: $j = 0, 1, 2, \cdots$, be an essentially infinite series of complex numbers. Let $X^{(K)}(j)$ denote the sequence of N $Z(j)$'s which starts with the Kth $Z(j)$. Thus, $X^{(0)}(j)$ is the sequence $Z(0), Z(1), \cdots, Z(N-1)$. Let $A^{(K)}(n)$ be the corresponding finite Fourier transform. It is sometimes required to compute the transform of $X^{(K)}(j)$ and then update it in the sense of computing the transform of $X^{(K+1)}(j)$, $X^{(K+2)}(j), \cdots$, i.e., compute successively $A^{(K)}(n)$, $A^{(K+1)}(n), \cdots$. This is easily accomplished as shown by Theorem 10.

Theorem 10

If

$$X^{(K)}(j) \leftrightarrow A^{(K)}(n),$$

then

$$X^{(K+1)}(j) \leftrightarrow W_N^n \left[A^{(K)}(n) + \frac{\{Z(K+N) - Z(K)\}}{N} \right].$$

Proof: Let

$$Y^{(K)}(j) = X^{(K)}(j) + \{Z(N+K) - Z(K)\}\delta(j);$$

then

$$X^{(K+1)}(j) = Y^{(K)}(j+1).$$

From Theorems 1 and 5, the transform of $Y^K(j)$ is

$$A^{(K)}(n) + \frac{Z(N+K) - Z(K)}{N}.$$

The result then follows from Theorem 4. Q.E.D.

VI. The Fast Fourier Transform Algorithm

The results of the previous sections will be used now to give an elementary derivation of the fast Fourier transform algorithm [2].

Suppose that we have a sequence $X(j)$ $(j = 0, \cdots, N-1)$, that $N = rs$, and that we wish to take its finite Fourier transform. If we use the formula

$$A(n) = \frac{1}{N} \sum_{j=0}^{N-1} X(j) W_N^{-nj}, \qquad (22)$$

we will require N^2 operations in the computation, where by an operation we mean a (complex) multiplication and addition. However, if we let $X_k(j)$ be a sequence s units long obtained by taking the kth, $r+k$th, $2r+k$th, etc., value of $X(j)$, i.e.,

$$X_k(j) = X(jr + k) \qquad (j = 0, \cdots, s-1),$$
$$(k = 0, 1, \cdots, r-1),$$

then we can express $X(j)$ as

$$X(j) = \sum_{k=0}^{r-1} \text{Stretch}_r \{j - k; X_k\}.$$

Now if we let $A_k(n)$ be the finite Fourier transform of $X_k(j)$, we have from Theorems 1, 4, and 8,

$$A(n) = \frac{1}{r} \sum_{k=0}^{r-1} A_k(n) W_N^{-kn} \qquad (23)$$
$$(n = 0, \cdots, N-1).$$

Now each $A_k(n)$ requires s^2 operations and there are r of them so that they require altogether rs^2 operations. Once we have the $A_k(n)$, then we have, from (23), that to calculate each element of the sequence $A(n)$ we require r operations. Since there are rs of these $A(n)$, their calculation requires r^2s operations. Hence, all told, $rs^2 + r^2s = N(r+s)$ operations are required. Finally, $N(r+s) < N^2 = N(rs)$ and we see that this latter method of calculating $A(n)$ is more economical. This procedure repeated on r and s until prime factors are reached is the fast Fourier transform algorithm. A history of the algorithm is given by Cooley, Lewis, and Welch [3]. Greater detail can be found in [1].

Acknowledgment

The authors would like to acknowledge their general indebtedness to H. Helms, C. Rader, G. Sande, and T. Stockham for information received informally on this subject, and to Prof. T. J. Schoenberg for the reference to his work. In addition, we would like to acknowledge indebtedness to papers of Sande [5], Stockham [6], and Gentleman and Sande [4], where portions of the material in our paper have appeared.

References

[1] G-AE Subcommittee on Measurement Concepts, "What is the fast Fourier transform ?," *IEEE Trans. Audio and Electroacoustics*, vol. AU-15, pp. 45–55, June 1967.

[2] J. W. Cooley and J. W. Tukey, "An algorithm for the machine calculation of complex Fourier series," *Math. Comp.*, vol. 19, pp. 297–301, 1965.

[3] J. W. Cooley, P. A. W. Lewis, and P. D. Welch, "Historical notes on the fast Fourier transform," *IEEE Trans. Audio and Electroacoustics*, vol. AU-15, pp. 76–79, June 1967.

[4] W. M. Gentleman and G. Sande, "Fast Fourier transforms for fun and profit," *1966 Fall Joint Computer Conf., AFIPS Proc.*, vol. 29. Washington, D. C.: Spartan, 1966, pp. 563–578.

[5] G. Sande, "On an alternative method of calculating covariance functions," Princeton University, Princeton, N. J., unpublished Tech. Note, 1965.

[6] T. G. Stockham, Jr., "High speed convolution and correlation," *1966 Spring Joint Computer Conf., AFIPS Proc.*, vol. 28. Washington, D. C.: Spartan, 1966, pp. 229–233.

[7] W. Rudin, *Fourier Analysis on Groups.* New York: Interscience, 1962.

[8] I. J. Schoenberg, "The finite Fourier series and elementary geometry," *Am. Math. Monthly*, vol. 57, no. 6, June–July 1950

Historical Notes on the Fast Fourier Transform

JAMES W. COOLEY, PETER A. W. LEWIS, AND PETER D. WELCH

***Abstract*—The fast Fourier transform algorithm has a long and interesting history that has only recently been appreciated. In this paper, the contributions of many investigators are described and placed in historical perspective.**

Historical Remarks

THE FAST FOURIER transform (FFT) algorithm is a method for computing the finite Fourier transform of a series of N (complex) data points in approximately $N \log_2 N$ operations. The algorithm has a fascinating history. When it was described by Cooley and Tukey [1] in 1965 it was regarded as new by many knowledgeable people who believed Fourier analysis to be a process requiring something proportional to N^2 operations with a proportionality factor which could be reduced by using the symmetries of the trigonometric functions. Computer programs using the N^2-operation methods were, in fact, using up hundreds of hours of machine time. However, in response to the Cooley–Tukey paper, Rudnick [5], of Scripps Institution of Oceanography, La Jolla, Calif., described his computer program which also takes a number of operations proportional to $N \log_2 N$ and is based on a method published by Danielson and Lanczos [2]. It is interesting that the Danielson-Lanczos paper described the use of the method in X-ray scattering problems, an area where, for many years after 1942, the calculations of Fourier transforms presented a formidable bottleneck to researchers who were unaware of this efficient method. Danielson and Lanczos refer to Runge [6], [7] for the source of their method. These papers and the lecture notes of Runge and König [8] describe the procedure in terms of sine-cosine series. The greatest emphasis, however, was on the computational economy that could be derived from the *symmetries* of the sine and cosine functions. In a relatively short section of Runge and König [8] it was shown how one could use the *periodicity* of the sine-cosine functions to obtain a $2N$-point Fourier analysis from two N-point analyses with only slightly more than N operations. Going the other way, if the series to be transformed is of length N and N is a power of 2, the series can be split into $\log_2 N$ subseries and this doubling algorithm can be applied to compute the finite Fourier transform in $\log_2 N$ doublings. The number of computations in the resulting successive doubling algorithm is therefore proportional to $N \log_2 N$ rather than N^2. The use of symmetries only reduces the proportionality factor while the successive doubling algorithm replaces N^2 by $N \log N$. This distinction was not important for the values of N used in the days of Runge and König. However, when the advent of computing machinery made calculations with large N possible, and the $N \log N$ methods should have been thoroughly exploited, they were apparently overlooked, even though they had been published by well-read and well-referenced authors.

The fast Fourier transform algorithm of Cooley and Tukey [1] is more general in that it is applicable when N is composite and not necessarily a power of 2. Thus, if two factors of N are used, so that $N = r \times s$, the data is, in effect, put in an r-column, s-row rectangular array, and a two-dimensional transform is performed with a phase-shifting operation intervening between the transformations in the two dimensions. This results in $N(r+s)$ operations instead of N^2. By selecting N to be highly composite, substantial savings result. For the very favorable situation when N is equal to a power of 2, the Cooley–Tukey method is essentially the successive doubling algorithm mentioned above and takes $N \log_2 N$ operations.

The 23-year hiatus in the use of the algorithm seemed quite remarkable, and prompted us to inquire of Prof. L. H. Thomas at the IBM Watson Scientific Computing Laboratory, New York City, N. Y., as to whether he was familiar with the successive doubling algorithm for computing Fourier series, and knew of any occasions when it had been used. It turned out that Prof. Thomas had spent three months in 1948 doing calculations of Fourier series on a tabulating machine, using what he referred to as the "Stumpff method of subseries." The algorithm described by Thomas [10] was thought at first to be essentially the same as the fast Fourier transform algorithm of Cooley and Tukey since it also achieved its economy by performing one-dimensional Fourier analysis by doing multidimensional Fourier analysis. However, the algorithms are different for the following reasons: 1) in the Thomas algorithm the factors of N must be mutually prime; 2) in the Thomas algorithm the calculation is precisely multidimensional Fourier analysis with no intervening phase shifts or "twiddle factors" as they have been called; and 3) the

Manuscript received January 26, 1967; revised March 6, 1967.
The authors are with the IBM Research Center, Yorktown Heights, N. Y.

Reprinted from *IEEE Trans. Audio Electroacoust.*, vol. AU-15, pp. 76–79, June 1967.

correspondences between the one-dimensional index and the multidimensional indexes in the two algorithms are quite different. The Thomas or "prime factor" algorithm is described in detail and compared with the fast Fourier transform algorithm in the next section.[1] It can be extremely useful when used in combination with the fast Fourier transform algorithm.

Several other calculations have been reported in the literature and in private communications which use one or the other of the two algorithms.

Another line of development has since led to the Thomas algorithm in its full generality. This comes from work in the analysis and design of experiments. Let $A(k_0, k_1, \cdots, k_{m-1})$ be, for example, a crop yield when a level k_i of treatment i, which may be an amount of fertilizer, is used. Yates [11] considered the case where $k_i=0$ or 1, meaning treatment i is or is not used. This yields $N=2^m$ values of crop yields and, to get all possible differences between all possible averages, one would, in principle, have to compute N linear combinations of all of the A's. This would require N^2 operations. Yates devised a scheme whereby one computed a new array of N sums and differences of pairs of the A's. The process was repeated on the new array with pairs selected in a different order. This was done $m=\log_2 N$ times, meaning he did the calculation in $N \log_2 N$ operations instead of N^2.

Good [4] noted that the Yates method could be regarded as m-dimensional Fourier analysis with only two points in each direction and that the procedure could be generalized to one for an arbitrary number of points in each direction. Then Good showed that if N is composite, with mutually prime factors, i.e., $N=r_1r_{,2}, \cdots, r_m$, one could do a one-dimensional Fourier analysis of N points by doing m-dimensional Fourier analysis on an m-dimensional, $r_1 \times r_2 \times, \cdots, \times r_m$, array of points. With these ideas put together and developed, Good's paper contains the full generalization of the Thomas prime factor algorithm.

The Prime Factor Algorithm

As mentioned in the previous section, the algorithm used by Thomas and described later by Good has been mistakenly said to be equivalent to the fast Fourier transform algorithm of Cooley and Tukey. It is important to distinguish between these two algorithms since each has its particular advantages which can be exploited in appropriate circumstances.

The differences will be illustrated by considering the calculation of a Fourier series using two factors of N.

TABLE I

Correspondence between One- and Two-Dimensional Indexing in the Arbitrary Factor Algorithm for the Case $r=8$, $s=3$, and $N=24$

$n=sn_1+n_0=3n_1+n_0$

	n_1: 0	1	2	3	4	5	6	7
n_0 = 0	0	3	6	9	12	15	18	21
1	1	4	7	10	13	16	19	22
2	2	5	8	11	14	17	20	23

$j=rj_1+j_0=8j_1+j_0$

	j_0: 0	1	2	3	4	5	6	7
j_1 = 0	0	1	2	3	4	5	6	7
1	8	9	10	11	12	13	14	15
2	16	17	18	19	20	21	22	23

The Fourier series is

$$X(j) = \sum_{n=0}^{N-1} A(n) W_N^{jn} \tag{1}$$

where $W_N=e^{2\pi i/N}$. Consider first the fast Fourier transform algorithm. We assume $N=r\cdot s$, and define a one-to-one mapping between the integers j, $0 \le j < N$, and the pairs of integers (j_1, j_0), $0 \le j_0 < r$, $0 \le j_1 < s$, by the relation

$$j = j_1 r + j_0. \tag{2}$$

Similarly, we let

$$n = n_1 s + n_0, \tag{3}$$

where

$$0 \le n < N, \qquad 0 \le n_0 < s, \qquad 0 \le n_1 < r.$$

This enables us to refer to $A(n)$ and $X(j)$ as though they were two-dimensional arrays and permits us to do the Fourier analysis in two steps

$$A_1(j_0, n_0) = \sum_{n_1=0}^{r-1} A(n_1, n_0) W_r^{j_0 n_1} \tag{4}$$

$$X(j_1, j_0) = \sum_{n_0=0}^{s-1} A_1(j_0, n_0) W_s^{j_1 n_0} W_N^{j_0 n_0}. \tag{5}$$

Table I shows where $A(n)$ and $X(j)$ are placed in the two-dimensional arrays indexed by (n_1, n_0) and (j_1, j_0), respectively, for $r=8$ and $s=3$. For this case, (4) consists of three eight-term Fourier series, one for each row of the n table. Then, if j_0 is taken to be the column index of the results, $A_1(j_0, n_0)$, (5) describes eight Fourier series of three terms each on the columns of the array of $A_1(j_0, n_0) W_N^{j_0 n_0}$. The factor $W^{j_0 n_0}$, referred to as the "twiddle factor" by Gentleman and Sande [3], is usually combined with either the $W_r^{j_0 n_1}$ factor in (4) or the $W_s^{j_1 n_0}$ factor in (5).

[1] Actually, Stumpff [9] gave only a doubling and a tripling algorithm and suggested (see Stumpff [9] p. 442, line 11) that the reader generalize to obtain the method for factors of N other than 2 or 3. Thomas made a further assumption (assuming that the index called r by Stumpff was equal to N/s where $s=2$ or 3) which led to his algorithm. Without this assumption, Stumpff's description leads to the Cooley–Tukey algorithm.

For the Thomas prime factor algorithm, one must require that r and s be mutually prime. In this case, different mappings of the one-dimensional arrays into two-dimensional arrays are used. These are also one-to-one mappings and are defined as follows. Let

$$n \equiv rn_0 + sn_1 \pmod{N} \qquad (0 \leq n < N) \tag{6}$$

and

$$j_0 \equiv j \pmod{r} \qquad (0 \leq j_0 < r)$$
$$j_1 \equiv j \pmod{s} \qquad (0 \leq j_1 < s). \tag{7}$$

Then the expression of j, in terms of j_0 and j_1, is a solution of the "Chinese remainder problem" and is given by

$$j \equiv s \cdot s_r j_0 + r \cdot r_s j_1 \pmod{N} \qquad (0 \leq j < N) \tag{8}$$

where s_r and r_s are solutions of

$$s \cdot s_r \equiv 1 \pmod{r} \qquad s_r < r$$
$$r \cdot r_s \equiv 1 \pmod{s} \qquad r_s < s,$$

respectively. Substituting (6) and (8) and using (7) gives

$$W_N^{jn} = W_N^{jn_0 r} W_N^{jn_1 s} = W_s^{jn} W_r^{jn_1} = W_s^{j_1 n_0} W_r^{j_0 n_1}$$

which enables one to write the Fourier series (1) in the form

$$A_1(j_0, n_0) = \sum_{n_1=0}^{r-1} A(n_1, n_0) W_r^{j_0 n_1} \tag{9}$$

$$X(j_1, j_0) = \sum_{n_0=0}^{s-1} A_1(j_0, n_0) W_s^{j_1 n_0}. \tag{10}$$

As in the fast Fourier transform algorithm, this is a two-dimensional Fourier transform. The essential difference is that the "twiddle factor" $W_N^{j_0 n_0}$ does not appear in (10) and the correspondence between one- and two-dimensional indexing is different. The presence of the "twiddle factor" does not introduce any more computation, but it does increase programming complexity slightly. To illustrate better how the indexing in the two algorithms differs, the mappings of n and j for the Thomas prime factor algorithm are given in Table II for comparison with the indexing described in Table I.

The prime factor algorithm can be programmed very easily in a source language like FORTRAN and, therefore, can be used efficiently with a subroutine designed for a number of terms equal to a power of two. For example, if r is a power of 2 and s is any odd number, the subseries (9) can be computed by the power of 2 subroutine.

TABLE II

CORRESPONDENCE BETWEEN ONE- AND TWO-DIMENSIONAL INDEXING IN THE PRIME FACTOR ALGORITHM FOR THE CASE $r=8$, $s=3$, AND $N=24$

$n \equiv rn_0 + sn_1 \equiv 8n_0 + 3n_1 \pmod{24,\ 0 \leq n < N}$

		n_1							
		0	1	2	3	4	5	6	7
n_0	0	0	3	6	9	12	15	18	21
	1	8	11	14	17	20	23	2	5
	2	16	19	22	1	4	7	10	13

$j \equiv s \cdot s_r j_0 + r \cdot r_s j_1 = 9j_0 + 16j_1 \pmod{24,\ 0 \leq j < N}$

		j_0							
		0	1	2	3	4	5	6	7
j_1	0	0	9	18	3	12	21	6	15
	1	16	1	10	19	4	13	22	7
	2	8	17	2	11	20	5	14	23

REFERENCES

[1] J. W. Cooley and J. W. Tukey, "An algorithm for the machine calculation of complex Fourier series," *Math. of Comput.*, vol. 19, pp. 297–301, April 1965.

[2] G. C. Danielson and C. Lanczos, "Some improvements in practical Fourier analysis and their application to X-ray scattering from liquids," *J. Franklin Inst.*, vol. 233, pp. 365–380 and 435–452, April 1942.

[3] W. M. Gentleman and G. Sande, "Fast Fourier transforms for fun and profit," *1966 Fall Joint Computer Conf., AFIPS Proc.*, vol. 29. Washington, D. C.: Spartan, 1966, pp. 563–578.

[4] I. J. Good, "The interaction algorithm and practical Fourier analysis," *J. Roy. Statist. Soc.*, ser. B, vol. 20, pp. 361–372, 1958; Addendum, vol. 22, 1960, pp. 372–375. (MR 21 1674; MR 23 A4231.)

[5] P. Rudnick, "Note on the calculation of Fourier series," *Math. of Comput.*, vol. 20, pp. 429–430, July 1966.

[6] C. Runge, *Zeit. für Math. und Physik*, vol. 48, p. 443, 1903.

[7] C. Runge, *Zeit. für Math. und Physik*, vol. 53, p. 117, 1905.

[8] C. Runge and H. König, "Die Grundlehren der Mathematischen Wissenschaften," *Vorlesungen über Numerisches Rechnen*, vol. 11. Berlin: Julius Springer, 1924.

[9] K. Stumpff, *Tafeln und Aufgaben zur Harmonischen Analyse und Periodogrammrechnung.* Berlin: Julius Springer, 1939.

[10] L. H. Thomas, "Using a computer to solve problems in physics," *Application of Digital Computers.* Boston, Mass.: Ginn, 1963.

[11] F. Yates, *The Design and Analysis of Factorial Experiments.* Harpenden: Imperial Bureau of Soil Science.

A Method for Computing the Fast Fourier Transform with Auxiliary Memory and Limited High-Speed Storage

RICHARD C. SINGLETON, SENIOR MEMBER, IEEE

Abstract—**A method is given for computing the fast Fourier transform of arbitrarily large size using auxiliary memory files, such as magnetic tape or disk, for data storage. Four data files are used, two in and two out. A multivariate complex Fourier transform of $n=2^m$ data points is computed in m passes of the data, and the transformed result is permuted to normal order by $m-1$ additional passes. With buffered input–output, computing can be overlapped with reading and writing of data. Computing time is proportional to $n \log_2 n$. The method can be used with as few as three files, but file passing for permutation is reduced by using six or eight files. With eight files, the optimum number for a radix 2 transform, the transform is computed in m passes without need for additional permutation passes.**

An ALGOL procedure for computing the complex Fourier transform with four, six, or eight files is listed, and timing and accuracy test results are given. This procedure allows an arbitrary number of variables, each dimension a power of 2.

Introduction

THE FAST FOURIER transform is a method of computing the finite complex Fourier transform

$$\alpha_k = \frac{1}{n}\sum_{j=0}^{n-1} x_j \exp(i2\pi jk/n) \quad \text{for } j = 0, 1, \cdots, n-1$$

or the inverse transform

$$x_j = \sum_{k=0}^{n-1} \alpha_k \exp(-i2\pi jk/n) \quad \text{for } j = 0, 1, \cdots, n-1.$$

The key idea of the method, that of factoring n as $n_1, n_2, \cdots, n_m$, then decomposing the transform into n/n_j transforms of size n_j for $j=1, 2, \cdots, m$, was first proposed in 1942 by Danielson and Lanczos [1], [2]. Since the labor of computing a single transform of dimension n is of the order of n^2, the decomposition gives a considerable saving when n is not a prime, reducing computing to the order of $n(n_1+n_2+\cdots+n_m)$. Good [3] in 1958 formulated the fast Fourier transform for complex-valued and multivariate data. Cooley and Tukey [4] programmed Good's method with improvements, and their 1965 paper touched off the current surge of research on fast Fourier transform methods [5], [6] and applications [7], [8].

In this paper, we first give some general comments on programs for the fast Fourier transform, then show a method for computing radix 2 transforms of arbitrarily large size on any computer having four auxiliary files available for data storage. This method also has possible application in the design of special-purpose hardware for parallel computation of the fast transform. A B5500 ALGOL procedure for computing a multivariate Fourier transform using four, six, or eight auxiliary files is listed in the Appendix; results of testing this procedure for time and accuracy are also given.

Programs for the Fast Fourier Transform

The fast Fourier transform can be computed for any size n, but since the number of arithmetic operations is proportional to $n\sum n_j$ for $n=\Pi n_j$, we avoid large factors if possible. An ALGOL procedure for the general case of arbitrary factors is available [9] and other programs are known to exist [6].

The special case of $n=2^m$ has received considerable attention for reasons of efficiency and ease of programming. A version of Cooley's original program, written in a mixture of FORTRAN II and FAP and available through SHARE [10], computes complex Fourier transforms with up to three variables

$$\alpha_{jkl} = \frac{1}{n_1 n_2 n_3}\sum_{p=0}^{n_1-1}\sum_{q=0}^{n_2-1}\sum_{r=0}^{n_3-1} x_{pqr} \cdot \exp(i2\pi(pj/n_1 + qk/n_2 + rl/n_3)),$$

for the $n_1n_2n_3=2^m$ case. The author has written ALGOL procedures for the univariate case [11], [12]. A univariate transform can be used to compute multivariate transforms if desired; for example, a bivariate transform of data stored in a two-dimensional array is computed by first transforming the rows one by one, then the columns of the array. Other programs for $n=2^m$ have been written by Gentleman and Sande [6]. They favor factors of 4, with a final factor of 2 if m is odd, and claim a doubling of efficiency by this approach. The author's experience contradicts their claim; when computing with all the data in high-speed storage, a good

Manuscript received December 29, 1966; revised February 14, 1967. This work was supported by the Stanford Research Institute, out of Research and Development funds.

The author is with the Stanford Research Institute, Menlo Park, Calif.

Reprinted from *IEEE Trans. Audio Electroacoust.*, vol. AU-15, pp. 91–97, June 1967.

radix 2 program is nearly as efficient as a radix 4 plus 2 program and is simpler. In computing the fast transform, a number of possibilities exist for reducing time by adding coding. However, the trade-off between efficiency and program size needs further study.

For a fast Fourier transform of size $n=2^m$, $n/4+1$ distinct sine function values, $\sin(2\pi k/n)$ for $k=0, 1, \cdots, n/4$, are needed during the computation. In the Cooley program the sine values are tabled, while in the author's programs sine and cosine values are generated as needed. Tabling gives a moderate increase in efficiency but requires additional storage space. The difference between the two methods is small if trigonometric function values are generated recursively by a difference equation.

The fast Fourier transform can be done in place, with the results of each transform of size n_j replacing the values used to compute the transform. In this case a complex transform of size n requires $2n$ data storage locations, with each complex quantity taking two locations, plus 2 $\max_j n_j$ temporary storage locations. If n is a power of 2 and if the number of words of high-speed storage in the computer is also a power of 2, then n can be no larger than one quarter of the memory size for a complex Fourier transform computed without auxiliary memory. Thus, Cooley's program will compute a transform of size 8192 on an IBM 7094 with 32 768 words of core (with a computing time of about 0.13 minute).

The speed of the fast Fourier transform leads us to search for methods of computing transforms larger than can be accommodated within high-speed storage. On a system with virtual memory, we can organize the sequence of computations so as to reduce time lost through overlay of data storage [9], [11]; using this approach the author has computed complex transforms of size up to 65 536 on a Burroughs B5500 with 32 768 words of core. Another approach suggested by Cooley [8] and in greater detail by Gentleman and Sande [6], is that of breaking the problem up into transforms that will fit within memory, storing each result in a separate auxiliary memory file, then combining results at the end. In the next section we propose still another approach, a method of computing the fast Fourier transform using four auxiliary files and requiring negligible high-speed storage for data.

An Algorithm for Use of Auxiliary Memory

Here, we assume the availability of four serial-organized auxiliary memory files, such as magnetic tapes or serial disk files. The reader will note, however, that three tape units can be used at the cost of changing two reels per pass or of added tape copying.

In this method the $n=2^m$ complex data points are first written on two of the files, with the first $n/2$ complex values in sequence on the first file and the remaining $n/2$ on the second. The fast Fourier transform is computed while copying the data back and forth between the two pairs of files. As with the fast transform done in place within high-speed memory, this method also leaves the transformed result in reverse binary order, requiring a permutation to restore it to normal order. Again, this permutation can either follow the transform or precede it, using in the later case a fast transform procedure operating on data stored initially in reverse binary order. We consider both approaches.

If the data are originally in normal order, we read a pair of complex data elements, one from each input file, compute a two-by-two transform and write the two resulting complex values on the first output file. After $n/2$ entries have been written we switch to writing on the second output file. The two-by-two transform computed on the kth pass, where $k=1, 2, \cdots, m$, is

$$y_{2j} = x_j + x_{j+n/2}$$
$$y_{2j+1} = x_j - x_{j+n/2} \exp\left(i\pi(j \div 2^{k-1})/2^{m-k}\right)$$

for $j=0, 1, \cdots (n/2)-1$, where $\div$ represents integer division without remainder. The successive passes are identical, except for interchanging input and output files and using a different sequence of trigonometric function values. The sequence of steps for the $n=8$ case is shown in Figs. 1 and 2, where

$$c^j = \exp(i2\pi j/n).$$

We see that the result is in reverse binary order.

This method of computing the fast Fourier transform offers interesting possibilities for parallel computation. As with other methods, the two-by-two transforms at each step are independent and can be computed in parallel. However, this method has the additional property of having the same schedule of data access and storage at each of the m steps of an $n=2^m$ transform.[1] If we copy back and forth between two data arrays, we need only a single fixed set of data transmission paths in each direction for computing the transform, plus an additional set of data transmission paths in one direction to permute the final result to normal order. We must, of course, also provide a way of changing the set of trigonometric function values used as multipliers of the data at each step in the transform.

If the data are originally in reverse binary—for example, as the result of a previous transform—we reverse the steps outlined above. In each pass we take pairs of complex entries from the first input file until it is exhausted, then from the second input file, compute a two-by-two transform with each pair, and write one result on each of the two output files. The two-by-two transform computed on the kth pass, where $k=1, 2, \cdots, m$, is

[1] M. Pease of Stanford Research Institute independently noted the parallel computation possibilities of this arrangement of the fast Fourier transform, and is exploring the idea further.

α_0	α_4	$(\alpha_0+\alpha_4)$	$(\alpha_2+\alpha_6)$
α_1	α_5	$(\alpha_0-\alpha_4)$	$c^2(\alpha_2-\alpha_6)$
α_2	α_6	$(\alpha_1+\alpha_5)$	$(\alpha_3+\alpha_7)$
α_3	α_7	$c^1(\alpha_1-\alpha_5)$	$c^3(\alpha_3-\alpha_7)$
Input to Step One		*Input to Step Two*	

$[(\alpha_0+\alpha_4)+(\alpha_2+\alpha_6)]$	$[(\alpha_1+\alpha_5)+(\alpha_3+\alpha_7)]$
$[(\alpha_0+\alpha_4)-(\alpha_2+\alpha_6)]$	$c^2[(\alpha_1+\alpha_5)-(\alpha_3+\alpha_7)]$
$[(\alpha_0-\alpha_4)+c^2(\alpha_2-\alpha_6)]$	$[c^1(\alpha_1-\alpha_5)+c^3(\alpha_3-\alpha_7)]$
$[(\alpha_0-\alpha_4)-c^2(\alpha_2-\alpha_6)]$	$c^2[c^1(\alpha_1-\alpha_5)-c^3(\alpha_3-\alpha_7)]$

Input to Step Three

Fig. 1. Auxiliary memory input files for $N=8$, data in normal order.

x_0	x_1	x_0+x_4	x_0-x_4
x_4	x_5	x_2+x_6	x_2-x_6
x_2	x_3	x_1+x_5	x_1-x_5
x_6	x_7	x_3+x_7	x_3-x_7
Input to Step One		*Input to Step Two*	

$(x_0+x_4)+(x_2+x_6)$	$(x_0+x_4)-(x_2+x_6)$
$(x_1+x_5)+(x_3+x_7)$	$(x_1+x_5)-(x_3+x_7)$
$(x_0-x_4)+c^2(x_2-x_6)$	$(x_0-x_4)-c^2(x_2-x_6)$
$(x_1-x_5)+c^2(x_3-x_7)$	$(x_1-x_5)-c^2(x_3-x_7)$

Input to Step Three

Fig. 3. Auxiliary memory input files for $N=8$, data in reverse binary order.

$$[(\alpha_0+\alpha_4)+(\alpha_2+\alpha_6)]+[(\alpha_1+\alpha_5)+(\alpha_3+\alpha_7)]=x_0$$
$$[(\alpha_0+\alpha_4)+(\alpha_2+\alpha_6)]-[(\alpha_1+\alpha_5)+(\alpha_3+\alpha_7)]=x_4$$
$$[(\alpha_0+\alpha_4)-(\alpha_2+\alpha_6)]+c^2[(\alpha_1+\alpha_5)-(\alpha_3+\alpha_7)]=x_2$$
$$[(\alpha_0+\alpha_4)-(\alpha_2+\alpha_6)]-c^2[(\alpha_1+\alpha_5)-(\alpha_3+\alpha_7)]=x_6$$

First Output File at Step Three

$$[(\alpha_0-\alpha_4)+c^2(\alpha_2-\alpha_6)]+[c^1(\alpha_1-\alpha_5)+c^3(\alpha_3-\alpha_7)]=x_1$$
$$[(\alpha_0-\alpha_4)+c^2(\alpha_2-\alpha_6)]-[c^1(\alpha_1-\alpha_5)+c^3(\alpha_3-\alpha_7)]=x_5$$
$$[(\alpha_0-\alpha_4)-c^2(\alpha_2-\alpha_6)]+c^2[c^1(\alpha_1-\alpha_5)-c^3(\alpha_3-\alpha_7)]=x_3$$
$$[(\alpha_0-\alpha_4)-c^2(\alpha_2-\alpha_6)]-c^2[c^1(\alpha_1-\alpha_5)-c^3(\alpha_3-\alpha_7)]=x_7$$

Second Output File at Step Three

Fig. 2. Transform output files for $N=8$, data in normal order.

$$[(x_0+x_4)+(x_2+x_6)]+[(x_1+x_5)+(x_3+x_7)]=\alpha_0$$
$$[(x_0-x_4)+c^2(x_2-x_6)]+c^1[(x_1-x_5)+c^2(x_3-x_7)]=\alpha_1$$
$$[(x_0+x_4)-(x_2+x_6)]+c^2[(x_1+x_5)-(x_3+x_7)]=\alpha_2$$
$$[(x_0-x_4)-c^2(x_2-x_6)]+c^3[(x_1-x_5)-c^2(x_3-x_7)]=\alpha_3$$

First Output File at Step Three

$$[(x_0+x_4)+(x_2+x_6)]-[(x_1+x_5)+(x_3+x_7)]=\alpha_4$$
$$[(x_0-x_4)+c^2(x_2-x_6)]-c^1[(x_1-x_5)+c^2(x_3-x_7)]=\alpha_5$$
$$[(x_0+x_4)-(x_2+x_6)]-c^2[(x_1+x_5)-(x_3+x_7)]=\alpha_6$$
$$[(x_0-x_4)-c^2(x_2-x_6)]-c^3[(x_1-x_5)-c^2(x_3-x_7)]=\alpha_7$$

Second Output File at Step Three

Fig. 4. Transform output files for $N=8$, data in reverse binary order.

$$y_j = x_{2j} + x_{2j+1}\exp\left(i\pi(j \div 2^{m-k})/2^{k-1}\right)$$

$$y_{j+n/2} = x_{2j} - x_{2j+1}\exp\left(i\pi(j \div 2^{m-k})/2^{k-1}\right)$$

for $j=0, 1, \cdots, (n/2)1$. The successive passes are identical except for using a different sequence of trigonometric function values. The sequence of steps for the $n=8$ case is shown in Figs. 3 and 4. We see that the final result is in normal order.

To reorder the complex data entries in reverse binary order, we require $m-1$ passes of the file. On the first pass, we read complex entries alternately from the first and second input files and write them on the first output file until it is filled with $n/2$ items, and then write them on the second output file. On the second pass, we copy two entries from the first input file, then two from the second, and so forth, until the first output file is filled, then continue with the second output file. On the kth pass, we copy 2^{k-1} entries from the first input file alternately with 2^{k-1} entries from the second. On completion of the $(m-1)$th pass, the complex data entries are in reverse binary order. The permutation matrix of this reordering is symmetric; thus, applying the procedure a second time will restore the initial ordering. We can also permute between normal and reverse binary order by following the above steps in inverse order.

If, instead of having the assumed initial ordering, the data are written with the real components in the first file and the imaginary components in the second file, we add one step at the beginning of the reordering for a total of m steps. On this step we alternately read single real and imaginary values from the two files and write the complex pairs on the first output file until it is filled, then on the second output file.

This method has been tested on a Burroughs B5500 computer, using both serial disk and magnetic tapes for auxiliary memory. Computing times were comparable with those for other fast Fourier transform programs on the same computer; however, input–output channel times were about twice the computing time. In the next section, we consider ways of reducing input–output time to achieve a better time balance.

Modification to Reduce Input–Output

In the previous section we separated the computation into a permutation phase and a transform phase. In doing this we have essentially followed the plan of computing the transform in place. Although the data are also permuted during each of the m transform steps, this permutation has a cycle length of m, giving the same order at the end as at the beginning.

The first modification in the algorithm is to start with the data in normal order, and alternate computing and permutation steps in such a way that the final result is in normal order. This change by itself gives no speed-up in the transform, but if we then read and write the data in records of 2^r items, where $1 \leq r < m-1$, we can conveniently combine computing and permuting during the final $r+2$ steps, thus eliminating $r+1$ permutation passes. The total number of passes of the file is then $2m-r-2$, except that if $r=m-1$ the transform is completed in m passes. The transform is done in two phases, the first with $m-r$ computing passes and $m-r-2$ permutation passes, and the second with r computing passes.

In this method we start with the first $n/2$ complex entries in normal order in one input file and the remaining $n/2$ entries in a second input file. During the first phase a record is read from each of the two input files and stored in an array. Complex pairs of entries are then selected in sequence from the two arrays, a two-by-two transform computed, and the results stored in place of the original entries. No permutation is done. When computing with one pair of records is completed, we write one record on each output file and then refill the input arrays as before. At the end of the computing pass the files are rewound and a permutation pass is done. In the permutation pass, groups of records are read from the first input file, and alternately written on the first and second output files. When the first input file is exhausted the second is similarly recopied. On the first permutation pass, groups of 2^{m-r-2} records are copied, i.e., one half of the first input file is written on the first output file and the other half on the second, and similarly with the second input file. On the second permutation pass, the group size is one quarter of the size of each input file. On the final permutation pass, groups of two records each are alternately written on the two output files until the first input file is exhausted, then from the second input file. Logically, the next permutation would alternate single records; but we avoid a separate pass for this permutation by altering the input procedure on the next succeeding computing pass, reading pairs of records from the first input file into the two data arrays, computing the transform, then writing the records as before on the two output files. After the first input file is exhausted, we go on with the second.

We then enter the second phase of the computation for the final r steps of the transform. In each of the r passes of the file the input file is read one record at a time, from the beginning of the first input file to the end of the second. Thus, we can if necessary reduce the number of files to three in this phase by copying the second output file onto the end of the first output file at the end of each pass. In this phase three data arrays of size equal to the record size are used in the computer, one to hold an input record and two more to hold records being built up for the two output files. The permutation in the kth step, where $m-r+1 \leq k \leq m$, is done by selecting pairs of entries 2^{m-k} apart from the input record to compute each two-by-two transform, then storing the pair of results in sequence, one each in the two arrays associated with the output files. Using the locations the results are stored in as the location index, the two-by-two transforms computed in both the first and second phase are

$$y_j = x_j + x_{j+n/2} \exp\left(i\pi(j \div 2^{m-k})/2^{k-1}\right)$$
$$y_{j+n/2} = x_j - x_{j+n/2} \exp\left(i\pi(j \div 2^{m-k})/2^{k-1}\right)$$

for $k=1, 2, \cdots, m$ and $j=0, 1, \cdots, (n/2)-1$.

A B5500 ALGOL procedure for computing the fast transform in this way is listed in the Appendix, along with a driver program for testing the procedure. The reader familiar with ALGOL-60 should be able to trace in detail the steps of the algorithm, keeping in mind the following differences in symbols:

B5500 ALGOL	ALGOL-60
DIV	$\div$
*	$\uparrow$
AND	$\wedge$
OR	$\vee$
NOT	$\urcorner$

The procedure READ (F, K, A [*]) reads a record of K words from file F into array A. The procedure WRITE is similar. The program as listed is set up for magnetic tape files, but only the four file declarations need be changed to use disk files.

Computing and input–output channel times were measured for transforms of 2^9, 2^{10}, $\cdots$, 2^{16} complex data points, using a buffer and array size of 512 words, and disk files for auxiliary memory. These times, shown in Fig. 5, increase somewhat faster than linearly with the number of data points. For larger problems, extrapolation beyond $n=2^{16}$ at a rate of $n \log_2 n$ should give good predictions of times. When magnetic tape files are used instead of disk, times increase by a small amount. At $n=2^{12}$ the ratio of computing time is 1.05 and the ratio of input–output channel times is 1.10, but these ratios decline gradually with increasing problem size. At $n=2^{16}$ the ratios for computing and input–output channel time are 1.02 and 1.05.

Another modification to reduce input–output channel time is to add two output files, using two files in and four files out, and copy two of the output files onto the ends of the other two at the end of each transform step during the first phase. This step replaces the permutation step and takes only half as long as the permutation.

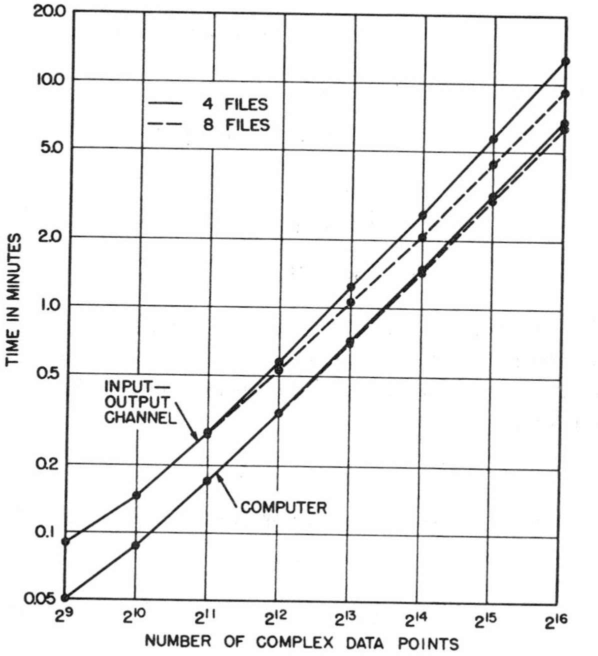

Fig. 5. Computer and input–output channel times for the fast Fourier transform.

As a further possibility, eight files can be used, four in and four out, and the permutation steps during the first phase avoided entirely. In this case the transform is computed in m passes of the files. The transform procedure in the Appendix runs with four, six, or eight files during the first phase, and closes the additional files, and runs with four during the second phase. Times for eight files are also shown in Fig. 5; times for six files lie about halfway between those for four and eight. For larger problems, the $n=16$ times listed in Table I can be extrapolated at an $n \log_2 n$ rate. Single variable transforms were computed during these timing tests. Multivariable transforms run slightly faster because of the fewer number of trigonometric function values and the smaller amount of indexing.

TABLE I

FAST FOURIER TRANSFORM TIMES FOR $N=2^{16}$

	Four Files	Six Files (time in minutes)	Eight Files
Computing	6.75	6.54	6.37
I–O Channel	12.71	10.60	9.25

Whatever the number of files used in the computation, the data are stored originally in two files, zero and two or one and three, and the output appears in one of these two pairs of files. In each pass the input file numbers are of one parity, odd or even, and the output files are of the other parity; the roles switch between passes. In the program listed in the Appendix, the switch file declaration is used as shown for four files, and is changed to

```
SWITCH FILE FT = FFT0, FFT1, FFT2, FFT3,
                 FFT4, FFT4, FFT5, FFT5
```

for six files, and to

```
SWITCH FILE FT = FFT0, FFT1, FFT2, FFT3,
                 FFT4, FFT5, FFT6, FFT7
```

for eight files. The value of the variable NF in the procedure call is also changed to agree with the number of files to be used.

The number of variables in the transform is given by the value of the parameter NV, and the log base 2 of the jth variable is given by the array entry DIM_j for $j=1, 2, \cdots, \mathrm{NV}$. If, for example, $\mathrm{NV}=3$, and the three dimensions are $n_1=2^{m_1}$, $n_2=2^{m_2}$, and $n_3=2^{m_3}$, where $n=2^{m_1+m_2+m_3}$ is the total number of complex data points, we set $\mathrm{DIM}_1=m_1$, $\mathrm{DIM}_2=m_2$, and $\mathrm{DIM}_3=m_3$, and store the data in locations

$$k_1 + k_2 n_1 + k_3 n_1 n_2 \qquad \text{for } 0 \leq k_j \leq n_j - 1,$$

viewing the input files as a single sequence from the beginning of the first to the end of the second. The transformed result is indexed in the same way. Since the number of variables is not limited, this procedure can be used to calculate interactions in a 2^m factorial experiment, as described by Good [3]; in this case $\mathrm{NV}=m$ and $\mathrm{DIM}_j=1$ for $j=1, 2, \cdots, m$.

To compute the single-variate transform of $n=2^m$ complex data values x_j for $j=0, 1, \cdots, n-1$, we set $\mathrm{NV}=1$ and $\mathrm{DIM}_1=m$. If negexp is false, then the transform

$$\alpha_k = \sum_{j=0}^{n-1} x_j \exp(i2\pi jk/n) \quad \text{for } k = 0, 1, \cdots, n-1$$

is computed; and if negexp is true,

$$\alpha_k = \sum_{j=0}^{n-1} x_j \exp(-i2\pi jk/n) \quad \text{for } k = 0, 1, \cdots, n-1$$

is computed. One followed by the other, in either order, gives n times the original data, except for roundoff errors.

With the radix 2 transform, no advantage is gained by using more than eight files. If, on the other hand, we consider using a radix 4 transform with eight files, four in and four out, a transform of $n=2^{2m}$ data points still takes $2m+1$ passes, including an initial pass to divide the data into four files and a final pass to consolidate the file, since a permutation step is needed between each transform step. To gain full advantage from a radix 4 transform we need 32 files, 16 in and 16 out; with this number of files an $n=4^m$ transform is done in $m+2$ passes, including an initial pass to divide the file and a final pass to consolidate. This plan looks practical for the B5500 computer, but has not been tried. In general, to compute a radix p transform without reordering between transform steps we need $2p^2$ files, p^2 in and p^2 out; however, only p of the input files need be open at any one time.

Accuracy

The arithmetic operations used in the fast Fourier transform procedure given here are unaltered by changes in the number of files. In testing for accuracy, we generated as data random normal deviates with zero mean and unit standard deviation. The transform results were first compared with those from another well-tested fast transform program to make sure the correct transform was being computed. In subsequent tests, the procedure was used to transform, then inverse transform, the data set, giving in theory n times the original data as a result. The squared differences of the original and $1/n$ times the result values were summed and divided by the sum of squares of the original values. The square root of this ratio, i.e., the ratio of rms error to rms data, was used as a measure of accuracy. Typical results are listed in Table II, with three trials for each transform size. In the floating point number representation used in the B5500 computer, an error in the final bit of a number x gives a difference of 14.6×10^{-12} for $1 < |x| < 8$, as is true for about one third of our data, and a difference of 1.8×10^{-12} for $\frac{1}{8} < |x| < 1$, as is true of most of the remainder of our data, except for about 10 percent with $|x| < \frac{1}{8}$. By comparison, the observed errors appear satisfactorily small. Single variable transforms were used for these tests; however, limited tests with three variables show even lower errors, as should be expected from the fact that fewer iterative extrapolations of trigonometric function values are needed.

TABLE II
Accuracy Tests of Single Variable Transform

Number of Data Points	First Trial	Second Trial	Third Trial
	(rms error/rms data in units of 10^{-12})		
2^9	19.9	19.9	18.9
2^{10}	21.8	21.5	22.0
2^{11}	26.2	25.2	25.1
2^{12}	40.4	40.5	40.8
2^{13}	65.5	66.5	65.8
2^{14}	106.2	105.3	106.3
2^{15}	109.9	109.5	109.7
2^{16}	153.8	153.8	154.2

In the procedure listed in the Appendix, cosine and sine function values are computed during the first phase using the second difference relations

$$\begin{aligned} C_{k+1} &= R \times \cos(k\theta) + C_k \\ \cos((k+1)\theta) &= \cos(k\theta) + C_{k+1}, \quad \text{and} \\ S_{k+1} &= R \times \sin(k\theta) + S_k \\ \sin((k+1)\theta) &= \sin(k\theta) + S_{k+1}, \end{aligned}$$

where the constant multiplier is

$$R = -4 \times \sin^2(\theta/2)$$

and the initial values are

$$\begin{aligned} C_0 &= 2 \times \sin^2(\theta/2), \\ S_0 &= \sin(\theta), \\ \cos(0) &= 1, \quad \text{and} \\ \sin(0) &= 0. \end{aligned}$$

This method gives very good accuracy. However, transform errors can be reduced by about one half (one binary place) by use of the following slightly slower method:

$$\begin{aligned} TEMP &= \cos(k\theta) \\ \cos((k+1)\theta) &= [CS \times \cos(k\theta) - SD \times \sin(k\theta)] \\ &\quad + \cos(k\theta) \\ \sin((k+1)\theta) &= [CS \times \sin(k\theta) + SD \times TEMP] \\ &\quad + \sin(k\theta), \end{aligned}$$

where the constant multipliers are

$$\begin{aligned} CS &= -2 \times \sin^2(\theta/2) \quad \text{and} \\ SD &= \sin(\theta). \end{aligned}$$

Trigonometric values are computed in this way during the second phase of the transform procedure listed in the Appendix. This method gives transform results that are nearly as accurate as when the computer's library trigonometric function procedure is used to compute all values.

Conclusions

We have shown that the fast Fourier transform of $n = 2^m$ complex data points can be computed using auxiliary memory for data storage with only a relatively small amount of high-speed memory required for temporary storage in addition to program. Using four auxiliary files, at most $2m-1$ passes of the data are required to compute the transform and restore the result to normal order. The logic of the transform calculation in this method may also find application in building special hardware for parallel computation of the fast Fourier transform.

Although the fast transform can be computed with as few as three auxiliary files, added efficiency results from increasing the number of files from four to six or eight. With eight auxiliary memory files the transform is computed in m passes of the data, with the final result in normal order. For the radix 2 transform there is no advantage gained by increasing the number of files beyond eight.

Appendix

A B5500 ALGOL Procedure for the Fast Fourier Transform with Auxiliary Memory

```
BEGIN COMMENT TEST AUXILIARY MEMORY TRANSFORM;
DEFINE KBF= 512#;
FILE FFT0 (2,512); FILE FFT1 (2,512);
FILE FFT2 (2,512); FILE FFT3 (2,512);
SWITCH FILE FT := FFT0,FFT1,FFT2,FFT3;
INTEGER NA,NB,NC,ND;
SAVE ARRAY AFA,AFB,AFC[0:KBF-1];

PROCEDURE FASTFOURIERAUX(NV,DIM,NEGEXP,NF);
    VALUE NV,NEGEXP,NF; INTEGER NV,NF;
    BOOLEAN NEGEXP; ARRAY DIM[0];
COMMENT NV IS THE NUMBER OF VARIABLES.
    DIM[J] IS THE LOG BASE 2 OF THE DIMENSION OF
        VARIABLE J.  THE INPUT FILE LOCATION OF THE
        COMPLEX DATA ENTRY WITH INDEX (K1,K2,...KNV) IS
        K1+K2×2*DIM[1]+...+KNV×2*(DIM[1]+...+DIM[NV-1]),
        CONSIDERING THE FILE AS A SINGLE SEQUENCE OF
        ENTRIES.
    NEGEXP=TRUE GIVES A NEGATIVE SIGN IN THE
        EXPONENTIAL AND FALSE GIVES A POSITIVE SIGN.
    NF IS THE NUMBER OF FILES, 4,6, OR 8, TO BE USED;
BEGIN INTEGER J,JJ,JK,K,KK,N,SPAN,KC,KD,KS,KR,NR,NS;
    REAL AB,BB,RE,IM,CN,SN,RAD,CD,SD,R,RS;
    BOOLEAN FIRST;
    LABEL L,L2,L3,L4,L5,L6,L7;
    PROCEDURE REWINDFILES;
    BEGIN FOR NB := 0,1,2,3 DO REWIND(FT[NB]);
        NA := NC; NC := 1-NC; ND := NC+2;
        NB := NA+2;
    END REWINDFILES;
    N := DIM[1]; J := JJ := K := KC := KR := 0;
    FOR JK := 2 STEP 1 UNTIL NV DO N := DIM[JK]+N;
    SPAN := N := 2*N; RAD := 6.28318530718/(4×N);
    IF NEGEXP THEN RAD := -RAD; JK := DIM[NV];
    NC := 1-NA; ND := NC+4; NB := NA+2;
    NR := JJ := N DIV (KBF+KBF); NS := N DIV KBF;
    IF JJ<2 OR NF<6 THEN
    BEGIN JJ := N; NF := 4; ND := NC+2 END ELSE
    IF JJ=2 OR NF<8 THEN NF := 6 ELSE NF := 8;
    IF NR=0 THEN FIRST := TRUE ELSE FIRST := FALSE;
    KD := KBF DIV 2; KS := N; GO TO L2;
    COMMENT THE FOLLOWING SECTION REORDERS RECORDS
        BETWEEN STEPS OF THE FIRST PHASE OF THE
        TRANSFORM WHEN 4 FILES ARE USED;
L:  READ(FT[NA],KBF,AFA[*]);
    WRITE(FT[NC],KBF,AFA[*]); KR := KR+1;
    K := K+KBF; IF K<KS THEN GO TO L; K := 0;
    IF NC<2 THEN NC := NC+2 ELSE NC := NC-2;
    IF KR<NS THEN GO TO L; KR := 0;
    IF NA<2 THEN
    BEGIN NA := NA+2; GO TO L END;
    REWINDFILES;
    COMMENT FIRST PHASE OF TRANSFORM STARTS HERE;
L2: CN := 1; SN := 0; SD := RS; RS := SIN(RAD×SPAN);
    R := -(RS+RS)*2; CD := -0.5×R;
    SPAN := SPAN DIV 2; KK := 0;
L3: READ(FT[NA],KBF,AFA[*]);
    READ(FT[NB],KBF,AFB[*]);
    FOR J := 0, J+1 WHILE J<KBF DO
    BEGIN AB := AFB[J]; BB := AFB[J+1];
        RE := CN×AB-SN×BB; IM := SN×AB+CN×BB;
        AFB[J] := AFA[J]-RE; AFA[J] := AFA[J]+RE;
        J := J+1;
        AFB[J] := AFAEJ]-IM; AFA[J] := AFA[J]+IM;
    END;
    KC := KC+KD; IF KC=SPAN THEN
    BEGIN KC := 0;
        COMMENT COMPUTE THE NEXT COS AND SIN VALUES;
        CD := R×CN+CD; CN := CN+CD;
        SD := R×SN+SD; SN := SN+SD;
    END;
    WRITE(FT[NC],KBF,AFA[*]);
    WRITE(FT[ND],KBF,AFB[*]);
    KK := KK+1; IF KK=JJ THEN
    BEGIN KK := 0; IF NC<2 THEN NC := NC+2 ELSE
        NC := NC-2; ND := NC+4;
    END;
    KR := KR+1; IF KR<NR THEN GO TO L3; KR := 0;
    FIRST := NOT FIRST; IF FIRST THEN
    BEGIN IF KS=N THEN GO TO L3 ELSE
        IF KS=KBF THEN NB := NA := NA+2 ELSE
        IF NF=8 THEN
        BEGIN NA := NA+4; NB := NB+4 END;
        GO TO L3;
    END;
    IF NF=8 AND KS≠N THEN
    BEGIN IF JJ=2 OR JJ=N THEN
        BEGIN CLOSE(FT[NA]); CLOSE(FT[NB]) END ELSE
        BEGIN REWIND(FT[NA]); REWIND(FT[NB]) END;
    END;
    IF JJ<N THEN
    BEGIN REWIND(FT[NC+4]); REWIND(FT[NC+6]);
        IF JJ>2 THEN JJ := JJ DIV 2 ELSE
        BEGIN NF := 8; JJ := N END;
    END;
    COMMENT THE FOLLOWING SECTION REORDERS RECORDS
        BETWEEN STEPS OF THE FIRST PHASE OF THE
        TRANSFORM WHEN 6 FILES ARE USED;
    IF NF=6 THEN FOR NB := NC, NC+2 DO
    BEGIN ND := NB+4;
L4:     READ(FT[ND],KBF,AFA[*]);
        WRITE(FT[NB],KBF,AFA[*]); KR := KR+1;
        IF KR<NR THEN GO TO L4; KR := 0;
        REWIND(FT[ND]);
    END;
    REWINDFILES; IF JJ<N THEN ND := NC+4;
    JK := JK-1; IF JK=0 THEN
    BEGIN NV := NV-1; JK := DIM[NV]; SPAN := N END;
    KS := KS DIV 2; IF KS>KBF THEN
    BEGIN IF NF≥6 THEN GO TO L2 ELSE GO TO L END
    ELSE IF KS=KBF THEN
    BEGIN NB := NA; NF := 4; GO TO L2 END; J := 0;
    COMMENT SECOND PHASE OF TRANSFORM STARTS HERE;
L5: CN := 1; SN := 0; SD := RS; RS := SIN(RAD×SPAN);
    CD := -2×RS*2;
    SPAN := SPAN DIV 2; KS := KD; KD := KD DIV 2;
L6: READ(FT[NA],KBF,AFC[*]); KK := 0; K := KS;
L7: AB := AFC[KK+KS]; BB := AFC[KK+KS+1];
    RE := CN×AB-SN×BB; IM := SN×AB+CN×BB;
    AB := AFC[KK]; BB := AFC[KK+1]; KK := KK+2;
    AFA[J] := AB+RE; AFB[J] := AB-RE; J := J+1;
    AFA[J] := BB+IM; AFB[J] := BB-IM; J := J+1;
    IF KK<K THEN GO TO L7;
    KC := KC+KD; IF KC=SPAN THEN
    BEGIN KC := 0; AB := CN;
        COMMENT COMPUTE THE NEXT COS AND SIN VALUES;
        CN := (CD×CN-SD×SN)+CN;
        SN := (CD×SN+SD×AB)+SN;
    END;
    KK := KK+KS; IF KK<KBF THEN
    BEGIN K := KK+KS; GO TO L7 END;
    IF J=KBF THEN
    BEGIN WRITE(FT[NC],KBF,AFA[*]);
        WRITE(FT[ND],KBF,AFB[*]); J := 0;
    END;
    KR := KR+1; IF KR<NS THEN GO TO L6; KR := 0;
    IF NA<2 THEN
    BEGIN NA := NA+2; GO TO L6 END;
    REWINDFILES;
    JK := JK-1; IF JK>0 THEN GO TO L5; SPAN := N;
    NV := NV-1; JK := DIM[NV]; IF NV>0 THEN GO TO L5;
END FASTFOURIERAUX;

COMMENT DRIVER PROGRAM STARTS HERE;
INTEGER J,JJ,K,M,N,NV,LIM,RDM,RDN;
INTEGER ARRAY D[0:6];
REAL SS1,SS2,SS3,SS4,AA,BB;
FILE LP 4(1,15);
PROCEDURE NORMAL(RDM,RX,RY);
    INTEGER RDM; REAL RX,RY;
COMMENT COMPUTES A PAIR RX,RY OF RANDOM NORMAL
    DEVIATES WITH MEAN = 0, S.D. = 1.
    CHOOSE INITIAL VALUE OF RDM ODD AND <2*27;
BEGIN REAL R; LABEL LR;
LR: RDM := 3589×RDM; RDM := RDM-(RDM DIV 134217728)×134217728;
    RX := (RDM-67108864)/67108864;
    RDM := 3589×RDM; RDM := RDM-(RDM DIV 134217728)×134217728;
    RY := (RDM-67108864)/67108864;
    R := RX*2+RY*2; IF R≥1.0 THEN GO TO LR;
    R := SQRT(-2×LN(R)/R); RX := RX×R; RY := RY×R;
END NORMAL;
RDN := 41365;
FOR M := 9 STEP 1 UNTIL 12 DO BEGIN
WRITE(LP[DBL],<"M =",I2>,M);
N := 2*M; LIM := N DIV KBF; JJ := KBF-1;

NV := 1; D[1] := M;
NA := 0; NB := 2; NC := 1; ND := 3;
RDM := RDN; SS3 := SS4 := 0;
WRITE(LP[DBL],<"RDM =",I10>,RDM);
COMMENT WRITE DATA FILES;
FOR J := 1 STEP 1 UNTIL LIM DO
BEGIN FOR K := 0 STEP 1 UNTIL JJ DO
    BEGIN NORMAL(RDM,AA,BB); AFA[K] := AA;
        AFB[K] := BB; SS3 := AA*2+SS3; SS4 := BB*2+SS4
    END;
    WRITE(FT[NA],KBF,AFA[*]);
    WRITE(FT[NB],KBF,AFB[*]);
END;
REWIND(FT[NA]); REWIND(FT[NB]);
FASTFOURIERAUX(NV,D,FALSE,4);
FASTFOURIERAUX(NV,D,TRUE,4);
RDM := RDN; SS1 := SS2 := 0;
FOR J := 1 STEP 1 UNTIL LIM DO
BEGIN READ(FT[NA],KBF,AFA[*]);
    READ(FT[NB],KBF,AFB[*]);
    FOR K := 0 STEP 1 UNTIL JJ DO
    BEGIN NORMAL(RDM,AA,BB);
        SS1 := (AFA[K]/N-AA)*2+SS1;
        SS2 := (AFB[K]/N-BB)*2+SS2;
    END;
END;
REWIND(FT[NA]); REWIND(FT[NB]);
WRITE(LP[DBL],<3E16.6>,SQRT(SS1/SS3),SQRT(SS2/SS4),
    SQRT((SS1+SS2)/(SS3+SS4)));
RDN := RDM; WRITE(LP[PAGE]);
END;
END.
```

References

[1] G. C. Danielson and C. Lanczos, "Some improvements in practical Fourier analysis and their application to X-ray scattering from liquids," *J. Franklin Institute*, vol. 233, pp. 365–380 and pp. 435–452, April 1942.

[2] P. Rudnick, "Note on the calculation of Fourier series," *Math. of Comput.*, vol. 20, pp. 429–430, July 1966.

[3] I. J. Good, "The interaction algorithm and practical Fourier series," *J. Royal Statist. Soc.*, ser. B, vol. 20, pp. 361–372, 1958; Addendum, vol. 22, pp. 372–375, 1960.

[4] J. W. Cooley and J. W. Tukey, "An algorithm for the machine calculation of complex Fourier series," *Math. of Comput.*, vol. 19, pp. 297–301, April 1965.

[5] T. G. Stockham, "High-speed convolution and correlation," *1966 Spring Joint Computer Conf., AFIPS Proc.*, vol. 28. Washington, D. C.: Spartan, 1966, pp. 229–233.

[6] W. M. Gentleman and G. Sande, "Fast Fourier transforms—for fun and profit," *1966 Fall Joint Computer Conf., AFIPS Proc.*, vol. 29. Washington, D. C.: Spartan, 1966, pp. 563–578.

[7] R. C. Singleton and T. C. Poulter, "Spectral analysis of the call of the male killer whale," this issue, p. 104–113.

[8] L. E. Alsop and A. A. Nowroozi, "Fast Fourier analysis," *J. of Geophys. Res.*, vol. 71, p. 5482–5483, November 15, 1966.

[9] R. C. Singleton, "An ALGOL procedure for the fast Fourier transform with arbitrary factors," SRI Project 181531-132, Stanford Research Institute, Menlo Park, Calif., December 1966 (Defense Doc. Ctr. AD-643 997).

[10] J. W. Cooley, "Harmonic Analysis Complex Fourier Series," SHARE Program Library SDA 3425, February 7, 1966.

[11] R. C. Singleton, "ALGOL procedures for the fast Fourier transform," SRI Project 181531-132, Stanford Research Institute, Menlo Park, Calif., November 1966 (Defense Doc. Ctr. AD-643 996).

[12] R. C. Singleton, "An ALGOL convolution procedure based on the fast Fourier transform," SRI Project 181531-132, Stanford Research Institute, Menlo Park, Calif., January 1967 (Defense Doc. Ctr. AD-646 628).

THE FAST FOURIER TRANSFORM ALGORITHM: PROGRAMMING CONSIDERATIONS IN THE CALCULATION OF SINE, COSINE AND LAPLACE TRANSFORMS†

J. W. COOLEY, P. A. W. LEWIS‡ AND P. D. WELCH

IBM Corporation, Thomas J. Watson Research Center, P.O. Box 218, Yorktown Heights, N.Y., U.S.A.

(*Received* 22 *August* 1969)

In the organization of programming packages for computing Fourier and Laplace transforms, it is useful, both for conceptual understanding and for operational efficiency to consider the discrete *complex* Fourier transform as a kind of nucleus around which programming for special applications is performed. An advantage of these procedures is that the basic complex Fourier transform algorithm is systematic and can relatively easily be implemented in efficient subroutines, micro-programs and special hardware devices. Once this is done, programming for special properties of the data can efficiently be left to the user to implement on a general purpose computer.

The problem of establishing the correspondence between the discrete transforms and the continuous functions with which one is usually dealing is described. The application of these results and the above-mentioned subroutines to the calculation and inversion of Laplace transforms is given with formulas and empirical results displaying the effect of optimal parameters on computational efficiency and accuracy.

1. INTRODUCTION

The fast Fourier transform (FFT) method is a computational algorithm which, on its rediscovery [1, 2, 3] greatly increased the speed with which Fourier transforms can be computed on digital devices. As a consequence, digital application of Fourier methods has been widely utilized and can be expected to be used even more as the FFT method is incorporated in efficient computer subroutines, microprograms, and even special hardware. Therefore, it seems useful to investigate further the essential properties of the discrete Fourier transform, its correspondence with integral transforms, and various algorithms for using the discrete Fourier transform in special circumstances.

The discrete complex Fourier series is a one-to-one mapping of any sequence $A(n)$, $n = 0, 1, 2, \ldots, N-1$, of N complex numbers onto another sequence defined by

$$X(j) = \sum_{n=0}^{N-1} A(n)\, W_N^{nj}, \quad j = 0, 1, \ldots, N-1, \tag{1}$$

where $i = \sqrt{-1}$ and where $W_N = \exp(2\pi i/N)$ is the principal Nth complex root of unity. This is also referred to as the inverse discrete Fourier transform (IDFT) since, in the literature, the formula for $A(n)$ in terms of $X(j)$,

$$A(n) = \frac{1}{N} \sum_{j=0}^{N-1} X(j)\, W_N^{-nj}, \tag{2}$$

† Presented at the lecture series on "Applications and methods of random data analysis" on 8 to 11 July 1969, at Southampton University.

‡ At present on leave at Department of Mathematics, Imperial College of Science and Technology, London, England.

Reprinted with permission from *J. Sound Vib.*, vol. 12, pp. 315–337, July 1970.

is known as the discrete Fourier transform (DFT). A direct calculation of (1) as an accumulated sum of products for each j would take N^2 operations. The fast Fourier transform method (FFT) is an algorithm for computing (1) or (2) in $N \log N$ operations, where "operation" means a complex multiplication and addition. The FFT algorithm has been described and programmed, in most cases, as a calculation of the operation defined by (1) on complex numbers. However, in actual practice, there is a wide variety of special conditions which one actually wants. For example, the data may be real rather than truly complex, and it may be even or odd so that cosine or sine transforms may be used. Of course, these transforms can be done directly with the complex DFT, but this would be inefficient since it would require redundant storage and computation. The purpose of some of the algorithms given below will be to avoid this. It will also be shown here how one can, by suitably manipulating the data, compute Laplace transforms in terms of the DFT. One could, of course, use the basic ideas of the FFT method and write special programs for each of these special cases. However, the logically simple and systematic nature of the FFT method for complex data makes it more practical to program it in a very efficient subroutine or special hardware and to obtain the above-mentioned special transforms by manipulating the data and using the complex FFT subroutine. The purpose of section 2 of this paper is to derive some of the special algorithms for doing this when the data is real and even or odd.

In section 3, a correspondence between the DFT and the Fourier integral transform will be shown. This gives some insight into the errors introduced by the discrete sampling of a function for only a finite interval of time. This will be used to develop an error analysis and procedures for selecting optimal parameters.

Finally, in section 4, it will be shown how one can use the DFT, computed by the FFT method, to invert Laplace transforms. Two methods are described: one of these computes Laguerre expansion coefficients; the other computes point values of the solution. For these, the data being supplied to the FFT subroutines contains some redundancy which can be used to reduce the computation. An additional parameter for the user to choose appears in the Laplace transform problem. This is c, the real part of the transform variable. It is shown here how the discussion of aliasing in section 3 leads to formulas for an optimal choice of c.

2. THE CALCULATION OF SINE AND COSINE TRANSFORMS OF REAL OR IMAGINARY DATA

The purpose of this section is to show how a subroutine or special hardware unit capable of computing the discrete complex Fourier series

$$X(j) = \sum_{n=0}^{N-1} A(n)\, W_N^{nj} \tag{3}$$

can, with some pre- and post-processing of the data, compute a variety of special transforms. The variety actually consists in using various types of redundancy in the data to reduce the size of the complex array on which the subroutine is to perform.

Before proceeding, some definitions and theorems will be summarized without proofs. Further details which are useful in computing special cases, have been published elsewhere [5].

Definition. If the sequence $A(n)$, $n = 0, 1, \ldots, N-1$ is the DFT of $X(j)$, we write

$$X(j) \leftrightarrow A(n). \tag{4}$$

Functions related to each other by an integral Fourier transform will also be connected by the double-headed arrow. In this case the function on the right is the Fourier integral transform of the one on the left. The sequence appearing on the right in (4) is usually a function of frequency and will be referred to as the discrete Fourier transform (DFT) of the sequence on the left. The sequence appearing on the left in (4) is usually a function of time, space, or simply of a sequential index and will be referred to here as a Fourier series or as the inverse discrete Fourier transform (IDFT) of the sequence on the right. It may also be referred to as "data".

An important property, used in the following sections, is that the indices of $X(j)$ and $A(n)$ are to be interpreted modulo N. Therefore, $X(-j)$ with $0 \leqslant j < N-1$ is understood to mean $X(N-j)$.

A sequence is even if

$$X(j) = X(-j), \tag{5a}$$

and odd if

$$X(j) = -X(-j). \tag{5b}$$

A sequence is conjugate even or conjugate odd if

$$X(j) = \tilde{X}(-j) \quad \text{or} \quad X(j) = -\tilde{X}(j), \tag{6}$$

respectively, where "~" denotes the complex conjugate.

Some useful theorems follow.

Theorem 1. The DFT is linear. If

$$\begin{aligned} X_1(j) &\leftrightarrow A_1(n), \quad n, j = 0, 1, \ldots, N-1, \\ X_2(j) &\leftrightarrow A_2(n), \end{aligned} \tag{7}$$

and

$$X(j) = aX_1(j) + bX_2(j),$$

then

$$A(n) = aA_1(n) + bA_2(n),$$

where $X(j) \leftrightarrow A(n)$.

Theorem 2.

$$X(-j) \leftrightarrow A(-n). \tag{8}$$

Corollary 2.1. $X(j)$ is even/odd if and only if $A(n)$ is even/odd.

Theorem 3.

$$\begin{aligned} \tilde{X}(j) &\leftrightarrow \tilde{A}(-n), \\ \tilde{X}(-j) &\leftrightarrow \tilde{A}(n). \end{aligned} \tag{9}$$

Corollary 3.1. $X(j)$ is real/imaginary if and only if $A(n)$ is conjugate even/odd, respectively; i.e. $A(n) = \pm\tilde{A}(-n)$. $A(n)$ is real-imaginary if and only if $X(j)$ is even/odd, respectively; i.e. $X(j) = \pm\tilde{X}(-j)$.

Corollary 3.2. $X(j)$ is real and even if and only if $A(n)$ is real and even.
$X(j)$ is real and odd if and only if $A(n)$ is imaginary and odd.
$X(j)$ is imaginary and even if and only if $A(n)$ is imaginary and even.
$X(j)$ is imaginary and odd if and only if $A(n)$ real and odd.

Theorem 4.

$$X(j-k) = W_N^{-nk} A(n), \tag{10a}$$

$$W_N^{mj} X(j) = A(n-m). \tag{10b}$$

Theorem 5.

$$\delta(j) \leftrightarrow 1/N, \tag{11a}$$

$$1 \leftrightarrow \delta(n), \tag{11b}$$

where $\delta(0) = 1$, $\delta(j) = 0$ for $j \neq 0$.

Theorem 6.

$$X(0) = \sum_{n=0}^{N-1} A(n), \tag{12}$$

$$A(0) = 1/N \sum_{j=0}^{N-1} X(j). \tag{13}$$

Theorem 7. Convolution theorem. If

$$X_1(j) \leftrightarrow A_1(n),$$
$$X_2(j) \leftrightarrow A_2(n),$$

then

$$1/N \sum_{k=0}^{N-1} X_1(k)\, X_2(j-k) \leftrightarrow A_1(n)\, A_2(n), \tag{14}$$

and

$$X_1(j)\, X_2(j) \leftrightarrow \sum_{m=0}^{N-1} A_1(m)\, A_2(n-m). \tag{15}$$

Corollary 7.1.

$$1/N \sum_{j=0}^{N-1} |X(j)|^2 = \sum_{n=0}^{N-1} |A(n)|^2. \tag{16}$$

Theorem 8. If

$$\text{stretch}_K\{j: X\} = X(0), 0, \ldots, 0, X(1), 0, \ldots, 0, X(2), \ldots, X(N-1), 0, \ldots, 0, \tag{17}$$

where $K-1$ zeros have been inserted between the $X(j)$'s, then

$$\text{stretch}_K\{j; X\} \leftrightarrow A(n)/K, \quad n = 0, 1, \ldots, NK-1. \tag{18}$$

Theorem 9. If

$$\text{sample}_K\{j, X\} = X(0), X(K), X(2K), \ldots, X(N-K), \tag{19}$$

then

$$\text{sample}_K\{j, X\} \leftrightarrow \sum_{m=0}^{K-1} A\left(n + m\frac{N}{K}\right). \tag{20}$$

Let us imagine that we have a Fourier series subroutine whose input is the complex sequence $A(n)$, $n = 0, 1, \ldots, N-1$ and whose output is the Fourier series (or IDFT),

$$X(j) = \sum_{n=0}^{N-1} A(n)\, W_N^{jn}, \quad j = 0, 1, \ldots, N-1. \tag{21}$$

We now describe procedures all of which are based on this single Fourier series subroutine which, we assume, embodies the fast Fourier transform algorithm.

2.1. PROCEDURE 1. THE FOURIER TRANSFORM

It can easily be seen that letting $\tilde{X}(n)/N$ replace $A(n)$ as input and $\tilde{A}(j)$ replace $X(j)$ as output, the above subroutine will be capable of computing the discrete Fourier transform (DFT)

$$A(n) = 1/N \sum_{j=0}^{N-1} X(j)\, W_N^{-nj}. \tag{22}$$

2.2. PROCEDURE 2. THE FOURIER TRANSFORM OF TWO SETS OF REAL DATA IN ONE PASS THROUGH A DFT SUBROUTINE

Using the linearity property, we see that if $X_1(j)$ and $X_2(j)$ are real sequences such that

$$X_1(j) \leftrightarrow A_1(n),$$
$$X_2(j) \leftrightarrow A_2(n),$$

and

$$X(j) = X_1(j) + iX_2(j), \tag{23}$$

then $X(j)$ has the transform

$$A(n) = A_1(n) + iA_2(n). \tag{24}$$

Replacing n by $N - n$, taking complex conjugates of both sides, and applying corollary 3.1 gives

$$\tilde{A}(N - n) = A_1(n) - iA_2(n). \tag{25}$$

Solving (24) and (25) for $A_1(n)$ and $A_2(n)$, we get

$$A_1(n) = \tfrac{1}{2}[\tilde{A}(-n) + A(n)], \qquad A_2(n) = \frac{i}{2}[\tilde{A}(-n) - A(n)]. \tag{26}$$

Hence, procedure 2 is to

(a) form $X(j)$ as defined by (23);
(b) compute $A(n)$ by means of the DFT subroutine;
(c) compute $A_1(n)$ and $A_2(n)$ according to (26)

for $n = 0, 1, 2, \ldots, N/2$.

2.3. PROCEDURE 3. DOUBLING ALGORITHM—COMPUTING THE FOURIER TRANSFORM OF $2N$ POINTS FROM THE TRANSFORMS OF TWO N-POINT SEQUENCES

Given the $2N$ data points $Y(j)$ with $Y(j) \leftrightarrow C(n)$, $n, j = 0, 1, \ldots, 2N - 1$, suppose the two N-point sequences

$$X_1(j) = Y(2j), \qquad X_2(j) = Y(2j + 1), \quad j = 0, 1, 2, \ldots, N - 1, \tag{27}$$

have the N-point transforms $A_1(n)$ and $A_2(n)$,

$$X_1(j) \leftrightarrow A_1(n), \qquad X_2(j) \leftrightarrow A_2(n). \tag{28}$$

Separating even- and odd-indexed points in the series for $C(n)$, we have

$$C(n) = \frac{1}{2N} \sum_{j=0}^{2N-1} Y(j)\, W_{2N}^{-nj} = \frac{1}{2N}\left\{\sum_{j=0}^{N-1} Y(2j)\, W_{2N}^{-2nj} + \sum_{j=0}^{N-1} Y(2j+1)\, W_{2N}^{-2nj-n}\right\}. \tag{29}$$

Since

$$W_{2N}^2 = W_N,$$

$$C(n) = \frac{1}{2N}\left\{\sum_{j=0}^{N-1} Y(2j)\, W_N^{-nj} + \sum_{j=0}^{N-1} Y(2j+1)\, W_N^{-nj}\, W_{2N}^{-n}\right\}. \tag{30}$$

The sums appearing in (30) define $A_1(n)$ and $A_2(n)$, so (30) can be written

$$C(n) = \tfrac{1}{2}\{A_1(n) + A_2(n)\, W_{2N}^{-n}\}. \tag{31}$$

Substituting $N + n$ for n, and using the fact that $W_{2N}^N = -1$, we get

$$C(N + n) = \tfrac{1}{2}\{A_1(n) - A_2(n)\, W_{2N}^{-n}\}. \tag{32}$$

Hence, procedure 3 is as follows:

(a) compute the two N-point DFT's of the sequences $X_1(j) = X(2j)$ and $X_2(j) = X(2j+1)$;
(b) apply (31) and (32) to the resulting transforms.

This procedure, when iterated upon to produce successive doublings up to N, the total number of points, is the FFT algorithm with radix 2.

2.4. PROCEDURE 4. CALCULATION OF THE DFT OF REAL DATA

The transform of a single sequence of $2N$ real data points can be done by means of one DFT of N complex points by using procedure 2 and procedure 3. First, assuming $Y(j)$, $= 0, 1, \ldots, 2N-1$ to be the real sequence, let

$$Y(j) \leftrightarrow C(n).$$

Then, form the N-point sequences,

$$\begin{aligned} X_1(j) &= Y(2j), \\ X_2(j) &= Y(2j+1), \\ X(j) &= X_1(j) + X_2(j)\,i. \end{aligned} \tag{33}$$

Compute the N-point Fourier transform of $X(j)$ and use procedure 2, equation (26), obtain the transforms of the two real sequences $X_1(j)$, $X_2(j)$ in terms of the DFT of $X(j)$. Equation (26) is repeated here,

$$A_1(n) = \tfrac{1}{2}[\tilde{A}(-n) + A(n)], \tag{34}$$

$$A_2(n) = \frac{i}{2}[\tilde{A}(-n) - A(n)].$$

Now, having the transforms of the even- and odd-indexed points of $Y(j)$, procedure 3, the doubling algorithm (31), tells how to obtain the transform of the full array $Y(j)$ by using formula

$$C(n) = \tfrac{1}{2}\{A_1(n) + A_2(n)\, W_{2N}^{-n}\}. \tag{35}$$

Note that the upper half of the $C(n)$ array is redundant; i.e. $Y(j)$ real implies that $C(n) = \tilde{C}(2N-n)$. Since $X_1(j)$ and $X_2(j)$ are real, $A_1(n)$ and $A_2(n)$ are conjugate even. Replacing N by $N+n$ in (35), we get

$$C(N+n) = \tfrac{1}{2}\{A_1(n) - A_2(n)\, W_{2N}^{-n}\}. \tag{36}$$

It is efficient to use (35) and (36) for $n = 0, 1, 2, \ldots, N/2$.

It can easily be shown that the $C(n)$'s in

$$Y(j) = \sum_{n=0}^{2N-1} C(n)\, W_{2N}^{nj}, \quad j = 0, 1, \ldots, 2N-1 \tag{37}$$

can be identified with the sine-cosine coefficients of the series

$$Y(j) = \tfrac{1}{2}a(0) + \sum_{n=1}^{N-1}\left\{a(n)\cos\frac{\pi jn}{N} + b(n)\sin\frac{\pi jn}{N}\right\} + \tfrac{1}{2}(-1)^j a(N), \tag{38}$$

as follows:

$$\begin{aligned} a(n) &= 2\,\mathrm{Re}\, C(n), \quad \text{for } 0 \leqslant n \leqslant N; \\ b(n) &= -2\,\mathrm{Im}\, C(n), \quad \text{for } 0 < n < N. \end{aligned} \tag{39}$$

Hence, procedure 4 is as follows:

(a) let the $2N$-point array $Y(j)$ be put into a complex N-point array $X(j)$ as defined by (33); in many computer programs, this means one does nothing since complex arrays are stored so that real and imaginary parts are already in alternating locations;
(b) compute the N-point DFT of $X(j)$;
(c) apply equation (34) for $n = 0, 1, \ldots, N/2$ to get $A_1(n)$ and $A_2(n)$;
(d) apply equations (35) and (36) for $n = 0, 1, \ldots, N/2$ to get $C(n)$ for $n = 0, 1, \ldots, N$.

Note that the $2N$ real data points $Y(j)$ are transformed to $N+1$ complex frequency values $C(n)$, $n = 0, 1, 2, \ldots, N$. Since $C(0)$ and $C(N)$ are real and since the remaining $C(n)$'s are complex, the sequence $C(n)$ contains $2N$ independent real numbers. One must remember that $C(n)$ is defined for $0 \leqslant n < 2N$, but that due to the property $C(n) = \tilde{C}(-n)$, values for $N < n < 2N$ are redundant and need not be computed or stored.

2.5. PROCEDURE 5. THE CALCULATION OF FOURIER SERIES FOR REAL DATA

This is performed by reversing procedure 4. To derive the formulas, solve (35) and (36) for $A_1(n)$ and $A_2(n)$ in terms of $C(n)$ and $C(N+n)$:

$$A_1(n) = C(n) + C(N+n), \tag{40}$$

$$A_2(n) = [C(n) - C(N+n)]\, W_{2N}^{n}. \tag{41}$$

Then, solving (34), one has

$$A(n) = A_1(n) + iA_2(n), \tag{42}$$

$$\tilde{A}(N-n) = A_1(n) - iA_2(n). \tag{43}$$

Hence, procedure 5 is as follows:

(a) generate $A_1(n)$ and $A_2(n)$ for $n = 0, 1, \ldots, N/2$ according to (40) and (41);
(b) generate $A(n)$ by using (42) and (43) with $n = 0, 1, \ldots, N/2$;
(c) compute the inverse DFT of the N-element complex array $A(n)$; the result will be the sequence $X(j)$ whose real and imaginary parts are the real elements of $Y(j)$ as defined by (33).

Procedures 4 and 5 also permit us to transform a conjugate even data sequence $X(j)$ to a real frequency sequence $A(n)$ and *vice versa*. This is easily seen to be accomplished by letting $\tilde{X}(n)/N$ replace $A(n)$ and $\tilde{A}(j)$ replace $X(j)$ in procedures 4 and 5, thereby switching the roles of "frequency" and "data".

2.6. PROCEDURE 6. THE CALCULATION OF COSINE SERIES FOR REAL DATA

It can easily be demonstrated that $Y(j)$ being real and even implies that its IDFT will be a cosine series. Thus, if $Y(j), j = 0, 1, 2, \ldots, 2N-1$, is real and $Y(j) = Y(2N-j)$,

$$Y(j) = \sum_{n=0}^{2N-1} C(n)\, W_{2N}^{nj},$$

or

$$Y(j) = \tfrac{1}{2}a(0) + \sum_{n=1}^{N-1} a(n) \cos\frac{\pi j n}{N} + \tfrac{1}{2}(-1)^j a(N), \tag{44}$$

where the $a(n)$'s are real and

$$a(n) = 2C(n).$$

It also follows that the DFT of $Y(j)$ can be expressed as a cosine series:

$$a(n) = \frac{2}{N}\left[\tfrac{1}{2} Y(0) + \sum_{j=1}^{N-1} Y(j) \cos \frac{\pi j n}{N} + \tfrac{1}{2}(-1)^j\, Y(N)\right]. \tag{45}$$

We derive the procedure for computing a cosine transform of real even data $Y(j)$, $j = 0, 1, 2, \ldots, 2N-1$, where

$$Y(2N-j) = Y(j).$$

(Actually only $Y(0), \ldots, Y(N)$ need be given.) First define the complex sequence

$$X(j) = Y(2j) + [Y(2j+1) - Y(2j-1)]\,i \tag{46}$$

for $j = 0, 1, 2, \ldots, N-1$. (Only terms with $j = 0, 1, \ldots, N/2$ have to be formed.) This is a complex conjugate even sequence and, therefore, its transform, $A(n)$, must be real. Procedure 5 gives an efficient way to transform a conjugate even sequence to a real sequence. We had in mind that the former was a frequency function and the latter was data. Here these must reverse roles. Letting the conjugate even sequence $\tilde{X}(j)/N$ be the input to procedure 5, the output will be $A(n)$, the real DFT of $X(j)$.

Having $A(n)$, procedure 2 can be used to obtain the transforms of the real and imaginary parts of $X(j)$. Let us define

$$Y(2j) \leftrightarrow A_1(n), \tag{47}$$

$$Y(2j+1) \leftrightarrow A_2(n),$$

$$Y'(2j) = Y(2j+1) - Y(2j-1) \leftrightarrow A_2'(n). \tag{48}$$

From procedure 2, equation (26),

$$A_1(n) = \tfrac{1}{2}[A(n) + A(-n)], \tag{49}$$

$$A_2'(n) = \frac{1}{2i}[A(n) - A(-n)], \tag{50}$$

where, it is to be remembered, $A(n)$ is real. From theorem 4, we derive the fact that

$$Y(2j-1) \leftrightarrow W_N^{-n} A_2(n),$$

and, therefore,

$$A_2'(n) = A_2(n) - W_N^{-n} A_2(n), \tag{51}$$

giving

$$A_2(n) = A_2'(n)/(1 - W_N^{-n}). \tag{52}$$

A special calculation must be made for $n = 0$ and $n = N$. For this, we must compute

$$A_2(0) = \frac{1}{N} \sum_{j=0}^{N-1} Y(2j+1). \tag{53}$$

Finally, procedure 3, the doubling algorithm, takes us from $A_1(n)$ and $A_2(n)$ to $C(n)$, the DFT of $Y(j)$. Substituting (49) and (51) in equation (31) of procedure 3, we get

$$a(n) = 2C(n) = \tfrac{1}{2}\{[A(n) + A(-n)] - [A(n) - A(-n)]/[2 \sin(\pi n/N)]\} \tag{54}$$

for $n = 1, 2, \ldots, N-1$. For $n = 0$ and $n = N$ we must use

$$C(0) = \tfrac{1}{2}\{A_1(0) + A_2(0)\}, \tag{55}$$

$$C(N) = \tfrac{1}{2}\{A_1(0) - A_2(0)\}.$$

Summarizing, procedure 6 for the cosine transform is:

(a) given the real even sequence $Y(j), j = 0, 1, \ldots, 2N-1$, define $X(j)$ according to (46);
(b) let $\tilde{X}(j)/N$ be the input to procedure 5; the output will be $\tilde{A}(n) = A(n)$;
(c) compute $C(n)$ using (54) for $n = 1, 2, \ldots, N-1$;

for $n = 0$ and $n = N$, let $A_1(0) = A(0)$ and compute $A_2(0)$ with (53); then, use (55) to obtain $C(0)$ and $C(N)$; finally, let $a(n) = 2C(n)$ for $n = 0, 1, \ldots, N$.

2.7. PROCEDURE 7. THE CALCULATION OF SINE SERIES FOR REAL DATA

It can be easily demonstrated that if $Y(j), j = 0, 1, \ldots, 2N-1$ is real and odd, it is expressible as a sine series,

$$Y(j) = \sum_{n=0}^{2N-1} C(n)\, W_{2N}^{nj} = \sum_{n=1}^{N-1} b(n) \sin(\pi nj/N), \tag{56}$$

where the $b(n)$'s are real and

$$b(n) = 2iC(n).$$

Note that one must have $Y(0) = Y(N) = 0$. It also follows that

$$b(n) = \frac{2}{N} \sum_{j=1}^{N-1} Y(j) \sin(\pi nj/N). \tag{57}$$

For the calculation, we make use of the fact that if $Y(j)$ is real and odd, then $iY(j)$ is conjugate even and corollary 3.1 then implies that its transform is real. Thus,

$$Y(j) \leftrightarrow C(n) = b(n)/2i$$

and

$$+iY(j) \leftrightarrow +iC(n) = \tfrac{1}{2}b(n).$$

Therefore, letting

$$X(j) = -[Y(2j+1) - Y(2j-1)] + Y(2j)i, \tag{58}$$

and, following a derivation similar to that for equation (54) in procedure 6, one arrives at formula (59) below for the coefficients of the cosine series.

Summarizing, procedure 7 is as follows:

(a) given the values $Y(j), j = 1, \ldots, N-1$ of a real odd sequence $Y(j), j = 0, 1, \ldots, 2N-1$, form $X(j)$ according to (58);
(b) let $\tilde{X}(j)/N$ be input to procedure 5; the output will be $\tilde{A}(n) = A(n)$, where $X(j) \leftrightarrow A(n)$;
(c) compute

$$b(n) = 2iC(n) = \tfrac{1}{2}\{[A(n) - A(-n)] - [A(n) + A(-n)]/[2\sin(\pi n/N)]\} \tag{59}$$

for $n = 1, 2, \ldots, N-1$.

Use of procedures 6 and 7 yields a fourfold decrease in computation and storage requirements since only half the real input data arrays need be supplied. (Actually, $N/2 + 1$ data for the cosine transform and $N/2 - 1$ for the sine transform.) The complex data $X(j)$ to be transformed contains only $N/2$ terms rather than the $2N$ required if one supplies the full array $Y(j)$ to a complex DFT subroutine.

3. THE CORRESPONDENCE BETWEEN DISCRETE AND INTEGRAL FOURIER TRANSFORMS

In the application of digital computers to Fourier methods with continuous functions, one must necessarily treat a discrete set of sampled values over a finite interval of time or distance. Furthermore, one has effects due to "rounding" errors resulting from representing and computing all quantities with finite numbers of digits. *Having the theory of discrete Fourier transforms and a set of subroutines to perform the transforms described in the previous section, one is presented with the problem of knowing how to sample the given data and prepare a sequence as input to a subroutine, and then how to interpret the output of the subroutine in terms of the desired solution.*

The present section briefly reviews a theorem [4] which relates the Fourier integral transform to the DFT in such a way as to give insight into the effects of sampling rates and intervals. From this, one hopes to be able to estimate optimal parameters and produce results with prescribed error bounds.

Consider the integral Fourier transform of a function $x(t)$:

$$a(f) = \int_{-\infty}^{\infty} x(t)\,\mathrm{e}^{-2\pi i f t}\,\mathrm{d}t \tag{60}$$

and its inverse

$$x(t) = \int_{-\infty}^{\infty} a(f)\,\mathrm{e}^{2\pi i f t}\,\mathrm{d}f. \tag{61}$$

This relationship will be denoted

$$x(t) \leftrightarrow a(f). \tag{62}$$

To observe the effect of sampling at finite intervals, express (61), evaluated at the points $x_j = j.\Delta t, j = 0, \pm 1, \pm 2, \ldots$, with $F = 1/\Delta t$,

$$x(t_j) = \int_{-\infty}^{\infty} a(f)\,\mathrm{e}^{2\pi i j f/F}\,\mathrm{d}f = \sum_{k=-\infty}^{\infty} \int_{kF}^{(k+1)F} a(f)\,\mathrm{e}^{2\pi i j f/F}\,\mathrm{d}f. \tag{63}$$

The exponential function $\mathrm{e}^{2\pi i j f/F}$ is periodic in f with period F, so this can be written

$$x(j\Delta t) = \int_{0}^{F} a_p(f)\,\mathrm{e}^{2\pi i j f/F}\,\mathrm{d}f, \tag{64}$$

where

$$a_p(f) = \sum_{k=-\infty}^{\infty} a(f + kF). \tag{65}$$

Thus, knowing $x(t)$ only as sampling points, the best one can do about obtaining $a(f)$ is to compute $a_p(f)$. The latter, it is noted, differs from $a(f)$ by the sum of the $a(f)$'s displaced by all multiples of F. This error is referred to as "aliasing" and, if $a(f)$ is monotonically decreasing as a function of $|f|$, there will be an error of roughly 100% at $f = F/2$, the Nyquist frequency.

Since $a_p(f)$ is periodic with period F, Fourier's theorem says it has a series expansion in powers of $\mathrm{e}^{-2\pi i f/F}$ whose coefficients are given by $(1/F)x(j.\Delta t)$. Thus,

$$a_p(f) = \frac{1}{F} \sum_{j=-\infty}^{\infty} x(j.\Delta t)\,\mathrm{e}^{-2\pi i f j/F}. \tag{66}$$

To consider the effect of sampling in the frequency domain at points $f_n = n.\Delta f$, $n = 0, \pm 1, \pm 2, \ldots$, where $\Delta f = 1/T$, write (66),

$$a_p(n.\Delta f) = \frac{1}{F} \sum_{j=-\infty}^{\infty} x(j.\Delta t)\, e^{-2\pi i n j/TF}. \tag{67}$$

Taking $T.F = N$ to be an integer, we use the fact that $e^{-2\pi i j n/N}$ is a periodic function of j, with period N, to put (67) in the form

$$a_p(n.\Delta f) = \frac{1}{F} \sum_{j=0}^{N-1} x_p(j\Delta t)\, e^{-2\pi i n j/N}, \tag{68}$$

where

$$x_p(j\Delta t) = \sum_{k=-\infty}^{\infty} x((j + kN)\Delta t). \tag{69}$$

The sum in (69) gives the sampled values of

$$x_p(t) = \sum_{k=-\infty}^{\infty} x(t + kT), \tag{70}$$

a periodic function of t, with period T, which is formed from $x(t)$ in exactly the same way as $a_p(f)$ is formed from $a(f)$. Substituting $\Delta t = T/N$ for $1/F$, (68) can be written

$$a_p(n.\Delta f) = \frac{1}{N} \sum_{j=0}^{N-1} T.x_p(j\Delta t)\, e^{-2\pi i n j/N}. \tag{71}$$

This is of the same form as the DFT defined by (2) if we let

$$X(j) = Tx_p(j\Delta t), \tag{72}$$

$$A(n) = a_p(n.\Delta f). \tag{73}$$

This "proves" the central theorem of this section. (A rigorous proof under rigorous conditions is beyond the scope of this paper.)

Theorem 10. If

$$x(t) \leftrightarrow a(f), \tag{74}$$

then

$$T.x_p(j.\Delta t) \leftrightarrow a_p(n.\Delta f), \quad j, n = 0, 1, 2, \ldots, N-1, \tag{75}$$

where $N = 1/\Delta t.\Delta f$.

An example of the use of this theorem is as follows. Suppose $x(t)$ is given and one wishes to compute $a(f)$, where it is assumed that $x(t)$ and $a(f)$ are related by (60) and that the integrals (60) and (61) exist. One may select F, a frequency interval, and N, a number of points. The parameters,

$$\Delta f = F/N,$$

$$T = 1/\Delta f,$$

$$\Delta t = 1/F,$$

are all determined. One then generates sampled values, $x_p(t_j)$ at $t_j = j.\Delta t, j = 0, 1, \ldots, N-1$ and lets $X(j) = Tx_p(t_j)$ be the input sequence to a DFT subroutine. The computed result will be $A(n) = a_p(f_n)$, the sampled values of $a_p(f)$ at sampling points $f_n = n.\Delta f$, where $n = 0, 1, \ldots, N-1$.

As an example, the results of a calculation of the Fourier transform of the function

$$x(t) = e^{-t} \quad \text{for } t > 0,$$
$$x(t) = 0 \quad \text{for } t < 0, \tag{76}$$

is shown in Figure 1. The correct result is

$$a(f) = 1/(1 + 2\pi i f). \tag{77}$$

With $F = 2$ and $N = 16$, the remaining parameters are $T = 8$, $\Delta t = 1/2$, and $\Delta f = 1/8$. The point values of $x(t)$, used for the input vector, are indicated by the dots on the upper curve. Aliasing is negligible, since $x(t)$ is insignificant at $t = 8$.

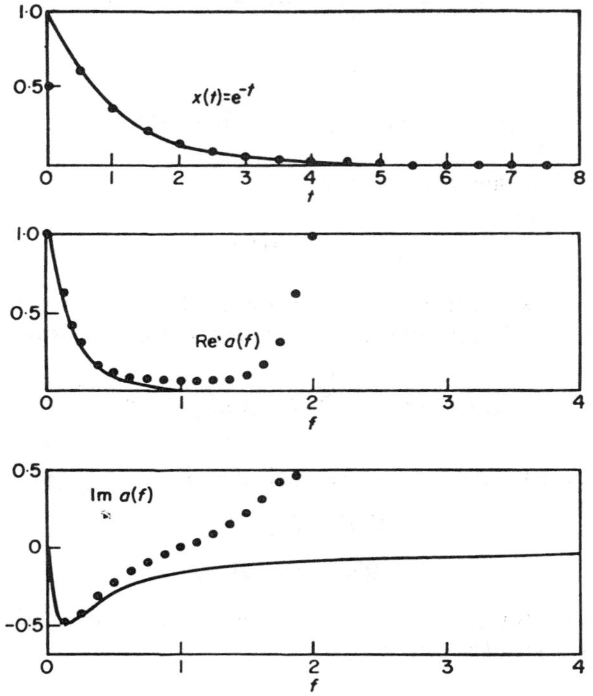

Figure 1. The calculation of the discrete Fourier transform of the function $x(t) = 0$ for $t < 0$, $x(t) = e^{-t}$ for $t > 0$ using $T = 8$, $N = 16$, $\Delta t = \frac{1}{2}$, $\Delta f = \frac{1}{8}$ and $F = 4$. Here $a(f) = 1/(1 + 2\pi i f)$.

The two lower solid curves are the real and imaginary parts of $a(f)$, and the computed results are indicated by the dots. For small f, the dots are quite close to the curve, but they deviate for larger f. At f near $F = 2$, the dots trace out a curvc looking like the negative half of the $a(f)$ curve. This is the aliasing present in the $a_p(f)$ function. It is clear what remedy one can use. It is merely to take a larger F, meaning a smaller Δt. In Figure 2, the same calculation, with $N = 32$ is shown. Here, $F = 4$ so that aliasing is reduced and a reasonable approximation to $a(f)$ is obtained over a wider range of f.

The procedure for computing $a(f)$ when given $x(t)$ can be derived from the above example. However, a word of caution about the formation of $a_p(f)$ is in order. A direct summation of

$$a_p(f) = \sum_{k=-\infty}^{\infty} a(f + kF) \tag{78}$$

may well fail to converge for a function like

$$a(f) = 1/(1 + 2\pi i f). \tag{79}$$

More will be said about this in the next section where Laplace transforms are discussed.

The coefficients of the Fourier series

$$x(t) = \sum_{n=-\infty}^{\infty} C(n)\,\mathrm{e}^{2\pi i n t/T}, \tag{80}$$

where $x(t)$ is defined for $0 < t < T$, can be obtained in terms of the DFT of the sequence $x(t_j)$, $t_j = j.\Delta t, j = 0, 1, \ldots, N-1, \Delta t = T/N$. This can be derived by appealing to theorem 10, but it may be simpler to obtain the results as follows. Consider

$$x(t_j) = \sum_{n=-\infty}^{\infty} C(n)\,\mathrm{e}^{2\pi i n j/N}. \tag{81}$$

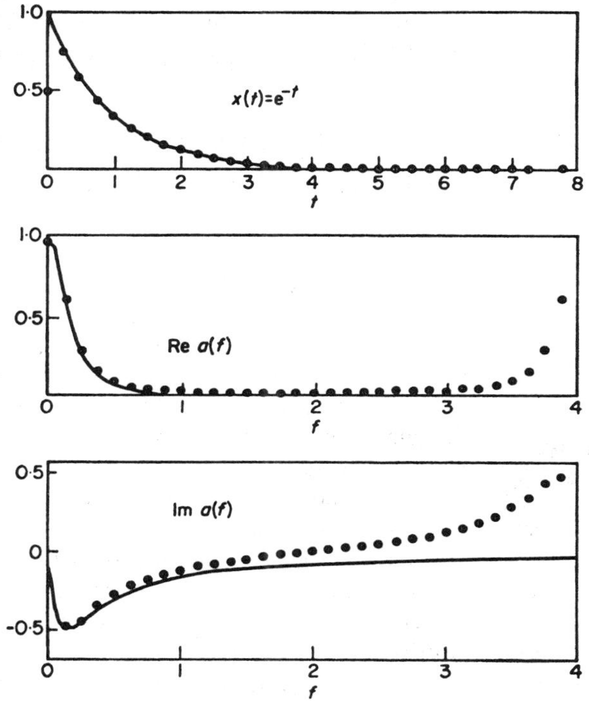

Figure 2. The same function and parameters as Figure 1 except that $N = 32$. This halves Δt (i.e. $\Delta t = \frac{1}{4}$) and doubles F (i.e. $F = 4$) so that aliasing errors in the solution are reduced.

Since $\mathrm{e}^{2\pi i j/N}$ is a periodic function of j with period N, the sum (81) can be expressed

$$x(t_j) = \sum_{n=0}^{N-1} \sum_{k=-\infty}^{\infty} C(n + kN)\,\mathrm{e}^{2\pi i n j/N},$$

giving

$$x(t_j) = \sum_{n=0}^{N-1} C_p(n)\,\mathrm{e}^{2\pi i n j/N}, \tag{82}$$

where $C_p(n)$ is the now familiar periodized function

$$C_p(n) = \sum_{k=-\infty}^{\infty} C(n + kN). \tag{83}$$

Consequently, we obtain the following theorem.

Theorem 11. Given T and N, and letting $t_j = j.\Delta t$, $\Delta t = T/N$, we have

$$x(t_j) \leftrightarrow C_p(n). \tag{84}$$

It is seen again that sampling in one domain causes aliasing in the other; sampling $x(t)$ at N points in the fixed interval $0 < t < T$ gives

$$C_p(n) = C(n) + \text{error}, \tag{85}$$

where

$$\text{error} = \sum_{k=1}^{\infty} [C(n + kN) + C(n - kN)]. \tag{86}$$

Thus, there will be aliasing at $n = k.N$ for all $k \neq 0$. Two procedures are then suggested: (i) when computing $C(n)$ from $x(t)$, one must take N so large that the error (86) is insignificant; (ii) when computing $x(t_j)$ from known values of $C(n)$, one first takes N to be the number of points desired, and then computes $C_p(n)$ according to equation (83). Then one computes its IDFT.

4. THE CALCULATION OF LAPLACE TRANSFORMS

The problem of inverting the Laplace transform, i.e. of determining a time function $g(t)$ from its Laplace transform,

$$G(s) = \int_0^{\infty} g(t)\,\mathrm{e}^{-st}\,\mathrm{d}t, \tag{87}$$

can often be solved analytically by applying a partial fraction expansion or an integration along some contour in the complex s-plane. When this proves to be too difficult or impossible due to the complexity of the formula for $G(s)$, or when $G(s)$ is known only in terms of its numerical values in the complex plane, numerical methods may be necessary. The inherent nature of the problem makes it impossible to suggest a unified approach to the problem. Rather, it happens that methods must be selected which only apply to some classes of functions $G(s)$, but not all. (For general references, see references 6 and 7.)

There have been two fairly general numerical approaches to the problem. One is to expand the unknown function as a series in a complete set of orthogonal functions. In the case where Laguerre functions are used, it has been shown by Weeks [8] that the expansion coefficients can be expressed as inverse Fourier transforms. Then Wing [9] modified the technique by using the FFT method to compute the inverse Fourier transform. This procedure is, of course, restricted in usefulness to functions which are representable as rapidly convergent series of Laguerre functions. A more general method has been suggested by Dubner and Abate [10]. They expressed the Laplace transform as a Fourier cosine transform whose inverse is expressible as a Fourier cosine series. Both methods will be described here as applications of theorems 10 and 11. The error analysis follows in terms of the aliasing mentioned there. Dubner and Abate neglected to mention the effect of rounding in choosing an optimal parameter c, to be defined below. Here, a formula for an optimal c is derived which balances aliasing and rounding errors.

In both of the methods described here, it is necessary to use values of $G(s)$ with the complex variable $s = c + i\omega$ on a vertical line in the complex plane. This is necessary so that one can use the periodicity in ω of the exponential function.

The method devised by Papoulis [6] uses an expansion

$$g(t) = \mathrm{e}^{-t} \sum_{n=0}^{\infty} C(n) L_n(t) \tag{88}$$

in Laguerre functions $L_n(t)$. Substituting this in (87), one gets

$$G(s) = \sum_{n=0}^{\infty} C(n) \frac{s^n}{(s+1)^{n+1}}, \tag{89}$$

for which Papoulis gives simple recursion formulas for the $C(n)$'s in terms of the values of the derivatives of $G(s)$ at $s = 0$.

Weeks' [8] modification of this consisted in letting $s = c + i\omega$ and using the expansion

$$g(t) = e^{(c-1/2T)t} \sum_{n=0}^{\infty} C(n) L_n(t/T). \tag{90}$$

This gives

$$G(s) = \sum_{n=0}^{\infty} C(n) \frac{(\omega i - T/2)^n}{(\omega i + T/2)^{n+1}}. \tag{91}$$

The factor $(\omega i - T/2)/(\omega i + T/2)$ has modulus 1 so one can define θ such that

$$e^{i\theta} = \frac{\omega i - T/2}{\omega i + T/2}, \tag{92}$$

and obtain a Fourier series

$$(\omega i + T/2)\, G(c + i\omega) = \sum_{n=0}^{\infty} C(n)\, e^{i\theta n}. \tag{93}$$

Solving (92) one gets ω in terms of θ;

$$\omega = (T/2) \cot \theta/2. \tag{94}$$

O. Wing [9] introduced the FFT algorithm to this method to solve (93) for the $C(n)$'s in terms of the values of the left side of (93).

Letting

$$x(\theta) = (\omega i + T/2)\, G(c + i\omega), \tag{95}$$

where ω is defined by (94), we note first that $x(\theta)$ is periodic in θ and that theorem 11 applies. However, the Fourier transform of $x(\theta)$, which is the sequence $C(n)$, is going to have aliasing about $n = N$, where N is the number of θ-intervals used. It can be seen that $x(\theta) = \bar{x}(-\theta)$ implies that real coefficients $C_p(n)$ will result. Weeks [8] used the even function $x_e(\theta) = x(\theta) + x(-\theta)$ in order to express the transform as a cosine transform of real sequences. This would permit one to use the cosine-transform-of-real-data form of the FFT algorithm as given in procedure 6 of section 2, but would require doubling N to keep aliasing errors within limits; therefore, procedure 3 giving the transform of conjugate symmetric data to real data would be recommended here. Next, there is the additional problem of choosing the optimal parameters T and c. Weeks [8] gives, as an empirical rule, the optimal parameter T by the formula

$$T = \frac{t_{max}}{N}, \tag{96}$$

where t_{max} is the maximum value of t for which a solution is desired. For the choice of optimal c, he suggests, on the basis of empirical evidence, the value

$$c = \max\left(c_0 + \frac{1}{t_{max}}, 0\right), \tag{97}$$

where c_0 is the real part of the rightmost pole of $G(s)$ in the complex plane.

To summarize, the procedure for using the FFT method to obtain values of an inverse Fourier transform in terms of an expansion in Laguerre functions is as follows.

(1) Select a scale factor T, a constant c, and an N.

(2) Let

$$\Delta\theta = 2\pi/N \tag{98}$$

and let

$$\theta_j = j.\Delta\theta, \quad j = 0, 1, 2, \ldots, N-1.$$

(3) Compute

$$\omega_j = (T/2)\cot\theta_j/2, \quad j = 0, 1, \ldots, N-1. \tag{99}$$

(4) Compute

$$X_j = (\omega_j i + T/2)\,G(c + i\omega_j) \quad \text{for } j = 0, 1, \ldots, N-1. \tag{100}$$

(5) Go to the FFT algorithm to solve,

$$X_j = \sum_{n=0}^{N-1} C_p(n)\exp(2\pi inj/N) \tag{101}$$

for the $C_p(n)$'s, $n = 0, 1, 2, \ldots, N-1$.

(6) For each value of t at which the inverse Laplace transform is desired, assume that $C(n) \approx C_p(n)$ for $n = 0, 1, \ldots, N-1$ and evaluate the series,

$$g(t) = e^{(c-T/2)t}\sum_{n=0}^{N-1} C(n)L_n(t/T). \tag{102}$$

The second method for inverting Laplace transforms is almost a direct application of theorem 10. In fact, in examples 1 and 2 of section 3, it was mentioned that the function being calculated could be regarded as a Laplace transform.

As before, we write the Laplace transform of a real function of time $g(t)$ as

$$G(s) = \int_0^\infty g(t)\,e^{-st}\,dt, \tag{103}$$

and let $s = c + i\omega$, where c and ω are real. Defining f by $\omega = 2\pi f$, we have

$$G(c + 2\pi if) = \int_0^\infty g(t)\,e^{-ct}\,e^{-2\pi ift}\,dt. \tag{104}$$

This is an integral Fourier transform pair where the functions involved are

$$a(f) = G(c + 2\pi if) \tag{105}$$

and

$$\begin{aligned} x(t) &= g(t)\,e^{-ct} && \text{if } t > 0, \\ &= 0 && \text{if } t < 0. \end{aligned} \tag{106}$$

The entire procedure for computing one of these functions when knowing values of the other function is as given before in section 3, examples 1 and 2. Aliasing and error considerations are the same. However, there is one new parameter c, which may be chosen by the user to optimize the calculation. Theoretically, c need only be taken larger than c_0, the maximum of the real parts of the poles of $G(s)$. However, there is a finite range of values of c which are useful in terms of numerical accuracy and computational effort. To see that the range of useful c-values is finite, we may observe first that the effect of large c is to attenuate $x(t)$, having the favorable effect of diminishing the aliasing in $x(t)$. The unfavorable consequence of using a large c is that for t also large, multiplying $x(t)$ by e^{ct} to get $g(t)$ increases the effect of rounding errors in $x(t)$.

Before making a quantitative analysis of this error, we will first describe another source of error. This is in the calculation of a $a_p(f)$, the periodized $a(f)$ function. For example, consider the common situation where the given Laplace transform behaves asymptotically like $1/s$ for large s. In this case, the infinite sum in the evaluation of $a_p(f)$ may have unbounded errors as more terms are taken. To deal with this problem, a slight variation on the above procedure can be made.

To do this, substitute (106) into (103) to obtain $a(f)$ as a Fourier transform

$$a(f) = \int_0^\infty x(t)\,\mathrm{e}^{-2\pi i f t}\,\mathrm{d}t. \tag{107}$$

Using the fact that $x(t)$ is real, the conjugate of $a(f)$ is

$$\widetilde{a(f)} = \int_0^\infty x(t)\,\mathrm{e}^{2\pi i f t}\,\mathrm{d}t, \tag{108}$$

which, added to $a(f)$, gives

$$\operatorname{Re} a(f) = \int_0^\infty x(t)\cos 2\pi f t\,\mathrm{d}t. \tag{109}$$

Subtracting, we can also get

$$\operatorname{Im} a(f) = \int_0^\infty x(t)\sin 2\pi f t\,\mathrm{d}t. \tag{110}$$

Procedures 5 and 6 of section 2 describe how one can compute the cosine and sine transforms.

Manipulation of these integrals yields the complex integral transforms in the forms

$$\operatorname{Re} a(f) = \int_{-\infty}^\infty x_e(t)\,\mathrm{e}^{-2\pi i f t}\,\mathrm{d}t, \tag{111}$$

$$\operatorname{Im} a(f) = \int_{-\infty}^\infty x_0(t)\,\mathrm{e}^{-2\pi i f t}\,\mathrm{d}t, \tag{112}$$

where

$$x_e(t) = \tfrac{1}{2}[x(t) + x(-t)] \tag{113}$$

and

$$x_0(t) = \tfrac{1}{2}[x(t) - x(-t)] \tag{114}$$

are the even and odd parts of $x(t)$, respectively.

Examples 1 and 2 of section 3 treated the case where $c = 1$ in the function

$$G(s) = 1/s, \quad s = c + i\omega, \quad \omega = 2\pi f.$$

This can be written

$$G(s) = \frac{c}{c^2 + \omega^2} - i\frac{\omega}{c^2 + \omega^2}, \tag{115}$$

giving

$$\operatorname{Re} a(f) = \frac{c}{c^2 + (2\pi f)^2} \tag{116}$$

and

$$\operatorname{Im} a(f) = -\frac{2\pi f}{c^2 + (2\pi f)^2}.$$

The sum

$$\operatorname{Re} a_p(f) = \sum_{k=-\infty}^{\infty} \operatorname{Re} a(f + kF) \tag{117}$$

$$= \sum_{k=-\infty}^{\infty} \frac{1}{c^2 + (2\pi)^2(f + kF)^2}$$

does converge and is a real symmetric function of f. Now, applying theorem 10, the inverse discrete Fourier transform of the sequence

$$A(n) = \operatorname{Re} a_p(n.\Delta f) \tag{118}$$

is the sequence

$$X(j) = T.x_{ep}(j\Delta t), \tag{119}$$

where

$$x_{ep}(t) = \sum_{k=-\infty}^{\infty} x_e(t+kT) \tag{120}$$

is the periodized $x_e(t)$.

The first error we will consider is that due to the truncation of the infinite series for $\operatorname{Re} a_p(f)$ by including only terms in a finite range $-K \leqslant k \leqslant K-1$. The effect of this is to replace $a(f)$ by the truncated function

$$a^*(f) = \beta(f;\bar{F}).a(f), \tag{121}$$

where $\bar{F} = KF$ and

$$\beta(f;\bar{F}) = 1 \quad \text{if } |f| < \bar{F}, \qquad = 0 \quad \text{otherwise.} \tag{122}$$

The inverse Fourier transform of the product of $a(f)$ and $\beta(f;\bar{F})$ is the convolution of their inverse transforms [4]. Thus, defining $x^*(t)$ by

$$\operatorname{Re} a^*(f) = \int_{-\infty}^{\infty} a_e^*(t)\, e^{-2\pi i f t}\, dt, \tag{123}$$

we have

$$x_e^*(t) = \int_{-\infty}^{\infty} x_e(t'-t)\, 2\bar{F} \operatorname{dif}(2\bar{F}t')\, dt', \tag{124}$$

where

$$\operatorname{dif} Z = \frac{\sin \pi Z}{\pi Z}. \tag{125}$$

The effect, therefore, of simply truncating the series is to convolve the desired inverse transform $x(t)$ with $2\bar{F} \times \operatorname{dif}(2\bar{F}t)$. This is a function which approaches the $\delta(t)$ function as $\bar{F} \rightarrow \infty$. Other "windows" on the Fourier transform than $\beta(f;\bar{F})$ are possible. A practical procedure is to limit errors by computing the infinite series (117) until the terms become insignificant or until one can estimate the tails of the series by some asymptotic formula. The procedure one should use depends on the behavior of the function $a(f)$ and how it is described; i.e., a functional expression, or numerical values obtained by experiment. Another observation to be made regarding this error is that the error in $x(t)$ due to truncation of the series may be expected to look like the frequency spectrum of the sum of the neglected terms of the series for $a_p(f)$. For example, $\operatorname{Re} 1/s$ is fairly close to being a constant function of f for K large and

$$KF \leqslant f < (K+1)F. \tag{126}$$

Hence, the greatest error will be in $x(0)$.

Since the manner of computing $a_p(f)$ and its effect upon error is so problem-dependent, it will not be discussed further here. The following analysis will be carried out as though the errors due to the calculation of $a_p(f)$ are negligible.

We will next consider the error due to aliasing in $x(t)$ and how it affects the optimal choice of parameters. Since the function $x_e(t)$ is even, the periodized function $x_{ep}(t)$ will be redundant in its upper half with $x_{ep}(t) = x_{ep}(T-t)$ for $T/2 \leqslant t < T$. Thus, we consider only $0 \leqslant t < T/2$. We first assume that T is sufficiently large so that aliasing in $x_p(t)$ in the interval $0 \leqslant t < T/2$ is due only to the $k = -1$ term of the series for $x_{ep}(t)$. Thus,

$$x_{ep}(t) \approx x_e(t) + x_e(t - T), \tag{127}$$

which, in the interval $0 \leqslant t < T/2$, is

$$x_{ep}(t) \approx \tfrac{1}{2}[x(t) + x(T-t)]. \tag{128}$$

When we multiply this by $2e^{ct}$, we get

$$\begin{aligned} 2e^{+ct} x_{ep}(t) &\approx e^{ct} x(t) + e^{ct} x(T-t) \\ &= g(t) + e^{-c(T-2t)} g(T-t). \end{aligned} \tag{129}$$

Thus, the aliasing error in the computed result is

$$E_a(t) = e^{-c(T-2t)} g(T-t). \tag{130}$$

This is a decreasing function of c and, with a given error criterion, ϵ, can be used to obtain a lower bound on c. On the other hand, an upper bound is imposed on c by the fact that the computed $x(t)$, with its rounding error $r(t)$, will have to be multiplied by e^{ct} to obtain $g(t)$. Thus, the rounding error in $g(t)$ will be

$$E_r(t) = e^{ct} r(t), \tag{131}$$

an increasing function of c. The optimal c, therefore, can be taken to be the value at which the maxima of the two errors, on an interval $0 \leqslant t < \tau$, are equal. Let

$$\epsilon_1 = \text{Max}\,|E_a(t)| = e^{-c(T-2\tau)} \hat{g}, \tag{132}$$

$$\epsilon_2 = \text{Max}\,|E_r(t)| = e^{c\tau} \hat{r}, \tag{133}$$

where

$$\hat{g} = \text{Max}\, g(T-t)$$

and

$$\hat{r} = \text{Max}\, r(t),$$

and where all maxima refer to the interval $0 < t < \tau$. Equating (132) and (133), we solve for c, getting

$$c = \frac{\ln(\hat{g}/\hat{r})}{T - \tau}. \tag{134}$$

Substituting this expression for c in (103) we obtain

$$\epsilon = \epsilon_1 + \epsilon_2 = 2\epsilon_1 = 2\epsilon_2 = 2\hat{g}^{\tau/(T-\tau)} \hat{r}^{(T-2\tau)/(T-\tau)}. \tag{135}$$

This error assumes the values,

$$\epsilon = 2\hat{r} \qquad \text{at } \tau = 0, \tag{136}$$

$$\epsilon = 2\hat{g}^{1/3} \hat{r}^{2/3} \qquad \text{at } \tau = T/4, \tag{137}$$

$$\epsilon = 2\hat{g} \qquad \text{at } \tau = T/2. \tag{138}$$

This shows that at $T/2$, the error is as large as $g(t)$ itself for all c. At smaller t-values, the error contains two factors: $\hat{g}$, arising from aliasing; and $\hat{r}$, arising from rounding. Therefore, the error criterion should be applied to a smaller range of t. A reasonable choice is to optimize the solution in the range $0 < t < T/4$. For this, one can take the c given by (134) for $\tau = T/4$,

$$c = (4/3T)\ln(\hat{g}/\hat{r}), \tag{139}$$

to obtain a solution with an error less than

$$\epsilon = 2\hat{g}^{1/3}\hat{r}^{2/3}. \tag{140}$$

As an application of formulas (139) and (140), let us first consider the problem in section 2 which treated the Laplace transform pair

$$G(s) = 1/s, \qquad g(t) = u(t), \tag{141}$$

where here, and in what follows, $u(t)$ is the step function

$$u(t) = \begin{cases} 0 & t < 0 \\ 0{\cdot}5 & t = 0 \\ 1 & t > 0. \end{cases} \tag{142}$$

In section 3 it was mentioned that this problem is equivalent to computing the Laplace transform of $u(t)$ with $s = 1 + i\omega$.

The numerical examples chosen here are designed to display the effects of the two sources of error. First, to determine an optimal c, we use the empirical result that errors in the inverse Fourier transform, $X(j)$, in single precision on the IBM 360 mod 67 are bounded by $N \times 10^{-6}$. The computed IDFT gives values of

$$X(j) = Tx_e(t_j) = T/2 . x(t_j). \tag{143}$$

Consequently, the error in $x(t)$ must be bounded by

$$\hat{r} \approx (2N/T)\,10^{-6} = 2F . 10^{-6}. \tag{144}$$

The parameters of the second part of the example in section 3 are $T = 8$, $N = 32$. These determine the parameters $\Delta t = 1/4$, $F = 4$, $\Delta f = 1/8$. In this example, $\hat{g} = 1$, and (144) gives $\hat{r} \approx 8 \times 10^{-6}$, resulting in the estimate of ϵ, from (140),

$$\epsilon = 2(8 \times 10^{-6})^{2/3} = 8 \times 10^{-4}. \tag{145}$$

From equation (139), we obtain

$$c = (4/3T) . \ln(10^6/8) = 15{\cdot}7/T, \tag{146}$$

which, with $T = 8$ gives $c \approx 2$.

Thus, for $G(s) = 1/s$, using $c = 2$, $t = 8$, and $N = 32$ should give the inverse Laplace transform in the range $0 < t < T/4 = 2$ with an error of no more than eight in the fourth digit.

To display further the effect of the choice of c, numerical calculations were carried out with the following transform pairs:

$$g(t) = u(t-4) \leftrightarrow G(s) = \exp(-4s)/s; \tag{147}$$

$$g(t) = J_0(t) \quad \leftrightarrow G(s) = 1/\sqrt{1+s^2}; \tag{148}$$

$$g(t) = t \quad \leftrightarrow G(s) = 1/s^2. \tag{149}$$

All three cases are run with $T = 32$ and $N = 32$. These determine the values

$$\Delta t = 1, \quad F = 1, \quad \Delta f = 1/32 \tag{150}$$

and, from (139),

$$\hat{r} = 2F \times 10^{-6} = 2 \times 10^{-6}.$$

For the first function (147), we have $\hat{g} = 1$ and obtain, from (139),

$$c = (4/3T)\ln(\hat{g}/\hat{r}) = 0{\cdot}55, \tag{151}$$

and, from (140),

$$\epsilon = 2\hat{g}^{1/3}\hat{r}^{2/3} = 2{\cdot}6 \times 10^{-4}. \tag{152}$$

The errors in the calculated values are given in Table 1 for $c = 0{\cdot}4$, $0{\cdot}55$, and $0{\cdot}7$. The best results are obtained with the predicted value of $c = 0{\cdot}55$; for the other values of c, errors accumulate for large t.

TABLE 1

Errors in the calculation of $g(t) = u(t-4)$, *where* $G(s) = \exp(-4s)/s$, *using* $T = 32$, $N = 32$, *and* $c = 0{\cdot}4$, $0{\cdot}55$, *and* $0{\cdot}7$

		|error| × 10⁴		
t	$g(t)$	$c = 0{\cdot}4$	$c = 0{\cdot}55$	$c = 0{\cdot}7$
0	0	0·4	0·2	0·1
1	0	0·6	0·3	0·2
2	0	1·1	0·7	0·5
3	0	2·7	2·0	1·9
4	0·5	1·2	0·0	0·5
5	1	2·9	5·2	6·5
6	1	0·3	4·1	6·0
7	1	4·7	3·9	7·1
8	1	13·7	3·6	10·7

Equations (152) and (151) predict an error bound of $\epsilon = 2{\cdot}6 \times 10^{-4}$ at the optimal $c = 0{\cdot}55$. Values of $G(s)$ with f up to $F = 160$ were used.

For the second function (148), the aliasing error will be smaller since $J_0(t)$ is a decreasing function of t. The bound $|J_0(t)| < \sqrt{2/\pi t} \approx 0{\cdot}8/\sqrt{t}$ is applied to the range $T/2 \leqslant t < T$ giving

$$\hat{g} = 0{\cdot}8/\sqrt{T/2} = 0{\cdot}2.$$

This gives

$$c = (4/96)\ln(0{\cdot}2) + 0{\cdot}55 = 0{\cdot}48 \tag{153}$$

and

$$\epsilon = 0{\cdot}2^{1/3} \times 2{\cdot}6 \times 10^{-4} = 1{\cdot}4 \times 10^{-4}. \tag{154}$$

The computed values of $g(t)$ in this case are seen in Table 2 to lie within the predicted error bound for the optimal c for all t except $t = 0$. The discrepancy at $t = 0$ is due to the truncation of the series for $a_p(f)$. The error in $a_p(f)$ is almost a constant function of f and, therefore affects almost exclusively, the value of $g(t)$ at $t = 0$ and has little effect on other values of t.

In the third case, (149), the fact that $g(t)$ is increasing means that aliasing will be more serious and a larger c will be required to attenuate it. Taking the bound $\hat{g} = 32$, we have

$$c = (4/96)\ln(32) + 0{\cdot}55 = 0{\cdot}79 \tag{155}$$

and

$$\epsilon = 32^{1/3} \times 2{\cdot}6 \times 10^{-4} = 17{\cdot}2 \times 10^{-4}. \tag{156}$$

The table of errors for $g(t) = t$ are given in Table 3. Here, the results obtained with $c = 0{\cdot}6$ are a little better than those obtained with $c = 0{\cdot}8$ although the predicted optimal value is

$c = 0{\cdot}79$. This is probably because the errors due to rounding are subject to random fluctuations and the value of c is not too critical within ranges of values as large as 0·2.

TABLE 2

Errors in the calculation of $g(t) = J_0(t)$, *where* $G(s) = 1/\sqrt{1 + s^2}$ *using* $T = 32$, $N = 32$, and $c = 0{\cdot}3, 0{\cdot}5, 0{\cdot}7$

		|error| × 10^4		
t	$g(t)$	$c = 0{\cdot}3$	$c = 0{\cdot}5$	$c = 0{\cdot}7$
0	1·00000	15·1	25·4	35·5
1	0·76520	0·0	0·1	0·1
2	0·22389	0·2	0·1	0·1
3	−0·26005	0·7	0·1	0·2
4	−0·39715	0·5	0·1	0·4
5	−0·17760	1·1	0·2	0·9
6	0·15064	4·0	0·3	1·8
7	0·30008	4·3	0·5	3·8
8	0·17165	4·8	1·3	7·6

Equations (154) and (153) predict an error bound of $\epsilon = 1{\cdot}4 \times 10^{-4}$ at the optimal $c = 0{\cdot}48$. Values of $G(s)$ with f up to $F = 20$ were used.

TABLE 3

Errors in the calculation of $g(t) = t$, *where* $G(s) = 1/s^2$ *using* $T = 32$, $N = 32$, *and* $c = 0{\cdot}4, 0{\cdot}6, 0{\cdot}8$ *and* $1{\cdot}0$

		|error| × 10^4			
t	$g(t)$	$c = 0{\cdot}4$	$c = 0{\cdot}6$	$c = 0{\cdot}8$	$c = 1{\cdot}0$
0	0	52·5	50·7	50·7	50·7
1	1	2·9	0·0	0·0	0·1
2	2	5·1	0·1	0·1	0·3
3	3	9·9	0·2	0·3	0·8
4	4	20·1	0·5	0·8	2·1
5	5	41·9	1·1	1·7	6·0
6	6	88·6	3·0	4·4	16·3
7	7	188·3	7·5	10·5	42·9
8	8	400·9	21·0	24·6	116·9

Equations (156) and (155) predict an error bound of $\epsilon = 17{\cdot}2 \times 10^4$ at the optimal $c = 0{\cdot}79$. Values of $G(s)$ with f up to $F = 20$ were used.

5. SUMMARY

The fast Fourier transform method (FFT) is an algorithm for computing, systematically and rapidly, the complex Fourier series. Programs using the basic idea of this algorithm for special situations where the data is real and/or cosine or sine transforms are required can be written. Instead, however, it is suggested here that a highly optimized subroutine or special hardware device should be made to compute the complex Fourier series only. This can be used with pre- and post-processing of the data to compute any of a variety of transforms, including cosine and sine transforms.

With a set of subroutines capable of computing sine, cosine, and Fourier transforms, one is then able, in a fairly straightforward manner, to compute and invert Laplace transforms. However, the method has three factors which limit its accuracy: one is due to the necessity of truncating the infinite sum (78) which defines $a_p(f)$; another results from rounding and depends on the number of significant digits carried by the computer being used, and the third comes from the aliasing due to the use of sampled values of function. The present analysis deals with each of these problems and results in a formula for selecting an optimal value of the parameter $c = \mathrm{Re}(s)$, where s is the Laplace transform variable.

It is hoped that the results obtained will permit computer programs automatically to select optimal parameters and operate efficiently on a wide class of functions. Such programs would provide important subroutines in larger programs for such problems as circuit analysis and the solution of systems of differential equations.

REFERENCES

1. J. W. Cooley and J. W. Tukey 1965 *Maths Comput.* **19**, 297. An algorithm for the machine calculation of complex Fourier series.
2. W. P. Cochran, J. W. Cooley, D. L. Favin, H. D. Helms, R. A. Kaenal, W. W. Lang, G. C. Maling Jr, D. E. Nelson, C. M. Rader and P. D. Welch 1967 *Proc. IEEE* **55**, 1664. What is the fast Fourier transform?
3. J. W. Cooley, P. A. W. Lewis and P. D. Welch 1967 *Proc. IEEE* **55**, 1675. Historical notes on the fast Fourier transform.
4. J. W. Cooley, P. A. W. Lewis and P. D. Welch 1967 *IEEE Trans. Audio Electroacoust.* **AU-15**, 79. Application of the fast Fourier transform to computation of Fourier integrals, Fourier series, and convolution integrals.
5. J. W. Cooley, P. A. W. Lewis and P. D. Welch 1969 *IBM Watson Research Center, Yorktown Heights, New York, RC 2285*. The discrete Fourier transform. Also published in amended form, 1969 *IEEE Trans Audio Electroacoust.* **AU-17**, 77. The discrete Fourier Transform?
6. A. Papoulis 1962 *The Fourier Integral and its Applications*. New York: McGraw-Hill.
7. R. Bellman, R. Kalaba and J. Lockett 1966 *Numerical Inversion of the Laplace Transform*. New York: American Elsevier.
8. W. T. Weeks 1966 *J. Ass. comput. Mach.* **13**, 419. Numerical inversion of Laplace transforms.
9. O. Wing 1967 *Archs electron. Comput.* **2**, 153. An efficient method of numerical inversion of Laplace transforms.
10. H. Dubner and J. Abate 1968 *J. Ass. comput. Mach.* **15**, 115. Numerical inversion of Laplace transforms.

An Algorithm for Computing the Mixed Radix Fast Fourier Transform

RICHARD C. SINGLETON, Senior Member, IEEE
Stanford Research Institute
Menlo Park, Calif. 94025

Abstract

This paper presents an algorithm for computing the fast Fourier transform, based on a method proposed by Cooley and Tukey. As in their algorithm, the dimension n of the transform is factored (if possible), and n/p elementary transforms of dimension p are computed for each factor p of n. An improved method of computing a transform step corresponding to an odd factor of n is given; with this method, the number of complex multiplications for an elementary transform of dimension p is reduced from $(p-1)^2$ to $(p-1)^2/4$ for odd p. The fast Fourier transform, when computed in place, requires a final permutation step to arrange the results in normal order. This algorithm includes an efficient method for permuting the results in place. The algorithm is described mathematically and illustrated by a FORTRAN subroutine.

Introduction

The fast Fourier transform (FFT) algorithm is an efficient method for computing the transformation

$$\alpha_k = \sum_{j=0}^{n-1} x_j \exp(i2\pi jk/n) \qquad (1)$$

for $k=0, 1, \cdots, n-1$, where $\{x_j\}$ and $\{\alpha_k\}$ are both complex-valued. The basic idea of the current form of the fast Fourier transform algorithm, that of factoring n,

$$n = \prod_{i=1}^{m} n_i,$$

and then decomposing the transform into m steps with n/n_i transformations of size n_i within each step, is that of Cooley and Tukey [1]. Most subsequent authors have directed their attention to the special case of $n=2^m$. Explanation and programming are simpler for $n=2^m$ than for the general case, and the restricted choice of values of n is adequate for a majority of applications. There are, however, some applications in which a wider choice of values of n is needed. The author has encountered this need in spectral analysis of speech and economic time series data.

Gentleman and Sande [2] have extended the development of the general case and describe possible variations in organizing the algorithm. They mention the existence of a mixed radix FFT program written by Sande. Available mixed radix programs include one in ALGOL by Singleton [3] and another in FORTRAN by Brenner [4]. A FORTRAN program based on the algorithm discussed here is included in Appendix I; this program was compared with Brenner's on several computers (CDC 6400, CDC 6600, IBM 360/67, and Burroughs B5500) and found to be significantly faster.

The Mixed Radix FFT

The complex Fourier transform (1) can be expressed as a matrix multiplication

$$\alpha = Tx,$$

where T is an $n\times n$ matrix of complex exponentials

$$t_{jk} = \exp(i2\pi jk/n).$$

In decomposing the matrix T, we use the factoring of Sande [2], rather than the original factoring of Cooley [1]. However, if the data are first permuted to digit-reversed order and then transformed, Cooley's factoring leads to an equally efficient algorithm.

In computing the fast Fourier transform, we factor T as

$$T = PF_mF_{m-1} \cdots F_2F_1,$$

where F_i is the transform step corresponding to the factor n_i of n and P is a permutation matrix. The matrix F_i has only n_i nonzero elements in each row and column and

Manuscript received December 2, 1968.

This work was supported by Stanford Research Institute, out of Research and Development funds.

Reprinted from *IEEE Trans. Audio Electroacoust.*, vol. AU-17, pp. 93–103, June 1969.

can be partitioned into n/n_i square submatrices of dimension n_i; it is this partition and the resulting reduction in multiplications that is the basis for the FFT algorithm. The matrices F_i can be further factored to yield

$$F_i = R_i T_i,$$

where R_i is a diagonal matrix of rotation factors (called twiddle factors by Gentleman and Sande [2]) and T_i can be partitioned into n/n_i identical square submatrices, each the matrix of a complex Fourier transform of dimension n_i. Although it might appear that this step increases the number of complex multiplications, it in fact enables us to exploit trigonometric function symmetries and multipliers of simple form (e.g., $e^{i\pi}$, $e^{i\pi/2}$, and $e^{i\pi/4}$) in computing T_i that more than compensate for the fewer than n multiplications in applying the rotation R_i. This point will be discussed further in later sections.

The permutation P is required because the transformed result is initially in digit-reversed order, i.e., the Fourier coefficient α_j, with

$$j = j_m n_{m-1} n_{m-2} \cdots n_1 + \cdots + j_2 n_1 + j_1,$$

is found in location

$$j' = j_1 n_2 n_3 \cdots n_m + j_2 n_3 n_4 \cdots n_m + \cdots + j_m.$$

As mentioned previously by the author [5], the permutation may be performed in place by pair interchanges if n is factored so that

$$n_i = n_{m-i}$$

for $i<n-i$. In this case, we can count j in natural order and j' in digit-reversed order, then exchange α_j and $\alpha_{j'}$ if $j<j'$. This method is a generalization of a well-known method for reordering the radix-2 FFT result.

Before computing the Fourier transform, we first decompose n into its prime factors. The square factors are arranged symmetrically about the factors of the square-free portion of n. Thus $n=270$ is factored as

$$3 \times 2 \times 3 \times 5 \times 3.$$

Then the permutation P is factored into two steps,

$$P = P_2 P_1.$$

The permutation P_1 is associated with the square factors of n and is done by pair interchanges as described above, except that the digits of n corresponding to the square-free factors are held constant and the digits of the square factors are exchanged symmetrically. Thus if

$$n = n_1 n_2 n_3 n_4 n_5 n_6 n_7$$

with $n_1=n_7$, $n_2=n_6$, and n_3, n_4, and n_5 are relatively prime, we interchange

$$j = j_7 n_6 n_5 \cdots n_1 + j_6 n_5 n_4 \cdots n_1 + j_5 n_4 n_3 n_2 n_1 + j_4 n_3 n_2 n_1 + j_3 n_2 n_1 + j_2 n_1 + j_1$$

and

$$j' = j_1 n_6 n_5 \cdots n_1 + j_2 n_5 n_4 \cdots n_1 + j_5 n_4 n_3 n_2 n_1 + j_4 n_3 n_2 n_1 + j_3 n_2 n_1 + j_5 n_1 + j_6.$$

The permutation P_1 in this case leaves each result element in its correct segment of length n/n_1n_2, grouped in subsequences of n_1n_2 consecutive elements. The permutation P_2 then completes the reordering by permuting the $n_3n_4n_5$ subsequences within each segment of length n/n_1n_2. In the FORTRAN subroutine given in Appendix I, P_2 is done by first determining the permutation cycles for digit reversal of the digits corresponding to the square-free factors, then permuting the data following these cycles. The permutation can be done using as few as two elements of temporary storage to hold a single complex result, but the program uses its available array space to permute the subsequences of length n_1n_2 if possible.

The Rotation Factor

In the previous section, we described the factoring of the transform step F_i, corresponding to a factor n_i of n, into a product R_iT_i of a matrix T_i of n/n_i identical Fourier transforms of dimension n_i and a diagonal rotation factor matrix R_i. Here we specify the elements of R_i for the Sande version of the FFT.

The rotation factor R_i following the transform step T_i has diagonal elements

$$r_j = \exp\left\{ i\frac{2\pi}{kk}(j \bmod k)\left[\frac{j \bmod kk}{k}\right]\right\}$$

for $j=0, 1, \cdots, n-1$ where

$$k = n/n_1n_2 \cdots n_i \quad \text{and} \quad kk = n_ik,$$

and the square brackets [ ] denote the greatest integer $\leq$ the enclosed quantity. The rotation factors multiplying each transform of dimension n_i within T_i have angles

$$0, \theta, 2\theta, \cdots (n_i - 1)\theta,$$

where θ may differ from one transform to another. No multiplication is needed for the zero angle, thus there are at most

$$n(n_i - 1)/n_i$$

complex multiplications to apply the rotation factor following the transform step T_i. In addition, $\theta=0$ for $(j \bmod k)=0$ or $(j \bmod kk)=0$, allowing the number of complex multiplications to be reduced by

$$(n_1 - 1) + n_1(n_2 - 1) + n_1n_2(n_3 - 1) + \cdots + n_1n_2 \cdots n_{m-1}(n_m - 1) = n - 1.$$

We note that the number of rotation factor multiplications is independent of the order of arrangement of the factors of n. The final rotation factor R_m has $\theta=0$ for all elements, and thus is omitted.

Counting Complex Multiplications

A complex multiplication, requiring four real multiplications and two real additions, is a relatively slow operation on most computers.[1] To a first approximation, the speed of an FFT algorithm is proportional to the number of complex multiplications used. The number of times we index through the data array is, however, an important secondary factor.

Using the results of the previous section, the number of complex multiplications for the rotation factors R_i is

$$\sum_{i=1}^{m} \frac{n(n_i - 1)}{n_i} - (n - 1),$$

assuming we avoid multiplication for all rotations of zero angle. To this number we must add the multiplications for the transform steps T_i.

For n a power of 2, we note that a complex Fourier transform of dimension 2 or 4 can be computed without multiplication and that a transform of dimension 8 requires only two real multiplications, equivalent to one-half a complex multiplication. Going one step further, a transform of dimension 16, computed as two factors of 4, requires the equivalent of six complex multiplications. Combining these results with the number of rotation factor multiplications and assuming that $n=2^m$ is a power of the radix, the total number of complex multiplications is as follows:

Radix	Number of Complex Multiplications
2	$mn/2-(n-1)$
4	$3mn/8-(n-1)$
8	$mn/3-(n-1)$
16	$21mn/64-(n-1)$

These results have been given previously by Bergland [6]. The savings for 16 over 8 is small, considering the added complexity of the algorithm. As Bergland points out, radix 8, with provision for an additional factor of 4 or 2, is a good choice for an efficient FFT program for powers of 2. For the mixed radix FFT, we transform with factors of 4 whenever possible, but also provide for factors of 2.

We now consider the number of complex multiplications for a radix-p transform of $n=p^m$ complex data values, where p is an odd prime. While at first it might appear that an elementary transform of dimension p requires $(p-1)^2$ complex multiplications, we show in the next section that $(p-1)^2$ real multiplications suffice, equivalent to $(p-1)^2/4$ complex multiplications. This result holds, in fact, for any odd value of p. Thus the transform steps for $n=p^m$ require the equivalent of

$$\frac{mn(p-1)^2}{4p}$$

complex multiplications. Adding the

$$\frac{mn(p-1)}{p} - (n-1)$$

multiplications for rotation factors, we obtain a total of

$$\frac{mn(p-1)(p+3)}{4p} - (n-1)$$

complex multiplications for a radix-p transform. Since p is assumed here to be an odd prime, we have no rotations with θ an integer multiple of $\pi/4$ to reduce further the number of complex multiplications.

The ratio of the number of complex multiplications to $n \log_2 n$ can serve as a measure of relative efficiency for the mixed radix FFT. The results of this section, neglecting the reduction by $n-1$ for $\theta=0$, yield the following comparison:

Radix	Relative Efficiency
2	0.500
4	0.375
8	0.333
16	0.328
3	0.631
5	0.689
7	0.763
11	0.920
13	0.998
17	1.151
19	1.227
23	1.374

The general term for an odd prime p is

$$\frac{(p-1)(p+3)}{4p \log_2 (p)}.$$

Decomposition of a Complex Fourier Transform

In the previous section, we promised to show that a complex transform of dimension p, for p odd, can be computed with $(p-1)^2$ real multiplications. Consider the complex transform

$$\begin{aligned} a_k + ib_k &= \sum_{j=0}^{p-1} (x_j + iy_j) \left\{ \cos\left(\frac{2\pi jk}{p}\right) + i \sin\left(\frac{2\pi jk}{p}\right) \right\} \\ &= x_0 + \sum_{j=1}^{p-1} x_j \cos\left(\frac{2\pi jk}{p}\right) - \sum_{j=1}^{p-1} y_j \sin\left(\frac{2\pi jk}{p}\right) \\ &\quad + i\left\{ y_0 + \sum_{j=1}^{p-1} y_j \cos\left(\frac{2\pi jk}{p}\right) + \sum_{j=1}^{p-1} x_j \sin\left(\frac{2\pi jk}{p}\right) \right\} \end{aligned}$$

[1] G. Golub (private communication) has pointed out that a complex multiplication can alternatively be done with three real multiplications and five real additions, as indicated by the following:

$$(a+ib)\cdot(c+id) = [(a+b)\cdot(c-d)+a\cdot d-b\cdot c]+i[a\cdot d+b\cdot c].$$

This method does not appear advantageous for FORTRAN coding, as the number of statements is increased from two to four.

$$= x_0 + \sum_{j=1}^{(p-1)/2} (x_j + x_{p-j}) \cos\left(\frac{2\pi jk}{p}\right)$$
$$- \sum_{j=1}^{(p-1)/2} (y_j - y_{p-j}) \sin\left(\frac{2\pi jk}{p}\right)$$
$$+ i\left\{y_0 + \sum_{j=1}^{(p-1)/2} (y_j + y_{p-j}) \cos\left(\frac{2\pi jk}{p}\right)\right.$$
$$\left. + \sum_{j=1}^{(p-1)/2} (x_j - x_{p-j}) \sin\left(\frac{2\pi jk}{p}\right)\right\}$$

for $k=0, 1, \cdots, p-1$. We note first that

$$a_0 + ib_0 = \sum_{j=0}^{p-1} (x_j + iy_j)$$

is computed without multiplications. The remaining Fourier coefficients can be expressed as

$$a_k = a_k^+ - a_k^-$$
$$a_{p-k} = a_k^+ + a_k^-$$
$$b_k = b_k^+ + b_k^-$$
$$b_{p-k} = b_k^+ - b_k^-$$

for $k=1, 2, \cdots, (p-1)/2$, where

$$a_k^+ = x_0 + \sum_{j=1}^{(p-1)/2} (x_j + x_{p-j}) \cos\left(\frac{2\pi jk}{p}\right)$$
$$a_k^- = \sum_{j=1}^{(p-1)/2} (y_j - y_{p-j}) \sin\left(\frac{2\pi jk}{p}\right)$$
$$b_k^+ = y_0 + \sum_{j=1}^{(p-1)/2} (y_j + y_{p-j}) \cos\left(\frac{2\pi jk}{p}\right)$$
$$b_k^- = \sum_{j=1}^{(p-1)/2} (x_j - x_{p-j}) \sin\left(\frac{2\pi jk}{p}\right).$$

Altogether there are $2(p-1)$ series to sum, each with $(p-1)/2$ multiplications, for a total of $(p-1)^2$ real multiplications.

For p an odd prime and for fixed j, the multipliers

$$\cos(2\pi jk/p) \qquad \text{for } k = 1, 2, \cdots (p-1)/2$$

have no duplications of magnitude, thus no further reduction in multiplications appears possible.[2] The same condition holds for the multipliers

$$\sin(2\pi jk/p) \qquad \text{for } k = 1, 2, \cdots (p-1)/2.$$

[2] C. M. Rader (private communication) has proposed an alternative decomposition of a Fourier transform of dimension 5, using the equivalent of 3 complex multiplications (12 real multiplications) instead of the 4 complex multiplications used in the algorithm described in this paper. In Appendix III we give a FORTRAN coding of Rader's method. When substituted in subroutine FFT (Appendix I), times were unchanged on the CDC 6600 computer and improved by about 5 percent for radix-5 transforms on the CDC 6400 computer (the 6400 has a relatively slower multiply operation). Rader's method looks advantageous for coding in machine language on a computer having multiple arithmetic registers available for temporary storage of intermediate results.

For even values of p, a decomposition similar to the above yields $4(p/2-1)$ series to sum, each with $(p/2-1)$ multiplications. Thus a complex Fourier transform for p even can be computed with at most $(p-2)^2$ real multiplications. For $p>2$, we know that this result can be improved. Combining results for the odd and even cases, we can state that a Fourier transform of dimension p can be computed with the equivalent of

$$\left[\frac{p-1}{2}\right]^2$$

or fewer complex multiplications, where the square brackets [ ] denote the largest integer value $\leq$ the enclosed quantity.

A Method for Computing Trigonometric Function Values

The trigonometric function values used in the fast Fourier transform can all be represented in terms of integer powers of

$$\exp(i2\pi/n),$$

the nth root of unity. Since we often use a sequence of equally spaced values on the unit circle, it is useful to have accurate methods of generating them by complex multiplication, rather than by repeated use of the library sine and cosine functions. For very short sequences, we use the simple method

$$\xi_{k+1} = \xi_k \exp(i\theta),$$

where

$$\xi_0 = 1$$

and $\{\xi_k\}$ is the sequence of computed values $\exp(ik\theta)$. This method suffers, however, from rapid accumulation of round-off errors. A better method, proposed by the author in an earlier paper [5], is to use the difference equation

$$\xi_{k+1} = \xi_k + \eta\xi_k,$$

where the multiplier

$$\eta = \exp(i\theta) - 1$$
$$= 2i \sin(\theta/2) \exp(i\theta/2)$$
$$= -2\sin^2(\theta/2) + i\sin(\theta)$$

decreases in magnitude with decreasing θ. This method gives good accurcay on a computer using rounded floating-point arithmetic (e.g., the Burroughs B5500). However, with truncated arithmetic (as on the IBM 360/67), the value of ξ_k tends to spiral inward from the unit circle with increasing k.

In Table I, we show the accumulated errors from extrapolating to $\pi/2$ in 2^k increments, using rounded arithmetic (machine language) and truncated arithmetic (FORTRAN) on a CDC 6400 computer; identical initial values, from the library sine and cosine functions, were used in computing the results in each of the three pairs of columns. In examining the second pair of columns, we find that the angle after 2^k extrapolation steps is very close to $\pi/2$, but that the magnitude has shrunk through truncation. To

TABLE I

Extrapolated Values of cos $\pi/2$ and sin $\pi/2-1$ on a CDC 6400 Computer, Using Rounded and Truncated Arithmetic Operations (Values in Units of 10^{-14})

Number of Extrapolations	Rounded Arithmetic Without Correction		Truncated Arithmetic Without Correction		Truncated Arithmetic With Correction	
	cos $\pi/2$	sin $\pi/2-1$	cos $\pi/2$	sin $\pi/2-1$	cos $\pi/2$	sin $\pi/2-1$
2^4	2.6	0.0	2.9	−3.6	2.8	−0.4
2^5	3.0	−0.7	4.2	−8.2	3.9	−0.4
2^6	2.5	0.0	3.8	−13.1	4.1	−0.4
2^7	5.1	−0.7	4.0	−26.3	3.6	−0.4
2^8	2.7	−0.7	4.5	−51.9	4.5	−0.4
2^9	4.9	−1.1	5.6	−104.4	5.2	−0.4
2^{10}	8.8	−1.1	5.2	−213.2	4.7	−0.4
2^{11}	14.2	2.1	2.8	−426.0	0.0	−0.4
2^{12}	13.2	−0.4	−1.1	−866.2	−5.4	−0.4
2^{13}	−0.2	−0.4	4.7	−1705.7	4.2	−0.4
2^{14}	−7.4	1.4	−9.6	−3440.1	1.2	−0.4
2^{15}	−10.3	−1.8	−9.4	−6841.5	36.9	−0.4
2^{16}	19.8	2.8	4.4	−13707.1	−33.3	−0.4
2^{17}	18.7	5.7	−8.9	−27416.7	3.5	−0.4

compensate for this shrinkage, we modify the above method to restore the extrapolated value to unit magnitude. We first compute a trial value

$$\gamma_k = \xi_k + \eta\xi_k$$

where

$$\eta = -2\sin^2(\theta/2) + i\sin(\theta)$$

and

$$\xi_0 = 1,$$

then multiply by a scalar

$$\xi_{k+1} = \delta_k\gamma_k,$$

where

$$\delta_k \approx \frac{1}{\sqrt{\gamma_k\gamma_k^*}},$$

to obtain the new value. Since $\gamma_k\gamma_k^*$ is very close to 1, we can avoid the library square-root function and use the approximation

$$\delta_k = \frac{1}{2}\left(\frac{1}{\gamma_k\gamma_k^*} + 1\right).$$

Or if division is more costly than multiplication, we can alternatively use the approximation

$$\delta_k = \tfrac{1}{2}(3 - \gamma_k\gamma_k^*).$$

On the CDC 6400 computer, both approximations give the results shown in the third pair of columns of Table I. This rescaling of magnitude uses four real multiplications and a divide (or five real multiplications) in addition to the four real multiplications to compute the trial value of γ_k. However, on most computers, these calculations will take less time than computing the values using the library trigonometric function.

The added step of rescaling the extrapolated trigonometric function values to the unit circle can also be used when computing with rounded arithmetic, but the gain in accuracy is small. The subroutines in Appendixes I and II include comment cards indicating the changes to remove the rescaling. On the other hand, the number of multiplications may be reduced by one when using truncated arithmetic, through using the overcorrection multiplier

$$\delta_k = 2 - \gamma_k\gamma_k^*.$$

In this case, the truncation bias stabilizes a method that mathematically borders on instability. On the CDC 6400 computer, this multiplier gives comparable accuracy to the multiplier suggested above.

A FORTRAN Subroutine for the Mixed Radix FFT

In Appendix I, we list a FORTRAN subroutine for computing the mixed radix FFT or its inverse, using the algorithm described above. This subroutine computes either a single-variate complex Fourier transform or the calculation for one variate of a multivariate transform.

To compute a single-variate transform (1) of n data values,

$$\text{CALL FFT}(A, B, n, n, n, 1).$$

The "inverse" transform

$$x_j = \sum_{k=0}^{n-1} \alpha_k \exp(-i2\pi jk/n)$$

is computed by

$$\text{CALL FFT}(A, B, n, n, n, -1).$$

Scaling is left to the user. The two calls in succession give the transformation

$$T^*Tx = nx,$$

i.e., n times the original values, except for round-off errors. The arrays A and B originally hold the real and imaginary components of the data, indexed from 1 to n;

the data values are replaced by the complex Fourier coefficients. Thus the real component of α_k is found in $A(k+1)$, and the imaginary component in $B(k+1)$, for $k=0, 1, \cdots, n-1$.

The difference between the transform and inverse calculation is primarily one of changing the sign of a variable holding the value 2π. The one additional change is to follow an alternative path within the radix-4 section of the program, using the angle $-\pi/2$ rather than $\pi/2$.

The use of the subroutine for multivariate transforms is described in the comments at the beginning of the program. To compute a bivariate transform on data stored in rectangular arrays A and B, the subroutine is called once to transform the columns and again to transform the rows. A multivariate transform is essentially a single-variate transform with modified indexing.

The subroutine as listed permits a maximum prime factor of 23, using four arrays of this dimension. The dimension of these arrays may be reduced to 1 if n contains no prime factors greater than 5. An array NP(209) is used in permuting the results to normal order; the present value permits a maximum of 210 for the product of the square-free factors of n. If n contains at most one square-free factor, the dimension of this array can be reduced to $j+1$, where j is the maximum number of prime factors of n. A sixth arrray NFAC(11) holds the factors of n. This is ample for any transform that can be done on a computer with core storage for 2^{17} real values (2^{16} complex values);

$$52\,488 = 2 \times 3^4 \times 2 \times 3^4 \times 2$$

is the only number $<2^{16}$ with as many as 11 factors, given the factoring used in this algorithm. The existing array dimensions do not permit unrestricted choice of n, but they rule out only a very small percentage of the possible values.

The transform portion of the subroutine includes sections for factors of 2, 3, 4, and 5, as well as a general section for odd factors. The sections for 2 and 4 include multiplication of each result value by the rotation factor; combining the two steps gives about a 10 percent speed improvement over using the general rotation factor section in the program, due to reduced indexing. The sections for 3 and 5 are similar to the general odd factors section, and they improve speed substantially for these factors by reducing indexing operations. The odd factors section is used for odd primes >5, but can handle any odd factor. The rotation factor section works for any factor but is used only for odd factors.

The permutation for square factors of n contains special code for single-variate transforms, since less indexing is required. However, the permutation for multivariate transforms also works on single-variate transforms.

The author has previously published an ALGOL procedure [3] of the same name and with a similar function. One significant difference between the two algorithms is that the ALGOL one is organized for computing large transforms on a virtual core memory system (e.g., the Burroughs B5500 computer). This constraint leads to a small loss in efficiency compared with the present algorithm. In the ALGOL algorithm, the Cooley version of the FFT algorithm is used, with a simulated recursive structure; rotation factor multiplication is included within the transform phase, requiring two additional arrays with dimension equal to the largest prime factor in n. The transform method for odd factors is like that used here. The permutation for square factors of n also has a simulated recursive structure, with one level of "recursion" for each square factor in n; in the present algorithm, this permutation is consolidated into a single step. The permutation for square-free factors is identical in both algorithms. The ALGOL algorithm contains a number of dynamic arrays, which is an obstacle to translation to FORTRAN. On the other hand, the FORTRAN subroutine given here can easily be translated to ALGOL, with the addition of dynamic upper bounds on all arrays other than NFAC; in making this translation, it would be desirable to modify the data indexing to go from 0 to $n-1$ to correspond with the mathematical notation.

Timing and Accuracy

The subroutine FFT was tested for time and accuracy on a CDC 6400 computer at Stanford Research Institute. The results are shown in Table II. The times are central processor times, which are measured with 0.002 second resolution; the times measured on successive runs rarely differ by more than 0.002 to 0.004 second. Furthermore, calling the subroutine with $n=2$ yields a timing result of 0 or 0.002 second; thus the time is apparently measured with negligible bias.

The data used in the trials were random normal deviates with a mean of zero and a standard deviation of one (i.e., an expected rms value of one). The subroutine was called twice:

$$\text{CALL FFT}(A, B, n, n, n, 1)$$
$$\text{CALL FFT}(A, B, n, n, n, -1);$$

then the result was scaled by $1/n$. The squared deviations from the original data values were summed, the real and imaginary quantities separately, then divided by n and square roots taken to yield an rms error value. The two values were in all cases comparable in magnitude, and an average is reported in Table II.

The measured times were normalized in two ways, first by dividing by

$$n \sum_{i=1}^{m} n_i,$$

and second by dividing by

$$n \log_2(n).$$

To a first approximation, computing time for the mixed radix FFT is proportional to n times the sum of the factors of n, and we observe in the present case that a proportionality constant of 25 μs gives a fair fit to this model.

TABLE II

Timing and Accuracy Tests of Subroutine FFT on a CDC 6400 Computer

Factoring of n	Time (seconds)	Time $n\sum n_i$ (μs)	Time $n \log_2 n$ (μs)	rms Error ($\times 10^{-13}$)
$512=4^2\times2\times4^2$	0.188	20.4	40.8	1.1
$1024=4^2\times4\times4^2$	0.398	19.4	38.9	1.2
$2048=4^2\times2\times2\times2\times4^2$	0.928	20.6	41.2	1.4
$4096=4^3\times4^3$	1.864	19.0	37.9	1.5
$2187=3^3\times3\times3^3$	1.494	32.5	61.6	1.6
$3125=5^2\times5\times5^2$	1.898	24.3	52.3	2.3
$2401=7^2\times7^2$	2.310	34.4	85.7	2.6
$1331=11\times11\times11$	1.324	30.1	95.9	2.5
$2197=13\times13\times13$	2.478	28.9	101.6	3.5
$289=17\times17$	0.272	27.7	115.1	2.5
$361=19\times19$	0.372	27.1	121.3	3.2
$529=23\times23$	0.636	26.1	132.9	3.5
$1000=2\times5\times2\times5\times2\times5$	0.546	26.0	54.8	1.6
$2000=4\times5\times5\times5\times4$	1.042	22.6	47.5	1.7

On the basis of counting complex multiplications, we would expect a decline in this proportionality constant with increasing radix; a decline is observed for odd primes > 5. Factors of 5 or less are of course favored by special coding in the program. The second normalized time value places all times on a comparable scale, allowing one to assess the relative efficiency of using values of n other than powers of 2; these results follow closely the relative efficiency values derived in an earlier section by counting complex multiplications, except that radix 5 is substantially better than predicted.

In Table III, we list the numbers up to 100 000 containing no prime factor greater than 5 to aid the user in selecting efficient values of n.

When compared with Brenner's FORTRAN subroutine [6] on the CDC 6400 computer, FFT was about 8 percent faster for radix 2, about 50 percent faster for radix 3 and 5, and about 22 percent faster for odd prime radix ≥ 7. Brenner's subroutine also requires working storage array space equal to that used for data when computing other than radix-2 transforms.

The FORTRAN style in the subroutine FFT was designed to simplify hand compiling into assembly language for the CDC 6600 to gain improved efficiency. Times on the CDC 6600 for the assembly language version are approximately 1/10 of those shown in Table II. The register arrangement of the CDC 6600 is well suited to the radix-2 FFT; the author has written a subroutine occupying 59 words of storage on this machine, including the constants used to generate all needed trigonometric function values, that computes a complex FFT for $n=1024$ in 42 ms.

Transforming Real Data

As others have pointed out previously, a single-variate Fourier transform of $2n$ real data values can be computed by use of a complex Fourier transform of dimension n. In Appendix II, we include a FORTRAN subroutine REALTR,

TABLE III

Numbers $\leq$ 100 000 Containing No Prime Factor Greater Than 5

2	288	1920	6750	19440	48000
3	300	1944	6912	19683	48600
4	320	2000	7200	20000	49152
5	324	2025	7290	20250	50000
6	360	2048	7500	20480	50625
8	375	2160	7680	20736	51200
9	384	2187	7776	21600	51840
10	400	2250	8000	21870	52488
12	405	2304	8100	22500	54000
15	432	2400	8192	23040	54675
16	450	2430	8640	23328	55296
18	480	2500	8748	24000	56250
20	486	2560	9000	24300	57600
24	500	2592	9216	24576	58320
25	512	2700	9375	25000	59049
27	540	2880	9600	25600	60000
30	576	2916	9720	25920	60750
32	600	3000	10000	26244	61440
36	625	3072	10125	27000	62208
40	640	3125	10240	27648	62500
45	648	3200	10368	28125	64000
48	675	3240	10800	28800	64800
50	720	3375	10935	29160	65536
54	729	3456	11250	30000	65610
60	750	3600	11520	30375	67500
64	768	3645	11664	30720	69120
72	800	3750	12000	31104	69984
75	810	3840	12150	31250	72000
80	864	3888	12288	32000	72900
81	900	4000	12500	32400	73728
90	960	4050	12800	32768	75000
96	972	4096	12960	32805	76800
100	1000	4320	13122	33750	77760
108	1024	4374	13500	34560	78125
120	1080	4500	13824	34992	78732
125	1125	4608	14400	36000	80000
128	1152	4800	14580	36450	81000
135	1200	4860	15000	36864	81920
144	1215	5000	15360	37500	82944
150	1250	5120	15552	38400	84375
160	1280	5184	15625	38880	86400
162	1296	5400	16000	39366	87480
180	1350	5625	16200	40000	90000
192	1440	5760	16384	40500	91125
200	1458	5832	16875	40960	92160
216	1500	6000	17280	41472	93312
225	1536	6075	17496	43200	93750
240	1600	6144	18000	43740	96000
243	1620	6250	18225	45000	97200
250	1728	6400	18432	46080	98304
256	1800	6480	18750	46656	98415
270	1875	6561	19200	46875	100000

similar to an ALGOL procedure REALTRAN given elsewhere [7] by the author.

The real data values are stored alternately in the arrays A and B,

$$A(1), B(1), A(2), B(2), \cdots A(n), B(n),$$

then we

$$\text{CALL FFT}(A, B, n, n, n, 1)$$

$$\text{CALL REALTR}\ (A, B, n, 1).$$

After scaling by $0.5/n$, the results in A and B are the Fourier cosine and sine coefficients, i.e.,

$$a_k = A(k+1)$$

$$b_k = B(k+1)$$

for $k=0, 1, \cdots, n$, with $b=b_n=0$. The inverse operation,

$$\text{CALL REALTR}\ (A, B, n, -1)$$

$$\text{CALL FFT}(A, B, n, n, n, -1),$$

after scaling by 1/2, evaluates the Fourier series and leaves the time domain values stored

$$A(1), B(1), A(2), B(2) \cdots A(n), B(n)$$

as originally.

The subroutine REALTR, called with ISN = 1, separates the complex transforms of the even- and odd-numbered data values, using the fact that the transform of real data has the complex conjugate symmetry

$$\alpha_{n-k} = \alpha_k^*$$

for $k = 1, 2, \cdots, n-1$, then performs a final radix-2 step to complete the transform for the $2n$ real values. If called with ISN = −1, the inverse operation is performed. The pair of calls

CALL REALTR $(A, B, n, 1)$

CALL REALTR $(A, B, n, -1)$

return the original values multiplied by 4, except for round-off errors.

Time on the CDC 6400 for $n = 1000$ is 0.100 second, and for $n = 2000$, 0.200 second. Time for REALTR is a linear function of n for other numbers of data values. The rms error for the above pair of calls of REALTR was 1.6×10^{-14} for both $n = 1000$ and $n = 2000$.

Conclusion

We have described an efficient algorithm for computing the mixed radix fast Fourier transform and have illustrated this algorithm by a FORTRAN subroutine FFT for computing multivariate transforms. The principal means of improving efficiency is the reduction in the number of complex multiplications for an odd prime factor p of n to approximately

$$n(p-1)(p+3)/4p.$$

The algorithm also permutes the result in place by pair interchanges for the square factors of n, using additional temporary storage during permutation only when n has two or more square-free factors.

A second subroutine REALTR for completing the transform of $2n$ real values is given, allowing efficient use of a complex transform of dimension n for the major portion of the computing in this case.

By use of these two subroutines, Fourier transforms can be computed for many possible values of n, with nearly as good efficiency as for n a power of 2. This expanded range of values has been found useful by the author in speech and economic time series analysis work.

Before Cooley and Tukey's paper [1], Good [8] presented a fast Fourier transform method based on decomposing n into mutually prime factors. This algorithm uses a more complicated indexing scheme than the Cooley–Tukey algorithm, but avoids the rotation factor multiplications. While the restriction to mutually prime factors is an obstacle to general use of Good's algorithm, we note that it could have been used here to transform the square-free factors of n. This alternative has not been tried, but the potential gain, if any, appears small.

References

[1] J. W. Cooley and J. W. Tukey, "An algorithm for the machine calculation of complex Fourier series," *Math. Comp.*, vol. 19, pp. 297–301, April 1965.

[2] W. M. Gentleman and G. Sande, "Fast Fourier transforms for fun and profit," *1966 Fall Joint Computer Conf., AFIPS Proc.*, vol. 29. Washington, D. C.: Spartan, 1966, pp. 563–578.

[3] R. C. Singleton, "An ALGOL procedure for the fast Fourier transform with arbitrary factors," *Commun ACM*, vol. 11, pp. 776–779, Algorithm 339, November 1968.

[4] N. M. Brenner, "Three FORTRAN programs that perform the Cooley–Tukey Fourier transform," M.I.T. Lincoln Lab., Lexington, Mass., Tech. Note 1967-2, July 1967.

[5] R. C. Singleton, "On computing the fast Fourier transform," *Commun. ACM*, vol. 10, pp. 647–654, October 1967.

[6] G. D. Bergland, "A fast Fourier transform algorithm using base 8 iterations," *Math. Comp.*, vol. 22, pp. 275–279, April 1968.

[7] R. C. Singleton, "ALGOL procedures for the fast Fourier transform," *Commun. ACM*, vol. 11, pp. 773–776, Algorithm 338, November 1968.

[8] I. J. Good, "The interaction algorithm and practical Fourier series," *J. Roy. Stat. Soc.*, ser. B, vol. 20, pp. 361–372, 1958; Addendum, vol. 22, pp. 372–375, 1960.

Appendix I

FORTRAN Subroutine FFT, for Computing the Mixed Radix Fourier Transform

```
      SUBROUTINE FFT(A,B,NTOT,N,NSPAN,ISN)
C  MULTIVARIATE COMPLEX FOURIER TRANSFORM, COMPUTED IN PLACE
C    USING MIXED-RADIX FAST FOURIER TRANSFORM ALGORITHM.
C  BY R. C. SINGLETON, STANFORD RESEARCH INSTITUTE, OCT. 1968
C  ARRAYS A AND B ORIGINALLY HOLD THE REAL AND IMAGINARY
C    COMPONENTS OF THE DATA, AND RETURN THE REAL AND
C    IMAGINARY COMPONENTS OF THE RESULTING FOURIER COEFFICIENTS.
C  MULTIVARIATE DATA IS INDEXED ACCORDING TO THE FORTRAN
C    ARRAY ELEMENT SUCCESSOR FUNCTION, WITHOUT LIMIT
C    ON THE NUMBER OF IMPLIED MULTIPLE SUBSCRIPTS.
C    THE SUBROUTINE IS CALLED ONCE FOR EACH VARIATE.
C    THE CALLS FOR A MULTIVARIATE TRANSFORM MAY BE IN ANY ORDER.
C  NTOT IS THE TOTAL NUMBER OF COMPLEX DATA VALUES.
C  N IS THE DIMENSION OF THE CURRENT VARIABLE.
C  NSPAN/N IS THE SPACING OF CONSECUTIVE DATA VALUES
C    WHILE INDEXING THE CURRENT VARIABLE.
C  THE SIGN OF ISN DETERMINES THE SIGN OF THE COMPLEX
C    EXPONENTIAL, AND THE MAGNITUDE OF ISN IS NORMALLY ONE.
C  A TRI-VARIATE TRANSFORM WITH A(N1,N2,N3), B(N1,N2,N3)
C    IS COMPUTED BY
C      CALL FFT(A,B,N1*N2*N3,N1,N1,1)
C      CALL FFT(A,B,N1*N2*N3,N2,N1*N2,1)
C      CALL FFT(A,B,N1*N2*N3,N3,N1*N2*N3,1)
C  FOR A SINGLE-VARIATE TRANSFORM,
C    NTOT = N = NSPAN = (NUMBER OF COMPLEX DATA VALUES), F.G.
C      CALL FFT(A,B,N,N,N,1)
C  THE DATA MAY ALTERNATIVELY BE STORED IN A SINGLE COMPLEX
C    ARRAY A, THEN THE MAGNITUDE OF ISN CHANGED TO TWO TO
C    GIVE THE CORRECT INDEXING INCREMENT AND A(2) USED TO
C    PASS THE INITIAL ADDRESS FOR THE SEQUENCE OF IMAGINARY
C    VALUES, E.G.
C      CALL FFT(A,A(2),NTOT,N,NSPAN,2)
C  ARRAYS AT(MAXF), CK(MAXF), BT(MAXF), SK(MAXF), AND NP(MAXP)
C    ARE USED FOR TEMPORARY STORAGE.  IF THE AVAILABLE STORAGE
C    IS INSUFFICIENT, THE PROGRAM IS TERMINATED BY A STOP.
C    MAXF MUST BE .GE. THE MAXIMUM PRIME FACTOR OF N.
C    MAXP MUST BE .GT. THE NUMBER OF PRIME FACTORS OF N.
C    IN ADDITION, IF THE SQUARE-FREE PORTION K OF N HAS TWO OR
C    MORE PRIME FACTORS, THEN MAXP MUST BE .GE. K-1.
      DIMENSION A(1),B(1)
C  ARRAY STORAGE IN NFAC FOR A MAXIMUM OF 11 FACTORS OF N.
C  IF N HAS MORE THAN ONE SQUARE-FREE FACTOR, THE PRODUCT OF THE
C    SQUARE-FREE FACTORS MUST BE .LE. 210
      DIMENSION NFAC(11),NP(209)
C  ARRAY STORAGE FOR MAXIMUM PRIME FACTOR OF 23
      DIMENSION AT(23),CK(23),BT(23),SK(23)
      EQUIVALENCE (I,II)
C  THE FOLLOWING TWO CONSTANTS SHOULD AGREE WITH THE ARRAY DIMENSIONS.
      MAXF=23
      MAXP=209
      IF(N .LT. 2) RETURN
      INC=ISN
      RAD=8.0*ATAN(1.0)
      S72=RAD/5.0
      C72=COS(S72)
      S72=SIN(S72)
      S120=SQRT(0.75)
      IF(ISN .GE. 0) GO TO 10
      S72=-S72
      S120=-S120
      RAD=-RAD
      INC=-INC
   10 NT=INC*NTOT
      KS=INC*NSPAN
      KSPAN=KS
      NN=NT-INC
```

```
      JC=KS/N
      RADF=RAC*FLCAT(JC)*0.5
      I=0
      JF=0
C  DETERMINE THE FACTCRS CF N
      M=0
      K=N
      GC TC 20
   15 M=M+1
      NFAC(M)=4
      K=K/16
   20 IF(K-(K/16)*16 .EC. 0) GO TO 15
      J=3
      JJ=9
      GO TC 30
   25 M=M+1
      NFAC(M)=J
      K=K/JJ
   30 IF(MCD(K,JJ) .EQ. 0) GC TO 25
      J=J+2
      JJ=J**2
      IF(JJ .LE. K) GC TC 30
      IF(K .GT. 4) GC TC 40
      KT=M
      NFAC(M+1)=K
      IF(K .NE. 1) M=M+1
      GO TC 80
   40 IF(K-(K/4)*4 .NE. 0) GC TC 50
      M=M+1
      NFAC(M)=2
      K=K/4
   50 KT=M
      J=2
   60 IF(MCD(K,J) .NE. 0) GC TC 70
      M=M+1
      NFAC(M)=J
      K=K/J
   70 J=((J+1)/2)*2+1
      IF(J .LE. K) GC TC 60
   80 IF(KT .EC. 0) GC TC 100
      J=KT
   90 M=M+1
      NFAC(M)=NFAC(J)
      J=J-1
      IF(J .NE. 0) GO TO 90
C  COMPUTE FOURIER TRANSFORM
  100 SD=RADF/FLOAT(KSPAN)
      CD=2.0*SIN(SD)**2
      SD=SIN(SD+SD)
      KK=1
      I=I+1
      IF(NFAC(I) .NE. 2) GO TO 400
C  TRANSFCRM FOR FACTOR OF 2 (INCLUDING ROTATION FACTOR)
      KSPAN=KSPAN/2
      K1=KSPAN+2
  210 K2=KK+KSPAN
      AK=A(K2)
      BK=B(K2)
      A(K2)=A(KK)-AK
      B(K2)=B(KK)-BK
      A(KK)=A(KK)+AK
      B(KK)=B(KK)+BK
      KK=K2+KSPAN
      IF(KK .LE. NN) GO TO 210
      KK=KK-NN
      IF(KK .LE. JC) GC TO 210
      IF(KK .GT. KSPAN) GO TC 800
  220 C1=1.0-CD
      S1=SD
  230 K2=KK+KSPAN
      AK=A(KK)-A(K2)
      BK=B(KK)-B(K2)
      A(KK)=A(KK)+A(K2)
      B(KK)=B(KK)+B(K2)
      A(K2)=C1*AK-S1*BK
      B(K2)=S1*AK+C1*BK
      KK=K2+KSPAN
      IF(KK .LT. NT) GO TO 230
      K2=KK-NT
      C1=-C1
      KK=K1-K2
      IF(KK .GT. K2) GO TO 230
      AK=C1-(CD*C1+SD*S1)
      S1=(SD*C1-CD*S1)+S1
C  THE FOLLCWING THREE STATEMENTS COMPENSATE FOR TRUNCATION
C    ERROR.  IF RCUNDED ARITHMETIC IS USED, SUBSTITUTE
C     C1=AK
      C1=0.5/(AK**2+S1**2)+0.5
      S1=C1*S1
      C1=C1*AK
      KK=KK+JC
      IF(KK .LT. K2) GO TO 230
      K1=K1+INC+INC
      KK=(K1-KSPAN)/2+JC
      IF(KK .LF. JC+JC) GO TO 220
      GO TO 100
C  TRANSFCRM FCR FACTCR OF 3 (OPTIONAL CODE)
  320 K1=KK+KSPAN
      K2=K1+KSPAN
      AK=A(KK)
      BK=B(KK)
      AJ=A(K1)+A(K2)
      BJ=B(K1)+B(K2)
      A(KK)=AK+AJ
      B(KK)=BK+BJ
      AK=-0.5*AJ+AK
      BK=-0.5*BJ+BK
      AJ=(A(K1)-A(K2))*S120
      BJ=(B(K1)-B(K2))*S120
      A(K1)=AK-BJ
      B(K1)=BK+AJ
      A(K2)=AK+BJ
      B(K2)=BK-AJ
      KK=K2+KSPAN
      IF(KK .LT. NN) GC TO 320
      KK=KK-NN
      IF(KK .LE. KSPAN) GO TO 320
      GC TC 700
C  TRANSFCRM FOR FACTOR CF 4
  400 IF(NFAC(I) .NE. 4) GC TO 6C0
      KSPNN=KSPAN
      KSPAN=KSPAN/4
  410 C1=1.0
      S1=0
  420 K1=KK+KSPAN
      K2=K1+KSPAN
      K3=K2+KSPAN
      AKP=A(KK)+A(K2)
      AKM=A(KK)-A(K2)
      AJP=A(K1)+A(K3)
      AJM=A(K1)-A(K3)
      A(KK)=AKP+AJP
      AJP=AKP-AJP
      BKP=B(KK)+B(K2)
      BKM=B(KK)-B(K2)
      BJP=B(K1)+B(K3)
      BJM=B(K1)-B(K3)
      B(KK)=BKP+BJP
      BJP=BKP-BJP
      IF(ISN .LT. 0) GC TO 450
      AKP=AKM-BJM
      AKM=AKM+BJM
      BKP=BKM+AJM
      BKM=BKM-AJM
      IF(S1 .EQ. 0.0) GO TO 460
  430 A(K1)=AKP*C1-BKP*S1
      B(K1)=AKP*S1+BKP*C1
      A(K2)=AJP*C2-BJP*S2
      B(K2)=AJP*S2+BJP*C2
      A(K3)=AKM*C3-BKM*S3
      B(K3)=AKM*S3+BKM*C3
      KK=K3+KSPAN
      IF(KK .LE. NT) GC TO 420
  440 C2=C1-(CD*C1+SD*S1)
      S1=(SD*C1-CD*S1)+S1
C  THE FOLLOWING THREE STATEMENTS COMPENSATE FOR TRUNCATION
C    ERROR.  IF ROUNDED ARITHMETIC IS USED, SUBSTITUTE
C     C1=C2
      C1=0.5/(C2**2+S1**2)+0.5
      S1=C1*S1
      C1=C1*C2
      C2=C1**2-S1**2
      S2=2.0*C1*S1
      C3=C2*C1-S2*S1
      S3=C2*S1+S2*C1
      KK=KK-NT+JC
      IF(KK .LE. KSPAN) GO TO 420
      KK=KK-KSPAN+INC
      IF(KK .LE. JC) GC TO 410
      IF(KSPAN .EQ. JC) GO TO 800
      GC TO 100
  450 AKP=AKM+BJM
      AKM=AKM-BJM
      BKP=BKM-AJM
      BKM=BKM+AJM
      IF(S1 .NE. 0.0) GO TO 430
  460 A(K1)=AKP
      B(K1)=BKP
      A(K2)=AJP
      B(K2)=BJP
      A(K3)=AKM
      B(K3)=BKM
      KK=K3+KSPAN
      IF(KK .LE. NT) GO TC 420
      GO TO 440
C  TRANSFCRM FOR FACTCR CF 5 (CPTIONAL CODE)
  510 C2=C72**2-S72**2
      S2=2.0*C72*S72
  520 K1=KK+KSPAN
      K2=K1+KSPAN
      K3=K2+KSPAN
      K4=K3+KSPAN
      AKP=A(K1)+A(K4)
      AKM=A(K1)-A(K4)
      BKP=B(K1)+B(K4)
      BKM=B(K1)-B(K4)
      AJP=A(K2)+A(K3)
      AJM=A(K2)-A(K3)
      BJP=B(K2)+B(K3)
      BJM=B(K2)-B(K3)
      AA=A(KK)
      BB=B(KK)
      A(KK)=AA+AKP+AJP
      B(KK)=BB+BKP+BJP
      AK=AKP*C72+AJP*C2+AA
      BK=BKP*C72+BJP*C2+BB
      AJ=AKM*S72+AJM*S2
      BJ=BKM*S72+BJM*S2
      A(K1)=AK-BJ
      A(K4)=AK+BJ
      B(K1)=BK+AJ
      B(K4)=BK-AJ
      AK=AKP*C2+AJP*C72+AA
      BK=BKP*C2+BJP*C72+BB
      AJ=AKM*S2-AJM*S72
      BJ=BKM*S2-BJM*S72
      A(K2)=AK-BJ
      A(K3)=AK+BJ
      B(K2)=BK+AJ
      B(K3)=BK-AJ
      KK=K4+KSPAN
      IF(KK .LT. NN) GO TO 520
      KK=KK-NN
      IF(KK .LE. KSPAN) GO TO 520
      GO TO 700
C  TRANSFCRM FOR ODD FACTORS
  600 K=NFAC(I)
      KSPNN=KSPAN
      KSPAN=KSPAN/K
      IF(K .EQ. 3) GO TC 320
      IF(K .EQ. 5) GO TO 510
      IF(K .EC. JF) GO TO 640
      JF=K
      S1=RAC/FLOAT(K)
      C1=CCS(S1)
      S1=SIN(S1)
      IF(JF .GT. MAXF) GC TO 998
      CK(JF)=1.0
      SK(JF)=0.0
      J=1
  630 CK(J)=CK(K)*C1+SK(K)*S1
      SK(J)=CK(K)*S1-SK(K)*C1
      K=K-1
      CK(K)=CK(J)
      SK(K)=-SK(J)
      J=J+1
      IF(J .LT. K) GO TC 630
```

```
  640 K1=KK
      K2=KK+KSPNN
      AA=A(KK)
      BB=B(KK)
      AK=AA
      BK=BB
      J=1
      K1=K1+KSPAN
  650 K2=K2-KSPAN
      J=J+1
      AT(J)=A(K1)+A(K2)
      AK=AT(J)+AK
      BT(J)=B(K1)+B(K2)
      BK=BT(J)+BK
      J=J+1
      AT(J)=A(K1)-A(K2)
      BT(J)=B(K1)-B(K2)
      K1=K1+KSPAN
      IF(K1 .LT. K2) GO TO 650
      A(KK)=AK
      B(KK)=BK
      K1=KK
      K2=KK+KSPNN
      J=1
  660 K1=K1+KSPAN
      K2=K2-KSPAN
      JJ=J
      AK=AA
      BK=BB
      AJ=0.0
      BJ=0.0
      K=1
  670 K=K+1
      AK=AT(K)*CK(JJ)+AK
      BK=BT(K)*CK(JJ)+BK
      K=K+1
      AJ=AT(K)*SK(JJ)+AJ
      BJ=BT(K)*SK(JJ)+BJ
      JJ=JJ+J
      IF(JJ .GT. JF) JJ=JJ-JF
      IF(K .LT. JF) GO TO 670
      K=JF-J
      A(K1)=AK-BJ
      B(K1)=BK+AJ
      A(K2)=AK+BJ
      B(K2)=BK-AJ
      J=J+1
      IF(J .LT. K) GO TO 660
      KK=KK+KSPNN
      IF(KK .LE. NN) GO TO 640
      KK=KK-NN
      IF(KK .LE. KSPAN) GO TO 640
C  MULTIPLY BY ROTATION FACTOR (EXCEPT FOR FACTORS OF 2 AND 4)
  700 IF(I .EQ. M) GO TO 800
      KK=JC+1
  710 C2=1.0-CD
      S1=SD
  720 C1=C2
      S2=S1
      KK=KK+KSPAN
  730 AK=A(KK)
      A(KK)=C2*AK-S2*B(KK)
      B(KK)=S2*AK+C2*B(KK)
      KK=KK+KSPNN
      IF(KK .LE. NT) GO TO 730
      AK=S1*S2
      S2=S1*C2+C1*S2
      C2=C1*C2-AK
      KK=KK-NT+KSPAN
      IF(KK .LE. KSPNN) GO TO 730
      C2=C1-(CD*C1+SD*S1)
      S1=S1+(SD*C1-CD*S1)
C  THE FOLLOWING THREE STATEMENTS COMPENSATE FOR TRUNCATION
C    ERROR.  IF ROUNDED ARITHMETIC IS USED, THEY MAY
C    BE DELETED.
      C1=0.5/(C2**2+S1**2)+0.5
      S1=C1*S1
      C2=C1*C2
      KK=KK-KSPNN+JC
      IF(KK .LE. KSPAN) GO TO 720
      KK=KK-KSPAN+JC+INC
      IF(KK .LE. JC+JC) GO TO 710
      GO TO 100
C  PERMUTE THE RESULTS TO NORMAL ORDER---DONE IN TWO STAGES
C  PERMUTATION FOR SQUARE FACTORS OF N
  800 NP(1)=KS
      IF(KT .EQ. 0) GO TO 890
      K=KT+KT+1
      IF(M .LT. K) K=K-1
      J=1
      NP(K+1)=JC
  810 NP(J+1)=NP(J)/NFAC(J)
      NP(K)=NP(K+1)*NFAC(J)
      J=J+1
      K=K-1
      IF(J .LT. K) GO TO 810
      K3=NP(K+1)
      KSPAN=NP(2)
      KK=JC+1
      K2=KSPAN+1
      J=1
      IF(N .NE. NTOT) GO TO 850
C  PERMUTATION FOR SINGLE-VARIATE TRANSFORM (OPTIONAL CODE)
  820 AK=A(KK)
      A(KK)=A(K2)
      A(K2)=AK
      BK=B(KK)
      B(KK)=B(K2)
      B(K2)=BK
      KK=KK+INC
      K2=KSPAN+K2
      IF(K2 .LT. KS) GO TO 820
  830 K2=K2-NP(J)
      J=J+1
      K2=NP(J+1)+K2
      IF(K2 .GT. NP(J)) GO TO 830
      J=1
  840 IF(KK .LT. K2) GO TO 820
      KK=KK+INC
      K2=KSPAN+K2
      IF(K2 .LT. KS) GO TO 840
      IF(KK .LT. KS) GO TO 830
      JC=K3
      GO TO 890
   PERMUTATION FOR MULTIVARIATE TRANSFORM
  850 K=KK+JC
  860 AK=A(KK)
      A(KK)=A(K2)
      A(K2)=AK
      BK=B(KK)
      B(KK)=B(K2)
      B(K2)=BK
      KK=KK+INC
      K2=K2+INC
      IF(KK .LT. K) GO TO 860
      KK=KK+KS-JC
      K2=K2+KS-JC
      IF(KK .LT. NT) GO TO 850
      K2=K2-NT+KSPAN
      KK=KK-NT+JC
      IF(K2 .LT. KS) GO TO 850
  870 K2=K2-NP(J)
      J=J+1
      K2=NP(J+1)+K2
      IF(K2 .GT. NP(J)) GO TO 870
      J=1
  880 IF(KK .LT. K2) GO TO 850
      KK=KK+JC
      K2=KSPAN+K2
      IF(K2 .LT. KS) GO TO 880
      IF(KK .LT. KS) GO TO 870
      JC=K3
  890 IF(2*KT+1 .GE. M) RETURN
      KSPNN=NP(KT+1)
C  PERMUTATION FOR SQUARE-FREE FACTORS OF N
      J=M-KT
      NFAC(J+1)=1
  900 NFAC(J)=NFAC(J)*NFAC(J+1)
      J=J-1
      IF(J .NE. KT) GO TO 900
      KT=KT+1
      NN=NFAC(KT)-1
      IF(NN .GT. MAXP) GO TO 998
      JJ=0
      J=0
      GO TO 906
  902 JJ=JJ-K2
      K2=KK
      K=K+1
      KK=NFAC(K)
  904 JJ=KK+JJ
      IF(JJ .GE. K2) GO TO 902
      NP(J)=JJ
  906 K2=NFAC(KT)
      K=KT+1
      KK=NFAC(K)
      J=J+1
      IF(J .LE. NN) GO TO 904
C  DETERMINE THE PERMUTATION CYCLES OF LENGTH GREATER THAN 1
      J=0
      GO TO 914
  910 K=KK
      KK=NP(K)
      NP(K)=-KK
      IF(KK .NE. J) GO TO 910
      K3=KK
  914 J=J+1
      KK=NP(J)
      IF(KK .LT. 0) GO TO 914
      IF(KK .NE. J) GO TO 910
      NP(J)=-J
      IF(J .NE. NN) GO TO 914
      MAXF=INC*MAXF
C  REORDER A AND B, FOLLOWING THE PERMUTATION CYCLES
      GO TO 950
  924 J=J-1
      IF(NP(J) .LT. 0) GO TO 924
      JJ=JC
  926 KSPAN=JJ
      IF(JJ .GT. MAXF) KSPAN=MAXF
      JJ=JJ-KSPAN
      K=NP(J)
      KK=JC*K+II+JJ
      K1=KK+KSPAN
      K2=0
  928 K2=K2+1
      AT(K2)=A(K1)
      BT(K2)=B(K1)
      K1=K1-INC
      IF(K1 .NE. KK) GO TO 928
  932 K1=KK+KSPAN
      K2=K1-JC*(K+NP(K))
      K=-NP(K)
  936 A(K1)=A(K2)
      B(K1)=B(K2)
      K1=K1-INC
      K2=K2-INC
      IF(K1 .NE. KK) GO TO 936
      KK=K2
      IF(K .NE. J) GO TO 932
      K1=KK+KSPAN
      K2=0
  940 K2=K2+1
      A(K1)=AT(K2)
      B(K1)=BT(K2)
      K1=K1-INC
      IF(K1 .NE. KK) GO TO 940
      IF(JJ .NE. 0) GO TO 926
      IF(J .NE. 1) GO TO 924
  950 J=K3+1
      NT=NT-KSPNN
      II=NT-INC+1
      IF(NT .GE. 0) GO TO 924
      RETURN
C  ERROR FINISH, INSUFFICIENT ARRAY STORAGE
  998 ISN=0
C     PRINT 999

      STOP
  999 FORMAT(44H0ARRAY BOUNDS EXCEEDED WITHIN SUBROUTINE FFT)
      END
```

Appendix II

FORTRAN Subroutine REALTR, for Completing the Fourier Transform of 2*n* Real Values

```
      SUBROUTINE REALTR(A,B,N,ISN)
C  IF ISN=1, THIS SUBROUTINE COMPLETES THE FOURIER TRANSFORM
C    OF 2*N REAL DATA VALUES, WHERE THE ORIGINAL DATA VALUES ARE
C    STORED ALTERNATELY IN ARRAYS A AND B, AND ARE FIRST
C    TRANSFORMED BY A COMPLEX FOURIER TRANSFORM OF DIMENSION N.
C    THE COSINE COEFFICIENTS ARE IN A(1),A(2),...A(N+1) AND
C    THE SINE COEFFICIENTS ARE IN B(1),B(2),...B(N+1).
C    A TYPICAL CALLING SEQUENCE IS
C      CALL FFT(A,B,N,N,N,1)
C      CALL REALTR(A,B,N,1)
C    THE RESULTS SHOULD BE MULTIPLIED BY 0.5/N TO GIVE THE
C    USUAL SCALING OF COEFFICIENTS.
C  IF ISN=-1, THE INVERSE TRANSFORMATION IS DONE, THE FIRST STEP
C    IN EVALUATING A REAL FOURIER SERIES.
C    A TYPICAL CALLING SEQUENCE IS
C      CALL REALTR(A,B,N,-1)
C      CALL FFT(A,B,N,N,N,-1)
C    THE RESULTS SHOULD BE MULTIPLIED BY 0.5 TO GIVE THE USUAL
C    SCALING, AND THE TIME DOMAIN RESULTS ALTERNATE IN ARRAYS A
C    AND B, I.E. A(1),B(1),A(2),B(2),...A(N),B(N).
C  THE DATA MAY ALTERNATIVELY BE STORED IN A SINGLE COMPLEX
C    ARRAY A, THEN THE MAGNITUDE OF ISN CHANGED TO TWO TO
C    GIVE THE CORRECT INDEXING INCREMENT AND A(2) USED TO
C    PASS THE INITIAL ADDRESS FOR THE SEQUENCE OF IMAGINARY
C    VALUES, E.G.
C      CALL FFT(A,A(2),N,N,N,2)
C      CALL REALTR(A,A(2),N,2)
C    IN THIS CASE, THE COSINE AND SINE COEFFICIENTS ALTERNATE IN A.
C  BY R. C. SINGLETON, STANFORD RESEARCH INSTITUTE, OCT. 1968
      DIMENSION A(1),B(1)
      REAL IM
      INC=IABS(ISN)
      NK=N*INC+2
      NH=NK/2
      SD=2.0*ATAN(1.0)/FLOAT(N)
      CD=2.0*SIN(SD)**2
      SD=SIN(SD+SD)
      SN=0.0
      IF(ISN .LT. 0) GO TO 30
      CN=1.0
      A(NK-1)=A(1)
      B(NK-1)=B(1)
   10 DO 20 J=1,NH,INC
      K=NK-J
      AA=A(J)+A(K)
      AB=A(J)-A(K)
      BA=B(J)+B(K)
      BB=B(J)-B(K)
      RE=CN*BA+SN*AB
      IM=SN*BA-CN*AB
      B(K)=IM-BB
      B(J)=IM+BB
      A(K)=AA-RE
      A(J)=AA+RE
      AA=CN-(CD*CN+SD*SN)
      SN=(SD*CN-CD*SN)+SN
C  THE FOLLOWING THREE STATEMENTS COMPENSATE FOR TRUNCATION
C    ERROR.  IF ROUNDED ARITHMETIC IS USED, SUBSTITUTE
C  20 CN=AA
      CN=0.5/(AA**2+SN**2)+0.5
      SN=CN*SN
   20 CN=CN*AA
      RETURN
   30 CN=-1.0
      SD=-SD
      GO TO 10
      END
```

Appendix III

Rader's Radix-5 Method, for Possible Substitution in Subroutine FFT in Appendix I

```
C  TRANSFORM FOR FACTOR OF 5 (OPTIONAL CODE),
C    USING METHOD DUE TO C. M. RADER
  510 C2=0.25*SQRT(5.0)
      S2=2.0*C72*S72
  520 K1=KK+KSPAN
      K2=K1+KSPAN
      K3=K2+KSPAN
      K4=K3+KSPAN
      AKP=A(K1)+A(K4)
      AKM=A(K1)-A(K4)
      BKP=B(K1)+B(K4)
      BKM=B(K1)-B(K4)
      AJP=A(K2)+A(K3)
      AJM=A(K2)-A(K3)
      BJP=B(K2)+B(K3)
      BJM=B(K2)-B(K3)
      AK=AKP+AJP
      AJP=(AKP-AJP)*C2
      BK=BKP+BJP
      BJP=(BKP-BJP)*C2
      AKP=A(KK)-0.25*AK
      A(KK)=A(KK)+AK
      BKP=B(KK)-0.25*BK
      B(KK)=B(KK)+BK
      AK=AKP+AJP
      AJP=AKP-AJP
      BK=BKP+BJP
      BJP=BKP-BJP
      AKP=AKM*S72+AJM*S2
      AKM=AKM*S2-AJM*S72
      BKP=BKM*S72+BJM*S2
      BKM=BKM*S2-BJM*S72
      A(K1)=AK-BKP
      A(K4)=AK+BKP
      B(K1)=BK+AKP
      B(K4)=BK-AKP
      A(K2)=AJP-BKM
      A(K3)=AJP+BKM
      B(K2)=BJP+AKM
      B(K3)=BJP-AKM
      KK=K4+KSPAN
      IF(KK .LT. NN) GO TO 520
      KK=KK-NN
      IF(KK .LE. KSPAN) GO TO 520
      GO TO 700
```

An Adaptation of the Fast Fourier Transform for Parallel Processing

MARSHALL C. PEASE

Stanford Research Institute, Menlo Park, California*

ABSTRACT. A modified version of the Fast Fourier Transform is developed and described. This version is well adapted for use in a special-purpose computer designed for the purpose. It is shown that only three operators are needed. One operator replaces successive pairs of data points by their sums and differences. The second operator performs a fixed permutation which is an ideal shuffle of the data. The third operator permits the multiplication of a selected subset of the data by a common complex multiplier.

If, as seems reasonable, the slowest operation is the complex multiplications required, then, for reasonably sized data sets—e.g. 512 complex numbers—parallelization by the method developed should allow an increase of speed over the serial use of the Fast Fourier Transform by about two orders of magnitude.

It is suggested that a machine to realize the speed improvement indicated is quite feasible.

The analysis is based on the use of the Kronecker product of matrices. It is suggested that this form is of general use in the development and classification of various modifications and extensions of the algorithm.

KEY WORDS AND PHRASES: transforms, Fourier transforms, Fourier series, parallel processors, parallel execution, simultaneous processors, convolution, power spectrum, Fast Fourier Transform

CR CATEGORIES: 5.19

1. *Introduction*

The Fast Fourier Transform and its variations have become the principal methods for the machine calculation of the Fourier series of time-sampled signals or other sequences of data. The basic algorithm has a long complicated history, as described by Cooley, Lewis and Welch [1], although general recognition of its computational power dates from the paper of Cooley and Tukey [2]. Elaborations led to applications to other purposes, by Stockham[3], Bingham, Godfrey, and Tukey [4], Gentleman and Sande [5], Singleton [6], Cooley, Lewis and Welch [7], and Helms [8].

The present work is concerned with the development of a variation of the algorithm that is better adapted to parallel processing in a special purpose machine. Our primary objective was to see what capabilities would be needed in such a machine.

This work may also be of value as an aid in understanding the algorithm, at least for those who are used to thinking in terms of matrix operators, and as a formalism within which other variations can be described and investigated.

For the purpose of parallel processing in a special purpose machine, we require that the operation be organized in a few levels, each of which involves a set of

* Computer Techniques Laboratory. This work was sponsored in part by the Office of Naval Research, Information Systems Branch, under Contract Nonr 4833(00).

elementary operations that can be done simultaneously. Preferably each level should involve only a single type of elementary process, so that the entire level can be initiated by a single command. Also there should be as few as possible distinct types of elementary operations, so that the parallel capability required shall be as simple as possible. For example, in the original algorithm each "pass" involves a different grouping of the data into pairs. We would like, if possible, to use only a single pattern of pairings which could then be "wired in."

On the other hand, in a special purpose machine—for example, one designed to be part of an autocorrelator, or a digital filtering system—the number of data points can be taken as fixed. We can then reasonably assume that this number is a power of 2, $N = 2^n$. This is fortunate since clearly we cannot expect to be able to use only a small number of types of operations in the general case, when we can only assume N to be a composite number. We therefore assume throughout that the number of data to be processed is $N = 2^n$.

2. *Finite Fourier Transform*

First, let us very briefly review what is meant by the Finite Fourier Transform. A more complete description can be found in the report of the G-AE Subcommittee on Measurement Concepts [9].

We consider a possibly complex-valued curve $f(t)$ over some interval of time that we can normalize to the unit interval $0 \leq t < 1$. We sample the curve at N points equally spaced along the interval:

$$\begin{aligned} t_0 &= 0, \\ t_i &= i/N, \qquad 0 \leq i < (N-1), \\ f_i &= f(t_i). \end{aligned}$$

The Fourier coefficients are obtained as

$$g_r = \frac{1}{N} \sum_{s=0}^{N-1} (\exp 2\pi jrs/N) f_s. \tag{1}$$

If the time scale is normalized so that $f(t)$ is periodic with period 1, then g_r is the complex amplitude of the component at frequency r of the usual Fourier series. If outside the given interval $f(t)$ vanishes, then g_r may be taken as a discrete approximation of the complex-valued Fourier integral. In either case frequencies higher than $r = (N - 1)$ cannot be computed meaningfully with the sampling rate specified.

The sets (f_s) and (g_s) can now be written as column vectors[1]:

$$\mathbf{f} = \text{col}\,(f_0, f_1 \cdots f_{N-1}), \qquad \mathbf{g} = \text{col}\,(g_0, g_1 \cdots g_{N-1}).$$

We define the matrix $\mathbf{T}$, whose coefficients are $(\mathbf{T})_{rs} = \exp\,(2\pi jrs/N)$. Then eq. (1) can be written as

$$\mathbf{g} = \frac{1}{N}\mathbf{Tf}. \tag{2}$$

[1] Lower-case boldface letters indicate vectors, and capital boldface letters indicate matrices.

The matrix **T** is called the Finite Fourier Transform. Writing w as the principal Nth root of unity, the coefficients of **T** are given by

$$(\mathbf{T})_{rs} = w^{rs}. \tag{3}$$

To simplify notation, we preserve only the exponent of w. That is, we write k in place of w^k. Then **T** can be written as

$$\mathbf{T} = \begin{bmatrix} 0 & 0 & 0 & 0 & \cdots & 0 \\ 0 & 1 & 2 & 3 & \cdots & N-1 \\ 0 & 2 & 4 & 6 & \cdots & 2(N-1) \\ 0 & 3 & 6 & 9 & \cdots & 3(N-1) \\ \cdot & \cdot & \cdot & \cdot & \cdot & \cdot \\ 0 & N-1 & 2(N-1) & 3(N-1) & \cdots & (N-1)^2 \end{bmatrix} \tag{4}$$

In this notation, multiplication of entries becomes addition. The expression for **T** can be further reduced by noting that each term may be replaced by its principal value (between 0 and $N - 1$), modulo N.

3. *The Fast Fourier Transform*

The transformation used by Cooley and Tukey accomplishes the same thing as **T**, except that the output is obtained in permuted order—specifically in digit-reversed order, where the rows are numbered from 0 to $(N - 1)$, as expressed in the base of the prime factor being used (in the present case, in binary notation). The transformation to this form is given, in general, by $\mathbf{T}' = \mathbf{QT}$, where **Q** is the appropriate permutation matrix.

We can obtain **T′** directly from **T** by permuting the rows of **T** as specified. For example, for $N = 8$, after reducing all terms modulo 8, we find that **T** in reduced exponent notation is given by

$$\mathbf{T} = \begin{bmatrix} 0 & 0 & 0 & 0 & 0 & 0 & 0 & 0 \\ 0 & 1 & 2 & 3 & 4 & 5 & 6 & 7 \\ 0 & 2 & 4 & 6 & 0 & 2 & 4 & 6 \\ 0 & 3 & 6 & 1 & 4 & 7 & 2 & 5 \\ 0 & 4 & 0 & 4 & 0 & 4 & 0 & 4 \\ 0 & 5 & 2 & 7 & 4 & 1 & 6 & 3 \\ 0 & 6 & 4 & 2 & 0 & 6 & 4 & 2 \\ 0 & 7 & 6 & 5 & 4 & 3 & 2 & 1 \end{bmatrix} \quad \begin{matrix} 0\ 0\ 0 \\ 0\ 0\ 1 \\ 0\ 1\ 0 \\ 0\ 1\ 1 \\ 1\ 0\ 0 \\ 1\ 0\ 1 \\ 1\ 1\ 0 \\ 1\ 1\ 1 \end{matrix}$$

where the binary row designation appears on the right. Then the modified transform is given by $\mathbf{T}'$ as follows:

$$\mathbf{T}' = \begin{bmatrix} 0 & 0 & 0 & 0 & 0 & 0 & 0 & 0 \\ 0 & 4 & 0 & 4 & 0 & 4 & 0 & 4 \\ 0 & 2 & 4 & 6 & 0 & 2 & 4 & 6 \\ 0 & 6 & 4 & 2 & 0 & 6 & 4 & 2 \\ 0 & 1 & 2 & 3 & 4 & 5 & 6 & 7 \\ 0 & 5 & 2 & 7 & 4 & 1 & 6 & 3 \\ 0 & 3 & 6 & 1 & 4 & 7 & 2 & 5 \\ 0 & 7 & 6 & 5 & 4 & 3 & 2 & 1 \end{bmatrix} \quad \begin{matrix} 0\;0\;0 \\ 1\;0\;0 \\ 0\;1\;0 \\ 1\;1\;0 \\ 0\;0\;1 \\ 1\;0\;1 \\ 0\;1\;1 \\ 1\;1\;1 \end{matrix} \tag{5}$$

The key to the Fast Fourier Transform algorithm lies in the fact that $\mathbf{T}'_N$ (when $N = 2^n$) can be first partitioned and then factored:

$$\begin{aligned} \mathbf{T}'_N &= \begin{bmatrix} \mathbf{T}'_{N/2} & \mathbf{T}'_{N/2} \\ (\mathbf{T}'_{N/2})\mathbf{K} & -(\mathbf{T}'_{N/2})\mathbf{K} \end{bmatrix} \\ &= \begin{bmatrix} \mathbf{T}'_{N/2} & \phi \\ \phi & \mathbf{T}'_{N/2} \end{bmatrix} \cdot \begin{bmatrix} \mathbf{I} & \mathbf{I} \\ \mathbf{K} & -\mathbf{K} \end{bmatrix} \\ &= \begin{bmatrix} \mathbf{T}'_{N/2} & \phi \\ \phi & \mathbf{T}'_{N/2} \end{bmatrix} \cdot \begin{bmatrix} \mathbf{I} & \phi \\ \phi & \mathbf{K} \end{bmatrix} \cdot \begin{bmatrix} \mathbf{I} & \mathbf{I} \\ \mathbf{I} & -\mathbf{I} \end{bmatrix}, \end{aligned} \tag{6}$$

where, in exponent notation, $\mathbf{K} = \text{diag}\ (0\,1\,2\,3\,\cdots)$ and ϕ indicates the null matrix of appropriate dimension.

The minus sign that occurs in eq. (6) applies to the *values* of the submatrices that follow it, and not to the exponent-coefficients that may actually be used. Since w is the Nth root of unity, $w^{N/2}$ is the square root of unity or the value (-1). Hence the minus sign can be translated to exponent notation by adding $N/2$ to the exponent-coefficients involved.

It should also be noted that $\mathbf{T}'_N$ is written in terms of w, the principle Nth root of unity. $\mathbf{T}'_{N/2}$ would normally be written in terms of the principle $(N/2)$-nd root, which is w^2. Hence, if eq. (6) is actually written out, the entries in $\mathbf{T}'_{N/2}$ will be twice those that would occur were it written in terms of the $(N/2)$-nd root.

The algorithm can now be iterated by applying the factorization of eq. (6) to each occurrence of $\mathbf{T}'_{N/2}$, and so on. If we carry the process to completion, we obtain the Fast Fourier Transform.

For the purposes of parallelization, the process given suffers from the defect that a coefficient in a given location is combined with coefficients in different locations each time the sum and difference are formed. For example, if we carry eq. (6) to the second stage of iteration, we get

$$\mathbf{T}'_N = \begin{bmatrix} \mathbf{T}'_{N/4} & \phi & \phi & \phi \\ \phi & \mathbf{T}'_{N/4} & \phi & \phi \\ \phi & \phi & \mathbf{T}'_{N/4} & \phi \\ \phi & \phi & \phi & \mathbf{T}'_{N/4} \end{bmatrix}$$

$$= \begin{bmatrix} \mathbf{I} & \phi & \phi & \phi \\ \phi & \mathbf{K}' & \phi & \phi \\ \phi & \phi & \mathbf{I} & \phi \\ \phi & \phi & \phi & \mathbf{K}' \end{bmatrix} \begin{bmatrix} \mathbf{I} & \mathbf{I} & \phi & \phi \\ \mathbf{I} & -\mathbf{I} & \phi & \phi \\ \phi & \phi & \mathbf{I} & \mathbf{I} \\ \phi & \phi & \mathbf{I} & -\mathbf{I} \end{bmatrix} \begin{bmatrix} \mathbf{I} & \phi \\ \phi & \mathbf{K} \end{bmatrix} \begin{bmatrix} \mathbf{I} & \mathbf{I} \\ \mathbf{I} & -\mathbf{I} \end{bmatrix}.$$

The last factor, which is the first operator applied to the data, forms the sums and differences of coefficients in the first half with corresponding terms in the second. The third factor forms the sums and differerices of the coefficients in the first quarter with corresponding terms in the second, and of the coefficients in the third quarter with those in the fourth. This prevents use of "wired-in" data paths, or else it requires an excessive number of such paths.

We therefore ask if the algorithm can be modified so as to avoid this difficulty.

4. *Modified Fast Fourier Transform*

We observe that the recursion formula of eq. (6) can be written in the form

$$\mathbf{T}'_N = (\mathbf{T}'_{N/2} \times \mathbf{I}_2)\, \mathbf{D}_N (\mathbf{I}_{N/2} \times \mathbf{T}'_2), \tag{7}$$

where $\mathbf{D}_N$ is the $N \times N$ diagonal matrix, quasidiag $(\mathbf{I}\ \mathbf{K})$, and $\mathbf{T}'_2$ is given by

$$\mathbf{T}'_2 = \begin{pmatrix} 0 & 0 \\ 0 & \frac{1}{2}N \end{pmatrix} = \begin{pmatrix} 1 & 1 \\ 1 & -1 \end{pmatrix} \tag{8}$$

where in the latter matrix the coefficients are the actual values, not exponents.

The multiplication sign in eq. (7) indicates the direct, or Kronecker, product [10] of the submatrices, indexed first on the indices of the first component and then on the indices of the second. For example, if $\mathbf{A} = (a_{ij})$, $\mathbf{B} = (b_{kh})$, with all indices being either 0 or 1, then

$$\mathbf{A} \times \mathbf{B} = \begin{pmatrix} a_{00}b_{00} & a_{01}b_{00} & a_{00}b_{01} & a_{01}b_{01} \\ a_{10}b_{00} & a_{11}b_{00} & a_{10}b_{01} & a_{11}b_{01} \\ a_{00}b_{10} & a_{01}b_{10} & a_{00}b_{11} & a_{01}b_{11} \\ a_{10}b_{10} & a_{11}b_{10} & a_{10}b_{11} & a_{11}b_{11} \end{pmatrix}.$$

The Kronecker product can be combined with matrix multiplication through the formula,

$$(\mathbf{A} \times \mathbf{B})(\mathbf{C} \times \mathbf{D}) = (\mathbf{AC}) \times (\mathbf{BD}),$$

provided that the dimensions of the matrices are compatible. In particular, we can write

$$(\mathbf{AB} \times \mathbf{I}) = (\mathbf{AB}) \times \mathbf{I}^2 = (\mathbf{A} \times \mathbf{I})(\mathbf{B} \times \mathbf{I})$$

or, more generally, if **A**, **B**, **C**, $\cdots$ are all $k \times k$ matrices, then

$$(\mathbf{ABC} \cdots) \times \mathbf{I} = (\mathbf{A} \times \mathbf{I})(\mathbf{B} \times \mathbf{I})(\mathbf{C} \times \mathbf{I}) \cdots .$$

Using this formula, we can apply eq. (7) iteratively and obtain, when $N = 2^n$,

$$\begin{aligned} \mathbf{T}_N' = {} & (\mathbf{T}_2' \times \mathbf{I}_2 \times \cdots \times \mathbf{I}_2) \cdot \mathbf{Q}_1 \\ & \cdot (\mathbf{I}_2 \times \mathbf{T}_2' \times \mathbf{I}_2 \times \cdots \times \mathbf{I}_2) \cdot \mathbf{Q}_2 \\ & \cdot (\mathbf{I}_2 \times \mathbf{I}_2 \times \mathbf{T}_2' \times \cdots \times \mathbf{I}_2) \cdot \mathbf{Q}_3 \\ & \quad \vdots \\ & \cdot (\mathbf{I}_2 \times \mathbf{I}_2 \times \cdots \times \mathbf{T}_2'), \end{aligned} \tag{9}$$

where each of the cross-products includes n factors, of which $(n - 1)$ are the 2×2 identity and one is $\mathbf{T}_2'$, and the $\mathbf{Q}_i$ are diagonal matrices. For example, for $N = 8$, $\mathbf{T}_8' = (\mathbf{T}_4' \times \mathbf{I}_2)\, \mathbf{D}_8 (\mathbf{I}_4 \times \mathbf{T}_2')$, and, since $\mathbf{I}_4 = \mathbf{I}_2 \times \mathbf{I}_2$ and

$$\mathbf{T}_4' = (\mathbf{T}_2' \times \mathbf{I}_2)\, \mathbf{D}_4 (\mathbf{I}_2 \times \mathbf{T}_2'),$$

we find that

$$\begin{aligned} \mathbf{T}_8' &= [\{(\mathbf{T}_2' \times \mathbf{I}_2)\, \mathbf{D}_4 (\mathbf{I}_2 \times \mathbf{T}_2')\} \times \mathbf{I}_2]\, \mathbf{D}_8 (\mathbf{I}_2 \times \mathbf{I}_2 \times \mathbf{T}_2') \\ &= \{(\mathbf{T}_2' \times \mathbf{I}_2 \times \mathbf{I}_2)(\mathbf{D}_4 \times \mathbf{I}_2)(\mathbf{I}_2 \times \mathbf{T}_2' \times \mathbf{I}_2)\}\, \mathbf{D}_8 (\mathbf{I}_2 \times \mathbf{I}_2 \times \mathbf{T}_2') \\ &= (\mathbf{T}_2' \times \mathbf{I}_2 \times \mathbf{I}_2)\, \mathbf{Q}_1 (\mathbf{I}_2 \times \mathbf{T}_2' \times \mathbf{I}_2)\, \mathbf{Q}_2 (\mathbf{I}_2 \times \mathbf{I}_2 \times \mathbf{T}_2'), \end{aligned}$$

where $\mathbf{Q}_1 = \mathbf{D}_4 \times \mathbf{I}_2$, $\mathbf{Q}_2 = \mathbf{D}_8$.

For the purpose of parallelization the difficulty now with eq. (9) is that $\mathbf{T}_2'$ occurs in different locations in the various factors. We therefore ask if there is a permutation operator, **P**, such that

$$\mathbf{P}(\mathbf{T}_2' \times \mathbf{A}_{N/2})\mathbf{P}^{-1} = \mathbf{A}_{N/2} \times \mathbf{T}_2'. \tag{10}$$

If so, we can write each of the product terms in eq. (9) in terms of a single one of them, say the first:

$$\begin{aligned} \mathbf{C} &= \mathbf{T}_2' \times \mathbf{I}_2 \times \mathbf{I}_2 \times \cdots \times \mathbf{I}_2 \\ &= \mathbf{T}_2' \times \mathbf{I}_{N/2}. \end{aligned} \tag{11}$$

By considering the behavior of the product operators on an arbitrary vector, we can find that a possible **P** is the "ideal shuffle," in which

$$\begin{aligned} \mathbf{P} \cdot \mathrm{col}\ & (x_0, x_1, \cdots, x_{\frac{1}{2}N-1}, x_{N/2} \cdots x_{N-1}) \\ & = \mathrm{col}\ (x_0, x_{N/2}, x_1, x_{\frac{1}{2}N+1}, \cdots, x_{\frac{1}{2}N-1}, x_{N-1}) \end{aligned} \tag{12}$$

or, for $N = 8$, for example,

$$\mathbf{P} = \begin{bmatrix} 1 & 0 & 0 & 0 & 0 & 0 & 0 & 0 \\ 0 & 0 & 0 & 0 & 1 & 0 & 0 & 0 \\ 0 & 1 & 0 & 0 & 0 & 0 & 0 & 0 \\ 0 & 0 & 0 & 0 & 0 & 1 & 0 & 0 \\ 0 & 0 & 1 & 0 & 0 & 0 & 0 & 0 \\ 0 & 0 & 0 & 0 & 0 & 0 & 1 & 0 \\ 0 & 0 & 0 & 1 & 0 & 0 & 0 & 0 \\ 0 & 0 & 0 & 0 & 0 & 0 & 0 & 1 \end{bmatrix}, \tag{13}$$

where the entries, in this case, are the actual values, not exponents.

(We could also use the ideal shuffle which does not keep the first coefficient fixed. In special circumstances other permutations are also possible. Our concern here, however, is to find one such **P** operator, not to define the range of possible permutations.)

With this **P**,

$$\mathbf{I}_2 \times \mathbf{T}_2' \times \mathbf{I}_{N/4} = \mathbf{P}(\mathbf{T}_2' \times \mathbf{I}_{n/2})\,\mathbf{P}^{-1} = \mathbf{PCP}^{-1},$$
$$\mathbf{I}_4 \times \mathbf{T}_2' \times \mathbf{I}_{N/8} = \mathbf{P}^2\mathbf{CP}^{-2},$$
$$\text{etc.,}$$

where **C** is defined in eq. (11). Then eq. (9) becomes

$$\mathbf{T}_N' = \mathbf{CE}_1\mathbf{PCP}^{-1}\mathbf{E}_2\mathbf{P}^2\mathbf{CP}^{-2}\mathbf{E}_3 \cdots \mathbf{P}^{n-1}\mathbf{CP}^{-(n-1)}. \tag{14}$$

$\mathbf{E}_i'$ and $\mathbf{E}_i''$ can now be defined by

$$\mathbf{E}_i = \mathbf{P}^{(i-1)}\mathbf{E}_i'\mathbf{P}^{-(i-1)} = \mathbf{P}^i\mathbf{E}_i''\mathbf{P}^{-i},$$

which permit us to pass the diagonal operator through the permutations. The operators $\mathbf{E}_i'$ and $\mathbf{E}_i''$ will also be diagonal, but with the values on the diagonal permuted. Using these transformations and noting that, since $\mathbf{P}^n = \mathbf{I}$, $\mathbf{P}^{-(n-1)} = \mathbf{P}$, eq. (14) becomes either

$$\mathbf{T}_N' = \mathbf{CE}_1'\mathbf{PCE}_2'\mathbf{PCE}_3' \ldots \mathbf{PCP} \tag{15}$$

or

$$\mathbf{T}_N' = \mathbf{CPE}_1''\mathbf{CPE}_2''\mathbf{CP} \ldots \mathbf{CP}. \tag{16}$$

The $\mathbf{E}_i'$ and $\mathbf{E}_i''$ matrices are diagonal and have the effect of multiplying selected elements of the data set by various powers of w. We can, if we like, factor each of these matrices into diagonal operators, each of which multiplies a selected subset of the data set by a common multiplier which is a power of w. For example, for

$N = 16$, $\mathbf{E}_1'$ is given by

$$\mathbf{E}_1' = \text{diag}\ (0\,0\,0\,4\,0\,2\,0\,6\,0\,1\,0\,5\,0\,3\,0\,7) = \mathbf{F}_1'\,\mathbf{F}_2'\,\mathbf{F}_4',$$

where

$$\mathbf{F}_1' = \text{diag}\ (0\,0\,0\,0\,0\,0\,0\,0\,0\,1\,0\,1\,0\,1\,0\,1),$$
$$\mathbf{F}_2' = \text{diag}\ (0\,0\,0\,0\,0\,2\,0\,2\,0\,0\,0\,0\,0\,2\,0\,2),$$
$$\mathbf{F}_4' = \text{diag}\ (0\,0\,0\,4\,0\,0\,0\,4\,0\,0\,0\,4\,0\,0\,0\,4),$$

in exponent notation. In particular, each $\mathbf{E}_i'$ or $\mathbf{E}_i''$ can be factored into the product of $\mathbf{F}_{(2^i)}'$ or $\mathbf{F}_{(2^i)}''$, where each $\mathbf{F}'$ or $\mathbf{F}''$ multiplies exactly $N/4$ of the data by w^{2^i}.

Whether or not this factorization of $\mathbf{E}_i'$ or $\mathbf{E}_i''$ is advantageous depends upon the particular capabilities available in the processor. If we can store $(n - 1)$ vectors internally and can multiply two vectors term by term in parallel, we can implement the $\mathbf{E}_i'$ or $\mathbf{E}_i''$ operations directly. If, on the other hand, internal storage is limited, so that the multiplying vector has to be loaded as used, it is probably better to use the factorization into the **F**-operators. Then we need only introduce the particular w^k and the enabling or disabling signals to designate the subset to be multiplied.

The alternative factorizations for $N = 16$ are given by

$$\mathbf{F}_{(4)}'' = \text{diag}\ (0\,0\,0\,0\,0\,0\,0\,0\,0\,4\,0\,4\,0\,4\,0\,4),$$
$$\mathbf{F}_{(2)}'' = \text{diag}\ (0\,0\,0\,0\,0\,0\,0\,0\,0\,0\,2\,2\,0\,0\,2\,2),$$
$$\mathbf{F}_{(1)}'' = \text{diag}\ (0\,0\,0\,0\,0\,0\,0\,0\,0\,0\,0\,0\,1\,1\,1\,1).$$

With these factorizations of the **E**-operators, eqs. (15) and (16) become

$$\mathbf{T}_{16}' = \mathbf{C}\mathbf{F}_{(4)}'\,\mathbf{F}_{(2)}'\,\mathbf{F}_{(1)}'\,\mathbf{P}\mathbf{C}\mathbf{F}_{(4)}'\,\mathbf{F}_{(2)}'\,\mathbf{P}\mathbf{C}\mathbf{F}_{(4)}'\,\mathbf{P}\mathbf{C}\mathbf{P}, \tag{17}$$

$$\mathbf{T}_{16}' = \mathbf{C}\mathbf{P}\mathbf{F}_{(4)}''\,\mathbf{F}_{(2)}''\,\mathbf{F}_{(1)}''\,\mathbf{C}\mathbf{P}\mathbf{F}_{(4)}''\,\mathbf{F}_{(2)}''\,\mathbf{C}\mathbf{P}\mathbf{F}_{(4)}''\,\mathbf{C}\mathbf{P}. \tag{18}$$

Since diagonal matrices commute, we can, of course, permute each subsequence of **F** operators as convenient.

Equation (18) was the first discovered. Equation (17) was found as a result of conversations with Jack Goldberg. It has the advantage, for some purposes, that adjacent data points are never involved in a given $\mathbf{F}_{(i)}'$.

The general rules for the **F**-matrices are as follows: We number the positions along the diagonal from 0 to $N - 1$, and express these numbers in binary form. Then, in $\mathbf{F}_{(2^m)}'$, the exponent-coefficients are 2^m in those positions in which the $(m + 1)$-st and the last digits are 1. All other exponent-coefficients are 0. For example, for $N = 16$, $\mathbf{F}_{(4)}'$ has the exponent-coefficient 4 in the 3rd, 7th, 11th, and 15th positions, for which the binary representations are 0 0 1 1, 0 1 1 1, 1 0 1 1, and 1 1 1 1, which are the numbers whose 3rd and last digits are 1. Similarly, in general, the $\mathbf{F}_{(2^m)}''$-matrix has the exponent-coefficient 2^m in the positions for which, in the binary representation, the first and $(m + 2)$-nd digits are 1.

The final results, given in eqs. (15) or (16), can be written as

$$\mathbf{T}_{N=2^n}' = \prod_{m=1}^{n} (\mathbf{C}\,\mathbf{E}_m'\,\mathbf{P}) \tag{19}$$

$$= \prod_{m=1}^{n} (\mathbf{C}\,\mathbf{P}\,\mathbf{E}_m''), \tag{20}$$

where $\mathbf{E}'_m$ and $\mathbf{E}''_m$ are the diagonal matrices defined before, and $\mathbf{E}'_n = \mathbf{E}''_n = \mathbf{I}$. In a parallel processor, each factor represents one "pass," a pass being defined as a single sequence of parallel operations involving, in eq. (19), first a permutation of the data set; second, the multiplication of selected elements of the data set by appropriate powers of w; and finally the replacement of the data set by the sums and differences of adjacent elements.

Equations (15) and (16), or (19) and (20), or, in the fully factored form, eqs. (17) and (18), and their generalization to arbitrary n, are the modified form of the Fast Fourier Transform that we seek.

5. *Generalizations and Variations*

There are some variations of the procedure that are worth noting.

(a) *Radix 4—Binary Reversal.* It is not necessary to factor $\mathbf{T}'$ all the way to factors involving $\mathbf{T}'_2$. It may, for example, be convenient to stop at $\mathbf{T}'_4$ since

$$\mathbf{T}'_4 = \begin{bmatrix} 0 & 0 & 0 & 0 \\ 0 & 8 & 0 & 8 \\ 0 & 4 & 8 & 12 \\ 0 & 12 & 8 & 4 \end{bmatrix} = \cdot \begin{bmatrix} 1 & 1 & 1 & 1 \\ 1 & -1 & 1 & -1 \\ 1 & j & -1 & -j \\ 1 & -j & -1 & j \end{bmatrix},$$

the latter being in terms of actual values, not exponents. $\mathbf{T}'_4$ then involves quite simple linear combinations of quadruples of elements of the data set. If this can be conveniently implemented in a parallel processor, its use in place of the C-operator will halve the number of passes required. The diagonal operators must then be redetermined. We can find, for example, that

$$\mathbf{T}'_{16} = (\mathbf{T}'_4 \times \mathbf{I}_4)\, \mathbf{Q}'(\mathbf{I}_4 \times \mathbf{T}'_4),$$

where

$$\mathbf{Q}' = \text{diag}\ (0\,0\,0\,0\,0\,2\,4\,6\,0\,1\,2\,3\,0\,3\,6\,9).$$

We can then use a four-fold interleave permutation in place of $\mathbf{P}$ to permute $\mathbf{I}_4 \times \mathbf{T}'_4$ into $\mathbf{T}'_4 \times \mathbf{I}_4$, as before. Presumably, this process can be generalized for the case where N is any even power of 2, although we have not worked out the details. We would also expect to be able to perform a similar factorization in terms of $\mathbf{T}'_8$, $\mathbf{T}'_{16}$, etc., if this were needed.

(b) *Full Radix 4.* As an alternative to the procedure used in (a) we can label the rows of $\mathbf{T}$ to the base 4 instead of 2, reverse the digits as before, and compute a $\mathbf{T}''$ as the new transform operator. We then obtain, for example,

$$\mathbf{T}''_{16} = (\mathbf{T}''_4 \times \mathbf{I}_4)\, \mathbf{Q}''\, (\mathbf{I}_4 \times \mathbf{T}''_4),$$

where now

$$\mathbf{T}_4'' = \begin{bmatrix} 0 & 0 & 0 & 0 \\ 0 & 4 & 8 & 12 \\ 0 & 8 & 0 & 8 \\ 0 & 12 & 8 & 4 \end{bmatrix} = \begin{bmatrix} 1 & 1 & 1 & 1 \\ 1 & j & -1 & -j \\ 1 & -1 & 1 & -1 \\ 1 & -j & -1 & j \end{bmatrix}$$

and

$$\mathbf{Q}'' = \text{diag}\ (0\,0\,0\,0\,0\,1\,2\,3\,0\,2\,4\,6\,0\,3\,6\,9).$$

Again, we can use in place of **P** the four-fold interleaving operator to permute the factors in the Kronecker product.

(c) *Arbitrary Radix.* If $N = p^n$, where p is any prime number, a similar process can be used. Nevertheless, with the possible exception of $p = 3$ there appears to be little advantage for parallelization in using such a transform, since the basic operator $\mathbf{T}_p'$ or $\mathbf{T}_{p^k}'$ becomes quite complicated.

(d) *Composite N.* The process can be generalized for the case where N is any composite number. To illustrate by a simple example, consider $N = 6$. Then

$$\mathbf{T} = \begin{bmatrix} 0 & 0 & 0 & 0 & 0 & 0 \\ 0 & 1 & 2 & 3 & 4 & 5 \\ 0 & 2 & 4 & 0 & 2 & 4 \\ 0 & 3 & 0 & 3 & 0 & 3 \\ 0 & 4 & 2 & 0 & 4 & 2 \\ 0 & 5 & 4 & 3 & 2 & 1 \end{bmatrix} \quad \begin{array}{cc} 0 & 0 \\ 0 & 1 \\ 1 & 0 \\ 1 & 1 \\ 2 & 0 \\ 2 & 1 \end{array} \quad \begin{array}{cc} 0 & 0 \\ 0 & 1 \\ 0 & 2 \\ 1 & 0 \\ 1 & 1 \\ 1 & 2 \end{array}$$

Now we have two possible labelings of the rows. Either the pth row can be labeled by $p = 3i + j,\ \ i = 0, 1;\ \ j = 0, 1, 2;$ or $p = 2i + j,\ \ i = 0, 1, 2,\ \ j = 0, 1,$ as indicated alongside the matrix. Reversing the digits of the first designation, and making the pth row the qth, where $q = 2j + i$, we obtain

$$\mathbf{T}_6' = \begin{bmatrix} 0 & 0 & 0 & 0 & 0 & 0 \\ 0 & 2 & 4 & 0 & 2 & 4 \\ 0 & 4 & 2 & 0 & 4 & 2 \\ 0 & 1 & 2 & 3 & 4 & 5 \\ 0 & 3 & 0 & 3 & 0 & 3 \\ 0 & 5 & 4 & 3 & 2 & 1 \end{bmatrix} = (\mathbf{T}_3' \times \mathbf{I}_2)\ \mathbf{Q}'(\mathbf{I}_3 \times \mathbf{T}_2'),\quad \mathbf{Q}' = \text{diag}\ (0\,0\,0\,0\,1\,2).$$

Using digit reversal on the second designation, we obtain

$$\mathbf{T}_6'' = \begin{bmatrix} 0 & 0 & 0 & 0 & 0 & 0 \\ 0 & 3 & 0 & 3 & 0 & 3 \\ 0 & 1 & 2 & 3 & 4 & 5 \\ 0 & 4 & 2 & 0 & 4 & 2 \\ 0 & 2 & 4 & 0 & 2 & 4 \\ 0 & 5 & 4 & 3 & 2 & 1 \end{bmatrix} = (\mathbf{T}_2'' \times \mathbf{I}_3)\, \mathbf{Q}''(\mathbf{I}_2 \times \mathbf{T}_3''), \qquad \mathbf{Q}'' = (0\,0\,0\,1\,0\,2).$$

We can, in either case, again find a permutation operator that will permute the order of the factors in the Kronecker products. In this situation, however, it does not appear to be of advantage to do so, since we do not, by this means, obtain a repeated factor. For example, the shuffle operator **P** takes $(\mathbf{I}_3 \times \mathbf{T}_2')$ into $\mathbf{T}_2' \times \mathbf{I}_3$ in $\mathbf{T}_6'$, but this is not the same as the first factor, $(\mathbf{T}_3' \times \mathbf{I}_2)$. This variation therefore seems to be mainly of academic interest.

Where the factorization of N is into relatively prime factors, as in the example, the factorization of $\mathbf{T}_N'$ is the "prime factor algorithm," which is sometimes confused with the Fast Fourier Transform [9].

6. *Conclusions*

With eqs. (17) and (18), we have expressed the transform $\mathbf{T}_N'$ ($N = 2^n$) using only the operators **C**, **P**, and a set of $(n - 1)$ operators $\mathbf{E}_{(i)}'$ or $\mathbf{E}_{(1)}''$, or $\mathbf{F}_{(2^i)}'$ or $\mathbf{F}_{(2^i)}''$. The operator **C** forms the sum and difference of pairs of successive components of a vector, and puts the computed values in place of the coefficients involved. The operator **P** permutes the components of the vector by a fixed permutation. The various $\mathbf{E}'$ or $\mathbf{E}''$ multiply each element of the data set by an appropriate power of w. The $\mathbf{F}_{(2^i)}'$ and $\mathbf{F}_{(2^i)}''$ multiply fixed subsets of the data set by the common multiplier w^{2^i}.

The transform, in the forms given in eqs. (17) or (18), could be implemented in a special purpose computer having the following capabilities:

(1) For i even:

$$a_i \leftarrow a_i + a_{i+1}\,, \qquad a_{i+1} \leftarrow a_i - a_{i+1}\,,$$

the computations to be done in parallel for successive pairs of data and in parallel for the real and imaginary parts.

(2) The parallel shift of data in accordance with **P**—i.e., an ideal shuffle. This is to be done separately but simultaneously for the real and imaginary parts, and can be conveniently done by a "wired-in" shift network involving bit-serial, word-parallel transfers.

(3) The ability to multiply in parallel a subset of the data set by a common multiplier. The subset can be selected by a stored "masking vector" as part of the control system.

Alternatively, if we use eqs. (15), (16), (19), or (20), in place of (3) we require

(3′) The ability to multiply in parallel each element of the data-set by a stored multiplier.

These capabilities should be relatively easy to implement in a special purpose machine. Hence it should be quite possible to build a special purpose computer using any of the given factorizations of the transform.

It is reasonable to suppose that the slowest operations of those required will be the complex multiplications required by the $\mathbf{E}'_{(i)}$ or $\mathbf{E}''_{(i)}$ operators. There are $\frac{1}{2}n(n-1)$ of these in the factorization of eqs. (17) or (18). If these can all be done in parallel, the resultant speed should be very high.

For example, consider a machine to form the Fourier transform of 512 complex data points. The direct computation of the Fourier transform requires $N^2 = 262{,}144$ complex multiplications. The serial implementation of the Fast Fourier Transform requires $N \log_2 N = nN = 4608$ complex multiplications. The machine suggested here requires only 36 sets of parallel complex multiplications with a common multiplier.

For data sets of the order indicated, the serial use of the Fast Fourier Transform is about two orders of magnitude faster than the direct transform. We would expect a special purpose machine implementing the factorization given here to be about another two orders of magnitude faster than the serial use of the Fast Fourier Transform.

We have also noted the possibility of other variations. For parallelization the most interesting of these is the possibility of using a factorization into $\mathbf{T}'_4$ or $\mathbf{T}''_4$, in the symbology of Section 5, i.e., a 4-step process. This cuts in half the number of passes required, but at some expense in terms of the operations needed per pass. Whether or not this would lead to a net gain in speed and, if so, at what cost in additional complexity in a special purpose machine, could be determined only as a result of detailed design work. If such a machine were built, this is a question that should be considered.

We conclude that if sufficient need exists it would be possible to build a special purpose parallel computer to calculate the Fourier transforms of large sets of data at extremely high speed.

ACKNOWLEDGMENT. The author is indebted to R. C. Singleton and W. H. Kautz of Stanford Research Institute for helpful discussions on this problem.

REFERENCES

1. COOLEY, J. W., LEWIS, P. A. W., AND WELCH, P. D. Historical notes on the Fast Fourier Transform. *IEEE Trans. AU-15* (June 1967), 76–79.
2. —— AND TUKEY, J. W. An algorithm for the machine calculation of complex Fourier series. *Math. Comput. 19*, 90 (April 1965), 297–301.
3. STOCKHAM, T. G., JR. High-speed convolution and correlation. Proc. AFIPS 1966 Spring Joint Comput. Conf., Vol. 28, pp. 229–233.
4. BINGHAM, C., GODFREY, M. D., AND TUKEY, J. W. Modern techniques of power spectrum estimation. (Unpublished.)
5. GENTLEMAN, W. M., AND SANDE, G. Fast Fourier Transforms—for fun and profit. Proc. AFIPS 1966 Fall Joint Comput. Conf., Vol. 29, pp. 563–578.
6. SINGLETON, R. C. A method for computing the Fast Fourier Transform with auxiliary memory and limited high-speed storage. *IEEE Trans. AU-15* (June 1967), 91–97.
7. COOLEY, J. W., LEWIS, P. A. W., AND WELCH, P. D. Application of the Fast Fourier Transform to computation of Fourier integrals, Fourier series, and convolution integrals. *IEEE Trans. AU-15* (June 1967), 79–84.
8. HELMS, H. D. Fast Fourier Transform method of computing difference equations and simulating filters. *IEEE Trans. AU-15* (June 1967), 85–90.
9. G-AE SUBCOMMITTEE ON MEASUREMENT CONCEPTS. What is the Fast Fourier Transform? *IEEE Trans. AU-15* (June 1967), 45–55.
10. PEASE, M. C. *Methods of Matrix Algebra*. Academic Press, New York, 1965, Ch. XIV.

A Linear Filtering Approach to the Computation of Discrete Fourier Transform

LEO I. BLUESTEIN, Member, IEEE
Electronic Systems Laboratory
General Telephone and Electronics Laboratories, Inc.
Waltham, Mass. 02154

Abstract

It is shown in this paper that the discrete equivalent of a chirp filter is needed to implement the computation of the discrete Fourier transform (DFT) as a linear filtering process. We show further that the chirp filter should not be realized as a transversal filter in a wide range of cases; use instead of the conventional FFT permits the computation of the DFT in a time proportional to $N \log_2 N$ for any N, N being the number of points in the array that is transformed. Another proposed implementation of the chirp filter requires N to be a perfect square. The number of operations required for this algorithm is proportional to $N^{3/2}$.

Introduction

There is currently a good deal of interest [1], [2] in a technique known as the fast Fourier transform (FFT), which is a method for rapidly computing the discrete Fourier transform of a time series (discrete data samples). This transform is the array of N numbers, A_n, $n=0, 1, \cdots, N-1$ which are defined by the relation

$$A_n = \sum_{k=0}^{N-1} X_k \exp(-2\pi jnk/N) \tag{1}$$

where X_k is the kth sample of a time series consisting of N (possibly complex) samples and $j=\sqrt{-1}$. Quite obviously, computing the A_n in a brute force manner requires N^2 operations,[1] a number which is too large to make the use of the discrete transform attractive. The fast Fourier transform discovered by Cooley and Tukey [3], on the other hand, requires about $N \log_2 N$ operations (if N is chosen to be a power of 2) a saving which is so great when N is large that it makes the use of the discrete transform attractive in such fields as digital filtering, computation of power spectra and autocorrelation functions and the like. If N is not chosen optimally, that is, if N is a composite number of the form

$$\prod_{i=1}^{s} p_i^{k_i}$$

where the p_i are prime numbers, the number of operations required by the Cooley–Tukey algorithm is proportional to

$$N \cdot \sum_{i=1}^{s} k_i p_i.$$

These results may be derived by viewing the computation of the DFT as a multistage operation on the array of X's.

We view the computational problem here rather as one of linear filtering. It is first shown that the discrete version of a chirp filter is an essential part of a linear filtering procedure which converts the X array to the A array. We then show that it is usually inadvisable to realize the chirp filter in transversal filter form since little saving in the number of operations would result. If instead the FFT is used to realize this filter, it is possible to compute the DFT in a time proportional to $N \log_2 N$ for any N. We have also devised another synthesis procedure for the chirp filter when N is a perfect square; in this case the number of computations is proportional to $N^{3/2}$.

An Algorithm Suggested by Chirp Filtering

This work was motivated by the observation that chirp filtering a waveform is nearly equivalent to taking its Fourier transform. For the sake of argument, assume that

[1] We usually take an operation to mean multiplication by a complex number plus all attendant computations, since multiplication of this sort on a computer requires an inordinate amount of time.

Manuscript received October 15, 1968; revised August 20, 1970.

Reprinted from *IEEE Trans. Audio Electroacoust.*, vol. AU-18, pp. 451–455, Dec. 1970.

a waveform $f(t)$ is applied to the input of a chirp filter. Because a chirp filter's delay characteristic is proportional to frequency, the low frequencies contained in $f(t)$ appear in the filter's output sooner than components at the high frequencies. This is then roughly a Fourier analysis of the waveform $f(t)$. Hence it appears reasonable to try to use a sampled equivalent of a chirp filter to generate the discrete Fourier transform.

For analog waveforms, a chirp filter is one whose frequency response is the bandpass equivalent of $\exp(-jf^2)$ over a range of frequencies. The impulsive response of this filter is proportional to $\exp[j(\pi t)^2]$. Now let us enter the realm of sampled systems. Motivated by the above discussion, we consider the effect of a sampled filter whose impulsive response is

$$h_r = \exp(+j\pi r^2/N) \tag{2}$$

on a sampled time function whose values are y_n. The y_n are assumed to be nonzero only for $0 \leq n \leq N-1$. Use of the convolution summation shows that the output of the filter at time $N+n$, u_{N+n} is

$$u_{N+n} = \sum_{k=0}^{N-1} y_k e^{j\pi(N+n-k)^2/N} \tag{3}$$

or

$$u_{N+n} e^{-j\pi n^2/N - j\pi N} = \sum_{k=0}^{N-1} (y_k e^{j\pi k^2/N}) e^{-2\pi jnk/N}. \tag{4}$$

Therefore, if we force

$$X_k = y_k e^{+j\pi k^2/N} \tag{5}$$

a comparison of (1) and (5) shows that

$$A_n = u_{N+n} e^{-j\pi n^2/N - j\pi N}. \tag{6}$$

Therefore, as Fig. 1 shows, the problem reduces to one of efficiently realizing a sampled chirp filter, whose impulsive response is given by (2). Moreover, we need only match (2) for $2N-1 \geq r \geq 0$. In addition, it is easily seen that the function $e^{-j\pi n^2/N}$ can be computed recursively to form the quantities y_k and A_k from X_k and u_{N+k}, respectively. This recursion is shown in Fig. 2 and depends on the observation that

$$e^{-j\pi(n+1)^2/N} = e^{-j\pi n^2/N} e^{-j2\pi n/N} e^{-j\pi/N} \tag{7}$$

In Fig. 2 the lower loop computes $e^{-j2\pi n/N}$; the upper loop performs the remaining operations indicated by (7). In using this recursion one must take care to avoid accumulation of round-off error.

Chirp Filter Synthesis Procedures

A. Transversal Filter Techniques

It is readily apparent that the sampled chirp filter essential to our algorithm may be synthesized as a transversal filter, i.e., as the sampled equivalent of a tapped delay line as shown in Fig. 3. The tap gains in this realization are of the form $e^{-j\pi r^2/N}$, $r=0, 1, 2, \cdots, N-1$. It is also clear that the number of distinct multipliers required in this realization can be reduced by observing that for certain values of r, say r_1 and r_2,

$$e^{-j\pi r_1^2/N} = \pm e^{-j\pi r_2^2/N} \tag{8}$$

or

$$e^{-j\pi r_1^2/N} = \pm je^{-j\pi r_2^2/N}. \tag{9}$$

If (8) holds, the quantity at the r_2th tap is multiplied by ± 1, added to the number on the r_1th tap and the sum multiplied by $e^{-j\pi r_1^2/N}$; if (9) holds, the r_2th quantity is multiplied by $\pm j$ (interchange of real and imaginary parts followed by possible negation) and summed as before.

The appropriate question now is, "How many multipliers are actually needed?" This question can be answered by the following considerations. If (8) holds,

$$r_1^2 = r_2^2 \bmod N \tag{10}$$

because (8) implies that

$$e^{-j\pi r_1^2/N} = e^{-j\pi(r_2^2+\alpha N)/N} \qquad \alpha = 0 \quad \text{or} \quad 1. \tag{11}$$

If (9) holds,

$$e^{-j\pi r_1^2/N} = e^{-j\pi(r_2^2+\beta(N/2))/N} \qquad \beta = +1 \quad \text{or} \quad -1. \tag{12}$$

Since $r_2^2+\beta(N/2)$ must be an integer, it follows that $N/2$ must be an integer, i.e., N is even. Thus (9) implies

$$r_1^2 = r_2^2 \bmod N/2. \tag{13}$$

If N is odd, we may disregard (9) and therefore (13) and concentrate on (10). The number of multipliers required is the number of distinct values of a such that

$$x^2 = a \bmod N \tag{14}$$

has a solution among the values $x=1, 2, 3, \cdots, N-1$. When this is so, and a and N are relatively prime, a is referred to as a quadratic residue of N.

If N is even, (10) implies (13). By the same considerations as above, we find that when N is even, the number of required multipliers is the number of distinct values of a, such that

$$x^2 = a \bmod N/2 \tag{15}$$

has a solution. Again, if a and $N/2$ are relatively prime, a is referred to as the quadratic residue of $N/2$.

Although we have not been able to find a complete solution to these questions, the following partial results may be obtained:

1) The number of distinct nonzero values of a satisfying (14) is equal to or greater than the largest integer less than $\sqrt{N}$, since values of x less than or equal to $\sqrt{N}$, when squared, generate distinct values of a. A similar consideration holds for (15).

2) When N is prime, a is always a quadratic residue. It then follows from number theory that the number of multipliers required is $(N-1)/2$.

3) When $N=2^{\alpha+1}$, the number of distinct values of a satisfying (15) can be computed by means of the following considerations. We divide the integers less than $N/2$ into

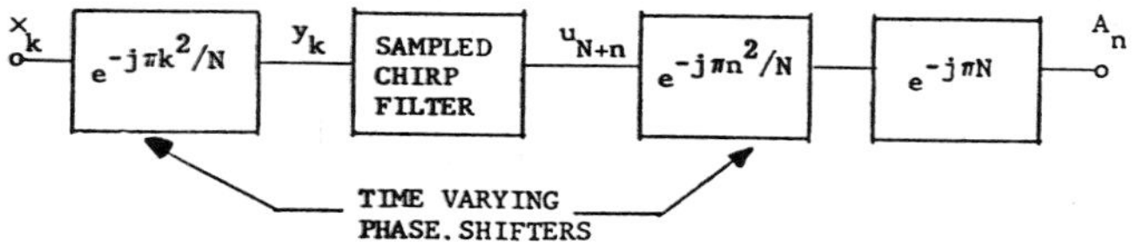

Fig 1. A Fourier transformer.

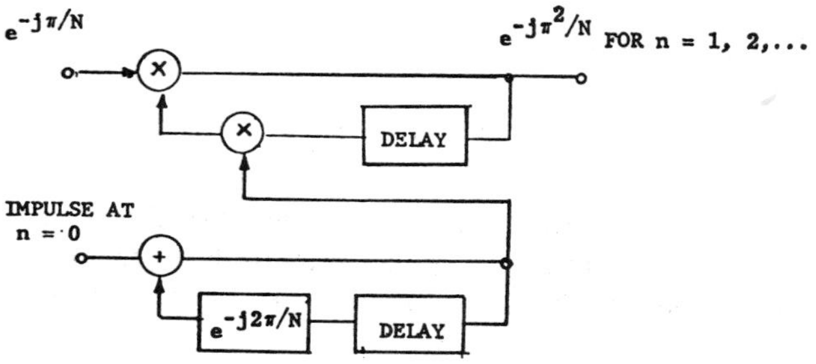

Fig. 2. A recursive method for generating $\exp(-j\pi n^2/N)$.

Fig. 3. A tapped delay-line version of a sampled chirp filter. $W \equiv e^{j\pi/N}$. Multiplications take place after the y_k's are read in.

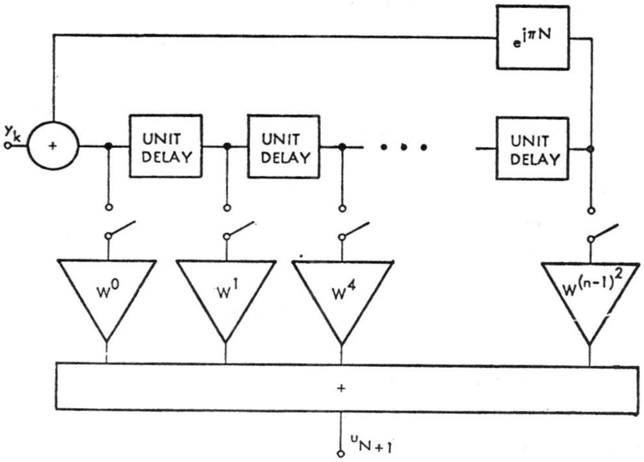

two sets, one comprised of odd numbers and one made up of the remaining even ones. The quadratic residues of 2^α are derived from the set of odd integers; these number $2^{\alpha-3}$ (if $\alpha \geq 3$).[2] Since each odd integer corresponds to some quadratic residue of 2^α, only the set of even integers need now be considered. We note that if an even value of x generates a distinct value of a, this value must be a multiple of 4. These values can therefore be found by taking the distinct values of a_1 which satisfy

$$x^2 = a_1 \bmod 2^{\alpha-2} \tag{16}$$

[2] This follows from two problem exercises on p. 39 of MacDuffee [4] and a result given on p. 94 by Vinagradov [5]. These are, respectively:

"Ex. 3. If $x^2 \equiv 1 \bmod n$ has exactly k solutions, and if $(c, n) = 1$, then $x^2 \equiv c \bmod n$ has no solution or exactly k solutions.
"Ex. 4. If c is a quadratic residue modulo n, and if $x^2 \equiv c \bmod n$ has k roots, there are exactly $\varphi(n)/k$ quadratic residues of n."

and

"The necessary conditions for the solvability of the congruence $x^2 \equiv a (\bmod 2^\alpha)$; $(a, 2) = 1$ are: $a \equiv 1 (\bmod 4)$ for $\alpha = 2$, $a \equiv 1 (\bmod 8)$ for $\alpha \geq 3$. If these conditions are satisfied, then the number of solutions is: 1 for $\alpha = 1$; 2 for $a = 2$; 4 for $\alpha \geq 3$."

The number of residue classes relatively prime to 2^α, $\varphi(2^\alpha)$, is just the number of odd integers less than 2^α and so is equal to $2^{\alpha-1}$. Since k (see Ex. 3) is 4, the number of quadratic residues is therefore $2^{\alpha-1}/4 = 2^{\alpha-3}$.

and multiplying them by 4. Once again we divide the integers less than $2^{\alpha-2}$ into odd and even groups; the group of odd numbers yield $2^{\alpha-5}$ distinct values of a_1 (if $\alpha - 2 \geq 3$). The remaining group of even integers will generate distinct values of a_1 corresponding to distinct values of a_2 which satisfy the new relation

$$x^2 = a_2 \bmod 2^{\alpha-4}. \tag{17}$$

The process continues until we have the relation

$$x^2 = a \bmod 4 \tag{18}$$

or

$$x^2 = a \bmod 2. \tag{19}$$

Equation (18) has only one nontrivial value of a which satisfies it, $a = 1$. Equation (19) has only a trivial value of a satisfying it, which does not require another multiplier. Thus, the number of distinct multipliers required to achieve a chirp response in transversal filter is

$$2^{\alpha-3} + 2^{\alpha-5} + 2^{\alpha-7} + \cdots \beta \tag{20}$$

where

$$\beta = \begin{cases} 0 & \alpha \text{ even} \\ 1 & \alpha \text{ odd.} \end{cases} \tag{21}$$

For large α, (20) is very nearly

$$2^{\alpha-3}\left(\frac{4}{3}\right) = \frac{2^{\alpha+1}}{12} \tag{22}$$

so that the number of multipliers needed is of the order of $N=2^{\alpha+1}$; combining terms prior to multiplying them only reduces the number of required multiplications by a factor of 12.

Thus, the transversal filter approach is not very fruitful for large N. The first observation implies that the best that one can hope for using this technique is a computational procedure with the number of computations proportional to $N^{3/2}$. The other two observations show that even this is not achievable for N a prime or a power of 2.

B. FFT Techniques

However, it should be observed[3] that we may realize *any* filter by means of the fast Fourier transform technique as suggested by Stockham [6]. In this method both the input to a filter and the filter's impulsive response are properly augmented by zeros and fast Fourier transformed. The results are multiplied together and the inverse transform applied. The number of operations is of the order of 3 $N'\log_2 N'$ where N' is both the length of the augmented impulsive response and the augmented input. The number N' is selected for this case to be the smallest power of 2 greater than or equal to $2N-1$. Thus, if a fast Fourier transform algorithm requiring N to be a power of 2 is available, it can easily be converted to one which accepts any N. The number of operations for the converted program will also be proportional to $N \log N$. This idea has been extended further by Rabiner, Shafer, and Rader [7].

C. Recursive Techniques

In this section we show that it is possible to achieve the performance promised by observation (1) using recursive techniques. We shall assume in what follows that N is a perfect square, i.e., that

$$N = m^2. \tag{23}$$

We will assume that the impulsive response given by (2) is zero for $r>2N-1$. The z transform, $H(z)$, of the impulsive response h_r is

$$H(z) = \sum_{r=0}^{2N-1} e^{j\pi r^2/m^2} z^{-r}. \tag{24}$$

Let us write

$$r = t + im \tag{25}$$

where

[3] This elegant observation is due to C. Rader of the M.I.T. Lincoln Laboratory.

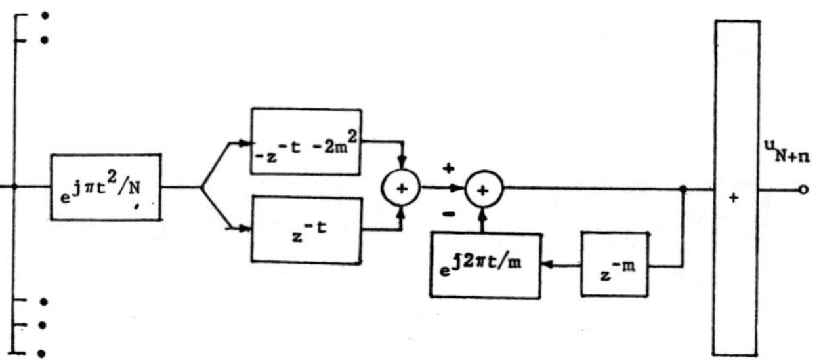

Fig. 4. A realization of $H(z)$.

$$t \leq m - 1. \qquad i \leq 2m - 1. \tag{26}$$

It then follows that

$$H(z) = \sum_{t=0}^{m-1} z^{-t} e^{j\pi t^2/N} \sum_{i=0}^{2m-1} e^{j2\pi ti/m} z^{-im} e^{j\pi i^2}. \tag{27}$$

but since

$$e^{j\pi i^2} = \begin{cases} -1 & i \text{ odd} \\ +1 & i \text{ even} \end{cases} \tag{28}$$

it follows that

$$e^{j\pi i^2} = (-1)^i \tag{29}$$

and

$$\begin{aligned} H(z) &= \sum_{t=0}^{m-1} z^{-t} e^{j\pi t^2/N} \sum_{i=0}^{2m-1} \left[-e^{j2\pi tm} z^{-m}\right]^i \\ &= \sum_{t=0}^{m-1} z^{-t} e^{j\pi t^2/N} \frac{1 - z^{-2m^2}}{1 + e^{j2\pi t/m} z^{-m}}. \end{aligned} \tag{30}$$

Thus, $H(z)$ may be realized by a bank of m filters as shown in Fig. 4. (A box labeled z^{-x} represents a delay of x units.) Since the impulsive response after $2m^2-1$ units does not concern us, we may open circuit the link connected to the box labeled z^{t-2m^2}. The number of operations required is about $3N\,m$, since we need only count operations which take place during the crucial interval between 0 and $2N-1$. This number is of the same form (within a multiplicative constant) as that required by the Cooley–Tukey algorithm for $N=m^2$ and m a prime.

Additional Comments

We have shown that by viewing the computation of the DFT as a linear filtering problem one is led naturally into algorithms which depend crucially upon the discrete version of a chirp filter. For arrays of large size, realizing the chirp filter by means of the FFT is appropriate, i.e., linear filtering cannot replace the FFT but can only extend its usefulness. For arrays of moderate size, and for wired program machines, algorithms based on the realizations shown in either Fig. 1 or Fig. 4 may be preferable for reasons of simplicity. Our results would tend to indicate that of these choices, the synthesis given in Fig. 4 is probably more desirable since asymptotically with N it behaves as well as any chirp transversal filter can ever be expected to perform. For special situations, and when hardware considerations are important, this tentative conclusion may not hold true.

References

[1] J. W. Cochran, J. W. Cooley, D. L. Favin, H. D. Helms, R. A. Kaenel, W. W. Lang, G. C. Maling, Jr., D. E. Nelson, C. M. Rader, and P. D. Welch, "What is the fast Fourier transform?" *Proc. IEEE*, vol. 55, pp. 1664–1674, October 1967.

[2] E. O. Brigham and R. E. Morrow, "The fast Fourier transform," *IEEE Spectrum*, vol. 4, pp. 63–70, December 1967.

[3] J. W. Cooley and J. W. Tukey, "An algorithm for the machine calculation of complex Fourier series," *Math. Comput.*, vol. 19, pp. 297–301, April 1965.

[4] C. C. MacDuffee, *An Introduction to Abstract Algebra.* New York: Wiley, 1940.

[5] I. M. Vinagradov, *Elements of Number Theory.* New York: Dover, 1958.

[6] T. G. Stockham, "High speed convolution and correlation," *1966 Spring Joint Comput. Conf., AFIPS Proc.*, vol. 28. Washington, D. C.: Spartan, 1966, pp. 229–233.

[7] L. R. Rabiner, R. W. Schafer, and C. M. Rader, "The chirp z-transform algorithm and its application," *Bell Syst. Tech. J.*, vol. 48, no. 5, pp. 1249–1292, May–June 1969.

The Chirp z-Transform Algorithm

L. R. RABINER, Member, IEEE
R. W. SCHAFER, Member, IEEE
Bell Telephone Laboratories, Inc.
Murray Hill, N. J.
C. M. RADER, Member, IEEE
Lincoln Laboratory[1]
Massachusetts Institute of Technology
Lexington, Mass.

Abstract

A computational algorithm for numerically evaluating the z-transform of a sequence of N samples is discussed. This algorithm has been named the chirp z-transform (CZT) algorithm. Using the CZT algorithm one can efficiently evaluate the z-transform at M points in the z-plane which lie on circular or spiral contours beginning at any arbitrary point in the z-plane. The angular spacing of the points is an arbitrary constant, and M and N are arbitrary integers.

The algorithm is based on the fact that the values of the z-transform on a circular or spiral contour can be expressed as a discrete convolution. Thus one can use well-known high-speed convolution techniques to evaluate the transform efficiently. For M and N moderately large, the computation time is roughly proportional to $(N+M)\log_2(N+M)$ as opposed to being proportional to $N \cdot M$ for direct evaluation of the z-transform at M points.

I. Introduction

In dealing with sampled data the z-transform plays the role which is played by the Laplace transform in continuous time systems. One example of its application is spectrum analysis. We shall see that the computation of sampled z-transforms, which has been greatly facilitated by the fast Fourier transform (FFT) [1], [2] algorithm, is still further facilitated by the chirp z-transform (CZT) algorithm to be described in this paper.

Manuscript received December 23, 1968; revised January 16, 1969.

This is a condensed version of a paper published in the *Bell System Technical Journal*, May 1969.

[1] Operated with support from the U. S. Air Force.

The z-transform of a sequence of numbers x_n is defined as

$$X(z) = \sum_{n=-\infty}^{\infty} x_n z^{-n}, \tag{1}$$

a function of the complex variable z. In general, both x_n and $X(z)$ could be complex. It is assumed that the sum on the right side of (1) converges for at least some values of z. We restrict ourselves to the z-transform of sequences with only a finite number N of nonzero points. In this case, we can rewrite (1) without loss of generality as

$$X(z) = \sum_{n=0}^{N-1} x_n z^{-n} \tag{2}$$

where the sum in (2) converges for all z except $z=0$.

Equations (1) and (2) are like the defining expressions for the Laplace transform of a train of equally spaced impulses of magnitudes x_n. Let the spacing of the impulses be T and let the train of impulses be $\sum_n x_n\delta(t-nT)$. Then the Laplace transform is $\sum_n x_n e^{-snT}$ which is the same as $X(z)$ if we let

$$z = e^{sT}. \tag{3}$$

If we are dealing with sampled waveforms the relation between the original waveform and the train of impulses is well understood in terms of the phenomenon of aliasing. Thus the z-transform of the sequence of samples of a time waveform is representative of the Laplace transform of the original waveform in a way which is well understood. The Laplace transform of a train of impulses repeats its values taken in a horizontal strip of the s-plane of width $2\pi/T$ in every other strip parallel to it. The z-transform maps each such strip into the entire z-plane, or conversely, the entire z-plane corresponds to any horizontal strip of the s-plane, e.g., the region $-\infty<\sigma<\infty$, $-\pi/T\leq\omega<\pi/T$, where $s=\sigma+j\omega$. In the same correspondence, the $j\omega$ axis of the s-plane, along which we generally equate the Laplace transform with the Fourier transform, is the unit circle in the z-plane, and the origin of the s-plane corresponds to $z=1$. The interior of the z-plane unit circle corresponds to the left half of the s-plane, and the exterior corresponds to the right half plane. Straight lines in the s-plane correspond to circles or spirals in the z-plane. Fig. 1 shows the correspondence of a contour in the s-plane to a contour in the z-plane. To evaluate the Laplace transform of the impulse train along the linear contour is to evaluate the z-transform of the sequence along the spiral contour.

Reprinted from *IEEE Trans. Audio Electroacoust.*, vol. AU-17, pp. 86–92, June 1969.

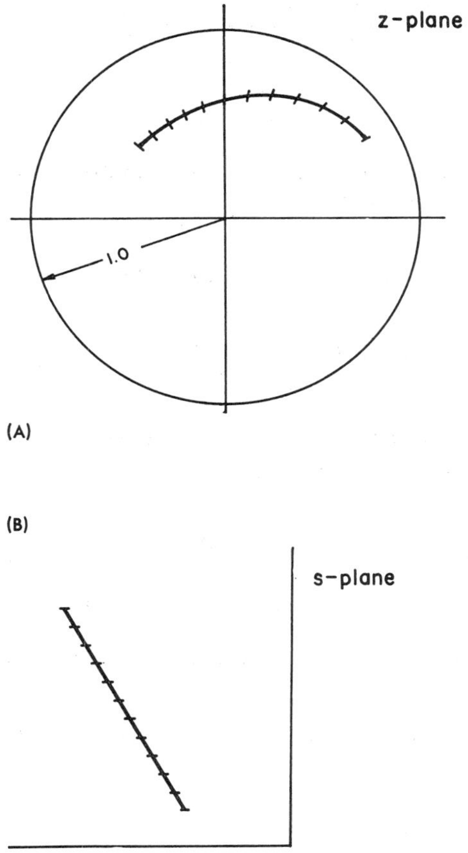

Fig. 1. The correspondence of (A) a z-plane contour to (B) an s-plane contour through the relation $z = e^{sT}$.

Values of the z-transform are usually computed along the path corresponding to the $j\omega$ axis, namely the unit circle. This gives the discrete equivalent of the Fourier transform and has many applications including the estimation of spectra, filtering, interpolation, and correlation. The applications of computing z-transforms off the unit circle are fewer, but one is presented elsewhere [6], namely the enhancement of spectral resonances in systems for which one has some foreknowledge of the locations of poles and zeroes.

Just as we can only compute (2) for a finite set of samples, so we can only compute (2) at a finite number of points, say z_k.

$$X_k = X(z_k) = \sum_{n=0}^{N-1} x_n z_k^{-n}. \qquad (4)$$

The special case which has received the most attention is the set of points equally spaced around the unit circle,

$$z_k = \exp(j2\pi k/N), \qquad k = 0, 1, \cdots, N-1 \qquad (5)$$

for which

$$X_k = \sum_{n=0}^{N-1} x_n \exp(-j2\pi k/N), \quad k = 0, 1, \cdots, N-1. \quad (6)$$

Equation (6) is called the discrete Fourier transform (DFT). The reader may easily verify that, in (5), other values of k merely repeat the same N values of z_k, which are the Nth roots of unity. The discrete Fourier transform has assumed considerable importance, partly because of its nice properties, but mainly because since 1965 it has become widely known that the computation of (6) can be achieved, not in the N^2 complex multiplications and additions called for by direct application of (6), but in something of the order of $N \log_2 N$ operations if N is a power of two, or $N \sum_i m_i$ operations if the integers m_i are the prime factors of N. Any algorithm which accomplishes this is called an FFT. Much of the importance of the FFT is that DFT may be used as a stepping stone to computing lagged products such as convolutions, autocorrelations, and cross convolutions more rapidly than before [3], [4]. The DFT has, however, some limitations which can be eliminated using the CZT algorithm which we will describe. We shall investigate the computation of the z-transform on a more general contour, of the form

$$z_k = AW^{-k}, \qquad k = 0, 1, \cdots, M-1 \qquad (7a)$$

where M is an arbitrary integer and both A and W are arbitrary complex numbers of the form

$$A = A_0 e^{j2\pi\theta_0} \qquad (7b)$$

and

$$W = W_0 e^{j2\pi\phi_0}. \qquad (7c)$$

(See Fig. 2.) The case $A = 1$, $M = N$, and $W = \exp(-j2\pi/N)$ corresponds to the DFT. The general z-plane contour begins with the point $z = A$ and, depending on the value of W, spirals in or out with respect to the origin. If $W_0 = 1$, the contour is an arc of a circle. The angular spacing of the samples is $2\pi\phi_0$. The equivalent s-plane contour begins with the point

$$s_0 = \sigma_0 + j\omega_0 = \frac{1}{T} \ln A \qquad (8)$$

and the general point on the s-plane contour is

$$s_k = s_0 + k(\Delta\sigma + j\Delta\omega) = \frac{1}{T}(\ln A - k \ln W), \qquad k = 0, 1, \cdots, M-1. \qquad (9)$$

Since A and W are arbitrary complex numbers we see that the points s_k lie on an arbitrary straight line segment of arbitrary length and sampling density. Clearly the contour indicated in (7a) is not the most general contour but it is considerably more general than that for which the DFT applies. In Fig. 2, an example of this more general contour is shown in both the z-plane and the s-plane.

To compute the z-transform along this more general contour would seem to require NM multiplications and additions as the special symmetries of $\exp(j2\pi k/N)$ which are exploited in the derivation of the FFT are absent in the more general case. However, we shall see that by using the sequence $W^{n^2/2}$ in various roles we can apply the FFT to the computation of the z-transform along the contour of (7a). Since for $W_0 = 1$, the sequence $W^{n^2/2}$ is a complex sinusoid of linearly increasing frequency, and since a

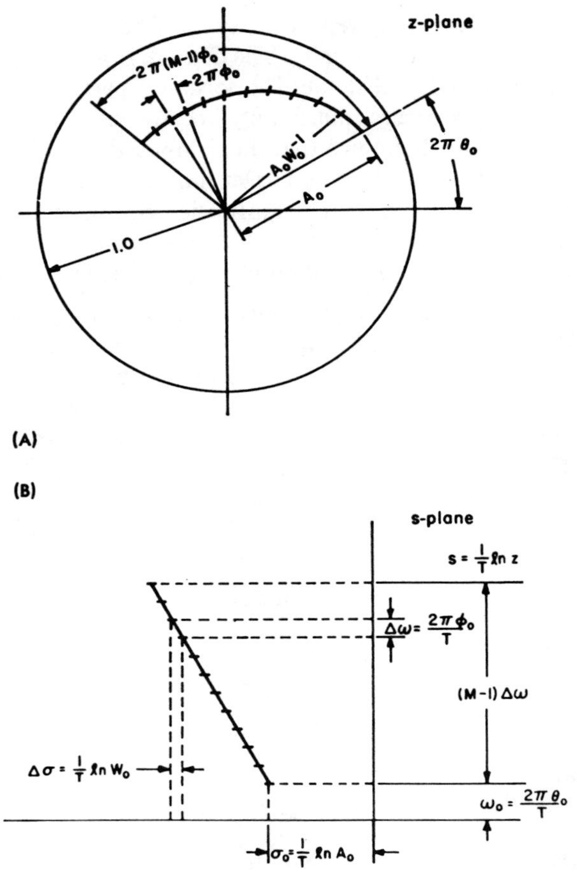

Fig. 2. An illustration of the independent parameters of the CZT algorithm. (A) How the z-transform is evaluated on a spiral contour starting at the point $z=A$. (B) The corresponding straight line contour and independent parameters in the s-plane.

similar waveform used in some radar systems has the picturesque name "chirp," we call the algorithm we are about to present the chirp z-transform (CZT). Since the CZT permits computing the z-transform on a more general tour than the FFT permits it is more flexible than the FFT, although it is also considerably slower. The additional freedoms offered by the CZT include the following:

1) The number of time samples does not have to equal the number of samples of the z-transform.

2) Neither M nor N need be a composite integer.

3) The angular spacing of the z_k is arbitrary.

4) The contour need not be a circle but can spiral in or out with respect to the origin. In addition, the point z_0 is arbitrary, but this is also the case with the FFT if the samples x_n are multiplied by z_0^{-n} before transforming.

II. Derivation of the CZT

Along the contour of (7a), (4) becomes

$$X_k = \sum_{n=0}^{N-1} x_n A^{-n} W^{nk}, \qquad k = 0, 1, \cdots, M-1 \tag{10}$$

which, at first appearance, seems to require NM complex multiplications and additions, as we have already observed. But, let us use the ingenious substitution, due to Bluestein [5],

$$nk = \frac{n^2 + k^2 - (k-n)^2}{2} \tag{11}$$

for the exponent of W in (10). This produces an apparently more unwieldly expression

$$X_k = \sum_{n=0}^{N-1} x_n A^{-n} W^{(n^2/2)} W^{(k^2/2)} W^{-(k-n)^2/2}, \qquad k = 0, 1, \cdots, M-1; \tag{12}$$

but, in fact, (12) can be thought of as a three-step process consisting of:

1) forming a new sequence y_n by weighting the x_n according to the equation

$$y_n = x_n A^{-n} W^{n^2/2}, \quad n = 0, 1, \cdots, N-1; \tag{13}$$

2) convolving y_n with the sequence v_n defined as

$$v_n = W^{-n^2/2} \tag{14}$$

to give a sequence g_k,

$$g_k = \sum_{n=0}^{N-1} y_n v_{k-n}, \quad k = 0, 1, \cdots, M-1; \tag{15}$$

3) multiplying g_k by $W^{k^2/2}$ to give X_k,

$$X_k = g_k W^{k^2/2}, \quad k = 0, 1, \cdots, M-1. \tag{16}$$

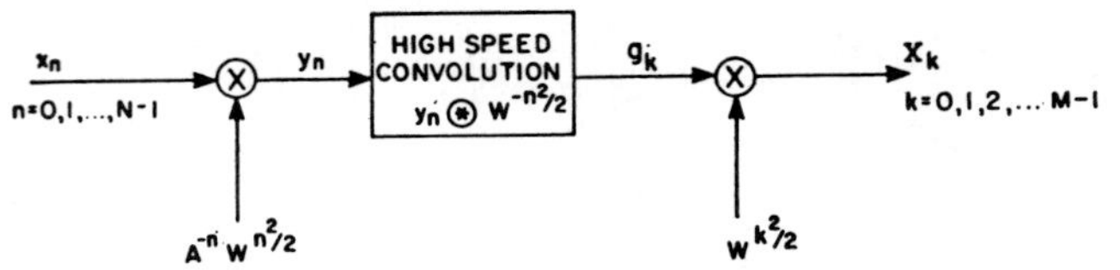

Fig. 3. An illustration of the steps involved in computing values of the z-transform using the CZT algorithm.

The three-step process is illustrated in Fig. 3. Steps 1 and 3 require N and M multiplications, respectively, and step 2 is a convolution which may be computed by the high-speed technique disclosed by Stockham [3], based on the use of the FFT. Step 2 is the major part of the computational effort and requires a time roughly proportional to $(N+M) \log (N+M)$.

Bluestein employed the substitution of (11) to convert a DFT to a convolution as in Fig. 3. The linear system to which the convolution is equivalent can be called a chirp filter which is, in fact, also sometimes used to resolve a spectrum. Bluestein [5] showed that for N a perfect square, the chirp filter could be synthesized recursively with $\sqrt{N}$ multipliers and the computation of a DFT could then be proportional to $N^{3/2}$.

The flexibility and speed of the CZT algorithm are related to the flexibility and speed of the method of high-speed convolution using the FFT. The reader should recall that the product of the DFT's of two sequences is the DFT of the circular convolution of the two sequences and, therefore, a circular convolution is computable as two DFT's, the multiplication of two arrays of complex numbers, and an inverse discrete Fourier transform (IDFT), which can also be computed by the FFT. Ordinary convolutions can be computed as circular convolutions by appending zeroes to the end of one or both sequences so that the correct numerical answers for the ordinary convolution can result from a circular convolution.

We shall now summarize the details of the CZT algorithm on the assumption that an already existing FFT program (or special-purpose machine) is available to compute DFT's and IDFT's.

Begin with a waveform in the form of N samples x_n and seek M samples of X_k where A and W have also been chosen.

1) Choose L, the smallest integer greater than or equal to $N+M-1$ which is also compatible with our high-speed FFT program. For most users this will mean L is a power of two. Note that while many FFT programs will work for arbitrary L, they are not equally efficient for all L. At the very least, L should be highly composite.

2) Form an L point sequence y_n from x_n by

$$y_n = \begin{cases} A^{-n}W^{n^2/2}x_n & n=0, 1, 2, \cdots, N-1 \\ 0 & n=N, N+1, \cdots, L-1. \end{cases} \tag{17}$$

3) Compute the L point DFT of y_n by the FFT. Call this Y_r, $r=0, 1, \cdots, L-1$.

4) Define an L point sequence v_n by the relation

$$v_n = \begin{cases} W^{-n^2/2} & 0 \le n \le M-1 \\ W^{-(L-n)^2/2} & L-N+1 \le n < L \\ \text{arbitrary} & \text{other } n, \text{ if any.} \end{cases} \tag{18}$$

Of course, if L is exactly equal to $M+N-1$, the region in which v_n is arbitrary will not exist. If the region does exist an obvious possibility is to increase M, the desired number of points of the z-transform we compute, until the region does not exist.

Note that v_n could be cut into two with a cut between $n=M-1$ and $n=L-N+1$ and if the two pieces were abutted together differently, the resulting sequence would be a slice out of the indefinite length sequence $W^{-n^2/2}$. This is illustrated in Fig. 4. The sequence v_n is defined the way it is in order to force the circular convolution to give us the desired numerical results of an ordinary convolution.

5) Compute the DFT of v_n and call it V_r, $r=0, 1, \cdots, L-1$.

6) Multiply V_r and Y_r point by point, giving G_r:

$$G_r = V_r Y_r, \quad r = 0, 1, \cdots, L-1.$$

7) Compute the L point IDFT g_k, of G_r.

8) Multiply g_k by $W^{k^2/2}$ to give the desired X_k:

$$X_k = W^{k^2/2} g_k, \quad k = 0, 1, 2, \cdots, M-1.$$

The g_k for $k \ge M$ are discarded.

Fig. 4 represents typical waveforms (magnitudes shown, phase omitted) involved in each step of the process.

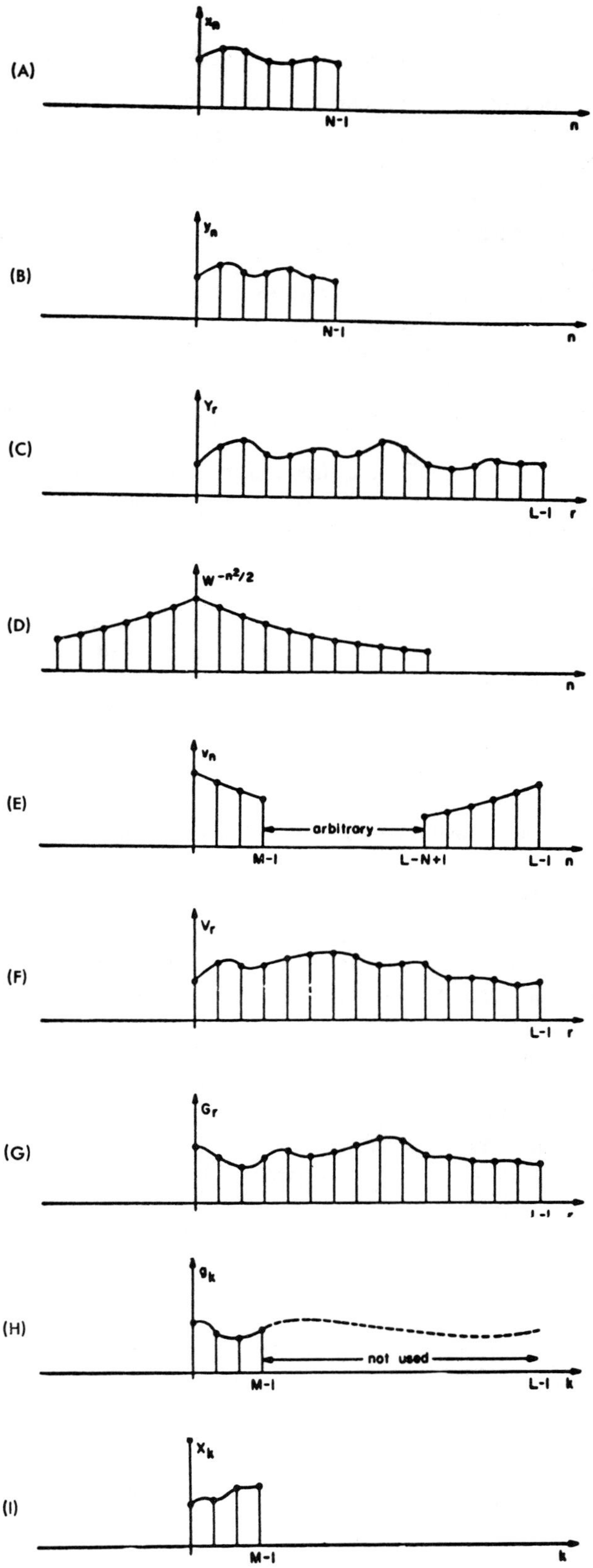

Fig. 4. Schematic representation of the various sequences involved in the CZT algorithm. (A) The input sequence x_n with N values. (B) The weighted input sequence $y_n = A^{-n}W^{n^2/2}x_n$. (C) The DFT of y_n. (D) The values of the indefinite sequence $W^{-n^2/2}$. (E) The sequence v_n formed appropriately from segments of $W^{-n^2/2}$. (F) The DFT of v_n. (G) The product $G_r = Y_r \cdot V_r$. (H) The IDFT of G_r. (I) The desired M values of the z-transform.

III. Fine Points of the Computation

Operation Count and Timing Considerations

An operation count can be made, roughly, from the eight steps just presented. We will give it step by step because there are, of course, many possible variations to be considered.

1) We assume that step 1, choosing L, is a negligible operation.

2) Forming y_n from x_n requires N complex multiplications, not counting the generation of the constants $A^{-n}W^{n^2/2}$. The constants may be prestored, computed as needed, or generated recursively as needed. The recursive computation would require two complex multiplications per point.

3) An L point DFT requires a time $k_{FFT}L \log_2 L$ for L a power of two, and a very simple FFT program. More complicated (but faster) programs have more complicated computing time formulas.

4), 5) v_n is computed for either M or N points, whichever is greater. The symmetry in $W^{-n^2/2}$ permits the other values of v_n to be obtained without computation. Again, v_n can be computed recursively. The FFT takes the same time as that in step 3. If the same contour is used for many sets of data, V_r need only be computed once, and stored.

6) This step requires L complex multiplications.

7) This is another FFT and requires the same time as step 3.

8) This step requires M complex multiplications.

As the number of samples of x_n or X_k grow large, the computation time for the CZT grows asymptotically as something proportional to $L \log_2 L$. This is the same sort of asymptotic dependence of the FFT, but the constant of proportionality is bigger for the CZT because two or three FFT's are required instead of one, and because L is greater than N or M. Still, the CZT is faster than the direct computation of (10) even for relatively modest values of M and N, of the order of 50.

Reduction in Storage

The CZT can be put into a more useful form for computation by redefining the substitution of (11) to read

$$nk = \frac{(n - N_0)^2 + k^2 - (k - n + N_0)^2 + 2N_0k}{2}.$$

Equation (12) can now be rewritten as

$$X_k = W^{k^2/2}W^{N_0k} \sum_{n=0}^{N-1} x_n A^{-n} W^{(n-N_0)^2/2} W^{-(k-n+N_0)^2/2}.$$

The form of the new equation is similar to (12) in that the input data x_n are pre-weighted by a complex sequence $(A^{-n}W^{(n-N_0)^2/2})$, convolved with a second sequence $(W^{-(n-N_0)^2/2})$, and post-weighted by a third sequence $(W^{k^2/2}W^{N_0k})$ to compute the output sequence X_k. However, there are differences in the detailed procedures for realizing the CZT. The input data x_n can be thought of as having been shifted by N_0 samples to the left; e.g., x_0 is weighted by $W^{N_0^2/2}$ instead of W_0. The region over which $W^{-n^2/2}$ must be formed, in order to obtain correct results from the convolution, is

$$-N + 1 + N_0 \leq n \leq M - 1 + N_0.$$

By choosing $N_0=(N-M)/2$ it can be seen that the limits over which $W^{-n^2/2}$ is evaluated are symmetric; i.e., $W^{-n^2/2}$ is a symmetric function in both its real and imaginary parts. (It follows thus that the transform of $W^{-n^2/2}$ is also symmetric in both its real and imaginary parts.) It can be shown that using this special value of N_0, only $(L/2+1)$ points of $W^{-n^2/2}$ need be calculated and stored and these $(L/2+1)$ complex points can be transformed using an $L/2$ point transform.[2] Hence the total storage required for the transform of $W^{-n^2/2}$ is $L+2$ locations.

The only other modifications to the detailed procedures for evaluating the CZT presented in Section II of this paper are: 1) following the L point IDFT of step 7, the data of array g_k must be rotated to the left by N_0 locations; and 2) the weighting factor of the g_k is $W^{k^2/2}W^{N_0k}$ rather than $W^{k^2/2}$. The additional factor W^{N_0k} represents a data shift of N_0 samples to the right, thus compensating the initial shift and keeping the effective positions of the data invariant to the value of N_0 used.

An estimate of the storage required to perform the CZT can now be made. Assuming that the entire process is to take place in core, storage is required for V_r which takes $L+2$ locations; for y_n, which takes $2L$ locations; and perhaps for some other quantities which we wish to save, e.g., the input, or values of $W^{+n^2/2}$ or $A^{-n}W^{n^2/2}$.

Additional Considerations

Since the CZT permits $M \neq N$, it is possible that occasions will arise where $M \gg N$ or $N \gg M$. In these cases, if the smaller number is small enough, the direct method of (10) is called for. However, if even the smaller number is large it may be appropriate to use the methods of sectioning described by Stockham [3]. Either the lap-save or lap-add methods may be used. Sectioning may also be used when problems too big to be handled in core memory arise. We have not actually encountered any of these problems and have not programmed the CZT with provision for sectioning.

Since the contour for the CZT is a straight line segment in the s-plane, it is apparent that repeated application of the CZT can compute the z-transform along a contour which is piecewise spiral in the z-plane or piecewise linear in the s-plane.

Let us briefly consider the CZT algorithm for the case of z_k all on the unit circle. This means that the z-transform is like a Fourier transform. Unlike the DFT, which by definition gives N points of transform for N points of

[2] The technique for transforming two real symmetric L point sequences using one $L/2$ point FFT was demonstrated by J. Cooley at the FFT Workshop, Arden House, October 1968. A summary of this technique is presented in the Appendix.

data, the CZT does not require $M=N$. Furthermore, the z_k need not stretch over the entire unit circle but can be equally spaced along an arc. Let us assume, however, that we are really interested in computing the N point DFT of N data points. Still the CZT permits us to choose any value of N, highly composite, somewhat composite, or even prime, without strongly affecting the computation time. An important application of the CZT may be computing DFT's when N is not a power of two and when the program or special-purpose device available for computing DFT's by FFT is limited to the case of N a power of two.

There is also no reason why the CZT cannot be extended to the case of transforms in two or more dimensions with similar considerations. The two-dimensional DFT becomes a two-dimensional convolution which is computable by FFT techniques.

We caution the reader to note that for the ordinary FFT the starting point of the contour is still arbitrary; merely multiply the waveform x_n by A^{-n} before using the FFT, and the first point on the contour is effectively moved from $z=1$ to $z=A$. However, the contour is still restricted to a circle concentric with the origin. The angular spacing of z_k for the FFT can also be controlled to some extent by appending zeroes to the end of x_n before computing the DFT (to decrease the angular spacing of the z_k) or by choosing only P of the N points x_n and adding together all the x_n for which the n are congruent modulo P; i.e., wrapping the waveform around a cylinder and adding together the pieces which overlap (to increase the angular spacing).

IV. Limitations

One limitation in using the CZT algorithm to evaluate the z-transform off the unit circle stems from the fact that we may be required to compute $W_0^{\pm n^2/2}$ for large n. If W_0 differs very much from 1.0, $W_0^{\pm n^2/2}$ can become very large or very small when n becomes large. (We require a large n when either M or N become large, since we need to evaluate $W^{n^2/2}$ for n in the range $-N<n<M$.) For example, if $W_0=e^{-.25/1000}\approx 0.999749$, and $n=1000$, $W_0^{\pm n^2/2}=e^{\pm 125}$ which exceeds the single precision floating point capability of most computers by a large amount. Hence the tails of the functions $W^{\pm n^2/2}$ can be greatly in error, thus causing the tails of the convolution (the high frequency terms) to be grossly inaccurate. The low frequency terms of the convolution will also be slightly in error but these errors are negligible in general.

The limitation on contour distance in or out from the unit circle is again due to computation of $W^{\pm n^2/2}$. As W_0 deviates significantly from 1.0, the number of points for which $W^{\pm n^2/2}$ can be accurately computed decreases. It is of importance to stress, however, that for $W_0=1$, there is no limitation of this type since $W^{\pm n^2/2}$ is always of magnitude 1.

The other main limitation on the CZT algorithm stems from the fact that two L point, and one $L/2$ point, FFT's must be evaluated where L is the smallest convenient integer greater than $N+M-1$ as mentioned previously. We need one FFT and $2L$ storage locations for the transform of $x_nA^{-n}W^{n^2/2}$; one FFT and $L+2$ storage locations for the transform $W^{-n^2/2}$; and one FFT for the inverse transform of the product of these two transforms. We do not know a way of computing the transform of $W^{-n^2/2}$ either recursively or by a specific formula (except in some trivial cases). Thus we must compute this transform and store it in an extra $L+2$ storage locations. Of course, if many transforms are to be done with the same value of L, we need not compute the transform of $W^{-n^2/2}$ each time.

We can compute the quantities $A^{-n}W^{n^2/2}$ recursively as they are needed to save computation and storage. This is easily seen from the fact that

$$A^{-(n+1)}W^{(n+1)^2/2} = (A^{-n}W^{n^2/2})\cdot W^nW^{1/2}A^{-1}. \quad (19)$$

If we define

$$C_n = A^{-n}W^{n^2/2} \quad (20)$$

and

$$D_n = W^nW^{1/2}A^{-1} \quad (21)$$

then

$$D_{n+1} = W\cdot D_n \quad (22)$$

and

$$C_{n+1} = C_n\cdot D_n. \quad (23)$$

Setting $A=1$ in (19) to (23) provides an algorithm for the coefficients required for the output sequence. A similar recursion formula can be obtained for generating the sequence $A^{-n}W^{(n-N_0)^2/2}$. The user is cautioned that recursive computation of these coefficients may be a major source of numerical error, especially when $W_0\approx 1$, or $\phi_0\approx 0$.

V. Summary

A computational algorithm for numerically evaluating the z-transform of a sequence of N time samples was presented. This algorithm, entitled the chirp z-transform algorithm, enables the evaluation of the z-transform at M equi-angularly spaced points on contours which spiral in or out (circles being a special case) from an arbitrary starting point in the z-plane. In the s-plane the equivalent contour is an arbitrary straight line.

The CZT algorithm has great flexibility in that neither N or M need be composite numbers; the output point spacing is arbitrary; the contour is fairly general and N need not be the same as M. The flexibility of the CZT algorithm is due to being able to express the z-transform on the above contours as a convolution, permitting the use of well-known high-speed convolution techniques to evaluate the convolution.

Applications of the CZT algorithm include enhancement of poles for use in spectral analysis; high resolution, narrowband frequency analysis; and time interpolation of data from one sampling rate to any other sampling rate. These applications are explained in detail elsewhere [6]. The CZT algorithm also permits use of a radix-2 FFT program or device to compute the DFT of an arbitrary number of samples. Examples illustrating how the CZT algorithm is used in specific cases are included elsewhere [6]. It is anticipated that other applications of the CZT algorithm will be found.

Appendix

The purpose of this Appendix is to show how the FFT's of two real, symmetric L point sequences can be obtained using one $L/2$ point FFT.

Let x_n and y_n be two real, symmetric L point sequences with corresponding DFT's X_k and Y_k. By definition,

$$\begin{aligned} x_n &= x_{L-n} \\ y_n &= y_{L-n} \end{aligned} \qquad n = 0, 1, 2, \cdots, L-1,$$

and it is easily shown that X_k and Y_k are real, symmetric L point sequences, so that

$$\begin{aligned} X_k &= X_{L-k} \\ Y_k &= Y_{L-k} \end{aligned} \qquad k = 0, 1, 2, \cdots, L-1.$$

Define a complex $L/2$ point sequence u_n whose real and imaginary parts are

$$\left.\begin{aligned} \text{Re}\,[u_n] &= x_{2n} - y_{2n+1} + y_{2n-1} \\ \text{Im}\,[u_n] &= y_{2n} + x_{2n+1} - x_{2n-1} \end{aligned}\right\} \quad n = 0, 1, \cdots, L/2-1.$$

The $L/2$ point DFT of u_n is denoted U_k and is calculated by the FFT. The values of X_k and Y_k may be computed from U_k using the relations

$$X_k = \tfrac{1}{2}\{\text{Re}\,[U_k] + \text{Re}\,[U_{L/2-k}]\} - \frac{1}{4 \sin \frac{2\pi}{L} k}\{\text{Re}\,[U_k] - \text{Re}\,[U_{L/2-k}]\}$$

$$Y_k = \tfrac{1}{2}\{\text{Im}\,[U_k] + \text{Im}\,[U_{L/2-k}]\} - \frac{1}{4 \sin \frac{2\pi}{L} k}\{\text{Im}\,[U_k] - \text{Im}\,[U_{L/2-k}]\}$$

for $k = 1, 2, \cdots, L/2-1$. The remaining values of X_k and Y_k are obtained from the relations

$$X_0 = \sum_{n=0}^{L-1} x_n$$

$$Y_0 = \sum_{n=0}^{L-1} y_n$$

$$X_{L/2} = \sum_{n=0}^{L-1} x_n(-1)^n$$

$$Y_{L/2} = \sum_{n=0}^{L-1} y_n(-1)^n.$$

References

[1] J. W. Cooley and J. W. Tukey, "An algorithm foı the machine calculation of complex Fourier series," *Math. Comp.*, vol. 19, pp. 297–301, 1965.

[2] G-AE Subcommittee on Measurement Concepts, "What is the fast Fourier transform?" *IEEE Trans. Audio and Electroacoustics*, vol. AU-15, pp. 45–55, June 1967.

[3] T. G. Stockham, Jr., "High speed convolution and correlation," *1966 Spring Joint Computer Conf., AFIPS Proc.*, vol. 28. Washington, D. C.: Spartan, 1966, pp. 229–233.

[4] H. D. Helms, "Fast Fourier transform method of computing difference equations and simulating filters," *IEEE Trans. Audio and Electroacoustics*, vol. AU-15, pp. 85–90, June 1967.

[5] L. I. Bluestein, "A linear filtering approach to the computation of the discrete Fourier transform," *1968 NEREM Rec.*, pp. 218–219.

[6] L. R. Rabiner, R. W. Schafer, and C. M. Rader, "The chirp z-transform algorithm and its applications," *Bell Sys. Tech. J.*, vol. 48, pp. 1249–1292, May 1969.

Discrete Fourier Transforms When the Number of Data Samples Is Prime

***Abstract*—The discrete Fourier transform of a sequence of N points, where N is a prime number, is shown to be essentially a circular correlation. This can be recognized by rearranging the members of the sequence and the transform according to a rule involving a primitive root of N. This observation permits the discrete Fourier transform to be computed by means of a fast Fourier transform algorithm, with the associated increase in speed, even though N is prime.**

To compute the discrete Fourier transform of a long sequence of data samples, one of the fast Fourier transform (FFT) algorithms should be used if possible.[1],[2] The limitation common to all of these algorithms is that the number of data samples, N, must be highly composite. In this letter we show how FFT techniques can be applied to the computation of a discrete Fourier transform when N is prime.

I. Notation and Definitions

We basically restrict our attention to the discrete Fourier transform. Let the data to be transformed be a sequence of N numbers $\{a_i\}$, $i=0, 1, \cdots, N-1$, where the braces indicate the entire sequence, while the symbol a_i represents only the ith member of the sequence. The discrete Fourier transform (DFT) is the sequence $\{A_k\}$, $k=0, 1, \cdots, N-1$, whose members are given by

$$A_k = \sum_{i=0}^{N-1} a_i \exp(-j(2\pi/N)ik). \tag{1}$$

The values of A_k are samples of the z-transform of the finite length sequence $\{a_i\}$ at N points equally spaced on the unit circle.

We need a notation for an integer modulo N. For brevity we will indicate this by superfluous parentheses:

$$((X)) = X \text{ modulo } N. \tag{2}$$

If N is prime there is some number g, not necessarily unique, such that there is a one-to-one mapping of the integers $i=1, \cdots, N-1$ to the integers $j=1, 2, \cdots, N-1$, given by

$$j = ((g^i)). \tag{3}$$

For example, let $N=7$ and $g=3$. The table below gives the mapping of i onto j.

i	1	2	3	4	5	6
j	3	2	6	4	5	1

In number theory, g is called a primitive root of N. A table of primitive roots of primes less than 10^4 is given by Abramowitz and Stegun.[3]

II. Computation of $\{A_k\}$ when N Is Prime

In (1) we have an expression for A_k for all k. The expression for A_0 is particularly simple,

$$A_0 = \sum_{i=0}^{N-1} a_i, \tag{4}$$

and is to be computed directly. For the other A_k we observe that a_0 is not to be multiplied, and we choose to add it last into the summation. We are left with the sequence $\{A_k - a_0\}$, $k=1, 2, \cdots, N-1$ to compute, given by

$$A_k - a_0 = \sum_{i=1}^{N-1} a_i \exp\left(-j\frac{2\pi}{N}ik\right). \tag{5}$$

We permute the terms in the summation, and change the order of the equations via transformations

$$\begin{aligned} i &\to ((g^i)) \\ k &\to ((g^k)) \end{aligned} \tag{6}$$

and, noting that $((g^{N-1})) = ((g^0))$, we can see that

$$(A_{((g^k))} - a_0) = \sum_{i=0}^{N-1} a_{((g^i))} \exp\left(-j\frac{2\pi}{N}g^{(i+k)}\right). \tag{7}$$

We are now able to recognize that the sequence $\{A_{((g^k))} - a_0\}$ is the circular correlation of the sequence $\{a_{((g^i))}\}$ and the sequence $\{\exp(-j(2\pi/N)g^i)\}$. But Stockham has shown how circular (or ordinary) correlation functions can be computed with a greatly reduced number of operations by making use of FFT algorithms.[4] There are two ways we can proceed. Since N is prime, $N-1$ must be composite. Suppose that it is highly composite. Then the $N-1$ point circular correlation (7) may be recognized as the inverse DFT of the product of the DFT of $\{a_{((g^{-i}))}\}$ and the DFT of $\{\exp(-j(2\pi/N)g^i\}$.[1] All the DFT operations called for are performed by an FFT algorithm.

$$\{A_{((g^k))} - a_0\} = \text{DFT}^{-1}\left\{(\text{DFT}\{a_{((g^{-i}))}\})\left(\text{DFT}\left\{\exp\left(-j\frac{2\pi}{N}g^i\right)\right\}\right)\right\}. \tag{8}$$

Proceeding in this first way we will be successful only if $N-1$ is highly composite. If $N-1$ is only modestly composite, as with $N=563$, the savings of the FFT algorithm will be overcome by the fact that more than one DFT must be computed. However, there is another way we can proceed which is not subject to these limitations. The second method is based on the observation that a circular correlation or convolution where the number of points is not highly composite can be computed as a part of a circular convolution with a larger number of points. Letting N' be any highly composite integer greater than $2N-4$, we create an N' point sequence $\{b_i\}$ by inserting $(N'-N+1)$ zeros between the zeroth and first points of $\{a_{((g^{-i}))}\}$ and we create a second N' point sequence $\{c_i\}$ by periodically repeating the $N-1$ point sequence $\{\exp(-j(2\pi/N)g^i)\}$ until N' points are present. Then the inverse DFT of the product of the DFTs of $\{b_i\}$ and $\{c_i\}$ contains $\{A_{((g^k))} - a_0\}$ as a subsequence—the first $N-1$ points. Since N' can be chosen to be highly composite, even a power of two, an FFT algorithm can be used to compute the DFTs.

Using either technique, about one-third of the computation can be saved if the transform of $\{\exp(-j(2\pi/N)g^i)\}$ is precomputed. One method requires a computation proportional to $(N-1)$ times the sum of the factors of $(N-1)$ whereas the second method requires a computation proportional to $N' \log N'$. Furthermore, the summation called for in (4) and the addition of a_0 to each other A_k can each be performed with negligible additional computation by operating on intermediate quantities available when the correlation is done by FFT techniques.

III. Conclusions

While the restriction that N be a highly composite number for FFT techniques to be useful has not proved severe, it is interesting to know that it can be removed. On the other hand, the recognition that a DFT can be expressed as a convolution may be useful in itself, as this implies that a single network with fixed parameters can compute all the points of a DFT. It is expected that such diverse applications as radar beam forming and modem design may profitably use this result.

Charles M. Rader
M.I.T. Lincoln Lab.[2]
Lexington, Mass. 02173

References

[1] J. W. Cooley and J. W. Tukey, "An algorithm for the machine computation of complex Fourier series," *Math. Comput.*, vol. 19, pp. 297–301, April 1965.

[2] "What is the fast Fourier transform?" *Proc. IEEE*, vol. 55, pp. 1664–1674, October 1967.

[3] M. Abramowitz and I. Stegun, *Handbook of Mathematical Functions*. New York: Dover, 1965, p. 827 and pp. 864–869.

[4] T. G. Stockham, "High speed convolution and correlation," *1966 Spring Joint Computer Conf., AFIPS Proc.*, vol. 28. Washington, D.C.: Spartan, 1966, pp. 229–233.

[1] It may be shown that this DFT has magnitude $\sqrt{N}$ at all frequencies but the zeroth, where it is -1.

[2] Operated with support from the U. S. Air Force.

Manuscript received March 4, 1968.

Reprinted from *Proc. IEEE*, vol. 56, pp. 1107–1108, June 1968.

HIGH-SPEED CONVOLUTION AND CORRELATION*

Thomas G. Stockham, Jr.
Massachusetts Institute of Technology, Project MAC
Cambridge, Massachusetts

INTRODUCTION

Cooley and Tukey[1] have disclosed a procedure for synthesizing and analyzing Fourier series for discrete periodic complex functions.† For functions of period N, where N is a power of 2, computation times are proportional to $N \log_2 N$ as expressed in Eq. (0).

$$T_{ct} = k_{ct} N \log_2 N \tag{0}$$

where k_{ct} is the constant of proportionality. For one realization for the IBM 7094, k_{ct} has been measured at 60 μsec. Normally the times required are proportional to N^2. For $N = 1000$ speed-up factors in the order of 50 have been realized! Eq. (1b) synthesizes the Fourier series in question. The complex Fourier coefficients are given by the analysis equation, Eq. (1a).

$$F(k) = \sum_{j=0}^{N-1} f(j) w^{-jk} \tag{1a}$$

$$f(j) = \frac{1}{N} \sum_{k=0}^{N-1} F(k) w^{jk} \tag{1b}$$

where $w = e^{2\pi i/N}$, the principal Nth root of unity. The functions f and F are said to form a discrete periodic complex transform pair. Both functions are of period N since

$$F(k) = F(k + cN) \tag{2a}$$

and

$$f(j) = f(j + cN) \tag{2b}$$

TRANSFORM PRODUCTS

Consider two functions g and h and their transforms G and H. Let G and H be multiplied to form the function C according to Eq. (3),

$$C(k) = G(k) \times H(k) \tag{3}$$

and consider the inverse transform $c(j)$. $c(j)$ is given by Eq. (4)

$$c(j) = \frac{1}{N} \sum_{J=0}^{N-1} g(J)h(j-J) = \frac{1}{N} \sum_{J=0}^{N-1} h(J)g(j-J) \tag{4}$$

as a sum of lagged products where the lags are performed circularly. Those values that are shifted from one end of the summation interval are circulated into the other.

The time required to compute $c(j)$ from either form of Eq. (4) is proportional to N^2. If one computes the transforms of g and h, performs the multiplication of Eq. (3), and then computes the inverse

*Work reported herein was supported (in part) by Project MAC, an M.I.T. research program sponsored by the Advanced Research Projects Agency, Department of Defense, under Office of Naval Research Contract Number Nonr-4102(01).

†To be able to use this procedure the period must be a highly composite number.

Reprinted with permission from *1966 Spring Joint Computer Conf., AFIPS Conf. Proc.*, vol. 28, pp. 229–233, 1966.

transform of C, one requires a time given by Eq. (5)

$$T_{\text{circ}} = 3\, k_{ct} N \log_2 N + k_m N$$
$$= k_{\text{circ}} N(\log_2 N + \mu) \quad (5)$$

where $k_{\text{circ}} = 3k_{ct}$, $\mu = k_m/k_{\text{circ}}$, and $k_m N$ is the time required to compute Eq. (3). Of course this assumes N is a power of 2. Similar savings would be possible provided N is a highly composite number.

APERIODIC CONVOLUTION

The circular lagged product discussed above can be alternately regarded as a convolution of periodic functions of equal period. Through suitable modification a periodic convolution can be used to compute an aperiodic convolution when each aperiodic function has zero value everywhere outside some single finite aperture.

Let the functions be called $d(j)$ and $s(j)$. Let the larger finite aperture contain M discrete points and let the smaller contain N discrete points. The result of convolving these functions can be obtained from the result of circularly convolving suitable augmented functions. Let these augmented functions be periodic of period L, where L is the smallest power of 2 greater than or equal to $M + N$. Let them be called $da(j)$ and $sa(j)$ respectively, and be formed as indicated by Eq. (6).

$$\begin{aligned} fa(j) &= f(j + j_0) && 0 \le j \le M - 1 \\ &= 0 && M \le j \le L - 1 \\ &= fa(j + nL) && \text{otherwise} \end{aligned} \quad (6)$$

where j_0 symbolizes the first point in the aperture of the function in question. The intervals of zero values permit the two functions to be totally non-overlapped for at least one lagged product even though the lag is a circular one. Thus, while the result is itself a periodic function, each period is an exact replica of the desired aperiodic result.

The time required to compute this result is given in Eq. (7).

$$T_{\text{aper}} = k_{\text{circ}} L(\log_2 L + \mu) \quad (7)$$

where $M + N \le L < 2(M + N)$. For this case, while L must be adjusted to a power of 2 so that the high-speed Fourier transform can be applied, no restrictions are placed upon the values of either M or N.

SECTIONING

Let us assume that M is the aperture of $d(j)$ and N is that of $s(j)$. In situations where M is considerably larger than N, the procedure may be further streamlined by sectioning $d(j)$ into pieces each of which contains P discrete points where $P + N = L$, a power of 2. We require K sections where

$$K = \text{least integer} \ge M/P \quad (8)$$

Let the ith section of $d(j)$ be called $d_i(j)$. Each section is convolved aperiodically with $s(j)$ according to the discussion of the previous section, through the periodic convolution of the augmented sections, $da_i(j)$ and $sa(j)$.

Each result section, $r_i(j)$, has length $L = P + N$ and must be additively overlapped with its neighbors to form the composite result $r(j)$ which will be of length

$$KP + N \ge M + N \quad (9a)$$

If $r_i(j)$ is regarded as an aperiodic function with zero value for arguments outside the range $0 \le j \le L - 1$, these overlapped additions may be expressed as

$$r(j) = \sum_{i=0}^{K-1} r_i(j - iP) \quad j = 0, 1, \ldots KP + N - 1 \quad (9b)$$

Each overlap margin has width N and there are $K - 1$ of them.

The time required for this aperiodic sectioned convolution is given in Eq. (10).

$$\begin{aligned} T_{\text{sect}} &= k_{ct}(P + N)\log_2(P + N) \\ &\quad + 2Kk_{ct}(P + N)\log_2(P + N) \\ &\quad + Kk_{\text{aux}}(P + N) \\ &= k_{ct}(2K + 1)(P + N)\log_2(P + N) \\ &\quad + Kk_{\text{aux}}(P + N) \\ &\approx k_{ct}(2K + 1)(P + N)[\log_2(P + N) + \mu'] \end{aligned} \quad (10)$$

where $\mu' = k_{\text{aux}}/2k_{ct}$. $Kk_{\text{aux}}(P + N)$ is the time required to complete auxiliary processes. These processes involve the multiplications of Eq. (3), the formation of the augmented sections $da_i(j)$, and the formation of $r(j)$ from the result sections $r_i(j)$. For the author's realization in which core memory was used for the secondary storage of input and output data, μ' was measured to be 1.5, which gives $k_{\text{aux}} = 3k_{ct} \approx 300\ \mu\text{sec}$. If slower forms of auxiliary storage were employed, this figure would be enlarged slightly.

For a specific pair of values M and N, P should be chosen to minimize T_{sect}. Since $P + N$ must be a power of 2, it is a simple matter to evaluate Eq. (10) for a few values of P that are compatible with this constraint and select the optimum choice. The size of available memory will place an additional constraint on how large $P + N$ may be allowed to become. Memory allocation considerations degrade the benefits of these methods when N becomes too large. In extreme cases one is forced to split the kernel, $s(j)$, into packets, each of which is considered separately. The results corresponding to all packets are then added together after each has been shifted by a suitable number of packet widths. For the author's realization N must be limited to occupy about ⅛ of the memory not used for the program or for the secondary storage of input/output data. For larger N, packets would be required.

COMBINATION OF SECTIONS IN PAIRS

If both functions to be convolved are real instead of complex, further time savings over Eq. (10) can be made by combining adjacent even and odd subscripted sections $da_i(j)$ into complex composites. Let even subscripted $da_i(j)$ be used as real parts and odd subscripted $da_{i+1}(j)$ be used as imaginary parts. Such a complex composite can then be transformed through the application of Eqs. (1a), (3), and (1b) to produce a complex composite result section. The desired even and odd subscripted result sections $r_i(j)$ and $r_{i+1}(j)$ are respectively the real and imaginary parts of that complex result section.

This device reduces the time required to perform the convolution by approximately a factor of 2. More precisely it modifies K by changing Eq. (8) to

$$K = \text{least integer} \geq M/2P \tag{11}$$

For very large numbers of sections, K, Eq. (10) can be simplified to a form involving M explicitly instead of implicitly through K. That form is given in Eq. (12)

$$T_{\text{fast}} \approx k_{ct} M((P + N)/P)[\log_2 (P + N) + \mu'] \tag{12}$$

Since it makes no sense to choose $P < N$, for simple estimates of an approximate computation time we can write

$$T_{\text{fast}} \approx 2k_{ct}M[\log_2 N + \mu' + 1] \tag{13}$$

EMPIRICAL TIMES

The process for combined-sectioned-aperiodic convolution of real functions described above was implemented in the MAD language on the IBM 7094 Computer. Comparisons were made with a MAD language realization of a standard sum of lagged products for N = 16, 24, 32, 48, 64, 96, 128, 192, and 256. In each case M was selected to cause Eq. (11) to be fulfilled with the equal sign. This step favors the fast method by avoiding edge effects. However, P was not selected according to the optimization method described above (under "Sectioning Convolution"), but rather by selecting L as large as possible under the constraint.

$$\ln L \geq P/N \tag{14}$$

This choice can favor the standard method.

Table 1 compares for various N the actual computation times required in seconds as well as times in milliseconds per unit lag. Values of M, K, and L are also given.

Relative speed factors are shown in Table 2.

ACCURACY

The accuracy of the computational procedure described above is expected to be as good or better

Table 1. Comparative Convolution Times for Various N

N	16	24	32	48	64	96	128	192	256
M	192	208	384	416	768	832	1536	1664	3584
K	2	1	2	1	2	1	2	1	1
L	64	128	128	256	256	512	512	1024	2048
					Time in seconds				
T_{standard}	0.2	0.31	0.8	1.25	3.0	5.0	12	20	48
T_{fast}	0.3	0.4	0.6	0.8	1.3	1.8	3.0	3.8	8.0
					Time in milliseconds per unit lag				
$T_{\text{standard}/M}$	1.0	1.4	2.0	3.0	3.9	6.0	7.8	12.0	13.3
$T_{\text{fast}/M}$	1.5	1.9	1.5	1.9	1.6	2.1	1.9	2.2	2.2

Table 2. Speed Factors for Various N

N	16	24	32	48	64	96	128	192	256	512	1024	2048	4096
Speed factor	$\frac{2}{3}$	$\frac{3}{4}$	$\frac{4}{3}$	1.5	2.3	2.8	4.0	5.2	6	13*	24*	44*	80*

*Estimated values.

than that obtainable by summing products. Specific investigations of the accuracy of the program used to accumulate the data of Tables 1 and 2 are in process at the time of this writing. The above expectations are fostered by accuracy measurements made for floating-point data on the Cooley-Tukey procedure and a standard Fourier procedure. Since the standard Fourier procedure computes summed products, its accuracy characteristics are similar to those of a standard convolution which also computes summed products. Cases involving functions of period 64 and 256 were measured and it was discovered that two Cooley-Tukey transforms in cascade produced respectively as much, and half as much, error as a single standard Fourier transform. This data implies that the procedures disclosed here may yield more accurate results than standard methods with increasing relative accuracy for larger N.

APPLICATIONS

Today the major applications for the computation of lagged products are digital signal processing and spectral analysis.

Digital signal processing, or digital filtering as it is sometimes called, is often accomplished through the use of suitable difference equation techniques. For difference equations characterized by only a few parameters, computations may be performed in times short compared to those required for a standard lagged product or the method described here. However, in some cases, the desired filter characteristics are too complex to permit realization by a sufficiently simple difference equation. The most notable cases are those requiring high frequency selectivity coupled with short-duration impulse response and those in which the impulse response is found through physical measurements. In these situations it is desirable to employ the techniques described here either alone or cascaded with difference equation filters.

The standard methods for performing spectral analysis[2] involves the computation of lagged products of the form

$$F(j) = \sum_{J=0}^{N-j-1} x(J)y(J+j) \qquad (15)$$

which, in turn, after weighting by so-called spectral windows are Fourier transformed into power spectrum estimates. Speed advantages can be gained when Eq. (15) is evaluated in a manner similar to that outlined above (under "Aperiodic Convolution") except that in this case L is only required to exceed $N + \Omega$ where Ω is the number of lags to be considered. This relaxed requirement on L is possible because it is not necessary to avoid the effect of performing the lags circularly for all L lags but rather for only Ω of them. An additional constraint is that Ω be larger than a multiple of $\log_2 L$. The usual practice is to evaluate Eq. (15) for a number of lags equal to a substantial fraction of N. Since the typical situation involves values of N in the hundreds and thousands, the associated savings may be appreciable for this application.

Digital spatial filtering is becoming an increasingly important subject.[3,4] The principles discussed here are easily extended to the computation of lagged products across two or more dimensions. Time savings depend on the total number of data points contained within the entire data space in question, and they depend on this number in a manner similar to that characterizing the one-dimension case.

ACKNOWLEDGMENTS

The author is indebted to Charles M. Rader of the MIT Lincoln Laboratory for his ideas concerning the Cooley-Tukey algorithm and to Alan V. Oppenheim of the Electrical Engineering Department, MIT, for suggesting that high-speed convolutions might be realized through the utilization of that algorithm. During the preparation of this work the author became aware of the related independent efforts of Howard D. Helms, Bell Telephone Laboratories, and Gordon Sande, Jr., Princeton University.

REFERENCES

1. J. W. Cooley and J. W. Tukey, "An Algorithm for the Machine Calculation of Complex Fourier Series," *Mathematics of Computation*, vol. 19, no. 90, pp. 297–301, (Apr. 1965).

2. R. B. Blackman and J. W. Tukey, *The Measurement of Power Spectra*, Dover Publications, New York, 1959; also *Bell System Technical Journal*, Jan. and Mar. 1958.

3. T. S. Huang and O. J. Tretiak, "Research in Picture Processing," *Optical and Electro-Optical Information Processing*, J. Tippett et al, eds., MIT Press, Cambridge, Mass., 1965, Chap. 3.

4. T. S. Huang, "PCM Picture Transmission," *IEEE Spectrum*, vol. 2, no. 12, pp. 57–63 (Dec. 1965).

The Use of Fast Fourier Transform for the Estimation of Power Spectra: A Method Based on Time Averaging Over Short, Modified Periodograms

PETER D. WELCH

Abstract—**The use of the fast Fourier transform in power spectrum analysis is described. Principal advantages of this method are a reduction in the number of computations and in required core storage, and convenient application in nonstationarity tests. The method involves sectioning the record and averaging modified periodograms of the sections.**

Introduction

THIS PAPER outlines a method for the application of the fast Fourier transform algorithm to the estimation of power spectra, which involves sectioning the record, taking modified periodograms of these sections, and averaging these modified periodograms. In many instances this method involves fewer computations than other methods. Moreover, it involves the transformation of sequences which are shorter than the whole record which is an advantage when computations are to be performed on a machine with limited core storage. Finally, it directly yields a potential resolution in the time dimension which is useful for testing and measuring nonstationarity. As will be pointed out, it is closely related to the method of complex demodulation described by Bingham, Godfrey, and Tukey.[1]

The Method

Let $X(j)$, $j=0, \cdots, N-1$ be a sample from a stationary, second-order stochastic sequence. Assume for simplicity that $E(X)=0$. Let $X(j)$ have spectral density $P(f)$, $|f| \leq \frac{1}{2}$. We take segments, possibly overlapping, of length L with the starting points of these segments D units apart. Let $X_1(j)$, $j=0, \cdots, L-1$ be the first such segment. Then

$$X_1(j) = X(j) \qquad j = 0, \cdots, L-1.$$

Similarly,

$$X_2(j) = X(j+D) \qquad j = 0, \cdots, L-1,$$

and finally

$$X_K(j) = X(j+(K-1)D) \quad j = 0, \cdots, L-1.$$

Manuscript received February 28, 1967.
The author is with the IBM Research Center, Yorktown Heights, N. Y.
[1] C. Bingham, M. D. Godfrey, and J. W. Tukey, "Modern techniques of power spectrum estimation," this issue, p. 56–66.

We suppose we have K such segments; $X_1(j), \cdots, X_K(j)$, and that they cover the entire record, i.e., that $(K-1)D+L=N$. This segmenting is illustrated in Fig. 1.

The method of estimation is as follows. For each segment of length L we calculate a modified periodogram. That is, we select a data window $W(j)$, $j=0, \cdots, L-1$, and form the sequences $X_1(j)W(j), \cdots, X_K(j)W(j)$. We then take the finite Fourier transforms $A_1(n), \cdots, A_K(n)$ of these sequences. Here

$$A_k(n) = \frac{1}{L}\sum_{j=0}^{L-1} X_k(j)W(j)e^{-2kijn/L}$$

and $i=(-1)^{1/2}$. Finally, we obtain the K modified periodograms

$$I_k(f_n) = \frac{L}{U}\,|A_k(n)|^2 \qquad k = 1, 2, \cdots, K,$$

where

$$f_n = \frac{n}{L} \qquad n = 0, \cdots, L/2$$

and

$$U = \frac{1}{L}\sum_{j=0}^{L-1} W^2(j).$$

The spectral estimate is the average of these periodograms, i.e.,

$$\hat{P}(f_n) = \frac{1}{K}\sum_{k=1}^{K} I_k(f_n).$$

Now one can show that

$$E\{\hat{P}(f_n)\} = \int_{-1/2}^{1/2} h(f)P(f-f_n)df$$

where

$$h(f) = \frac{1}{LU}\left|\sum_{j=0}^{L-1} W(j)e^{2\pi ifj}\right|^2$$

and

$$\int_{-1/2}^{1/2} h(f)df = 1.$$

Reprinted from *IEEE Trans. Audio Electroacoust.*, vol. AU-15, pp. 70–73, June 1967.

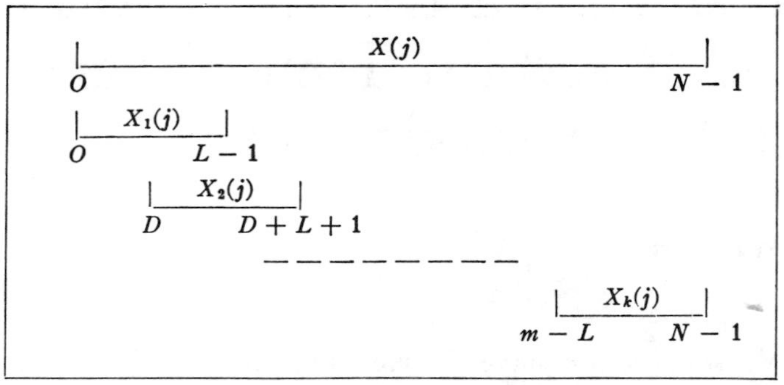

Fig. 1. Illustration of record segmentation.

Hence, we have a spectral estimator $\hat{P}(f)$ with a resultant spectral window whose area is unity and whose width is of the order of $1/L$.

Choice of Data Windows

We suggest two reasonable choices for the data window $W(j)$; one of them has the *shape* $1-t^2$: $-1 \leq t \leq 1$ and gives a spectral window which, when the two are normalized to have the same half-power width, is very close in shape to the hanning or cosine arch spectral window; the other data window has the *shape* $1-|t|$: $-1 \leq t \leq 1$ and gives the Parzen spectral window. The actual functions for a particular segment length L are

$$W_1(j) = 1 - \left(\frac{j - \frac{L-1}{2}}{\frac{L+1}{2}} \right)^2$$

and

$$W_2(j) = 1 - \left| \frac{j - \frac{L-1}{2}}{\frac{L+1}{2}} \right| \qquad j = 0, 1, \cdots, L-1.$$

The resultant spectral windows corresponding to these data windows are given approximately by

$$h_1(f) \approx \frac{1}{LU}\left\{\frac{2}{\pi^2(L+1)f^2}\left[\frac{\sin\{(L+1)\pi f\}}{(L+1)\pi f} - \cos\{(L+1)\pi f\}\right]\right\}^2$$

$$h_2(f) \approx \frac{1}{LU}\left\{\frac{(L+1)}{2}\,\frac{\sin^2\{(L+1)\pi f/2\}}{\{(L+1)\pi f/2\}^2}\right\}^2.$$

In the preceding approximations L is a scale parameter. In changing L we change the shape of $h_1(f)$ and $h_2(f)$ only in stretching or shrinking the horizontal dimension. For $h_1(f)$ the half-power width is

$$\Delta_1 f \approx \frac{(1.16)}{L+1}.$$

For $h_2(f)$ the half-power width is

$$\Delta_2 f \approx \frac{(1.28)}{L+1}.$$

The Variances of the Estimates

As developed above our estimator is given by

$$\hat{P}(f_n) = \frac{1}{K}\sum_{k=1}^{K} I_k(f_n), \qquad (n = 0, 1, \cdots, L/2).$$

Now, if we let

$$d(j) = \text{Covariance}\,\{I_k(f_n),\, I_{k+j}(f_n)\}$$

then it is easily shown that

$$\text{Var}\,\{\hat{P}(f_n)\} = \frac{1}{K}\left\{d(0) + 2\sum_{j=1}^{K-1}\frac{K-j}{K}\,d(j)\right\}.$$

Further, if

$$\rho(j) = \text{Correlation}\,\{I_k(f_n),\, I_{k+j}(f_n)\} = \frac{d(j)}{d(0)}$$

then,

$$\text{Var}\,\{\hat{P}(f_n)\} = \frac{d(0)}{K}\left\{1 + 2\sum_{j=1}^{K-1}\frac{K-j}{K}\,\rho(j)\right\} = \frac{\text{Var}\,\{I_k(f_n)\}}{K}\left\{1 + 2\sum_{j=1}^{K-1}\frac{K-j}{K}\,\rho(j)\right\}.$$

Assume now that $X(j)$ is a sample from a Gaussian process and assume that $P(f)$ is flat over the passband of our estimator. Then we can show[2] that

$$\text{Var}\,\{I_k(f_n)\} = P^2(f_n).$$

Further, under the above assumptions and assuming that $h(f-f_n) = 0$ for $f<0$ and $f>\frac{1}{2}$ we can show[3] that

$$\rho(j) = \left[\sum_{k=0}^{L-1} W(k)W(k+jD)\right]^2 \left[\sum_{k=0}^{L-1} W^2(k)\right]^2.$$

Hence, we have the following result which enables us to estimate the variances of $\hat{P}(f_n)$ when f_n is not close to 0 or $\frac{1}{2}$.

Result: If $X(j)$ is a sample from a Gaussian process, and $P(f)$ is flat over the passband of the estimator, and $h(f-f_n) = 0$ for $f<0$ and $f>\frac{1}{2}$, then

$$\text{Var}\,\{\hat{P}(f_n)\} = \frac{P^2(f_n)}{K}\left\{1 + 2\sum_{j=1}^{K-1}\frac{K-j}{K}\,\rho(j)\right\}$$

[2] P. D. Welch, "A direct digital method of power spectrum estimation," *IBM J. Res. and Dev.*, vol. 5, pp. 141–156, April 1961.

[3] In Welch[2] we obtained the variance spectrum of $I_k(f_n)$ considered as a function of time. The above result is obtained by taking the Fourier transform of this spectrum.

where

$$\rho(j) = \left[\sum_{k=0}^{L-1} W(k)W(k+jD)\right]^2 \Big/ \left[\sum_{k=0}^{L-1} W^2(k)\right]^2.$$

For estimating the spectrum of $P(f_n)$ at 0 and $\frac{1}{2}$ the variance is twice as great, as given by the following result:

Result: If $X(j)$ is a sample from a Gaussian process and $P(f)$ is flat over the passband of the estimator, then

$$\text{Var}\{\hat{P}(0 \text{ or } 1/2)\} = \frac{2P^2(0 \text{ or } 1/2)}{K}\left\{1 + 2\sum_{=1}^{K-1} \frac{K-j}{K}\rho(j)\right\}$$

where $\rho(j)$ is as defined above.

In the above results note that $\rho(j) \geq 0$ and that $\rho(j) = 0$ if $D \geq L$. Hence, if we average over K segments the best we can do is obtain a reduction of the variance by a factor $1/K$. Further, this $1/K$ reduction can be achieved (under these conditions) if we have nonoverlapping segments. Hence, if the total number of points N can be made sufficiently large the computationally most efficient procedure for achieving any desired variance is to have nonoverlapping segments, i.e., to let $D = L$. In this case we have

$$\text{Var}\{\hat{P}(f_n)\} = \frac{P^2(f_n)}{K} = \frac{P^2(f_n)L}{N}.$$

Further, under these conditions $E\{\hat{P}(f_n)\} = E\{I_K(f_n)\} = P(f_n)$ and, hence,

$$\frac{E^2\{P(f_n)\}}{\text{Var}\{P(f_n)\}} = K$$

and the equivalent degrees of freedom of the approximating chi-square distribution is given by

$$\text{E.D.F.}\{\hat{P}(f_n)\} = 2K.$$

If the total number of points N cannot be made arbitrarily large, and we wish to get a near maximum reduction in the variance out of a fixed number of points then a reasonable procedure is to overlap the segments by one half their length, i.e., to let $D = L/2$. In this case, if we use $W_1(j)$ as the data window we get $\rho(1) \approx 1/9$ and $\rho(j) = 0$ for $j > 1$. Letting $\hat{P}_1(f_n)$ be the estimate, we have

$$\text{Var}\{\hat{P}_1(f_n)\} = \frac{P^2(f_n)}{K}\left\{1 + \frac{2}{9} - \frac{2}{9K}\right\} \approx \frac{11P^2(f_n)}{9K}.$$

The factor 11/9, compared with the factor 1.0 for nonoverlapped segments, inflates the variance. However, an overall reduction in variance for fixed record length is achieved because of the difference in the value of K. For nonoverlapped segments we have $K = N/L$; for the overlapping discussed here

$$K = \frac{N}{L/2} - 1 = \frac{2N}{L} - 1 \approx \frac{2N}{L}.$$

Therefore, for fixed N and L the overall reduction in variance achieved by this overlapping is by a factor of 11/18. Now again $E\{\hat{P}_1(f_n)\} = P(f_n)$ and, hence,

$$\frac{E^2\{\hat{P}_1(f_n)\}}{\text{Var}\{\hat{P}_1(f_n)\}} \approx \frac{9K}{11} \approx \frac{18N}{11L}.$$

Finally,

$$\Delta_1(f) \approx \frac{(1.16)}{L+1} \approx \frac{7}{6L}.$$

Thus,

$$\frac{E^2\{\hat{P}_1(f_n)\}}{\text{Var}\{\hat{P}_1(f_n)\}} \approx 1.4N\Delta f$$

and the equivalent degrees of freedom of the approximating chi-square distribution is

$$\text{E.D.F.}\{\hat{P}_1(f_n)\} \approx 2.8N\Delta f.$$

Similarly, if we use $W_2(j)$ as our data window we get $\rho(1) = 1/16$ and $\rho(j) = 0$, $j > 1$. Letting $\hat{P}_2(f_n)$ be the estimate in this case we get, by following the above steps and using the result $\Delta_2 f = (1.28)/(L+1)$, that the equivalent degrees of freedom is again approximately

$$\text{E.D.F.}\{\hat{P}_2(f_n)\} \approx 2.8N\Delta f.$$

Thus, both $W_1(j)$ and $W_2(j)$ yield roughly the same variance when adjusted to have windows of equal half power width. Finally, we should point out that the above variances need to be doubled and the equivalent degrees of freedom halved for the points $f_n = 0$ and $\frac{1}{2}$.

Details in the Application of the Fast Fourier Transform Algorithm

Our estimator $\hat{P}(f_n)$ is given by

$$\hat{P}(f_n) = \frac{1}{K}\sum_{k=1}^{K} I_k(f_n) = \frac{L}{UK}\sum_{k=1}^{K} |A_k(n)|^2,$$

where L is the length of the segments, and K is the number of segments into which the record is broken, and

$$U = \frac{1}{L}\sum_{j=0}^{L-1} W^2(j).$$

We will first discuss how the complex algorithm can be used to obtain the summation $\sum_{k=1}^{K} |A_k(n)|^2$ two terms at a time with $K/2$ [or $(K+1)/2$, if K is odd] rather than K transforms. Suppose K is even and let

$$\left.\begin{aligned} Y_1(j) &= X_1(j)W(j) + iX_2(j)W(j) \\ &\vdots \\ Y_{K/2}(j) &= X_{K-1}(j)W(j) + iX_K(j)W(j) \end{aligned}\right\} j = 0, \cdots, L-1.$$

Let $B_k(n)$ be the transform of $Y_k(j)$. Then, by the linearity property of the finite Fourier transform

$$B_k(n) = A_{2k-1}(n) + iA_{2k}(n).$$

Further,

$$\begin{aligned} B_k(N-n) &= A_{2k-1}(N-n) + iA_{2k}(N-n) \\ &= \widetilde{A_{2k-1}(n) + iA_{2k}(n)}. \end{aligned}$$

Now,

$$| B_k(n) |^2 = (A_{2k-1}(n) + iA_{2k}(n))\widetilde{(A_{2k-1}(n)} - iA_{2k}(n))$$

$$| B_k(N-n) |^2 = (A_{2k-1}(n) - iA_{2k}(n))\widetilde{(A_{2k-1}(n)} + iA_{2k}(n)).$$

These equations yield, with some algebra,

$$| B_k(n) |^2 + | B_k(N-n) |^2 = 2(| A_{2k-1}(n) |^2 + | A_k(n) |^2).$$

Hence, finally,

$$\hat{P}(f_n) = \frac{L}{2UK} \sum_{k=1}^{K/2} (| B_k(n) |^2 + | B_k(N-n) |^2).$$

If K is odd this procedure can be extended in an obvious fashion by defining $Y_{(K+1)/2}(j) = X_K(j)$ and summing from 1 to $(K+1)/2$.

A second observation on the actual application of the algorithm concerns the bit-inverting. If the algorithm is applied as described here, and one is especially concerned with computation time, then the bit-inverting could be postponed until after the summation. Thus, instead of bit-inverting $K/2$ times, one would only have to bit-invert once.

Computation Time

The time required to perform a finite Fourier transform on a sequence of length L is approximately $k'L \log_2 L$ where k' is a constant which depends upon the program and type of computer. Hence, if we overlap segments by an amount $L/2$ we require an amount of computing time (performing two transforms simultaneously) approximately equal to

$$\left(\frac{1}{2}\right)\left(\frac{N}{L/2}\right) k'L \log_2 L = k'N \log_2 L,$$

plus the amount of time required to premultiply by the data window and average. If we only consider the time required for the Fourier transformation this compares with approximately $k'N(\log_2 N)/2$ for the smoothing of the periodogram. Hence, if $L < (N)^{1/2}$ it requires less computing time than the smoothing of the periodogram.

Relation of This Method to Complex Demodulation

It is appropriate to mention here the process of complex demodulation and its relation to this method of spectral estimation. Complex demodulation is discussed in Tukey,[4] Godfrey,[5] and Bingham, Godfrey, and Tukey.[1] The functions $A_k(n)e^{-2\pi ikD/L}$ considered as functions of k are complex demodulates sampled at the sampling period D. In this case the demodulating function is $e^{-2\pi i f n j}$. A phase coherency from sample to sample is retained in the complex demodulates. This phase is lost in estimating the spectrum and, hence, as a method of estimating spectra, complex demodulation is identical to the method of this section. However, additional information can be obtained from the time variation of the phase of the demodulates.

The Spacing of the Spectral Estimates

This method yields estimates spaced $1/L$ units apart. If more finely spaced estimates are desired zeros can be added to the sequences $X_k(j)W(j)$ before taking the transforms. If L' zeros are added giving time sequences $L+L'=M$ long and we let $A_k'(n)$ be the finite Fourier transforms of these extended sequences, i.e.,

$$A_k'(n) = \frac{1}{M} \sum_{j=0}^{L-1} X_k(j)W(j)e^{-2\pi ijn/M}$$

then the modified periodogram is given by

$$I_k(f_n) = \frac{M^2}{LU} | A_k'(n) |^2$$

where

$$f_n = \frac{n}{M} \qquad n = 0, 1, \cdots, M/2.$$

Everything proceeds exactly as earlier except that we have estimates spaced at intervals of $1/M$ rather than $1/L$.

Estimation of Cross Spectra

Let $X(j)$, $j=0, \cdots, N-1$, and $Y(j)$, $j=0, \cdots, N-1$, be samples from two second-order stochastic sequences. This method can be extended in a straightforward manner to the estimation of the cross spectrum, $P_{xy}(f)$. In exactly the same fashion each sample is divided in K segments of length L. Call these segments $X_1(j), \cdots, X_K(j)$ and $Y_1(j), \cdots, Y_K(j)$. Modified cross periodograms are calculated for each pair of segments $X_k(j)$, $Y_k(j)$, and the average of these modified cross periodograms constitutes the estimate $\hat{P}_{xy}(f_n)$. The spectral window is the same as is obtained using this method for the estimation of the spectrum.

[4] J. W. Tukey, "Discussion, emphasizing the connection between analysis of variance and spectrum analysis," *Technometrics*, vol. 3, pp. 191–219, May 1961.

[5] M. D. Godfrey, "An exploratory study of the bispectrum of economic time series," *Applied Statistics*, vol. 14, pp. 48–69, January 1965.

An Improved Algorithm for High Speed Autocorrelation with Applications to Spectral Estimation

CHARLES M. RADER, Member, IEEE
Lincoln Laboratory
Massachusetts Institute of Technology
Lexington, Mass. 02173

Abstract

A common application of the method of high speed convolution and correlation is the computation of autocorrelation functions, most commonly used in the estimation of power spectra. In this case the number of lags for which the autocorrelation function must be computed is small compared to the length of the data sequence available. The classic paper by Stockham, revealing the method of high speed convolution and correlation, also discloses a number of improvements in the method for the case where only a small number of lag values are desired, and for the case where a data sequence is extremely long. In this paper, the special case of autocorrelation is further examined. An important simplification is noted, based on the linearity of the discrete Fourier transform, and the circular shifting properties of discrete Fourier transforms. The techniques disclosed here should be especially important in real-time estimation of power spectra, in instances where the data sequence is essentially unterminated.

An autocorrelation function can be computed by means of a fast Fourier transform [1] algorithm (FFT), as first suggested by Stockham [2]. The technique uses the fact that the product of the discrete Fourier transforms (DFT) of any two sequences is equal to the DFT of the circular convolution of the two sequences. A circular convolution is made to yield results arithmetically identical to those of an ordinary convolution by inserting zeros at the end of one, or both, of the original sequences prior to computing the DFTs. The algorithm may be considered to be high speed because the DFTs are computed with FFT speed. In many instances, and the most notable of these is the estimation of power spectra, the data for which the autocorrelation function is desired forms a very long, perhaps indefinitely long sequence; also the autocorrelation function is desired for only a number of lag values which is but a small fraction of the length of the data sequence [4], [5]. These two situations combine to make the direct application of Stockham's algorithm impractical, first because it calls for computing DFTs of impractically long sequences, and second, because better methods are available. These better methods were also suggested by Stockham, and developed for computing convolutions. They can also be applied to computing correlations, but there are special considerations which apply to autocorrelation which permit still further simplification. Such further simplification is the purpose of this paper.

Manuscript received June 12, 1970.

This work was sponsored by the Department of the Air Force.

In what follows, all DFTs are of M point sequences. M may be chosen to be a power of 2 for efficiency. At any rate, we will require that it be even. The data for which the autocorrelation is desired will be called $x(n)$, and will be available for $n=0, 1, 2, \cdots$, perhaps of indefinite length. We will follow the convention that n is a time index, k is a frequency index, and m is a lag index.

In Fig. 1, we see a possible sequence $x(n)$, plotted as a continuous function for convenience. The vertical separation markers are $M/2$ points apart. Let us form a series of subsequences $\{x_i(n)\}$ as also shown in Fig. 1.

$$x_i(n) = \begin{cases} x(n + iM/2) & 0 \leqq n < M/2 \\ 0 & M/2 \leqq n < M, \\ & i = 0, 1, 2, \cdots. \end{cases} \quad (1)$$

Also, for now, let us form a second series of sequences, $\{y_i(n)\}$, (See Fig. 1). These y_i sequences will not be ultimately needed, but are a pedagogical device.

$$y_i(n) = x(n + iM/2) \qquad 0 \leqq n < M, \quad i = 0, 1, 2, \cdots. \quad (2)$$

If we compute the DFTs of $\{x_i(n)\}$ and, for now, of $\{y_i(n)\}$,

$$\begin{aligned} X_i(k) &= \text{DFT } \{x_i(n)\} \\ Y_i(k) &= \text{DFT } \{y_i(n)\} \end{aligned} \quad (3)$$

and form the product

$$W_i(k) = X_i^*(k)\, Y_i(k) \quad (4)$$

we have the DFT, $W_i(k)$ of the sequence

Reprinted from *IEEE Trans. Audio Electroacoust.*, vol. AU-18, pp. 439–441, Dec. 1970.

$$w_i(m) = \sum_{n=0}^{(M/2)-1} x^*(n + iM/2)x(n + iM/2 + m),$$
$$m = 0, 1, \cdots, M/2. \quad (5)$$

Because of the circular nature of the correlation, the values of $w_i(m)$ are useless for $m > M/2$, and are not given by the above equation.

Let us define $R_x(m)$, our desired autocorrelation function, as

$$R_x(m) = (1/N) \sum_{n=0}^{N-1} x^*(n)x(n + m) \quad (6)$$

where N is the number of terms we can afford to sum in practice. Except for the factor $(1/N)$ it should be apparent how $R_x(m)$ can be formed by summing $w_i(m)$. In fact, let $\{Z_i(m)\}$ be

$$z_i(m) = \sum_{j=0}^{i} w_j(m)$$
$$= \sum_{j=0}^{i} \sum_{n=0}^{(M/2)-1} x^*(n + jM/2)x(n + jM/2 + m)$$
$$z_i(m) = \sum_{n=0}^{(i+1)M/2-1} x^*(n)x(n + m). \quad (7)$$

When $i = (2N/M) - 1$, $z_i(m)$ is $N \times R_x(m)$. The preceding equations demonstrate how an autocorrelation function can be obtained by the use of M point DFTs even though the data sequence $x(n)$ could be indefinitely long. For each subsequence $\{x_i(n)\}$ used, the procedure described here uses three DFTs.

It is simple to see how to eliminate some work in this process almost immediately. The sum indicated in (7) can be carried out in the frequency domain,

$$Z_i(k) = \sum_{j=0}^{i} W_j(k) = Z_{i-1}(k) + W_i(k) \quad (8)$$

and the return to the lag value domain need be made only once, for $i = (2N/M) - 1$. This means that to compute $R_x(m)$ for lag values $m = 0, 1, \cdots, M/2$ requires, by the method as developed above, two DFTs of M points each, per $M/2$ points in the data sequence used, plus an additional computational load at the beginning and end of the process.

The next step in the simplification, which is believed to be novel, is to note that the computation of $Y_i(k)$ can be simply made without use of $\{y_i(n)\}$—in fact

$$Y_i(k) = X_i(k) + (-1)^k X_{i+1}(k). \quad (9)$$

This is justified by observing that multiplying a DFT by $(-1)^k$ is equivalent to rotating the corresponding time function by $M/2$ positions [3]. But such rotation would make the nonzero values of $\{x_{i+1}(n)\}$ exactly overlap the zero values of $\{x_i(n)\}$, and vice versa, so that the sum, in either the time domain or the frequency domain is identical to $\{y_i(n)\}$, or $Y_i(k)$. We note with considerable satisfaction that $\{y_i(n)\}$ need never be formed at all, and its transform can be obtained with only M additions or subtractions, given that $X_{i+1}(k)$ is about to be formed as

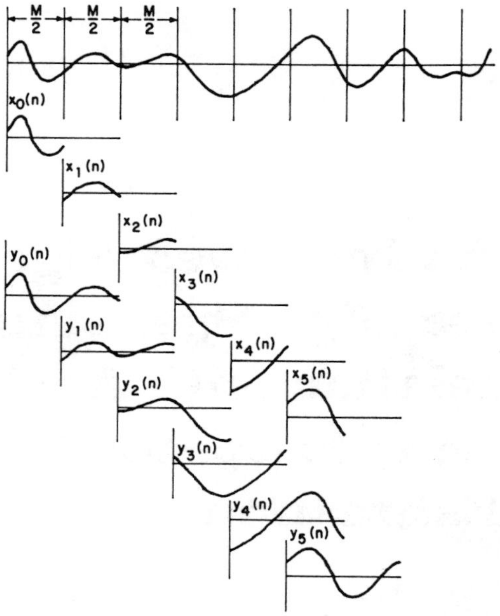

Fig. 1.

the transform of the next section of $M/2$ points anyway. We summarize the steps of the computation of $R_x(m)$, incorporating this latest trick, below:

1) form $\{x_0(n)\}$ and compute its transform $X_0(k)$. Clear out $Z_0(k)$;
2) for $i = 0, 1, \ldots, (2N/M) - 2$
   a) form $\{x_{i+1}(n)\}$ and compute $X_{i+1}(k)$;
   b) compute

$$Z_{i+1}(k) = Z_i(k) + X_i^*(k)[X_i(k) + (-1)^k X_{i+1}(k)];$$

3) compute
   $R_x(m) = (1/N)\mathrm{IDFT}\{Z_{(2N/M)-1}(k)\}$
   keeping only the first $(M/2)+1$ values.

Some special comments seem to be in order at this time. For most high speed convolutions and correlations, it is possible to avoid the phenomenon of bit reversal which is associated with computing the DFT by FFT techniques in place, that is, within the same array of memory as the initial data to be transformed. To go from time domain to frequency domain, one uses an FFT algorithm which accepts normally ordered input data, and produces bit-reversed ordered frequency coefficients. To go back to the time domain (here the lag domain) one uses an FFT algorithm which accepts bit-reversed ordered frequency coefficients and produces normally ordered output. To use this trick in the procedure given above, we must consider the effect of the bit reversed order on the quantity $(-1)^k$ in (9). This term will be $+1$ for the first $M/2$ frequency coefficients in the bit reversed array, and -1 for the last $M/2$ frequency coefficients in the bit reversed array. These two groups represent the even and odd numbered points in a normally ordered array.

Secondly we note that if $x(n)$ is real we may compute the M point DFTs of two sections at once by making use of the properties of odd and even symmetry in the transforms of pure real or pure imaginary sequences. This use of symmetry on bit reversed data will probably be as complicated as unsorting, so as to negate the comment about $(-1)^k$ in the previous paragraph. These same symmetries may be used to store only one half of $Z_i(k)$ and to halve the number of multiplications performed in (4).

In what we have done so far, the number of lag values for which $R_x(m)$ was computed was rigidly tied to the section length chosen. However, additional lags could be computed by correlating $\{x_i(n)\}$ with $\{y_{i+1}(n)\}$, with $\{y_{i+2}(n)\}$, etc. We can form the transform of any $\{y_{i+p}(n)\}$ without using much computational effort by the use of (9) suitably restated, i.e.,

$$Y_{i+p}(k) = X_{i+p}(k) + (-1)^k X_{i+p+1}(k). \quad (10)$$

Lag values $pM/2 \leq m \leq (p+1)M/2$ are then obtained by accumulating

$$Z_{i+1}{}^p(k) = Z_i{}^p(k) + X_i^*(k)[X_{i+p}(k) + (-1)^k X_{i+p+1}(k)]. \quad (11)$$

The only extra DFTs required for extra lags are the extra DFTs at the beginning of the process and the extra DFTs at the end of the process to account for the extra answers. Note that there is one common lag value between the values of $R_x(m)$ obtained by inverse transforming successive subarrays $Z_i{}^p$ and $Z_i{}^{p+1}$. Possibly these overlapping values are of some use as a check. The important point to make note of is that in a real-time application, the number of operations which need to be performed per input datum, exclusive of beginning and ending situation, is a very weak function of the number of lags desired. In fact, for a data sequence $x(n)$ which is indefinitely long, a computer could compute $R_x(m)$ for μ lags in real time if it could compute one real M point DFT and μ complex products and sums in the time in which $M/2$ data are sampled, i.e., if the computation time T_{comp} per sample is less than the sampling interval.

$$\begin{aligned} T_{\text{comp}} &= \frac{2}{M}(k_{\text{fft}} M \log_2 M + k_{\text{cma}}\mu) \\ &= 2k_{\text{fft}} \log_2 M + 2k_{\text{cma}}\mu/M \end{aligned} \quad (12)$$

where k_{fft} is a proportionality constant associated with the computation of a DFT of a real valued sequence, and k_{cma} is a proportionality constant associated with a complex multiply time, a complex add time, and whatever bookkeeping is required. Proper choice of M will minimize this T_{comp}. Analytically, the minimum based on a zero value of the derivative is

$$M = \mu k_{\text{cma}} \ln (2)/k_{\text{fft}} \quad (13)$$

but since we are assuming that M is a power of 2, this value will not be strictly correct. It will probably suffice to try the values of M just above and just below the analytical expression to find the power of 2 which produces the optimum T_{comp} for a given number of lags.

We know of no reason why the technique disclosed here should not be just as useful in estimating multidimensional autocorrelations as in estimating one-dimensional autocorrelations. In the particular instance of spectral estimation, it is known that the convergence of a spectral estimate is poor unless the measured autocorrelation function is multiplied by a window function whose effective lag width is a small fraction of the number of data samples used in the estimate. On the other hand, the requirement for good spectral resolution will lead to the necessity to measure the autocorrelation function for many lags. Therefore, the technique we have just explored should be admirably suited to the computation of the autocorrelation estimate needed in estimation of a power spectrum.

References

[1] J. W. Cooley and J. W. Tukey, "An algorithm for the machine computation of complex Fourier series," *Math. Comput.*, vol. 19, pp. 297–301, April 1965.

[2] T. G. Stockham, Jr., "High speed convolution and correlation," in *1966 Spring Joint Comput. Conf., AFIPS Proc.*, vol. 28. Wasington, D. C.: Spartan, 1966, pp. 229–233.

[3] B. Gold and C. M. Rader, *Digital Processing of Signals.* New York: McGraw-Hill, 1969, pp. 162–168.

[4] R. B. Blackman and J. W. Tukey, *The Measurement of Power Spectra.* New York: Dover, 1958.

[5] G. M. Jenkins and D. G. Watts, *Spectral Analysis and Its Application.* San Francisco: Holden Day, Inc., 1968, pp. 209–309.

Computation of Spectra with Unequal Resolution Using the Fast Fourier Transform

Abstract—The discrete Fourier transform of a sequence, which can be computed using the fast Fourier transform algorithm, represents samples of the z transform equally spaced around the unit circle. In this letter, a technique is discussed and illustrated for transforming a sequence to a new sequence whose discrete Fourier transform is equal to samples of the z transform of the original sequence at unequally spaced angles around the unit circle.

In many applications we are concerned with the problem of computing samples of the z transform of a sequence on the unit circle. To obtain samples *equally* spaced around the unit circle, the most efficient procedure is to compute the discrete Fourier transform (DFT) using the fast Fourier transform (FFT) algorithm.[1] If we are interested in obtaining samples equally spaced within a particular region of the unit circle, then one efficient procedure consists of using the chirp z-transform algorithm.[2] Often we would like to obtain samples that are unequally spaced—corresponding, for example, to a constant Q spectral analysis of the original sequence. An algorithm for accomplishing this with an efficiency similar to that achievable with the FFT algorithm is not known. One procedure sometimes used is to evaluate the samples explicitly at the desired frequencies. Another procedure used is to add equally spaced frequency samples in bands. A related alternative procedure corresponds to implementing a spectral analysis of the sequence with a recursive or nonrecursive filter bank. This letter is directed toward a procedure that perhaps is slightly more efficient than the alternative just mentioned, and may also have some advantages when considering hardware implementation of a spectral analysis with nonuniform resolution.

The procedure consists of transforming the original sequence to a new sequence having the property that its DFT is equal to samples of the z transform of the original sequence at unequally spaced angles around the unit circle.[3] Letting $f(n)$ represent the original sequence, and $g(k)$ represent the transformed sequence, we consider linear transformations between $f(n)$ and $g(k)$ corresponding to expanding $f(n)$ in terms of a set of linearly independent sequences $\psi_k(n)$ so that

$$f(n) = \sum_{k=-\infty}^{+\infty} g(k)\psi_k(n). \tag{1}$$

The basic property that we would like this transformation to have is that the z transform of the sequence $f(n)$ and the z transform of the sequence $g(k)$ are related by a change of variables, so that on the unit circle, if

$$G(e^{j\omega}) = \sum_{k=-\infty}^{+\infty} g(k)e^{-j\omega k}$$

and

$$F(e^{j\Omega}) = \sum_{n=-\infty}^{+\infty} f(n)e^{-j\Omega n}$$

then

$$\omega = \theta(\Omega)$$

so that

$$G(e^{j\theta(\Omega)}) = F(e^{j\Omega}). \tag{2}$$

Consequently, what is required is that the Fourier transform of $f(n)$ and the Fourier transform of $g(k)$ be related by a distortion of the frequency axis. It can be shown that the requirement placed on the set of functions $\psi_k(n)$ such that (2) is satisfied is that

$$\Psi_k(e^{j\Omega}) = e^{-jk\theta(\Omega)} \tag{3}$$

where

$$\Psi_k(e^{j\Omega}) = \sum^{+\infty} \psi_k(n)e^{-j\Omega n}.$$

Therefore, the functions $\psi_k(n)$ must have an all-pass characteristic, that is, their z transform on the unit circle must have unity magnitude independent of frequency. With these conditions satisfied, the relationship between the frequency variable ω corresponding to the Fourier transform of the new sequence $g(k)$ and the frequency variable Ω corresponding to the Fourier transform of the original sequence $f(n)$ is that

$$\omega = \theta(\Omega). \tag{4}$$

If we restrict the mapping from Ω to ω to be such that when Ω changes by 2π then ω changes by 2π, and if we require that the z transform $\Psi_k(z)$ of the functions $\psi_k(n)$ be rational functions of z, then $\Psi_k(z)$ must be of the form

$$\Psi_k(z) = \left(\frac{z^{-1} - a}{1 - az^{-1}}\right)^k. \tag{5}$$

As required, the magnitude of $\Psi_k(z)$ for z on the unit circle is unity, and the phase factor $\theta(\Omega)$ is given by

$$\omega = \theta(\Omega) = \tan^{-1}\left[\frac{(1 - a^2)\sin\Omega}{(1 + a^2)\cos\Omega - 2a}\right]. \tag{6}$$

It can be shown that the inverse relation is

$$\Omega = \theta^{-1}(\omega) = \tan^{-1}\left[\frac{(1 - a^2)\sin\omega}{(1 + a^2)\cos\omega + 2a}\right] \tag{7}$$

which corresponds to replacing a by $-a$ in (6). Consequently, (6) specifies the relationship between the new frequency variable ω and the original

Manuscript received June 11, 1970; revised July 30, 1970. A. Oppenheim and D. Johnson were supported in part by the Joint Services Electronics Programs under Contract DA-28-043-AMC-02536(E) and by the U. S. Air Force Cambridge Research Laboratories, Office of Aerospace Research, under Contract F19628-69-C-0044. K. Steiglitz was supported in part by the U. S. Army Research Office, Durham, N. C., under Contract DAHC04-69-C-0012.

[1] J. W. Cooley and J. W. Tukey, "An algorithm for the machine calculation of complex Fourier series," *Math. Comput.*, vol. 19, pp. 297–301, April 1965.

[2] L. Rabiner, R. Schafer, and C. Rader, "The chirp z-transform algorithm and its applications," *Bell Syst. Tech. J.*, vol. 48, pp. 1249–1292, May–June 1969.

[3] A. Oppenheim and D. Johnson, "Discrete representations of analog signals," M.I.T. Res. Lab. Electron., Cambridge, Mass., Quart. Progr. Rep. 97, pp. 185–190, April 15, 1970.

Reprinted from *Proc. IEEE*, vol. 59, pp. 299–301, Feb. 1971.

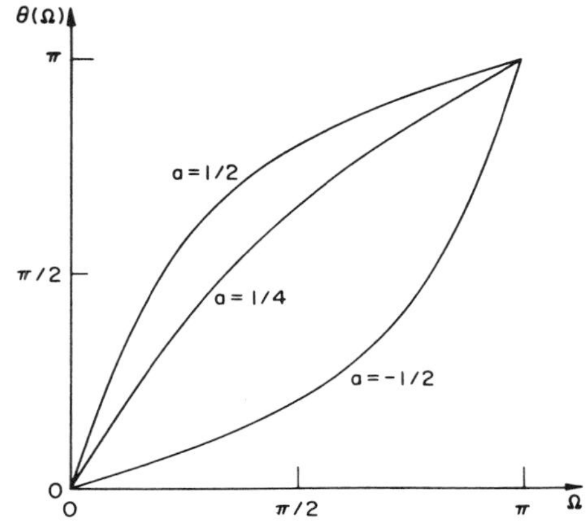

Fig. 1. Distortion of frequency for several values of a.

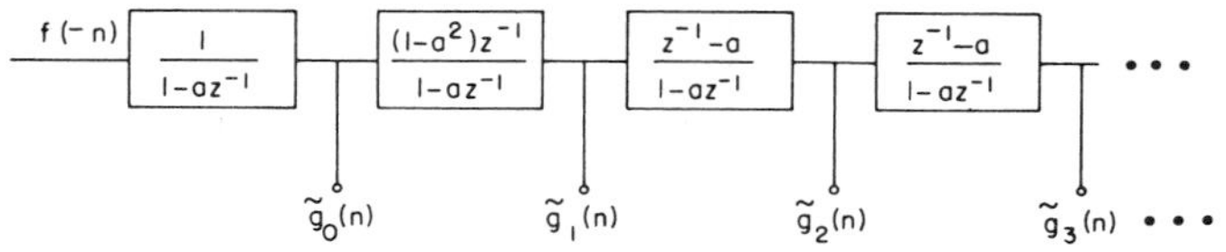

Fig. 2. All-pass network used to implement a distortion of the frequency axis.

frequency variable Ω. Since a computation of the DFT of the new sequence $g(k)$ corresponds to sampling uniformly in ω, the frequency samples obtained will correspond to nonuniform sampling in the original frequency variable Ω. Fig. 1 shows the function $\theta(\Omega)$ for several values of the parameter a.

For the remainder of the discussion, we will assume that $f(n)$ and $g(k)$ are both zero for $n<0$. It is straightforward to modify the discussion to incorporate the more general case. For the particular functions $\psi_k(n)$ defined by (5), it can be shown that

$$\sum_{n=0}^{\infty} n\psi_r(n)\psi_k(n) = k, \qquad r = k$$
$$= 0, \qquad r \neq k.$$

Consequently, this set of functions $\psi_k(n)$ is orthogonal with a weighting function of n and a normalizing constant of k. From this relation together with (1), the sequence $g(k)$ is specified by

$$g(k) = \frac{1}{k}\sum_{n=0}^{\infty} n\psi_k(n)f(n), \qquad k = 1, 2, \cdots \tag{8a}$$

$$g(0) = \sum_{n=0}^{\infty} f(n)a^n. \tag{8b}$$

One implementation of (8) consists of passing the sequence $f(-n)$ through the linear shift-invariant network shown in Fig. 2. With the outputs in the network of Fig. 2 designated as $\tilde{g}_k(n)$, the sequence $g(k)$ is related to $\tilde{g}_k(n)$ by

$$g(k) = \tilde{g}_k(0).$$

To obtain the samples of the spectrum of the original sequence on the distorted frequency scale, the DFT of the sequence $g(k)$ is computed. From the curves in Fig. 1, with the parameter a real and between 0 and 1, the effect on the spectrum is to sample with higher resolution at low frequencies and lower resolution at higher frequencies. If, instead, a is negative, between 0 and -1, then the effect is the reverse, that is, the spectrum is sampled with greater resolution at higher frequencies than at low frequencies. It is also straightforward to implement the network in Fig. 2 with a complex value for a. This would then result in high spectral resolution at some intermediate frequency.

In general, if the sequence $f(n)$ is of finite duration, the sequence $g(k)$ will be of infinite duration. Consequently, in order to compute the DFT

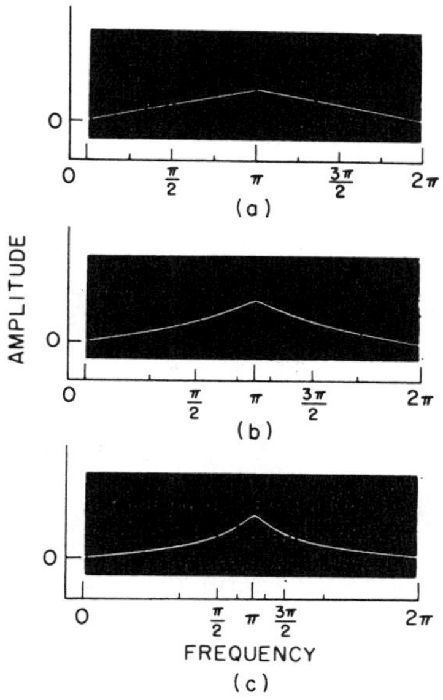

Fig. 3. Example of frequency distortion. (a) Original spectrum. (b) Distorted spectrum with $a=1/4$. (c) Distorted spectrum with $a=1/2$.

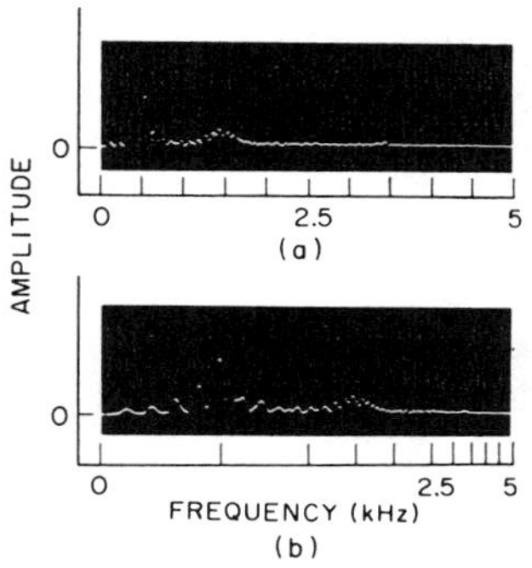

Fig. 4. Example of frequency distortion on a sample of speech. (a) Original spectrum. (b) Distorted spectrum with $a=1/2$.

of $g(k)$, an appropriate finite-duration window must be applied. Applying such a window corresponds to "smearing" the spectrum of the sequence before obtaining the spectral samples. The spectral window applied to the transform of $g(k)$ maintains constant width in ω which corresponds to smearing the spectrum of $f(n)$ with unequal bandwidth. From (6) it follows that for $a>0$, the bandwidth of the spectral window increases with frequency in the same way that the spacing of the spectral samples increases with frequency. This is similar (but not identical) to the situation where a spectral analysis is implemented by implementing a constant Q filter bank, in which case both the spacing of the spectral samples and the bandwidth of the filter increase with increasing frequency.

Examples of spectral analysis implemented on the basis of this discussion are illustrated in Figs. 3 and 4. Fig. 3(a)–(c) corresponds to the DFT of an original sequence and the transformed sequences for $a=\frac{1}{4}$ and $a=\frac{1}{2}$. Similarly, Fig. 4(a) and (b) corresponds to the DFT of an original and a transformed segment of a speech waveform. In all of these spectra, the transform size was 512 points.

ALAN OPPENHEIM
DON JOHNSON
Dep. of Elec. Eng. and Res. Lab. of Electron.
Mass. Inst. Tech.
Cambridge, Mass. 02139
KENNETH STEIGLITZ
Dep. of Elec. Eng.
Princeton University
Princeton, N. J. 08540

A Bound on the Output of a Circular Convolution with Application to Digital Filtering

ALAN V. OPPENHEIM, Member, IEEE
CLIFFORD WEINSTEIN, Student Member, IEEE
Lincoln Laboratory
Massachusetts Institute of Technology
Lexington, Mass. 02173

Abstract

When implementing a digital filter, it is important to utilize in the design a bound or estimate of the largest output value which will be obtained. Such a bound is particularly useful when fixed point arithmetic is to be used since it assists in determining register lengths necessary to prevent overflow. In this paper we consider the class of digital filters which have an impulse response of finite duration and are implemented by means of circular convolutions performed using the discrete Fourier transform. A least upper bound is obtained for the maximum possible output of a circular convolution for the general case of complex input sequences. For the case of real input sequences, a lower bound on the least upper bound is obtained. The use of these results in the implementation of this class of digital filters is discussed.

I. Introduction

When implementing a digital filter, either in hardware or on a computer, it is important to utilize in the design a bound or estimate of the largest output value which will be obtained. Such a bound is particularly useful when fixed point arithmetic is to be used, since it assists in determining register lengths necessary to prevent overflow. This paper considers the class of digital filters which have an impulse response of finite duration and are implemented by means of convolution sums performed using the discrete Fourier transform (DFT). The output samples of such a filter are obtained from the results of N-point circular convolutions of the filter impulse response (kernel) with sections of the input. These circular convolutions are obtained by computing the DFT of the input section, multiplying by the DFT of the impulse response, and inverse transforming the result. Stockham [1] has discussed procedures for utilizing the results of these circular convolutions to perform linear convolutions, rationales for choosing the transform length N, and speed advantages to be gained by using the fast Fourier transform (FFT) to implement the DFT. We concern ourselves here only with bounding the output of the N-point circular convolutions.

Manuscript received November 26, 1968.
This work was sponsored by the U. S. Air Force.

II. Problem Statement

According to the above discussion, we would like to determine an upper bound on the maximum modulus of an output value that can result from an N-point circular convolution. With $\{x_n\}$ denoting the input sequence, $\{h_n\}$ denoting the kernel, and $\{y_n\}$ denoting the output sequence, we have

$$y_n = \sum_{k=0}^{N-1} x_k h_{(n-k) \bmod N} \quad n = 0, 1, \cdots, N-1 \tag{1}$$

where it is understood that, in general, each of the three sequences may be complex. The circular convolution is accomplished by forming the product

$$Y_k = H_k X_k \tag{2}$$

where

$$X_k = \frac{1}{N} \sum_{n=0}^{N-1} x_n W^{nk} \quad k = 0, 1, \cdots, N-1 \tag{3}$$

$$Y_k = \frac{1}{N} \sum_{n=0}^{N-1} y_n W^{nk} \quad k = 0, 1, \cdots, N-1 \tag{4}$$

$$H_k = \sum_{n=0}^{N-1} h_n W^{nk} \quad k = 0, 1, \cdots, N-1 \tag{5}$$

with W defined as $W = \exp[j2\pi/N]$.

For convenience in notation, we imagine the computation to be carried out on fixed point fractions. Thus we bound the input values so that

$$|x_n| \leq 1. \tag{6}$$

By virtue of (3) we are then assured that

$$|X_k| \leq 1$$

Reprinted from *IEEE Trans. Audio Electroacoust.*, vol. AU-17, pp. 120–124, June 1969.

so that the values of X_k do not overflow in the fixed point word.

In the typical cases, the sequence h_n is known and, consequently, so is the sequence H_k. Therefore it is not necessary to continually evaluate (5); that is, the sequence H_k is computed, normalized, and stored in advance. Thus it is reasonable to only apply a normalization to H_k and not to h_n, so that we require[1]

$$|H_k| \le 1. \tag{7}$$

A normalization of the transform of the kernel so that the maximum modulus is unity allows maximum energy transfer through the filter, consistent with the requirement that Y_k does not overflow the register length.

Our objective is to obtain an upper bound on $|y_n|$ for all sequences $\{x_n\}$ and $\{H_k\}$ consistent with (6) and (7). This bound will specify, for example, the scaling factor to be applied in computing the inverse of (4) to guarantee that no value of y_n overflows the fixed point word. The following results will be obtained.

Result A: With the above constraints, the result of the N-point circular convolution of (1) is bounded by

$$|y_n| \le \sqrt{N}.$$

Result B: In the general case where $\{x_n\}$ and $\{h_n\}$ are allowed to be complex, the bound in Result A is a least upper bound. This will be shown by demonstrating a sequence that can achieve the bound.

Result C: If we restrict $\{x_n\}$ and/or $\{h_n\}$ to be real valued, the bound of Result A is no longer a least upper bound for every N. However, the least upper bound $\beta(N)$ is itself bounded by

$$\frac{\sqrt{N}}{2} \le \beta(N) \le \sqrt{N}.$$

III. Derivation of Results

Proof of Result A

Parseval's relation requires that

$$\sum_{n=0}^{N-1} |y_n|^2 = N \sum_{k=0}^{N-1} |Y_k|^2 \tag{8}$$

and

$$\sum_{n=0}^{N-1} |x_n|^2 = N \sum_{k=0}^{N-1} |X_k|^2. \tag{9}$$

Substituting (2) into (8) and using (7),

$$\sum_{n=0}^{N-1} |y_n|^2 \le N \sum_{k=0}^{N-1} |X_k|^2, \tag{10}$$

or, using (9),

$$\sum_{n=0}^{N-1} |y_n|^2 \le \sum_{n=0}^{N-1} |x_n|^2 \tag{11}$$

[1] The restrictions of (6) and (7) do not impose any loss of generality, and are introduced only for convenience. The bounds to be derived on max $|y_n|$ can be interpreted in a more general sense as bounds on the ratio max $|y_n|/\{\max|x_n| \max|H_k|\}$.

with equality if and only if $|H_k| = 1$. However, (6) requires that

$$\sum_{n=0}^{N-1} |x_n|^2 \le N \tag{12}$$

with equality if and only if $|x_n| = 1$. Combining (11) and (12),

$$\sum_{n=0}^{N-1} |y_n|^2 \le N. \tag{13}$$

But

$$|y_n|^2 \le \sum_{n=0}^{N-1} |y_n|^2 \tag{14}$$

and therefore

$$|y_n| \le \sqrt{N}. \tag{15}$$

Proof of Result B

To show that $\sqrt{N}$ is a least upper bound on $|y_n|$, we review the conditions for equality in the inequalities used above. We observe that for equality to be satisfied in (15), it must be satisfied in (11), (12), and (14), requiring that

1) $|H_k| = 1$
2) $|x_n| = 1$
3) Any output sequence $\{y_n\}$ which has a point whose modulus is equal to $\sqrt{N}$ can contain only one nonzero point.

The third requirement can be rephrased as a requirement on the input sequence and on the sequence H_k. Specifically, if the output sequence contains only one nonzero point then Y_k for this sequence must be of the form

$$Y_k = AW^{n_0 k} = |A| \exp\left[j\left[\frac{2\pi}{N} n_0 k + \rho\right]\right]$$

where ρ is a real constant and n_0 is an integer so that, from (2),

$$H_k X_k = |A| \exp\left[j\left[\frac{2\pi}{N} n_0 k + \rho\right]\right]. \tag{16}$$

We can express H_k and X_k as

$$H_k = e^{j\eta_k}$$

and

$$X_k = |X_k| e^{j\theta_k}$$

where we have used the fact that $|H_k| = 1$. For (16) to be satisfied, then

$$|X_k| = |A| \tag{17}$$

and

$$\eta_k = -\theta_k + \frac{2\pi}{N} n_0 k + \rho. \tag{18}$$

Therefore, requirement 3) can be replaced by the requirement that:

3′) $|X_k|$ = constant and the phase of H_k be chosen to satisfy (18).

As an additional observation, we note that for *any* input sequence $\{x_n\}$,

$$|y_n| \leq \sum_{k=0}^{N-1} |H_k| \, |X_k|$$

with equality for some value of n if and only if $|H_k|=1$ and the phase of H_k is chosen on the basis of (18). Therefore, for any $\{x_n\}$ the output modulus is maximized when H_k is chosen in this manner. This maximum value will only equal $\sqrt{N}$, however, if, in addition, $|x_n|=1$ and $|X_k|=\text{constant}$.

For N even, a sequence having the property that $|x_n|=1$ and $|X_k|=\text{constant}$ is (see Appendix) the sequence

$$x_n = \exp\left[j\frac{\pi n^2}{N}\right] = W^{n^2/2}. \tag{19}$$

For N odd, a sequence with $|x_n|=1$ and $|X_k|=\text{constant}$ is[2] (see Appendix)

$$x_n = \exp\left[j\frac{2\pi n^2}{N}\right] = W^{n^2}. \tag{20}$$

Using one of these sequences as the input, and choosing $H_k=e^{j\eta_k}$, with η_k given by (18), equality in (15) can be achieved for any N. Thus the bound given in Result A is a least upper bound.

Proof of Result C

Consider first the case where $\{x_n\}$ is restricted to be real. It can be verified by consideration of all possibilities that for $N=2$ and $N=3$, no real sequence exists for which both the sequence and its transform have constant modulus. Therefore, for these values of N at least, (15) does not provide a least upper bound, since requirements 2) and 3′) cannot be satisfied simultaneously. Note, however, that for $N=4$ the real sequence $\{x_n\}=\{1, 1, -1, 1\}$ satisfies 2) and 3′), and therefore for $N=4$, (15) is a least upper bound.

If $\{h_n\}$ is required to be real (with no such restriction on $\{x_n\}$), then one can verify for $N=2$ that if $\{x_n\}$ is chosen to satisfy 2) and 3′), then the phase of H_k cannot be chosen to satisfy 3′), and thus (15) is not a least upper bound for this case.

To show that $\beta(N)\geq\sqrt{N}/2$ for $\{x_n\}$ and/or $\{h_n\}$ restricted to be real valued, it suffices to show that $\beta(N)\geq\sqrt{N}/2$ for both $\{x_n\}$ and $\{h_n\}$ real valued. This we will demonstrate only for the case where N is even, since the argument for N odd is identical.

Consider the complex sequence

$$f_n = \exp[j\pi n^2/N] = \cos\frac{\pi n^2}{N} + j\sin\frac{\pi n^2}{N}$$

with DFT denoted by

$$F_k = R_k + jI_k$$

where R_k and I_k are real valued and (see Appendix)

$$R_k^2 + I_k^2 = \frac{1}{N}\cdot \tag{21}$$

Since $\exp[j\pi n^2/N]$ is an even function of n, i.e.,

$$\exp[j\pi n^2/N] = \exp[j\pi(N-n)^2/N],$$

R_k is the DFT of $\cos(\pi n^2/N)$ and I_k is the DFT of $\sin(\pi n^2/N)$. Now, if we choose

$$x_n = \cos\left(\frac{\pi n^2}{N}\right)$$

and

$$H_k = \begin{cases} 1 & R_k > 0 \\ -1 & R_k \leq 0, \end{cases}$$

then

$$y_0 = \sum_{k=0}^{N-1} |R_k|. \tag{22}$$

Similarly, if we choose $x_n'=\sin(\pi n^2/N)$, then we can choose $\{H_k\}$ in such a way that

$$y_0' = \sum_{k=0}^{N-1} |I_k|. \tag{23}$$

We note that since $\{x_n\}$ and $\{x_n'\}$ are both real, the values y_0 and y_0' will be obtained with $\{H_k\}$ having even magnitude and odd phase, corresponding to real $\{h_n\}$. Now, if β is the least upper bound for $|y_n|$, then

$$\beta \geq y_0 \tag{24a}$$

$$\beta \geq y_0' \tag{24b}$$

and, from (21),

$$|R_k| \leq \frac{1}{\sqrt{N}}$$

$$|I_k| \leq \frac{1}{\sqrt{N}}$$

and, hence,

$$|R_k| \geq \sqrt{N}\,|R_k|^2 \tag{25a}$$

$$|I_k| \geq \sqrt{N}\,|I_k|^2. \tag{25b}$$

Combining (22), (23), (24), and (25),

$$\beta \geq \sum_{k=0}^{N-1} |R_k| \geq \sqrt{N}\sum_{k=0}^{N-1} |R_k|^2 \tag{26a}$$

$$\beta \geq \sum_{k=0}^{N-1} |I_k| \geq \sqrt{N}\sum_{k=0}^{N-1} |I_k|^2. \tag{26b}$$

Adding (26a) and (26b) and using (21),

$$2\beta \geq \sqrt{N}$$

or

$$\beta \geq \frac{\sqrt{N}}{2}\cdot \tag{27}$$

Since we argued previously that $\beta\leq\sqrt{N}$, Result C is proved.

[2] The sequences (19) and (20) were suggested to the authors by C. M. Rader of the M.I.T. Lincoln Laboratory.

IV. Discussion

The bound obtained in the previous sections can be utilized in several ways. If the DFT computation is carried out using a block floating-point strategy so that arrays are rescaled only when overflows occur, then a final rescaling must be carried out after each section is processed so that it is compatible with the results from previous sections. For general input and filter characteristics, the final rescaling can be chosen based on the bounds given here to insure that the output will not exceed the available register length.

The use of block floating-point computation requires the incorporation of an overflow test. In some cases we may wish instead to incorporate scaling in the computation in such a way that we are guaranteed never to overflow. For example, when we realize the DFT with a power of two algorithm, overflows in the FFT computation of $\{X_k\}$ will be prevented by including a scaling of $\frac{1}{2}$ at each stage, since the maximum modulus of an array in the computation is nondecreasing and increases by at most a factor of two as we proceed from one stage to the next [2]. With this scaling, the bound derived in this paper guarantees that with a power of two computation, scaling is not required in more than half the arrays in the inverse FFT computation. Therefore, including a scaling of $\frac{1}{2}$ in the first half of the stages in the inverse FFT will guarantee that there are no overflows in the remainder of the computation. The fact that $\beta \geq \sqrt{N}/2$ indicates that if we restrict ourselves to only real input data, at most one rescaling could be eliminated for some values of N.

The bounds derived and method of scaling mentioned above apply to the general case; that is, except for the normalization of (7), they do not depend on the filter characteristics. This is useful when we wish to fix the scaling strategy without reference to any particular filter. For specific filter characteristics, the bound can be reduced. Specifically, it can be verified from (1) and (6) that in terms of $\{h_n\}$

$$|y_n| \leq \sum_{l=0}^{M-1} |h_l| \tag{28}$$

where M denotes the length of the impulse response. This is a least upper bound since a sequence $\{x_n\}$ can be selected which will result in this value in the output. This will be significantly lower than the bound represented in (15) if, for example, the filter is very narrow band, or if the kernel has many points with zero value.

Appendix

We wish to demonstrate that for N even, the sequence

$$x_n = \exp\left[j\frac{\pi n^2}{N}\right] \qquad \begin{array}{l} n = 0, 1, \cdots, N-1 \\ N \text{ even} \end{array} \tag{29}$$

has a discrete Fourier transform with constant modulus and that for N odd, the sequence

$$x_n = \exp\left[j\frac{2\pi n^2}{N}\right] \qquad \begin{array}{l} n = 0, 1, \cdots, N-1 \\ N \text{ odd} \end{array} \tag{30}$$

has a discrete Fourier transform with constant modulus. We consider first the case of (29). Letting X_k denote the DFT of x_n,

$$X_k = \frac{1}{N}\sum_{n=0}^{N-1} \exp\left[j\frac{\pi n^2}{N}\right] \exp\left[j\frac{2\pi nk}{N}\right]$$

or

$$X_k = \frac{1}{N}\exp\left[-j\frac{\pi k^2}{N}\right] \sum_{n=0}^{N-1} \exp\left[j\pi(n+k)^2\right]. \tag{31}$$

We wish to show first that

$$\sum_{n=0}^{N-1} \exp\left[j\frac{\pi}{N}(n+k)^2\right]$$

is a constant. It is easily verified by a substitution of variables that

$$\sum_{n=0}^{2N-1} \exp\left[j\pi(n+k)^2/N\right] = \text{constant} \triangleq B. \tag{32}$$

But

$$\begin{aligned} &\sum_{n=0}^{2N-1} \exp\left[j\pi(n+k)^2/N\right] \\ &= \sum_{n=0}^{N-1} \exp\left[j\pi(n+k)^2/N\right] + \sum_{n=N}^{2N-1} \exp\left[j\pi(n+k)^2/N\right] \\ &= \sum_{n=0}^{N-1} \exp\left[j\pi(n+k)^2/N\right] \\ &\quad + \sum_{n=0}^{N-1} \exp\left[j\pi(n+k)^2/N\right] \exp\left[j\pi N\right] \end{aligned}$$

or, since N is even,

$$\sum_{n=0}^{2N-1} \exp\left[j\pi(n+k)^2/N\right] = 2\sum_{n=0}^{N-1} \exp\left[j\pi(n+k)^2/N\right]. \tag{33}$$

Combining (31), (32), and (33),

$$X_k = \frac{1}{N}\cdot B \cdot \exp\left[-j\pi k^2/N\right].$$

To determine the modulus of B, Parseval's relation requires that

$$\sum_{n=0}^{N-1} |x_n|^2 = N\sum_{n=0}^{N-1} |X_k|^2$$

or

$$N = |B|^2.$$

Therefore

$$|B| = \sqrt{N}$$

or

$$|X_k| = \frac{1}{\sqrt{N}}.$$

It can be verified by example (try $N=3$) that the sequence of (29) does not have a DFT with constant modulus if N is odd.

Consider next the sequence of (30). We will show that X_k has constant modulus by showing that the circular autocorrelation of x_n, which we denote by c_n, is nonzero only at $n=0$. Specifically, consider

$$c_n = \sum_{r=0}^{N-1} x_r \overset{*}{x}_{(n+r)\bmod N}$$

$$= \sum_{r=0}^{N-1} \exp\left[j\frac{2\pi r^2}{N}\right]\exp\left[-j\frac{2\pi[(n+r)^2]\bmod N}{N}\right].$$

Now,

$$\exp\left[-j\frac{2\pi[(n+r)^2]\bmod N}{N}\right] = \exp\left[-j\frac{2\pi(n+r)^2}{N}\right].$$

Therefore,

$$c_n = \sum_{r=0}^{N-1} \exp\left[j\frac{2\pi r^2}{N}\right]\exp\left[-j\frac{2\pi(r+n)^2}{N}\right]$$

$$= \sum_{r=0}^{N-1} \exp\left[-j\frac{2\pi n^2}{N}\right]\exp\left[-j\frac{4\pi rn}{N}\right]$$

$$= \exp\left[-j\frac{2\pi n^2}{N}\right]\sum_{r=0}^{N-1} \exp\left[-j\frac{4\pi rn}{N}\right].$$

But

$$\sum_{r=0}^{N-1} \exp\left[-j\frac{4\pi rn}{N}\right] = \begin{cases} 1 & n=0 \\ 0 & n\neq 0, \quad N \text{ odd} \\ 1 & n=\frac{N}{2}, \quad N \text{ even} \\ 0 & n\neq 0,\ n\neq\frac{N}{2}, \quad N \text{ even.} \end{cases}$$

Since we are considering the case of N odd,

$$c_n = \begin{cases} 1 & n=0 \\ 0 & n\neq 0. \end{cases}$$

Since $|X_k|$ is constant, we may again use Parseval's theorem to show that $|X_k| = 1/\sqrt{N}$.

References

[1] T. G. Stockham, "High speed convolution and correlation," *1966 Spring Joint Computer Conf., AFIPS Proc.*, vol. 28. Washington, D. C.: Spartan, 1966, pp. 229–233.

[2] P. D. Welch, "A fixed-point fast Fourier transform error analysis," this issue, pp. 151–157.

Fast Fourier Transform Hardware Implementations–An Overview

GLENN D. BERGLAND, Member, IEEE
Bell Telephone Laboratories, Inc.
Whippany, N. J. 07981

Abstract

This discussion served as an introduction to the Hardware Implementations Session of the IEEE Workshop on Fast Fourier Transform Processing. It introduces the problems associated with implementing the FFT algorithm in hardware and provides a frame of reference for characterizing specific implementations. Many of the design options applicable to an FFT processor are described, and a brief comparison of several machine organizations is given.

Manuscript received February 14, 1969.

Introduction

Software implementations of the Cooley–Tukey fast Fourier transform (FFT) algorithm [1] have in many cases reduced the time required to perform Fourier analysis by nearly two orders of magnitude. Even greater gains can be realized through special-purpose hardware designed specifically for performing the FFT algorithm. In order to design this hardware one should examine:

1) The reasons for building special-purpose hardware;
2) The options that should be considered;
3) The tradeoffs that must be made;
4) Other considerations.

Special-Purpose Hardware

Most applications for special-purpose FFT processors result from signal processing problems which have an inherent real-time constraint. Examples include digital vocoding, synthetic-aperture radar mapping, sonar signal processing, radar signal processing, and digital filtering. In these examples, a processing rate slower than real-time would overload the system with input data or lead to worthless results.

Other applications involve off-line processing where the volume of data makes processing impractical unless a dedicated machine is used. Studies in radio astronomy and crystallography have involved Fourier analysis taking nearly a month to perform on a general-purpose computer. In several cases, the use of special-purpose hardware could reduce this time to less than a day and make a corresponding reduction in cost.

Experience with the FFT processor built by Bell Telephone Laboratories [2] indicates that the cost reduction resulting from special-purpose hardware is nearly as great as the reduction which came with the Cooley–Tukey algorithm. The FFT signal processing system costs 5 times less per hour than a large general-purpose computer while performing the fast Fourier transform algorithm 20 times faster. Thus, on the FFT part of the processing, a 100 to 1 cost saving is possible. As a result of this reduction in cost, people who had not even heard of the fast Fourier transform three years ago are now finding that they cannot get along without it.

Options

The many and varied forms of the FFT algorithm have been described at length in the literature [3]–[11]. The execution times and memory requirements of software implementations of these algorithms can be evaluated rather conveniently. The criteria which apply to evaluating hardware implementations, however, are not as easily specified.

Reprinted from *IEEE Trans. Audio Electroacoust.*, vol. AU-17, pp. 104–108, June 1969.

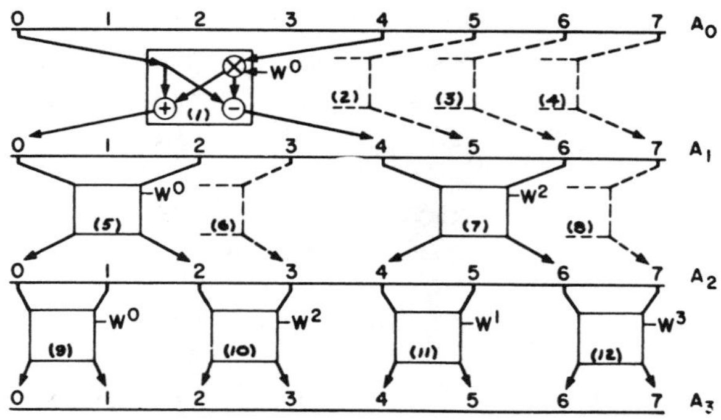

Fig. 1. Fast Fourier transform flow diagram for $N=8$.

The first design choice often made concerns constraining the number of data points to be analyzed to being a power of 2 (i.e., $N=2^m$ for $m=0, 1, 2, \cdots$). This represents a tradeoff of versatility for cost and performance. This choice is not as limiting as it was previously. The convolutional form of the FFT [11] can be used to find the discrete Fourier transform for any value of N even though the FFT processor performing the convolutions requires that N be a power of 2.

If the input time series consists of real numbers, the second option which should be considered involves the use of a real-input algorithm [10] or a modified complex-input algorithm [12]. By exercising this option, a two-to-one improvement in performance and a two-to-one reduction in storage can be achieved.

While a radix-8 algorithm may be near optimum for a software implementation, the simplicity of the radix-4 and radix-2 algorithms is a considerable advantage when dealing with hardware. Since the cost is proportional to the number of options included, the use of only one basic operation in the radix-2 algorithm in many cases offsets the additional computation required.

The organization of an FFT processor is usually dictated by the performance and cost requirements and the technology assumed. Four families of machine organizations, which have appeared in some form in the literature, will be described and characterized.

The Sequential Processor

The first hardware implementation considered involves implementing the basic operation shown in Fig. 1, in hardware [13]. This basic operation (i.e., one complex multiplication followed by an addition and a subtraction) can be applied sequentially to the 12 sets of data shown in the diagram. The same memory can be used to store the input data, the intermediate results, and the resulting Fourier coefficients. Since only one basic operation is involved and the accessing pattern is very regular, the amount of hardware involved can be relatively small. A simplified block diagram of the resulting processor is shown in Fig. 2.

Fig. 2. The functional block diagram of a sequential fast Fourier transform processor.

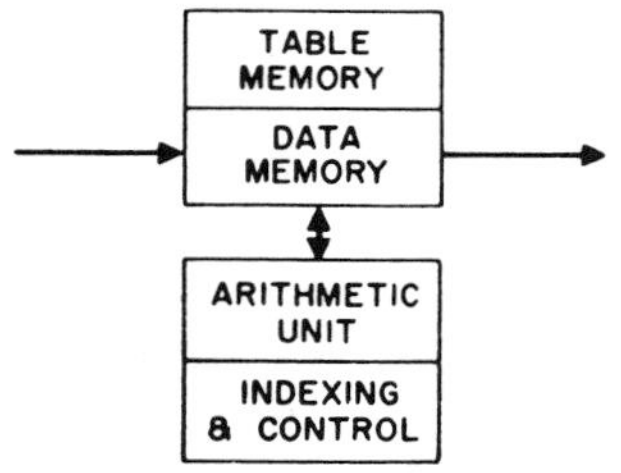

This organization is similar to that of a small general-purpose computer except that the table memory, data memory, arithmetic unit, and control unit can usually all operate concurrently. Since the processor operates on batches of data, a real-time environment would usually dictate that some buffering precede or be incorporated into the data memory.

For this discussion, the sequential processor will be characterized as having: 1) one arithmetic unit, 2) $(N/2) \log_2 N$ operations performed sequentially, and 3) an execution time of $B(N/2) \log_2 N$ μs where B is the time required for performing one basic operation. The reordering of the Fourier coefficients can be done either in place or during I/O.

The Cascade Processor

To improve the performance of the processor, parallelism can be introduced into the flow diagram shown in Fig. 1 [14], [15]. By using a separate arithmetic unit for each iteration, the throughput can be increased by a factor of $\log_2 N$. In the diagram of Fig. 1 this means that the first arithmetic unit performs the operations labeled 1 through 4, the second performs operations 5 through 8,

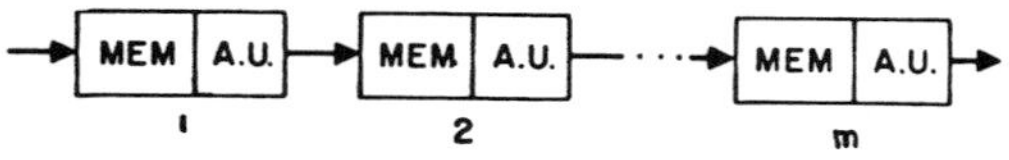

Fig. 3. The functional block diagram of a cascade processor.

Fig. 4. The functional block diagram of a parallel-iterative processor.

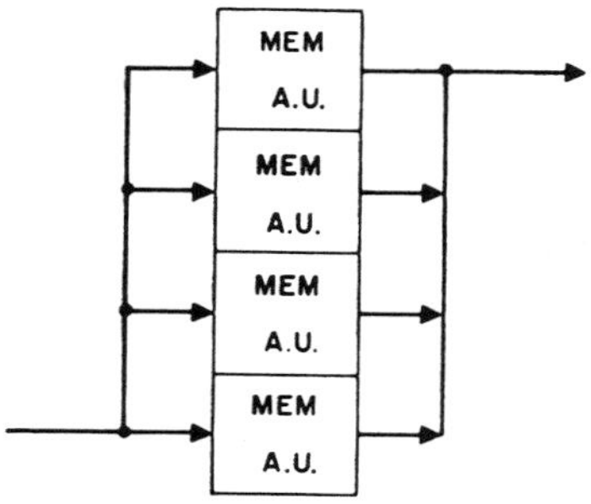

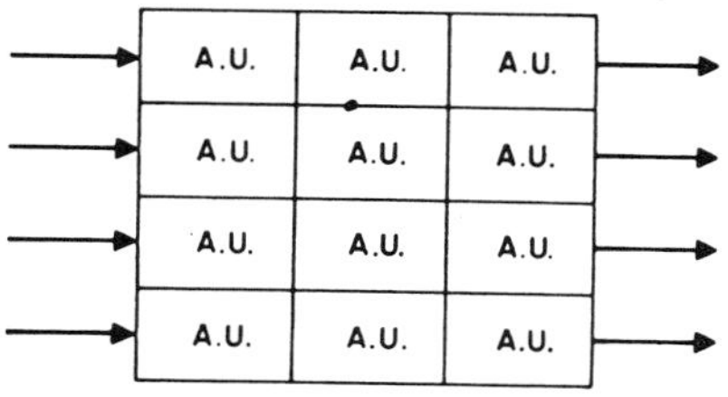

Fig. 5. The functional block diagram of an array processor.

and the third performs operations 9 through 12. A simplified block diagram of the resulting processor is shown in Fig. 3.

In this processor, the buffering required for processing a continuous stream of data is incorporated directly into the organization and takes the form of a delay line.

For this discussion, the cascade processor will be characterized as having: 1) m arithmetic units, 2) m iterations performed in parallel, 3) $N/2$ operations performed sequentially, 4) an execution time of $(B \cdot N)$ μs per record, and 5) buffering incorporated within the processor in the form of time delays.

Although this discussion is directed toward radix-2 algorithms, versions of this processor for radix-4 algorithms, real-valued algorithms, and arbitrary-radix algorithms should be apparent. It should also be apparent that this organization lends itself well to multichannel operation where sets of samples are interleaved. Although this adds to the amount of buffering within the processor, in many applications it allows one to realize this buffering via relatively inexpensive drum or disk storage used in the form of time delays. As recently shown by Sande [16], this same magnetic drum can also be used to perform the reordering.

The Parallel Iterative Processor

A third alternative for improving performance involves introducing parallelism within each iteration. By using four arithmetic units, the operations labeled 1 through 4 can be performed in parallel before performing operations 5 through 8 in parallel, etc. The processor performs the iterations sequentially, but performs all of the operations within each iteration in parallel. A simplified block diagram of the resulting processor is shown in Fig. 4.

In practice, this organization would often be combined with the sequential processor so that only the degree of parallelism actually needed is implemented. Without the use of large-scale integration, the cost of $N/2$ arithmetic units (as shown in Fig. 4) would usually be prohibitive.

The problem areas associated with this organization involve the communication between the arithmetic units and the generation of the required sine and cosine functions. Both of these problems, however, have been dealt with successfully [17], [18].

For this discussion, the parallel iterative processor will be characterized as having 1) $N/2$ arithmetic units, 2) $N/2$ operations performed in parallel, 3) m iterations performed sequentially, and 4) an execution time of $B(\log_2 N)$ μs.

The Array Analyzer

For completeness, a processor can be considered in which all 12 of the operations of Fig. 1 are performed in parallel [19]. By pipelining three different sets of data through this processor simultaneously, the effective execution time is simply the time required for performing one basic operation. A simplified block diagram is shown in Fig. 5 for the example of $N=8$.

For this discussion, the array analyzer will be characterized as having: 1) $(N/2)(\log_2 N)$ arithmetic units, 2) $(N/2)(\log_2 N)$ operations performed in parallel, and 3) an execution time of B μs. At this point in time, the cost of this approach severely limits its application.

Summary

By assuming the relatively fast processing rate of one μs per basic operation, the approximate processing rates of the different families of analyzers can be compared in Table I. The time saving of special-purpose hardware is apparent when these times are compared to the typical 500 000 μs execution time of a large general-purpose computer. Note that you can introduce as much parallelism into the fast Fourier transform algorithm as your problem demands.

TABLE I

Ambitious Processing Rates for $N=1024$ With $B=1\ \mu s$

Machine Organization	Arith. Units	Execution Time (μs)	Processing Rate (samples/s)
Sequential	1	5000	200 000
Cascade	10	1000	1 000 000*
Parallel Iterative	512	10	100 000 000
Array	5120	1	1 000 000 000

* Capable of doing two channels at this rate simultaneously.

Tradeoffs

In designing a special-purpose fast Fourier transform processor, there are many possible tradeoffs which can be made in terms of cost, speed, and accuracy. These tradeoffs are related specifically to the arithmetic unit, the memory, the control unit, and the algorithm being implemented.

As described previously, there is usually a cost penalty associated with allowing N to be other than a power of 2. If enough options are added, the control unit can rapidly overtake the arithmetic unit in complexity.

When versatility is more important than speed, it may be worth changing from a hard-wired control unit to a software or microprogrammed control unit.

The choice of a core memory, a semiconductor memory, or some combination of the two in a sequential processor will generally be dictated by the performance and cost requirements placed on the system. In the cascade processor, a plot of cost versus speed can be very erratic since the internal storage could be anything from magnetic drum storage to semiconductor shift registers.

The arithmetic units for each type of processor can vary considerably. The cost ratio between building five inexpensive arithmetic units and building one very expensive one can dictate the organization of the processor. One also has the option of building one fast real-input arithmetic unit, or four slower arithmetic units tied together to form a complex-input arithmetic unit. In some cases a combinatorial (or static) multiplier is required. In other cases the less expensive iterative multiplier is fast enough. The speed is also highly dependent on whether the numbers are represented in fixed point, floating point, or "poor man's floating point" (i.e., with an exponent common to the whole array).

A more complete list of design options is given in the FFT processor survey [20].

Other Considerations

In many cases, people tend to focus on only the FFT hardware since it is the best defined part of the system. Those parts of the problem which should not be overlooked are preprocessing, postprocessing, data reduction, and diagnostics.

The preprocessing function often involves applying an appropriate data window, forming redundant records, and buffering. Postprocessing frequently involves smoothing, interpolating, automatic gain controlling, convolving, and correlating.

Since every set of N numbers put into an FFT processor results in N Fourier coefficients coming out, any data reduction functions which can be defined are usually worth building into the hardware.

The problem of designing a set of diagnostic checks for the FFT processor is helped considerably when the processor is a computer attachment. Even then, however, the testing function should be given at least as much thought as the FFT algorithm.

Conclusions

Four families of FFT machine organizations have been defined which represent increasing degrees of parallelism, performance, and cost. Within each family, however, the cost and performance of two processors can still differ widely due to choices of arithmetic unit, memory, and control. It is clear that, given enough parallelism and enough money, FFT processing can be done digitally at rates less than one sample per second or as high as a billion samples per second.

References

[1] J. W. Cooley and J. W. Tukey, "An algorithm for the machine calculation of complex Fourier series," *Math. Comp.*, vol. 19, pp. 297–301, April 1965.

[2] R. Klahn, R. R. Shively, E. Gomez, and M. J. Gilmartin, "The time-saver: FFT hardware," *Electronics*, pp. 92–97, June 24, 1968.

[3] J. W. Cooley, "Complex finite Fourier transform subroutine," SHARE Doc. 3465, September 8, 1966.

[4] W. M. Gentleman and G. Sande, "Fast Fourier transforms for fun and profit," *1966 Fall Joint Computer Conf., AFIPS Proc.*, vol. 29. Washington, D. C.: Spartan, 1966, pp. 563–578.

[5] G. D. Bergland, "The fast Fourier transform recursive equations for arbitrary length records," *Math. Comp.*, vol. 21, pp. 236–238, April 1967.

[6] N. M. Brenner, "Three FORTRAN programs that perform the Cooley–Tukey Fourier transform," M.I.T. Lincoln Lab., Lexington, Mass., Tech. Note 1967-2, July 1967.
[7] R. C. Singleton, "On computing the fast Fourier transform," *Commun. ACM*, vol. 10, pp. 647–654, October 1967.
[8] G. D. Bergland, "A fast Fourier transform algorithm using base 8 iterations," *Math. Comp.*, vol. 22, pp. 275–279, April 1968.
[9] M. C. Pease, "An adaption of the fast Fourier transform for parallel processing," *J. ACM*, vol. 15, pp. 252–264, April 1968.
[10] G. D. Bergland, "A fast Fourier transform algorithm for real-valued series," *Commun. ACM*, vol. 11, pp. 703–710, October 1968.
[11] L. I. Bluestein, "A linear filtering approach to the computation of the discrete Fourier transform," *1968 NEREM Rec.*, pp. 218–219.
[12] J. W. Cooley, P. A. W. Lewis, and P. D. Welch, "The fast Fourier transform algorithm and its applications," IBM Research Paper RC-1743, February 1967.
[13] R. R. Shively, "A digital processor to generate spectra in real time," *IEEE Trans. Computers*, vol. C-17, pp. 485–491, May 1968.
[14] G. D. Bergland and H. W. Hale, "Digital real-time spectral analysis," *IEEE Trans. Electronic Computers*, vol. EC-16, pp. 180–185, April 1967.
[15] R. A. Smith, "A fast Fourier transform processor," Bell Telephone Labs., Inc., Whippany, N. J., 1967.
[16] G. Sande, University of Chicago, Chicago, Ill., private communication.
[17] M. C. Pease, III, and J. Goldberg, "Feasibility study of a special-purpose digital computer for on-line Fourier analysis," Advanced Research Projects Agency, Order 989, May 1967.
[18] G. D. Bergland and D. E. Wilson, "An FFT algorithm for a global, highly-parallel processor," this issue, pp. 125–127.
[19] R. B. McCullough, "A real-time digital spectrum analyzer," Stanford Electronics Labs., Stanford, Calif., Sci. Rept. 23, November 1967.
[20] G. D. Bergland, "Fast Fourier transform hardware implementations—a survey," this issue, pp. 109–119.

A Pipeline Fast Fourier Transform

HERBERT L. GROGINSKY, SENIOR MEMBER, IEEE, AND GEORGE A. WORKS

Abstract—This paper describes a novel structure for a hardwired fast Fourier transform (FFT) signal processor that promises to permit digital spectrum analysis to achieve throughput rates consistent with extremely wide-band radars. The technique is based on the use of serial storage for data and intermediate results and multiple arithmetic units each of which carries out a sparse Fourier transform. Details of the system are described for data sample sizes that are binary multiples, but the technique is applicable to any composite number.

Index Terms—Cascade Fourier transform, digital signal processor, Doppler radar, fast Fourier transform, radar–sonar signal processor, radix-two fast Fourier transform, real-time signal processor.

Introduction

THIS paper describes a novel structure for a hardwired FFT signal processor that promises to permit digital spectrum analysis to achieve throughput rates consistent with extremely wide-band radars.

The processor consists of a number of modular units connected in cascade through switches that direct the flow of information from memory to arithmetic units. The switching required to carry out the process is simple and is controlled by a binary counter. The processor is similar to the binary analyzer described by Bergland and Hale [1], but employs only N complex words of storage to compute the FFT of N complex data samples.[1] Bergland [2] has listed many alternative organizations of FFT processors. Recently O'Leary [10] has also proposed a similar structure.

We show that the Cooley–Tukey algorithm does a natural interleaving of data gathered by the time multiplexing of a number of independent channels, typical of radars and sonars. In this concept, the successive stages or iterations of the fundamental algorithm are each carried out in the separate cascaded modules. Using shift registers as digital delay lines permits new data to be entered into the processor while the processing of earlier data blocks is carried out. In effect the overall delay required is equal to the time required to gather the analysis sample block N in each of the separate channels. As the Nth complex data sample is loaded into the digital delay line, the first analysis frequency appears at the output. The output appears in precisely the same channel sequence as the data when they were loaded into the delay line. The output frequencies, however, appear in the scrambled sequence associated with the algorithm.

The control device, namely the binary counter, yields a digital number identifying both the channel number and the frequency currently being outputted. In addition, it specifies the instants at which the separate modules are to be

Manuscript received November 7, 1969; revised April 27, 1970. This work was supported by Raytheon research and development funding. A patent has been filed on the basic structure of this signal processor. This paper was presented at EASCON'69 (Electronics and Aerospace Systems Convention), Washington, D. C., October 27–29, 1969.

The authors are with the Raytheon Company, Sudbury, Mass.

[1] Although the processor described here is cascade in structure, we prefer the pipeline designation used by computer designers [11] because this structure permits direct application of pipeline arithmetic techniques.

Reprinted from *IEEE Trans. Comput.*, vol. C-19, pp. 1015–1019, Nov. 1970.

switched and a digital number identifying the sine/cosine values needed by each of the stages. This structure, although hardwired, does permit the flexible interchange of channels processed for data sample length per channel. Thus a system capable of processing N complex samples in a single channel is also capable of processing N/L samples in each of L channels provided L is a factor of N. The modular design of the device permits the duplicate arithmetic units to weight the input prior to the FFT operation and to present the output in magnitude. Furthermore, it allows computation of Fourier transforms at the rate at which new data can be inserted into the digital delay line. Fundamentally, the signal processing rate is independent of the data sample length.

A pipeline FFT configured to process radar data from a pulse Doppler tracking radar has been designed and tested. The system uses MOSFET shift registers as the digital delay lines, TTL in the arithmetic units and MOS LSI READ ONLY memory to store sine/cosine tables and filter shaping weight functions.

The system processes eight range channels taking 512 complex samples per channel. It is designed to obtain subclutter visibility of 60 dB and achieves this using 12-bit fixed-point internal operations in the arithmetic unit. The throughput rate achieved by the system is 128K samples per second.

Theory of Operation

In this section, the method of operation of the pipeline FFT is explained in terms of the fundamental mathematical operations that must be carried out. An extensive literature now exists [3] explaining the basic principles of the FFT. The discussion here emphasizes certain features of the analysis, permitting a hardwired realization of the algorithm to achieve the goals set forth in the introduction.

The discrete Fourier transform (DFT) is defined by

$$X_m = \sum_{n=0}^{N-1} x_n W_N^{mn} \tag{1}$$

where

$$W_N^z = e^{-j(2\pi z/N)},$$

x_n is a complex data sample, and X_m is the complex image of the data at frequency m/N.

Theory shows that when N is a composite number

$$N = \prod_{k=1}^{M} r_k \tag{2}$$

where the r_k are a set of integers (possibly with repeats), and (1) may be calculated iteratively in M stages as follows.

$$a_{\mu,\nu}^m = \sum_{l=0}^{r_m-1} W_{R_m}^{-\mu l} a_{p,q}^{m-1} \tag{3}$$

with

$$p = \mu \bmod R_{m-1}$$
$$q = \nu + lC_m$$

and the ranges

$$1 \le m \le M, \quad 0 \le \mu < R_m, \quad 0 \le \nu < C_m$$

where

$$R_m = \prod_{k=1}^{m} r_k$$
$$C_m = N/R_m$$
$$a_{\mu,0}^0 = x_n$$

and

$$\mu \bmod r \triangleq \mu - \text{greatest integer in } \mu/r.$$

In fact, when this is done, the number of calculations drops from N^2 complex operations (MULTIPLY and ADD) to $N(r_1+r_2+\cdots+r_M)$ such operations. This iterative process both reduces the amount of hardware required to realize the operation, and provides the basic pipeline structure permitting new calculations to proceed before the results of earlier calculations are completed. This form of the algorithm is known as the Cooley–Tukey version.

When $r_k=2$ for all k, the algorithm is conveniently summarized by the flow diagram shown in Fig. 1. In this figure, the input data enter the left-hand column and each successive column corresponds to a later stage of the iterative process. The coefficients indicated at the input then correspond to the index of the input data x_n and give its order in the data stream. The figures shown at each later stage indicate the coefficients of the rotation vector W_{2^m} that must be applied to the lower branch entering each node.

A number of important features of the algorithm may be seen by examining Fig. 1. First, we observe that each stage needs only the data generated from the preceding stage. Second, if each stage is processed in order of arrival, the first stage examines data points displaced by half the data length ($N/2$), the second by one quarter of the data length ($N/4$), etc. Third, if the data were available in a continuous stream, the first stage could be processing one block of data while the second stage processed the next earlier block and so on through all M stages. Fourth, the rotation vector required, W_{2^m}, has the same periodicity as the data displacement interval. Finally, we note that the output appears in the usual scrambled order of frequency associated with this version of the algorithm.

The significance of these remarks is that each of the stages may be realized with a basic component whose general form is shown in Fig. 2. Any m module alternately transfers blocks of 2^m data samples into the delay line and into the arithmetic unit. When the data block just fills the delay line, the arithmetic unit obtains a rotation vector (from a READ ONLY memory) and begins its operation. The next block of 2^m input data samples are sent to the arithmetic unit that now produces two complex outputs in response to the two complex inputs it receives. One of the outputs is immediately transferred to the next stage while the other output is sent to the delay line. Thus in the interim period when the delay line is filled with fresh input data, the contents of the

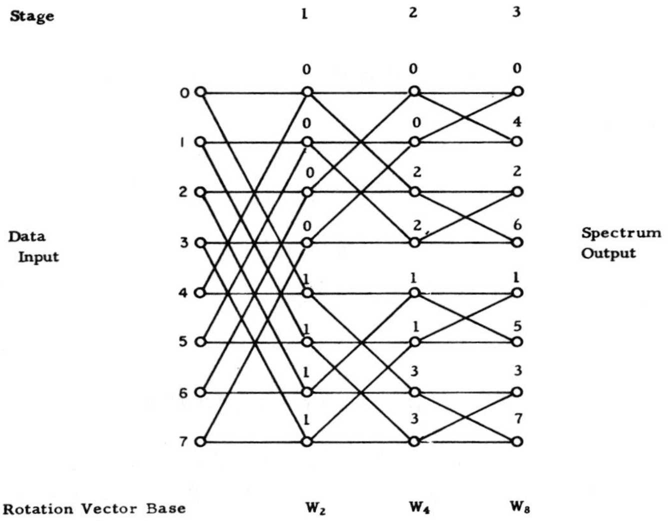

Fig. 1. Flow diagram of a Cooley–Tukey FFT.

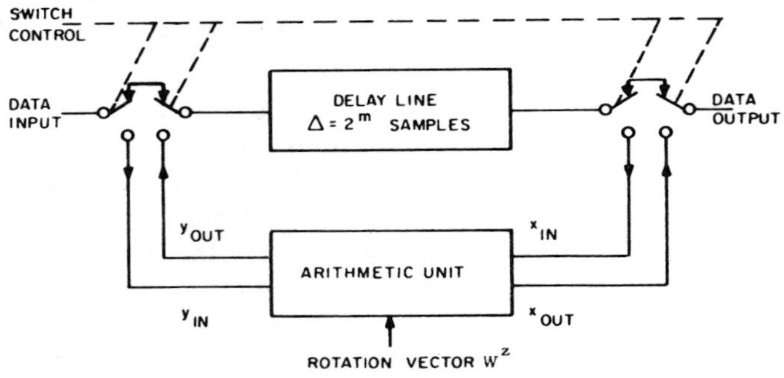

Fig. 2. Pipeline FFT m module.

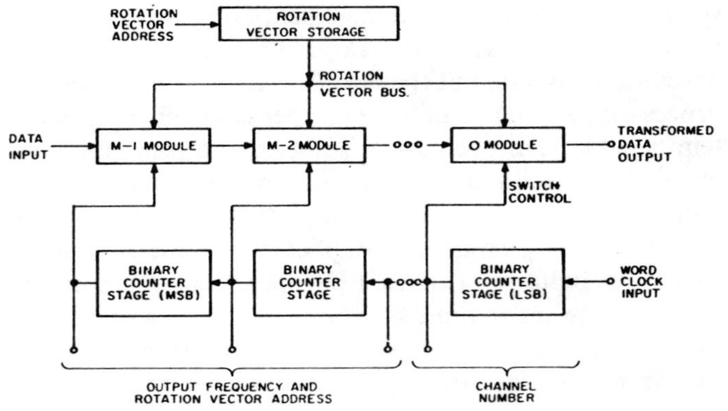

Fig. 3. Pipeline FFT processor.

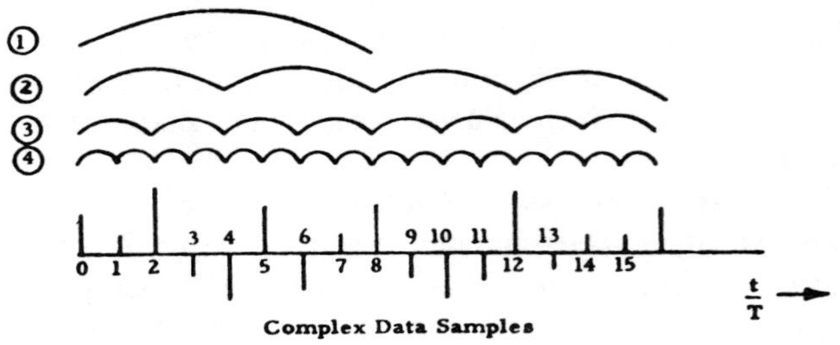

Fig. 4. FFT interleaving mechanism.

line containing the results of processing the earlier blocks are transferred to the next stage. The arithmetic unit, of course, computes the complex two-point transforms, shown below.

$$x_{\text{out}} = x_{\text{in}} + y_{\text{in}} W^z$$
$$y_{\text{out}} = x_{\text{in}} - y_{\text{in}} W^z. \tag{4}$$

This module design may be assembled into a system for computing DFTs in blocks of $N = 2^M$ samples as shown in Fig. 3. The rotation vector storage shown in the figure is a table of roots of unity, or sines and cosines, which is shared by all m modules. Fig. 1 shows that $N/2$ different rotation vectors are read to process one block of N samples. Indeed if they are produced in the order required for the last stage, the rotation vectors required for the earlier stages may be obtained by strobing this list at the proper instant in advance of its need. Thus the sines and cosines required for each stage may be obtained by providing a register for each arithmetic unit, all driven from a common bus. Note that exactly M arithmetic units and exactly $N-1$ complex data points of storage are needed in this system and that the first transform output is obtained immediately after the last data sample in the block of N is received.

The theory cited above leads to still another important observation, namely that the output at any intermediate stage of the process has a simple interpretation. Theory shows that

$$a^m_{\mu,\nu} = \sum_{k=0}^{R_m - 1} x_{\nu + kC_m} W^{\mu k}_{R_m}, \tag{5}$$

which is precisely the discrete Fourier transform of all groups of data points separated by an interval C_m. This may be better understood in reference to Fig. 4, which shows the natural FFT interleaving mechanism that results from the Cooley–Tukey algorithm. In terms of a sequential process, the output of the first stages results in $N/2$ independent two-point transforms; the output of the second stage yields $N/4$ independent four-point transforms, etc. Thus, if two independent streams of complex data were entered into the input interleaved with one another, the module 1 stage of the cascade processor would produce two independent DFTs of each data stream. The spectral component of each channel of data is outputted before the spectral frequency is changed. In particular for pulsed radar or sonar application, where the data for many range samples are received before a new sample may be taken, this system permits the data to be processed in order of arrival with no modification of the control circuits and without requiring the data to be reassembled first into consecutive (noninterleaved) data streams.

Equation (5) shows that the index μ, which is the rotation vector coefficient in (3), can also be regarded as the current frequency being outputted, namely μ/R_m. When these normalized frequencies (chosen to make R_m a unit period) are modified to account for the expanded sampling interval C_m, the current frequency being outputted may be regarded on absolute scales as the frequency $\mu/R_m C_m = \mu/N$. Thus, in Fig. 1, the coefficient given at every stage in the process indicates not only the coefficient of the rotation vector applied to the lower branch but the current frequency (in absolute terms) as well.

The control mechanism for this system is perhaps its most elegant feature. Fig. 1 shows clearly that a binary counter driven in synchronism with the data stream generates a signal designating the processing interval (i.e., the switch position) for each stage. However, if any output is taken at any intermediate stage, say at the $k+1$st stage, then the lower k bits of the counter, in normal order, give the channel number of the data currently being outputted, while the upper $M-k$ bits taken in bit-reversed order, give the frequency that is currently being outputted and, in addition, describes the address (θ) of the sine and cosines needed in the current computation of the $k+1$st stage. Thus, the control mechanism contains all the information needed to carry out the spectral analysis as well as to descramble the output data. In many applications, it is unnecessary to descramble the output data provided one can identify the frequency of any component.

Indeed the structure shown in Fig. 3 may be readily modified to calculate transforms when the data samples are given in scrambled order. This structure also permits the trade-off of channels processed for data length per channel by taking outputs at an intermediate stage. The modified structure produces the output sequence in natural order in both the channels and in time.

Computational Errors and Quantization

If the input signal to an FFT machine is obtained from an analog-to-digital converter, it must be sufficiently finely quantized so that quantization noise is uncorrelated to avoid distortion of the Fourier transform. Widrow [4] has shown that for signals in the presence of Gaussian noise, choosing the quantization grain or value of the least significant bit of the quantizer equal to three times the noise standard derivation provides a good approximation to this condition. The value of the most significant bit of the quantizer must be greater than or equal to one half the peak value of the input signal (without noise) to avoid peak limiting. The minimum number of bits Q required to represent a signal of peak value $\pm V_p$ with additive Gaussian noise of standard deviation σ_n is therefore given by

$$Q = \log_2(V_p/3\sigma_n) + 1. \tag{6}$$

One or more bits may be added to further reduce quantization noise.

The number of bits per data word required in an FFT processor is usually greater than the number of bits Q in the input word to reduce computational noise. The two-point transform (4), which is the building block of larger Fourier transforms, may be accomplished by four real multiplications and six real additions. Following each two-point transform, words may be truncated or rounded to maintain constant word size or allowed to grow.

Welch [5] has shown that rounding after each two-point transform leads to a relative rms output error ε, which is bounded above by

$$\varepsilon = (0.3)2^{-B+(M+3)/2}/\text{rms (input)} \tag{7}$$

for a transform of 2^M samples using B-bit arithmetic. The error may be reduced by allowing words to lengthen from stage to stage. This can be particularly attractive in a cascade machine because most of the storage in such a machine may be associated with the first few stages of the transform.

Alternate Configurations

The computing module circuitry may be configured in a number of different ways to meet different requirements. The arithmetic unit that computes two-point transforms may employ either four real multiplications and six real additions or three real multiplications and nine real additions. The multiplications may be performed either before data are stored in the shift register or after data emerge from the shift register. Goldstone [6] has shown that multipliers may be time-shared between adjacent modules by performing half of the required multiplications as data enter the shift register and the other half as data leave. Some of the total shift register delay may be incorporated in a pipeline multiplier for high-speed operation.

In order to make use of the speed possible in a pipeline processor, one word delays must be inserted between the computing modules. These delays permit each module to begin computation at the start of a word time rather than to wait for the preceding modules to compute the input it requires. These intermodule delays do not appreciably complicate the control circuitry of the processor, since they may be compensated by delaying the control and rotation vector inputs to the module by a delay equal to the total data delay.

A cascade processor requires no multipliers in the first two modules. The rotation vectors (W_2^0 and W_4^1) used in these modules may be implemented by, at most, a switching circuit to interchange the real and imaginary part of the data word and invert the sign of the resulting real part when rotation by W_4^1 is required. The multipliers may be implemented if desired and used for data windowing.

Data windowing, or multiplying input data samples by a data window function, is a technique used to change the frequency response of the equivalent N filters whose output is computed by the FFT. If an FFT of N samples taken at equal intervals T is computed, a data window that is exactly zero everywhere except over sampling interval of duration NT must be used. If no data windowing is intentionally performed, input data samples are weighted by a data window function that is unity over the sampling interval and zero elsewhere.

The frequency response of the equivalent FFT filters is given by the discrete Fourier transform of the data window function. The frequency response of an FFT filter when no data windowing is performed is therefore the DFT of a unit-amplitude pulse of duration NT, which (normalized) is

$$H(f) = \frac{\sin \pi NTf}{N \sin \pi Tf} \quad (8)$$

The relatively slow decrease in amplitude with increasing frequency of this function makes it undesirable for Fourier analyzer applications, in which data windowing is commonly used. Frequently used data window functions include the Hamming [7], Hanning [7], Dolph–Tchebyscheff [8], and Taylor [9] functions. The properties of these functions have been extensively described.

Implementation

The pipeline structure was used to implement a real-time FFT machine to provide spectral analysis for a tracking pulse-Doppler radar. This machine simultaneously processes eight channels of Doppler data at a sample rate of 16 000 complex samples per second per channel with a transform length of 512 complex samples and a quantization of 24 bits per complex sample. The FFT throughput rate is slightly over three million bits per second. A Taylor data window is employed to permit 60-dB subclutter visibility of high-speed targets.

The machine is composed of nine processing modules, a synchronizer, A–D converter, and display. The processing modules employ a total of 2600 TTL integrated circuits for switching and arithmetic functions and a total of 500 200-bit MOSFET shift registers for delay. Words are stored in bit-serial form in the shift registers, but arithmetic operations are performed in bit-parallel. Interstage delay is incorporated in the arithmetic units. Data windowing is performed by the "spare" multiplier in the first processing module. An eight complex sample input buffer following the A–D converter allows the FFT machine to operate at a constant word rate equal to eight times the 8–16 kHz radar PRF, independent of the ranges at which samples are collected. Fig. 5 shows a photograph of the complete breadboard FFT cascade processor.

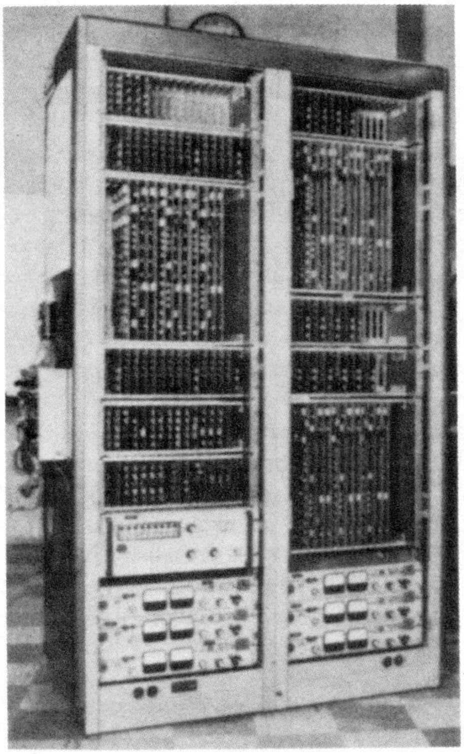

Fig. 5. Pipeline processor machine.

Conclusions

The pipeline FFT processor has proven to be an effective tool to meet the real-time spectral analysis requirements of many radar and sonar systems. Its structure permits sufficient paralleling of operations, such that the processing time is limited solely by the time it takes to collect the data. It is efficient in storage requirement since it requires only as much storage as the number of samples to be processed. It is simple to control because basically all of the control information can be generated by a binary counter. Furthermore the use of this counter guarantees proper synchronization of the control function with the data stream passing through the device.

Perhaps its greatest disadvantage is the scrambled order in which the output appears. In the radar/sonar applications, for which the device was designed, this fault was transparent. It does, however, make certain operations in the frequency domain, such as smoothing over frequency, somewhat more difficult. Descrambling is possible in a pipeline processor at the cost of additional memory less than the number of words in the processed data block.

The remarkable flexibility with which the structure may be reconfigured to trade channels for data sample length and to carry out the inverse transform functions is one of its most satisfying properties. Properly utilized, the device is able to carry out auto- and cross-correlation, block convolutions, and cross-spectral density calculations.

References

[1] G. D. Bergland and H. W. Hale, "Digital real-time spectral analysis," *IEEE Trans. Electronic Computers*, vol. EC-16, pp. 180–185, April 1967.

[2] G. D. Bergland, "Fast Fourier transform hardware implementations—An overview," *IEEE Trans. Audio Electroacoust.*, vol. AU-17, pp. 104–108, June 1969.

[3] ——, "A guided tour of the fast Fourier transform," *IEEE Spectrum*, vol. 6, pp. 41–52, July 1969.

[4] B. Widrow, "Statistical analysis of amplitude-quantized sampled-data systems," *AIEE Trans.*, vol. 79, pt. 2, pp. 555–567, January 1961.

[5] P. D. Welch, "A fixed-point fast Fourier transform error analysis," *IEEE Trans. Audio Electroacoust.*, vol. AU-17, pp. 151–157, June 1969.

[6] B. J. Goldstone, "Serial FFT—More efficient utilization of the multiplier," Raytheon Co. internal memo BFX-R-29, October 1968.

[7] R. B. Blackman and J. W. Tukey, *The Measurement of Power Spectra*. New York: Dover, 1958.

[8] C. L. Dolph, "A current distribution for broadside arrays which optimizes the relationship between beam width and side-lobe level," *Proc. IRE*, vol. 34, pp. 335–348, June 1946.

[9] T. T. Taylor, "Design of line-source antennas for narrow beamwidth and low sidelobes," *IRE Trans. Antennas Propag.*, vol. AP-3, pp. 16–28, January 1955.

[10] G. O'Leary, "A high-speed cascade fast Fourier transformer," presented at IEEE Arden House Workshop on Digital Filtering, January 1970.

[11] W. R. Graham, "The parallel, pipeline and conventional computer," *Datamation*, vol. 16, pp. 68–71, April 1970.

Part 3
Effects of Finite Word Lengths

Effect of Finite Word Length on the Accuracy of Digital Filters—A Review

BEDE LIU, MEMBER, IEEE

Invited Paper

Abstract—The accuracy of a digital filter is limited by the finite word length used in its implementation. Techniques have been developed to analyze this problem. Good agreement between the theoretical and experimental results has been reported. This paper discusses some of these accuracy problems and reviews some of the approaches used in investigating them. The calculation of the statistical mean-squared error at the output of the filter is discussed in detail.

I. Introduction

A DIGITAL filter operates on an input-sampled signal to produce an output-sampled signal by means of a computational algorithm. It can be simulated on a general-purpose computer or can be constructed with special-purpose digital hardware. These filters have found important applications in an increasing number of fields in science and engineering, and design techniques have been developed to achieve desired filter characteristics.

Despite the many advantages offered by digital filters, there is an inherent limitation on the accuracy of these filters due to the fact that all digital networks operate with only a finite number of bits. This paper attempts to present a survey and discussion of the effect of finite word length on the accuracy of digital filters.

The class of digital filters considered is specified by the linear constant-coefficient difference equation

$$w_n = \sum_{k=0}^{M} b_k x_{n-k} - \sum_{k=1}^{L} a_k w_{n-k} \tag{1}$$

where $\{x_n\}$ is the input sequence and $\{w_n\}$ is the output sequence. The three common sources of error due to finite word length are[1]

1) the quantization of the input signal $\{x_n\}$ into a set of discrete levels;
2) the representation of the filter coefficients a_k and b_k by a finite number of bits;
3) the accumulation of roundoff errors committed at arithmetic operations.

Manuscript received March 8, 1971; revised May 19, 1971 and July 1971. This work was supported by the National Science Foundation under Grant GK-1439 and by the Air Force Office of Scientific Research, Office of Aerospace Research, USAF, under Grant AFOSR-1333-67.

The author is with the Department of Electrical Engineering, Princeton University, Princeton, N. J. 08540.

[1] If a digital filter is used to process analog signals, there may be additional sources of error as a result of sampling of the input analog signal and the reconstruction of the output analog signal. See [1] and [2] for a discussion of these errors.

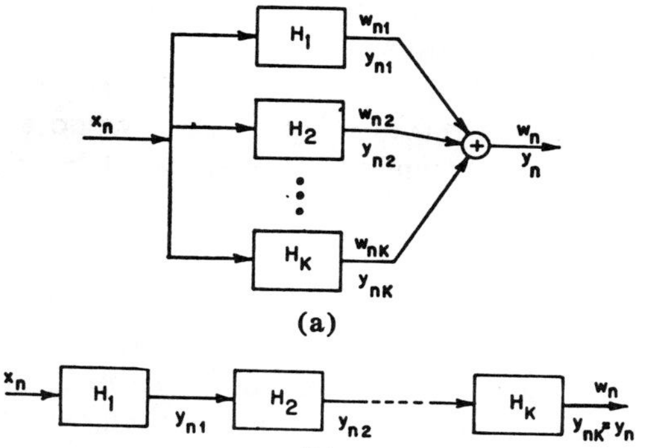

Fig. 1.

Because of these errors the actual output will be $\{y_n\}$ and will be, in general, different from the ideal output $\{w_n\}$. One may define

$$e_n = y_n - w_n \tag{2}$$

as the error at the output at the nth sample, and it is important for the designer and the user of a digital filter to be able to determine some measure of the error e_n.

In addition to the word length, the accuracy of a digital filter depends on two important factors: the form of realization and the type of arithmetic used.

There are three basic forms for realizing a digital filter [3].[2] If the output is calculated by using (1), the filter is said to be realized in the *direct form*. If the transfer function of the filter

$$H(z) = \frac{\sum_{k=0}^{M} b_k z^{-k}}{1 + \sum_{k=1}^{L} a_k z^{-k}} \tag{3}$$

is written as a partial fraction of first- and second-order terms, $H(z) = \sum_{i=1}^{K} H_i(z)$, the entire filter may be visualized as a parallel connection of the simpler filter $H_i(z)$ [Fig. 1(a)]. In this case, the filter is said to be realized in the *parallel form*. Similarly, $H(z)$ may be written as a product of first- and second-order factors, $H(z) = \prod_{i=1}^{K} H_i(z)$, and the filter may be visualized as a cascade of simpler filters, as shown in Fig.

[2] There is a fourth realization form, called the *canonical* form by Gold and Rader [4], which is regarded by Kaiser [3] as another way of realizing the direct form. It will not be discussed in this paper.

Reprinted from *IEEE Trans. Circuit Theory*, vol. CT-18, pp. 670–677, Nov. 1971.

1(b). The original filter is then said to be realized in the *cascade form*.

We shall study number representation by a finite number of bits in Section II, and discuss the effects of the three sources of error in Sections III, IV, and V.

II. Representation of Numbers with Fixed Number of Bits

A real number can be represented using a finite number of bits in either the fixed-point form or the floating-point form [5], [6]. The error introduced in such a representation is discussed in this section. Only binary arithmetic will be discussed. Consider first the fixed-point case. Suppose a number v which has been normalized so that $|v| \leq 1$ has the binary expansion[3]

$$v = -v_0 + \sum_{k=1}^{\infty} v_k 2^{-k}, \qquad v_k = 1 \text{ or } 0. \tag{4}$$

To approximate v by a "word" of only t bits, *rounding* or *chopping* is used. In rounding, a 1 or a 0 is first added to the tth bit v_{t-1} according to whether the $(t+1)$th bit v_t is 1 or 0. Then, only the first t bits of the result are kept. In chopping, those bits beyond the most significant t bits are simply dropped. Since the error introduced by chopping is more serious than that introduced by rounding because of a bias, chopping arithmetic is not commonly used. We shall only discuss rounding arithmetic in this paper, although modification for chopping arithmetic is rather straightforward in most cases.

Let $(v)_t$ be the t-bit representation of the number v. It is then clear that

$$-2^{-t} < v - (v)_t \leq 2^{-t} \tag{5}$$

if rounding is used. An error of approximation ϵ may be defined by

$$(v)_t = v + \epsilon \tag{6}$$

with $-2^{-t} \leq \epsilon < 2^{-t}$. The approximation of v by $(v)_t$ is identical to the quantization of the number v by a quantizer with uniform step size $q = 2^{-t+1}$, and the error ϵ is referred to as the "quantizing noise" [7].

When two t-bit fixed-point numbers are added, their sum would still have t bits, *provided* there is no overflow. Therefore, under this assumption of no overflow, fixed-point addition causes no error. On the other hand, the product of two t-bit numbers may have more than t bits. Thus rounding is needed if only t bits are to be kept. Let the actual computed product of two numbers v_1 and v_2 be denoted by $(v_1 v_2)_t$; then it is clear from the above discussion that

$$(v_1 v_2)_t = v_1 v_2 + \epsilon \tag{7}$$

where the error ϵ is bounded by $-2^{-t} \leq \epsilon < 2^{-t}$.

To study the accuracy of digital filters using the above results, the approach may be taken to regard the errors introduced by the quantization as small random quantities, statistically independent of one another and with a uniform probability distribution in the range $(-2^{-t}, 2^{-t})$ [7]–[9]. This viewpoint simplifies the analysis and enables useful theoretical results to be derived. The validity of such an approach lies ultimately in the experimental verification of these theoretical results. Computer simulation tends to confirm this statistical approach when the number of quantization levels, or equivalently the word length, is not too small.[4]

A floating-point number is written in the form (sgn) $2^a \cdot b$, where a is a binary integer called the exponent and b is a fraction between $\frac{1}{2}$ and 1 called the mantissa. The number of bits of the exponent determines the range of numbers that can be so represented, and the mantissa can usually take on the value 0. To represent a number v in the floating-point form with only a t-bit mantissa,[5] one first determines the smallest integer exceeding $\log_2|v|$, denoted by $[\log_2|v|]$. The binary expansion of the fraction $v/[\log_2|v|]$ is then rounded to t bits. Let $(v)_t$ denote the t-bit mantissa floating-point approximation of the number v; then it is clear that

$$(v)_t = v(1 + \epsilon) \tag{8}$$

where the relative error ϵ is bounded by $-2^{-t} \leq \epsilon < 2^{-t}$.

Unlike the fixed-point case, both addition and multiplication in floating-point can introduce roundoff error. Let $(v_1 + v_2)_t$ and $(v_1 \cdot v_2)_t$ denote, respectively, the actual computed sum and product of the two numbers v_1 and v_2; then [5], [6]

$$(v_1 + v_2)_t = (v_1 + v_2)(1 + \epsilon)$$
$$(v_1 \cdot v_2)_t = (v_1 \cdot v_2)(1 + \delta) \tag{9}$$

where the relative errors ϵ and δ are bounded by $-2^{-t} \leq \epsilon < 2^{-t}$, and $-2^{-t} \leq \delta < 2^{-t}$.

As in the fixed-point case, the approach has been taken to regard these errors as random quantities in analyzing the accuracy of floating-point digital filters. ϵ and δ are assumed to be uniformly distributed in their range $(-2^{-t}, 2^{-t})$. Again, the validity of the statistical approach lies with the experimental confirmation, and computer-simulation results suggest that the approach is justified when the word length is not too short.[4]

III. Effect of Input Quantization

The first place where quantization of the input signal may take place is at the analog-to-digital converter. Although in principle the input may be subject to a further quantization if the word length used in the digital filter is shorter than that of the quantizer at the analog-to-digital converter, this is seldom the case.

Suppose the quantizer has equal step size Δ; then the input to the actual filter is $\{x_n + e_n^Q\}$ where each e_n^Q is bounded by

[3] This is essentially the 2's complement representation.

[4] On talking with workers experienced in this field, one finds that the model is quite satisfactory at least down to eight bits.

[5] It should be noted that for fixed-point numbers t is the entire word length, and for floating-point numbers t is only the length of the mantissa.

$-(\Delta/2) \leq e_n^Q < (\Delta/2)$. Since the filter is linear, the output is the sum of two components, one due to the signal $\{x_n\}$ and the other to $\{e_n^Q\}$. The usual approach for treating the effect of input quantization is to regard $\{e_n^Q\}$ as white noise [7]–[11]. That is, it has zero mean and variance $\Delta^2/12$. Therefore, the steady-state output component due to $\{e_n^Q\}$ is a zero-mean wide-sense-stationary (w.s.s.) sequence with power spectral density given by

$$H(z)H\left(\frac{1}{z}\right)\frac{\Delta^2}{12} \tag{10}$$

where $H(z)$ is the transfer function of the filter [see (3)]. Here the effect on the output of coefficient inaccuracy and round-off accumulation has been ignored, since their effect on the response to $\{e_n^Q\}$ is much smaller than that due to the response to $\{x_n\}$.

To relate Δ to the word length of the digital filter, scaling of the input may need to be considered. For example, if the input has been scaled such that $|x_n| \leq 1$ and quantization is at the input of a fixed-point filter with a t-bit quantizer, then $\Delta = 2^{-t+1}$. Scaling is usually not important in floating-point filters. When it is used, the input signal spectrum and Δ^2 are scaled by the same factor.

The mean-squared value of the error at the output due to input quantization can be obtained by integrating the power spectral density given by (10). It is equal to[6]

$$\frac{1}{2\pi j}\oint H(z)H\left(\frac{1}{z}\right)\frac{\Delta^2}{12}\frac{dz}{z} \tag{11}$$

and can be evaluated, either numerically or algebraically, by a computer program or a table [12].

One can also bound the output component due to input quantization. It is easily seen that the output due to $\{e_n^Q\}$ is bounded in absolute value by $\sum_n |h_n| \Delta/2$, where $\{h_n\}$ is the impulse response of the filter. Although this bound can be approached with a particular input sequence, it is extremely unlikely for the $\{e_n^Q\}$ to take on these values.

IV. Effect of Coefficient Inaccuracy

As a result of the finite word length used in a digital filter, each coefficient is replaced by its t-bit representation according to (6) or (8) depending on whether fixed-point or floating-point arithmetic is used. That is, the coefficient a_k is replaced by $(a_k)_t$, which equals $a_k + \alpha_k$ for fixed-point or $a_k(1+\alpha_k)$ for floating-point, with α_k bounded in absolute value by 2^{-t}. Similarly, each b_k is replaced by $(b_k)_t$, which is either $b_k + \beta_k$ or $b_k(1+\beta_k)$. Therefore, the filter characteristics are changed. This problem can be approached in a number of ways. First, one can simply compute the frequency response of the actual filter with t-bit rounded coefficients, that is, by using the actual transfer function

$$[H(z)]_t = \frac{\sum_{k=0}^{M} (b_k)_t z^{-k}}{1 + \sum_{k=1}^{L} (a_k)_t z^{-k}}. \tag{12}$$

The result can then be compared with the ideal response for the original design. For a certain bandstop filter, calculations [13] show that in addition to a greatly increased transition-region width between stop- and passbands, the minimum in-band rejection deteriorates from 75 to less than 50 dB when the word length is reduced from 40 to 12 bits.

One can also calculate the movements of the poles and zeros of the transfer function due to coefficient rounding and then apply network sensitivity theory to study the changes in the filter response [3], [4]. Suppose the poles of $H(z)$ are at z_i, $i = 1, 2, \cdots, L$, and the poles of $[H(z)]_t$ are at $z_i + \Delta z_i$. Then it can be shown readily that

$$\Delta z_i = \sum_{k=1}^{L} \frac{z_i^{k+1}}{\prod_{\substack{m=1\\ m\neq i}}^{L}\left(1 - \frac{z_i}{z_m}\right)} \Delta a_k \tag{13}$$

where Δa_k is the change in the coefficient a_k which is either α_k or $a_k\alpha_k$. Similar results can be obtained for the movement of the zeros. From these movements, the change in the overall filter response can be studied.

If a single number as a measure of the change is desired, an integrated squared deviation of the frequency response may be used such as

$$\frac{1}{2\pi j}\oint \left| H(z) - [H(z)]_t \right|^2 \frac{dz}{z} \tag{14}$$

where $H(z)$ is the transfer function given by (3) and $[H(z)]_t$ is the transfer function when each coefficient is replaced by its t-bit approximation [see (12)]. The meaning of such a measure is not always clear, since sometimes one may be mostly interested in a limited frequency range. By regarding the filter coefficient errors α_k and β_k as independent random variables, Knowles and Olcayto [13] have calculated the statistical average of the integrated squared frequency response error as defined by (14). However, since the error in each coefficient is fixed throughout the operation of the filter, the validity of the assumption of random coefficient error can be questioned when the order of the filter is low.

When the filter has high Q poles, that is, poles that are close to the unit circle in the z plane, instability may occur as a result of coefficient error. This problem can be quite serious when the sampling rate is relatively high, even for moderately low-order filters simulated on the computer in the direct form [14]. Kaiser [15] has shown that for an Lth-order low-pass filter operating at a sampling rate of $1/T$ with distinct poles at $e^{-p_k T}$, stability is guaranteed if the number

[6] Unless otherwise stated, contour integration in the z plane is taken around $|z| = 1$ counterclockwise.

of bits used m_b satisfies the inequality

$$m_b > \left[-\log_2 \left(\frac{5\sqrt{L}}{2^{L+2}} \prod_{k=1}^{L} p_k T \right) \right] \tag{15}$$

where the bracket denotes the smallest integer exceeding the quantity inside. It is possible to extend the result to include multiple poles and to derive similar results for filters of other than the low-pass type. Otnes and McNamee [16] have related the minimum number of bits needed to insure stability directly to the filter bandwidth for specific filters. Mantey [17] has studied a related problem of state-variable selection in digital systems and eigenvalue sensitivity and has applied the results to digital filters.

In general, the effect of coefficient inaccuracy is more pronounced for a high-order filter when it is realized in the direct form than when it is realized in the parallel or cascade form. As a rule, therefore, the parallel or cascade form should be used for high-order filters whenever possible. The saving in the number of coefficient bits can be quite substantial.

Since the filter characteristics are altered as a result of coefficient error, the output sequence will no longer be the ideal one specified by (1). It is possible to study the error at the output of the filter due to coefficient rounding. This is done in the next section in conjunction with roundoff accumulation.

V. Effect of Roundoff Accumulation

A. Fixed-Point Filters

We shall consider first the direct form of realization and then use the result to treat parallel and cascade forms. The development here follows largely that of Knowles and Edwards [10].

Direct Form: From (1) and (7) it is seen that the actual output sequence $\{y_n\}$ is given by

$$y_n = \sum_{k=0}^{M} (b_k)_t x_{n-k} - \sum_{k=1}^{L} (a_k)_t y_{n-k} + \epsilon_n \tag{16}$$

where $(a_k)_t$ and $(b_k)_t$ are t-bit fixed-point representations of the coefficients a_k and b_k, and ϵ_n denotes the roundoff error in the calculation of y_n. From (6) and the discussion of Section II, we have $(a_k)_t = a_k + \alpha_k$ and $(b_k)_t = b_k + \beta_k$ where α_k and β_k are the coefficient errors.

The error of the nth sample of the output is given by the difference between the actual output y_n and the ideal output w_n [see (2)]. From (1), (2), and (16) we have

$$e_n = -\sum_{k=1}^{L} a_k e_{n-k} + u_n \tag{17}$$

where

$$u_n = \sum_{k=0}^{M} \beta_k x_{n-k} - \sum_{k=1}^{L} \alpha_k w_{n-k} + \epsilon_n. \tag{18}$$

Suppose $\{x_n\}$ is zero mean and w.s.s. with autocorrelation function $R_{xx}(n)$ and power spectral density $\Phi_{xx}(z)$. Then $\{w_n\}$ is zero mean and w.s.s. with power spectral density $\Phi_{ww}(z)$ given by

$$\Phi_{ww}(z) = H(z) H\left(\frac{1}{z}\right) \Phi_{xx}(z). \tag{19}$$

It can be shown that $\{u_n\}$ is also zero mean and w.s.s. with the autocorrelation function given by

$$\Phi_{uu}(z) = [B(z) - H(z)A(z)] \left[B\left(\frac{1}{z}\right) - H\left(\frac{1}{z}\right) A\left(\frac{1}{z}\right) \right] \Phi_{xx}(z) + q^2(\mu + \nu) \tag{20}$$

where

$$A(z) = \sum_{k=1}^{L} \alpha_k z^{-k} \quad \text{and} \quad B(z) = \sum_{k=0}^{M} \beta_k z^{-k}. \tag{21}$$

$q^2 = 2^{-2t}/3$ is the variance of a random variable uniformly distributed in the interval $(-2^{-t}, 2^{-t})$ and μ and ν are, respectively, the number of b_k and a_k that are neither 1 nor 0. For simplicity, they may be taken to be $(M+1)$ and L, respectively. The error $\{e_n\}$ is zero mean and w.s.s. with

$$\Phi_{ee}(z) = \frac{1}{D(z)D(1/z)} \Phi_{uu}(z) = \frac{C(z)C(1/z)}{D(z)D(1/z)} \Phi_{xx}(z) + \frac{(\mu + \nu)q^2}{D(z)D(1/z)} \tag{22}$$

where $D(z)$ is the denominator of the transfer function given by

$$D(z) = 1 + \sum_{k=1}^{L} a_k z^{-k}$$

and

$$C(z) = B(z) - H(z)A(z). \tag{23}$$

The mean-squared value of e_n is then

$$E\{e_n^2\} = \frac{1}{2\pi j} \oint \Phi_{ee}(z) \frac{dz}{z}. \tag{24}$$

Suppose there is no coefficient rounding error; then the first term in (20) and (22) is absent and the result is that given by Knowles and Edwards [10], as to be expected. Suppose there is no roundoff error; then the second term of (22) is absent. Thus we see that the error at the output of the filter consists of two components; one is due to roundoff accumulation and the other to the rounding of the coefficients to t bits. The component due to roundoff accumulation is uncorrelated with both the input $\{x_n\}$ and the ideal output $\{w_n\}$. From (1), (17), and (18) we can arrive at the block diagram shown in Fig. 2, which will facilitate our discussion of the parallel and cascade realization forms. It is interesting to note that (22) can be written down almost by inspection of Fig. 2.

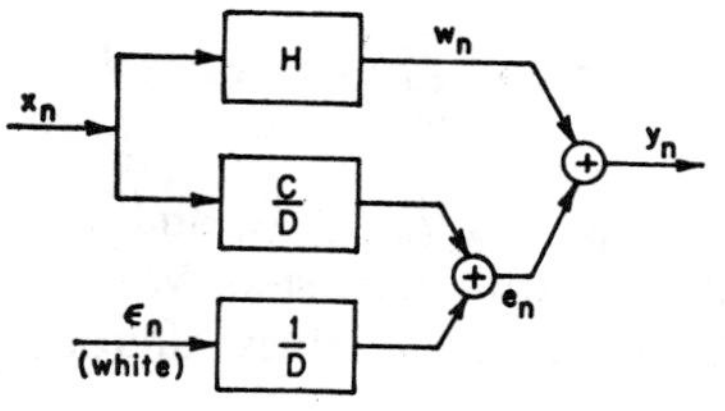

Fig. 2.

Parallel Form: For the parallel form of filter realization, $H(z)$ is written as

$$H(z) = \sum_{i=1}^{K} H_i(z) \tag{25}$$

where

$$H_i(z) = \frac{N_i(z)}{D_i(z)} = \frac{b_{0i} + b_{1i}/z}{1 + a_{1i}/z + a_{2i}/z^2}. \tag{26}$$

Equation (26) includes the possibility of a real pole or constant by setting $a_{2i}=b_{1i}=0$ or $a_{1i}=a_{2i}=b_{1i}=0$. The parallel form of implementation is shown in Fig. 1(a) where K intermediate outputs $\{w_{ni}\}$ $i=1, 2, \cdots, K$, are calculated from $\{x_n\}$ and then summed to form the final output $\{w_n\}$.

Suppose the actual coefficients for the ith branch are $(b_{0i})_t$, $(b_{1i})_t$, $(a_{1i})_t$, and $(a_{2i})_t$, which according to (6) are related to the ideal coefficients by $(b_{0i})_t=b_{0i}+\beta_{0i}$, $(b_{1i})_t=b_{1i}+\beta_{1i}$, $(a_{1i})_t=a_{1i}+\alpha_{1i}$, and $(a_{2i})_t=a_{2i}+\alpha_{2i}$. Let $\{y_{ni}\}$ be the actual output of the ith branch and $\{e_{ni}\}$ the error:

$$e_{ni} = y_{ni} - w_{ni}. \tag{27}$$

By using Fig. 2, we can draw a block diagram as shown in Fig. 3, from which one quickly arrives at an expression for the power spectral density of the output error $\{e_n\}$:

$$\Phi_{ee}(z) = \Phi_{xx}(z)\left[\sum_{i=1}^{K}\frac{C_i(z)}{D_i(z)}\right]\left[\sum_{i=1}^{K}\frac{C_i(1/z)}{D_i(1/z)}\right] + q^2\sum_{i=1}^{K}\frac{\mu_i+\nu_i}{D_i(z)D_i(1/z)} \tag{28}$$

where

$$C_i(z) = B_i(z) - H_i(z)A_i(z)$$
$$B_i(z) = \beta_{0i} + \beta_{1i}z^{-1}$$
$$A_i(z) = \alpha_{1i}z^{-1} + \alpha_{2i}z^{-2}. \tag{29}$$

μ_i and ν_i may both be taken as 2. The mean-squared value of e_n can be computed by using (24) and (28).

Cascade Form: To realize the digital filter in the cascade form, $H(z)$ is written as

$$H(z) = c\prod_{i=1}^{K} H_i(z) \tag{30}$$

where c is a constant which shall be taken as 1 for simplicity, and

$$H_i(z) = \frac{N_i(z)}{D_i(z)} = \frac{1 + b_{1i}/z + b_{2i}/z^2}{1 + a_{1i}/z + a_{2i}/z^2} \tag{31}$$

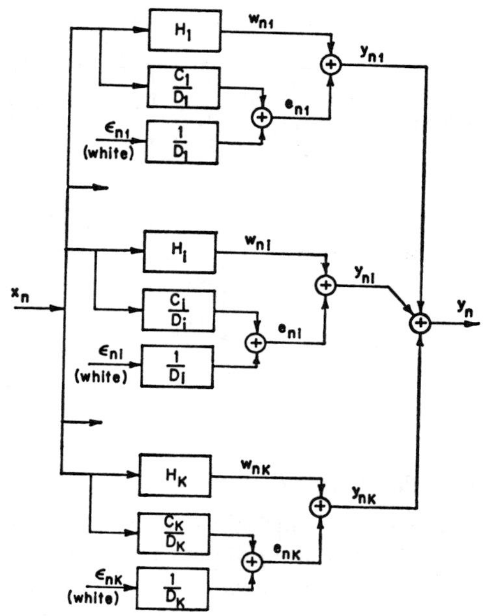

Fig. 3.

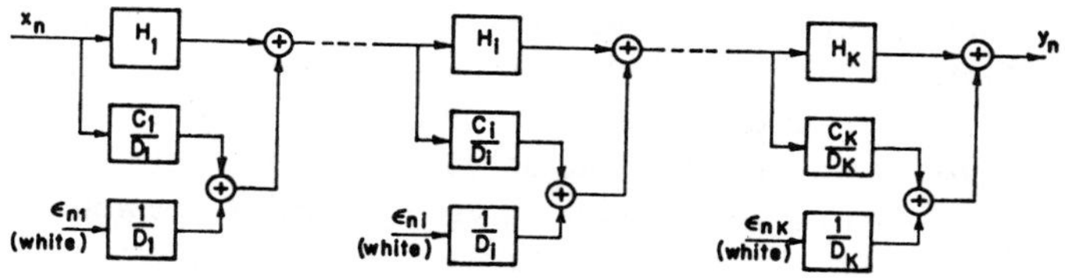

Fig. 4.

Notice that the numerator $N_i(z)$ is different from that in (26). Suppose $(b_{1i})_t$, $(b_{2i})_t$, $(a_{1i})_t$, and $(a_{2i})_t$ are the actual coefficients. Again we have $(b_{1i})_t=b_{1i}+\beta_{1i}$, $(b_{2i})_t=b_{2i}+\beta_{2i}$, $(a_{1i})_t=a_{1i}+\alpha_{1i}$, and $(a_{2i})_t=a_{2i}+\alpha_{2i}$. By using Fig. 2, we arrive at the block diagram shown in Fig. 4 where

$$C_i(z) = B_i(z) - H_i(z)A_i(z)$$
$$B_i(z) = \beta_{1i}z^{-1} + \beta_{2i}z^{-2}$$
$$A_i(z) = \alpha_{1i}z^{-1} + \alpha_{2i}z^{-2}. \tag{32}$$

The power spectral density of the actual output $\{y_n\}$ can be determined easily from Fig. 4 by neglecting terms involving fourth or higher powers of q. From the expression so obtained, the power spectral density of the ideal output $\{w_n\}$ is subtracted. The remaining part is the power spectral density of the error $\{e_n\}$. The result is

$$\Phi_{ee}(z) = \Phi_{xx}(z)\sum_{i=1}^{K}\frac{C_i(z)C_i(1/z)}{D_i(z)D_i(1/z)}\prod_{\substack{j=1\\ j\neq i}}^{K} H_j(z)H_j(1/z) + q^2\left[\frac{\mu_K+\nu_K}{D_K(z)D_K(1/z)} + \sum_{i=1}^{K-1}\frac{\mu_i+\nu_i}{D_i(z)D_i(1/z)}\prod_{j=i+1}^{K} H_j(z)H_j(1/z)\right]. \tag{33}$$

Both μ_i and ν_i can be taken as 2. The mean-squared value of e_n can be computed by using (24) and (33).

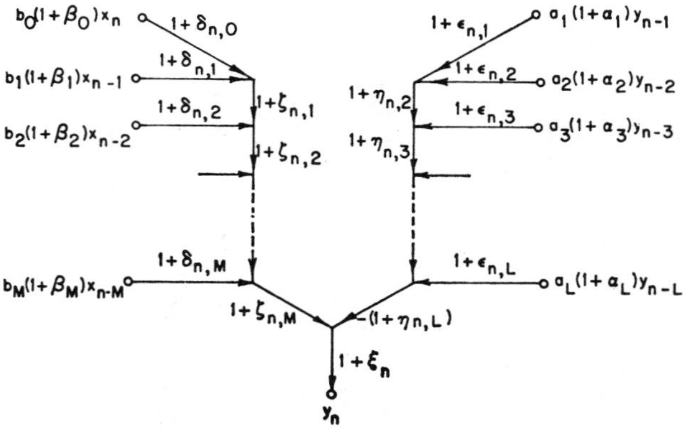

Fig. 5.

B. Floating-Point Filters

The treatment of roundoff accumulation for floating-point filters [18]–[21] is quite different from that for fixed-point filters, since the errors introduced are relative according to (8) and (9). This problem was first analyzed by Sandberg [18] who derived bounds on the time average of the squared error at the output of the filter. We shall discuss below the calculation of statistical mean-squared error at the output. Only the direct form of realization will be considered, since the extension to treat the other forms is straightforward.

According to (8), the actual filter coefficients are $a_k(1+\alpha_k)$ and $b_k(1+\beta_k)$ where α_k and β_k are bounded in absolute value by 2^{-t}. The actual computed output y_n is given by

$$y_n = \mathrm{fl}\left[\sum_{k=0}^{M} b_k(1+\beta_k)x_{n-k} - \sum_{k=1}^{L} a_k(1+\alpha_k)y_{n-k}\right] \qquad (34)$$

where fl[ ] denotes the actual computed result by floating-point arithmetic of the quantity inside the bracket. We assume the computation of (34) is carried out in the following order: the products $b_k(1+\beta_k)x_{n-k}$ and $a_k(1+\alpha_k)y_{n-k}$ are first formed; the two sums are then calculated; and finally the difference is taken to give y_n. Each of these arithmetic operations introduces a roundoff error which may be characterized by (9). Following Sandberg [18], we may draw the flowgraph shown in Fig. 5 which includes all the roundoff error in the calculation of y_n. For example, $\delta_{n,k}$ is introduced when the product of $b_k(1+\beta_k)x_{n-k}$ is formed, and $\zeta_{n,1}$ is introduced when the computed products of $b_0(1+\beta_0)x_n$ and $b_1(1+\beta_1)x_{n-1}$ are added. The actual output y_n is thus seen to be

$$y_n = \sum_{k=0}^{M} b_k(1+\beta_k)\theta_{n,k}x_{n-k} - \sum_{k=1}^{L} a_k(1+\alpha_k)\phi_{n,k}y_{n-k} \qquad (35)$$

where

$$\theta_{n,0} = (1+\xi_n)(1+\delta_{n,0})\prod_{i=1}^{M}(1+\zeta_{n,i})$$

$$\theta_{n,k} = (1+\xi_n)(1+\delta_{n,k})\prod_{i=k}^{M}(1+\zeta_{n,i}), \quad k=1,2,\cdots,M$$

$$\phi_{n,1} = (1+\xi_n)(1+\epsilon_{n,1})\prod_{i=2}^{L}(1+\eta_{n,i})$$

$$\phi_{n,k} = (1+\xi_n)(1+\epsilon_{n,k})\prod_{i=k}^{L}(1+\eta_{n,i}), \quad k=2,3,\cdots,L. \qquad (36)$$

The quantities ξ_n, $\epsilon_{n,k}$, $\delta_{n,k}$, $\zeta_{n,k}$, and $\eta_{n,k}$ are the errors introduced at each arithmetic step. We shall take them to be independent random variables uniformly distributed in $(-2^{-t}, 2^{-t})$.

By neglecting higher order terms, it can be shown from (1), (2), and (35) that the error $\{e_n\}$ satisfies the following equation:

$$\sum_{k=0}^{L} a_k e_{n-k} = u_n' + u_n'' \qquad (37)$$

where $a_0 = 1$ and

$$u_n' = \sum_{k=0}^{M} b_k\beta_k x_{n-k} - \sum_{k=1}^{L} a_k\alpha_k w_{n-k}$$

$$u_n'' = \sum_{k=0}^{M} b_k(\theta_{n,k}-1)x_{n-k} - \sum_{k=1}^{L} a_k(\phi_{n,k}-1)w_{n-k}. \qquad (38)$$

$\{u_n'\}$ is due to coefficient rounding, and $\{u_n''\}$ is due to roundoff accumulation. As before, we assume that the input $\{x_n\}$ is zero mean and w.s.s. It is seen that both components $\{u_n'\}$ and $\{u_n''\}$ have zero mean and are w.s.s., and they are uncorrelated, since $\theta_{n,k}$ and $\phi_{n,k}$ have mean equal to 1 and are independent of $\{x_n\}$ and $\{w_n\}$.

Therefore from (37), the error sequence $\{e_n\}$ is also zero mean and w.s.s. Its power spectral density is related to those of $\{u_n'\}$ and $\{u_n''\}$ through

$$\Phi_{ee}(z) = \frac{1}{D(z)D(1/z)}\left[\Phi_{u'u'}(z) + \Phi_{u''u''}(z)\right]. \qquad (39)$$

$\Phi_{u'u'}(z)$ can be calculated from (38). It is given by

$$\Phi_{u'u'}(z) = [B(z) - H(z)A(z)][B(1/z) - H(1/z)A(1/z)]\Phi_{xx}(z) \qquad (40)$$

where $H(z)$ is given by (3) and

$$A(z) = \sum_{k=1}^{L} a_k\alpha_k z^{-k}$$

$$B(z) = \sum_{k=0}^{M} b_k\beta_k z^{-k}. \qquad (41)$$

From (38) and (36) it can be shown that the sequence $\{u_n''\}$ is "white" with power spectral density given by

$$\Phi_{u''u''}(z) = \frac{q^2}{2\pi j}\oint\Bigg(F(z) + G(z)H(z)H\left(\frac{1}{z}\right) - N\left(\frac{1}{z}\right)[D(z)-1]H(z) - N(z)\left[D\left(\frac{1}{z}\right)-1\right]H\left(\frac{1}{z}\right)\Bigg)\Phi_{xx}(z)\frac{dz}{z} \qquad (42)$$

where $N(z)=\sum_{k=0}^{M} b_k z^{-k}$ is the numerator of the transfer function (3), and

$$F(z) = \sum_{k=0}^{M} \sum_{i=0}^{M} b_k b_i F_{k,i} z^{k-i}$$

$$G(z) = \sum_{k=1}^{L} \sum_{i=1}^{L} a_k a_i G_{k,i} z^{k-i}$$

$$F_{k,i} = \begin{cases} M + 2 - \max(k, i), & k \neq i \text{ or } k = i = 0 \\ M + 3 - k, & k = i \neq 0 \end{cases}$$

$$G_{k,i} = \begin{cases} L + 2 - \max(k, i), & k \neq i \text{ or } k = i = 1 \\ L + 3 - k, & k = i \neq 1. \end{cases} \tag{43}$$

Again, the mean-squared value of the error e_n can be calculated from $\Phi_{ee}(z)$ by using (24).

It is noted that in the absence of coefficient rounding, $\Phi_{u'u'}(z)=0$ and the result reduces to that given in [19] and [21] as expected. If there is no roundoff, then $\Phi_{u''u''}(z)=0$. Thus, the error $\{e_n\}$ consists of two uncorrelated components, one due to the coefficient inaccuracy and the other to the accumulation of roundoff errors.

The output-error spectrum depends on the input-signal spectrum, as can be seen from (40) and (42). It is possible to derive bounds on the mean-squared error that are independent of the input spectrum for filters with no zero on the unit circle $|z|=1$ [19], [21].

VI. Concluding Remarks

The accuracy of a digital filter is limited by the finite word length used in its implementation. For filters simulated on a general-purpose computer, the word length is usually fixed, and one is therefore interested in knowing whether a certain accuracy can be achieved. When a filter is constructed with digital hardware, one is also interested in determining the minimum word length needed for a specified performance accuracy.

These problems have been investigated extensively, and techniques have been developed to treat the effects of the three common sources of error: input quantization, coefficient inaccuracy, and roundoff accumulation. Good agreement between theory and the experimental results have been reported. We have briefly reviewed in this paper some of these problems and have discussed some of the analytical methods used in treating them. Particular emphasis has been placed in the calculation of the statistical mean-squared error at the output. It is seen that the error consists of three components, one due to each of the three sources of error.

Studies show that for a fixed word length, the accuracy achievable with a direct realization of a high-order filter is considerably less than that with either cascade or parallel realization of the same filter. Thus a number of studies have concentrated on first- and second-order filters which are the basic building blocks for higher order filters. The significance of analyzing the error for the direct realization form thus lies in the fact that the results obtained form the basis for the analysis of both the parallel and cascade forms of realization, as illustrated in Section V.

The calculation of the mean-squared error discussed in Section V is based on the assumption that the roundoff errors introduced are small and independent random variables. For floating-point filters, because of the moderately high number of bits commonly used, such an assumption is justified and the results have been confirmed by computer simulation. For fixed-point filters the validity of random multiplication error becomes somewhat questionable when the input signal is small, and it is important to examine the deterministic limit cycle behavior [see 2) below].

There are a number of important and interesting problems associated with the accuracy of digital filters that are not discussed in this paper. Among them, we mention the following.

1) Block floating-point arithmetic has been proposed for digital filter realization. The effect of roundoff accumulation for first- and second-order filters has been calculated and compared with filters using fixed- and floating-point arithmetics [22].

2) Sustained oscillations, or limit cycles, may appear at the output of the digital filter even when there is no applied input. This problem can usually be ignored in filters using floating-point arithmetic [18]. For fixed-point filters there are two distinct types of oscillation. The first is due to the overflow of registers with limited dynamic range. When the roundoff error is ignored, a second-order filter can be modeled by

$$y_n + f(a_1 y_{n-1} + a_2 y_{n-2}) = 0 \tag{44}$$

where for the 2's complement arithmetic, $f(u)$ is a periodic function, with $f(u)=u$ in $|u| \leq 1$, and repeats itself periodically beyond the interval $(-1, 1)$. The oscillation of such a model and the conditions on a_1 and a_2 to produce oscillation have been studied in detail. It can be shown [23] that by suitably modifying the function $f(u)$, oscillation can be eliminated. In particular, the use of a "saturation"-type arithmetic with $f(u)=\operatorname{sgn}(u)$ for $|u|>1$ would eliminate the oscillation. By adding an extra term λ_n to (44), this result can be generalized to include the effect of roundoff error and a possible further complication [24]. The second type of limit cycle is due to the rounding of multiplication. It has been referred to as the deadband effect and has received considerable treatment for first- and second-order filters [25]–[27].

3) An approach has been found to reduce the sensitivity with respect to coefficient inaccuracy for filters whose unit delay is too small relative to the reciprocal of the critical frequency [28]. A holding technique to reduce the number of coefficient bits has been reported [29].

4) The selection of specific configurations for digital-filter realization to achieve low-noise performance with a specified dynamic-range constraint has been studied [30]–[32].

5) The fast-Fourier transform (FFT) algorithm can be used for digital filtering [33], [34]. The error of FFT has been analyzed [35]–[38].

Acknowledgment

The author wishes to thank J. F. Kaiser, I. W. Sandberg, and an anonymous reviewer for many valuable suggestions.

References

[1] J. B. Thomas and B. Liu, "Error problems in sampling representations," *1964 IEEE Int. Conv. Rec.*, pt. 5, pp. 269–277.
[2] B. Liu and J. B. Thomas, "Error problems in the reconstruction of signals from sampled data," in *Proc. Nat. Electronics Conf.*, vol. 23, Oct. 1967, pp. 803–807.
[3] J. F. Kaiser, "Digital filters," in *System Analysis by Digital Computers*, F. F. Kuo and J. F. Kaiser, Eds. New York: Wiley, 1966, pp. 218–285.
[4] B. Gold and C. M. Rader, *Digital Processing of Signals.* New York: McGraw-Hill, 1969.
[5] J. H. Wilkinson, *Rounding Errors in Algebraic Processes.* Englewood Cliffs, N. J.: Prentice-Hall, 1963.
[6] G. Forsythe and C. B. Moler, *Computer Solution of Linear Algebraic Systems.* Englewood Cliffs, N. J.; Prentice-Hall, 1967.
[7] W. R. Bennett, "Spectra of quantized signals," *Bell Syst. Tech. J.*, vol. 27, July 1948, pp. 446–472.
[8] B. Widrow, "Statistical analysis of amplitude-quantized sampled-data systems," *AIEE Trans. Appl. Ind.*, vol. 79, Jan. 1961, pp. 555–568.
[9] J. Katzenelson, "On errors introduced by combined sampling and quantization," *IRE Trans. Automat. Contr.*, vol. AC-7, Apr. 1962, pp. 58–68. See also D. G. Watts and J. Katzenelson, "Discussion of 'On errors introduced by combined sampling and quantization,'" *IEEE Trans. Automat. Contr.* (Corresp.), vol. AC-8, Apr. 1963, pp. 187–188.
[10] J. B. Knowles and R. Edwards, "Effects of a finite-word-length computer in a sampled-data feedback system," *Proc. Inst. Elec. Eng.*, vol. 112, June 1965, pp. 1197–1207.
[11] B. Gold and C. M. Rader, "Effects of quantization noise in digital filters," in *Proc. AFIPS 1966 Spring Joint Computer Conf.*, vol. 28, 1966, pp. 213–219.
[12] K. J. Aström, E. I. Jury, and R. G. Agniel, "A numerical method for the evaluation of complex integrals," *IEEE Trans. Automat. Contr.* (Short Papers), vol. AC-13, Aug. 1970, pp. 468–471.
[13] J. B. Knowles and E. M. Olcayto, "Coefficient accuracy and digital filter response," *IEEE Trans. Circuit Theory*, vol. CT-15, Mar. 1968, pp. 31–41.
[14] C. S. Weaver *et al.*, "Digital filtering with applications to electrocardiogram processing," *IEEE Trans. Audio Electroacoust.*, vol. AU-16, Sept. 1968, pp. 350–391.
[15] J. F. Kaiser, "Some practical considerations in the realization of linear digital filters," in *Proc. 3rd Annu. Allerton Conf. Circuit and System Theory*, 1965, pp. 621–633.
[16] R. K. Otnes and L. P. McNamee, "Instability thresholds in digital filters due to coefficient rounding," *IEEE Trans. Audio Electroacoust.*, vol. AU-18, Dec. 1970, pp. 456–463.
[17] P. E. Mantey, "Eigenvalue sensitivity and state-variable selection," *IEEE Trans. Automat. Contr.*, vol. AC-13, June 1968, pp. 263–269.
[18] I. W. Sandberg, "Floating-point round-off accumulation in digital filter realization," *Bell Syst. Tech. J.*, vol. 46, Oct. 1967, pp. 1775–1791.
[19] T. Kaneko and B. Liu, "Round-off error of floating-point digital filters," in *Proc. 6th Annu. Allerton Conf. Circuit and System Theory*, Oct. 1968, pp. 219–227.
[20] C. Weinstein and A. V. Oppenheim, "A comparison of round-off noise in floating point and fixed point digital filter realizations," *Proc. IEEE* (Corresp.), vol. 57, June 1969, pp. 1181–1183.
[21] B. Liu and T. Kaneko, "Error analysis of digital filters realized with floating-point arithmetic," *Proc. IEEE*, vol. 57, Oct. 1969, pp. 1735–1747.
[22] A. V. Oppenheim, "Realization of digital filters using block-floating-point arithmetic," *IEEE Trans. Audio Electroacoust.*, vol. AU-18, June 1970, pp. 130–136.
[23] P. M. Ebert, J. E. Mazo, and M. G. Taylor, "Overflow oscillations in digital filters," *Bell Syst. Tech. J.*, vol. 48, Nov. 1968, pp. 3021–3030.
[24] I. W. Sandberg, "A theorem concerning limit cycles in digital filters," in *Proc. 7th Annu. Allerton Conf. Circuit and System Theory*, 1968, pp. 63–68.
[25] L. B. Jackson, "An analysis of limit cycles due to multiplication rounding in recursive digital filters," in *Proc. 7th Annu. Allerton Conf. Circuit and System Theory*, 1969, pp. 69–79.
[26] F. Bonzanigo, "Constant-input behavior of recursive digital filters," presented at IEEE Arden House Workshop in Digital Filtering, Harriman, N. Y., Jan, 1970.
[27] R. B. Blackman, *Linear Data-Smoothing and Prediction in Theory and Practice.* Reading, Mass.: Addison-Wesley, 1965, pp. 75–79.
[28] W. A. Gardner, "Reduction of sensitivity in sampled-data filters," *IEEE Trans. Circuit Theory* (Corresp.), vol. CT-17, Nov. 1970, pp. 660–663.
[29] R. M. Golden and S. A. White, "A holding technique to reduce number of bits in digital transfer functions," *IEEE Trans. Audio Electroacoust.*, vol. AU-16, Sept. 1968, pp. 433–436.
[30] R. Edwards, J. Bradley, and J. B. Knowles, "Comparison of noise performance of programming methods in the realization of digital filters," in *Proc. Symp. Computer Processing Communications*, PIB-MRI Symposia Series, vol. 19, 1969, pp. 295–311.
[31] L. B. Jackson, "On the interaction of roundoff noise and dynamic range in digital filters," *Bell Syst. Tech. J.*, vol. 49, Feb. 1970, pp. 159–184.
[32] ——, "Roundoff-noise analysis for fixed-point digital filters realized in cascade or parallel form," *IEEE Trans. Audio Electroacoust.*, vol. AU-18, June 1970, pp. 107–122.
[33] T. G. Stockham, "High speed convolution and correlation," in *Proc. AFIPS 1966 Spring Joint Computer Conf.*, vol. 28, 1966, pp. 229–233.
[34] H. D. Helms, "Fast Fourier transform method of computing difference equations and simulating filters," *IEEE Trans. Audio Electroacoust.*, vol. AU-15, June 1967, pp. 85–90.
[35] W. W. Gentleman and G. Sande, "Fast Fourier transform for fun and profit," in *Proc. AFIPS 1966 Fall Joint Computer Conf.*, 1966, vol. 29, pp. 563–578.
[36] P. D. Welch, "A fixed-point fast Fourier transform error analysis," *IEEE Trans. Audio Electroacoust.*, vol. AU-17, June 1969, pp. 151–157.
[37] C. J. Weinstein, "Roundoff noise in floating point fast Fourier transform computation," *IEEE Trans. Audio Electroacoust.*, vol. AU-17, Sept. 1969, pp. 209–215.
[38] T. Kaneko and B. Liu, "Accumulation of round-off error in fast Fourier transforms," *J. Ass. Comput. Mach.*, vol. 17, Oct. 1970, pp. 637–654.

SOME PRACTICAL CONSIDERATIONS IN THE REALIZATION OF LINEAR DIGITAL FILTERS

J. F. KAISER
Bell Telephone Laboratories, Incorporated, Murray Hill, New Jersey

ABSTRACT

The literature on sampled-data filters, although extensive on design methods, has not treated adequately the important problems connected with the actual realization of the obtained filters with finite arithmetic elements. Beginning with a review of the traditional design procedures a comparison is made between the different canonical realization forms and their related computational procedures. Special attention is directed to the problems of coefficient accuracy and of rounding and trucation effects. A simple expression is derived which yields an estimate of the required coefficient accuracy and which shows clearly the relationship of this accuracy to both sampling rate and filter complexity.

Reprinted from *Proc. 3rd Annu. Allerton Conf. Circuit System Theory,* pp. 621–633, 1965.

The high speed general purpose digital computer has become a powerful and widely used tool for the simulation of large and complex dynamic systems[1] and for the processing and reducing of large amounts of data by filter methods. The increased computational accuracy of the machines, the broader dynamic ranges in both amplitude and frequency of the system variables, and the increasing order or complexity of the dynamic systems themselves have made it necessary to take a much closer look at the computational and realization details of the designed digital filters. Many of the problems now coming to light were not noticed before[2,3,4,5,6,7] either because the filters were of low order with low (two or three decimal) accuracy requirements or because the sampling frequencies were comparable to the dynamic system and signal frequencies. An understanding of these computational problems and realization considerations is of vital interest to the users of the different digital filter design methods as their presence may often spell the success or failure of a particular application.

The problems to be treated in this paper relate to the numerical determination of the digital filter coefficients, to the stability of the digital filters themselves, to the precision of the arithmetic necessary to carry out the desired filtering operation, and to the choice of filter design methods. The choice of a satisfactory realization scheme is discussed in detail as it pertains to the previously mentioned problems.

Stability and Coefficient Accuracy

The two most widely used methods for the design of digital filters that approximate continuous linear filters with rational transfer characteristics are the bilinear z transformation[8] and the standard z transformation[2] methods.

The bilinear z transform is algebraic in nature and consists simply of a substitution or change of variable in the continuous filter transfer characteristic $H(s)$; i.e., the digital filter $H^*(z^{-1})$ is formed as

$$H^*(z^{-1}) = H(s) \Big|_{s \rightarrow \frac{2}{T}\frac{(1-z^{-1})}{(1+z^{-1})}} \qquad (1)$$

where

$$z = e^{sT} \qquad (2)$$

the unit advance operator.

The bilinear z transform simply maps the imaginary axis of the s-plane into the unit circle of z^{-1} plane with the left half of the s-plane mapping into the exterior of the unit circle in the z^{-1} plane. The mapping is one-to-one and thus unique. Thus if the transformation indicated by (1) is carried out exactly then $H^*(z^{-1})$ will be stable if $H(s)$ is stable and will be of precisely the same order. The bilinear z form can theoretically be applied directly to a rational transfer characteristic, $H(s)$, in either polynominal or factored form. It will be shown later which form is to be preferred.

The standard z transformation method consists first of obtaining a partial fraction expansion of $H(s)$ in its poles and then z transforming each partial fraction by making use of the transform pair

$$\frac{1}{s+a} \rightarrow \frac{T}{1-e^{-aT}z^{-1}} \tag{3}$$

or transform pairs derived therefrom. The mapping is again such that the left half of the s-plane maps into the exterior of the unit circle in the z^{-1} plane with the portion of the imaginary axis in the s-plane from $n\pi/T$ to $(n+2)\pi/T$, for n any integer, mapping into the unit circle. Thus the standard z transformation when applied to a stable transfer characteristic $H(s)$ always yields an $H^*(z^{-1})$ that is also stable provided the arithmetic is carried out with infinite precision.

For both transforms the resulting $H^*(z^{-1})$ for linear lumped continuous filters is of the general form

$$H^*(z^{-1}) = \frac{\sum_{k=0}^{n} a_k z^{-k}}{1 + \sum_{k=1}^{n} b_k z^{-k}} = \frac{N_d(z^{-1})}{D_d(z^{-1})} = \frac{N_d(z^{-1})}{\prod_{k=1,n} (1-z^{-1}/z_k)} \tag{4}$$

where b_0 has been set unity with no loss in generality. The question now arises as to what accuracy must the coefficients b_k be known to insure that the zeros of $D_d(z^{-1})$ all lie external to the unit circle, the requirement for a stable digital filter. First a crude bound will be established to be followed by a more refined evaluation of the coefficient sensitivity.

The polynomial $D_d(z^{-1})$ can be written in factored form as follows

$$D_d(z^{-1}) = \prod_{k=1}^{n} (1-z^{-1}/z_k) \tag{5}$$

For ease of presentation only simple poles are assumed for the basically low-pass transfer characteristic H(s), there being no difficulty in extending the analysis to the multiple order pole and non low-pass filter cases. If the standard z transform is used then $D_d(z^{-1})$ becomes

$$D_d(z^{-1}) = \prod_{k=1}^{n} (1-e^{p_k T} z^{-1}) \tag{6}$$

where p_k represents the kth pole of H(s) and may be complex. For the bilinear z transform there results

$$D_d(z^{-1}) = \prod_{k=1}^{n} \left[1 - \frac{(1+p_k T/2)}{(1-p_k T/2)} z^{-1} \right] \tag{7}$$

Assuming now that the sampling rate, 1/T, has been chosen to be at least twice the highest critical frequency in the H(s) and in the signals to be processed by the H(s), it is of interest to see how $D_d(z^{-1})$ behaves as the sampling rate is increased further.

The Nyquist limit constraint dictates that for the standard z transformation

$$\mathscr{I}[p_k T] \leq \pi \tag{8}$$

The critical frequencies are normalized with respect to half the sampling frequency as

$$\mu_k = \frac{p_k T}{\pi} = p_k/\omega_n \tag{9}$$

where ω_n is the Nyquist frequency or one-half the sampling frequency. Normally

$$|\mu_k| < 1. \tag{10}$$

Thus as the sampling frequency is increased the μ_k decrease from unity and approach zero. Then one can write for the standard z transform case

$$\left[1 - e^{p_k T} z^{-1}\right] \Longrightarrow \left[1 - (1+\mu_k \pi) z^{-1}\right] \quad \text{as } T \to 0 \tag{11}$$

and for the bilinear case

$$\left[1 - \frac{(1+p_k T/2)}{(1-p_k T/2)} z^{-1}\right] \Longrightarrow [1 - (1+\mu_k \pi) z^{-1}] \quad \text{as } T \to 0 \tag{12}$$

which illustrates that the two design methods yield essentially the same characteristic polynomials, $D_d(z^{-1})$, in the limit as T is made small.

Inspection of (11) and (12) show that the zeros of $D_d(z^{-1})$ tend to cluster about the point $z^{-1} = +1$ in the z^{-1} plane, i.e.,

$$z_k \doteq \frac{1}{1+\mu_k \pi} \approx 1 - \mu_k \pi \tag{13}$$

where for a stable system the μ_k has a negative real part. Now the filter $H^*(z^{-1})$ will become unstable if any of its poles move across the unit circle to the interior as a result of some perturbation or change in the coefficients b_j. To estimate the order of this effect one computes the change necessary to cause a zero of $D_d(z^{-1})$ to occur at the point $z^{-1} = 1$. From (5) and (13) there results

$$D_d(z^{-1}) \Big|_{z^{-1}=1} \doteq \prod_{k=1}^{n} (\mu_k \pi) = \prod_{k=1}^{n} (p_k T) \tag{14}$$

But

$$D_d(z^{-1}) \Big|_{z^{-1}=1} = 1 + \sum_{k=1}^{n} b_k z^{-k} \Big|_{z^{-1}=1} = 1 + \sum_{k=1}^{n} b_k \tag{15}$$

The right hand side of this expression is an important quantity and is therefore defined as

$$F_0 \triangleq 1 + \sum_{k=1}^{n} b_k \tag{16}$$

Thus by combining (14) and (15) it is immediately seen that if any of the b_k are changed by the amount given by (16) then the $D_d(z^{-1})$ will have a zero at $z^{-1} = 1$ and the filter $H^*(z^{-1})$ will thus have a singularity on the stability boundary. A zero of $D_d(z^{-1})$ at $z^{-1} = 1$ causes $H^*(z^{-1})$ to behave as if an integration were present in the H(s). Any further change in the magnitudes of any combination of the b_k in such a manner as to cause $D_d(z^{-1})\Big|_{z^{-1}=1}$ to change sign will result in an unstable filter, i.e., with some of the zeros of $D_d(z^{-1})$ lying inside the unit circle. Hence (14) is the desired crude bound on coefficient accuracy.

Equation (14) has a significant interpretation; it states that for small μ_k (large sampling rates) the bound on coefficient accuracy is dependent on both the order n of the filter and the sampling rate or normalized filter pole locations. Thus going from an nth order filter to a $(2n)^{th}$ order filter at the same normalized frequency will require approximately twice as many digits accuracy for the representation of the b_k. Similarly doubling the sampling rate for an nth order filter requires $n \times \log_{10} 2$ or $0.3 \times n$ additional decimal digits in the representation of each of the b_k.

Equation (15) has the interpretation that it represents the return difference[9] at zero frequency when the filter $H^*(z^{-1})$ is realized in direct form as shown in Fig. 1. This expression is also recognized as simply the reciprocal of Blackman's deadband factor[10] λ. Thus for complex filters with fairly large sampling rates the quantity F_0 will usually be very small. For example a fifth order Butterworth low pass filter with its break frequency at 1/10 the sampling frequency yields an $H^*(z^{-1})$ having $F_0 = 7.9 \times 10^{-5}$.

The coefficient accuarcy problem is somewhat further aggravated by the fact that as T is made smaller the b_k tend to approach in magnitude the binomial coefficients,[11] $\binom{n}{k}$, and tend to alternate in sign. Thus the evaluation of $D_d^*(z^{-1})$ involves the perennial computational problems associated with the differencing of large numbers. A better bound on coefficient accuracy is obtained by dividing F_0 as obtained from (16) by the magnitude of the largest b_k. The largest b_k is given approximately by

$$\max b_k \approx \binom{n}{[n/2]} \approx \frac{2^n}{\sqrt{n}} \sqrt{\frac{2}{\pi}}$$

$$\therefore \quad \max b_k \approx \frac{4}{5} * \frac{2^n}{\sqrt{n}} \qquad (17)$$

Hence from (14), (16), and (17) an absolute minimum bound on the number of decimal digits m_d required for representing the b_k is found as

$$m_d \geq 1 + \left[-\log_{10}\left(\frac{5\sqrt{n}}{2^{n+2}} \prod_{k=1}^{n} p_k T\right)\right] \tag{18}$$

where [x] denotes the "greatest integer in x."

While the foregoing analysis has yielded an easily computable absolute accuracy bound on the denominator coefficients of the recursive digital filter, the bound is not necessarily the best possible nor does it say anything about what happens to the zeros of $D_d(z^{-1})$ as small perturbations, less than F_0, are introduced in the values of the b_k. It is not enough to say that the digital filter $H^*(z^{-1})$ is simply stable; what is necessary is that the obtained digital filter have response characteristics close in some sense to those of the continuous filter it is approximating. This means that the sensitivity of the zeros of $D_d(z^{-1})$ to changes in the b_k must be determined. The most direct way to establish this relationship is to equate the two forms of the denominator of (4) and then to compute $\partial z_i/\partial b_k$. There results for filters with simple poles only

$$\frac{\partial z_i}{\partial b_k} = \frac{z_i^{k+1}}{\prod\limits_{\substack{k=1\\k\neq i}}^{n}\left(1 - \frac{z_i}{z_k}\right)} \tag{19}$$

from which the total differential change in any zero may be evaluated as

$$dz_i = \sum_{k=1}^{n} \frac{\partial z_i}{\partial b_k} db_k \tag{20}$$

These results extend directly to the multiple order pole case.[12]

Utilizing the fact that a pole, p_k, of $H(s)$ transforms to a zero of $D_d(z^{-1})$ at $e^{-p_k T}$ for the standard z transform and that for T such that $p_k T \ll 1$ the zero becomes approximately equal to $1 - p_k T$, as given previously by (13) the fractional change δ_i in a zero location z_i can be expressed in terms of the fractional change ε_k in a coefficient b_k of the polynomial $D_d(z^{-1})$. Using (19) there results

$$\varepsilon_k \approx \left[\prod_{\substack{\ell=1 \\ \ell \neq i}}^{n} (z_\ell - z_i)\right] \times \frac{\delta_i}{b_k} \tag{21}$$

where $\varepsilon_k = \frac{\Delta b_k}{b_k}$, $\delta_i = \frac{\Delta z_i}{z_i}$

and $|1 - z_i| \ll 1$, i.e., tightly clustered zeros have been assumed. For the purpose of quickly estimating the value required for ε_k, the product of the (n-1) factors can be approximated coarsely by the product (14) which is simply F_0. Thus (21) illustrates that the precision required for the representation of the b_k is increased by the factor $\log_{10}(1/\delta_i)$ over that given by (18).

Returning to (19) and (20) it is seen that the detailed changes in the positions of the zeros resulting from changes in the b_k are in general complex functions as the z_i may be complex. The changes in the b_k can occur as a result of imprecise arithmetic used in their computation or as a result of truncating or rounding the obtained b_k coefficients to a smaller number of significant digits. The qualitative evaluation of (19) can also be carried out by using the well developed ideas of the root locus.[13] For example the changes in location of all the zeros of $D_d(z^{-1})$ as a result of a change ρ_k in b_k are found from

$$\left(1 + \sum_{i=1}^{n} b_i z^{-i}\right) + \rho_k z^{-k} = 0 \tag{22}$$

or

$$\prod_{i=1}^{n}\left(1 - \frac{z^{-1}}{z_i}\right) + \rho_k z^{-k} = 0 \tag{23}$$

This has the appearance of the standard root locus problem for a single feedback loop having the loop transmission poles of the z_i, a kth order zero at the origin, and a loop gain factor of ρ_k. The parameter F_0 is simply the "gain" ρ_k required when the root locus passes through the point $z^{-1} = 1$. Thus all the techniques of the root locus method and the insight gained thereby can be brought to bear on the problem.

By viewing the coefficient sensitivity problem in terms of root loci the effects of both increasing filter order and especially increasing the sampling rate can be easily observed. Increasing the sampling rate tends to cluster the poles of $H^*(z^{-1})$ even more compactly about the point $z^{-1} = 1$ as Fig. 2 shows for a third order filter. As filter order increases so does the possible order k of the zero at the origin of the z^{-1} plane. All n branches of the root loci begin at the roots z_i; as ρ_k increases k branches converge on the kth order zero at the origin and n-k branches move off toward infinity with eventually radial symmetry. The angles the loci make as they leave the z_i are simply the angles given by evaluating (19) at each z_i. The value of ρ_k at which a branch of the locus first crosses the unit circle (the stability boundary) gives the measure of total variation that can be made in b_k and still keep the filter stable. Clearly the closer the roots z_i are to the unit circle initially the smaller will be the value of ρ_k necessary to move them to lie on the boundary. Thus by varying the ρ_k (the changes in b_k) the extent of the stability problem can be viewed.

The development up to this point assumed that the H(s) was basically low pass with simple poles. Extension to filters of high pass, bandpass, or bandstop types and with multiple order poles presents no real problems when viewed using the root locus idea. For example a digital version of a narrow bandpass filter with center frequency at ω_c would have its poles and zeros located as shown in Fig. 3 about the radial lines at $\pm\omega_c T$ radians. The coefficient sensitivity analysis proceeds in the same way as before except that now the points on the unit circle in the vicinity of $z^{-1} = \cos \omega_c T \pm j \sin \omega_c T$ replace the point $z^{-1} = 1$ in the stability computations. This is easily seen for the standard z transform where a continuous pole of H(s) at $p_k = \sigma_k + j\omega_k$ transforms to a zero of $D_d(z^{-1})$ at $e^{-p_k T}$, i.e., at $e^{-\sigma_k T} \angle \omega_k T$ when written in polar form. Thus it follows that it is primarily the smallness of the real parts of the filter poles of H(s) that cause the z^{-1} plane poles to be very near the unit circle and as a result to contribute measurably to the coefficient accuracy and the related sensitivity problems. It can be shown that expressions quite similar to (14) and (18) can be developed for digital bandpass and bandstop filters. The expressions differ primarily in that for an nth order filter only n/2 terms in the product $\prod^{n} p_k T$ will in general be small and thus contribute to this measure of sensitivity.

It is interesting to note that in the construction of continuous filters from RLC elements performance is limited primarily by the obtainable Q's of the inductors. The representation of the continuous filter by a digital filter is also strongly influenced by the Q's required of the filter section as it has been shown this directly establishes the number of digits needed to represent the digital filter coefficients.

In this section some of the relationships between filter order, pole locations, sampling frequency and digital filter coefficient accuracy have been established. The question then logically arises, how do these results affect the form chosen for realizing the digital filter? This question is discussed in the following section.

Realization Schemes

The three basic forms for realizing linear digital filters are the direct, the cascade and the parallel forms as shown in Fig. 4. As far as the stability question goes the two variations of the direct form, Fig. 4(a) and Fig. 4(b), are entirely equivalent with the configuration of Fig. 4(a) requiring fewer delay elements. The stability results developed in the previous section indicate clearly that the coefficient accuracy problem will be by far the most acute for the direct form realization. For any reasonably complex filter with steep transitions between pass and stop bands the use of the direct form should be avoided.

The choice between the utilization of either the cascade, Fig. 4(c), or parallel, Fig. 4(d), forms is not clear cut but depends somewhat on the initial form of the continuous filter and the transformation scheme to be used. In any case the denominator of H(s) must be known in factored form. If the parallel form is desired then a partial fraction expansion of H(s) must first be made. This is followed by a direct application of either (1) or (3) if the bilinear or standard z transforms are used respectively. For bandpass or bandstop structures the midfrequency gains of the individual parallel sections may vary considerably in magnitude introducing a small problem of the differencing of large numbers. This parallel form is perhaps the most widely used realization forms.

For cascade realization the bilinear z form requires that the numerator in addition to the denominator of H(s) be known in factored form. The splitting into simpler cascaded forms can then be done rather arbitrarily since the bilinear z operator has the property that

$$Z(G_1G_2) = Z(G_1)\cdot Z(G_2) \tag{24}$$

If the standard z transform is utilized, a partial fraction expansion must first be made followed by z transforming term by term. Then the fractions must be collapsed to yield an $N_d(z^{-1})$ which must then be factored to permit the cascade realization. This more involved procedure is necessary because the standard z transform does not possess the transform property given by (24).

The discussion up to this point has centered on satisfactory means for obtaining the digital filter coefficients required for the desired realization form. In actually using the digital filter to process data streams the performance of the filter will also be affected by the quantization of the data and by roundoff in the multiplication and addition operations. The recent paper by Knowles and Edwards[14] treats this aspect of the problem in some detail. Their results tend to indicate also that the parallel form realization exhibits slightly less performance error than the cascade form and that the direct form is definitely inferior to both the cascade and parallel forms.

Summary

After reviewing briefly two design procedures for digital filters expressions were derived estimating the accuracy required in the obtained filter coefficients. The expressions showed clearly the relationship of coefficient accuracy to filter complexity and sampling rate. These results also indicated which of the canonical realization forms are to be preferred.

BIBLIOGRAPHY

1. Golden, R. M., "Digital Computer Simulation of Communication Systems Using the Block Diagram Compiler: BLODIB, " Third Annual Allerton Conference on Circuit and System Theory, Monticello, Illinois, October 1965.

2. Ragazzini, J. R. and G. F. Franklin, "Sampled Data Control Systems," McGraw Hill, 1958.

3. Monroe, A. J., "Digital Processes for Sampled Data Systems," John Wiley, 1962.

4. Jury, E. I., "Sampled-Data Control Systems," John Wiley, 1958.

5. Tou, J. T., "Digital and Sampled-Data Control Systems," McGraw Hill, 1959.

6. Jury, E. I., "Theory and Application of the z-Transform Method," John Wiley, 1964.

7. Freeman, H., "Discrete Time Systems," John Wiley, 1965.

8. Kaiser, J. F., "Design Methods for Sampled-Data Filters," Proceedings First Allerton Conference on Circuit and System Theory, November 1963.

9. Bode, H. W., "Network Analysis and Feedback Amplifier Design," Van Nostrand, 1945, pp. 47-49.

10. Blackman, R. B., "Linear Data-Smoothing and Prediction in Theory and Practice," Addison-Wesley, 1965, p. 76.

11. Mansour, M., "Instability Criteria of Linear Discrete Systems," Automatica Vol. 2, n. 3, January 1965, pp. 167-178.

12. Maley, C. E., "The Effect of Parameters on the Roots of an Equation System," Computer Journal Vol. 4, 1961-2, pp. 62-63.

13. Truxal, J. G., "Automatic Feedback Control System Synthesis," McGraw Hill Book Co., Inc., New York 1955, pp. 223-250.

14. Knowles, J. B. and R. Edwards, "Effect of a Finite-Word Length Computer in a Sampled-Data Feedback System, Proc. IEE Vol. 112, No. 6 June 1965, pp. 1197-1207.

$$H_*^*(z^{-1}) = \frac{\sum\limits_{0}^{n} a_k z^{-k}}{1 + \sum\limits_{1}^{n} b_k z^{-k}}$$

Fig. 1

Fig. 2

x T=3
△ T=1

$\sqrt{\ } k^{th}$ order zero

Fig. 3

$\omega_c T$

$-\omega_c T$

4(a)

4(b)

4(c)

4(d)

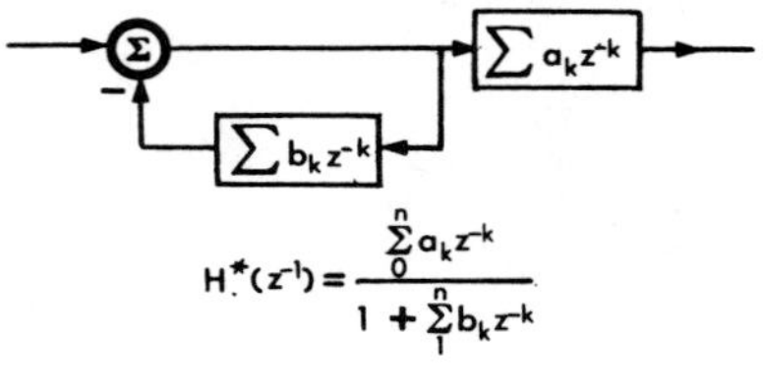

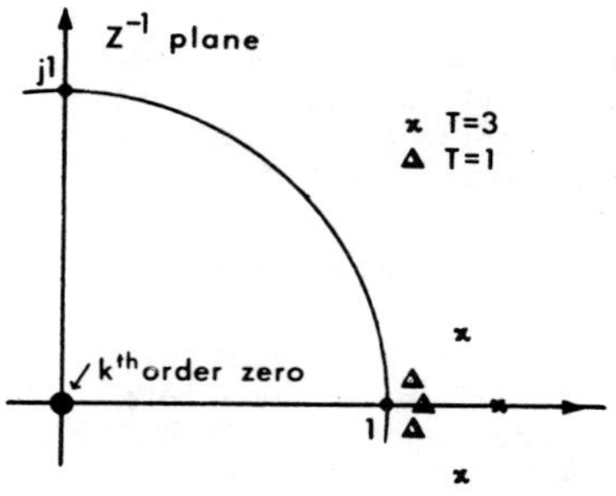

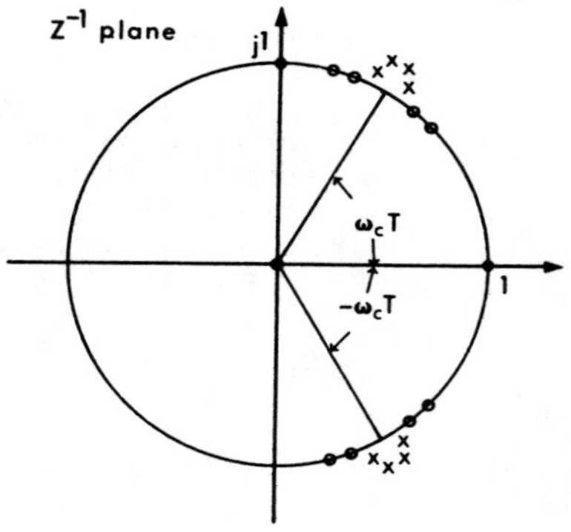

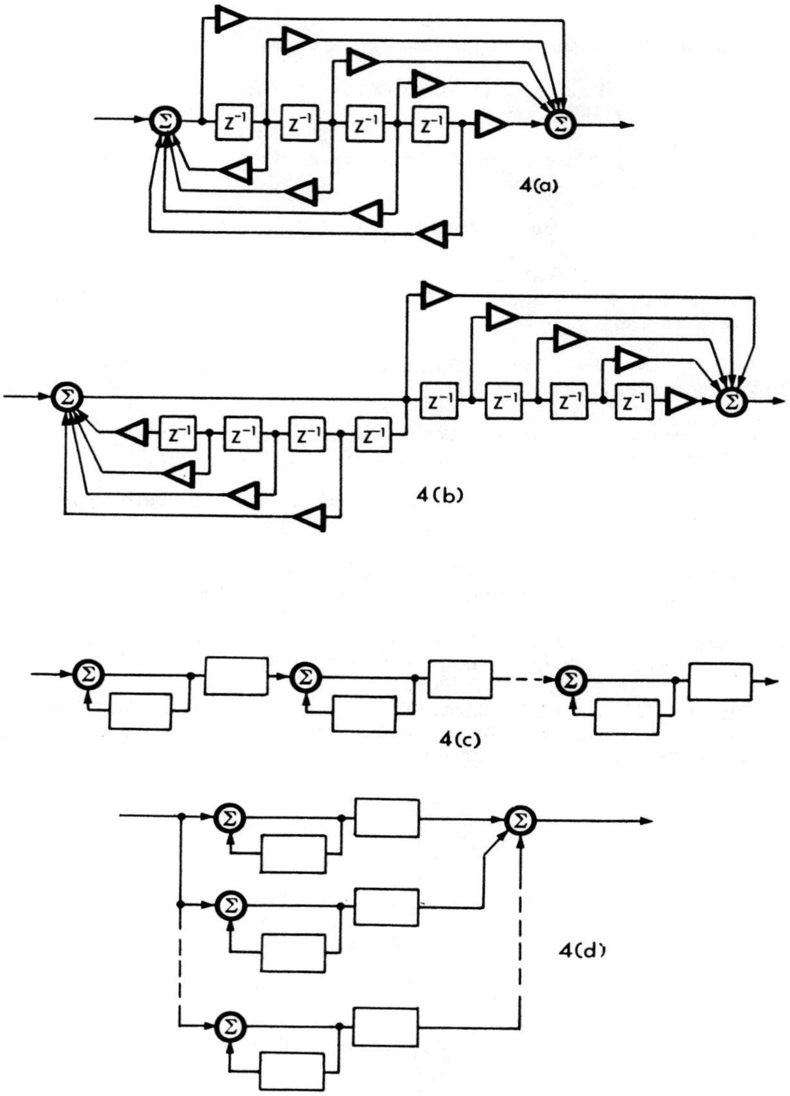

Fig. 4

Effects of Parameter Quantization on the Poles of a Digital Filter

Abstract—**A configuration is proposed which leads to a digital filter that is less sensitive to parameter quantization than are the standard configurations. A few examples illustrate the differences between the various realizations.**

Recursive digital-filter design techniques usually proceed on the assumption that the difference equations resulting from the design may be executed with effectively infinite-precision arithmetic. Knowles and Edwards[1] and Gold and Rader[2,3] have considered the effects of rounding of the multiplications in the difference equations, showing that this rounding introduces additive noise whose magnitude and spectral shape may be closely estimated by z-transform techniques. Kaiser[4,5] considered the effects of rounding the coefficients of the difference equations in terms of the resulting errors in the positions of the poles and zeros of the z-transfer function of the filter, and states that it is almost always preferable to realize an Nth-order digital filter as a cascade or parallel combination of first- and second-order digital filters in order to assure that small errors in the coefficients (due to rounding) do not result in unacceptably large errors in pole positions. In this letter, we analyze the errors in pole positions due to coefficient rounding for first- and second-order systems and propose an improved realization of a second-order system for which the pole positions are not as sensitive to coefficient errors. Since all digital filters can and should be realized as combinations of first- and second-order systems, the results presented here constitute, in a sense, a complete picture of the effects of parameter quantization on the pole positions of digital filters.

For a first-order equation

$$y(nT) = Ky(nT - T) + x(nT) \tag{1}$$

there is one coefficient, K, and the pole position is equal to the parameter value. Thus, if rounding causes the filter to be realized with $(K + \Delta K)$ then the error in the pole position is ΔK. The relative importance of such an error is greatest when K is nearly unity; the maximum gain of a one-pole filter is

$$\max\{H(z)\} = \frac{1}{1 - K}. \tag{2}$$

In the realization of a digital one-pole filter (as, for example, by any of the design methods in Rader and Gold[3]) with a bandwidth α at a sampling frequency of $1/T$, the ratio of bandwidth to sampling frequency (αT) is a useful index. As the sampling frequency increases, αT decreases and the pole of the filter moves toward the point $z = 1$. Therefore, a simulation of an analog filter may become less, rather than more, accurate as sampling rate is increased beyond a certain point.

The situation is more complicated for a second-order equation such as

$$y(nT) = Ky(nT - T) - Ly(nT - 2T) + x(nT) \tag{3}$$

which will be assumed to have complex conjugate poles at $re^{\pm j\theta}$ where

$$r = \sqrt{L} \quad \text{and} \quad \theta = \cos^{-1}(K/2\sqrt{L}). \tag{4}$$

If K and L are quantized such that in the realization of the filter they are in error by ΔK and ΔL, we can approximate the error in pole position by differentiating r and θ with respect to K and L, and applying Newton's approximation

$$\Delta v \approx \frac{\partial v}{\partial x}\Delta x + \frac{\partial v}{\partial y}\Delta y. \tag{5}$$

We have for the errors in r and θ

$$\Delta r \approx \Delta L/2r, \qquad \Delta\theta \approx (\Delta L/2r\tan\theta) - (\Delta K/2r\sin\theta). \tag{6}$$

The error in radius, for the interesting case of r near unity, is not very different from the case of the first-order equation, but the error in θ, for θ near 0, can be quite large. If the ratio of resonant frequency to sampling frequency is made small, as happens when the sampling rate is increased, the poles in the z plane approach $z = 1$, and the effects of the error in r and especially of the error in θ increase rapidly. Furthermore, the error in θ tends to be largest precisely when θ is smallest, so that the relative error in resonant frequency tends to behave like $1/\theta^2$.

To realize a second-order transfer function in which the pole locations are less sensitive to parameter errors, consider the simultaneous equations

$$\left.\begin{aligned} y_1(nT) &= Ky_1(nT - T) - Ly_2(nT - T) + Ax(nT) \\ y_2(nT) &= Ly_1(nT - T) + Ky_2(nT - T) + Bx(nT) \end{aligned}\right\} \tag{7}$$

with

$$K = r\cos\theta \quad \text{and} \quad L = r\sin\theta. \tag{8}$$

We can write the z-transforms of (7), after slight rearrangement, as

$$\left.\begin{aligned} Y_1(z)[1 - Kz^{-1}] + Y_2(z)[Lz^{-1}] &= AX(z) \\ Y_1(z)[-Lz^{-1}] + Y_2(z)[1 - Kz^{-1}] &= BX(z) \end{aligned}\right\} \tag{9}$$

so that

$$\begin{aligned} \frac{Y_1(z)}{X(z)} &= \frac{A(1 - Kz^{-1}) - BLz^{-1}}{1 - 2Kz^{-1} + (K^2 + L^2)z^{-2}}, \\ \frac{Y_2(z)}{X(z)} &= \frac{B(1 - Kz^{-1}) + ALz^{-1}}{1 - 2Kz^{-1} + (K^2 + L^2)z^{-2}}. \end{aligned} \tag{10}$$

These transfer functions each have poles at $re^{\pm j\theta}$ and a zero whose position can be specified by suitably choosing A and B.

Differentiating (8) with respect to L and K and applying (5), we obtain

$$\Delta r \approx (\cos\theta)\Delta K + (\sin\theta)\Delta L \tag{11}$$

$$\Delta\theta \approx \left(\frac{\cos\theta}{r}\right)\Delta L - \left(\frac{\sin\theta}{r}\right)\Delta K. \tag{12}$$

These errors in pole position are quite different than for (3) and for θ small, the error in θ is substantially less for (7) than for (3). Of course, (7) requires substantially more computation than (3), and it should therefore be resorted to only when the quantization of parameters is a problem.

Manuscript received February 23, 1967.

[1] J. B. Knowles and R. Edwards, "Effect of a finite-word-length computer in a sampled-feedback system," *Proc. IEE* (*London*), vol. 112, June 1965.

[2] B. Gold and C. M. Rader, "Effects of quantization noise in digital filters," *Proc. SJCC*, 1966.

[3] C. M. Rader and B. Gold, "Digital filter design techniques in the frequency domain," *Proc. IEEE*, vol. 55, pp. 149–171, February 1967.

[4] J. F. Kaiser, "Some practical considerations in the realization of linear digital filters," presented at the 3rd Allerton Conf., 1965.

[5] ——, private communication.

Reprinted from *Proc. IEEE*, vol. 55, pp. 688–689, May 1967.

Example

Consider a linear digital filter defined by the transfer function

$$H(z) = \frac{1.95772751 \times 10^{-6} z^{-1}}{1 - 1.99768072 z^{-1} + 0.99768268 z^{-2}}. \tag{13}$$

This filter has complex poles with radii $r = 0.99884$ and angles $\theta = \pm 7.838367 \times 10^{-4}$ radians, which parameters make this filter suitable for the simulation, at a 10-kHz sampling rate, of a volume unit (VU) meter.[5,6] If the coefficients of the denominator must be rounded to the nearest multiple of 2^{-18}, it can be shown[5] that the error $\Delta\theta$ in the pole angles is about 100 percent, an unacceptably large error.[7] Using an ingenious method Kaiser was able to reduce $\Delta\theta$, for the given sampling rate, to about five percent of θ.

If Kaiser had used the realization specified by (7), with $A = 1$ and $B = 0$, using $y_2(nT)$ as the output, the transfer function

$$H(z) = \frac{r \sin \theta z^{-1}}{(1 - re^{+j\theta} z^{-1})(1 - re^{-j\theta} z^{-1})} \tag{14}$$

would have been obtained (this is the desired transfer function except for a constant gain factor). Since the absolute maximum theoretical rounding errors, ΔL and ΔK, cannot exceed 2^{-19}, or about 2×10^{-6}, we see from (11) and (12) that $\Delta r < 2 \times 10^{-6}$, which is still acceptably small compared to $(1-r)$, and $\Delta\theta < 2 \times 10^{-6}$, which is only about 0.3 percent of θ. The realization via (7) leads to a VU meter filter design which works generally for a wide range of sampling rates, and needs no special compensation when the requirement of 18-bit arithmetic is imposed.

[6] L. L. Beranek, *Acoustic Measurements*. New York: Wiley, 1949, pp. 504–506.

[7] Equations (6) are good approximations to the error in pole position only when the sensitivity of pole position to changes in L and K at the "new" pole positions does not differ significantly from the sensitivity at the "old" pole positions. In this case, $\Delta K = -0.056 \times 10^{-6}$ and $\Delta L = 1.798 \times 10^{-6}$

$$\Delta r \approx 0.9 \times 10^{-6}$$

which is small enough compared to $(1-r)$ so that (6) remains valid. However, using (6) we get

$$\Delta\theta \approx 0.132 \times 10^{-2}$$

which is greater than θ itself. Therefore (6) does not give us a good estimate of the error $\Delta\theta$.

While (7) requires four multiplications per input sample, the number of coefficients determining the pole positions is only two, due to constraints. A generalization of (7), yielding a second-order transfer function, is

$$\left.\begin{aligned} y(nT) &= Ay(nT - T) + Bw(nT - T) + Ex(nT) \\ w(nT) &= Cy(nT - T) + Dw(nT - T) + Fx(nT) \end{aligned}\right\} \tag{15}$$

where

$$A + D = 2r\cos\theta \quad \text{and} \quad AD - BC = r^2 \tag{16}$$

If we impose constraints $A = D$ and $B = -C$, we arrive at (7). The constraints $C = 1$, $D = 0$ lead to (3). However, other, more flexible, constraints could have been chosen. For example, if $A = D$ is the only constraint, (16) becomes

$$r\cos\theta = A, \qquad r^2 = A^2 - BC \tag{17}$$

Another degree of freedom is now allowed for reduction of the quantization error. If the problem is to minimize the error in r, and if A is set equal to the quantized value nearest to that specified by (17), then the error in r can be computed each time a choice of B is made; this choice of B determines C [via (17)] and thus determines the quantization errors in both B and C. The error in r can then be found using the quantized values of A, B, and C, and can thus be empirically minimized; since such a computation is usually of interest only when the quantization errors are relatively large (which means a short computer-register length), it is reasonable to try all values of B.

The realization of a given digital filter affects the quantization noise caused by round-off errors of the multiplications performed.[1,2] We have computed the mean squared noise caused by round-off (using the techniques described by Gold and Rader[2]) for the realization defined by (7); the result obtained is $\sigma^2/(1-r^2)$, where $\sigma^2 = E_0^2/12$ if products are summed before rounding, and where E_0 is the quantization increment. For r close to unity, this reduces to $E_0^2/24\varepsilon$, where $\varepsilon = 1 - r$; the corresponding results obtained[2] were $E_0^2/48\varepsilon$ for the canonic realization of (3).

C. M. Rader
B. Gold
M.I.T. Lincoln Lab.[8]
Lexington, Mass.

[8] Operated with support from the U. S. Air Force.

On the Approximation Problem in the Design of Digital Filters with Limited Wordlength

The influence of coefficients having limited length on the frequency response of selective filters will be considered, using a specific bandpass as an example. It will be shown, that a certain minimum wordlength of the coefficients is necessary in order to satisfy a given tolerance scheme. In the particular case, the needed wordlength can be reduced by 3 bits, if an optimization procedure is used in the discrete parameter space.

Über das Approximationsproblem beim Entwurf von digitalen Filtern mit begrenzter Wortlänge

Der Einfluß von Koeffizienten begrenzter Wortlänge auf den Frequenzgang selektiver Filter wird an einem Beispiel demonstriert. Es wird gezeigt, daß eine gewisse Mindestwortlänge nötig ist, um einem gegebenen Toleranzschema zu genügen. Mit Hilfe einer Optimierung mit diskreten Parametern kann in dem betrachteten Beispiel die erforderliche Wortlänge wesentlich reduziert werden.

One of the most promising aspects of digital filters, if compared with continuous filters, is the fact that they can be implemented with any necessary accuracy by choosing the wordlengths of the coefficients and the state variables sufficiently large. On the other hand the fundamental property, that both wordlengths are limited, raises two important problems:

1. Due to the rounding in the arithmetic operations, especially in the multiplication, a noise of random character is introduced, which determines the signal-to-noise ratio of the system.
2. The limitation of the wordlength of the coefficients may lead to a system, which does not satisfy any more the original conditions to be met e.g. by the frequency response.

Since the cost of the implementation depends heavily on the wordlength, it is necessary to choose it as small as possible but large enough that the system just fulfills the requirements. As far as the first problem is concerned the noise can be reduced by choosing a suitable structure of the digital filter [1], [2]. This paper deals with the second problem. The design of a bandpass will be used as an example.

Let

$$H(z) = \frac{\sum_{\mu=0}^{m} b_\mu z^\mu}{\sum_{\nu=0}^{n} c_\nu z^\nu} = \frac{b_m}{c_n} \frac{\prod_{\mu=1}^{m} (z - z_{0\mu})}{\prod_{\nu=1}^{n} (z - z_{\infty\nu})} \quad \text{with} \quad z = e^{sT}$$

be the transfer function of the proposed digital filter. We consider the magnitude of this function on the unit circle $z = e^{j\omega T} = e^{j\Omega}$:

$$|H(\Omega)| = |H(z)|_{z = e^{j\Omega}}.$$

In the example considered here $|H(\Omega)|$ has to satisfy the tolerance scheme given in Fig. 1, described by the four frequencies Ω_{-P}, Ω_{+P}, Ω_{-S}, Ω_{+S} and the tolerated deviations δ_1 in the passband and δ_2 in the stopband. Since the tolerance scheme is given in normalized quantities, we introduce a normalized magnitude function $|H_n(\Omega)|$ by

$$|H_n(\Omega)| = 2 \frac{|H(\Omega)|}{\max |H(\Omega)| + \min |H(\Omega)|}$$

where $\max |H(\Omega)|$ is the maximum and $\min |H(\Omega)|$ the minimum of the magnitude inside the passband. With this expression we define a normalized error-function

$$\varepsilon(\Omega) = \begin{cases} \frac{1}{\delta_1} \left| 1 - |H_n(\Omega)| \right| & \Omega_{-P} \leq \Omega \leq \Omega_{+P} \quad \text{(passband)} \\ \frac{1}{\delta_2} |H_n(\Omega)| & 0 \leq \Omega \leq \Omega_{-S} \\ & \Omega_{+S} \leq \Omega \leq 180^\circ. \quad \text{(stopbands)} \end{cases}$$

Since there are no constraints in the ranges between passband and the stopbands, $\varepsilon(\Omega)$ is not defined in these ranges. In the following we consider $\max \varepsilon(\Omega)$ as a figure of merit. A solution with $\max \varepsilon \leq 1$ satisfies the tolerance scheme, while a set of parameters leading to a transfer function with $\max \varepsilon > 1$ cannot be accepted.

Using an optimization procedure for the minimization of $\max \varepsilon(\Omega)$ and coefficients with a wordlength of 36 bits, i.e. the full wordlength available in the computer, a system of 8th degree has been found as an acceptable solution with $\max \varepsilon_{36} = 0.526054$. The frequency response $|H_n(\Omega)|$ has been drawn with dotted lines in Fig. 1. Obviously there is some margin for a reduction of the wordlength.

For many reasons a cascade of subsystems of second order (third canonic form in [3]) has been found to be a suitable structure for realization [1], [2]. Restricting ourselves to the case given here, that all zeros are located on the unit circle, we write the transfer function as

$$H(z) = \prod_{\varkappa=1}^{n/2} H_\varkappa(z) = \prod_{\varkappa=1}^{n/2} \frac{b_{0\varkappa} z^2 + b_{1\varkappa} z + b_{0\varkappa}}{z^2 + c_{1\varkappa} z + c_{0\varkappa}}.$$

If we assume the minimum increment of the coefficients $b_{1\varkappa}$, $b_{0\varkappa}$, $c_{1\varkappa}$ and $c_{0\varkappa}$ to be $Q = 2^{-6}$ (i.e. in our case a wordlength of 6 bits for $|b_{1\varkappa}|$, $b_{0\varkappa}$, $|c_{1\varkappa}|$, and $c_{0\varkappa}$), we

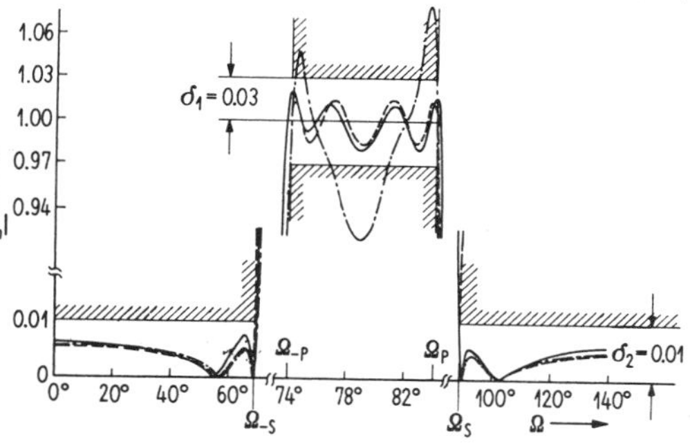

Fig. 1. To the influence of rounding and optimization on the frequency response;
— — — $|H_n(\Omega)|$ with $Q = 2^{-36}$,
—·—·— $|H_n(\Omega)|$ with $Q = 2^{-6}$,
——— $|H_n(\Omega)|$ with $Q = 2^{-6}$ after optimization.

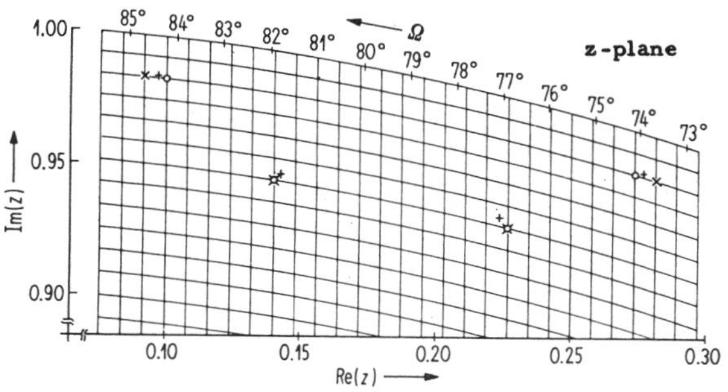

Fig. 2. Pol positions before and after rounding to $Q = 2^{-6}$ and after optimization;
pol positions: + solution with $k = 36$,
○ solution with $k = 6$ after rounding,
× solution with $k = 6$ after optimization.

Reprinted with permission from *Arch. Elek. Übertragung*, vol. 24, pp. 571–572, 1970.

get a grid of possible pole-zero positions in the unit circle. A part of it has been drawn in Fig. 2, where in addition the wanted positions of poles for the example considered here have been indicated. Rounding these coefficients to a minimum increment of $Q = 2^{-k}$ increases max ε as shown in Fig. 3. It is to be seen, that in some cases a smaller value of Q, i.e. a larger wordlength leads to a worth, i.e. larger value of max ε. The smallest value of k, leading to

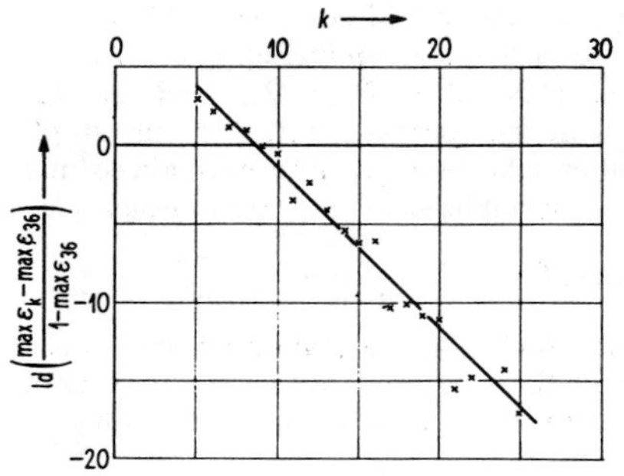

Fig. 3. The figure of merit max ε as a function of k.

a tolerable solution, if the coefficients are rounded, turns out to be $k = 9$. For $k = 6$ the not tolerable frequency response has been drawn in Fig. 1 and the pole locations are indicated in Fig. 2.

A modified Gauss-Seidel procedure has been used to minimize max ε in the discrete parameter space. This procedure starts at the point found by rounding the original coefficients. A detailed description will be given elsewhere [4]. In general it will find a set of parameters having a smaller max ε. But it cannot be guaranteed that the optimum to be found is the global and not a local one.

In the example treated here a minimum with max $\varepsilon = 0.897$ has been found for $k = 6$. Therefore, due to the optimization the wordlength of the coefficients can be reduced in this particular case by 3 bits. Of course the possible reduction depends heavily on the treated filter.

The function $|H_n(\Omega)|$ for this solution has been drawn in Fig. 1. In addition the pole and zero locations have been indicated in Fig. 2.

The computations have been done on the CD 3300 of the Computing Centre of the University Erlangen-Nürnberg.

(Received September 1st, 1970.)
Dipl.-Ing. E. AVENHAUS,
Prof. Dr. W. SCHÜSSLER
Institut für Nachrichtentechnik
der Universität
D-852 Erlangen, Egerlandstrasse 5

References

[1] EDWARDS, R., BRADLEY, J., and KNOWLES, J. B., Comparison of noise performances of programming methods in the realization of digital filters. Proc. Symposium on Computer Processing in Communications, XIX, PIB-MRI Symposia Series (1969), pp. 295–311.

[2] AVENHAUS, E., Zur Realisierung digitaler Filter mit günstigem Nutz-Störsignalverhältnis. Nachrichtentech. Z. 23 [1970], 217–219.

[3] SCHÜSSLER, W., Zur allgemeinen Theorie der Verzweigungsnetzwerke. AEÜ 22 [1968], 361–367.

[4] AVENHAUS, E., Zum Entwurf digitaler Filter mit minimaler Speicherwortlänge für Koeffizienten und Zustandsgrössen. Ausgewählte Arbeiten über Nachrichtensysteme, No. 13. University of Erlangen-Nürnberg, Germany, 1971. W. Schüssler, Ed.

On the Accuracy Problem in the Design of Nonrecursive Digital Filters

For the case of a nonrecursive digital filter with linear phase and equal ripple attenuation in the passband as well as in the stopband the influence of inaccurate coefficients will be considered. It will be shown, that a structure formed as a cascade of appropriate subsystems is superior to the direct realization, especially in the stopband.

Zum Genauigkeitsproblem beim Entwurf nichtrekursiver digitaler Filter

Der Einfluß der Rundung von Koeffizienten auf die Genauigkeit des Frequenzganges eines nichtrekursiven digitalen Filters linearer Phase wird für den Fall betrachtet, daß die Dämpfung des Filters im Durchlaßbereich den Wert Null und im Sperrbereich einen vorgeschriebenen Mindestwert gleichmäßig approximiert. Es wird gezeigt, daß eine Kaskadenanordnung geeignet gewählter Teilsysteme speziell im Sperrbereich ein wesentlich günstigeres Verhalten als eine direkte Realisierung zeigt

A nonrecursive filter with linear phase can be described by the transfer function

$$H(z) = \frac{1}{z^{2n}} \frac{1}{2} \sum_{\mu=0}^{n} d_\mu (z^{n+\mu} + z^{n-\mu}) = \frac{1}{z^{2n}} H_1(z) = \tag{1}$$

$$= \frac{1}{z^{2n}} \frac{d_n}{2} \prod_{\mu=1}^{n} (z - z_{0\mu})(z - z_{0\mu}^{-1}) \text{ with } |z_{0\mu}| \leq 1, \tag{2}$$

$$H(\Omega) = \frac{1}{e^{jn2\pi\Omega}} \sum_{\mu=0}^{n} d_\mu \cos \mu\, 2\pi\Omega = e^{-jn2\pi\Omega} H_0(\Omega). \tag{3}$$

Here $H_1(z)$ is a mirror-image-polynomial having zeros on the unit circle or reciprocal to it, while $H_0(z) = z^{-n} H_1(z)$.

In a recent paper [1] a new class of filters of this type has been introduced having equal ripple attenuation in the

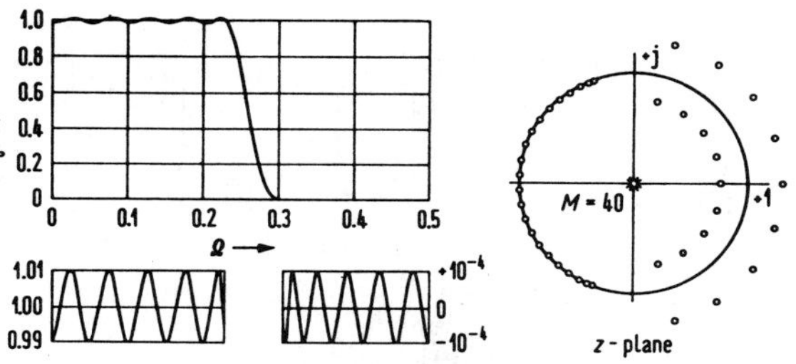

Fig. 1. Frequency response and zero-positions of the filter 40.01.001.9

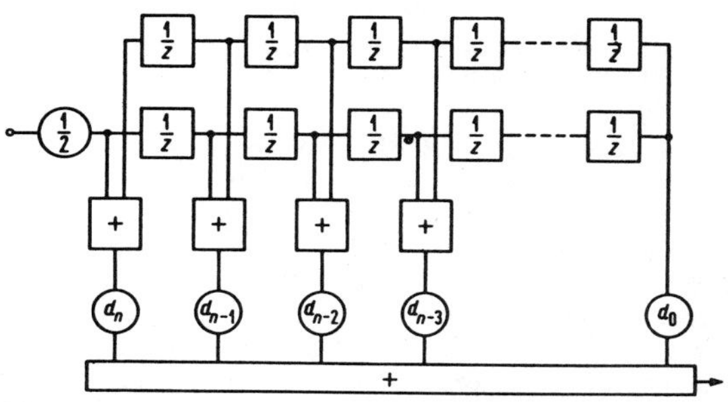

Fig. 2. Structure of a nonrecursive filter with linear phase.

Reprinted with permission from *Arch. Elek. Übertragung*, vol. 24, pp. 525–526, 1970.

passband as well as in the stopband. Fig. 1 shows the frequency response and the positions of the zeros $z_{0\mu}$ of one of these filters, having a tolerated deviation of 1% in the passband and an attenuation of at least 80 dB in the stopband. One possible structure for the realization is the well known direct form, which uses the coefficients d_μ of the polynomial given in eq. (1). Fig. 2 shows the structure for the case of the linear phase filter to be discussed here.

We consider the effect of inaccurate coefficients d_μ won as the result of rounding the original coefficients. Obviously we still get a mirror-image-polynomial and therefore a system with linear phase. But the frequency response $H_0(\Omega)$ will change. Fig. 3 shows this function in the passband and in the stopband for the same filter as in Fig. 1 after rounding the coefficients to 12 bit. It is to be seen that the behavior in the passband has not changed while the function in the stopband has been effected seriously. Obviously some of the zeros have been moved off the unit circle.

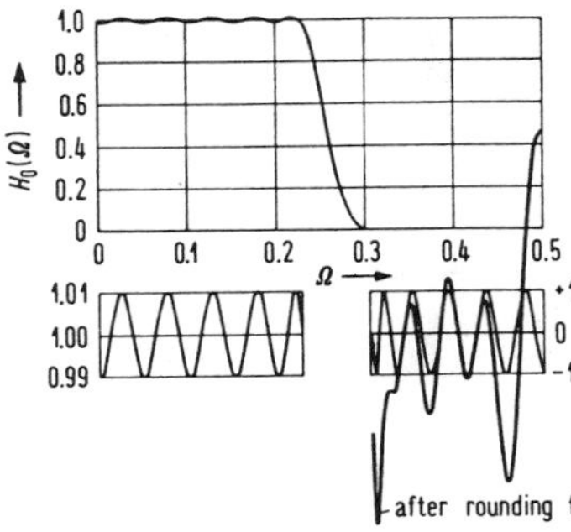

Fig. 3. Frequency response of the filter with rounded coefficients if realized in direct form.

It is well known that in the case of recursive filters the cascade-form is superior to the direct form at least in the case of selective filters with attenuation poles. For nonrecursive filters this form cannot be applied directly, since the structure has to be chosen in such a way that the linear phase property is guaranteed even after rounding. Beside the trivial case of polynomials of second degree with zeros on the unit circle a mirror-image-polynomial of fourth degree is the simplest one of interest:

$$H_{1\nu}(z) = z^4 + d_{1\nu} z^3 + d_{0\nu} z^2 + d_{1\nu} z + 1 . \tag{4}$$

Let $$z_{0\nu} = r_\nu \, \mathrm{e}^{\pm \mathrm{j} \psi_\nu} \quad \text{and} \quad \frac{1}{z_{0\nu}} = \frac{1}{r_\nu} \mathrm{e}^{\pm \mathrm{j} \psi_\nu}$$

be the zeros of $H_{1\nu}(z)$. $H_{1\nu}(z)$ can be written as

$$H_{1\nu}(z) = (z^2 - 2 r_\nu \cos \psi_\nu z + r_\nu^2) \frac{1}{r_\nu^2} (r_\nu^2 z^2 - 2 r_\nu \cos \psi_\nu z + 1) .$$

Here in both polynomials of second degree the same coefficients appear, which are simply related to the zeros. Fig. 4 shows the structure of one block of the cascade and the grid of possible zero-positions inside the unit circle, if an increment of 2^{-5} for both coefficients is prescribed. Unfortunately the structure needs four multipliers per block beside the one in front of the second section.

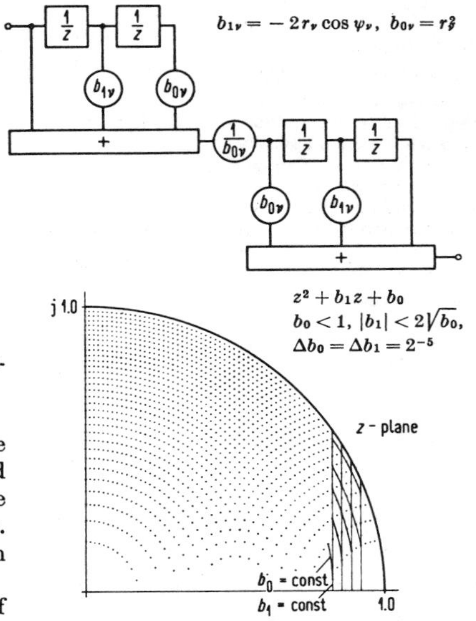

Fig. 4. A possible structure of a block of fourth order having linear phase and its grid of possible zero positions.

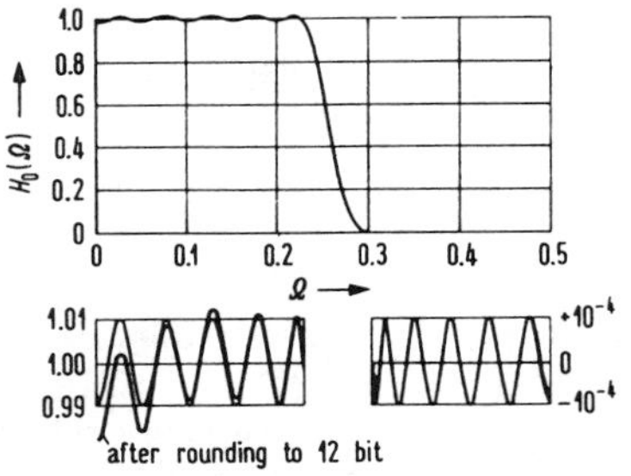

Fig. 6. Frequency response of the filter with rounded coefficients if realized in cascade form, using the structure of Fig. 5.

The direct realization of the polynomial (4) leeds to the structure of Fig. 5, which needs only two multipliers and a total of n multipliers as the direct form requires. In addition Fig. 5 shows the possible zero-positions in this case for different increments of d_0 and d_1 in different areas of the unit circle. It is to be seen that the grid of zero-positions becomes very dense close to the origin, while the density is very small close to the unit circle and especially close to the point $z = 1.0$. This explains the rather high influence of inaccurate coefficients in the passband, while the behavior in the stopband stays almost unaltered as to be seen in Fig. 6.

Further effort will be necessary to find a set of coefficients of limited length, satisfying a given tolerance scheme.

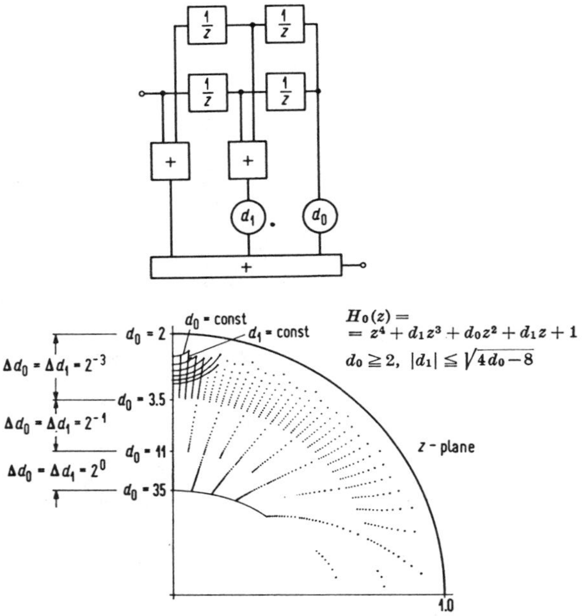

Fig. 5. Another structure of a block of fourth order having linear phase and its grid of possible zero positions.

(Received August 27, 1970.)

Dr. O. Herrmann,
Prof. Dr. W. Schüssler
Institut für Nachrichtentechnik
der Universität
D-852 Erlangen
Egerlandstrasse 5

Reference

[1] Herrmann, O., Design of nonrecursive digital filters with linear phase. Electron. Letters 6 [1970], 328–329.

On the Interaction of Roundoff Noise and Dynamic Range in Digital Filters*

By LELAND B. JACKSON

(Manuscript received October 22, 1969)

The interaction between the roundoff-noise output from a digital filter and the associated dynamic-range limitations is investigated for the case of uncorrelated rounding errors from sample to sample and from one error source to another. The required dynamic-range constraints are derived in terms of L_p norms of the input-signal spectrum and the transfer responses to selected nodes within the filter. The concept of "transpose configurations" is introduced and is found to be quite useful in digital-filter synthesis; for although such configurations have identical transfer functions, their roundoff-noise outputs and dynamic-range limitations can be quite different, in general. Two transpose configurations for the direct form of a digital filter are used to illustrate these results.

I. INTRODUCTION

With the rapid development of digital integrated circuits in the 1960's and the potential for large-scale integration (LSI) of these circuits in the 1970's, digital signal processing has become much more than a tool for the simulation of analog systems or a technique for the implementa-

* This paper is taken in part from a thesis submitted by Leland B. Jackson in partial fulfillment of the requirements for the degree of Doctor of Science in the Department of Electrical Engineering at Stevens Institute of Technology.[1]

tion of very complex and costly one-of-a-kind systems alone. The traditional advantages of digital systems, such as high accuracy, stable parameter values, and straight-forward realization, have been supplemented through the use of integrated circuits by the additional advantages of high reliability, small circuit size, and ever-decreasing cost. As a result, it now appears that many signal processing systems which have been in the exclusive domain of analog circuits may in the future be implemented using digital circuits; while other proposed systems which could not be implemented at all because of the practical limitations of analog circuits may now be realized with digital circuits.[2]

The key element in most of these new signal-processing systems is the digital filter. The term "digital filter" here denotes a time-invariant, discrete or sampled-data filter with finite accuracy in the representation of all data and parameter values.[3-5] That is, all data and parameters within the filter are "quantized" to a finite set of allowable values with, in general, some form of error being incurred as a result of the quantization process. Implicit in this quantization is a maximum value or set of maximum values for the magnitudes of these data and parameters which, in the case of the data, is usually referred to as the "dynamic range" of the filter.

Without the above quantization effects, linear discrete filters could be implemented exactly. Of course, one very significant feature of digital signal processing is that arbitrarily high accuracy can, in fact, be maintained once the initial analog-to-digital (A-D) conversion (if any) has taken place. However, there are still practical limitations to the accuracy of any physical system, and often it is desirable to minimize the accuracy of the implementation (while still satisfying the system specifications) in order to minimize the cost of the system. Hence, a thorough understanding of quantization errors in digital filters is quite important if the full potential of digital signal processing is ever to be realized.

II. QUANTIZATION ERRORS IN DIGITAL FILTERS

The specific sources of quantization error in the implementation and operation of a digital filter are as follows:

(*i*) The filter coefficients (multiplying constants) must be quantized to some finite number of digits (usually binary digits, or bits).

(*ii*) The input samples to the filter must also be quantized to a finite number of digits.

(*iii*) The products of the multiplications (of data by coefficients) within the filter must usually be rounded or truncated to a smaller number of digits.

(*iv*) When floating-point arithmetic is used, rounding or truncation must usually be performed before or after additions as well.

The first source of error above is deterministic and straightforward to analyze in that the filter characteristics must simply be recomputed to reflect the (small) changes in the filter coefficients due to quantizing.[6,7] However, the inclusion of coefficient quantization in the initial filter synthesis procedure in order to minimize (in some sense) the resulting filter complexity produces a complex problem in nonlinear integer programming which has only begun to be investigated.

The second source of error is often referred to as "quantization noise". It is inherent in any A-D conversion process and has been studied in great depth.[8] Hence, input quantization has not been included in our investigation, except as it relates to other error sources of interest.

The third and fourth error sources are similar to the second since they also involve quantization of the data, but they differ in two respects: (i) The data to be quantized is already digital in form, and (ii) the rounding or truncation of the data takes place at various points *within* the filter, not just at its input. To distinguish these sources of error from the input quantization noise, the resulting error processes will be referred to as "roundoff noise" (to be used generically, whether rounding or truncation is actually employed). Because of (ii), the roundoff noise is potentially much larger than the input quantization noise, and it is one of the principal factors which determine the complexity of the digital filter implementation, especially when special-purpose hardware is used.

There are three variables in the filter implementation which determine the level and character of the roundoff noise for a given input signal:

(*i*) the number of digits (bits) used to represent the data within the filter,

(*ii*) the "mode" of arithmetic employed (that is, fixed-point or floating-point), and

(*iii*) the circuit configuration of the digital filter. The number of digits in the data may be thought of as determining either the quantization step size or the dynamic range of the filter. We choose here the latter interpretation in order to have the same step size for all filters. Therefore, with this interpretation, the number of data digits does not affect the level of the roundoff noise directly, but rather it limits the maximum allowable signal level and hence the realizable signal-to-noise ratio. Data within the filter must, of course, be properly "scaled" if the maximum signal-to-noise ratio is to be maintained without exceeding the dynamic-range limitations. Among the principal results reported here are the determination of appropriate scaling for certain important classes of input signals and the calculation of the effect of this scaling on the output roundoff noise.

The output roundoff noise from a floating-point digital filter is usually (but not always) less than that from a fixed-point filter with the same total number of data digits because of the automatic scaling provided by floating-point arithmetic.[9,10] However, since floating-point arithmetic is significantly more complex and costly to implement, most special-purpose digital filters have been, and will probably continue to be, constructed with fixed-point hardware. Hence, we have considered only fixed-point digital filters in this work although much of the analysis could be adapted to floating-point filters. Oppenheim has recently proposed another interesting mode of arithmetic for digital filter implementation, called "block-floating-point", which provides a simplified form of automatic scaling of the filter data.[11] As would be expected, the performance of block-floating-point appears to lie somewhere between those of fixed-point and of floating-point.

The third variable in the implementation of a digital filter, that of circuit configuration, is the principal factor determining the character (spectrum) of the output roundoff noise and, along with mode of the arithmetic, ultimately determines the number of data digits required to satisfy the performance specifications. In fact, the key step in the synthesis of a digital filter is the selection of an appropriate configuration for the digital circuit. There are a multitude of equivalent circuit configurations for any given linear *discrete* filter (whose transfer function is expressible as a rational fraction in z); but in the implementation of the corresponding *digital* filter, these configurations are no longer equivalent, in general, because of the effects of coefficient quantization and roundoff noise. As noted previously, the effects of coefficient quantization are deterministic and can thus be accounted for exactly as a (typically small) change in the transfer function of the discrete filter. Therefore, assuming that the coefficients for the configurations under consideration have been (or can be) quantized satisfactorily, the choice between these configurations is then determined by the level and character of their output roundoff noise. As we will show, there can be very significant differences between the roundoff-noise outputs of otherwise equivalent digital filter configurations.

The content and complexity of any analysis of roundoff noise are determined to a large extent by the assumed correlation between roundoff errors. If these errors may be assumed to be uncorrelated from sample to sample and from multiplier (or other rounding point) to multiplier, then the roundoff-noise analysis is relatively straightforward, and the results are independent of the exact nature of the input signal to the filter. If, on the other hand, uncorrelated errors may not be assumed, then the analysis is much more complex, and the results are generally dependent on the particular input signal or class of input signals. This paper is concerned exclusively with the uncorrelated-error case because this assumption seems to be valid for most filters with input signals of reasonable amplitude and spectral content. Even in this case, the inclusion of the associated dynamic-range constraints makes the analysis reasonably involved and the corresponding synthesis problem quite complex.

Although the generic term "roundoff noise" has been used to include the case of truncation as well as rounding, we actually concentrate on the rounding case. As long as the assumption of uncorrelated errors can be made, our results are applicable to either case, with the error variance for truncation being four times that for rounding. However, as the input signals become less "random", the uncorrelated-error assumption tends to break down for truncation more readily than for rounding. Hence, additional care must be exercised in applying these results to the truncation case.

III. FILTER MODEL FOR UNCORRELATED-ROUNDOFF-NOISE ANALYSIS

The analyses appearing in the literature concerning roundoff noise in digital filters usually employ the simplifying and often reasonable assumption of uncorrelated roundoff errors from sample to sample and from one error source (multiplier or other rounding point) to another.[9,12,13] This assumption is based on the intuitively plausible and experimentally supported notion that for sufficiently large and dynamic signals within the filter, the small roundoff error made at one point in the network and/or in time should have little relationship to (that is, correlation with) the roundoff error made at any other point in the network and/or time. The advantage of assuming uncorrelated errors from one sample to another is that the noise injected into the filter by each rounding operation is then "white"; while the advantage of assuming uncorrelated error sources is that the output noise power spectrum may then be computed as simply the superposition of the (filtered) noise spectra due to the separate error sources.[12] Experimental results which support the validity of this assumption, even in the case of a single sinusoidal input, are presented in Ref. 1. In this section, we introduce the notation and develop the analysis pertaining to uncorrelated roundoff noise for later use in investigating the synthesis of digital filters.

Digital filter networks are composed of three basic elements: adders, constant multipliers, and delays. The interconnection of these elements into a particular network configuration is the key step in digital filter synthesis. For our purposes here, we need only consider the network as a directed graph, with the multipliers and delays being represented by graph branches. The branch interconnection points, or nodes, will be divided into two types: "summation nodes", which correspond to the adders and have multiple inputs and a single output, and "branch nodes", which correspond to simple "wired" interconnections that have a single input and one or more outputs.

A digital filter network may thus be represented as shown in Fig. 1. The input to and output from the filter at time $t = nT$ are denoted by $u(n)$ and $y(n)$, respectively. The corresponding output from the ith branch node is denoted by $v_i(n)$; while the roundoff error introduced into the filter at the jth summation node is denoted by $e_j(n)$. Since with fixed-point arithmetic, rounding is performed only after multiplications, non-zero roundoff errors are "input" to the filter only at those summation nodes which follow constant (non-integer) multiplier branches, as depicted in Fig. 2.

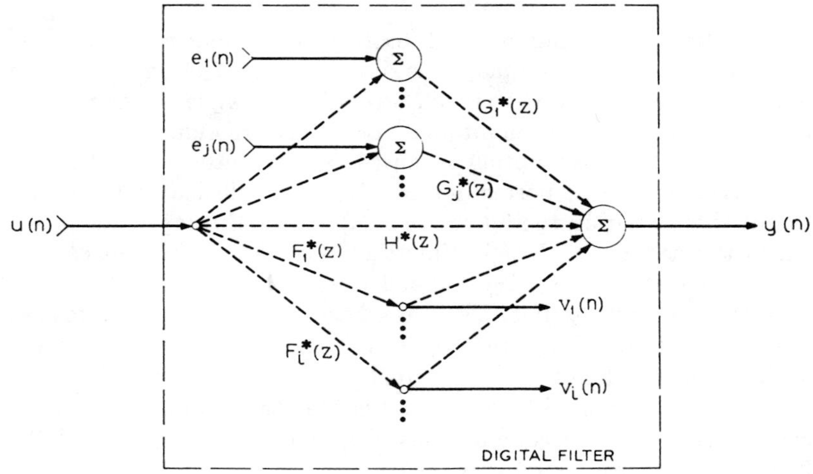

Fig. 1 — General digital filter model.

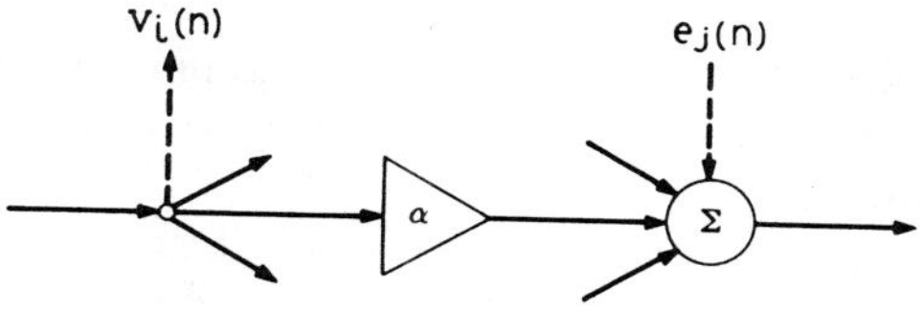

Fig. 2 — Constant multiplier with preceding branch node and succeeding summation node.

For a unit sample input to the filter at $t = 0$ and no rounding [that is, $u(0) = 1$, $u(n) = 0$ for $n \neq 0$, and $e_j(n) = 0$ for all j and n], the resulting output values $y(n)$ and $v_i(n)$ for all $n \geqq 0$ and all i are designated as $h(n)$ and $f_i(n)$, respectively. Alternatively, for a unit sample input to the jth summation node and zero inputs otherwise [that is, $e_j(0) = 1$, $e_j(n) = 0$ for $n \neq 0$, and $e_k(n) = u(n) = 0$ for all n and for $k \neq j$], the resulting output values $y(n)$ for all $n \geqq 0$ are denoted by $g_j(n)$. We thus have the following transfer functions of interest, expressed in z-transform form:

From filter input to output:

$$H^*(z) = \sum_{n=0}^{\infty} h(n) z^{-n}. \tag{1}$$

From filter input to i^{th} branch-node output:

$$F_i^*(z) = \sum_{n=0}^{\infty} f_i(n) z^{-n}. \tag{2}$$

From j^{th} summation-node input to filter output:

$$G_j^*(z) = \sum_{n=0}^{\infty} g_j(n) z^{-n}. \tag{3}$$

These transfer functions are indicated in Fig. 1.

The frequency responses (Fourier transforms) corresponding to the above transfer functions are given by [3–5]

$$H(\omega) = H^*(e^{i\omega T}), \tag{4}$$

$$F_i(\omega) = F_i^*(e^{i\omega T}), \tag{5}$$

$$G_k(\omega) = G_k^*(e^{i\omega T}). \tag{6}$$

This notation will be used throughout this paper. That is, for any z-transform $A^*(z)$ which converges for $|z| = 1$, the corresponding Fourier transform is given by

$$A(\omega) = A^*(e^{i\omega T}).$$

If scaling has been included in the filter design in order to satisfy certain dynamic-range constraints, then prime marks (′) are added to denote this fact [for example, $F_i'(\omega)$, $F_j^{*\prime}(z)$].

Each error source (rounding operation) within the filter is assumed to inject white noise of uniform power-spectral density N_0. Assuming uniformly distributed rounding errors with zero mean, the variance of the roundoff noise from each error source is given by[12,13]

$$\sigma_0^2 = \Delta^2/12 \tag{7}$$

where Δ is the spacing of the quantization steps (after rounding). To eliminate the sampling period T from certain expressions of interest, we now define $N_0 = \sigma_0^2$. Hence, the variance, or total average power, corresponding to an arbitrary power-density spectrum $N(\omega)$ with no DC component (which implies a zero-mean process) is given by†

$$\sigma^2 = \frac{1}{\omega_s} \int_0^{\omega_s} N(\omega)\, d\omega \tag{8}$$

where ω_s is the radian sampling frequency given by

$$\omega_s = 2\pi/T. \tag{9}$$

Assume now that k_j error sources input to the jth summation node. The spectral density of the roundoff error sequence $\{e_j(n)\}$ is then just $k_j N_0$ by our assumption of uncorrelated error sources. The total roundoff noise in the output of the filter thus has a power-density spectrum given by[12]

$$N_y(\omega) = \sigma_0^2 \sum_j k_j \mid G_j(\omega) \mid^2 \tag{10a}$$

where we have substituted σ_0^2 for N_0. If scaling has been included in the filter design, then the corresponding expression is just

$$N_y(\omega) = \sigma_0^2 \sum_j k_j' \mid G_j'(\omega) \mid^2 \tag{10b}$$

where $k_j' \geqq k_j$ to account for the additional scaling multipliers.

IV. DYNAMIC-RANGE CONSTRAINTS

The ultimate objective of the synthesis procedures to be investigated will be the minimization of some norm of $N_y(\omega)$ for a given quantization step size Δ, subject to certain "constraints". One constraint is that the

† This normalization of $N(\omega)$ is further motivated by the derivation in Section V leading to equation (30b).

specified transfer function $H^*(z)$ must be maintained. Another fundamental, but often overlooked, constraint is the finite dynamic range of the filter. Specifically, the signals $v_i(n)$ at certain branch nodes within the filter cannot be allowed to "overflow" (that is, exceed the dynamic-range limitations), at least not more than some small percentage of the time, in order to prevent severe distortion in the filter output.

Overflow constraints are required only at certain branch nodes in the digital circuit because it is only the inputs to the constant multipliers which cannot be allowed to overflow when several standard numbering systems are used (for example, one's- or two's-complement binary).[14] Specifically, in the summation of more than two numbers, if the magnitude of the correct total sum is small enough to allow its representation by the K available digits, then in these numbering systems the correct total sum will be obtained regardless of the order in which the numbers are added, even if an overflow occurs in one of the partial sums. Hence, those node outputs which correspond to partial sums comprising a larger total sum may be allowed to overflow, as long as the total sum is constrained not to overflow. This property also applies when one of the inputs to a summation node has overflowed as a result of a multiplication by a coefficient of magnitude greater than one.

Turning to the formulation of the required overflow constraints, we may easily derive an upper bound on the magnitude of the signals $v_i(n)$ for all possible input sequences $\{u(n)\}$, neglecting the (small) error signals $e_j(n)$. Assuming zero initial conditions in the filter and $e_j(n) = 0$ for all j and n, the ith branch-node output $v_i(n)$ is given by

$$v_i(n) = \sum_{k=0}^{\infty} f_i(k)u(n-k), \quad \text{all} \quad n. \tag{11}$$

Therefore, given that $u(n)$ is bounded in magnitude by some number M for all n, an upper bound on the magnitude of $v_i(n)$ is given by[15]

$$|v_i(n)| \leqq M \sum_{k=0}^{\infty} |f_i(k)|, \quad \text{all} \quad n. \tag{12}$$

Thus, if the node signal $v_i(n)$ is also to be bounded in magnitude by M for all possible input sequences, the associated scaling must ensure that

$$\sum_{k=0}^{\infty} |f_i'(k)| \leqq 1. \tag{13}$$

That (13) is not only a sufficient condition to rule out overflow for all possible input sequences $\{u(n)\}$, but also a necessary condition, is easily shown by letting $u(n) = \pm M$ for all n, with $\operatorname{sgn}[u(n_0 - k)] = \operatorname{sgn}[f_i(k)]$ for some $n = n_0$ and all $k \geqq 0$. Then from equation (11) we see that (12) is satisfied with equality in this case, and thus (13) is a necessary condition, as well.

The norm of $f_i'(k)$ employed in (13) is not very useful in practice because of the difficulty of evaluating the indicated summation in all but the simplest cases. Also, for large classes of input signals, (12) and thus (13) are overly pessimistic. Therefore, we now derive alternate conditions on (the transform of) the scaled unit-sample response $\{f_i'(n)\}$ which ensure that for certain classes of input signals, the corresponding branch-node output $v_i(n)$ cannot overflow. The derivation of these conditions for discrete systems closely parallels the corresponding derivation for continuous systems, as given by Papoulis.[16]

An alternate expression for equation (11) in terms of z-transforms is derived as follows: Consider an (absolutely summable) deterministic input sequence $\{u(n)\}$ possessing the z-transform

$$U^*(z) = \sum_{n=-\infty}^{\infty} u(n)z^{-n}, \qquad a < |z| < b, \tag{14}$$

for some $a < 1$ and $b > 1$. Stability requires that $F_i^*(z)$, defined in equation (2), exist for all $|z| > c$ for some $c < 1$. Hence, the z-transform of $\{v_i(n)\}$ is given by[3]

$$V_i^*(z) = F_i^*(z)U^*(z), \qquad d < |z| < b, \tag{15}$$

where $d = \max(a, c)$. The inverse transform of equation (15) is given by[3]

$$v_i(n) = \frac{1}{2\pi j}\oint_\Gamma V_i^*(z)z^{n-1}\,dz \tag{16}$$

where the contour of integration Γ is contained in the region of convergence $d < |z| < b$. Since $d < 1$ and $b > 1$, let Γ be the unit circle in the z plane ($|z| = 1$), and perform the change of variables $z = e^{j\omega T}$ in equation (16). Using equation (15), the resulting equation becomes

$$v_i(n) = \frac{1}{\omega_s}\int_0^{\omega_s} F_i(\omega)U(\omega)e^{jn\omega T}\,d\omega. \tag{17}$$

The conditions to be derived from equation (17) are most easily expressed in terms of L_p norms, defined for an arbitrary periodic function $A(\cdot)$ with period ω_s by[17]

$$\|A\|_p = \left[\frac{1}{\omega_s}\int_0^{\omega_s} |A(\omega)|^p\,d\omega\right]^{1/p} \tag{18a}$$

for each real $p \geqq 1$ such that

$$\int_0^{\omega_s} |A(\omega)|^p \, d\omega < \infty .$$

It can be shown[17] that for $A(\cdot)$ continuous, the limit of equation (18a) as $p \rightarrow \infty$ exists and is given by

$$\| A \|_\infty = \max_{0 \leq \omega \leq \omega_s} | A(\omega) | . \tag{18b}$$

Assume now that $| U(\omega) |$ is bounded from above by some number M(that is, $\| U \|_\infty \leqq M$). Then, from equation (17),

$$| v_i(n) | \leqq M \frac{1}{\omega_s} \int_0^{\omega_s} | F_i(\omega) | \, d\omega$$

or

$$| v_i(n) | \leqq \| F_i \|_1 \cdot \| U \|_\infty . \tag{19}$$

In exactly the same manner, we may also show that

$$| v_i(n) | \leqq \| F_i \|_\infty \cdot \| U \|_1 . \tag{20}$$

Applying the Schwarz inequality to equation (17), on the other hand, yields that

$$| v_i(n) |^2 \leqq \frac{1}{\omega_s^2} \int_0^{\omega_s} | F_i(\omega) |^2 \, d\omega \int_0^{\omega_s} | U(\nu) |^2 \, d\nu$$

or

$$| v_i(n) | \leqq \| F_i \|_2 \cdot \| U \|_2 . \tag{21}$$

Note that (19), (20), and (21) are all of the form

$$| v_i(n) | \leqq \| F_i \|_p \cdot \| U \|_q , \quad \left(\frac{1}{p} + \frac{1}{q} = 1 \right) \tag{22}$$

for $p, q = 1, 2$, and ∞. It can be shown[18] that (22) is true in general for all $p, q > 1$ satisfying $1/p + 1/q = 1$; and we have shown in (19) and (20) that if the L_∞ norms exist, then (22) holds for $p, q = 1$, as well. The general relation in (22) for all $p, q > 1$, is derived from Holder's inequality.

A simple, but important special case of (22) results from letting $F_i^*(z) = F_i(\omega) = 1$. Since $\| 1 \|_p = 1$ for all $p \geqq 1$, we then have simply

$$| u(n) | \leqq \| U \|_q , \quad \text{all} \quad q \geqq 1. \tag{23}$$

But since (23) holds for all sequences $\{u(n)\}$, it must also be true that

$$| v_i(n) | \leqq || V_i ||_r , \quad \text{all} \quad r \geqq 1.$$

This is, in fact, the basis of (22), for Holder's inequality actually states that

$$|| V_i' ||_1 \leqq || F_i ||_p \, || U ||_q , \qquad \left(\frac{1}{p} + \frac{1}{q} = 1\right).$$

Therefore, the real implication of (22) is that the mean absolute value of $V_i(\omega)$ is bounded by $|| F_i ||_p || U ||_q$, and this, in turn, provides a bound on $| v_i(n) |$.

Assume, therefore, that the input transform $U(\omega)$ satisfies $|| U ||_q \leqq M$ for some $q \geqq 1$. From (23) we immediately have that $| u(n) | \leqq M$ for all n. Then, if $| v_i(n) |$ is also to be bounded by M, (22) provides a sufficient condition on the scaling to ensure this, namely

$$|| F_i' ||_p \leqq 1, \qquad (|| U ||_q \leqq M) \tag{24}$$

for $p = q/(q - 1)$. Inequality (24) is the desired condition to replace the more general, but often less useful condition given by (13).

From an engineering viewpoint, the most significant values for p and q would seem to be 1, 2, and ∞. The case $p = 1$, $q = \infty$ requires that the input transform $U(\omega)$ be everywhere bounded in magnitude by M (that is, $|| U ||_\infty \leqq M$), in which case only the L_1 norm of the scaled transfer function $F_i'(\omega)$ need satisfy (24). For an input of finite energy $E = \sum_n u^2(n)$, Parseval's identity implies that $|| U ||_2^2 = E$, and thus with $M \geqq (E)^{\frac{1}{2}}$, (24) can be satisfied for $p = q = 2$.

The case of $p = \infty$, $q = 1$ in (24) implies the most stringent condition on $F_i'(\omega)$ because from equation (18) it is evident that

$$|| F_i' ||_p \leqq || F_i' ||_\infty \tag{25}$$

for all $p \geqq 1$. It is clear, for example, that for a sinusoidal input of amplitude $A \leqq M$ and arbitrary frequency ω_0, we must have $| F_i'(\omega) | \leqq 1$ for all ω (that is, $|| F_i' ||_\infty \leqq 1$) to ensure that $| v_i(n) | \leqq M$ for all n. However, a sinusoidal input sequence $\{u(n)\}$ is not absolutely summable, and thus $U^*(z)$ as defined in equation (14) does not exist in this case. This difficulty may be circumvented, as is common in Fourier analysis, by assuming a finite sequence of length N and then passing to the limit as $N \to \infty$. The resulting (Fourier) transform of $\{u(n)\}$ is of the form

$$U_0(\omega) = \frac{A}{2} e^{i\theta}[\delta(\omega - \omega_0) + \delta(\omega - \omega_s + \omega_0)], \qquad (0 \leqq \omega \leqq \omega_s) \tag{26}$$

where $\delta(\omega)$ is the familiar Dirac delta function defined by

$$\delta(\omega) = 0, \qquad \omega \neq 0,$$
$$\int_{-\infty}^{\infty} \delta(\omega)\, d\omega = 1. \tag{27}$$

$U_0(\omega)$ is, of course, periodic in ω with period ω_s. From equations (18a), (26), and (27), we immediately have that $\| U_0 \|_1 = A \leqq M$, and thus with $p = \infty$, (24) is applicable for sinusoidal input sequences, as expected.

V. RANDOM INPUT CASE

In the case of random input sequences, (24) is not directly applicable because the z-transform $U^*(z)$ is not defined. Similar conditions may be obtained, however, by considering the discrete autocorrelation function $\varphi(\cdot)$, defined for a (wide-sense) stationary sequence $\{w(n)\}$ by

$$\varphi_w(m) = E[w(n)w(n+m)] \tag{28}$$

where $E[\cdot]$ is the expected-value operator. A z-transform $\Phi_w^*(z)$ may be defined for the sequence $\{\varphi_w(m)\}$ as in equation (14) with an inverse transform as in (16). Assuming ergodicity and a zero mean ($E[w(n)] = 0$) for $\{w(n)\}$, we immediately have from equation (28) that the variance, or total average power, of $\{w(n)\}$ is given by

$$\varphi_w(0) = E[w^2(n)] = \sigma_w^2, \tag{29}$$

and from equation (16) we also have

$$\varphi_w(0) = \frac{1}{2\pi j} \oint_\Gamma \Phi_w^*(z) z^{-1}\, dz. \tag{30a}$$

Letting Γ be the unit circle ($z = e^{j\omega T}$), equations (29) and (30a) imply that

$$\sigma_w^2 = \frac{1}{\omega_s} \int_0^{\omega_s} \Phi_w(\omega)\, d\omega. \tag{30b}$$

Hence, from equation (8) we see that $\Phi_w(\omega)$ is just the power-density spectrum of the sequence $\{w(n)\}$.

For an input sequence $\{u(n)\}$ whose autocorrelation function has the z-transform $\Phi_u^*(z)$, it is well-known that the corresponding transform for the output $\{v_i(n)\}$ is given by

$$\Phi_{v_i}^*(z) = F_i^*(z) F_i^*(z^{-1}) \Phi_u^*(z) \tag{31a}$$

or

$$\Phi_{v_i}(\omega) = | F_i(\omega) |^2 \Phi_u(\omega). \tag{31b}$$

Equations (29) through (31) imply then that

$$\sigma_{v_i}^2 = \frac{1}{\omega_s} \int_0^{\omega_s} | F_i(\omega) |^2 \Phi_u(\omega) \, d\omega. \tag{32}$$

Since equation (32) is of the same basic form as (17), a derivation similar to that leading to (22) must yield the following relations for $p, q \geqq 1$:

$$\sigma_{v_i}^2 \leqq || F_i^2 ||_p \cdot || \Phi_u ||_q , \qquad \left(\frac{1}{p} + \frac{1}{q} = 1\right) \tag{33a}$$

or, from equation (17),

$$\sigma_{v_i}^2 \leqq || F_i ||_{2p}^2 \cdot || \Phi_u ||_q , \qquad \left(\frac{1}{p} + \frac{1}{q} = 1\right). \tag{33b}$$

Two cases of (33) are of particular interest, namely

$$\sigma_{v_i}^2 \leqq || F_i ||_2^2 \cdot || \Phi_u ||_\infty \tag{34}$$

and

$$\sigma_{v_i}^2 \leqq || F_i ||_\infty^2 \cdot || \Phi_u ||_1 . \tag{35}$$

In view of equation (25), we see that (34) implies the most stringent condition on the input spectrum $\Phi_u(\omega)$, whereas (35) yields the most stringent condition on the transfer function $F_i(\omega)$. From (34) and (30b), for example, we have that if the input power-density spectrum is "white" [that is, $\Phi_u(\omega) = \sigma_u^2$ for all ω], then $\sigma_{v_i}^2 \leqq || F_i ||_2^2 \sigma_u^2$. Hence, if the input sequence $\{u(n)\}$ is a Gaussian process,[19] the node output sequence $\{v_i(n)\}$ will overflow no more (in percentage of time) than does the input, provided only that

$$|| F_i' ||_2 \leqq 1. \tag{36}$$

The inequality in (35) requires, on the other hand, that for an input sinusoid of arbitrary amplitude and frequency, $F_i'(\omega)$ must satisfy

$$|| F_i' ||_\infty \leqq 1 \tag{37}$$

to ensure against overflow, as we have seen earlier from (24).

To summarize, dynamic-range constraints of the form

$$|| F_i' ||_p \leqq 1, \qquad p \geqq 1 \tag{38}$$

have been derived for both deterministic and random inputs, where

$F_i'(\omega)$ is the (scaled) transfer response from the filter input to the ith branch node and $\|\cdot\|_p$ denotes the L_p norm defined in equation (18). For a deterministic input with amplitude spectrum $U(\omega)$, (38) assumes that

$$\| U \|_q \leqq M, \qquad q = \frac{p}{p-1}, \tag{39}$$

where M is the maximum allowable signal amplitude. For a random input, on the other hand, the use of (38) requires appropriate conditions on $\| \Phi_u \|_r$, $r = p/(p-2)$ and $p \geqq 2$, where $\Phi_u(\omega)$ is the power-density spectrum of the input sequence.

The effect of (38) and (39) is to bound the mean absolute value of the amplitude spectrum at the ith branch node (that is, $\| V_i \|_1$) which, in turn, bounds the peak signal amplitude at that node. The use of (38) in conjunction with (33), however, bounds only the average power at the ith branch node, and thus the relationship between this average power and the peak signal amplitude at the node must also be determined in order to provide an effective dynamic-range constraint.

VI. TRANSPOSE SYSTEMS

In the evaluation of different circuit configurations for a given digital filter, a useful concept relating certain of these configurations is that of "transpose configurations". This relationship is a general property of linear graphs[20] and will be presented here in terms of a state-variable formulation.

The general state equations for a linear, time-invariant discrete system are given by[21]

$$\begin{aligned} \mathbf{x}(n+1) &= A\mathbf{x}(n) + B\mathbf{u}(n), \\ \mathbf{y}(n) &= C\mathbf{x}(n) + D\mathbf{u}(n) \end{aligned} \tag{40}$$

where $\mathbf{x}(n)$ is an N-dimensional vector describing the state of the system at time $t = nT$, $\mathbf{u}(n)$ is the corresponding J-dimensional input vector, $\mathbf{y}(n)$ is the corresponding I-dimensional output vector, and A, B, C, and D are fixed parameter matrices of the appropriate dimensions relating the input, state, and output vectors as given by equation (40). The $(N+I) \times (N+J)$ matrix S defined by

$$S = \begin{bmatrix} A & B \\ C & D \end{bmatrix} \tag{41}$$

provides a convenient single parameter matrix which describes the complete discrete system.

A transfer function matrix $\mathcal{H}_S^*(z)$ may be defined for the system (described by) S relating the input and output vector sequences $\{\mathbf{u}(n)\}$ and $\{\mathbf{y}(n)\}$ by

$$\mathbf{Y}^*(z) = \mathcal{H}_S^*(z)\mathbf{U}^*(z) \tag{42}$$

where $\mathbf{U}^*(z)$ and $\mathbf{Y}^*(z)$ are the vector z-transforms of $\{\mathbf{u}(n)\}$ and $\{\mathbf{y}(n)\}$, respectively. $\mathcal{H}_S^*(z)$ is readily shown to be given by[21]

$$\mathcal{H}_S^*(z) = C(zI - A)^{-1}B + D \tag{43}$$

where $(\cdot)^{-1}$ denotes the matrix inverse and I is the N-dimensional identity matrix.

Consider now a new system which is described by the parameter matrix S^t, that is,

$$S^t = \begin{bmatrix} A^t & C^t \\ B^t & D^t \end{bmatrix} \tag{44}$$

where $(\cdot)^t$ denotes the matrix transpose. From equations (41) and (43) it is easily seen that the transfer function matrix for the new system S^t is given by

$$\begin{aligned} \mathcal{H}_{S^t}^*(z) &= B^t(zI - A^t)^{-1}C^t + D^t \\ &= [\mathcal{H}_S^*(z)]^t. \end{aligned} \tag{45}$$

Thus, the transfer function matrix for the system S^t is simply the transpose of the transfer function matrix for the system S. That is, the element $H_{ij}^*(z)$ from $\mathcal{H}_S^*(z)$, which is the transfer function from the jth input to the ith output of system S, equals the element $H_{ji}^{*t}(z)$ from $\mathcal{H}_{S^t}^*(z)$, that is, the transfer function from the ith input to the jth output of S^t. Note also that while the system S has a total of J inputs and I outputs, the system S^t has I inputs and J outputs.

The concept of transpose systems will be particularly useful to us in conjunction with the digital-filter model introduced in Section III and depicted in Fig. 1. Defining the input and output vectors for the filter by

$$\mathbf{u}(n) = \begin{bmatrix} u(n) \\ e_1(n) \\ \vdots \\ e_J(n) \end{bmatrix} \quad \text{and} \quad \mathbf{y}(n) = \begin{bmatrix} y(n) \\ v_1(n) \\ \vdots \\ v_I(n) \end{bmatrix} \tag{46}$$

respectively, the transfer function matrix for the filter is given by

$$\mathcal{H}^*(z) = \begin{bmatrix} H^*(z) & G_1^*(z) & \cdots & G_J^*(z) \\ F_1^*(z) & \text{---} & \text{---} & \text{---} \\ \vdots & \text{---} & \text{---} & \text{---} \\ F_I^*(z) & \text{---} & \text{---} & \text{---} \end{bmatrix} \tag{47}$$

where the specific expressions for the elements in other than the first row and first column are unimportant for our purposes. By equation (45), the transfer function matrix for the corresponding transpose system is then simply

$$\mathcal{H}_t^*(z) = \begin{bmatrix} H^*(z) & F_1^*(z) & \cdots & F_I^*(z) \\ G_1^*(z) & \text{---} & \text{---} & \text{---} \\ \vdots & \text{---} & \text{---} & \text{---} \\ G_J^*(z) & \text{---} & \text{---} & \text{---} \end{bmatrix}. \tag{48}$$

Note, in particular, that the transfer function from input-1 to output-1 [that is, $H^*(z)$, the ideal transfer function from filter input to filter output] is the same for both systems.

As discussed more fully in Ref. 1, the circuit configuration realizing a given system S is not necessarily unique, and hence neither is the configuration for the transpose system S^t. However, given a particular configuration for the system S, a unique "transpose configuration", which realizes S^t, may be derived from the given configuration for S by simply reversing the direction of all branches in the given network! In particular, then, all delays and constant multipliers remain the same except for the change in direction. All summation nodes in the given configuration become branch nodes in the transpose configuration, and all branch nodes become summation nodes. Likewise, all inputs in the given configuration become outputs in the transpose configuration, and all outputs become inputs.†

That the transpose configuration defined above actually realizes the transpose system S^t is easily seen by considering the state equations in (40). The constant multiplier(s) corresponding to the element d_{ij} of the matrix D and relating the jth input and the ith output of the original configuration must relate the ith input and the jth output of the transpose

† Note that the transpose system S^t is fundamentally different from the "adjoint" system[22] because, although the signal flow is reversed in both, the transpose system does not run "backwards in time."

configuration, and thus $d_{ij} = d^t_{ji}$ for all i and j. The multiplier(s) corresponding to the element b_{ij} of B and relating the jth input and the ith state of the original configuration must, on the other hand, relate the ith state and the jth output of the transpose configuration, and thus $b_{ij} = c^t_{ji}$ for all i and j. Similarly, $c_{ij} = b^t_{ji}$ for all i and j. Finally, the multiplier(s) corresponding to a_{ij} and relating $x_j(n)$ and $x_i(n + 1)$ in the original configuration must, in the transpose configuration, relate $x_i(n)$ and $x_j(n + 1)$, and thus $a_{ij} = a^t_{ji}$ for all i and j. Therefore, the transpose configuration indeed realizes the system S^t.

VII. AN EXAMPLE: THE DIRECT FORM

To demonstrate the application of the results of the preceding sections, we now evaluate and compare the roundoff-noise outputs from two transpose configurations for a digital filter. The scaling required to satisfy the overflow constraints in (38) is derived, and the effect of this scaling on the output roundoff noise is determined.

The transfer function $H^*(z)$, defined in equation (1) and relating the input and output of the digital filter, may be expressed as a rational function in z of the form[3,4]

$$H^*(z) = \frac{\sum_{i=0}^{N} a_i z^{-i}}{1 + \sum_{i=1}^{N} b_i z^{-i}} = \frac{A^*(z)}{B^*(z)}. \tag{49}$$

Assuming that a_N and b_N are not both zero, N is referred to as the "order" of the filter. There are many different, but equivalent, forms in which equation (49) may be written, with a number of equivalent circuit configurations corresponding to each of these forms (at least two transpose configurations). Those forms such as equation (49) which require the minimum number of multiplications and additions in the general case (that is, $2N + 1$ and $2N$, respectively) are referred to as "canonical" forms. In general, however, it is necessary to add additional scaling multipliers to these canonical forms in order to satisfy the overflow constraints in (38).

The form of $H^*(z)$ given in equation (49) is often called the "direct form" of a digital filter. It has been pointed out by Kaiser[6] that use of the direct form is usually to be avoided because of the sensitivity of the roots of higher-order polynomials to small variations (that is, quantization errors) in the polynomial coefficients. The roundoff-noise outputs from the direct form can also be much larger than from other canonical forms.[15] Nevertheless, the direct form is of theoretical interest, and it provides a convenient illustration of our results. Similar investigations for the two canonical forms most commonly employed in practice—the cascade and parallel forms—are described in Ref. 1.

Two transpose configurations which implement the direct form with scaling are shown in Figs. 3 and 4. These configurations actually realize $H^*(z)$ in the form

$$H^*(z) = \frac{K_k' \sum_{i=0}^{N} {}_k a_i' z^{-i}}{1 + \sum_{i=1}^{N} b_i z^{-i}} \tag{50}$$

where ${}_k a_i' = a_i/K_k'$, and the additional scaling multipliers K_k', $k = 1, 2$, are required to satisfy (38) in the general case. The configuration in Fig. 3 will be designated as form 1 (that is, $k = 1$), and Fig. 4 as form 2 (that is, $k = 2$).

The branch nodes at which overflow constraints are required (because these signals input to multipliers) are indicated by (*). The dynamic-range limitations are obviously satisfied (by assumption) at the input to the filter, but for completeness, an overflow constraint is included there as indicated. The scaled transfer responses ${}_k F_i'(\omega)$ to these nodes are noted in Figs. 3 and 4, and the corresponding unscaled responses ${}_k F_i(\omega)$ apply, of course, when $K_k' = 1$.

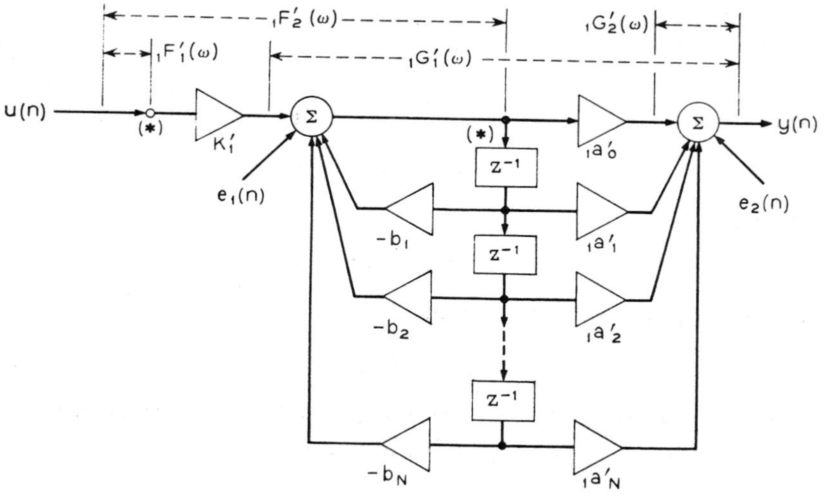

Fig. 3 — Direct form 1 with scaling.

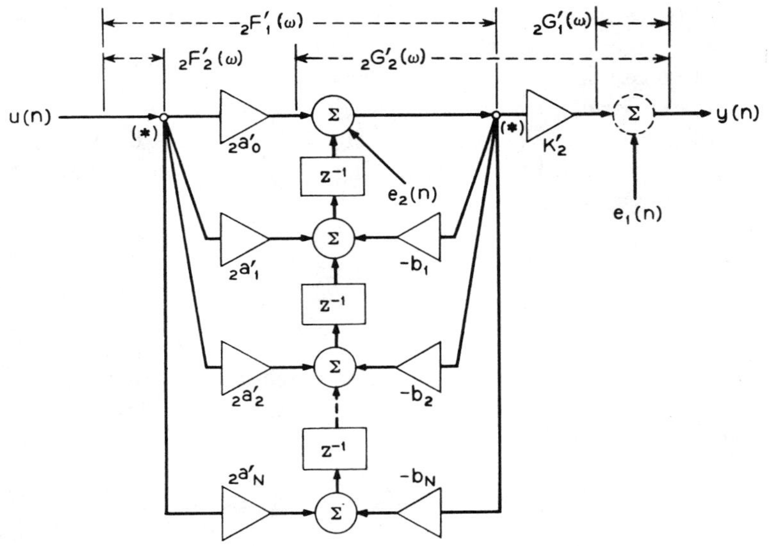

Fig. 4 — Direct form 2 with scaling.

It is intuitively clear that to preserve the greatest possible signal-to-noise ratio, the scaling should reduce the magnitude of ${}_kF'_i(\omega)$ no more than is necessary (or should increase it as much as possible, as the case may be). In other words, ${}_kF'_i(\omega)$ should satisfy

$$\| {}_kF'_i \|_p = 1. \tag{51}$$

This condition will be satisfied if the scaling factors ${}_ks_i$, defined by

$${}_kF'_i(\omega) = {}_ks_i \cdot {}_kF_i(\omega), \tag{52a}$$

are given by

$${}_ks_i = 1/\| {}_kF_i \|_p . \tag{52b}$$

It is readily seen from Figs. 3 and 4 that

$${}_1F'_1(\omega) = {}_2F'_2(\omega) = 1, \tag{53}$$

and hence equation (51) is automatically satisfied for these responses. Of more interest, however, are the responses

$${}_1F'_2(\omega) = \frac{K'_1}{B(\omega)} = K'_1 \, {}_1F_2(\omega) \tag{54}$$

and

$$ {}_2F_1'(\omega) = \frac{H(\omega)}{K_2'} = \frac{{}_2F_1(\omega)}{K_2'}. \tag{55} $$

From equations (52), (54), and (55), it follows that (51) is satisfied for these configurations if (and only if)

$$ K_1' = 1/\|\, 1/B \,\|_p \tag{56} $$

and

$$ K_2' = \|\, H \,\|_p \,. \tag{57} $$

The rounding-error inputs $e_i(n)$ are also shown in Figs. 3 and 4 along with the transfer responses ${}_kG_j'(\omega)$ from these inputs to the output of the filter. Note that in form 2 (Fig. 4) the error input $e_2(n)$ incorporates the roundoff errors from all of the multipliers except K_2' even though these error sources are separated by delays (z^{-1}). This is done for convenience and is possible because of the assumption of uncorrelated errors from sample to sample and source to source. The noise weights k_j' [see equation (10)] for form 1 are thus

$$ {}_1k_1' = {}_1k_2' = N + 1; \tag{58a} $$

while for form 2,

$$ {}_2k_1' = 1 \quad \text{and} \quad {}_2k_2' = 2N + 1. \tag{58b} $$

The indices i and j of the ${}_kF_i(\omega)$ and ${}_kG_j(\omega)$ have been assigned in such a way that forms 1 and 2 are related as in equations (47) and (48). That is, these unscaled responses satisfy the following equations:

$$ {}_1F_i(\omega) = {}_2G_i(\omega), \qquad i = 1, 2, \tag{59a} $$

$$ {}_1G_j(\omega) = {}_2F_j(\omega), \qquad j = 1, 2. \tag{59b} $$

Note that the scaled responses ${}_kF_i'(\omega)$ and ${}_kG_j'(\omega)$ are not related as in equation (59) because, in general, $K_1' \neq K_2'$. In particular,

$$ {}_1G_1'(\omega) = \frac{H(\omega)}{K_1'} = \left(\frac{K_2'}{K_1'}\right) {}_2F_1'(\omega); \tag{60} $$

while

$$ {}_2G_2'(\omega) = \frac{K_2'}{B(\omega)} = \left(\frac{K_2'}{K_1'}\right) {}_1F_2'(\omega). \tag{61} $$

However, we do have, as in equation (53), that

$$ {}_1G_2'(\omega) = {}_2G_1'(\omega) = 1. \tag{62} $$

From equations (10) and (53) through (62), the power-spectral densities of the roundoff-noise outputs from these two configurations are thus computed to be

$$ {}_1N_v(\omega) = \sigma_0^2(N + 1)\left\{1 + \left\| \frac{1}{B} \right\|_p^2 \mid H(\omega) \mid^2\right\} \tag{63a} $$

and

$$ {}_2N_v(\omega) = \sigma_0^2\left\{1 + (2N + 1) \mid\mid H \mid\mid_p^2 \left| \frac{1}{B(\omega)} \right|^2\right\}. \tag{63b} $$

The variances, or total average powers of the output roundoff noise from these configurations are then, from equations (8) and (18), simply

$$ \mid\mid {}_1N_v \mid\mid_1 = \sigma_0^2(N + 1)\left\{1 + \left\| \frac{1}{B} \right\|_p^2 \mid\mid H \mid\mid_2^2\right\} \tag{64a} $$

and

$$ \mid\mid {}_2N_v \mid\mid_1 = \sigma_0^2\left\{1 + (2N + 1) \mid\mid H \mid\mid_p^2 \left\| \frac{1}{B} \right\|_2^2\right\}. \tag{64b} $$

The peak noise densities $\mid\mid {}_kN_v \mid\mid_\infty$ are, on the other hand, bounded by

$$ \mid\mid {}_1N_v \mid\mid_\infty \leqq \sigma_0^2(N + 1)\left\{1 + \left\| \frac{1}{B} \right\|_p^2 \mid\mid H \mid\mid_\infty^2\right\} \tag{65a} $$

and

$$ \mid\mid {}_2N_v \mid\mid_\infty \leqq \sigma_0\left\{1 + (2N + 1) \mid\mid H \mid\mid_p^2 \left\| \frac{1}{B} \right\|_\infty^2\right\}. \tag{65b} $$

We now compare direct forms 1 and 2 on the basis of (64) and (65). Although comparisons based on bounds for $\mid\mid {}_kN_v \mid\mid_\infty$ as in (65) do not, of course, necessarily hold for $\mid\mid {}_kN_v \mid\mid_\infty$ itself, experimental results have indicated that such comparisons are quite effective qualitatively, and often quantitatively as well.[1] Consider first the expressions in equation (64) for $p = 2$ and in (65) for $p = \infty$ (that is, $\mid\mid N_v \mid\mid_r$, $r = 1, \infty$, for $p = r + 1$]. In these two cases, the only difference between the (a) and (b) expressions for forms 1 and 2, respectively, are the k_i', as given in equation (58). In particular, for $\mid\mid 1/B \mid\mid_p^2 \mid\mid H \mid\mid_p^2 \gg 1$ as is often the case, the $\mid\mid N_v \mid\mid_r$ for form 1 are approximately half, or 3 db less than, those for form 2. This result simply reflects the fact that only half of the noise sources in form 1 input at other than the filter output; whereas in form 2, all but one input within the filter. Hence, if the gains from these inputs to the output are large, form 1 is preferable to form 2 by up to 3 db.

For $p \neq r + 1$, however, the differences in the k_i' are of secondary importance compared with the potential differences due to the mixture of L_2 and L_∞ norms in (64) and (65). In particular, letting

$$\theta_{pq} = \left\| \frac{1}{B} \right\|_p^2 \| H \|_q^2 , \tag{66a}$$

we immediately see that if $\theta_{\infty 2} \gg \theta_{2\infty}$, then form 2 is better for $p = \infty$ while form 1 is better for $p = 2$. If, on the other hand, $\theta_{\infty 2} \ll \theta_{2\infty}$, then the opposite applies.

To gain insight into the above conditions, we rewrite equation (66a) as

$$\theta_{pq} = \left\| \frac{1}{B} \right\|_p^2 \left\| \frac{A}{B} \right\|_q^2 . \tag{66b}$$

It is then clear that the difference between $\theta_{\infty 2}$ and $\theta_{2\infty}$ is due entirely to the effect of $A(\omega)$ on the L_q norms of $A(\omega)/B(\omega)$ for $q = 2, \infty$ versus the corresponding norms of $1/B(\omega)$. In particular, $A(\omega)$ affects the L_∞ norm in $\theta_{2\infty}$. But the L_∞ norm of a function "concentrates" exclusively on the maximum absolute value of that function; whereas the L_2 norm of a function reflects the r.m.s. absolute value of that function over all argument values. Therefore, the effect of $A(\omega)$ in $\theta_{2\infty}$ results from the alteration of the maxima of $| 1/B(\omega) |$ in $| A(\omega)/B(\omega) |$; while in $\theta_{\infty 2}$, the effect concerns the difference between $| 1/B(\omega) |$ and $| A(\omega)/B(\omega) |$ over all ω.

Intuitively, one expects that the former effect is potentially much greater; that is, in many cases $A(\omega)$ should affect the L_∞ norm in $\theta_{2\infty}$ much more than the L_2 norm in $\theta_{\infty 2}$. In particular, if $| A(\omega) |$ significantly attenuates the maxima of $| 1/B(\omega) |$ [as in a band-rejection filter, for example], then $\theta_{2\infty}$ should be much smaller than $\theta_{\infty 2}$. In this case, form 2 should be used for $p = \infty$, and form 1 for $p = 2$. If, however, $| A(\omega) |$ does not provide such attenuation, then $| A(\omega) |$ must be relatively constant within the band(s) where $| 1/B(\omega) |$ is largest [by the nature of $A(\omega)$], and hence

$$\left\| \frac{A}{B} \right\|_q \approx | A(\omega_0) | \cdot \left\| \frac{1}{B} \right\|_q \tag{67}$$

where ω_0 is a frequency at or near a maximum of $| 1/B(\omega) |$. But then,

$$\theta_{pq} \approx | A(\omega_0) | \left\| \frac{1}{B} \right\|_p \left\| \frac{1}{B} \right\|_q \approx \theta_{qp} , \tag{68}$$

and the difference between direct forms 1 and 2 should be less in this case.

VIII. SUMMARY

The interaction between the roundoff-noise output from a digital filter and the associated dynamic-range limitations has been investigated for the case of uncorrelated rounding errors from sample to sample and from one error source to another. The spectrum of the output roundoff noise from fixed-point implementations was readily shown to be of the form

$$N_y(\omega) = \sigma_0^2 \sum_i k_i' \mid G_i'(\omega) \mid^2 \tag{69}$$

where the $G_i'(\omega)$ are scaled transfer responses from certain "summation nodes" in the digital circuit to the filter output. σ_0^2 is the variance of the rounding errors from each multiplier (or other rounding point), and the k_i' are integers indicating the number of error inputs to the respective summation nodes.

Defining $F_i'(\omega)$ to be the scaled transfer response from the input to the ith "branch node" at which a dynamic-range constraint is required, constraints of the form

$$\| F_i' \|_p \leqq 1 \tag{70}$$

for $p \geqq 1$ were then derived, where $\| F_i' \|_p$ is the L_p norm of the response $F_i'(\omega)$. The appropriate value of p is determined by assumed conditions on the spectra of the input signals to the filter. The effect of (70) is to bound the maximum signal amplitude (for deterministic inputs) or the maximum average power (for random inputs) at the ith branch node.

A state-variable description was employed to formulate the general concept of "transpose configurations" for a digital network and to illustrate the usefulness of this concept in digital-filter synthesis. A particularly important result is that for a given unscaled configuration with transpose responses $F_i(\omega)$ and $G_i(\omega)$, as described above, the responses $F_i^t(\omega)$ and $G_i^t(\omega)$ for the corresponding transpose configuration are given by

$$F_i^t(\omega) = G_i(\omega) \quad \text{and} \quad G_i^t(\omega) = F_i(\omega). \tag{71}$$

Hence, although the overall transfer functions for these two configurations are the same, their roundoff-noise outputs can be quite different, in general. The transpose configuration is obtained by simply reversing the direction of all branches in the given network configuration, and the poles and zeros of the network are thus realized in reverse order in the transpose configuration.

To illustrate these results, the roundoff-noise spectra $N_y(\omega)$ for two transpose configurations for the direct form of a digital filter were calculated and compared. The direct form should usually be avoided in practice,[6] but it is still of theoretical interest and provides a convenient example of our general approach. Using a very natural assignment of the indices i and j for the unscaled $F_i(\omega)$ and $G_j(\omega)$, equation (69) was shown to be of the form

$$N_v(\omega) = \sigma_0^2\left\{k'_{M+1} + \sum_{i=1}^{M} k'_i \, || F_i ||_p^2 \, | G_i(\omega) |^2\right\} \tag{72}$$

for these (scaled) configurations for the direct form, where M is the number of error inputs at other than the output of the filter. Hence, the variance, or total average power, of the output roundoff noise is simply

$$\sigma_v^2 = \sigma_0^2\left\{k'_{M+1} + \sum_{i=1}^{M} k'_i \, || F_i ||_p^2 \, || G_i ||_2^2\right\}; \tag{73}$$

while the peak spectral density $|| N_v ||_\infty$ is bounded by

$$|| N_v ||_\infty \leqq \sigma_0^2\left\{k'_{M+1} + \sum_{i=1}^{M} k'_i \, || F_i ||_p^2 \, || G_i ||_\infty^2\right\}. \tag{74}$$

Identical expressions to (72) through (74) can also be derived for the parallel and cascade forms of a digital filter.[1] The relationship between the noise outputs of corresponding transpose configurations is immediately indicated by (71) through (74) [although, in general, $k'_i \neq k''_i$].

REFERENCES

1. Jackson, L. B., *An Analysis of Roundoff Noise in Digital Filters,* Sc.D. Thesis, Stevens Institute of Technology, Hoboken, New Jersey (1969).
2. McDonald, H. S., "Impact of Large-Scale Integrated Circuits on Communication Equipment," *Proc. of the National Electronics Conf., 24* (December 1968), pp. 569–72.
3. Rader, C. M., and Gold, B., *Digital Processing of Signals,* New York: McGraw-Hill, 1969, pp. 1–130.
4. Kaiser, J. F., "Digital Filters," *System Analysis by Digital Computer,* New York: Wiley, 1966, pp. 218–85.
5. Rader, C. M., and Gold, B., "Digital Filter Design Techniques in the Frequency Domain," *Proc. IEEE, 55,* No. 2 (February 1967), pp. 149–71.
6. Kaiser, J. F., "Some Practical Considerations in the Realization of Linear Digital Filters," *Proc. Third Annual Allerton Conf. on Circuit and System Theory,* Monticello, Illinois, October 1965, pp. 621–33.
7. Knowles, J. B., and Olcayto, E. M., "Coefficient Accuracy and Digital Filter Response," *IEEE Trans. on Circuit Theory, CT-15,* No. 1 (March 1968), pp. 31–41.
8. Bennett, W. R., Spectra of Quantized Signals," B.S.T.J., *27,* No. 3 (July 1948), pp. 446–72.
9. Kaneko, T., and Liu, B., "Round-off Error of Floating-Point Digital Filters," *Proc. Sixth Annual Allerton Conf. on Circuit and System Theory,* Monticello, Illinois, October 1968, pp. 219–27.

10. Weinstein, C., and Oppenheim, A. V., "A Comparison of Roundoff Noise in Floating-Point and Fixed-Point Digital-Filter Realizations," *Proc. IEEE, 57,* No. 6 (June 1969), pp. 1181–3.
11. Oppenheim, A. V., "Block-Floating-Point Realization of Digital Filters," MIT Lincoln Laboratory, Technical Note 1969–19 (March 20, 1969).
12. Knowles, J. B., and Edwards, R., "Effects of a Finite-Word-Length Computer in a Sampled-Data Feedback System," *Proc. IEE, 112,* No. 6 (June 1965), pp. 1197–1207.
13. Gold, B., and Rader, C. M., "Effects of Quantization Noise in Digital Filters," *Proc. AFIPS,* 1966 SJCC, pp. 213–19.
14. Jackson, L. B., Kaiser, J. F., and McDonald, H. S., "An Approach to the Implementation of Digital Filters," *IEEE Trans. on Audio and Electroacoustics, AU-16,* No. 3 (September 1968), pp. 413–21.
15. Edwards, R., Bradley, J., and Knowles, J. B., "Comparison of Noise Performance of Programming Methods in the Realization of Digital Filters," *Proc. of the Symposium on Computer Processing in Communications, XIX,* PIB–MRI Symposia Series (1969).
16. Papoulis, A., "Limits on Bandlimited Signals," *Proc. IEEE, 55,* No. 10 (October 1967), pp. 1677–85.
17. Rice, J. R., *The Approximation of Functions,* Reading, Mass.: Addison-Wesley, 1964, pp. 4–10.
18. Bachman, G., and Naria, L., *Functional Analysis,* New York: Academic Press, 1966, pp. 110–11.
19. Davenport, W. B., Jr., and Root, W. L., *Random Signals and Noise,* New York: McGraw-Hill, 1958, pp. 154–7.
20. Mason, S. J., and Zimmerman, H. J., *Electronic Circuits, Signals and Systems,* New York: Wiley, 1960, pp. 122–3.
21. Freeman, H., *Discrete-Time Systems,* New York: Wiley, 1965, pp. 19–27.
22. Laning, J. H., Jr., and Battin, R. H., *Random Processes in Automatic Control,* New York: McGraw-Hill, 1956, pp. 239–43.

Roundoff-Noise Analysis for Fixed-Point Digital Filters Realized in Cascade or Parallel Form

LELAND B. JACKSON, Member, IEEE
Bell Telephone Laboratories, Inc.
Murray Hill, N. J.

Abstract

The roundoff-noise outputs from two transpose configurations, each for the cascade and parallel forms of a digital filter, are analyzed for the case of uncorrelated roundoff noise and fixed dynamic range. Corresponding transpose configurations are compared on the basis of the variance, or total average power, and the peak spectral density of the output roundoff noise. In addition to providing general computational techniques to be employed in choosing an appropriate configuration for the digital filter, these results also indicate useful "rules of thumb" relating to this choice of configuration. Included are indications of good (although not necessarily optimum) sequential orderings and pole–zero pairings for the second-order sections comprising the cascade form. Computational results are presented which indicate that the analysis is quite accurate and useful.

Introduction

Several analyses of the roundoff-noise output from a digital filter with fixed dynamic range (i.e., implemented using fixed-point arithmetic) have recently appeared in the literature [1]–[4]. Comparisons between the roundoff-noise outputs of different circuit configurations for a digital filter have been of particular interest because of the desire to maximize some measure of the output signal-to-noise ratio. A general framework for these comparisons has been provided by the author [2], and two configurations for the direct form of a digital filter have been analyzed and compared using this approach. In this paper the techniques in [2] are applied to two configurations each for the cascade and parallel forms, and the sequential ordering of the second-order sections comprising the cascade form is also investigated.

Manuscript received January 12, 1970.

This paper is taken in part from a thesis submitted by the author in partial fulfillment of the requirements for the degree of Doctor of Science to the Department of Electrical Engineering, Stevens Institute of Technology, Hoboken, N. J. [1].

The author is now with Rockland Systems Corp., Blauvelt, N. Y. 10913.

Background

The interaction between the roundoff-noise output from a digital filter and the associated dynamic-range limitations is investigated in [2] for the case of uncorrelated rounding errors from sample to sample and from one error source to another. The spectrum of the output roundoff noise from fixed-point implementations is readily shown to be of the form

$$N_y(\omega) = \sigma_0^2 \sum_j k_j' \,|\, G_j'(\omega) \,|^2 \tag{1}$$

where the $G_j'(\omega)$ are scaled transfer responses from certain "summation nodes" in the digital circuit to the filter output. σ_0^2 is the variance of the rounding errors from each multiplier (or other rounding point), and the k_j' are integers indicating the number of error inputs to the respective summation nodes.

Defining $F_i'(\omega)$ to be the scaled transfer response from the input to the ith "branch node" at which a dynamic-range constraint is required, constraints of the form

$$\|F_i'\|_p \leq 1 \tag{2}$$

for $p \geq 1$ are then derived, where $\|F_i'\|_p$ is the L_p norm of the response $F_i'(\omega)$ defined by

$$\|F_i'\|_p = \left[\frac{1}{\omega_s}\int_0^{\omega_s} |\, F_i'(\omega) \,|^p d\omega\right]^{1/p}. \tag{3}$$

It can be shown [5] that for $F_i'(\omega)$ continuous (as is always the case for digital filters), the limit of (3) as $p \to \infty$ exists and is given by

$$\|F_i'\|_\infty = \max_{0 \leq \omega \leq \omega_s} |\, F_i'(\omega) \,| \tag{4}$$

where ω_s is the radian sampling frequency $2\pi/T$. The appropriate value of p in (2) is determined by assumed conditions on the spectra of the input signals to the filter. The effect of (2) is to bound the maximum signal amplitude (for deterministic inputs) or the maximum average power (for random inputs) at the ith branch node.

Reprinted from *IEEE Trans. Audio Electroacoust.*, vol. AU-18, pp. 107–122, June 1970.

As implied above, the prime is used to indicate that the filter has been scaled to satisfy (2). It is intuitively clear that to preserve the greatest possible signal-to-noise ratio, this scaling should reduce the magnitude of $F_i'(\omega)$ no more than is required (or increase it as much as possible, as the case may be). In other words, $F_i'(\omega)$ should actually satisfy

$$\|F_i'\|_p = 1. \tag{5}$$

This condition will be satisfied if the scaling factors s_i, defined by

$$F_i'(\omega) = s_i F_i(\omega), \tag{6a}$$

are given by

$$s_i = 1/\|F_i\|_p. \tag{6b}$$

The unscaled responses $F_i(\omega)$ and $G_j(\omega)$, as well as the overall filter response $H(\omega)$, are depicted in Fig. 1 where

$$\begin{aligned} H(\omega) &= H^*(e^{j\omega T}) \\ F_i(\omega) &= F_i^*(e^{j\omega T}) \\ G_k(\omega) &= G_k^*(e^{j\omega T}). \end{aligned} \tag{7}$$

This notation will be used throughout this paper. That is, for any z transform $A^*(z)$ which converges for $|z| = 1$, the corresponding Fourier transform is given by

$$A(\omega) = A^*(e^{j\omega T}).$$

The sequences $\{e_j(n)\}$ in Fig. 1 represent the roundoff-noise "inputs" at the summation nodes; while the sequences $\{v_i(n)\}$ are the total "outputs" from those branch nodes at which dynamic-range restraints are required.

A state-variable description is then employed in [2] to formulate the general concept of "transpose configurations" for a digital network and to illustrate the usefulness of this concept in digital-filter synthesis. A particularly important result is that for a given unscaled configuration with transpose responses $F_i(\omega)$ and $G_j(\omega)$, as described above, the responses $F_i^t(\omega)$ and $G_j^t(\omega)$ for the corresponding transpose configuration are given by

$$F_i^t(\omega) = G_i(\omega),$$

and

$$G_j^t(\omega) = F_j(\omega). \tag{8}$$

Hence, although the overall transfer responses $H(\omega)$ for these two configurations are the same, their roundoff-noise outputs and/or dynamic-range limitations can be quite different, in general. The transpose configuration is obtained by simply reversing the direction of all branches in the given network configuration, and the poles and zeros of the network are thus realized in reverse order in the transpose configuration.

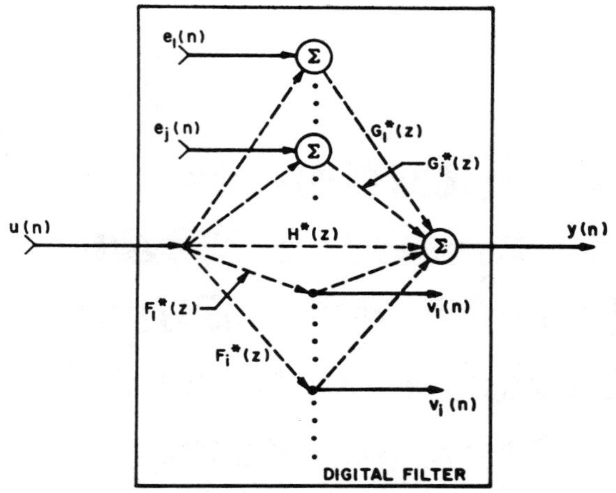

Fig. 1. General digital-filter model.

The Parallel Form

Two transpose configurations for the parallel form of a digital filter are shown in Figs. 2 and 3. The parallel form corresponds to an $H^*(z)$ of the general form [6], [7]

$$H^*(z) = \gamma_0 + \sum_{i=1}^{M} \frac{\gamma_{1i}z^{-1} + \gamma_{0i}}{\beta_{2i}z^{-2} + \beta_{1i}z^{-1} + 1}. \tag{9}$$

However, in order to satisfy (5) in the general case, additional scaling multipliers ρ_i have been added to the configurations in Figs. 2 and 3 (called forms $1P$ and $2P$, respectively). Hence, in form $1P$, $H^*(z)$ is actually realized as

$$H^*(z) = \gamma_0 + \sum_{i=1}^{M} \rho_{1i} \frac{{}_1\gamma_{1i}'z^{-1} + {}_1\gamma_{0i}'}{\beta_{2i}z^{-2} + \beta_{1i}z^{-1} + 1} \tag{10}$$

where the scaling multipliers ρ_{1i} are given by

$$\rho_{1i} = {}_{1P}s_i = 1/\|{}_{1P}F_i\|_p$$

and

$${}_1\gamma_{ki}' = {}_1\gamma_{ki}/\rho_{1i} \qquad k = 0, 1. \tag{11}$$

The transfer function in (10) obviously equals that in (9), but now the transfer responses $F_i'(\omega)$ [from the filter input to the nodes marked by (*)] also satisfy (5). The unscaled transfer functions ${}_{1P}F_i^*(z)$ are given by

$${}_{1P}F_i^*(z) = \frac{1}{\beta_i^*(z)} \qquad i = 1, 2, \cdots, M \tag{12a}$$

where

$$\beta_i^*(z) = \beta_{2i}z^{-2} + \beta_{1i}z^{-1} + 1,$$

and thus

$${}_{1P}F_i(\omega) = \frac{1}{\beta_i(\omega)} \qquad i = 1, 2, \cdots, M. \tag{12b}$$

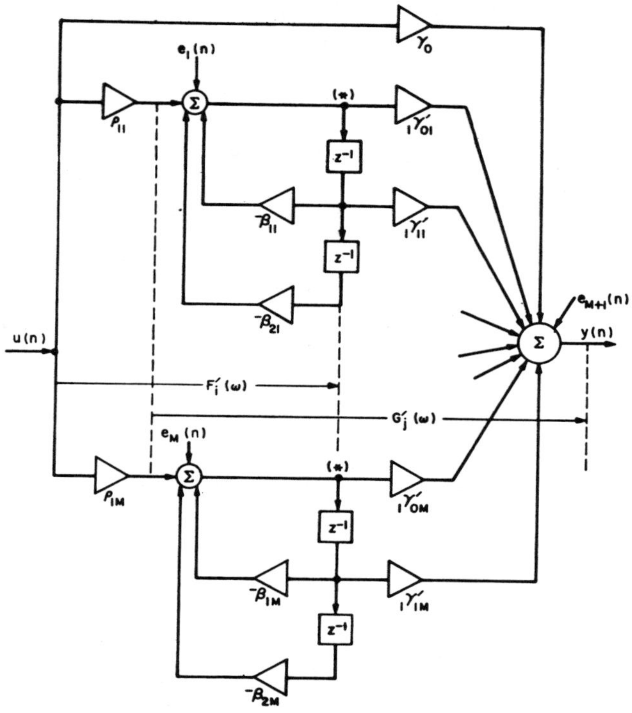

Fig. 2. Parallel form 1P with scaling.

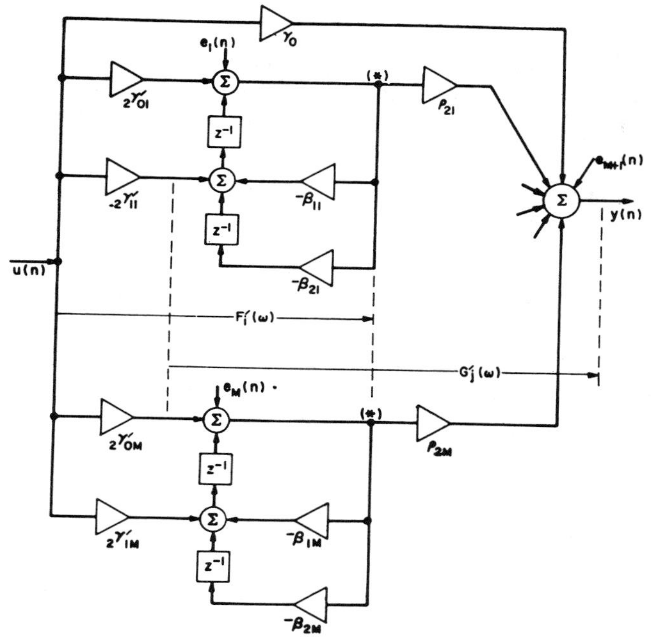

Fig. 3. Parallel form 2P with scaling.

The sealing for parallel from $2P$ is shown in Fig. 3. The transfer function $H^*(z)$ for this form is also given by (10), except that the scaling multipliers ρ_{1i} are replaced by

$$\rho_{2i} = 1/{}_{2P}s_i = \|{}_{2P}F_i\|_p \tag{13a}$$

and now, of course,

$${}_2\gamma_{ki}' = {}_2\gamma_{ki}/\rho_{2i} \qquad k = 0, 1. \tag{13b}$$

The transfer responses to the form-$2P$ branch nodes marked by (*) are

$${}_{2P}F_i'(\omega) = \frac{{}_2\gamma_i'(\omega)}{\beta_i(\omega)} = ({}_{2P}s_i)\,{}_{2P}F_i(\omega) \tag{14}$$

where

$$\gamma_i^{*\prime}(z) = \gamma_{1i}'z^{-1} + \gamma_{0i}'$$

and thus ${}_{2P}F_i'(\omega)$ indeed satisfies (5), as required.

Note that in form $1P$ the scaling is accomplished via the ρ_{1i} multipliers and is then compensated for in the γ_{ki}' multipliers; whereas in form $2P$, the scaling is performed in the γ_{ki}' multipliers and is then compensated for in the ρ_{2i} mutlipliers. Note also that scaling has increased the required number of constant multipliers in these parallel forms to $2N+M+1$ from $2N+1$, in the general case. We will find that this applies to the cascade forms as well. Practically speaking, however, it is assumed that in most cases scaling will be performed with only a few (extra) bits in the multipliers. In particular, scaling may be restricted to powers of 2 in a binary system, in which case the overall transfer function can be maintained exactly and no extra multiplication hardware is required (just shifting operations). In this case, (5) is not satisfied generally, but we can ensure that

$$\tfrac{1}{2} < \|F_i'\|_p \le 1. \tag{15}$$

Since such finite-accuracy considerations are not expected to influence significantly the noise comparisons to follow, we will assume for the purpose of analysis that ideal scaling is performed as indicated in (11) and (13), and thus that (5) is indeed satisfied.

The roundoff-noise output from each of the two parallel forms is readily analyzed, as one would expect, because this output noise is simply the sum of the noise produced by the individual second-order sections comprising the filter. Referring to Fig. 2, the transfer responses $G_j'(\omega)$ for form $1P$, corresponding to the roundoff-noise inputs $e_j(n)$, are easily seen to be

$$\begin{aligned} {}_{1P}G_j'(\omega) &= \frac{{}_1\gamma_j'(\omega)}{\beta_j(\omega)} \qquad i = 1, 2, \cdots, M \\ {}_{1P}G_{M+1}'(\omega) &= 1. \end{aligned} \tag{16}$$

Assuming that the coefficients β_{kj}, ${}_1\gamma_{kj}'$, and ρ_{1j} for all k and j are nonintegers, the weights k_j' corresponding to the $G_j'(\omega)$ are simply

$$
\begin{aligned}
{}_{1P}k_j' &= 3 \quad j = 1, 2, \cdots, M - 1 \\
{}_{1P}k_M' &= 3 - [2M - N] \\
{}_{1P}k_{M+1}' &= N + 1
\end{aligned} \tag{17}
$$

where [ ] denotes the "integer part of," and we have assumed that if the order (N) of the filter is odd, the Mth section is the one (degenerate) first-order section. As indicated previously, with all noninteger coefficients the k_j' for each of the configurations to be considered satisfy

$$\sum_j k_j' = 2N + M + 1. \tag{18}$$

Using the fact that $M = [(N+1)/2]$, (17) is readily shown to satisfy (18).

The effect of scaling on the transfer responses ${}_{1P}G_j'(\omega)$ is seen by substituting (11) into (16) to yield

$${}_{1P}G_j'(\omega) = \|{}_{1P}F_j\|_p \frac{\gamma_j(\omega)}{\beta_j(\omega)} \quad j = 1, 2, \cdots, M \tag{19a}$$

or, from (12),

$${}_{1P}G_j'(\omega) = \left\|\frac{1}{\beta_j}\right\|_p \frac{\gamma_j(\omega)}{\beta_j(\omega)} \quad j = 1, 2, \cdots, M. \tag{19b}$$

The total roundoff noise in the output of parallel form $1P$ is then, from (1),

$${}_{1P}N_y(\omega) = \sigma_0^2 \left\{(N+1) + \sum_{j=1}^{M} {}_{1P}k_j' \left\|\frac{1}{\beta_j}\right\|_p^2 \left|\frac{\gamma_j(\omega)}{\beta_j(\omega)}\right|^2\right\}. \tag{20}$$

Turning to parallel form $2P$, shown in Fig. 3, we note that only one noise input $e_j(n)$ has been shown for each second-order section, rather than two as would be indicated by the number of summation nodes per section. This is done for convenience, and it is possible because the two summation nodes are separated only by a delay and, therefore, actually comprise a single summation. The corresponding transfer responses $G_j'(\omega)$ are then given by

$${}_{2P}G_j'(\omega) = \frac{\rho_{2j}}{\beta_j(\omega)} \quad j = 1, 2, \cdots, M. \tag{21}$$

But from (13) through (15) this is just

$${}_{2P}G_j'(\omega) = \left\|\frac{\gamma_j}{\beta_j}\right\|_p \frac{1}{\beta_j(\omega)} \quad j = 1, 2, \cdots, M. \tag{22a}$$

Note the difference between (19b) and (22a). The transfer response for the input $e_{M+1}(n)$ is, of course, just

$${}_{2P}G_{M+1}'(\omega) = 1 \tag{22b}$$

as in (16).

The noise weights k_j' for form $2P$ are

$$
\begin{aligned}
{}_{2P}k_j' &= 4 \quad j = 1, 2, \cdots, M - 1 \\
{}_{2P}k_M' &= 4 - 2[2M - N] \\
{}_{2P}k_{M+1}' &= M + 1
\end{aligned} \tag{23}
$$

and again (23) is readily shown to satisfy (18). The total roundoff noise in the output of form $2P$ is then

$${}_{2P}N_y(\omega) = \sigma_0^2 \left\{(M+1) + \sum_{j=1}^{M} {}_{2P}k_j' \left\|\frac{\gamma_j}{\beta_j}\right\|_p^2 \left|\frac{1}{\beta_j(\omega)}\right|^2\right\}. \tag{24}$$

We now compare the output roundoff noise for form $1P$ given by (20) with that for form $2P$ given in (24). First note that if N is odd, the noise contribution of the one degenerate first-order section (the Mth section) is the same for both forms. This is seen by noting that, in this case, ${}_{1P}k_M' = {}_{2P}k_M' = 2$, while

$$
\begin{aligned}
({}_{1P}G_M'(\omega))_{N\text{ odd}} &= ({}_{2P}G_M'(\omega))_{N\text{ odd}} \\
&= \gamma_{0M} \left\|\frac{1}{\beta_M}\right\|_p \frac{1}{\beta_M(\omega)}.
\end{aligned} \tag{25}
$$

We next note that $N = M + [N/2]$ and hence that

$${}_{1P}k_{M+1}' = {}_{2P}k_{M+1}' + \left[\frac{N}{2}\right]. \tag{26}$$

Therefore, defining

$$N_1(\omega) = \sigma_0^2 \cdot \left\{(M+1) + 2[2M - N]\gamma_{0M}^2 \left\|\frac{1}{\beta_M}\right\|_p^2 \left|\frac{1}{\beta_M(\omega)}\right|^2\right\}, \tag{27}$$

we may rewrite the equations for $N_y(\omega)$ in (20) and (24) as follows:

$$
\begin{aligned}
{}_{1P}N_y(\omega) &= N_1(\omega) + \sigma_0^2 \sum_{j=1}^{[N/2]} \left(1 + 3\left\|\frac{1}{\beta_j}\right\|_p^2 \left|\frac{\gamma_j(\omega)}{\beta_j(\omega)}\right|^2\right) \\
{}_{2P}N_y(\omega) &= N_1(\omega) + \sigma_0^2 \sum_{j=1}^{[N/2]} 4 \left\|\frac{\gamma_j}{\beta_j}\right\|_p^2 \left|\frac{1}{\beta_j(\omega)}\right|^2.
\end{aligned} \tag{28}
$$

Consider now the variance, or total average power, of the output roundoff noise, given by

$$\sigma_y^2 = \|N_y\|_1 = \frac{1}{\omega_s} \int_0^{\omega_s} N_y(\omega) d\omega \tag{29}$$

(since $N_y(\omega)$ is a nonnegative real function). From (28) and the definition of $\| \ \|_2$ in (3), it follows that

$$\|{}_{1P}N_y\|_1 = \|N_1\|_1 + \sigma_0^2 \sum_{j=1}^{[N/2]} \left(1 + 3\left\|\frac{1}{\beta_j}\right\|_p^2 \left\|\frac{\gamma_j}{\beta_j}\right\|_2^2\right); \tag{30a}$$

whereas

$$\|{}_{2P}N_y\|_1 = \|N_1\|_1 + \sigma_0^2 \sum_{j=1}^{[N/2]} 4 \left\|\frac{\gamma_j}{\beta_j}\right\|_p^2 \left\|\frac{1}{\beta_j}\right\|_2^2. \tag{30b}$$

Hence, letting

$${}_{pq}\theta_j = \left\|\frac{1}{\beta_j}\right\|_p^2 \left\|\frac{\gamma_j}{\beta_j}\right\|_q^2, \tag{31}$$

we have the following simple result for the case $p = q = 2$:

$$\sum_{j=1}^{[N/2]} {}_{pq}\theta_j \begin{cases} < [N/2] \Rightarrow \text{form } 2P \\ > [N/2] \Rightarrow \text{form } 1P. \end{cases} \tag{32}$$

Note, however, that for $p = 2$, the output noise variances $\|N_y\|_1$ for these two forms can differ at most by a factor of

4/3, or 1.25 dB, and thus it really makes little difference in this case whether form $1P$ or form $2P$ is chosen.

The comparison of (30a) and (30b) in the case of $p=\infty$ is not so obvious as for $p=2$, but it is basically the same. Let ω_{0j} be a frequency at which the maximum of $|1/\beta_j(\omega)|$ is achieved; that is,

$$\left|\frac{1}{\beta_j(\omega_{0j})}\right| = \left\|\frac{1}{\beta_j}\right\|_\infty . \tag{33}$$

Because of the very peaked nature of $|1/\beta_j(\omega)|$ relative to $|\gamma_j(\omega)|$, we may then approximate the L_p norm of $\gamma_j(\omega)/\beta_j(\omega)$ by

$$\left\|\frac{\gamma_j}{\beta_j}\right\|_p \approx \gamma_j(\omega_{0j})\left\|\frac{1}{\beta_j}\right\|_p , \tag{34}$$

and with this approximation (31) becomes

$$_{pq}\theta_j \approx \gamma_j(\omega_{0j})\left\|\frac{1}{\beta_j}\right\|_p\left\|\frac{1}{\beta_j}\right\|_q . \tag{35a}$$

Hence, in this case

$$_{pq}\theta_j \approx {_{qp}\theta_j} \tag{35b}$$

and (32) again applies. As before, it really makes little difference which parallel form is chosen. However, experience has shown that usually $_{\infty 2}\theta_j > 1$, and hence form $1P$ is generally to be preferred over form $2P$ for $p=\infty$.

The other norm of $N_y(\omega)$ to be considered is the L_∞ norm, i.e., the maximum of $N_y(\omega)$ over all ω. Here we actually compare bounds on $\|N_y\|_\infty$ for forms $1P$ and $2P$, rather than expressions or approximations for $\|N_y\|_\infty$ itself. Although comparisons based on bounds for $\|N_y\|_\infty$ do not, of course, necessarily hold for $\|N_y\|_\infty$ itself, experimental results supporting the application of these comparisons to $\|N_y\|_\infty$ are presented in a later section.

Again employing $N_1(\omega)$ as defined by (27) in (20) and (24), the following bounds on the L_∞ norms of $N_y(\omega)$ for forms $1P$ and $2P$ are readily obtained:

$$\begin{aligned} \|{_{1P}N_y}\|_\infty &\le \|N_1\|_\infty + \sigma_0^2 \sum_{j=1}^{[N/2]} (1 + 3\,{_{p\infty}\theta_j}) \\ \|{_{2P}N_y}\|_\infty &\le \|N_1\|_\infty + \sigma_0^2 \sum_{j=1}^{[N/2]} 4\,{_{\infty p}\theta_j} \end{aligned} \tag{36}$$

where $_{pq}\theta_j$ is given by (31). Utilizing the approximation for $_{pq}\theta_j$ in (35), the conclusions contained in (32) apply in this case as well. As indicated previously, we usually find that $_{2\infty}\theta_j > 1$ and almost always that $_{\infty\infty}\theta_j > 1$.

Therefore, for all four cases considered (i.e., $\|N_y\|_r$ for $r=1$, ∞ and $p=2$, ∞) there is really little advantage to be gained by using form $1P$ over form $2P$ (or vice versa); but as a general "rule of thumb," form $1P$ is somewhat to be preferred. The situation is much more interesting in the case of the cascade form, where large differences are possible between the roundoff-noise outputs of the form-1 and form-2 configurations. In addition, there is the question of how to sequentially order the cascade-form sections.

The Cascade Form

The two most commonly employed (transpose) configurations for the cascade form of a digital filter are shown in Figs. 4 and 5 and are designated as forms $1D$ and $2D$, respectively. (The letter D indicates that each second-order section is realized in direct form, as opposed to other possible configurations [1].) The cascade form corresponds to an $H^*(z)$ of the general form [6], [7]

$$H^*(z) = a_0 \prod_{i=1}^{M} \frac{\alpha_{2i}z^{-2} + \alpha_{1i}z^{-1} + 1}{\beta_{2i}z^{-2} + \beta_{1i}z^{-1} + 1} . \tag{37}$$

However, in the scaled versions of forms lD, $l=1, 2$, $H^*(z)$ is realized as

$$H^*(z) = {_l\alpha_0'} \prod_{i=1}^{M} \frac{{_l\alpha_i^{*\prime}}(z)}{\beta_i^*(z)} \tag{38}$$

where for form $1D$ (letting $_{1D}s_{M+1} = a_0$)

$$_1\alpha_0' = {_{1D}s_1}$$

$$_1\alpha_i^{*\prime}(z) = \frac{_{1D}s_{i+1}}{_{1D}s_i}\,\alpha_i^*(z) \qquad i = 1, 2, \cdots, M;$$

while for form $2D$ (letting $_{2D}s_0 = 1$)

$$_2\alpha_i^{*\prime}(z) = \frac{_{2D}s_i}{_{2D}s_{i-1}}\,\alpha_i^*(z) \qquad i = 1, 2, \cdots, M$$

$$_2\alpha_0' = \frac{a_0}{_{2D}s_M} .$$

The corresponding transfer responses to the branch nodes marked by (*) are then

$$\begin{aligned} {_{1D}F_i'}(\omega) &= {_{1D}s_i}\,\frac{1}{\beta_i(\omega)} \prod_{j=1}^{i-1} \frac{\alpha_j(\omega)}{\beta_j(\omega)} \qquad i = 1, 2, \cdots, M \\ {_{2D}F_i'}(\omega) &= {_{2D}s_i} \prod_{j=1}^{i} \frac{\alpha_j(\omega)}{\beta_j(\omega)} \qquad i = 1, 2, \cdots, M \end{aligned} \tag{39}$$

where we define $\prod_{j=1}^{0}(\cdot) = 1$. Hence, from the definition of s_i in (6), the $_{lD}F_i'(\omega)$ in (39) satisfy (5), as required.

Scaling is accomplished in cascade forms $1D$ and $2D$ via the altered multipliers $_l\alpha_{ki}'$ *within* the second-order sections, rather than by inserting multipliers *between* the second-order sections, for the following reasons. In form $1D$ the insertion of additional multipliers between the second-order sections would require that an additional overflow constraint be added at the output of each section. In a large number of cases (although not in every case), these additional overflow constraints would significantly reduce the potential signal-to-noise ratio which could otherwise be realized in the filter. In form $2D$, on the other hand, it is readily apparent that even if the additional overflow constraints (now required at the inputs to the second-order sections) do not actually change the required scaling, the total rounding error in the outputs of the α_{1i} and α_{2i} multipliers must be greater with scaling (and rounding) between sections, rather than within them.

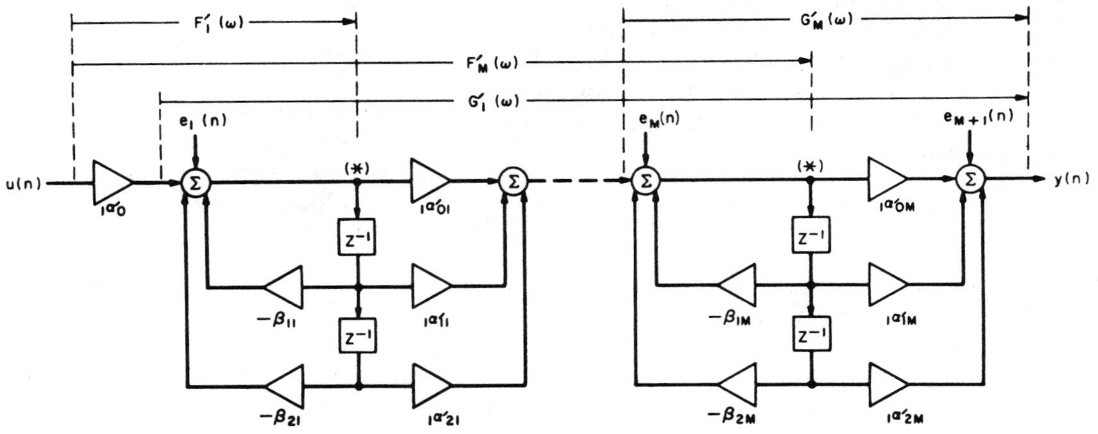

Fig. 4. Cascade form 1D with scaling.

Fig. 5. Cascade form 2D with scaling.

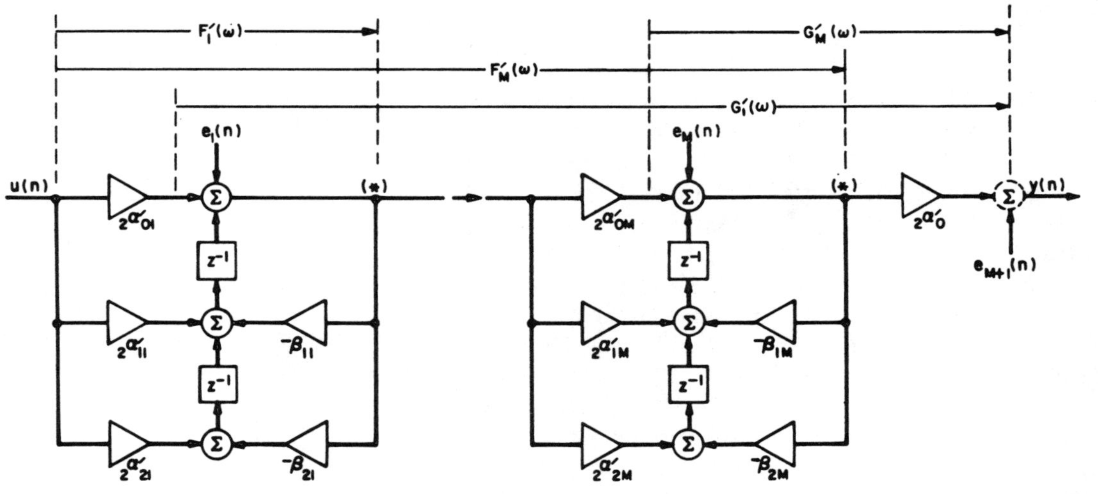

The transfer responses $G_j'(\omega)$ for form $1D$ are given by

$$
\begin{aligned}
{}_{1D}G_j'(\omega) &= \prod_{i=j}^{M} \frac{{}_1\alpha_i'(\omega)}{\beta_i(\omega)} \qquad j = 1, 2, \cdots, M \\
{}_{1D}G_{M+1}'(\omega) &= 1.
\end{aligned}
\tag{40}
$$

Assuming that the coefficients ${}_1\alpha_{ki}'$ and β_{ki} for all k and i are nonintegers, the corresponding weights k_j' for form $1D$ are

$$
\begin{aligned}
{}_{1D}k_1' &= 3 \\
{}_{1D}k_j' &= 5 \qquad j = 2, 3, \cdots, M-1 \\
{}_{1D}k_M' &= 5 - [2M - N] \\
{}_{1D}k_{M+1}' &= 3 - [2M - N]
\end{aligned}
\tag{41}
$$

where we have assumed, for the present, that if the order (N) of the filter is odd, the Mth section is the one first-order section. Equation (41) is readily shown to satisfy (18). The effect of scaling on the transfer responses ${}_{1D}G_j'(\omega)$ is seen by substituting the expressions for ${}_1\alpha_i'(\omega)$ in (38) into (40), which yields

$$
{}_{1D}G_j'(\omega) = \frac{a_0}{{}_{1D}s_j} \prod_{i=j}^{M} \frac{\alpha_i(\omega)}{\beta_i(\omega)} \qquad j = 1, 2, \cdots, M \tag{42}
$$

or, from (39) and (6),

$$
{}_{1D}G_j'(\omega) = a_0 \left\| \frac{1}{\beta_j} \prod_{i=1}^{j-1} \frac{\alpha_i}{\beta_i} \right\|_p \prod_{i=j}^{M} \frac{\alpha_i(\omega)}{\beta_i(\omega)} \qquad j = 1, 2, \cdots, M. \tag{43}
$$

In cascade form $2D$, shown in Fig. 5, we have again chosen to represent the roundoff-noise inputs to each second-order section by a single error source $e_j(n)$. The corresponding $G_j'(\omega)$ are then

$$
{}_{2D}G_j'(\omega) = {}_2\alpha_0' \frac{1}{\beta_j(\omega)} \prod_{i=j+1}^{M} \frac{{}_2\alpha_i'(\omega)}{\beta_i(\omega)} \qquad j = 1, 2, \cdots, M. \tag{44}
$$

But from (38) and (39) this becomes

$$ {}_{2D}G_j'(\omega) = a_0 \left\| \prod_{i=1}^{j} \frac{\alpha_i}{\beta_i} \right\|_p \frac{1}{\beta_j(\omega)} \prod_{i=j+1}^{M} \frac{\alpha_i(\omega)}{\beta_i(\omega)} \qquad j = 1, 2, \cdots, M. \tag{45a} $$

The transfer response for the input $e_{M+1}(n)$ is, of course, just

$$ {}_{2D}G_{M+1}'(\omega) = 1. \tag{45b} $$

The noise weights k_j' for form $2D$ are

$$ \begin{aligned} {}_{2D}k_j' &= 5 \qquad j = 1, 2, \cdots, M-1 \\ {}_{2D}k_M' &= 5 - 2[2M - N] \\ {}_{2D}k_{M+1}' &= 1. \end{aligned} \tag{46} $$

The variance, or total average power, of the output roundoff noise from forms $1D$ and $2D$ is derived from (1) and (42) through (46) to be

$$ \| {}_{1D}N_y \|_1 = \sigma_0^2 \left\{ {}_{1D}k_{M+1}' + a_0 \sum_{j=1}^{M} {}_{1D}k_j' \cdot \left\| \frac{1}{\beta_j} \prod_{i=1}^{j-1} \frac{\alpha_i}{\beta_i} \right\|_p^2 \left\| \prod_{i=j}^{M} \frac{\alpha_i}{\beta_i} \right\|_2^2 \right\} \tag{47a} $$

$$ \| {}_{2D}N_y \|_1 = \sigma_0^2 \left\{ 1 + a_0 \sum_{j=1}^{M} {}_{2D}k_j' \cdot \left\| \prod_{i=1}^{j} \frac{\alpha_i}{\beta_i} \right\|_p^2 \left\| \frac{1}{\beta_j} \prod_{i=j+1}^{M} \frac{\alpha_i}{\beta_i} \right\|_2^2 \right\}. \tag{47b} $$

There are two separate questions to be considered in the minimization of (47): first, which of these two forms is best; and second, what should be the sequential ordering of the M sections. We investigate the first question in this section, and the second in the next.

Cascade forms $1D$ and $2D$ are related via the concept of transpose configurations, as previously described. In particular, assuming that the $\alpha_i(\omega)$ and $\beta_i(\omega)$ appearing in (47) are the same for both forms, form $2D$ corresponds to the *section-by-section* transpose of form $1D$. Therefore, the sequential ordering of the individual sections in form $1D$ is not reversed in form $2D$. There is, of course, another form-2 configuration (denoted as form $2D_t$) which corresponds to the *overall* transpose of the given form-$1D$ configuration, and likewise, a form-1 configuration which corresponds to the overall transpose of the given form-$2D$ configuration. The sequential ordering of the individual sections of forms $1D$ and $2D$ is thus reversed in forms $1D_t$ and $2D_t$.

Since we will separately investigate the ordering of the sections of the cascade forms, we may here compare the noise output from form $1D$ with that from form $2D$ or $2D_t$. We will find it most convenient to consider form $2D_t$ in the case of $p=2$. Denoting the numerator and denominator factors of the D_t forms by $\alpha_i^t(\omega)$ and $\beta_i^t(\omega)$, respectively, we then note that

$$ \beta_i^t(\omega) = \beta_{M-i+1}(\omega) \qquad i = 1, 2, \cdots, M $$

and hence that

$$ \prod_{i=r}^{s} \frac{\alpha_i^t(\omega)}{\beta_i^t(\omega)} = \prod_{i=M-s+1}^{M-r+1} \frac{\alpha_i(\omega)}{\beta_i(\omega)} \qquad 1 \le r \le s \le M. \tag{48} $$

Therefore, letting $l = M - j + 1$, we may rewrite (47b) for form $2D_t$ as follows:

$$ \| {}_{2D_t}N_y \|_1 = \sigma_0^2 \left\{ 1 + a_0 \sum_{l=1}^{M} {}_{2D}k_l' \left\| \prod_{i=l}^{M} \frac{\alpha_i}{\beta_i} \right\|_p^2 \cdot \left\| \frac{1}{\beta_l} \prod_{i=1}^{l-1} \frac{\alpha_i}{\beta_i} \right\|_2^2 \right\} \tag{49} $$

where we have employed the fact that

$$ {}_{2D_t}k_{M-l+1}' = {}_{2D}k_l' \qquad l = 1, 2, \cdots, M. $$

Comparing $\|N_y\|_1$ for form $1D$ as given in (47a) with that for form $2D_t$ in (49), we see that these two expressions differ in only two respects: first, the functions appearing in the L_p and L_2 norms are reversed; and second, the k_j' correspond to forms $1D$ and $2D$, respectively. From (41) and (46), the k_j' for these two forms are related in the following way:

$$ \begin{aligned} {}_{1D}k_1' &= {}_{2D}k_1' - 2 \\ {}_{1D}k_j' &= {}_{2D}k_j' \qquad j = 2, 3, \cdots, M-1 \\ {}_{1D}k_M' &= {}_{2D}k_M' + [2M - N] \\ {}_{1D}k_{M+1}' &= {}_{2D}k_{M+1}' + 2 - [2M - N]. \end{aligned} \tag{50} $$

Therefore, for $p=2$, $\|N_y\|_1$ differs for forms $1D$ and $2D_t$ only by virtue of the k_j' for $j=1$, M, and $M+1$. And for N even, only k_1' and k_{M+1}' are different for these two forms. Hence, as was the case for the parallel form, it really makes little difference which form is used when $p=2$ (and we are considering $\|N_y\|_1$).

In the case of $p=\infty$, however, there can be a significant difference between $\|{}_{1D}N_y\|_1$ and $\|{}_{2D_t}N_y\|_1$ due to the reversal of the L_∞ and L_2 norms in (47a) and (49). Neglecting the small difference due to the different k_j' for $j=1$, $M+1$, and (for N odd) M, the choice between forms $1D$ and $2D_t$ is determined by the relative magnitudes of ${}_{pq}\theta_j$ and ${}_{qp}\theta_j$ for $p=\infty$ and $q=2$, where

$$ {}_{pq}\theta_j = \left\| \frac{1}{\beta_j} \prod_{i=1}^{j-1} \frac{\alpha_i}{\beta_i} \right\|_p^2 \left\| \prod_{i=j}^{M} \frac{\alpha_i}{\beta_i} \right\|_q^2. \tag{51} $$

Note that this definition for ${}_{pq}\theta_j$ in the case of the D forms is consistent with that given in (31) for the P forms because in both cases

$$ {}_{pq}\theta_j = \| {}_1F_j \|_p^2 \| {}_1G_j \|_q^2 = \| {}_{2t}G_j \|_p^2 \| {}_{2t}F_j \|_q^2. \tag{52} $$

In particular, then,

$$ \sum_{j=1}^{M} ({}_{pq}\theta_j - {}_{qp}\theta_j) \begin{array}{l} < 0 \Rightarrow \text{form } 1D \\ > 0 \Rightarrow \text{form } 2D_t \end{array} \tag{53} $$

for $p=\infty$ and $q=2$.

Although it was especially convenient to compare form $1D$ with form $2D_t$ in the case of $p=2$, another useful comparison can be made using form $2D$ when $p=\infty$. Noting that

$$\begin{aligned} {}_{2D}F_j(\omega) &= \alpha_j(\omega)\,{}_{1D}F_j(\omega) \\ {}_{1D}G_j(\omega) &= \alpha_j(\omega)\,{}_{2D}G_j(\omega) \end{aligned} \qquad j = 1, 2, \cdots, M$$

we may rewrite (47) in the following form:

$$\|{}_{1D}N_y\|_1 = \sigma_0^2 \Big\{ {}_{1D}k_{M+1}' + \sum_{j=1}^{M} {}_{1D}k_j' \cdot \|{}_{1D}F_j\|_p^2 \|(\alpha_j)\,{}_{2D}G_j\|_2^2 \Big\} \tag{54a}$$

$$\|{}_{2D}N_y\|_1 = \sigma_0^2 \Big\{ 1 + \sum_{j=1}^{M} {}_{2D}k_j' \cdot \|(\alpha_j)\,{}_{1D}F_j\|_p^2 \|{}_{2D}G_j\|_2^2 \Big\}. \tag{54b}$$

Hence, neglecting the difference in (54a) and (54b) due to the different k_j' for $j=1$, $M+1$, and (for N odd) M, the primary difference between these two expressions results from the occurrence of the extra factors $\alpha_j(\omega)$ multiplying ${}_{2D}G_j(\omega)$ in the former and ${}_{1D}F_j(\omega)$ in the latter for all j.

The implication of the extra factors $\alpha_j(\omega)$ in (54) for the case of $p=\infty$ is as follows. In (54a) the factors $\alpha_j(\omega)$ affect the L_2 norms, while in (54b) they affect the L_∞ norms. But the L_∞ norm of a function "concentrates" exclusively on the maxima of that function, whereas the L_2 norm of a function reflects the rms value of that function over all argument values. Therefore, the effect of the extra $\alpha_j(\omega)$ in (54b) results from the alteration of the maxima of ${}_{1D}F_j(\omega)$ in $\alpha_j(\omega)\,{}_{1D}F_j(\omega)$; while in (54a), the effect concerns the difference in ${}_{2D}G_j(\omega)$ and $\alpha_j(\omega)\,{}_{2D}G_j(\omega)$ over all ω.

Intuitively, one expects that the former effect is potentially much greater; that is, in some cases the extra $\alpha_j(\omega)$ should affect the L_∞ norms in (54b) much more than the L_2 norms in (54a). The experimental evidence presented later will support this conclusion. In particular, we will find that when the $\alpha_j(\omega)$ provide significant attenuation in the neighborhood of the maxima of the corresponding responses ${}_{1D}F_j(\omega)$, $\|N_y\|_1$ can be significantly less for form $2D$ than for form $1D$ (when $p=\infty$). On the other hand, when the $\alpha_j(\omega)$ do not provide such attenuation, there is not a great difference between the $\|N_y\|_1$ for these two forms.

Turning now to the consideration of $\|N_y\|_\infty$ (i.e., the maximum of $N_y(\omega)$ over all ω), the analysis goes through exactly as above, but with the L_2 norms being replaced by L_∞ norms. As in the case of the parallel forms, the comparisons are actually based on bounds for $\|N_y\|_\infty$, rather than on $\|N_y\|_\infty$ itself, with intuition and experimental evidence supporting this approach. By analogy with (54) and (49), these bounds on $\|N_y\|_\infty$ for forms $1D$, $2D$, and $2D_t$ are readily derived to be

$$\|{}_{1D}N_y\|_\infty \le \sigma_0^2 \Big\{ {}_{1D}k_{M+1}' + \sum_{j=1}^{M} {}_{1D}k_j' \cdot \|{}_{1D}F_j\|_p^2 \|(\alpha_j)\,{}_{2D}G_j\|_\infty^2 \Big\} \tag{55a}$$

	$\|N_y\|_r$	
$\|F_i'\|_p$	r = 1	r = ∞
p = 2	EITHER	FORM 1D
p = ∞	FORM 2D	EITHER

Fig. 6. General rules for selection of cascade-form configuration.

$$\|{}_{2D}N_y\|_\infty \le \sigma_0^2 \Big\{ 1 + \sum_{j=1}^{M} {}_{2D}k_j' \cdot \|(\alpha_j)\,{}_{1D}F_j\|_p^2 \|{}_{2D}G_j\|_\infty^2 \Big\} \tag{55b}$$

$$\|{}_{2D_t}N_y\|_\infty \le \sigma_0^2 \Big\{ 1 + \sum_{j=1}^{M} {}_{2D}k_j' \cdot \|(\alpha_j)\,{}_{2D}G_j\|_p^2 \|{}_{1D}F_j\|_\infty^2 \Big\} \tag{55c}$$

where, as before,

$${}_{1D}F_j(\omega) = \frac{1}{\beta_j(\omega)} \prod_{i=1}^{j-1} \frac{\alpha_i(\omega)}{\beta_i(\omega)}$$

and

$${}_{2D}G_j(\omega) = \frac{1}{\beta_j(\omega)} \prod_{i=j+1}^{M} \frac{\alpha_i(\omega)}{\beta_i(\omega)}.$$

Therefore, it is now the case of $p=\infty$ for which the only difference between these bounds for forms $1D$ and $2D_t$ lies in the k_j' for $j=1$, $M+1$, and (for N odd) M. And for $p=2$, the attenuation provided by the extra factors $\alpha_j(\omega)$ in (55a) and (55b) should, when significant, benefit from $1D$ over form $2D$.

The general results of this section are thus summarized as follows. Significant differences in $\|N_y\|_r$ for cascade forms $1D$ and $2D$ (or $2D_t$) can occur in the cases of $r=1$, $p=\infty$ and $r=\infty$, $p=2$. In the first case, form $2D$ tends to be superior; whereas in the latter case, form $1D$ tends to be superior. For $p=r+1$, $r=1$, ∞; however, there seems to be little difference between these two forms. A chart incorporating these conclusions is given in Fig. 6.

Sequential Ordering of Cacsade Sections

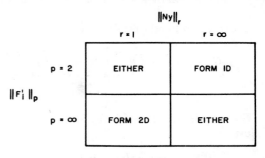

In this section we present a heuristic discussion of the sequential ordering of the second-order sections comprising the cascade D forms, based upon the analytical results of the preceding section. We must not only consider the ordering of given second-order factors $\alpha_i^*(z)/\beta_i^*(z)$, but also the pairing of each numerator factor $\alpha_i^*(z)$ with a denominator factor $\beta_i^*(z)$. An analytical technique to determine the best sequential ordering, other than a complete enumeration and evaluation of all possible permutations, has not yet been devised, but it is possible to state some general rules which can be quite helpful in the design of a given digital filter. In view of the fact that there are up to $M!$ possible pairings of the $\alpha_i^*(z)$ and $\beta_i^*(z)$ and $M!$ possible permutations of each set of $\alpha_i^*(z)/\beta_i^*(z)$, it is important to determine such general rules, where possible.

Consider the relations in (54) and (55) for $\|N_y\|_r$, $r=1$ and $r=\infty$, respectively. Each term of these expressions corresponds to the noise contribution from one section of the filter (i.e., the jth section). In each of these terms, the transfer response ${}_{1D}F_j(\omega)$ [containing the factors $\alpha_i(\omega)/\beta_i(\omega)$ for the $j-1$ sections preceding that section] appears as the argument of an L_p norm, while the response ${}_{2D}G_j(\omega)$ [containing the factors $\alpha_i(\omega)/\beta_i(\omega)$ for the $M-j$ sections following that section] appears as the argument of an L_{r+1} norm. This, of course, just reflects the fact that it is the sections preceding a given section which determine the overflow scaling for that section, while the succeeding sections filter the roundoff noise produced by that section.

Therefore, for $p \neq r+1$ the primary difference between different sequential orderings of the M sections most likely results from the different characteristics of the L_∞ and L_2 norms, as discussed in the preceding section. In particular, since the L_∞ norm is much more sensitive to the maxima of its argument function than the L_2 norm, one would expect that the preferred sequential ordering should minimize (in some sense) the peaked nature of those functions which appear as arguments of the L_∞ norms of (54) and (55). This would, in turn, indicate that the $\alpha_i(\omega)/\beta_i(\omega)$ should be ordered from "most peaked" to "least peaked", or vice versa, depending upon whether $r=\infty$ or $p=\infty$, respectively.

Since it is the L_2 and L_∞ norms with which we are concerned, a reasonable measure of the peakedness of the $\alpha_i(\omega)/\beta_i(\omega)$ for our purposes is simply

$$\mathcal{P}_i = \left\|\frac{\alpha_i}{\beta_i}\right\|_\infty \Big/ \left\|\frac{\alpha_i}{\beta_i}\right\|_2. \tag{56}$$

We would then expect from the above discussion that for $p \neq r=\infty$, a good (but not necessarily optimum) ordering for the M sections of the cascade form is provided by having $\mathcal{P}_i$ decrease (i.e., decreasing peakedness) with increasing i; while for $r \neq p=\infty$, $\mathcal{P}_i$ should increase with increasing i. Our experimental results support this general rule. Note that because of the numerator factors $\alpha_i(\omega)$, $\mathcal{P}_i$ is not necessarily proportional to the Q of the section, although this is often the case.

For $p=r+1$, however, the factors $\alpha_i(\omega)/\beta_i(\omega)$ for all $i \neq j$ appear in one L_p norm or another, and hence the output roundoff noise in this case should be less sensitive to the ordering of the M sections of the cascade form than for $p \neq r+1$. There will still, of course, be some effect due to the grouping of the $\alpha_i(\omega)/\beta_i(\omega)$ within these norms; but the primary effect is probably due, as before, to the attenuation provided by the extra factors $\alpha_j(\omega)$ which occur in the L_p norms for form $1D$ and the L_{r+1} norms for form $2D$. In particular, when the $\alpha_j(\omega)$ provide significant attenuation in the neighborhood of the peaks of the responses ${}_{1D}F_j(\omega)$ and ${}_{2D}G_j(\omega)$, this attenuation should be applied to the most peaked of these two sets of responses. This, in turn, implies (from the preceding discussion for $p \neq r+1$) that for $p=r+1$, a good ordering for the M sections of cascade form $1D$ should result from having $\mathcal{P}_i$ increase with increasing i; while for cascade form $2D$, $\mathcal{P}_i$ should decrease with increasing i. This and the preceding rules for sequential ordering are contained in the chart of Fig. 7.

It is instructive to note that something like the above rule for $p=r+1$ was to be expected from the results of the preceding section, where each form-$1D$ configuration was found to be comparable to the corresponding transpose configuration ($2D_t$) in this case. Note from Figs. 6 and 7, however, that this rule is *opposite* to that for $p \neq r+1$ since, for $p \neq r=\infty$, form $1D$ with decreasing $\mathcal{P}_i$ is indicated; while for $r \neq p=\infty$, form $2D$ with increasing $\mathcal{P}_i$ is indicated. An alternate, and perhaps clearer, presentation of these results is given in the charts of Fig. 8, which show the recommended combination(s) of form and ordering for each p, r pair.

$\|\cdot\|_p$ \ $\|Ny\|_r$	$r=1$	$r=\infty$
$p=2$	DECREASING $\mathcal{P}_i$ WITH FORM 2D INCREASING $\mathcal{P}_i$ WITH FORM 1D	DECREASING $\mathcal{P}_i$
$p=\infty$	INCREASING $\mathcal{P}_i$	DECREASING $\mathcal{P}_i$ WITH FORM 2D INCREASING $\mathcal{P}_i$ WITH FORM 1D

Fig. 7. General rules for sequential ordering of cascade-form sections.

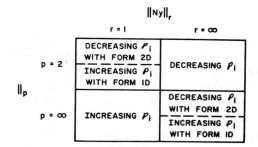

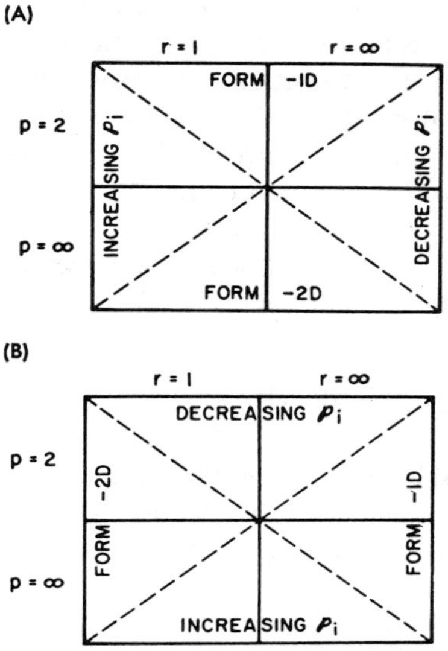

Fig. 8. Alternative description of general rules in Figs. 6 and 7 [use either (A) or (B)].

Having considered the sequential ordering of the section responses $\alpha_i(\omega)/\beta_i(\omega)$, we now address the question of how best to pair the numerator factors $\alpha_i(\omega)$ with the denominator factors $\beta_i(\omega)$ to produce these responses. There are, of course, many cases where this question does not arise including, for example, the cases of Butterworth or Chebyshev (type-1) low-pass or high-pass filters, where all zeros occur at $z=1$ or $z=-1$ (assuming that the bilinear z transform has been employed). In these cases the $\alpha_i(\omega)$ are equal for all i (except for one first-order factor when N is odd), and there is no choice to be made in pairing the $\alpha_i(\omega)$ and $\beta_i(\omega)$ (assuming that the first-order factors, if present, are paired together).

When the second-order factors $\alpha_i(\omega)$ are not equal, however, there is almost always a significant effect on the roundoff-noise output due to the pairing of these factors with the $\beta_i(\omega)$. The reason is apparent from (54) and (55), where the $\alpha_i(\omega)$ occur only in ratio with the corresponding $\beta_i(\omega)$, and the ratios $\alpha_i(\omega)/\beta_i(\omega)$ for all i appear in each term of the summations. Since the norms of $2M-1$ different combinations of the $\alpha_i(\omega)/\beta_i(\omega)$ occur in each relation for $\|N_y\|_r$ in (54) and (55), it is most reasonable to assume that the minimum $\|N_y\|_r$ will result when the individual $\alpha_i(\omega)/\beta_i(\omega)$ are minimized (in some sense). And since

$$\left\|\frac{\alpha_i}{\beta_i}\right\|_p \leq \left\|\frac{\alpha_i}{\beta_i}\right\|_\infty$$

for all $p \geq 1$, it is reasonable to minimize the L_∞ norms of the individual $\alpha_i(\omega)/\beta_i(\omega)$, insofar as this is possible.

Although the above discussion may seem inconclusive as it stands, the proper pairing of the $\alpha_i(\omega)$ and $\beta_i(\omega)$ is really quite evident from such considerations in most cases of practical interest. Consider, for example, the z-plane diagram of Fig. 9 for a sixth-order band-rejection filter. The dotted lines indicate the best pairing of the zeros corresponding to the $\alpha_i(\omega)$ with the poles corresponding to the $\beta_i(\omega)$. This pairing obviously minimizes the L_∞ norms of the $\alpha_i(\omega)/\beta_i(\omega)$ *for all i* and thus, most likely, the values of $\|\alpha_i/\beta_i\|_p$ for all i and p, as well. Almost as obvious are the pairings of Fig. 10 for a sixth-order elliptic bandpass filter. The two "higher-Q" pole pairs are combined with the nearest zero pairs, leaving the real zeros to be combined with the "lower-Q" pole pair. This should minimize the maximum $\|\alpha_i/\beta_i\|_\infty$ over i. Experimental evidence of the effect of proper pole–zero combination is presented in the next section.

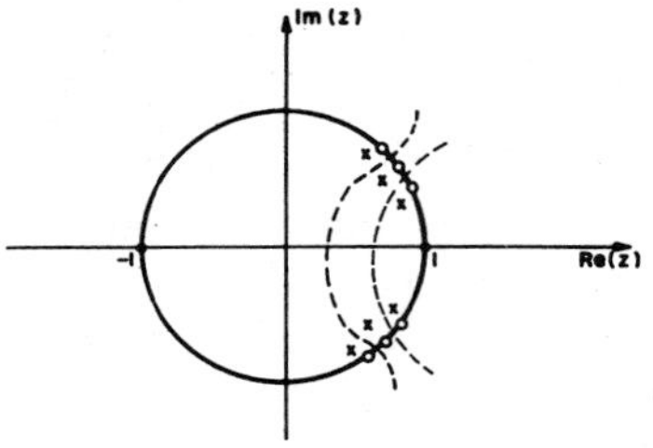

Fig. 9. Proper pairing of poles and zeros in cascade form for sixth-order band-rejection filter.

Fig. 10. Proper pairing of poles and zeros in cascade form for sixth-order bandpass filter.

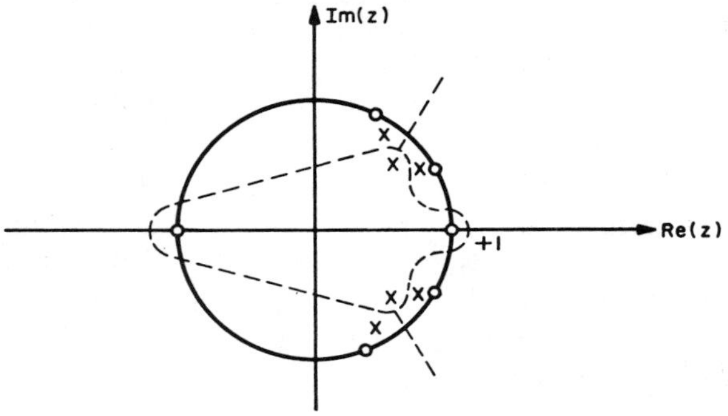

Computer Implementation: An Example

Two digital computer programs have been written to implement and test the analytical results of the preceding sections. These programs are written in FORTRAN IV and have been run successfully on the GE635 digital computer at Bell Telephone Laboratories, Inc., Murray Hill, N. J. The first is the Noise Analysis Program (NAP), which, given the transfer function $H^*(z)$ for a digital filter in either cascade or parallel form, computes the scaling required for that form to satisfy the overflow constraints in (5) (for $p=2$ or for $p=\infty$) and then predicts the resulting roundoff-noise spectrum $N_y(\omega)$ from (1). The other is the Filter Simulation Program (FSP), which simulates the (scaled) digital filter in the appropriate form using sinusoidal or white-noise input signals and estimates the power-density spectrum $N_y(\omega)$ of the output roundoff noise. In other words, NAP implements the synthesis and analysis procedures developed in this paper, and FSP tests the validity of these results for specific filters and input signals.

In both NAP and FSP the output noise density $N_y(\omega)$ is normalized with respect to $N_0=\sigma_0^2$ (i.e., the density or variance of the white noise from a single rounding operation). That is, instead of $N_y(\omega)$ we actually compute

$$\frac{N_y(\omega)}{N_0} = \sum_j k_j' \mid G_j'(\omega) \mid^2. \tag{57}$$

Hence $N_y(\omega)$ and norms thereof are always given in decibels relative to N_0. This normalization is very helpful in relating the results of NAP and FSP, for it eliminates any (direct) reference to the input signal level. In [1] NAP and FSP are described in greater detail, and several representative examples of their operation are given.

One of these examples is the sixth-order Chebyshev (type-2) band-rejection filter (BRF) described in Table I. The unscaled coefficients for both the cascade and parallel

TABLE I

Sixth-Order Chebyshev-2 Band-Rejection Filter

$\alpha_0 = 0.76091619$; $\gamma_0 = 1.3142206$

	Section 1	2	3	1′	2′
β_{2i}	0.90352914	0.84506679	0.75829007	0.90352914	0.84506679
β_{1i}	−1.7636952	−1.4427789	−1.5334490	−1.7636952	−1.4427789
α_{2i}	1.0	1.0	1.0	1.0	1.0
α_{1i}	−1.8118373	−1.6545862	−1.7442502	−1.6545862	−1.8118373
$\lVert\alpha_i/\beta_i\rVert_\infty$	1.65	1.37	1.137	5.717	2.694
$\lVert\alpha_i/\beta_i\rVert_2$	1.056	1.076	1.066	1.602	1.280
ρ_i	1.56	1.27	1.07	3.57	2.11
γ_{1i}	0.09494903	0.17123073	0.24439853	—	—
γ_{0i}	−0.10898306	−0.16408810	−0.28023324	—	—

Fig. 11. Overall frequency response of sixth-order Chebyshev-2 BRF in parallel form.

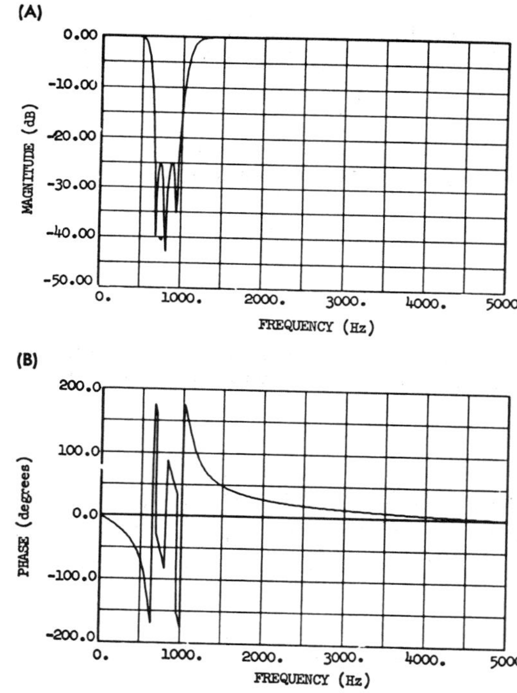

TABLE II

Sixth-Order Chebyshev-2 Band-Rejection Filter

	Variance ($r=1$) $p=\infty$ F1	$p=\infty$ F2	$p=2$ F1	$p=2$ F2	Peak Noise ($r=\infty$) $p=\infty$ F1	$p=\infty$ F2	$p=2$ F1	$p=2$ F2
Parallel (*P*) Forms	24.1*	25.2	13.8**	14.3	35.6*	37.0	23.0*	24.4
Cascade (*D*) Forms								
(123)	34.2	23.5**	22.1	21.0	34.8	34.8	22.7**	31.2
(321)	31.8	24.2	20.4*	23.6	35.0	36.0	23.0	35.7
(213)	33.6	23.5**	21.3	22.4	36.0	35.0	23.3	33.7
(312)	34.0	23.8	22.0	21.7	35.0	35.5	23.1	32.5
(231)	31.6	24.5	20.3*	23.8	35.0	37.0	23.3	36.0
(132)	34.4	23.7	22.3	20.8	35.0	34.5**	23.1	30.7

Note: $\lVert N_y\rVert_r/N_0$ in decibels.

forms of this filter are given in Table I, as well as the L_2 and L_∞ norms of the cascade-form responses $\alpha_i(\omega)/\beta_i(\omega)$ and the corresponding ratios $\mathcal{P}_i$. The effect of proper pole–zero pairing is also illustrated by reversing the pairings of Sections 1 and 2 to yield Sections 1′ and 2′, as described in Table I. The pairings in Sections 1, 2, and 3 are in accordance with the discussion contained in the preceding section, and illustrated in Fig. 9.

The frequency response (both magnitude and phase) of the BRF is shown in Fig. 11. The specifications for the filter are 2.26-dB passband ripple, 25-dB stopband attenuation, and a transition ratio of 0.53. The filter was designed from these specifications using a FORTRAN-IV computer program developed and written by R. M. Golden,[1] J. F. Kaiser, and E. J. Sitar of Bell Telephone Laboratories, Inc. The plots were generated on a Stromberg–Carlsen 4060 microfilm plotter using another subroutine due to Kaiser and Sitar.

Predictions of the output roundoff-noise spectra $N_y(\omega)$ were made by NAP for parallel forms $1P$ and $2P$ and all orderings of cascade forms $1D$ and $2D$ with $p=2$ and $p=\infty$. The results are summarized in Table II, where the predicted $\lVert N_y\rVert_r$ for $r=1, \infty$ are given in decibels relative to N_0 for all cases. The form-1 results are indicated by "F1" column headings, and form-2 by "F2."

The ordering (123) corresponds to decreasing $\mathcal{P}_i$, i.e., decreasing peakedness. Note from Table II that there is not much difference between the different cascade-form orderings, although for $p=r+1$ form $2D$ is somewhat better when Section 1 is first and form $1D$ is better when

[1] Now with Technology Service Corporation, Santa Monica, Calif.

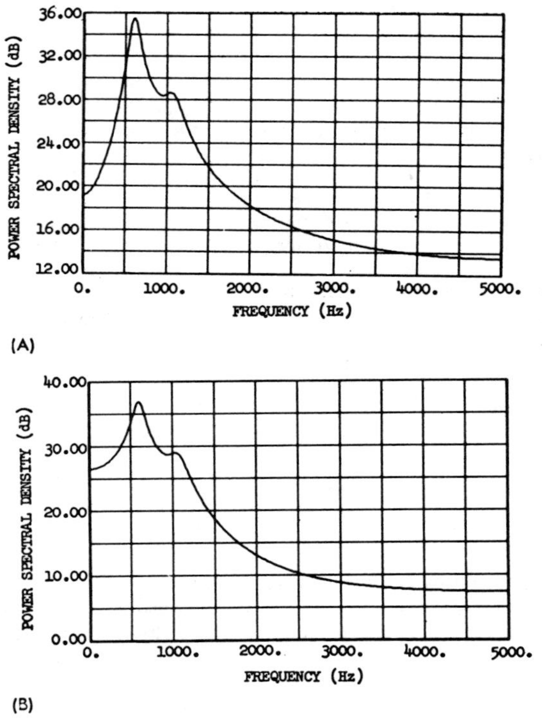

Fig. 12. Predicted roundoff-noise densities for sixth-order BRF in parallel forms 1P (A) and 2P (B) with $p=\infty$.

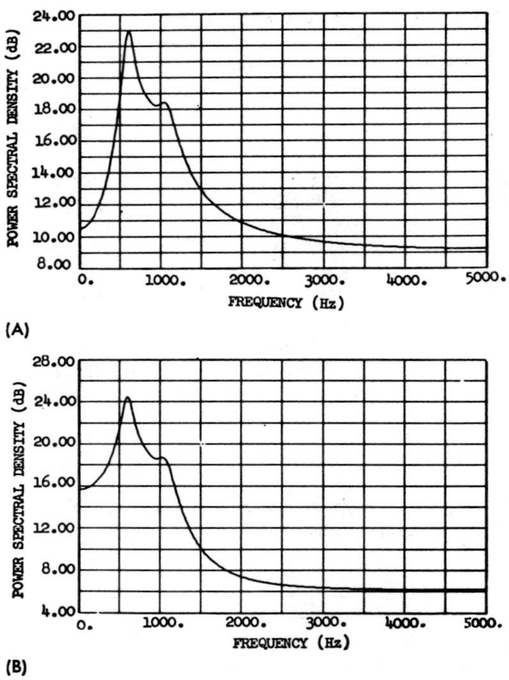

Fig. 13. Predicted roundoff-noise densities for sixth-order BRF in parallel forms 1P (A) and 2P (B) with $p=2$.

Fig. 14. Predicted roundoff-noise densities for sixth-order BRF in cascade forms 1D (A) and 2D (B) with $p=\infty$ (123 ordering).

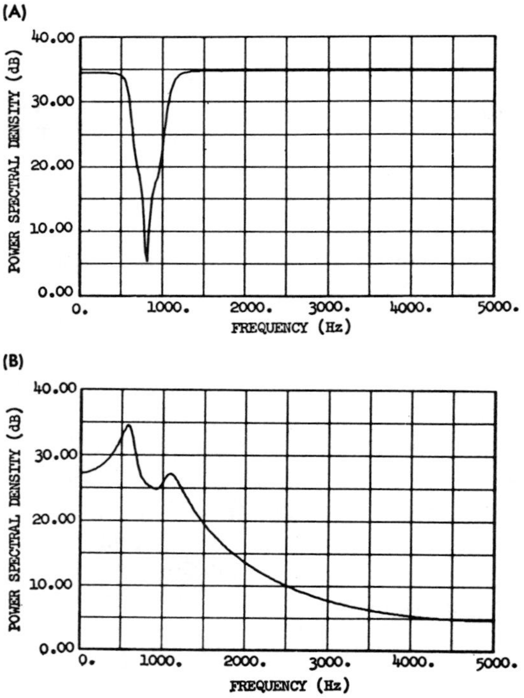

Fig. 15. Predicted roundoff-noise densities for sixth-order BRF in cascade forms 1D (A) and 2D (B) with $p=2$ (123 ordering).

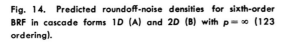

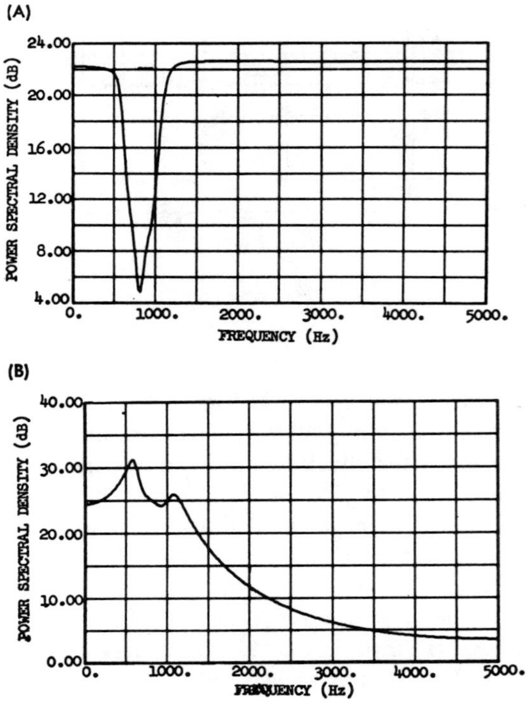

TABLE III
Sixth-Order Chebyshev-2 Band-Rejection Filter (Pairing Reversed in Sections 1 and 2)

Cascade (D) Forms	Variance ($r=1$)				Peak Noise ($r=\infty$)			
	$p=\infty$		$p=2$		$p=\infty$		$p=2$	
	F1	F2	F1	F2	F1	F2	F1	F2
(1′2′3)	36.5	28.3	23.9	20.6	39.4	37.7*	26.7*	29.0
(32′1′)	29.6	27.7*	19.9*	24.9	37.7*	40.4	28.1	37.6
(2′1′3)	30.7	27.6*	20.4	23.8	38.2	39.8	28.0	35.8
(31′2′)	35.8	28.2	23.5	20.9	39.5	38.4	27.1	30.6
(2′31′)	32.5	32.9	21.6	28.4	42.1	46.0	31.2	41.6
(1′32′)	41.1	31.3	27.9	22.0	45.8	42.1	32.6	32.9

Note: $\|N_y\|_r/N_0$ in decibels.

Section 1 is last, in agreement with the general rules in Fig. 7. This relative insensitivity to ordering is explained by the relatively small variation in $\mathcal{P}_i$ for Sections 1, 2, and 3, as given in Table I.

The largest differences in the predicted $\|N_y\|_r$ for the BRF occur between forms $1D$ and $2D$ when $p \neq r+1$ and between the P and D forms when $p=r+1=2$. Note from Table II that these differences are of the order of 7 to 12 dB. In agreement with Fig. 6, form $2D$ is superior for $r \neq p=\infty$, while form $1D$ is best when $p \neq r=\infty$. Parallel form $1P$ is comparable to the best cascade form for each p, r pair except $p=r+1=2$, where it ($1P$) is definitely superior.

The above differences in $\|N_y\|_r$ are readily apparent from the predicted densities in Figs. 12 through 15. Figs. 12 and 13 show the predicted $N_y(\omega)$ for the parallel forms, and Figs. 14 and 15 give the predicted $N_y(\omega)$ for the (123)-ordered cascade forms. Note that the shapes of the form-$2D$ densities are very similar to those for the P forms. In particular, the densities for form $1D$ resemble the wide-band response of the BRF itself, whereas those for forms $2D$, $1P$, and $2P$ are narrow-band in character and achieve their peak values in the BRF stopband.

With LI scaling ($p=\infty$) the peak values $\|N_y\|_\infty$ are comparable in every case, but because of the wide-band character of ${}_{1D}N_y(\omega)$, the total power $\|N_y\|_1$ for form $1D$ is much greater than for the other forms. For L2 scaling ($p=2$), however, the peak density for form $1D$ is much less than for form $2D$ although their total noise powers are very comparable. The peak density for parallel form $1P$ is almost the same as for form $1D$ when $p=2$, but because of its narrow-band character the total noise power for the parallel form is much less in this case.

Another implication of the wide-band response of the BRF is that the form-$2D$ noise densities are not much less for $p=2$ than for $p=\infty$, in contrast with other examples presented in [1]. The reason for this is that the transfer responses $F_i(\omega)$ are all wide-band in this case, and hence the scaling for wide-band inputs ($p=2$) is not much different from that for narrow-band inputs ($p=\infty$). For form $1D$, however, the $F_i(\omega)$ are not wide-band, and the noise densities are, therefore, much less for $p=2$ than for $p=\infty$ due to scaling.

As an example of the effect of proper pole–zero pairing, we now consider the corresponding results for the BRF when the pairings of the $\alpha_i(\omega)$ and $\beta_i(\omega)$ for Sections 1 and 2 are reversed to form Sections 1′ and 2′. The resulting $\|N_y\|_r$ are given in Table III for all orderings of the cascade forms. As before, the ordering (1′2′3) corresponds to decreasing $\mathcal{P}_i$, but now the variation in $\mathcal{P}_i$ is much greater. Note that the results in Table III are in complete agreement with the general rules in Figs. 6 and 7 concerning form and ordering.

Our main point is, however, that the lowest $\|N_y\|_r$ for each p, r pair in Table III is significantly greater (3 to 4 dB) than that for the cascade forms in Table II in every case except for $p=r+1=2$. And in the latter case, the results are comparable. Hence, proper pole–zero pairing is indeed important in the synthesis of this digital filter, as expected.

The results of four representative FSP simulations of the sixth-order BRF are presented in Table IV and Figs. 16 through 19. The cascade-form results correspond to the (123) ordering (with proper pole–zero pairing). The scaling appropriate to the input signal is employed (i.e., $p=\infty$ for sinusoidal inputs and $p=2$ for white noise). The input signal amplitudes, peak output roundoff errors, estimated error means, and peak signal levels at overflow-constrained branch nodes are listed in Table IV in quantization-step units (assuming rounding to the nearest integer). The estimated and predicted error variances are given in decibels relative to N_0. Parzen lag windows have been employed in each case.

Note first of all the excellent agreement between the estimated noise densities in Figs. 16 through 19 and the corresponding predictions in Figs. 12(A), 13(A), 14(B), and 15(A), respectively. For example, note that the peak densities and those at dc and 5000 Hz agree quite closely with their predicted values. The estimated error variances, or average noise powers, in these cases are all within 0.9

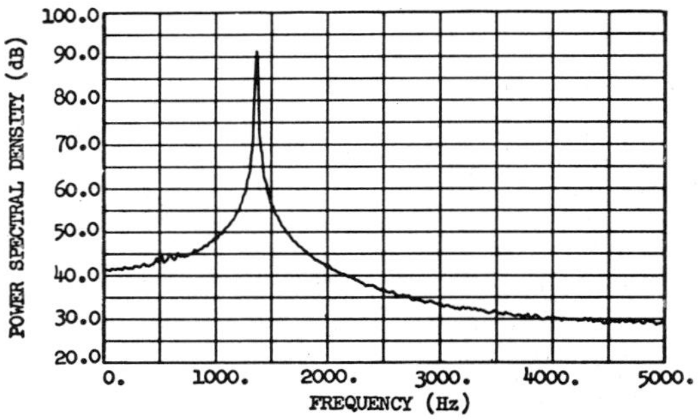

(A)

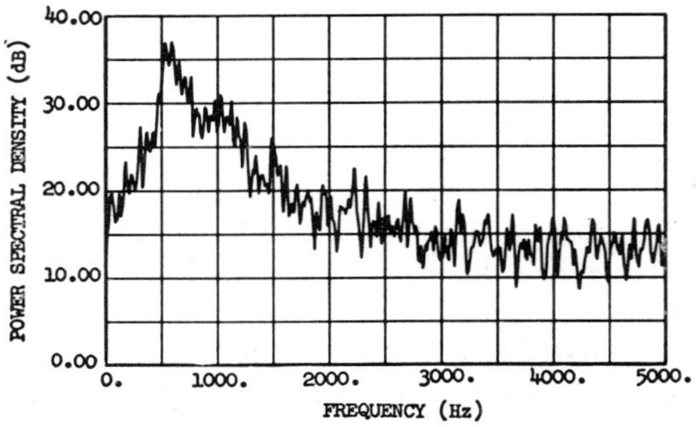

(B)

Fig. 16. Estimated densities of filter output and roundoff noise from simulation of sixth-order BRF in parallel from 1*P* with 1369-Hz input, $p=\infty$. (A) Actual filter output including roundoff error. (B) Roundoff error in actual filter output.

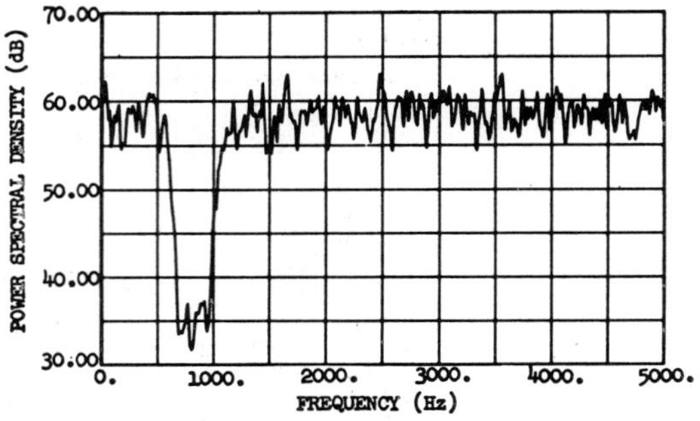

(A)

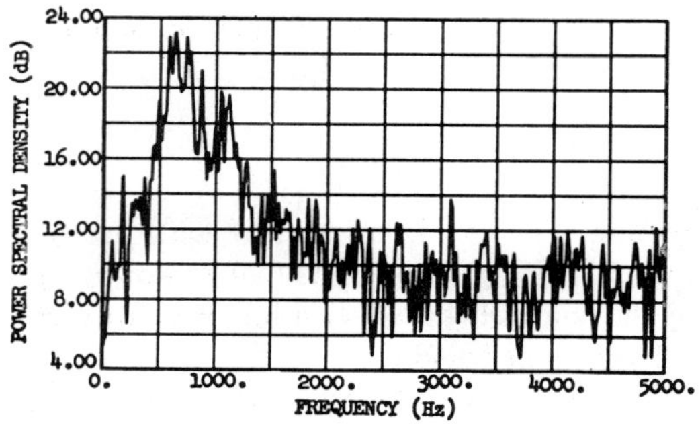

(B)

Fig. 17. Estimated densities of filter output and roundoff noise from simulation of sixth-order BRF in parallel form 1*P* with white-noise input, $p=2$. (A) Actual filter output including roundoff error. (B) Roundoff error in actual filter output.

TABLE IV

Simulation of Sixth-Order Chebyshev-2 Band-Rejection Filter (2048 Samples)

Scaling	Form	Input	Amplitude	Frequency or PF (ρ)	Peak Error	Error Mean	Error Variance (dB)	Predicted Variance (dB)	Peak Data
LI	1*P*	sine	1024	1369 Hz	15.54	−0.0253	24.16	24.1	372
L2	1*P*	noise	1024	4.0	4.545	0.0218	13.45	13.8	1076
LI	2*D*	sine	1024	1369 Hz	17.99	−0.284	24.42	23.5	1025
L2	1*D*	noise	1024	4.0	11.88	0.0707	22.05	22.1	909

dB of their predicted values. This agreement is perhaps surprising in the first and third cases because the input signal is a single sinusoid (at 1369 Hz) and significant correlation might be expected in the output roundoff error as a result. However, the correlation in the individual roundoff-error inputs seems to average out in the total output from the filter. Cases can, of course, be generated with significant correlation being apparent in the form of harmonics of the sinusoid [1], but these cases are the exception, not the rule.

There is also good agreement between our analytical results and other measured data. The estimated error means are very close to zero in all cases, and the peak data at overflow-constrained branch nodes are close to the input signal amplitude (1024). The one case (form 2*D*, LI) where the peak data exceeded 1024 (plus the peak error) resulted from the residual transient remaining in the data after the 200-sample initialization period. Otherwise, the overflow constraints appear to have been satisfied in all cases.

Summary

The roundoff-noise outputs from two transpose configurations each for the cascade and parallel forms of a digital filter have been analyzed and compared using the

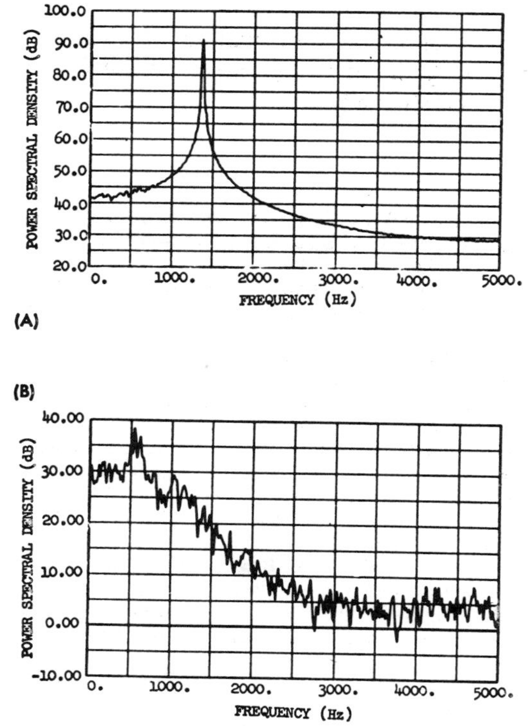

Fig. 18. Estimated densities of filter output and roundoff noise from simulation of sixth-order BRF in cascade form 2D with 1369-Hz input, $p=\infty$. (A) Actual filter output including roundoff error. (B) Roundoff errror in actual filter output.

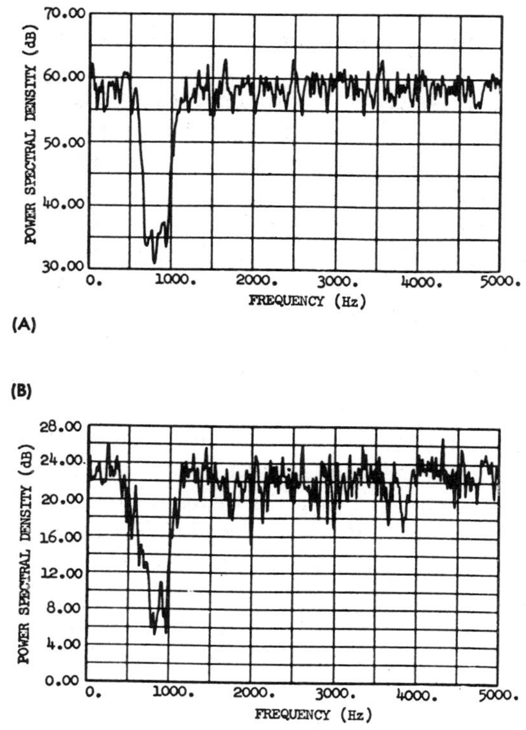

Fig. 19. Estimated densities of filter output and roundoff noise from simulation of sixth-order BRF in cascade form 1D with white-noise input, $p=2$. (A) Actual filter output including roundoff error. (B) Roundoff error in actual filter output.

techniques developed in [2]. The spectrum of the output roundoff noise from these four configurations has been shown to be of the form

$$N_y(\omega) = \sigma_0^2 \left\{ k_{M+1}' + \sum_{j=1}^{M} k_j' \|F_j\|_p^2 \, | G_j(\omega)|^2 \right\} \qquad (58)$$

where $F_i(\omega)$ is the (unscaled) transfer response from the input to the ith branch node, $G_j(\omega)$ is the (unscaled) transfer response from the jth summation node to the output, k_j' is the number of noise sources inputting (directly) to the jth summation node, M is the number of second-order sections comprising the cascade or parallel form, and σ_0^2 is the variance of the noise from each rounding operation. Hence, the variance, or total average power, of the output roundoff noise is simply

$$\sigma_y^2 = \sigma_0^2 \left\{ k_{M+1}' + \sum_{j=1}^{M} k_j' \|F_j\|_p^2 \|G_j\|_2^2 \right\}, \qquad (59)$$

while the peak spectral density $\|N_y\|_\infty$ is bounded by

$$\|N_y\|_\infty \le \sigma_0^2 \left\{ k_{M+1}' + \sum_{j=1}^{M} k_j' \|F_j\|_p^2 \|G_j\|_\infty^2 \right\}. \qquad (60)$$

Corresponding transpose configurations have been compared on the basis of (59) and (60), and the comparisons based on the bounds in (60), rather than on $\|N_y\|_\infty$ itself, have been justified by the excellent experimental agreement reported here and in [1]. A heuristic discussion of the sequential ordering and pole–zero pairing of the cascade-form sections has also been presented on the basis of (59) and (60). From these results, the following general "rules of thumb" may be stated.

1) Parallel form $1P$ is generally preferable to form $2P$, although the maximum difference in $\|N_y\|_r$ for these two forms (for a given p) is only of the order of 4/3, or 1.25 dB.

2) For $p=r+1$ there is little difference in $\|N_y\|_r$ between any given form-$1D$ configuration and the corresponding transpose configuration (form $2D_t$).

3) For $p \neq r+1$, however, cascade form $2D$ is generally superior to form $1D$ for $p=\infty$, and vice versa for $r=\infty$; and the difference can be quite significant.

4) Good (but not necessarily optimum) sequential orderings for the M sections of the cascade form are indicated in the charts of Figs. 7 and 8, where they are stated in terms of the variation in

$$\mathcal{P}_i = \left\| \frac{\alpha_i}{\beta_i} \right\|_\infty \Big/ \left\| \frac{\alpha_i}{\beta_i} \right\|_2$$

with increasing i.

5) The numerator factors $\alpha_i(\omega)$ and denominator factors $\beta_i(\omega)$ of the cascade D forms should be paired so

as to minimize the individual $\|\alpha_i/\beta_i\|_\infty$, insofar as this is possible.

The analytical results obtained do not provide a general rule relating to the choice between parallel form 1*P* and one of the cascade *D* forms. The experimental results do indicate, however, that form 1*P* is usually comparable to or somewhat better than the best *D* form(s). Therefore, from the viewpoint of roundoff noise alone, the simplest and most reliable choice of a digital-filter configuration would seem to be parallel form 1*P*. However, the cascade *D* forms often have other advantages which make them most desirable [8], especially when the zeros of $H^*(z)$ lie on the unit circle ($|z|=1$). In these cases, the *D*-form options of form, sequential ordering, and pole–zero pairing contained in (2) through (5) must be considered.

Finally, two computer programs written to implement and test the analytical results have been briefly described. A representative example has been included to illustrate the operation of these programs and, more importantly, to demonstrate the effectiveness of the analysis and synthesis procedures upon which they are based. These experimental results indicate that the uncorrelated-error analysis is quite accurate, even for sinusoidal inputs, unless the input is periodic in a multiple of the sampling period *T* and/or is of insufficient magnitude [1]. The general rules concerning circuit configuration have also been found to be very helpful and effective.

References

[1] L. B. Jackson, "An analysis of roundoff noise in digital filters," Sc.D. dissertation, Dept. of Elec. Engrg., Stevens Institute of Technology, Hoboken, N. J., 1969.

[2] ——, "On the interaction of roundoff noise and dynamic range in digital filters," *Bell Sys. Tech. J.*, vol. 49, no. 2, February 1970.

[3] R. Edwards, J. Bradley, and J. B. Knowles, "Comparison of noise performance of programming methods in the realization of digital filters," *Proc. Symp. on Computer Processing in Communications*, PIB-MRI Symposia Ser., vol. 19, 1969.

[4] C. Weinstein and A. V. Oppenheim, "A comparison of roundoff noise in floating point and fixed point digital filter realizations," *Proc. IEEE* (Letters), vol. 57, pp. 1181–1183, June 1969.

[5] J. R. Rice, *The Approximation of Functions.* Reading, Mass.: Addison-Wesley, 1964, pp. 4–10.

[6] C. M. Rader and B. Gold, *Digital Processing of Signals.* New York: McGraw-Hill, 1969.

[7] J. F. Kaiser, "Digital filters," in *System Analysis by Digital Computer*, F. F. Kuo and J. F. Kaiser, Eds. New York: Wiley, 1966, ch. 7, pp. 218–285.

[8] L. B. Jackson, J. F. Kaiser, and H. S. McDonald, "An approach to the implementation of digital filters," *IEEE Trans. Audio and Electroacoustics*, vol. AU-16, pp. 413–421, September 1968.

A Comparison of Roundoff Noise in Floating Point and Fixed Point Digital Filter Realizations

Abstract—A statistical model for roundoff noise in floating point digital filters, proposed by Kaneko and Liu, is tested experimentally for first- and second-order digital filters. Good agreement between theory and experiment is obtained. The model is used to specify a comparison between floating point and fixed point digital filter realizations on the basis of their output noise-to-signal ratio, and curves representing this comparison are presented. One can find values of the filter parameters at which the fixed and the floating point curves will cross, for equal total register lengths.

Recently, Kaneko and Liu[1] used a statistical model to predict theoretically the effect of roundoff noise in digital filters realized with floating point arithmetic. This letter is concerned with providing an experimental verification of the model, and the use of the model in specifying a quantitative comparison between fixed point and floating point realizations. We restrict attention to first- and second-order filters, both in the interest of simplicity and because more complicated digital filters are often constructed as combinations of first- and second-order filters.

First-Order Case

For a first-order filter of the form

$$w_n = aw_{n-1} + x_n, \tag{1}$$

where x_n is the input and w_n is the output, the computed output y_n is

$$y_n = [ay_{n-1}(1 + \varepsilon_n) + x_n](1 + \xi_n). \tag{2}$$

The random variables ε_n and ξ_n account for the roundoff errors due to the floating point multiply and add, respectively, and are bounded by

$$|\varepsilon_n| \le 2^{-t}, \qquad |\xi_n| \le 2^{-t}. \tag{3}$$

Following Kaneko and Liu, we define the error $e_n = y_n - w_n$, subtract (1) from (2), neglect second-order terms in e, ε, and ξ, and obtain a difference equation for the error e_n, as

$$e_n - ae_{n-1} = aw_{n-1}(\varepsilon_n + \xi_n) + x_n\xi_n = u_n. \tag{4}$$

Assuming that ε_n and ξ_n are independent from sample to sample (white), and that ε_n, ξ_n, and the signal x_n are mutually independent, u_n in (4) is white noise with variance dictated by the statistics of x_n and the variances σ_ε^2 and σ_ξ^2 of ε_n and ξ_n. The variance σ_e^2 of the output noise e_n is obtained easily from the variance σ_u^2 of u_n as

$$\sigma_e^2 = \sigma_u^2 \sum_{n=0}^{\infty} h_n^2 = \frac{1}{1 - a^2}\sigma_u^2 \tag{5}$$

where $h_n = a^n$ is the filter impulse response.

For example, if we assume that x_n is stationary white noise of variance σ_x^2, we obtain

$$\sigma_e^2 = \frac{\sigma_\xi^2 + a^2\sigma_\varepsilon^2}{(1 - a^2)^2}\sigma_x^2. \tag{6}$$

For the case of a high gain filter, with $a = 1 - \delta$, and δ small, (6) becomes

$$\sigma_e^2 = \frac{(\sigma_\varepsilon^2 + \sigma_\xi^2)\sigma_x^2}{4\delta^2}. \tag{6a}$$

If, instead, x_n is taken to be a sine wave of the form $A \sin(\omega_0 n + \phi)$ with ϕ uniformly distributed in $(0, 2\pi)$, then

$$\sigma_e^2 = \frac{A^2(\sigma_\xi^2 + a^2\sigma_\varepsilon^2)}{2(1 - a^2)(a^2 - 2a\cos\omega_0 + 1)}. \tag{7}$$

To test the model, σ_e^2 was measured experimentally for white noise and sine wave inputs. Each input was applied to a filter using a 27-bit mantissa, and also a filter with the same coefficient a, but using a shorter (e.g., 12-bit) mantissa in the computation. The outputs of the two filters were then subtracted, squared, and averaged over a sufficiently long period to obtain a stable estimate of σ_e^2. Kaneko and Liu assumed that ξ_n and ε_n were both uniformly distributed in $(-2^{-t}, 2^{-t})$ with variances $\sigma_\varepsilon^2 = \sigma_\xi^2 = \frac{1}{3}2^{-2t}$ Actual measurements of the noise due to a multiply and an add verified that ε_n and ξ_n have zero mean, but indicated that the variances

$$\sigma_\varepsilon^2 = \sigma_\xi^2 = (0.23)(2^{-2t}) \tag{8}$$

would better represent these noise sources. Using (1), (6), (7), and (8), we can compute the output noise-to-signal ratio for both white noise and sinusoidal inputs for the first-order case as

$$\frac{\sigma_e^2}{\sigma_w^2} = (0.23)2^{-2t}\left(\frac{1 + a^2}{1 - a^2}\right). \tag{9}$$

In Fig. 1, experimental curves for noise-to-signal ratio are compared with the theoretical curve of (9).

Manuscript received February 10, 1969. This work was sponsored by the U.S. Air Force.

[1] T. Kaneko and B. Liu, "Round-off error of floating-point digital filters," presented at the Sixth Annual Allerton Conf. on Circuit and System Theory, Allerton, Ill., October 1968. To be published in the *Proc. of the Conference.*

Reprinted from *Proc. IEEE*, vol. 57, pp. 1181–1183, June 1969.

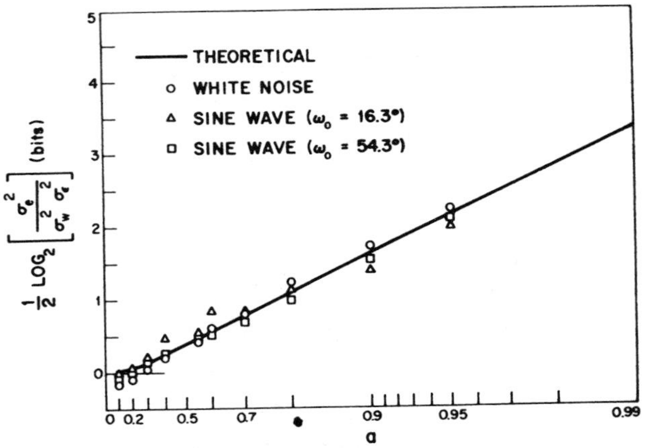

Fig. 1. Theoretical and experimental noise-to-signal ratio for a first-order filter, as a function of pole position. The noise-to-signal ratio is represented in bits.

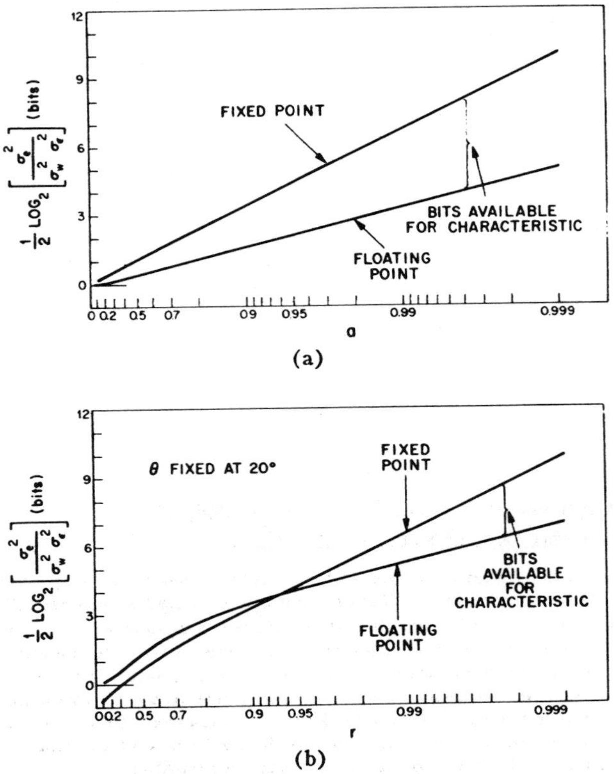

Fig. 2. Comparison of fixed point and floating point noise-to-signal ratios. (a) First-order filter. (b) Second-order filter, $\theta = 20°$.

TABLE I

Theoretical and Experimental Noise-to-Signal Ratio for a Second-Order Filter, as a Function of Pole Position

r	θ	$\frac{1}{2}\log_2\left[\frac{\sigma_e^2}{\sigma_w^2\sigma_\varepsilon^2}\right]$ (bits)			
		White Noise		Sine Wave	
		Theoretical	Experimental	Theoretical	Experimental
0.55	22.5	1.48	1.66	1.54	1.64
0.7	22.5	2.16	2.33	2.23	2.38
0.9	22.5	3.32	3.33	3.35	3.45
0.55	45.0	0.93	1.08	0.97	0.94
0.7	45.0	1.36	1.44	1.37	1.51
0.9	45.0	2.28	2.51	2.22	2.14
0.55	67.5	0.42	0.46	0.39	0.33
0.7	67.5	0.75	0.88	0.65	0.62
0.9	67.5	1.63	1.97	1.45	0.99

Second-Order Case

An analysis similar to the above can be carried out for a second-order filter of the form

$$w_n = -r^2 w_{n-2} + 2r\cos\theta w_{n-1} + x_n, \tag{10}$$

with a complex conjugate pole pair at $z = re^{\pm j\theta}$. Based on experimental verification of (8) in the first-order case, we assume here that the ε's and ζ's representing the errors in the second-order case have the same variance, as given by (8).

When x_n is stationary white noise, we obtain for the variance of the noise e_n,

$$\sigma_e^2 = \sigma_\varepsilon^2\sigma_x^2\left[G + G^2\left(3r^4 + 12r^2\cos^2\theta - 16\frac{r^4\cos^2\theta}{1+r^2}\right)\right] \tag{11}$$

where

$$G = \frac{\sigma_e^2}{\sigma_u^2} = \sum_{n=0}^{\infty} h_n^2 = \left(\frac{1+r^2}{1-r^2}\right)\left(\frac{1}{r^4 + 1 - 4r^2\cos^2\theta + 2r^2}\right). \tag{12}$$

For the case of a high gain filter, with $r = 1-\delta$, (11) becomes approximately

$$\sigma_e^2 = \sigma_\varepsilon^2\sigma_x^2\left(\frac{3 + 4\cos^2\theta}{16\delta^2\sin^2\theta}\right). \tag{13}$$

For the case of sinusoidal input, we obtain

$$\sigma_e^2 = A^2 G\sigma_\varepsilon^2\left[\tfrac{3}{2}r^4|H|^2 + 6r^2\cos^2\theta|H|^2 + \tfrac{1}{2} - 4r^3|H|^2\cos\theta\cos\omega_0 - r^2|H|\cos(\phi - 2\omega_0) + 2r|H|\cos\theta\cos(\phi - \omega_0)\right] \tag{14}$$

where $|H|$ and ϕ represent the magnitude and phase of the filter system function at the input frequency ω_0. In Table I, a comparison of theoretical and experimental values for output noise-to-signal ratio are displayed for a second-order filter.

Fixed Versus Floating Point Comparison

The statistical model of floating point roundoff noise proposed by Kaneko and Liu and one of fixed point roundoff noise as presented for example by Gold and Rader[2] provide the framework for comparing these two structures on the basis of the resulting noise-to-signal ratio. We consider only the case of white noise input.

For the fixed point case, the register length must be chosen sufficiently long so that the output cannot overflow the fixed point word. If h_n denotes the impulse response of the filter, then output w_n is bounded according to

$$\max(|w_n|) = \max(|x_n|)\sum_{n=0}^{\infty}|h_n|. \tag{15}$$

Interpreting the fixed point numbers as signed fractions, we require for no overflows that $|w_n| < 1$, restricting x_n to the range

$$-\frac{1}{\sum_{n=0}^{\infty}|h_n|} < x_n < +\frac{1}{\sum_{n=0}^{\infty}|h_n|}. \tag{16}$$

[2] B. Gold and C. M. Rader, "Effect of quantization noise in digital filters," *1966 Spring Joint Computer Conf., AFIPS Proc.*, vol. 28, Washington, D. C. Spartan, 1966, pp. 213–219.

With x_n white and uniformly distributed between the limits in (16), the resulting output noise-to-signal ratio for a first-order filter is

$$\frac{\sigma_e^2}{\sigma_w^2} = \frac{1}{4} 2^{-2t} \left(\sum_{n=0}^{\infty} |h_n| \right)^2 = \frac{1}{4} \frac{2^{-2t}}{(1-a)^2}, \qquad (17)$$

and for a second-order filter

$$\frac{\sigma_e^2}{\sigma_w^2} = \frac{1}{2} 2^{-2t} \left(\sum_{n=0}^{\infty} |h_n| \right)^2 = \frac{1}{2} 2^{-2t} \left(\frac{1}{\sin\theta} \sum_{n=0}^{\infty} r^n |\sin[(n+1)\theta]| \right)^2. \qquad (18)$$

The variance of the roundoff noise due to a multiplication is taken as $\frac{1}{12} 2^{-2t}$ with t denoting the fixed point register length.

For the case of floating point computation, the noise-to-signal ratio for the first-order filter is

$$\frac{\sigma_e^2}{\sigma_w^2} = (0.23) 2^{-2t} \frac{1+a^2}{1-a^2} \qquad (19)$$

where t is the number of bits in the mantissa. For the second-order filter, we have

$$\frac{\sigma_e^2}{\sigma_w^2} = (0.23) 2^{-2t} \left[1 + G\left(3r^4 + 12r^2 \cos^2\theta - 16 \frac{r^4 \cos^2\theta}{1+r^2} \right) \right]. \qquad (20)$$

For a comparison of floating and fixed point arithmetic in the case of a first-order filter, Fig. 2(a) presents curves of $\frac{1}{2}\log_2 (\sigma_e^2/\sigma_\varepsilon^2 \sigma_w^2)$ as determined from (8), (17), and (19). These curves represent a comparison of the rms noise-to-signal ratio for the two cases, in units of bits. In Fig. 2(b), a similar comparison is illustrated for the second-order case. For the purpose of the illustration, θ was kept fixed and only r varied.

Fig. 2(a) and (b) indicates that floating point arithmetic leads to a lower noise-to-signal ratio than fixed point if the floating point mantissa is equal in length to the fixed point word. We notice that for high gain filters, as a increases toward unity in the first-order case, and as r increases toward unity for θ fixed in the second-order case, the noise-to-signal ratio for fixed point increases faster than for floating point.

However, this comparison does not account for the number of bits needed for the characteristic in floating point. If c denotes the number of bits in the characteristic, this would be accounted for in Fig. 2 by numerically adding the constant c to the floating point data. This shift will cause the floating and fixed point curves to cross at a point where the noise-to-signal ratios are equal for equal total register lengths.

For the sake of the comparison, we provide just enough bits in the characteristic to allow the same dynamic range for both the floating and the fixed point filters. If t_{fx} denotes the fixed point word length, then the requirement of identical dynamic range requires that

$$c = \log_2 t_{fx}. \qquad (21)$$

Assuming for example that $t_{fx} = 16$ so that $c = 4$, crossover points in the noise-to-signal ratio will occur at $a = 0.996$ in the first-order case, and at $r = 0.99975$, $\theta = 20°$, in the second-order case depicted by Fig. 2(b).

CLIFFORD WEINSTEIN
ALAN V. OPPENHEIM
M.I.T. Lincoln Lab.
Lexington, Mass. 02173

Realization of Digital Filters Using Block-Floating-Point Arithmetic

ALAN V. OPPENHEIM, Member, IEEE
Massachusetts Institute of Technology
Lincoln Laboratory
Lexington, Mass. 02173
and
Department of Electrical Engineering, and
Research Laboratory of Electronics
Cambridge, Mass.

Abstract

Recently, statistical models for the effects of roundoff noise in fixed-point and floating-point realizations of digital filters have been proposed and verified, and a comparison between these realizations presented. In this paper a structure for implementing digital filters using block-floating-point arithmetic is proposed and a statistical analysis of the effects of roundoff noise is carried out. On the basis of this analysis, block-floating-point is compared to fixed-point and floating-point arithmetic with regard to roundoff noise effects.

Manuscript received August 8, 1969.

This work was sponsored by the Department of the Air Force.

Introduction

Recently, statistical models for the effects of roundoff noise in fixed-point and floating-point realizations of digital filters have been proposed and verified, and a comparison between these realizations has been suggested [1]–[3]. In general terms, the comparison revolves around the fact that while floating-point arithmetic has a larger dynamic range than fixed-point, the latter is more accurate when the full register length can be utilized. Because of the limited dynamic range of fixed-point arithmetic, for high-gain filters, the input signal must be attenuated to prevent overflow in the output. Thus, for sufficiently high gain, floating-point arithmetic leads to lower noise-to-signal ratio than fixed point. On the other hand, floating-point arithmetic implies a more complex hardware structure than fixed-point arithmetic.

An alternative realization, block-floating-point, has some of the advantages of both fixed point and floating point. In this paper a structure for implementing digital filters using block-floating-point arithmetic is proposed and a statistical analysis of the effects of roundoff noise presented. On the basis of this analysis, block-floating-point is compared to fixed-point and floating-point arithmetic with regard to roundoff noise effects.

A Structure for Block-Floating-Point Realization

In block-floating-point arithmetic the input and filter states (i.e., the outputs of the delay registers) are jointly normalized before the multiplications and adds are performed using fixed-point arithmetic. The scale factor obtained during the normalization is then applied to the final output to produce a fixed-point result. To illustrate, consider a first-order filter described by the difference equation

$$y_n = x_n + a_1 y_{n-1}. \tag{1}$$

For convenience we will treat all numbers as fixed-point fractions. To perform the computation in a block-floating-point manner, we define

$$A_n = \frac{1}{\text{IP}[\max\{|x_n|, |y_{n-1}|\}]} \tag{2}$$

where IP$[M]$ is used to denote the integer power of two such that $M < \text{IP}(M) \leq 2M$, i.e., with M written as $M = 2^m \cdot P$ with P between $\frac{1}{2}$ and 1, $\text{IP}(M) = 2^m$. For M a fraction, 2^m is less than or equal to unity so that A_n is greater than or equal to unity. Thus A_n represents the power-of-two scaling which will jointly normalize x_n and y_{n-1}. Thus with block-floating-point we can compute y_n as

$$y_n = \frac{1}{A_n}[A_n x_n + a_1 A_n y_{n-1}] \tag{3}$$

where the multiplications and addition in (3) are carried out in a fixed-point manner.

Reprinted from *IEEE Trans. Audio Electroacoust.*, vol. AU-18, pp. 130–136, June 1970.

Because of the recursive nature of the computation for a digital filter, it is advantageous to modify (3) as

$$\hat{y}_n = A_n x_n + a_1 \Delta_n w_{1n} \tag{4}$$

with

$$w_{1n} = A_{n-1} y_{n-1}$$

$$y_n = \frac{1}{A_n} \hat{y}_n$$

and

$$\Delta_n = A_n / A_{n-1}.$$

The difference between (3) and (4) is meant to imply that the number $A_n y_n$ rather than y_n is stored in the delay register of the filter. Because of (2), $A_n y_n$ is always more accurate (or as accurate) as y_n since multiplication by A_n corresponds to a left shift of the register.

A disadvantage with (4) is that y_{n-1} must be available to compute A_n, and Δ_n must then be obtained from A_n and A_{n-1}. An alternative is represented by the set of equations

$$\hat{y}_n = \Delta_n \hat{x}_n + a_1 \Delta_n w_{1n} \tag{5a}$$

with

$$\hat{x}_n = A_{n-1} x_n \tag{5b}$$

and

$$\Delta_n = A_n / A_{n-1} = \frac{1}{\text{IP}[\max\{|\hat{x}_n|, |w_{1n}|\}]}. \tag{5c}$$

In this case, we first scale x_n by A_{n-1} to form $\hat{x}_n$ and then determine the incremental scaling using (5c). As in (4), the scaled value $\hat{y}_n$ is stored in the delay register and the output value y_n is determined from $\hat{y}_n$. If we consider the general case of an Nth order filter of the form

$$y_n = x_n + a_1 y_{n-1} + a_2 y_{n-2} + \cdots + a_N y_{n-N},$$

then the block-floating-point realization corresponding to (5) and represented in the direct form is depicted in Fig. 1. For the general case,

$$\Delta_n = \frac{1}{\text{IP}[\max\{|\hat{x}_n|, |w_{1n}|, |w_{2n}|, \cdots, |w_{Nn}|\}]} \tag{6}$$

and

$$A_n = \frac{1}{\text{IP}[\max\{|x_n|, |y_{n-1}|, |y_{n-2}|, \cdots, |y_{n-N}|\}]} \tag{7}$$
$$= A_{n-1}\Delta_n.$$

As an additional consideration, we note that because of the block normalization, there is the possibility of overflow in the addition, which cannot be avoided by an attenuation of the input. This possibility of overflow can be avoided by decreasing the normalization constant A_n by a fixed amount. Thus we modify (6) and (7) as

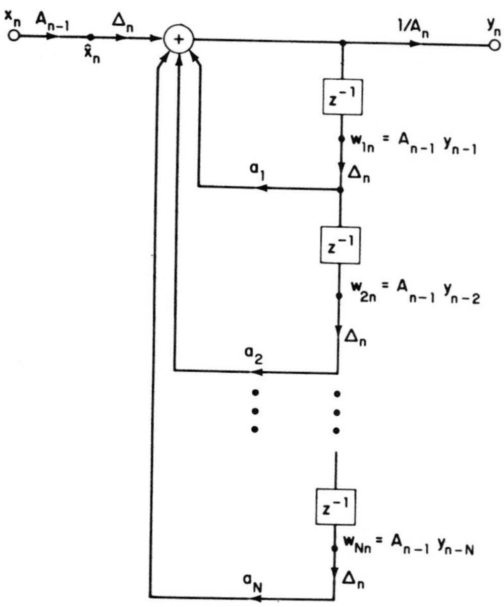

Fig. 1. Network for block-floating-point realization of an Nth-order filter.

$$\Delta_n = \frac{1}{\alpha\,\text{IP}[\max\{|\hat{x}_n|, |w_{1n}|, |w_{2n}|, \cdots, |w_{Nn}|\}]} \tag{6'}$$

and

$$A_n = \frac{1}{\alpha\,\text{IP}[\max\{|x_n|, |y_{n-1}|, |y_{n-2}|, \cdots, |y_{n-N}|\}]} \tag{7'}$$

where α is a constant that may be changed depending on the filter to be implemented. In a first-order filter, for example, α need never be greater than two.

The Effect of Roundoff Noise in Block-Floating-Point Filters

In evaluating the performance of the block-floating-point realization in the presence of roundoff noise, we will restrict attention to the implementation of (5) and Fig. 1 for the first- and second-order cases. We will assume that no roundoff occurs in the computation of $\hat{x}_n$ from x_n and the subsequent multiplication by Δ_n. Since A_{n-1} and $A_{n-1}\Delta_n$ are always nonnegative powers of two, that is, they always correspond to a positive scaling, the above assumption corresponds to allowing more bits in the representation of the intermediate variable $\hat{x}_n$. This is reasonable if we take the attitude that it is primarily in the variables used in the arithmetic computations that the register length is important.

For the first-order case, roundoff noise is introduced in the multiplication of w_{1n} by Δ_n, the multiplication by a_1, and the final multiplication by $1/A_n$. The effects of multiplier roundoff will be modeled by representing the roundoff by additive white noise sources. We consider, for convenience, the fixed-point numbers in the registers to represent signed fractions, with the register length excluding sign denoted by t bits. Each of the roundoff noise generators is assumed to be white, mutually independent and independent of the input, and to have a variance σ_ϵ^2 equal to $(1/12)\cdot 2^{-2t}$. The network for the first-order filter including the noise sources representing roundoff error is presented in Fig. 2(A). In Fig. 2(B) an equivalent representation is shown, where the noise sources are at the filter input. If we consider the input to be a stationary random signal, then the noise source ξ_n will be white stationary random noise with variance

$$\sigma_\xi^2 = \sigma_\epsilon^2(1 + a_1^2)k^2 \tag{8}$$

where k^2 denotes the expected value of $(1/A_n)^2$. Letting η_n denote the noise in the filter output due to the noise ξ_n, the variance of the output noise η_n will be

$$\sigma_\eta^2 = \sigma_\xi^2 \sum_{n=0}^{\infty} h_n^2 + \overline{\epsilon_{3n}^2} = \sigma_\epsilon^2\left[1 + \frac{1 + a_1^2}{1 - a_1^2} k^2\right]. \tag{9}$$

This result is derived by observing that in Fig. 2(B) the transmission from the noise source ξ_n to the output is that of a first-order filter with unit sample response h_n given by $h_n = a^n$. For the case of a second-order filter a similar procedure can be followed. Fig. 3(A) shows a second-order filter with the roundoff noise sources included. In Fig. 3(B) an equivalent representation is shown, where equivalent noise sources are introduced at the filter input. Again, considering the input to be a stationary random signal, then

$$\begin{aligned}\sigma_\xi^2 &= \overline{\left(\frac{1}{A_n}\right)^2}\sigma_\epsilon^2[4r^2\cos^2\theta + 2 + r^4] \\ &\quad + r^4\sigma_\epsilon^2\overline{\left(\frac{1}{A_{n-1}}\right)^2} \\ &= k^2\sigma_\epsilon^2[4r^2\cos^2\theta + 2 + 2r^4]\end{aligned} \tag{10}$$

where we assume that the mean-square values of $(1/A_n)$ and $(1/A_{n-1})$ are equal. Hence the variance of the output noise η_n is

$$\sigma_\eta^2 = \sigma_\epsilon^2 + k^2\sigma_\epsilon^2 G[4r^2\cos^2\theta + 2 + 2r^4] \tag{11}$$

where

$$G = \left(\frac{1 + r^2}{1 - r^2}\right)\frac{1}{1 + r^4 - 4r^2\cos^2\theta + 2r^2}\cdot \tag{12}$$

Experimental Verification

To verify the validity of (9) and (11) the values of k^2 were measured and the values of σ_ξ^2 computed from (9) and (11) using these measured values. These results were taken as the theoretical results since they incorporate the

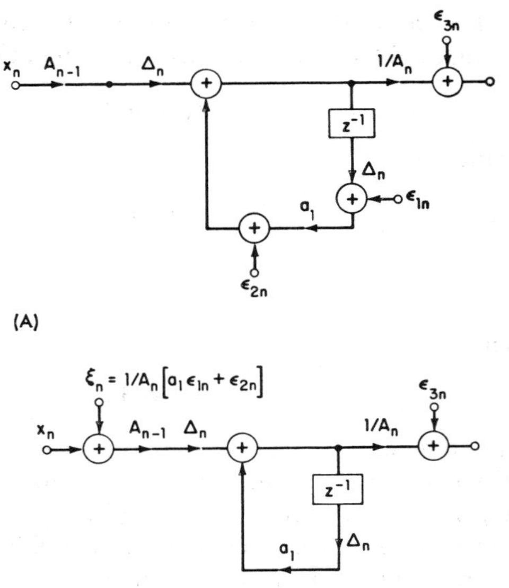

Fig. 2. (A) Noise model for first-order filter. (B) Equivalent noise model.

Fig. 3. (A) Noise model for second-order filter. (B) Equivalent noise model.

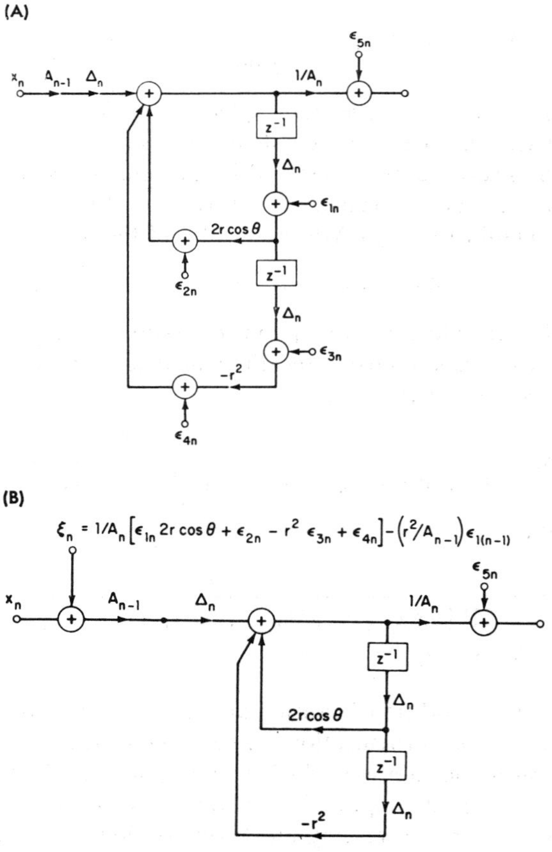

TABLE I

Measured Values of k^2 and Theoretical and Experimental Values of Output Noise Variance for a First-Order Filter with White Noise Input in the Range $|x_n| \leq 1/16$

a_1	k^2	(1/2) $\log_2[2^{2t}\sigma_\eta^2]$(bits) Theoretical	Experimental
0.1	0.0137	−1.7825	−1.7808
0.3	0.0161	−1.7787	−1.7737
0.4	0.0175	−1.7753	−1.7740
0.55	0.0200	−1.7760	−1.7640
0.7	0.0253	−1.7410	−1.7403
0.9	0.0612	−1.4613	−1.4457
0.95	0.1162	−0.9386	−0.9328

TABLE II

Measured Values of k^2 and Theoretical and Experimental Values of Output Noise Variance for a Second-Order Filter with White Noise Input in the Range $|x_n| \leq 1/128$

r	θ	k^2	(1/2) $\log_2[2^{2t}\sigma_\eta^2]$(bits) Theoretical	Experimental
0.55	22.5	0.00288	−1.774	−1.7539
0.55	45.0	0.00186	−1.7861	−1.7767
0.55	67.5	0.00151	−1.7893	−1.7822
0.7	22.5	0.00499	−1.7168	−1.6755
0.7	45.0	0.00248	−1.7780	−1.7661
0.7	67.5	0.00177	−1.7872	−1.7813
0.9	22.5	0.01667	−1.0268	−1.0078
0.9	45.0	0.00581	−1.6824	−1.6494
0.9	67.5	0.00385	−1.7576	−1.7242
0.95	22.5	0.03113	−0.2425	−0.2635
0.95	45.0	0.0110	−1.2369	−1.3760
0.95	67.5	0.00708	−1.6696	−1.5968
0.99	22.5	0.15527	2.0165	2.0199
0.99	45.0	0.04688	0.1605	0.3076
0.99	67.5	0.03125	−0.6065	−0.1395

assumptions of the model. The variance of the roundoff noise ξ_n was then measured experimentally. This was done by simulating the block-floating-point filter with a signed mantissa of 12 bits and comparing the output values with the output of an identical filter simulated with 36-bit fixed-point arithmetic. In all of these measurements the input was white noise with a uniform amplitude distribution. For the first-order filter, the value of α in (6′) and (7′) was taken as two. For the second-order filter, the value of α was taken as four.

In Table I, measured values of k^2 and the theoretical and experimental values of the variance of the roundoff noise for the first-order case are given. The input is white noise in the range $|x_n| \leq \frac{1}{16}$. In a similar manner, theoretical and experimental results for the second-order case are summarized in Table II. The input is white noise in the range $|x_n| \leq \frac{1}{128}$.

A Comparison of Block-Floating-Point, Floating-Point, and Fixed-Point Realizations

Using the model presented in the previous section, the block-floating-point realization of digital filters can be compared with fixed-point and floating-point realizations. The comparison to be presented here will be on the basis of the output noise-to-signal ratio when the input is a random signal with a flat spectrum, using results presented by Gold and Rader [1], Kaneko and Liu [2], and Weinstein and Oppenheim [3]. With σ_η^2 denoting the variance of the roundoff noise as it appears in the output, we have for the first-order filter

$$\text{fixed point: } \sigma_\eta^2 = \frac{1}{12}\cdot 2^{-2t}\,\frac{1}{1-a_1^2} \tag{13}$$

$$\text{floating point: } \sigma_\eta^2 = 0.23\cdot 2^{-2t}\,\frac{1+a_1^2}{1-a_1^2}\,\sigma_y^2 \tag{14}$$

and for the second-order filter

$$\text{fixed point: } \sigma_\eta^2 = \frac{1}{6}\cdot 2^{-2t}G \tag{15}$$

$$\text{floating point: } \sigma_\eta^2 = 0.23\cdot 2^{-2t}$$

$$\cdot\left[1 + G\left(3r^4 + 12r^2\cos^2\theta - \frac{16r^4\cos^2\theta}{1+r^2}\right)\right]\sigma_y^2 \tag{16}$$

where t is the number of bits in the mantissa, not including sign, σ_y^2 is the variance of the output signal, and G is given by (12). In the fixed-point case the output noise is independent of the output signal variance, and in the floating-point case the output noise is proportional to the output signal variance. The expression for block-floating-point noise has a term independent of the signal and a term which depends on the signal through the factor k^2. In both the fixed-point and block-floating-point cases, the dynamic range for the output is constrained by the register length. Consequently, as the filter gain increases, the input must be scaled down to prevent the output from overflowing the register length. Since the output is given by

$$y_n = \sum_{k=0}^{\infty} h_k x_{n-k},$$

then

$$|y_n| \leq \max(|x_n|)\sum_{k=0}^{\infty}|h_k|.$$

To insure that the output fits within a register length, we require that, with x_n and y_n interpreted as fractions,

$$|y_n| \leq 1$$

so that

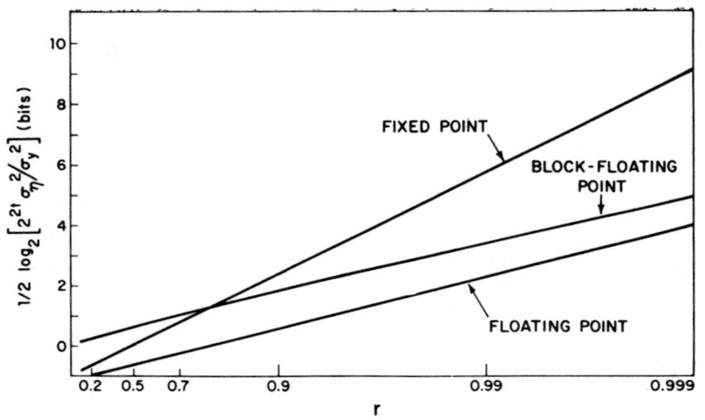

Fig. 4. Comparison of noise-to-signal ratios for first-order filter using fixed-point, floating-point, and block-floating-point arithmetic.

$$\frac{-1}{\sum_{k=0}^{\infty} |h_k|} \leq x_n \leq \frac{1}{\sum_{k=0}^{\infty} |h_k|}. \tag{17}$$

With this constraint on the input, we can then compute an output noise-to-signal ratio for fixed-point, floating-point, and block-floating realizations. Specifically, for the first-order case,

$$\left(\frac{\sigma_\eta^2}{\sigma_y^2}\right)_{\text{fixed point}} = \frac{1}{12} 2^{-2t} \cdot \frac{3}{(1-a_1)^2} \tag{18}$$

$$\left(\frac{\sigma_\eta^2}{\sigma_y^2}\right)_{\text{floating point}} = 0.23 \cdot 2^{-2} \frac{1+a_1^2}{1-a_1^2} \tag{19}$$

$$\left(\frac{\sigma_\eta^2}{\sigma_y^2}\right)_{\text{block floating}} = \frac{1}{12} \cdot 2^{-2t} \left[\frac{3(1-a_1^2)}{(1-a_1)^2} + 3(1+a_1^2)\bar{k}^2\right] \tag{20}$$

where $\bar{k}^2$ is the value for k^2 when x_n is uniformly distributed between plus and minus unity.[1]

In a similar manner, for the second-order case,

$$\left(\frac{\sigma_\eta^2}{\sigma_y^2}\right)_{\text{fixed point}} = \frac{1}{2} 2^{-2t} \left(\frac{1}{\sin\theta} \sum_{n=0}^{\infty} r^n \left| \sin[(n+1)\theta] \right| \right)^2 \tag{21}$$

[1] Equations (20) and (23) are expressed in terms of $\bar{k}^2$ rather than k^2 to facilitate a later approximation. In deriving these equations, k^2 is given as $\bar{k}^2$ times the square of the maximum value of the input as dictated by (17). Thus, although expressed in terms of a normalized value for k^2, (20) and (23) are consistent with the constraint that x_n be sufficiently small so that the output is less than unity.

$$\left(\frac{\sigma_\eta^2}{\sigma_y^2}\right)_{\text{floating point}} = (0.23) 2^{-2t} \cdot \left[1 + G\left(3r^4 + 12r^2\cos^2\theta - 16\frac{r^4\cos^2\theta}{1+r^2}\right)\right] \tag{22}$$

$$\left(\frac{\sigma_\eta^2}{\sigma_y^2}\right)_{\text{block floating}} = \frac{1}{12} 2^{-2t} \left\{\frac{3}{G}\left[\frac{1}{\sin\theta}\sum_{n=0}^{\infty} r^n \left|\sin(n+1)\theta\right|\right]^2 + 3\bar{k}^2(2 + 2r^4 + 4r^2\cos^2\theta)\right\}. \tag{23}$$

In Fig. 4, (18), (19), and (20) are compared. In Fig. 5, (21), (22), and (23) are compared. In these figures the noise-to-signal ratios are plotted in bits so that the difference between two of the curves reflects the number of bits that the mantissas should differ by to achieve the same noise-to-signal ratio. In each of the cases, the difference between floating-point and block-floating-point is approximately constant as the filter gain (or the proximity of the poles to the unit circle) increases. This difference is approximately one bit in the first-order case and two bits in the second-order case. In contrast, the fixed-point noise-to-signal ratio increases at a faster rate than floating-point or block-floating-point, and for low gain is better and for high gain is worse than block-floating-point.

In evaluating the comparison between fixed-point, floating-point, and block-floating-point filter realizations, it is important to note that Figs. 4 and 5 are based only on the mantissa length and do not reflect the additional bits needed to represent the characteristic in either floating-point or block-floating-point arithmetic.

An additional consideration which is not reflected in these curves is that in both fixed-point and block-floating-point the noise-to-signal ratio is computed on the assump-

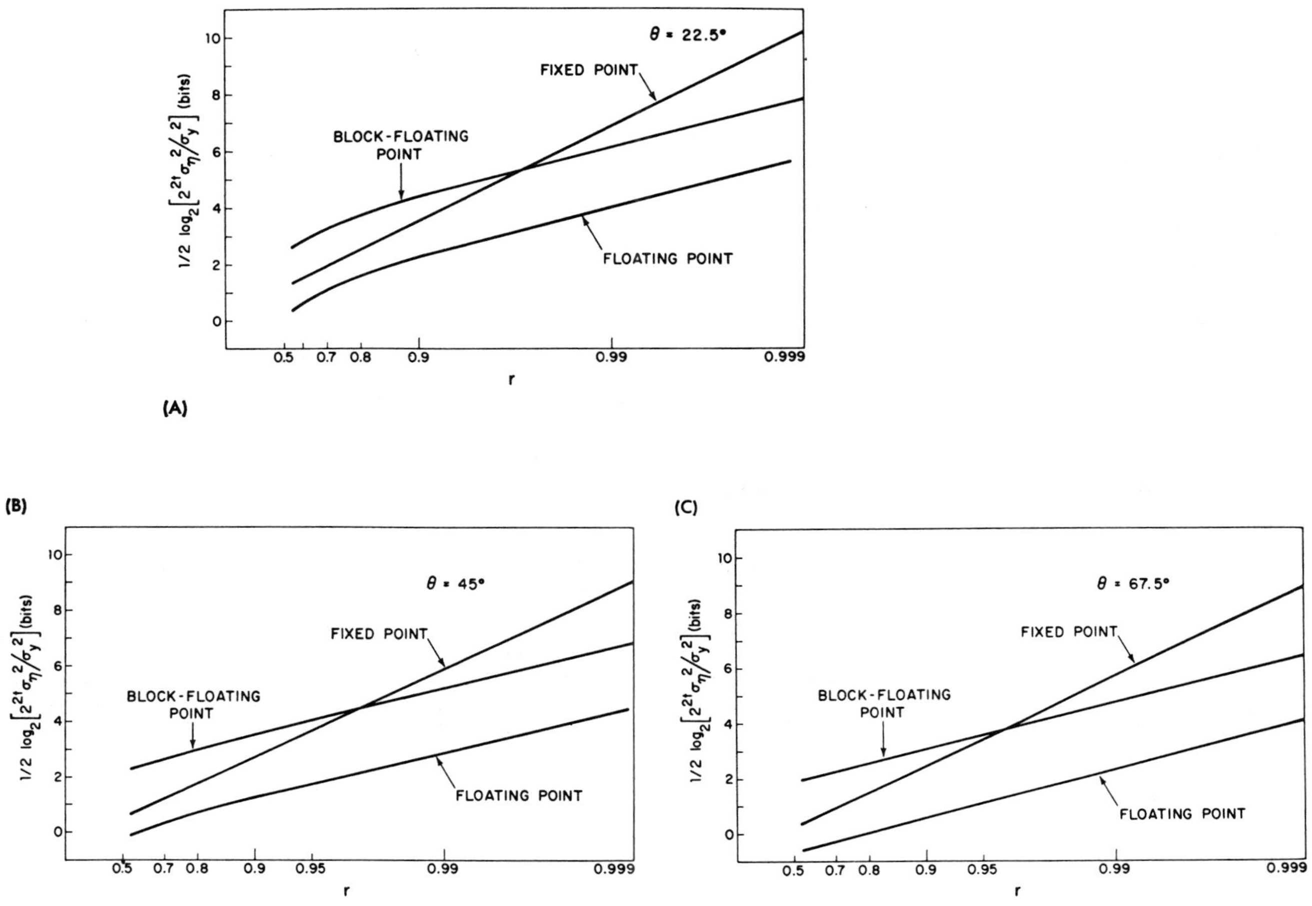

Fig. 5. Comparison of noise-to-signal ratios for second-order filter using fixed-point, floating-point, and block-floating-point arithmetic.

tion that the input signal is as large as possible consistent with the requirement that the output fit within the register length. If the input signal is in fact smaller than permitted, then the noise-to-signal ratio for the fixed-point case will be proportionately higher. For block-floating-point, as the input signal decreases, k^2 decreases, thus reducing the output noise. From (9) and (12) we observe that as the input signal decreases the output noise variance asymptotically approaches σ_ϵ^2.

For the case of high-gain filters, (18) through (23) can be approximated by asymptotic expressions which place in evidence the relationship between them. For the high-gain case, that is, for a_1 close to unity in the first-order filter and r close to unity and θ small in the second-order filter, we will assume that $|x_n|$ is always smaller than $|y_n|$ so that $(1/A_n) \cong 2|y_n|$ for the first-order filter and $(1/A_n) \cong 4|y_n|$ for the second-order filter. Then, if we consider y_n as a random variable with a symmetric probability density,

$$k^2 \cong \frac{4}{3}\,\frac{1}{1+a_1^2}$$

in (20) and

$$k^2 \cong \frac{16}{3} G$$

in (23).

Representing a as $1-\delta$ for the first-order case and r as $1-\delta$ for the second-order case, with δ small, we can approximate (18) through (20) as

$$2^{2t}\left(\frac{\sigma_\eta^2}{\sigma_y^2}\right)_{\text{fixed point}} \cong \frac{0.25}{\delta^2} \tag{24}$$

$$2^{2t}\left(\frac{\sigma_\eta^2}{\sigma_y^2}\right)_{\text{floating point}} \cong 0.23\,\frac{1}{\delta} \tag{25}$$

$$2^{2t}\left(\frac{\sigma_\eta^2}{\sigma_y^2}\right)_{\text{block floating}} \cong 0.83\,\frac{1}{\delta}\,. \tag{26}$$

For the second-order case we will want to bracket the expression

$$\frac{1}{\sin\theta}\sum_{n=0}^{\infty} r^n \,|\sin(n+1)\theta\,| .$$

This sum is the sum of the absolute values of the impulse response and as such is an upper bound on the filter output with a maximum input of unity. Consequently, it must be greater than or equal to the response of the second-order filter to a sinusoid of unity amplitude at the resonant frequency. This resonance response is given by $1/(1-r)(1+r^2-2r\cos 2\theta)^{1/2}$ or, with the high-gain approximation, $1/(2\delta\sin\theta)$. An upper bound is easily obtained on the sum as

$$\left(\frac{1}{\sin\theta}\sum_{n=0}^{\infty} r^n\right) \quad \text{or} \quad \frac{1}{\delta\sin\theta} .$$

Furthermore, for the high-gain case we approximate G as $G \cong 1/4\delta\sin^2\theta$. We can then write that

$$\frac{1}{8}\,\frac{1}{\delta^2\sin^2\theta} \le 2^{2t}\left(\frac{\sigma_\eta^2}{\sigma_y^2}\right)_{\text{fixed point}} \le \frac{1}{2}\,\frac{1}{\delta^2\sin^2\theta} \tag{27}$$

$$2^{2t}\left(\frac{\sigma_\eta^2}{\sigma_y^2}\right)_{\text{floating point}} = 0.23\left[1 + \frac{3+4\cos^2\theta}{4\delta\sin^2\theta}\right] \tag{28}$$

$$\frac{1}{\delta}\left(\frac{1}{4} + \frac{4(1+\cos^2\theta)}{3\sin^2\theta}\right) \le 2^{2t}\left(\frac{\sigma_\eta^2}{\sigma_y^2}\right)_{\text{block floating}} \le \frac{7+\cos^2\theta}{3\delta\sin^2\theta} . \tag{29}$$

We note that the behavior of these expressions as a function of δ is consistent with the results plotted in Figs. 4 and 5.

Acknowledgment

The author would like to thank Dr. C. Weinstein for his contributions to this work through many discussions and Mrs. A. Fried for her help in carrying out many of the computations involved.

References

[1] B. Gold and C. M. Rader, "Effect of quantization noise in digital filters," *1966 Spring Joint Computer Conf., AFIPS Proc.*, vol. 28. Washington, D. C.: Thompson, 1966, pp. 213–219.

[2] T. Kaneko and B. Liu, "Round-off error of floating-point digital fiiters," *Proc. 6th Ann. Allerton Conf. on Circuit and System Theory*, October 2–4, 1968.

[3] C. Weinstein and A. V. Oppenheim, "A comparison of roundoff noise in floating point and fixed point digital filter realizations," *Proc. IEEE* (Letters), vol. 57, pp. 1181–1183, June 1969.

AN ANALYSIS OF LIMIT CYCLES DUE TO MULTIPLICATION ROUNDING IN RECURSIVE DIGITAL (SUB)FILTERS*

LELAND B. JACKSON
Bell Telephone Laboratories, Inc., Murray Hill, New Jersey

ABSTRACT

An analysis of the roundoff limit cycles occuring in fixed-point implementations of second-order recursive digital (sub)filters is presented. A heuristic approach employing an "effective-value" linear model is used to derive certain second-order "deadbands" in which these limit cycles can exist. The results agree with experimental evidence.

INTRODUCTION

Methods for determining whether or not a linear time-invariant sampled-data system is stable are well-known, and they closely parallel the corresponding methods for linear time-invariant continuous systems.[1] However, the implementation of sampled-data systems with digital hardware implies inherent amplitude quantization and a finite dynamic range which introduce nonlinearities into the otherwise linear systems. These digital systems and, in particular, recursive digital filters can thus possess instabilities which are not explained by the usual linear analysis.

Stability in the bounded-input, bounded-output sense is an inherent property of most digital systems because of their finite dynamic range. On the other hand, such systems often do not possess asymptotic stability because periodic outputs, or "limit cycles," can be maintained with zero input to the system. Note that outputs which do not go to zero with zero input must be periodic in a deterministic digital system because of the finite number of internal states for the system.

Two different types of limit cycles can occur in recursive digital filters. The first type is essentially a result of the finite dynamic range as implemented in certain arithmetic systems such as two's-complement binary.[2] These "overflow" limit cycles are generally very large and undesirable, and hence it is important to eliminate the possibility that they can occur. This can always be done, in fact, since one sure way is to zero the internal states of the filter whenever an undesirable overflow occurs. (All overflows are not necessarily undesirable in digital filters as, for example, in the partial sums of more than two data samples).[3] Sandberg has recently shown that a "saturating" form of arithmetic also eliminates overflow limit cycles, permitting only the second type described below.[4]

The second type of limit cycle in recursive digital filters, and the principal subject of this paper, results from amplitude quantization within the filter or, in particular, from the rounding of multiplication products in a feedback loop. Such limit cycles were first noted by R. B. Blackman,[5] who referred to the amplitude intervals within which these limit cycles are

---

* This paper is taken in part from a thesis to be submitted in partial fulfillment of the requirements for the degree of Doctor of Science in the Department of Electrical Engineering at Stevens Institute of Technology.[7]

Reprinted from *Proc. 7th Annu. Allerton Conf. Circuit System Theory*, pp. 69-78, 1969.

confined as "deadbands." Blackman considered only first-order (DC) limit cycles, however, (in filters of arbitrary order) with rounding after the summation of the full-accuracy multiplication products. In this paper, we investigate both the first- and second-order deadbands produced in second-order digital (sub)filters as a result of rounding after multiplication. We consider only the zero-input case although results of the same general character can be derived for any constant input. Our investigation is also confined to fixed-point arithmetic implementation because roundoff limit cycles may generally be neglected with floating-point.[6]

COMPLEX-CONJUGATE EFFECTIVE POLES

For a first-order recursive digital (sub)filter with input sequence $\{x_n\}$ and output sequence $\{y_n\}$, the output sequence is assumed to be computed from

$$y_n = x_n - (\alpha y_{n-1})' \tag{1}$$

where ( )' denotes the operation of rounding to the nearest integer. The first-order deadband within which limit cycles can exist with zero input to the filter is then[5]

$$D_1 = [-k,k] \tag{2a}$$

where k is the largest integer satisfying

$$k \leq \frac{0.5}{1 - |\alpha|} \, . \tag{2b}$$

The limit cycles occurring within D_1 can be of only two forms: constant magnitude and sign for α negative, or constant magnitude with alternating signs for α positive.

Note that for all output values y_n within D_1, the multiplier α has an "effective value" (α') of ±1. That is,

$$(\alpha y_n)' = \alpha' y_n = \pm y_n \qquad \text{for all } y_n \in D_1 . \tag{3}$$

But for a linear discrete filter with transfer function $1/(1+\alpha' z^{-1})$, $\alpha' = \pm 1$ implies that the z-plane pole occurs at $z = \mp 1$. Therefore, we will say that the limit cycles occurring with D_1 are the result of an "effective pole" at $z = \pm 1$.

Now consider the second-order digital (sub)filter defined by

$$y_n = x_n - (\beta_1 y_{n-1})' - (\beta_2 y_{n-2})' . \tag{4}$$

With $\beta_1^2 \leq 4\beta_2$ (complex-conjugate poles) and without rounding, this filter would be stable if, and only if, $0 < \beta_2 < 1$. The case of $\beta_2 = 1$ (i.e., poles on the unit circle in the z plane) is sometimes referred to as being "marginally stable" because with zero input the filter will yield a sinusoidal (and hence nonincreasing) output. But considering a deadband such as that in (2) with $|\alpha|$ replaced by β_2, we see that for all y_n within this deadband, β_2 will have an effective value (β_2') of 1! That is, defining

$$D_2 = [-k,k] \tag{5a}$$

where k is the largest integer satisfying

$$k \leq \frac{0.5}{1-\beta_2}, \quad (0 < \beta_2 < 1), \tag{5b}$$

we then have, as in (3), that

$$(\beta_2 y_n)' = \beta_2' y_n = y_n \quad \text{for all } y_n \in D_2. \tag{6}$$

We thus hypothesize that second-order roundoff limit cycles are a result of an effective complex-conjugate pole pair on the unit circle in the z plane and, as such, are confined (approximately) to the deadband D_2 defined in (5). Although this is an heuristic approach and bounds on the second-order limit cycles have not been derived, extensive computer simulation has corroborated these results in all their essential detail. The only variations from the predicted behavior are the (small) expected deviations from a perfectly sinusoidal waveform due to the rounding errors in the $\beta_1 y_{n-1}$ multiplications. In particular, some of the observed limit cycles for $\beta_2 \geq 0.9$ have exceeded the limits of D_2 by ±1 (i.e., one quantization step), but never more.

Given the amplitude of any particular limit cycle within D_2, the oscillation frequency of the limit cycle can be controlled only by β_1 since the effective value of β_2 is always ±1. To gain insight into this mechanism, it is useful to consider a diagram of the parameter space (β_1, β_2 plane) for the second-order section as shown in Fig. 1. Here, within the (triangular) region of linear stability for the digital filter, the lines $\beta_2 = 0.5$, 0.75, 0.833, and 0.875 have been drawn corresponding to the minimum β_2 values which produce the deadbands $[-k,k]$, $k = 1, 2, 3$, and 4, respectively. In particular, no limit cycles resulting from complex effective poles are possible for $\beta_2 < 0.5$. The minimum β_2 values (β_{2k}) for the deadband $D_2 = [-k,k]$ are, in general, given by

$$\beta_{2k} = \frac{2k-1}{2k}, \quad k = 1,2,\ldots \tag{7a}$$

Hence, a sufficient condition for a deadband of $[-k,k]$ is simply

$$\beta_2 \geq \beta_{2k}. \tag{7b}$$

Above each horizontal line ($\beta_2 = \beta_{2k}$) in Fig. 1 are drawn the vertical lines $\beta_1 = m(1-\beta_{2k})$ for m odd. Each subregion of the β_1, β_2 plane delimited by these lines must have a common limit cycle behavior because all integers y_n in the corresponding deadband will yield the same rounded products for any β_1, β_2 pair in the subregion. Thus, for example, for $0.5 \leq \beta_2 < 0.75$ corresponding to a deadband of $[-1,1]$, only the subregions $|\beta_1| \geq 1.5$, $0.5 \leq |\beta_1| < 1.5$, and $|\beta_1| < 0.5$ can display distinctive limit cycle behavior because the integers (-1, 0, and +1) contained in this deadband yield the same rounded products for all β_1 values in each of these subregions.

For each of the β_1, β_2 subregions described above, the effective β_1 value (β_1') will be determined by the oscillation frequency of the limit cycle corresponding to that subregion. That is, for an oscillation frequency of f_0 and a Nyquist frequency of f_N (one-half the sampling rate), the effective β_1 value is defined by

$$\beta_1' \triangleq -2\cos(\Pi f_0/f_N). \tag{8}$$

Thus, to estimate f_0 for a given subregion, we can choose a suitable β_1' and compute the estimate from (8). For example, for the case of $0.5 \leq \beta_2 < 0.75$ discussed above, the effective values β_1' for the subregions $|\beta_1| \geq 1.5$, $0.5 \leq |\beta_1| < 1.5$, and $|\beta_1| < 0.5$ must obviously be just ±2, ±1, and 0, respectively. Hence, the oscillation frequencies (f_0) for these subregions are as follows: f_N ($\beta_1' = 2$); $2f_N/3$ ($\beta_1' = 1$); $f_N/2$ ($\beta_1' = 0$); $f_N/3$ ($\beta_1' = -1$); and DC, or $f_0 = 0$ ($\beta_1' = -2$).

The correct β_1' for each subregion is not generally as obvious as it was above, but as β_2 approaches 1, the accuracy of any guess for β_1' must become better and better. Thus, for sufficiently high-Q filters (how high depends upon the application), the approximation $\beta_1' \approx \beta_1$ should suffice to estimate the oscillation frequency f_0. The experimentally determined oscillation frequencies for all subregions having $\beta_2 < .875$ are listed in Tables 1A and 1B. The entries in these tables are for the parameter λ defined by $f_0 = \lambda f_N$. Note that λ must be a rational fraction because, by definition, a discrete limit cycle must be periodic with period NT, where T is the sampling period and N is an integer.

REAL EFFECTIVE POLES

The results in (1)-(3) are easily generalized to include first-order limit cycles in digital filters of arbitrary order with rounding after multiplication. Assuming that

$$y_n = x_n - \sum_{i=1}^{N} (\beta_i y_{n-i})', \tag{9}$$

constant (DC) limit cycles of amplitude k can occur if, and only if,

$$- \sum_{i=1}^{N} a^i (\beta_i k)' = k \tag{10}$$

for a = +1; while alternating-sign limit cycles result if, and only if, (10) is satisfied for a = -1. In particular, in second-order digital (sub)filters, we must have

$$- (\beta_2 k)' \pm (\beta_1 k)' = k \tag{11a}$$

or, from the definition of the corresponding effective values, simply

$$- \beta_2' \pm \beta_1' = 1. \tag{11b}$$

The lines $- \beta_2 \pm \beta_1 = 1$ form the "sides" of the triangular region of linear stability in Fig. 1, and for any β_1, β_2 pair on these boundaries, at least one of the corresponding z-plane poles must lie at $z = \pm 1$. We will thus say that limit cycles satisfying (11) are due to a real effective pole at $z = \pm 1$.

The subregions of the β_1, β_2 plane which satisfy (11) for a given value of k are readily determined. For k = 1, for example, we must have $\pm (\beta_2)' - (\beta_1)' = 1$, which yields the following subregions: $|\beta_1| \geq 0.5$, $|\beta_2| < 0.5$; $|\beta_1| \geq 1.5$, $\beta_2 \geq 0.5$; and $|\beta_1| < 0.5$, $\beta_2 \leq -0.5$. These subregions along with those for k = 2 and k = 4 are shown in Fig. 2. The cross-hatched region of Fig. 2 contains those β_1, β_2 points which do not satisfy (11) for any nonzero value of k.

A necessary condition on the β_1, β_2 values satisfying (11) for a given k is seen from Fig. 2 to be

$$|\beta_1| \geq \left(\frac{k-1}{k}\right) + \beta_2. \tag{12}$$

A dotted line in Fig. 2 shows this condition for k = 2. Note that for β_1, β_2 pairs not satisfying (12), the sufficient condition in (7) for a deadband of [-k,k] must also be necessary (remembering that our definition of the deadband D_2 in (5) neglects the possibility of slightly larger limit cycles due to rounding errors from the β_1 multiplication).

UNIFIED RESULTS FOR SECOND-ORDER SECTIONS

We now combine the results of the preceding sections for complex-conjugate and real effective poles, respectively, into a unified description of the limit-cycle behavior of second-order digital filter sections. First, combining Figs. 1 and 2 results in the overall description in Fig. 3. The cross-hatched region in Fig. 3 indicates those β_1, β_2 pairs for which no limit cycles are possible. The subregions corresponding to the deadbands [-1,1] [-2,2], and [-4,4] are also shown. Based upon our heuristic derivation and the supporting experimental results, we hypothesize that a diagram such as Fig. 3 depicts both the necessary and sufficient conditions for a deadband of [-k,k] for all k $\geq$ 1 and all β_1 and β_2.

In the case where the filter has complex-conjugate poles (i.e., $\beta_1^2 \leq 4\beta_2$), an alternate depiction of the limit-cycle conditions in terms of the (ideal) pole positions in the z-plane is sometimes more useful. To map the limit-cycle conditions from the β_1, β_2 plane to the z-plane, consider the mapping of the arbitrary line $\beta_2 = s\beta_1 + r$. The complex-conjugate poles of the filter are given by

$$z_p = -\frac{\beta_1}{2} \pm j\sqrt{\beta_2 - \left(\frac{\beta_1}{2}\right)^2} \tag{13}$$

By writing $z_p = x_p + jy_p$, substituting $\beta_2 = s\beta_1 + r$ into (13) and performing some algebra, we arrive at the relationship

$$y_p^2 + (x_p+s)^2 = r + s^2. \tag{14}$$

Thus, the locus of pole positions in the z plane corresponding to the line $\beta_2 = s\beta_1 + r$ for $\beta_1^2 \leq 4\beta_2$ is a circle centered at $z_p = -s$ with a radius of $\sqrt{r+s^2}$.

Applying the above mapping to various lines of interest in Fig. 3 produces the z-plane representation shown in Fig. 4. The necessary conditions for a deadband of [-k,k] are indicated by an N, and the sufficient conditions by an S. These conditions are shown for k = 1, 2, and 4, as before. The separate necessary conditions (circles centered at z = ±1) are taken from (12); while the separate sufficient conditions (vertical lines) are derived as follows: (7) and (11) are sufficient conditions for a deadband of [-k,k]. Therefore, if β_1 satisfies (11) for $\beta_2 = \beta_{2k}$, then either (7) or

(11) must be satisfied for any given value of β_2. The condition on β_1 ensuring that (11) is satisfied for $\beta_2 = \beta_{2k}$ is

$$|\beta_1| \geq \frac{4k-3}{2k} . \tag{15}$$

We now give mathematical expressions for the conditions depicted in Fig. 4 for complex-conjugate poles. First, the sufficient condition (7) becomes

$$|z_p| \geq \sqrt{\frac{2k-1}{2k}} . \tag{16}$$

For $y_p^2 + (|x_p| - 1/3)^2 \geq (2/3)^2$, (16) is also necessary. Otherwise, the necessary condition (12) becomes

$$y_p^2 + (|x_p| - 1)^2 \leq \frac{1}{k} ; \tag{17}$$

while the sufficient condition (15) implies that

$$|x_p| \geq \frac{4k-3}{4k} . \tag{18}$$

Finally, for $|z_p| < 1/\sqrt{2}$ and k = 1, the sufficient condition given by (18) is also necessary; that is, for $|z_p| < 1/\sqrt{2}$, a necessary and sufficient condition for a nonzero deadband is $|x_p| \geq \frac{1}{4}$.

CONCLUSIONS

A heuristic approach based upon an "effective value" linear model has been employed to derive the second-order deadbands in which limit cycles due to multiplication rounding in recursive digital filters are found to exist. Extensive computer simulations have been performed for representative β_1, β_2 pairs over the entire region of linear stability for a second-order filter, and these simulations have confirmed the predicted limit-cycle behavior in all cases. The only variations from the predicted results are the (small) expected deviations from a perfectly sinusoidal limit-cycle waveform due to rounding. In particular, some of the observed limit cycles (for $\beta_2 \geq 0.9$) have exceeded the limits of the computed deadband (D_2) by one quantization step, but never more.

Initial experimental evidence also indicated that these results are quite "stable" with respect to the introduction of small nonconstant inputs to the filter. That is, with very small nonconstant inputs, the spectrum of the output remains essentially the same as for the largest zero-input limit cycle. As the input is increased, the output remains largely unchanged (with perhaps some spectral broadening) until the ideal filter output approaches the magnitude of the largest zero-input limit cycle. Above this transition input level, the filter output spectrum approaches the ideal output spectrum more and more closely as the effect of rounding becomes relatively smaller and smaller.

The extension of these results to higher-order digital filters realized as either a parallel or cascade combination of second-order filter sections[3] is obvious in the former case, but not in the latter. In the parallel case, the individual second-order sections share a common input, and their outputs are summed to produce the filter output. Hence, the limit-cycle behavior of each section is independent of all other sections, and the output of the filter for zero input must be simply the sum of the limit-cycle outputs of the individual sections.

In the cascade case, however, each filter section derives its input from the preceding section, and thus, in the general case, only the first section in the cascade necessarily has zero input with zero input to the filter. The previously described "stability" of our limit-cycle results would suggest that perhaps with small (or zero) inputs the filter behaves essentially as though each section were producing its maximum zero-input limit cycle, or else filtering the output of the previous section, whichever is the larger output. The extent to which this conjecture is valid has yet to be determined.

REFERENCES

1. E. I. Jury, Sampled-Data Control Systems, (New York: John Wiley, 1958).

2. P. M. Ebert, J. E. Mazo, and M. G. Taylor, "Overflow Oscillations in Digital Filters," B.S.T.J., Vol. 48, No. 9 (November, 1969).

3. L. B. Jackson, J. F. Kaiser, and H. S. McDonald, "An Approach to the Implementation of Digital Filters," IEEE Trans. on Audio and Electroacoustics, Vol. AU-16, No. 3 (September, 1968), pp. 413-21.

4. I. W. Sandberg, "A Theorem Concerning Limit Cycles in Digital Filters," Proc. Seventh Annual Allerton Conf. on Circuit and System Theory, Monticello, Illinois (October, 1969).

5. R. B. Blackman, Linear Data-Smoothing and Prediction in Theory and Practice, (Reading, Mass.: Addison-Wesley, 1965), pp. 75-9.

6. I. W. Sandberg, "Floating-Point-Roundoff Accumulation in Digital-Filter Realizations," B.S.T.J., Vol. 46, No. 8 (October, 1967), pp. 1775-91.

7. L. B. Jackson, An Analysis of Roundoff Noise in Digital Filters, Sc.D. Thesis, Stevens Institute of Technology, Hoboken, New Jersey (1969).

TABLE 1A (λ: $f_0 = \lambda f_N$)

$\beta_1 \leq$	$\beta_2 \geq 0.5$ [-1,1]	$\beta_2 \geq 0.75$ [-2,2]	$\beta_2 \geq 0.833$ [-3,3]
0.	1/2	1/2	1/2
-0.167	1/2	1/2	5/11
-0.25	1/2	3/7	4/9
-0.5	1/3	2/5	2/5
-0.75	1/3	1/3	3/8
-0.833	1/3	1/3	1/3
-1.167	1/3	1/3	2/7
-1.25	1/3	1/4	1/4
-1.5	0	1/5	1/5
-1.75	-	0	1/7
-1.833	-	-	0

TABLE 1B (λ: $f_0 = \lambda f_N$)

$\beta_1 \geq$	$\beta_2 \geq 0.5$ [-1,1]	$\beta_2 \geq 0.75$ [-2,2]	$\beta_2 \geq 0.833$ [-3,3]
0.	1/2	1/2	1/2
0.167	1/2	1/2	6/11
0.25	1/2	4/7	5/9
0.5	2/3	3/5	3/5
0.75	2/3	2/3	5/8
0.833	2/3	2/3	2/3
1.167	2/3	2/3	5/7
1.25	2/3	3/4	3/4
1.5	1	4/5	4/5
1.75	-	1	6/7
1.833	-	-	1

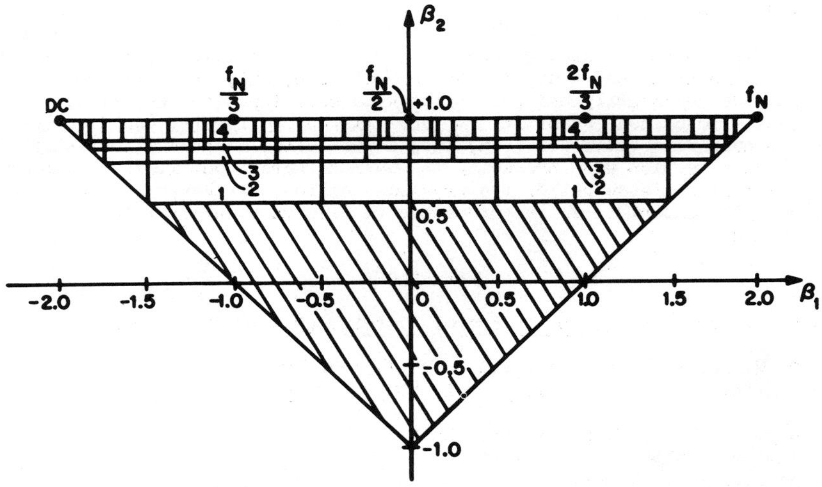

Fig. 1 - Parameter Space for Complex-Conjugate Effective Poles.

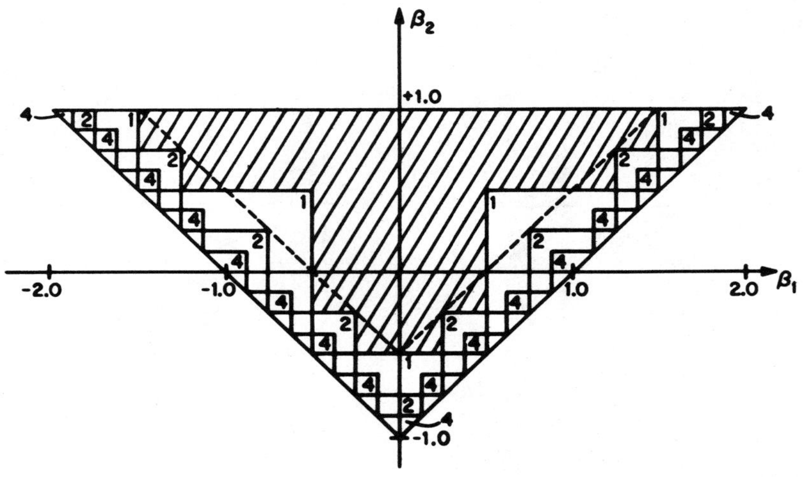

Fig. 2 - Parameter Space for Real Effective Poles.

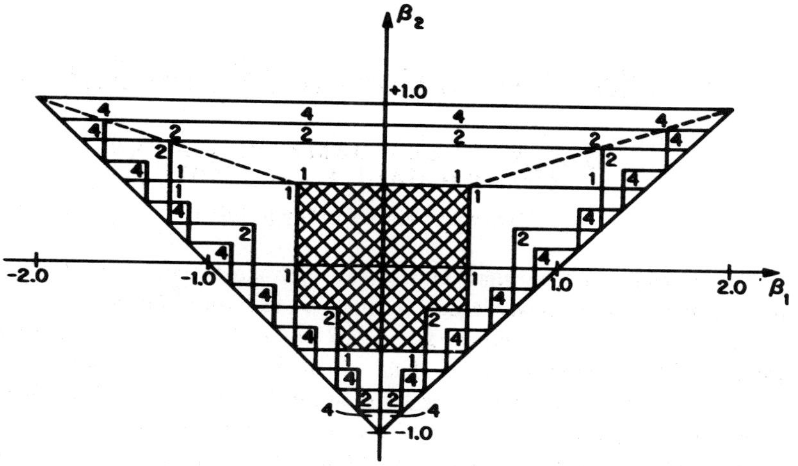

Fig. 3 - All Deadband Subregions (Figs. 1 & 2 combined).

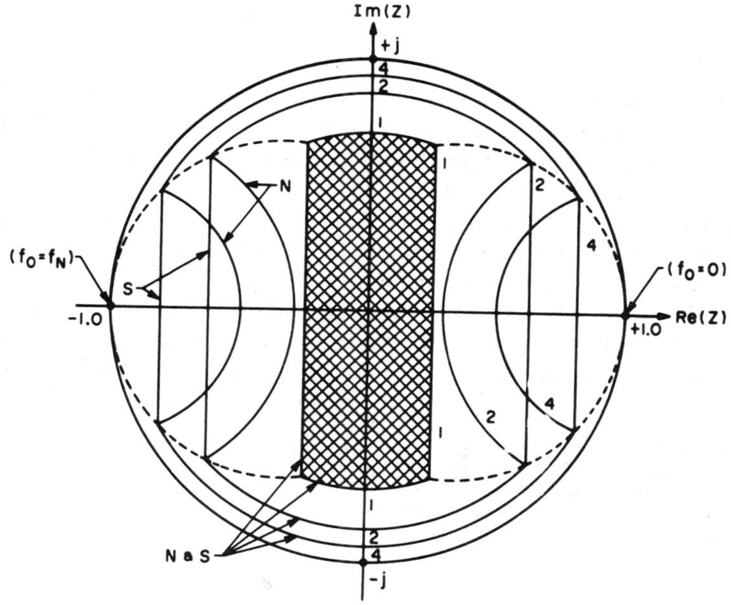

Fig. 4 - Mapping of Fig. 3 to z plane for $\beta_1^2 \leq 4\beta_2$.

Overflow Oscillations in Digital Filters

By P. M. EBERT, JAMES E. MAZO,
AND MICHAEL G. TAYLOR

(Manuscript received May 9, 1969)

The cascade and parallel realizations of an arbitrary digital filter are both formed using second order sections as building blocks. This simple recursive filter is commonly implemented using 2's complement arithmetic for the addition operation. Overflow can then occur at the adder and the resulting nonlinearity causes self-oscillations in the filter. The character of the resulting oscillations for the second order section are here analyzed in some detail. A simple necessary and sufficient condition on the feedback tap gains to insure stability, even with the presence of the nonlinearity, is given although for many desired designs this will be too restrictive. A second question studied is the effect of modifying the "arithmetic" in order to quench the oscillations. In particular it is proven that if the 2's complement adder is modified so that it "saturates" when overflow occurs, then no self-oscillations will be present.

I. INTRODUCTION

A digital filter using idealized operations can easily be designed to be stable.[1] Nevertheless, in actual implementations, the output of such a stable filter can display large oscillations even when no input is present.* A known cause of this phenomenon is the fact that the digital filter realization of the required addition operation can cause overflow, thereby creating a severe nonlinearity.† Our purpose here is twofold. The first is to give a somewhat detailed analysis of the character of the oscillations when the filter is a simple second order recursive section with two feedback taps. This unit is the fundamental building block for the cascade and the parallel realization of digital filters, and as such is worthy of some scrutiny.[2] A simple conclusion which one can draw from

* To the best of our knowledge, these oscillations were first observed and diagnosed by L. B. Jackson of Bell Telephone Laboratories.

† In the present work rounding errors in multiplication or storage are neglected and therefore so are the little-understood oscillations attendant upon these nonlinearities.

the analysis is that the design of many useful filters requires using values of feedback coefficients such that the threat of oscillations is always present (with 2's complement arithmetic). Optimum solutions that cope with this state of affairs are still unknown. Some recent proposals include observing when overflow at the adder is to occur and then taking appropriate action. Our second purpose, then, is to discuss the effectiveness of some of these ideas, and to give a proof that modifying 2's complement arithmetic so that the adder "saturates" is an effective way to eliminate the oscillations. Questions of how this nonlinearity will affect the desired outputs from a particular ensemble of input signals are not yet answered however, and perhaps for some applications other solutions need be considered.

II. PROBLEM FORMULATION AND GENERAL DISCUSSION

As explained in the introduction, this paper deals primarily with the simple structure shown in Fig. 1. The outputs of the registers, which are storage elements with one unit of delay, are multiplied by coefficients a and b respectively, fed back, and "added" to the input in the accumulator. No round-off error is considered either in multiplication or storage, but overflow of the accumulator is not neglected. In other words, the accumulator will perform as a true adder if the sum of its inputs is in some range; otherwise a nonlinear behavior is observed.

Figure 2 shows the instantaneous input-output characteristic $f(v)$ of the device motivated by using 2's complement arithmetic. It is also important to note that there is no memory of the accumulator for past outputs; that is, the device is zeroed after the generation of each output.

If we let $x(t)$ be the input signal to the device, $y(t)$ the output, and

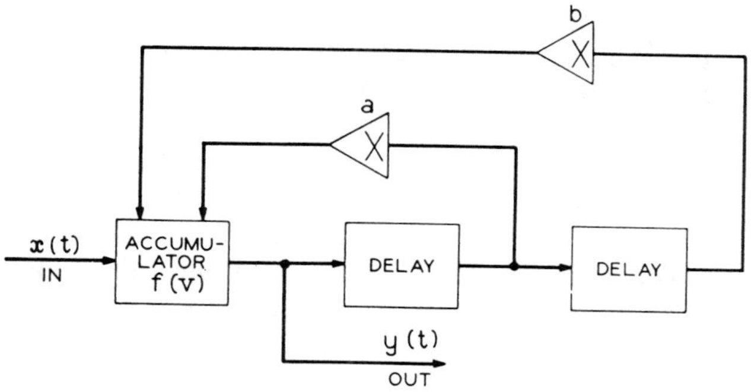

Fig. 1 — Basic configuration for the digital filter $\cdot y_{k+2} = f[ay_{k+1} + by_k + x_{k+2}]$.

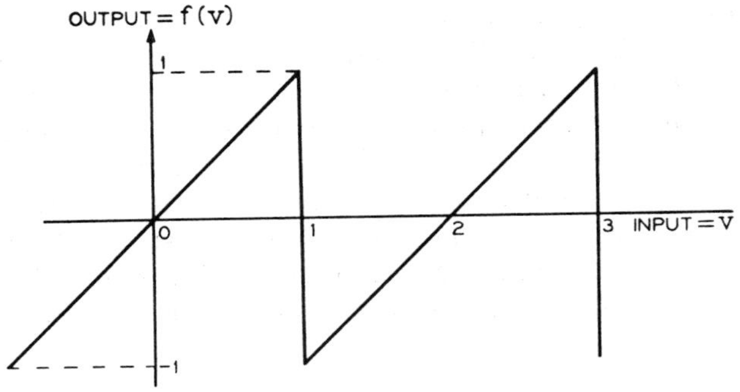

Fig. 2 — Instantaneous transfer function of the accumulator.

$f(\cdot)$ the nonlinear characteristic of the accumulator, we have the basic equation

$$y(t + 2) = f[ay(t + 1) + by(t) + x(t + 2)]. \tag{1}$$

We shall be concerned with the self-sustaining oscillations of the device that are observed even when no input is present [$x(t) = 0$], and when linear theory would predict the device to be stable.

By making this linear approximation $f(v) = v$, the linearized version of equation (1) becomes, with no driving term in the equation,

$$y(t + 2) - ay(t + 1) - by(t) = 0. \tag{2}$$

The roots of the characteristic equation for equation (2) are

$$\rho_{1,2} = \frac{a \pm (a^2 + 4b)^{\frac{1}{2}}}{2} \tag{3}$$

and the region of linear stability corresponds to the requirement that $|\rho_i| < 1$. This region is depicted as a subset of the a–b plane in Fig. 3. One has $|\rho_i| < 1$ if and only if one is within the large triangle shown in Fig. 3. For this situation any solution of (2) will damp out to zero after a sufficient period of time. Now note that (2) is not necessarily a valid reduction of (1) even when $x(t) = 0$. The output, by choice of f, has been assumed to be constrained to be less than unity, but this is not sufficient to guarantee that the argument of the function f is less than unity. For this to be the case we require

$$|ay(t + 1) + by(t)| < 1. \tag{4}$$

Since $|y(t)| < 1$, equation (4) will always be satisfied provided that

$$|a| + |b| < 1. \tag{5}$$

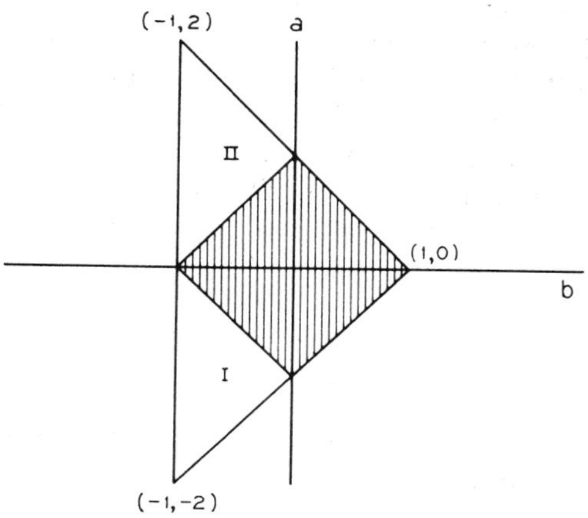

Fig. 3—Some interesting regions in the "space" of feedback tap weights. The hatching indicates stability even with the nonlinearity.

The subset of the a–b plane for which (5) is true is shown in Fig. 3 with vertical hatching, and is a subset of the region of linear stability. It is shown in this Section that if (5) is not satisfied there always exist self-sustained oscillations of the digital filter and hence (5) is both a necessary and sufficient condition for absence of self-sustained oscillations.* One way to avoid the oscillations in question is simply to impose the requirement (5). This trick has its limitations, however, for it clearly restricts design capabilities. The region of the s-plane which is shaded in Fig. 4 shows the allowable pole positions. Roughly speaking, one concludes that there are desirable filter characteristics that can be realized with this restriction and there are desirable characteristics that cannot.

It is not our purpose here to outline those applications for which (5) will not be restrictive; we proceed to sketch the situation when $|a| + |b| > 1$ and the threat of oscillation is present. Sections III and IV contain, we believe, a novel and interesting mathematical treatment of the general problem of classifying the self-oscillations of the nonlinear difference equation (1). However, for the user of digital filters a simple proof of the $|a| + |b| > 1$ being sufficient for threat of oscillations is of more immediate interest. After reading the simple proof of this fact given next in the present section, such a reader may wish to proceed directly to Section V.

Consider the possibility of undriven nonlinear operation giving a dc

* I. W. Sandberg has informed the authors that the necessity and sufficiency of (5) holding for absence of oscillations has also been obtained jointly by him and L. B. Jackson.

output, that is, $y_k \equiv y$ for all k. Equation (1), with $x(t) = 0$ becomes $y = f[(a + b)y]$. Assuming for definitness that $y > 0$, we can easily see from Fig. 2 that the above equation will be true if $(a + b)y = y - 2$, which implies $y = 2/(1 - a - b)$. One can show (see discussion following equation 17), that this y will have magnitude < 1 provided only that the tap values a and b lie in the region labeled I in Fig. 3. Thus a consistent dc oscillation is always possible for all (a, b) pairs in this region. Next consider the possibility of a period 2 oscillation. This amounts to finding a consistent solution to $y = f[(b - a)y]$. Proceeding as before we obtain

$$y = \frac{2}{1 + a - b}.$$

Thus y_k will be given by $(-1)^k y$, and will have magnitude less than unity if the (a, b) pair lies anywhere in region II of Fig. 3.

III. FURTHER ANALYSIS OF THE OSCILLATIONS

To analyze equation (1) in greater detail, it is very convenient to write it in the form similar to (2),

$$y(t + 2) - ay(t + 1) - by(t) = \sum_n a_n u(t + 2 - n), \tag{6}$$

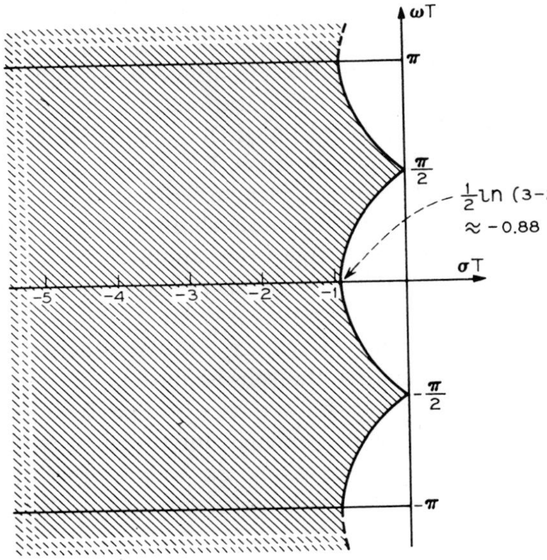

Fig. 4 — Pole locations in the s-plane (shaded region) realizable under the constraint that $|a| + |b| < 1$.

where $u(t)$ is a square pulse of unit height that one may conveniently think of as lasting from $t = 0$ until $t = 1$. This, of course, means that one interprets the solution of (6) to be a piecewise constant function like the actual output of the digital filter. For mathematical manipulations it is sometimes desirable to also interpret (6) as a difference equation, defined only for integer t. In this case one would write that $u(t - n) = \delta_{tn}$ where δ_{tn} is the familiar Kroneker symbol.

The point of the right side of (6) is simply to keep $|f(v)| < 1$ regardless of what value v has. From Fig. 2 we see that if $|v| < 1$, this added term is not needed and we take $a_n = 0$. If $1 < v < 3$ then we take $a_n = -2$, and if $-3 < v < -1$ we take $a_n = +2$. Since we have that $|y(t)| < 1$ and that linear stability (see Fig. 3) implies $|a| < 2$, $|b| < 1$, we need not consider further values of $|v|$. Thus in (6) $a_n = 0, \pm 2$ depending on whether or not $v(t) \equiv ay(t + 1) + by(t)$ crosses the lines $v = \pm 1$. It will be convenient to have a word for such crossings; we shall call them "clicks", borrowing a favorite word from FM theory. Then $a_n = 0, \pm 2$ depending on whether or not a click does not, or does, occur.

Note if one knew what the click sequence $\{a_n\}$ was, one could solve (6) simply by using the clicks to be the driving term for a linear equation. We are mainly interested in describing the self-sustained steady state oscillations of arbitrary period N. Hence initial conditions will play no essential role for us, for while they determine which oscillating mode appears as $t \rightarrow \infty$, they play no role in describing the modes. Our procedure will be as follows:

(*i*) Assume a click sequence of period N;

$$\begin{aligned} &a_0, a_1, a_2, \cdots, a_{N-1} \\ &a_{lN+k} = a_k . \qquad l = 0, 1, \cdots \\ &\qquad\qquad\qquad 0 \leqq k < N - 1. \end{aligned} \tag{7}$$

(*ii*) Using the assumed $\{a_n\}$, find the steady state solution of (6). However, only solutions that have $|y(t)| < 1$ for all t are allowed. (*iii*) Check that this steady state solution actually generates the assumed click sequence.

In carrying out the above program for some simple cases we observed that step *iii* never seemed to yield anything new. Indeed, surprising as it seems at first glance, step *iii* never has to be carried out. If one obtains a solution with $|y(t)| < 1$, this solution is consistent. That is, it automatically generates the assumed click sequence. The proof is simple.

One calculates the argument of the function f from (6):

$$ay(t + 1) + by(t) = y(t + 2) - \sum a_n u(t + 2 - n). \tag{8}$$

We have a click at time $t + 2 = m$ if $| ay(m - 2) + by(m - 1) | > 1$. From (8),

$$| ay(m - 2) + by(m - 1) | = | y(m) - a_m |. \tag{9}$$

Note then if in (9) $a_m = 0$, then $| ay(m - 2) + by(m - 1) | = | y(m) | < 1$; thus if there is no click at a particular time in the assumed click sequence the "solution" will not generate one. Next assume $a_m = +2$; then

$$ay(m - 2) + by(m - 1) = y(m) - 2 < -1, \tag{10}$$

where we use $| y(t) | < 1$ again. Equation (10) says if a positive click is present in the assumed click sequence then the solution obtained from the linear equation (6), given by this click sequence, will reproduce the positive click. Obviously the same argument holds for a negative click, $a_m = -2$, and the proof of this point is complete.

The steady-state solution of our fundamental equation (6) for an arbitrary click sequence $\{a_m\}$ of period N is derived in the appendix. If we define

$$A_{N-1}\left(\frac{1}{z}\right) \equiv \sum_{n=0}^{N-1} a_n z^{-n} \tag{11}$$

and

$$D(z) \equiv z^2 - az - b, \tag{12}$$

and let r_i, $i = 1, \cdots, N$, be the N Nth roots of unity, then the (periodic) output values are given by

$$y_k = \frac{1}{N} \sum_{i=1}^{N} \frac{A_{N-1}\left(\frac{1}{r_i}\right)}{D(r_i)} r_i^k . \tag{13}$$

The above expression gives the $\{y_k\}$ output sequence for any click sequence. We emphasize, however, that it is only a solution corresponding to a self-sustained oscillation of the digital filter if we have $| y_k | < 1$, all k. Whether or not this is true depends on the particular click sequence assumed.

Another form of the solution can be obtained by manipulation of (13). To write this down, define

$$b_n^{(k)} \equiv (\bar{a}_{k-1-n} + \bar{a}_{k-1-n+N})/2, \tag{14}$$

where we understand $\bar{a}_j \equiv 0$ if j does not lie between 0 and $N - 1$, inclusive, and $\bar{a}_j \equiv a_j$ if it does. One of the $\bar{a}$'s in (14) will thus always be zero and $b_n^{(k)}$ has values of ± 1, 0. The other form of the solution is then

$$y_k = \frac{2}{\rho_1 - \rho_2} \sum_{n=0}^{N-1} b_n^{(k)} \left[\frac{\rho_1^n}{1 - \rho_1^N} - \frac{\rho_2^n}{1 - \rho_2^N} \right] \qquad k = 0, 1, \cdots, N-1 \tag{15}$$

where ρ_i are given in (3).

In (15) we have N vectors of dimension N, namely the $\{b_n^{(k)}\}$ $k = 0, 1, 2, \cdots, N - 1$. Note from (14), however, that they are all cyclic permutations of one another. Hence we may refer to the b vector, $\mathbf{b}$, of a solution, understanding that the $\mathbf{b}$ and all its cyclic permutations generate a solution in the sense of (15). Note that a cyclic permutation of the y_k has no real significance here; it simply changes the origin of time.

An interesting property of the solutions which we have written down follows from the fact that if we transform the point (a, b) in the ab-plane into another point by

$$\begin{aligned} a &\rightarrow a' = -a \\ b &\rightarrow b' = b \end{aligned} \tag{16a}$$

then under this transformation

$$\begin{aligned} \rho_1 &\rightarrow \rho_1' = -\rho_2 \\ \rho_2 &\rightarrow \rho_2' = -\rho_1 . \end{aligned} \tag{16b}$$

The property is this: Let N be an even integer and let $\mathbf{b} = (b_0, b_1, \cdots, b_{N-1})$ be a click vector generating a solution at point (a, b). Then the vector $\mathbf{b}' = (b_0, -b_1, b_2, -b_3, \cdots, b_{N-1})$ generates a solution at reflected point $(-a, b)$. The proof is simple. Note from (15),

$$\begin{aligned} y'^{(k)} &= \frac{2}{\rho_1' - \rho_2'} \sum_n b_n'^{(k)} \left[\frac{\rho_1'^n}{1 - \rho_1'^N} - \frac{\rho_2'^n}{1 - \rho_2'^N} \right] \\ &= \frac{2}{-\rho_2 + \rho_1} \sum_n (-1)^{k+n} b_n \left[\frac{(-\rho_2)^n}{1 - \rho_2^N} - \frac{(-\rho_1)^n}{1 - \rho_1^N} \right] = (-1)^k y^{(k)} . \end{aligned}$$

Hence if $| y^{(k)} | < 1$ then $| y'^{(k)} | < 1$. Note that the proof also supplies the value for $y'^{(k)}$ in terms of $y^{(k)}$. This theorem will be used later to generate new solutions from old ones.

Before leaving this general discussion in favor of exhibiting some solutions in the next section, we list a few more observations related to the click vector **b**. The click vector **b**, whose only allowed component values are ± 1, 0, completely characterizes the associated oscillation. Clearly there can then only be a finite number of oscillations of given period N. This number is upper bounded by 3^N, but will generally be much less. Also note that a cyclic permutation of the components of b cyclically permutates the output values y^k, and this latter is merely a shift in time. The permutated values are not physically distinct.

Also note that if we perform $\mathbf{b} \to -\mathbf{b}$ then $\mathbf{y} \to -\mathbf{y}$, and a solution of opposite sign is obtained. While this may often be distinguishable from the first solution, it is trivially related to it. Finally if one were to count the number **b** vectors of dimension N that yield new information, one would wish to exclude subperiods of N. Thus if $(+, 0, 0)$ is an generating **b** vector for period 3, $(+, 0, 0, +, 0, 0)$ generates a period 6 oscillation but this is not new information. We have not solved the problem of counting how many of the 3^N vectors are left after we impose the requirements of cyclic shifts, sign changes, and subperiods. At any rate, it is essential to test the ones that remain to check that they generate allowed solutions, $|y^k| < 1$.

IV. SOME EXPLICIT PERIODS AND REGIONS OF OSCILLATION

Now for a few explicit solutions. Consider the possibility of a dc "oscillation", namely, set $N = 1$. The only nontrivial click vector is $\mathbf{b} = (+)$. The solution is more immediate if we use (13). We have

$$y = \frac{2}{1 - a - b} \tag{17}$$

for the dc value of output. For what values of a and b within the triangle of Fig. 3 will we have $|y| < 1$? We require

$$|1 - a - b| > 2 \tag{18}$$

which is equivalent to either

$$1 - a - b > 2 \tag{19a}$$

or

$$-1 + a + b > 2. \tag{19b}$$

Inequality (19a) (coupled with the linear stability requirement) defines the triangle labeled "I" in Fig. 3, while (19b) is outside the stability region and needs no further consideration. Thus any portion of the region $a < 0$ that we have not excluded from oscillations has now been shown to have them. They are of period 1; other period oscillations may (and do) occur in this region.

At this point it is amusing to use an earlier remark on the possibility of generating new solutions from an even period one by "reflection". Letting $N = 2$, the click vector $\mathbf{b} = (+, +)$ certainly generates a period 2 oscillation (albeit one with subperiods) in region I. Then the click vector $\mathbf{b} = (+, -)$ generates something really new: a period 2 oscillation in the region labeled II in Fig. 3. The amplitudes of the output are

$$y^{(k)} = (-1)^k \frac{2}{1 + a - b}, \qquad a > 0. \tag{20}$$

One more possibility of a click vector exists for period 2, and that is $\mathbf{b} = (+, 0)$. From (13) we write for possible output values

$$\begin{aligned} y_0 &= \frac{1}{1 - a - b} + \frac{1}{1 + a - b} \\ y_1 &= \frac{1}{1 - a - b} - \frac{1}{1 + a - b}. \end{aligned} \tag{21}$$

After a little uninteresting analysis one can conclude that we cannot have $|y_0| < 1$, $|y_1| < 1$ in (21) for any allowed values of a and b. Thus there are no other period 2 oscillations.

On to period 3. Now there are four click vectors which must be considered. These are $(+00)$, $(++0)$, $(+-0)$, $(++-)$. Even in this case an exhaustive check that the "solutions" generated are legitimate ones is trying. Therefore, we resort to a trick; we look for periods which may exist in the immediate neighborhood of the point $(a = 0, b = 1)$. This means $\rho_1 = i$, $\rho_2 = -i$. In this immediate neighborhood $\rho_2 = \rho_1^*$, and (15) reads

$$y = \frac{2}{\operatorname{Im} z} \operatorname{Im} \sum_{n=0}^{N-1} \frac{b_n z^n}{1 - z^N}, \tag{22}$$

where we have let $z = \rho_1$. Letting $N = 3$, $z = i$ gives

$$\begin{aligned} y_0 &= -b_0 + b_1 + b_2 \\ y_1 &= -b_1 + b_2 + b_0 \\ y_2 &= -b_2 + b_0 + b_1 . \end{aligned} \tag{23}$$

We now require $y_k = \pm 1$ as a test for the click vector $\mathbf{b}$. We see that only $(+00)$ qualifies as possibly yielding a solution in the neighborhood of $(a = 0, b = -1)$. A computer study shows that indeed the solution extends into the interior of the triangle and the region found is shown in Fig. 5. This immediately implies existence of the period 6 oscillation generated by $(+00-00)$ in the reflected region. Similarly, a period 5 oscillation region (with the concomitant period 10) generated by $(+0000)$ is shown in Fig. 6.

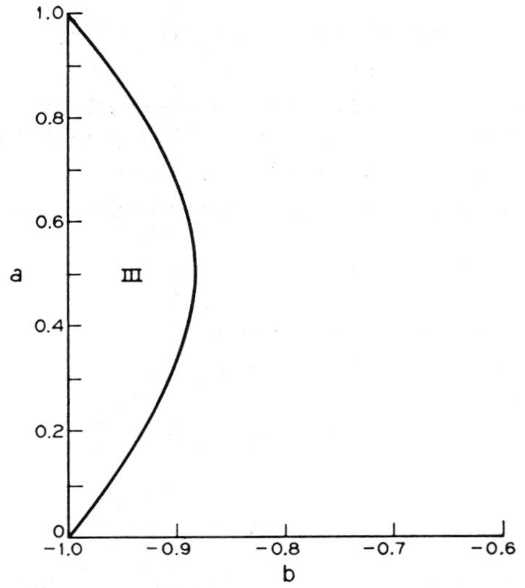

Fig. 5 — A region for period 3 oscillations.

It is very tempting to conjecture that the point ($a = 0$, $b = -1$) is a boundary point of any allowed region of oscillation. If this is true, a procedure like that used above may eliminate some otherwise very respectable **b** vectors from consideration. Note that for $N = 2$, $b = (+, 0)$ satisfies the required condition at $\rho_1 = i$, but we have shown this

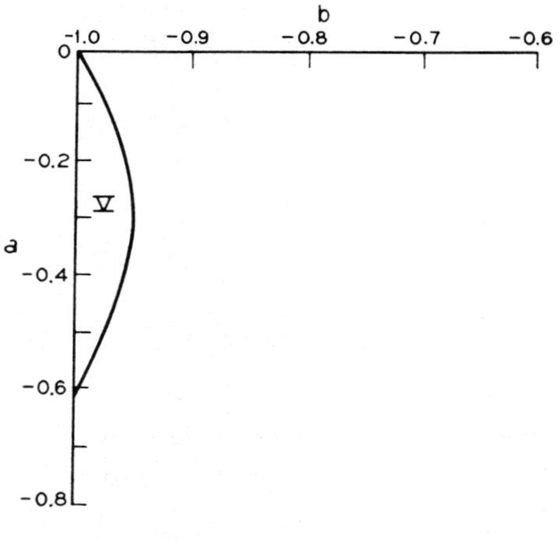

Fig. 6 — A region for period 5 oscillations.

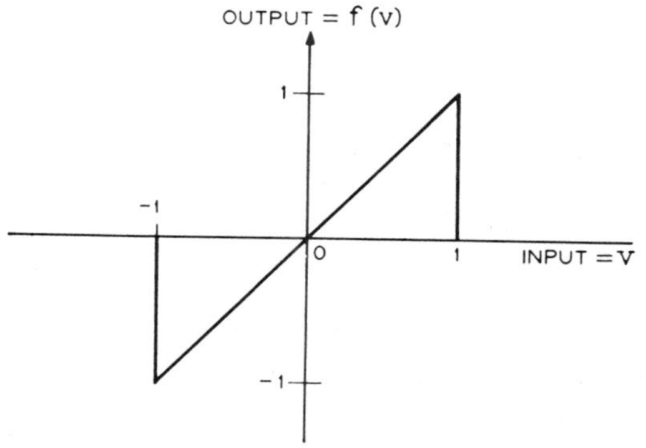

Fig. 7 — Zeroing arithmetic, shown above, also gives rise to oscillations.

is not extendable into the interior of the triangle. Hence existence at $z = i$ does not guarantee an allowed solution.

V. STABILITY WITH A MODIFIED ARITHMETIC

In an attempt to eliminate these oscillations, proposals have been made which rely on detecting overflow. One such suggestion dictates that when overflow occurs, the adder is directed to shift out zero. For ref-ereace we call this zeroing arithmetic. The effective transfer function of the adder for zeroing arithmetic is given in Fig. 7. However, it can be shown by numerical example that such a procedure still leads to oscil-lations. Another possibility, "saturation arithmetic," is displayed in Fig. 8. Here a one (with the appropriate sign) is put out when overflow is detected. The remaining portion of this paper is devoted to proving that saturation arithmetic leads to stable operation whenever linear theory would predict it to be so.

To begin, we suppose for the moment that we ignore the fact that the digitally implemented adder is nonlinear. Then the second-order linear difference equation which governs the behavior of the undriven system has solutions y_k which may be described as follows:

Case 1: Complex roots for characteristic equation

$$y_k = \text{Re } K_0 \exp(-\alpha k), \quad K_0 \text{ and } \alpha \text{ complex}, \quad \text{Re } \alpha > 0. \qquad k = 0, 1, 2, \cdots . \tag{24}$$

Case 2: Real but unequal roots

$$y_k = K_1 \exp(-\alpha k) + K_2 \exp(-\beta k). \quad K_i \text{ real}; \quad \alpha > 0, \quad \beta > 0. \tag{25}$$

Case 3: Real and equal roots

$$y_k = [K_1 + K_2 k] \exp(-\alpha k). \quad K_i \text{ real}; \quad \alpha > 0. \tag{26}$$

Using this information, coupled with knowledge of y_j and y_{j+1} for some j, it is easy to give a bound on the magnitudes of all future ($k \geqq j$) values of the output and to show this value goes to zero with increasing j. This is just another way to say that the solutions go to zero for the linear case. In the nonlinear case we cannot exclude the situation that some y_{k+1} will exceed unity and the nonlinearity will be operative. For saturation arithmetic the offending value must be set to unity if, for example, $y_{k+1} > +1$. We can, for conceptual purposes, regard this as a "squeezing" of the output from a value greater than unity down to the value one which is performed in a continuous fashion. The crux of the proof now comes in showing that the partial derivative of our bound (on future outputs) with respect to the most recent output y_{k+1} has, for saturation arithmetic, the same sign as y_{k+1}. Hence decreasing a value that is too large in magnitude will decrease the bound as well, and it will go to zero at least as fast as it does for the linear case.

To show how the above outline works, consider first the linear case with complex roots. From the form of the solution

$$y_k = \text{Re}\, K_0 \exp(-\alpha k), \quad \text{Re}\, \alpha > 0, \quad k = 0, 1, 2, \cdots,$$

it is clear that if we define

$$B_0 = |K_0|^2 \tag{27}$$

then $y_k^2 \leqq B_0$ for all $k \geqq 0$. We now express B_0 in terms of the values y_0, y_1 which are initially stored in the shift registers to yield

$$B_0 = y_0^2 + \frac{[y_1 - y_0 \text{ Re} \exp(-\alpha)]^2}{[\text{Im} \exp(-\alpha)]^2}. \tag{28}$$

This suggests that one define the more general set of numbers

$$B_j = y_j^2 + \frac{[y_{j+1} - y_j \text{ Re} \exp(-\alpha)]^2}{[\text{Im} \exp(-\alpha)]^2}. \tag{29}$$

Clearly, from the way that B_j is defined, we have that

$$y_k = \text{Re}\, K_j \exp[-\alpha(k - j)], \qquad k \geqq j \tag{30}$$

where K_j is some appropriate complex number that satisfies

$$B_j = |K_j|^2. \tag{31}$$

From (30), the additional inequality that $y_k^2 \leqq B_j$ for all $k \geqq j$ follows.

Furthermore, one can see by comparing (30) and (24) that

$$|K_j|^2 = |K_0|^2 \, |\exp(-\alpha j)|^2. \tag{32}$$

Hence, since the real part of α is positive, B_j goes monotonically to zero with increasing j.

To generalize the above arguments to a nonlinear situation of interest,* consider the following equation which follows from (29):

$$\frac{\partial B_j}{\partial y_{j+1}} = \frac{2}{[\text{Im} \exp(-\alpha)]^2} [y_{j+1} - y_j \text{ Re} \exp(-\alpha)]. \tag{33}$$

Now imagine B_{j-1} has been calculated from values stored in the registers. From *linear* theory we predict $y_{j+1}^{(L)}$ and $B_j^{(L)} \leqq B_{j-1} \exp(-2\alpha)$, by (32). Now if the $y_{j+1}^{(L)}$ generated by the linear equation were too large, say, then decreasing it to unity would, according to (33), *decrease* the bound B_j if we knew that

$$y_{j+1} - y_j \text{ Re} [\exp(-\alpha)] \geqq 0 \quad \text{for} \quad y_{j+1}^{(L)} \geqq y_{j+1} \geqq y_{j+1}^{(C)} \tag{34}$$

where $y_{j+1}^{(L)}$ is the linear prediction for y_{j+1} and $y_{j+1}^{(C)}$ is the correct value for the nonlinear circuit resulting from "squeezing" $y_{j+1}^{(L)}$ down. Since $|y_j| \leqq 1$ and Re exp $(-\alpha) < 1$, (34) is always true for saturation arithmetic (see Fig. 8) because $y_{j+1}^{(C)} = +1$ (assuming $y_{j+1}^{(L)} > +1$) and (34) can never swing negative. Similar things happen, of course, if $y_{j+1} < -1$. Thus the bound decreases at least as fast as for the linear case (which is exponential) and stability is assured. For zeroing arithmetic $y_{j+1}^{(C)} = 0$, and thus the appropriate sign for (34) cannot be guaranteed which is in satisfying agreement with the known instability for this case.

For the next case of real but unequal roots, we now have reference to equation (25) and define our initial bound as

$$B_0 = 2(K_1^2 + K_2^2)$$
$$= 2 \frac{[y_1 - \exp(-\alpha)y_0]^2 + [y_1 - \exp(-\beta)y_0]^2}{[\exp(-\alpha) - \exp(-\beta)]^2}. \tag{35}$$

The remaining details are too similar to those of the preceding case to warrant recording again; stability for saturation arithmetic holds here as well.

The last case to discuss occurs when we have real and equal roots.

---

* B_j calculated from (29) is a bound on future outputs for the nonlinear as well as the linear case. If $B_j \leq 1$ the two cases coincide, while of $B_j > 1$ the conclusion follows equally trivially since $|y_k| \leq 1$ for the nonlinear situation.

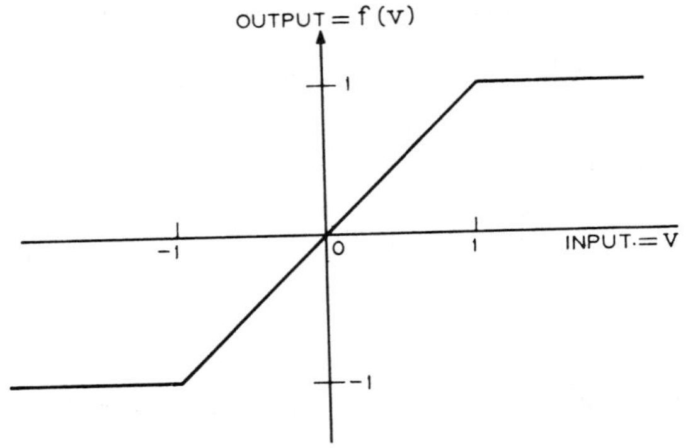

Fig. 8 — The above nonlinearity corresponds to saturation arithmetic and leads to stable behavior.

This situation, represented for the linear equation by equation (26), is more difficult to treat than the previous ones. The analog of (27) and (35) now is

$$B_0 = \max \begin{cases} 4K_1^2 \\ \dfrac{4K_2^2}{\alpha^2} \end{cases}. \tag{36}$$

That (36) yields a bound follows from the facts that (for $t \geqq 0$)

$$\begin{aligned} y_k^2 &\leqq \max_t \, [(K_1 + K_2 t) \exp(-\alpha t)]^2 \\ &\leqq 2 \max_t \, [K_1^2 + K_2^2 t^2] \exp(-2\alpha t) \\ &\leqq 4 \max \begin{cases} \max_t K_1^2 \exp(-2\alpha t) \\ \max_t K_2^2 t^2 \exp(-2\alpha t) \end{cases} \\ &= 4 \max \begin{cases} K_1^2 \\ \dfrac{K_2^2 \exp(-2)}{\alpha^2} \end{cases} \\ &\leqq 4 \max \begin{cases} K_1^2 \\ \dfrac{K_2^2}{\alpha^2} \end{cases} \end{aligned}$$

Since

$$K_1^2 = y_0^2$$
$$\frac{K_2^2}{\alpha^2} = \frac{(y_1 \exp \alpha - y_0)^2}{\alpha^2}, \tag{37}$$

we define our general bound as

$$B_j = 4 \max \begin{cases} y_j^2 \\ \dfrac{(y_{j+1} \exp \alpha - y_j)^2}{\alpha^2}. \end{cases} \tag{38}$$

Using the solution $y_j = (K_1 + K_2 j) \exp(-\alpha j)$, we see that

$$\theta_j \equiv \frac{(y_{j+1} \exp \alpha - y_j)^2}{\alpha^2} \tag{39}$$

decreases by the multiplicative factor $\exp(-2\alpha)$ for every unit increase of j. Further, suppose that $B_j = 4y_j^2$ for some j. That is, suppose

$$\frac{(y_{j+1} \exp \alpha - y_j)^2}{\alpha^2} < y_j^2 . \tag{40}$$

This implies

$$y_{j+1}^2 < y_j^2(1 + \alpha)^2 \exp(-2\alpha), \tag{41}$$

and so if next time $B_{j+1} = 4y_{j+1}^2$, then we have decreased by $(1 + \alpha)^2 \exp(-2\alpha) < 1$. On the other hand, if at the next step we have to choose $B_{j+1} = 4\theta_{j+1}$, we see

$$\frac{B_{j+1}}{B_j} = \frac{\theta_{j+1}}{y_j^2} \leqq \frac{\theta_{j+1}}{\theta_j} \leqq \exp(-2\alpha). \tag{42}$$

Likewise if we go from $4\theta_j$ to $4\theta_{j+1}$ we decrease by $\exp(-2\alpha)$. Finally, a "transition" from $4\theta_j$ as a bound to $4y_{j+1}^2$ decreases the bound by a multiplicative factor of $(1 + \alpha)^2 \exp(-2\alpha)$. To see this we note that, by assumption,

$$B_j = \frac{4[y_{j+1} \exp \alpha - y_j]^2}{\alpha^2} \geqq 4y_j^2 . \tag{43}$$

Using the left-hand equality in (43) implies

$$| y_{j+1} | \exp \alpha \leqq \frac{\alpha(B_j)^{\frac{1}{2}}}{2} + | y_j |. \tag{44}$$

while $B_i \geqq 4y_i^2$ yields

$$| y_i | \leqq \frac{(B_i)^{\frac{1}{2}}}{2}. \tag{45}$$

Using (45) in (44) then allows us to deduce that

$$B_{i+1} = 4y_{i+1}^2 \leqq (1 + \alpha)^2 \exp(-2\alpha)B_i \tag{46}$$

as was claimed. To extend these arguments to the nonlinear case we again observe that

$$\frac{\partial B_i}{\partial y_{i+1}} \geqq 0 \tag{47}$$

for saturation arithmetic.

VI. GENERALIZATIONS TO OTHER STABLE NONLINEARITIES

Aside from the three nonlinearities already mentioned, there does not appear to be immediate engineering interest in seeing which other nonlinearities will or will not give rise to stable behavior of the filter. Having come this far, however, it is hard to resist asking if the method of proof we have used, or some slight extension of it, does suggest other nonlinearities for which stability will hold. The extension we consider is not to require

$$\frac{\partial B_i}{\partial y^{i+1}} \geqq 0$$

all during the "squeezing" operation, but merely that

$$B_i^L - B_i^C \geqq 0, \tag{48}$$

where B_i^L is the value of the bound using linear theory and B_i^C is the "correct" value. An inspection of the previous proofs shows that this is equivalent to

$$(y_{i+1}^L - ay_i)^2 - (y_{i+1}^C - ay_i)^2 > 0 \tag{49}$$

for all real a such that $| a | < 1$.

A little manipulation reduces (49) to

$$(y_{k+1}^L - y_{k+1}^C)(y_{k+1}^L + y_{k+1}^C - 2ay_k) \geqq 0. \tag{50}$$

Assuming $y_{k+1}^L > 0$, the first term in (50) to be nonnegative, and $| y_k | \leqq 1$, makes it apparent that

$$y_{k+1}^L + y_{k+1}^C \geqq 2 \tag{51}$$

is sufficient. The "stable nonlinearities" deduced from this kind of reasoning are outlined in Fig. 9. Thus any nonlinearity whose graph coincides with the identity function on the interval $[-1, 1]$ and whose remaining portions lie in the closed shaded region of Fig. 9 will be stable. The function in these regions need not be continuous and need not obey $f(-u) = -f(u)$.

An even higher degree of generality is achieved when we realize that nothing in our proofs required the nonlinearity $f(u)$ to be the same for successive values of the parameter k. This is tantamount to allowing the nonlinearity to be random in the following manner. Suppose a value of $y_{k+1}^L > 1$ has been predicted from linear theory (see Fig. 9). The perpendicular P to the v axis through y_{k+1}^L intersects the shaded region shown in Fig. 9 along a line segment. Choose randomly from this line segment the "value" of the nonlinearity to give y_{k+1}^C. The discussion in this Section shows that the solutions of the difference equation

$$y_{k+2} = f[ay_{k+1} + by_k] \tag{52}$$

which has the stochastic nonlinearity just described will be stable whenever the linear version has stable solutions.

APPENDIX

Derivation of the Steady-State Solution

We obtain the steady-state solution of our fundamental equation (6) using z-transforms. Recall that if one has a bounded sequence of number $\{a_n\}$, the z-transform is defined by

$$f(z) = \sum_{n=0}^{\infty} a_n z^{-n} \tag{53}$$

where (53) converges and is analytic outside the unit circle, $|z| > 1$. It is easy to show that if $\{a_n\}$ is periodic of period N, that is if $a_{N+n} = a_n$, then (53) becomes

$$f(z) = \frac{A_{N-1}\left(\frac{1}{z}\right)}{1 - z^{-N}} \tag{54}$$

where A_{N-1} is the polynomial of degree $(N - 1)$ in $1/z$ given by

$$A_{N-1}\left(\frac{1}{z}\right) = \sum_{n=0}^{N-1} a_n z^{-n}. \tag{55}$$

The N poles of $f(z)$ at the N roots of unity are apparent from (12), and there are no other poles.

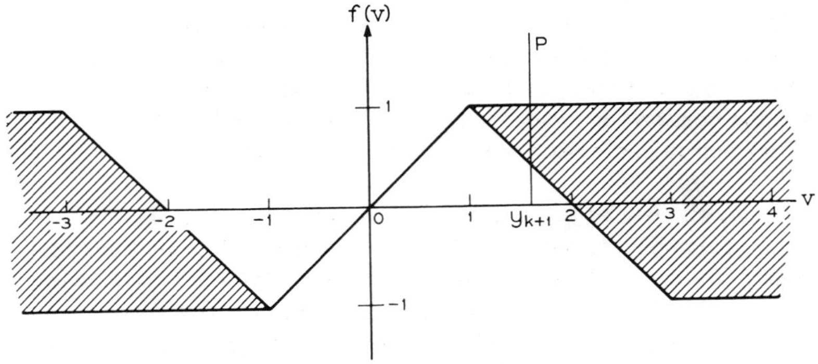

Fig. 9 — Any nonlinearity whose graph coincides with the identity function on the interval [−1, +1] and whose remaining portions lie in the (closed) shaded region will be stable. The possibility of generalizing this to a stochastic nonlinearity is also noted in the text.

Denoting by $Y(z)$ the z-transform of $y(t)$ *excluding* the additive terms involving initial conditions (since these will damp out because of linear stability) we have from (6) that

$$Y(z) = \frac{A_{N-1}\left(\frac{1}{z}\right)}{(z^2 - az - b)(1 - z^{-N})}. \tag{56}$$

The z-transform of the steady-state solution $\hat{Y}(z)$ must still be extracted from $Y(z)$. Since the unit circle $|z| = 1$ corresponds to the frequency axis if one were using Fourier transforms, we know, by analogy, the state steady-state portion of (56) will be the pole-terms. Let r_i, $i = 1, \cdots, N$ be the N Nth roots of unity and define

$$Q_i^{N-1}\left(\frac{1}{z}\right) \equiv \sum_{k=0}^{N-1} \left(\frac{1}{r_i}\right)^{N-1-k} \left(\frac{1}{z}\right)^k = \frac{1 - z^{-N}}{\frac{1}{r_i} - \frac{1}{z}}. \tag{57}$$

Note (57) implies

$$Q_i^{N-1}\left(\frac{1}{r_i}\right) = Nr_i . \tag{58}$$

Then from (56)–(58) we have

$$\hat{Y}(z) = \sum_{i=1}^{N} \frac{A_{N-1}\left(\frac{1}{r_i}\right)}{\left(\frac{1}{r_i} - \frac{1}{z}\right) \cdot Nr_i \cdot D(r_i)}, \tag{59}$$

where we have let

$$D(z) = z^2 - az - b. \tag{60}$$

Using (57) once more, the steady-state solution (59) may be written

$$\hat{Y}(z) = \frac{1}{1 - z^{-N}} \cdot \frac{1}{N} \sum_{i=1}^{N} \frac{A_{N-1}\left(\frac{1}{r_i}\right) Q_i^{N-1}\left(\frac{1}{z}\right)}{r_i D(r_i)}. \tag{61}$$

Referring back to the discussion at the beginning of this section, we see that (61) is the z-transform of a sequence $\{y_k\}$ of period N where

$$y_k = \text{coefficient of } z^{-k} \text{ in } \left\{ \frac{1}{N} \sum_{i=1}^{N} \frac{A_{N-1}\left(\frac{1}{r_i}\right) Q_i^{N-1}\left(\frac{1}{z}\right)}{r_i D(r_i)} \right\}$$

$$k = 0, 1, \cdots, N - 1. \tag{62}$$

Using (57) in (62) we obtain

$$y_k = \frac{1}{N} \sum_{i=1}^{N} \frac{A_{N-1}\left(\frac{1}{r_i}\right)}{D(r_i)} r_i^k, \tag{63}$$

where, in writing (63), we have used the fact that $r_i^N = 1$. Expression (63) thus gives the $\{y_k\}$ sequence for any click sequence. It is a solution corresponding to a self-sustained oscillation of the digital filter only if we have $| y_k | < 1$, all k.

Two sums appear in (63). The explicit one shown is the sum over the roots of unity; the hidden one is the polynomial $A_{N-1}(1/r_i)$. We will exhibit another form of solution (63) by explicitly doing the sum over the N roots. We begin by writing

$$A_{N-1}\left(\frac{1}{r_i}\right) = 2 \sum_{l=0}^{N-1} \frac{p_l}{r_i^l}, \qquad p_l = \pm 1, 0. \tag{64}$$

Thus p_l are the coefficients, except for the factor of 2, of the polynomial $A_{N-1}(z)$. We also write, by factoring $D(z)$ and expanding in partial fractions,

$$\frac{1}{D(z)} = \frac{1}{(z - \rho_1)(z - \rho_2)} \doteq \frac{1}{\rho_1 - \rho_2} \left[\frac{1}{z - \rho_1} - \frac{1}{z - \rho_2} \right]. \tag{65}$$

Now note that if z is such a number than $z^N = 1$, we have (since $| \rho | < 1$ and $| z | = 1$)

$$\frac{1}{z-\rho} = \frac{1}{z}\sum_{n=0}^{\infty}\left(\frac{\rho}{z}\right)^n. \tag{66}$$

Let us look at the sum of the $n = 0, N, 2N$, etc., terms in the right side of (66), that is

$$1 + \frac{\rho^N}{z^N} + \frac{\rho^{2N}}{z^{2N}} + \frac{\rho^{3N}}{z^{3N}} + \cdots$$
$$= 1 + \rho^N + \rho^{2N} + \rho^{3N} + \cdots = \frac{1}{1-\rho^N}. \tag{67}$$

Treating the sum of terms

$$n = 1, N + 1, 2N + 1, \cdots$$
$$n = 2, N + 2, 2N + 2, \cdots$$

$$n = N - 1, N + (N - 1), 2N + (N - 1), \cdots$$

similarly, we have

$$\frac{1}{z-\rho} = \frac{1}{z}\cdot\frac{1}{1-\rho^N}\left[1 + \frac{\rho}{z} + \frac{\rho^2}{z^2} + \cdots + \frac{\rho^{N-1}}{z^{N-1}}\right]. \tag{68}$$

Finally letting $z = 1/r_i$ gives

$$\frac{1}{\frac{1}{r_i} - \rho} = \frac{r_i}{1-\rho^N}\sum_{n=0}^{N-1}[\rho r_i]^n. \tag{69}$$

Using (65) and (64) in (63) yields

$$y_k = \frac{1}{\rho_1 - \rho_2}\cdot\frac{2}{N}\sum_i r_i^k\left(\sum_{l=0}^{N-1}\frac{p_l}{r_i^l}\right)$$
$$\cdot\left[\frac{1}{r_i}\sum_{n=0}^{N-1}\frac{1}{r_i^n}\left(\frac{\rho_1^n}{1-\rho_1^n} - \frac{\rho_2^n}{1-\rho_2^n}\right)\right]. \tag{70}$$

Two sums in (70) are immediately done. First look at the sum over the roots of unity. This involves observing that

$$\sum_i r_i^{k-l-1-n} = \begin{cases} N & \text{if} \quad k - l - 1 - n \equiv 0 \mod N, \\ 0 & \text{otherwise.}\end{cases} \tag{71}$$

The congruence indicated in (71) can only be satisfied here if $l = k -$

$1 - n$ or if $l = k - 1 - n + N$. Thus it is useful to define

$$2b_n^{(k)} \equiv \bar{a}_{k-1-n} + \bar{a}_{k-1-n+N}\,, \tag{72}$$

where we understand $\bar{a}_j \equiv 0$ if j does not lie between 0 and $N - 1$, inclusive, and $\bar{a}_j \equiv a_j$ if it does. One of the $\bar{a}$'s in (72) will thus always be zero and $b_n^{(k)}$ has values, like the p's, of $\pm 1, 0$. Using the discussion above surrounding equations (71) and (72) we perform next the sum over l and write another form of the solution:

$$y_k = \frac{2}{\rho_1 - \rho_2} \sum_{n=0}^{N-1} b_n^{(k)} \left[\frac{\rho_1^n}{1 - \rho_1^N} - \frac{\rho_2^n}{1 - \rho_2^N} \right]$$

$$k = 0, 1, \cdots, N - 1. \tag{73}$$

REFERENCES

1. Rader, C. M., Gold, B., "Digital Filter Design Techniques in the Frequency Domain," Proc. IEEE, *55*, No. 2 (February 1967), pp. 149–171.
2. Jackson, L. B., Kaiser, J. F., and McDonald, H. S., "An Approach to the Implementation of Digital Filters," IEEE Trans. Audio and Electroacoustics, *AV-16*, No. 3 (September 1968), pp. 413–421.

A Fixed-Point Fast Fourier Transform Error Analysis

PETER D. WELCH, Member, IEEE
IBM Watson Research Center
Yorktown Heights, N. Y.

Abstract

This paper contains an analysis of the fixed-point accuracy of the power of two, fast Fourier transform algorithm. This analysis leads to approximate upper and lower bounds on the root-mean-square error. Also included are the results of some accuracy experiments on a simulated fixed-point machine and their comparison with the error upper bound.

Manuscript received February 26, 1969; revised April 9, 1969.

This work was supported in part by the Advanced Research Projects Agency, Dept. of Defense, Contract AF 19-67-C-0198.

I. Introduction

In many situations there is interest in implementing the fast Fourier transform using fixed-point arithmetic. In this case the effect of the word size on the accuracy of the calculation is of obvious importance both with regard to the design of special-purpose machines and with regard to the accuracy attainable from existing machines. This paper contains an analysis of the fixed-point accuracy of the power of two, fast Fourier transform (FFT) algorithm. This analysis leads to approximate upper and lower bounds on the root-mean-square error. Also included are the results of some accuracy experiments on a simulated fixed-point machine and their comparison with the error upper bound.

II. The Finite Fourier Transform

If $X(j)$, $j=0, 1, \cdots, N-1$, is a sequence of complex numbers, then the finite Fourier transform of $X(j)$ is the sequence

$$A(n) = (1/N) \sum_{j=0}^{N-1} X(j) \exp -2\pi ijn/N \qquad n = 0, 1, \cdots, N-1. \tag{1}$$

The inverse transform is

$$X(j) = \sum_{n=0}^{N-1} A(n) \exp 2\pi ijn/N. \tag{2}$$

In both of the above equations, $i=(-1)^{1/2}$. We will be considering a fixed-point calculation of these transforms using the fast Fourier transform algorithm [1], [2]. In connection with (1), we will consider the calculation of $NA(n)$ from $X(j)$. N^{-1} would then be included as an overall scale factor at the end. Now considering the calculation of $NA(n)$ from $X(j)$ or $X(j)$ from $A(n)$, Parseval's theorem states:

$$\sum_{j=0}^{N-1} | X(j) |^2 = N \sum_{n=0}^{N-1} | A(n) |^2$$

or

$$\sum_{n=0}^{N-1} | NA(n) |^2 = N \sum_{j=0}^{N-1} | X(j) |^2 \tag{3}$$

and we see that the mean-square value of the result is N times the mean-square value of the initial sequence. This fact will be used below.

III. The Inner Loop of the Fast Fourier Transform Algorithm: Step-by-Step Scaling

The inner loop of the power of two FFT algorithm operates on two complex numbers from the sequence. It takes these two numbers and produces two new complex numbers which replace the original ones in the sequence.

Reprinted from *IEEE Trans. Audio Electroacoust.*, vol. AU-17, pp. 151–157, June 1969.

Let $X_m(i)$ and $X_m(j)$ be the original complex numbers. Then, the new pair $X_{m+1}(i)$, $X_{m+1}(j)$ are given by

$$\begin{aligned} X_{m+1}(i) &= X_m(i) + X_m(j)W \\ X_{m+1}(j) &= X_m(i) - X_m(j)W \end{aligned} \tag{4}$$

where W is a complex root of unity. If we write these equations out in terms of their real and imaginary parts, we get

$$\begin{aligned} \text{Re}\,\{X_{m+1}(i)\} &= \text{Re}\,\{X_m(i)\} + \text{Re}\,\{X_m(i)\}\ \text{Re}\,\{W\} \\ &\quad - \text{Im}\,\{X_m(i)\}\ \text{Im}\,\{W\} \\ &\quad \vdots \\ \text{Im}\,\{X_{m+1}(j)\} &= \text{Im}\,\{X_m(i)\} - \text{Re}\,\{X_m(i)\}\ \text{Im}\,\{W\} \\ &\quad - \text{Im}\,\{X_m(j)\}\ \text{Re}\,\{W\}. \end{aligned} \tag{5}$$

At each stage the algorithm goes through the entire sequence of N numbers in this fashion, two at a time. If $N=2^M$, then the number of such stages in the computation is M.

As we move from stage to stage through the calculation, the magnitudes of the numbers in the sequence generally increase which means that it can be kept properly scaled by right shifts. Consider first the root-mean square of the complex numbers. From (4) we have

$$\left[\frac{|X_{m+1}(i)|^2 + |X_{m+1}(j)|^2}{2}\right]^{1/2} = \sqrt{2}\left[\frac{|X_m(i)|^2 + |X_m(j)|^2}{2}\right]^{1/2}. \tag{6}$$

Hence, in the root-mean-square sense, the numbers (both real and complex) are increasing by $\sqrt{2}$ at each stage. Consider next the maximum modulus of the complex numbers. From (4) one can easily show that

$$\begin{aligned} \max\,\{|X_m(i)|,\ |X_m(j)|\} &\\ \le \max\,\{|X_{m+1}(i)|,\ |X_{m+1}(j)|\} & \\ \le 2\max\,\{|X_m(i)|,\ |X_m(j)|\}. & \end{aligned} \tag{7}$$

Hence the maximum modulus of the array of complex numbers is nondecreasing.

In what follows, we will assume that the numbers are scaled so that the binary point lies at the extreme left. With this assumption the relationships among the numbers is as shown in Fig. 1. The outside square gives the region of possible values, $\text{Re}\{X_m(i)\}<1$ and $\text{Im}\{X_m(i)\}<1$. The circle inscribed in this square gives the region $|X_m(i)|<1$. The inside square gives the region $\text{Re}\{X_m(i)\}<1/2$, $\text{Im}\{X_m(i)\}<1/2$. Finally, the circle inscribed in this latter square gives the region $|X_m(i)|<1/2$. Now if $X_m(i)$ and $X_m(j)$ are inside the smaller circle, then (7) tells us that $X_{m+1}(i)$ and $X_{m+1}(j)$ will be inside the larger circle and hence not result in an overflow. Consequently, if we control the sequence at the mth stage so that $|X_m(i)|<1/2$, we are certain we will have no overflow at the $m+1$st stage. However, if $X_m(i)$ and $X_m(j)$ are inside the smaller square, then it is possible for $X_{m+1}(i)$ or $X_{m+1}(j)$ to be outside the larger square and hence result in an overflow. Consequently, we cannot control the sequence to prevent overflow by keeping the absolute values of the real and imaginary parts less than one-half. Furthermore, the maximum absolute value of the real and imaginary parts can increase by more than a factor of two and hence a simple right shift is not a sufficient correction.

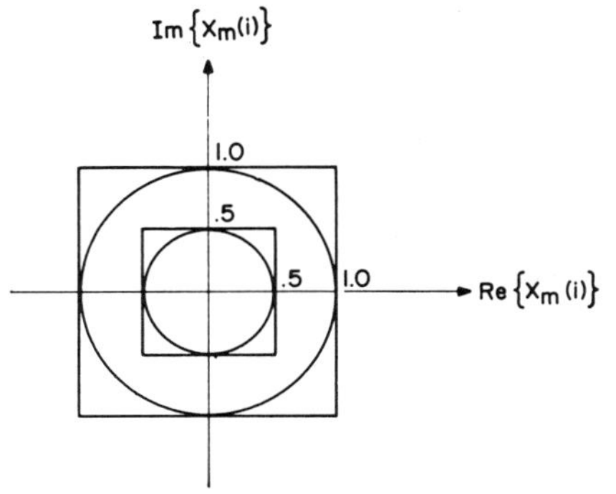

Fig. 1. Regions important to the rescaling of the sequence.

The above results and observations suggest a number of alternative ways of keeping the array properly scaled. The three that seem most reasonable are the following.

1) *Shifting Right One Bit At Every Iteration:* If the initial sequence, $X_0(i)$, is scaled so that $|X_0(i)|<1/2$ for all i and if there is a right shift of one bit after every iteration (excluding the last) then there will be no overflows.

2) *Controlling the Sequence so that* $|X_m(i)|<1/2$*:* Again assume the initial sequence is scaled so that $|X_0(i)|<1/2$ for all i. Then at each iteration we check $|X_m(i)|$ and if it is greater than one half for any i we shift right one bit before each calculation throughout the next iteration.

3) *Testing for an Overflow:* In this case the initial sequence is scaled so that $\text{Re}\{X_0(i)\}<1$ and $\text{Im}\{X_0(i)\}<1$. Whenever an overflow occurs in an iteration the entire sequence (part of which will be new results, part of which will be entries yet to be processed) is shifted right by one bit and the iteration is continued at the point at which the overflow occurred. In this case there could be two overflows during an iteration.

The first alternative is the simplest, but the least accurate. Since it is not generally necessary to rescale the sequence at each iteration, there is an unnecessary loss in accuracy. The second alternative is also not as accurate as possible because one less than the total number of bits available is being used for the representation of the sequence. This alternative also requires the computation of the modulus of every member of the sequence at each iteration. The third alternative is the most accurate. It has the disadvantage that one must process through the sequence an additional time whenever there is an overflow. The indexing for this processing is, however, straightforward. It would not be the complex indexing required for the algorithm. In comparing the speed of the second and third alternatives one would be comparing the speed of two overflow tests, two loads, two stores, and a transfer with that of the calculation or approximation of the modulus and a test of its magnitude. This comparison would depend greatly upon the particular machine and the particular approximation to the magnitude function.

A modification of the second alternative was adopted by Shively [3]. In this modification, if $|X_m(i)| > 1/2$, the right shift was made *after* each calculation in the next iteration. Provision was made for possible overflow. We will give an error analysis of the third alternative below. A microcoding performance study of this third alternative for the IBM 360/40 can be found in [4]. Although this error analysis applies to the third alternative it can be easily modified to apply to the second. In addition, the upper bound given applies directly to the first alternative. The analysis can also be modified for the power of four algorithm.

IV. A Fixed-Point Error Analysis

A. Introduction

We will assume, in this analysis, that the inputs [i.e., the real and imaginary parts of $X(j)$ or $A(n)$] are represented by B bits plus a sign. We assume the binary point lies to the left of the leftmost bit. We showed earlier that the magnitudes of the members of the sequence would generally increase as we moved from stage to stage in calculation. Hence, the method of operation is to test for overflow within the inner loop. If there is no overflow, the calculation proceeds as usual. If there is an overflow, then the two inputs producing the overflow are shifted right until there is no overflow. The amount of the shift is recorded (it will be either one or two bits) and the entire sequence is shifted right this same amount. In this scheme, we shift not only those elements we have already calculated but also those yet to be done. The total number of shifts is accumulated and the power of two, raised to the negative of this total number of shifts, constitutes an overall scale factor to be applied to the final sequence.

There are two operations which produce errors which are propagated through the calculation:

1) When two B bit numbers are multiplied together a $2B$ bit product results. If this product is rounded to B bits, an error whose variance is

$$\Delta_1^2 = 2^{-2B}/12 \tag{8}$$

is created. This error has a standard deviation of

$$\Delta_1 = 2^{-B}/\sqrt{12} \approx 0.3(2^{-B}). \tag{9}$$

2) When two B bit numbers are added together and there is an overflow, then the sum must be shifted right and a bit lost. If this bit is a zero, there is no error. If it is a one, there is an error of $\pm 2^{-B}$ depending upon whether the number is positive or negative. The variance of this error (it is unbiased assuming there are an equal number of positive and negative numbers) is

$$\Delta_2^2 = 2^{-2B}/2. \tag{10}$$

It has a standard deviation

$$\Delta_2 = 2^{-B-1/2} \approx 0.7(2^{-B}). \tag{11}$$

In addition, we will consider the effects of the propagation of errors present in the initial sequence. The variance of these errors we designate by δ^2. In the simplest case these errors would be the quantization errors resulting from the A/D conversion of an analog signal.

B. Upper Bound Analysis

In this section, we give an upper bound analysis of the ratio of the rms error to the rms of the completed transform. This upper bound is obtained by assuming that during each step of the calculation there is an overflow and a need to rescale. We let $X_k(j)$ be a typical real element at the kth stage (i.e., the real or imaginary part of a complex element) and let

$$\begin{aligned} V(X_k) &= \overline{\text{variance}\,\{X_k(j)\}} \\ &= \frac{1}{N}\sum_{j=0}^{N-1} \text{variance}\,\{X_k(j)\}. \end{aligned} \tag{12}$$

(This notation, a bar over the symbol indicating an average over the sequence, will be carried throughout the paper.) We will, in what follows, replace Δ_2^2 by $6\Delta_1^2$. We will also let $\Delta^2 = \Delta_1^2$.

Since the first stage gives an overflow, the original data must be rescaled or truncated by one bit. Hence,

$$V(X_0) = 6\Delta^2 + \delta^2. \tag{13}$$

In going from the original data to the results of the first stage, $W=1$ and, hence, there is no multiplication and we either add or subtract. Further, we assume that the next stage will result in an overflow and hence we will have to rescale. This gives

$$\begin{aligned} V(X_1) &= 2V(X_0) + 4\cdot 6\Delta^2 \\ V(X_1) &= 2(6\Delta^2) + 2\delta^2 + 4\cdot 6\Delta^2. \end{aligned} \tag{14}$$

In these expressions and this entire discussion, we are assuming all errors to be independent and, hence, that the variance of the sum is the sum of the variances. Going from the first stage to the second stage, we have $W=(-1)^{1/2}$ and again there are only additions and subtractions. Thus, with the rescaling,

$$\begin{aligned} V(X_2) &= 2V(X_1) + 4^2\cdot 6\Delta^2 \\ &= 2^2(6\Delta^2) + 2^2\delta^2 + 2(4\cdot 6\Delta^2) + 4^2\cdot 6\Delta^2. \end{aligned} \tag{15}$$

In going from the second stage to the third stage, we have multiplications and we have them in all subsequent stages. In generating the third stage, half the inner loops have multiplications. Consider the first equation of (5). All the other equations are identical in terms of error propagation. Remember that $X_3(i)$ is complex:

$$\text{Re}\{X_3(i)\} = \text{Re}\{X_2(i)\} + \text{Re}\{X_2(j)\}\,\text{Re}\{W\} - \text{Im}\{X_2(j)\}\,\text{Im}\{W\}. \tag{16}$$

Equation (16) yields, with rounding to B bits after the addition and with rescaling,

$$\begin{aligned} V'(X_3) &= V(X_2) + [\overline{\text{Re}^2\{X_2(j)\}} + \overline{\text{Im}^2\{X_2(j)\}}]V(W) \\ &\quad + [\text{Re}^2(W) + \text{Im}^2(W)]V(X_2) \\ &\quad + (4^3\Delta^2) + 4^3\cdot 6\Delta^2 \\ &= V(X_2) + \overline{|X_2(j)|^2}\Delta^2 + V(X_2) + (4^3\Delta^2) + 4^3\cdot 6\Delta^2. \end{aligned} \tag{17}$$

In (17), the first term is the variance of the first term of (16). The second and third terms of (17) are the variance of the full $2B$ bit products given by the second and third terms of (16). The fourth term of (17) is the result of rounding after the addition. The fifth term is the rescaling term. Finally, we saw in (6) that the average modulus squared of the complex numbers is increasing by a factor of 2 every stage. Hence, if we let K equal the average modulus squared of the initial array, i.e.,

$$K = \overline{|X_0(j)|^2} = \frac{1}{N}\sum_{j=0}^{N-1} |X_0(j)|^2,$$

then we have

$$V'(X_3) = 2V(X_2) + 2^2K\Delta^2 + 4^3\Delta^2 + 4^3\cdot 6\Delta^2. \tag{18}$$

Equation (18) would be correct for $V(X_3)$ if all the inner loops involved multiplications. However, at this stage only half of them do and, hence,

$$\begin{aligned} V(X_3) &= 2V(X_2) + 2K\Delta^2 + 4^3\Delta^2/2 + 4^3\cdot 6\Delta^2 \\ &= 2^3(6\Delta^2) + 2^3\delta^2 + 2^2(4\cdot 6\Delta^2) + 2(4^2\cdot 6\Delta^2) \\ &\quad + 4^3\cdot 6\Delta^2 + 2K\Delta^2 + 4^3\Delta^2/2. \end{aligned} \tag{19}$$

In the next stage, three quarters of the inner loops require multiplications and these multiplications get progressively more numerous as the stages increase. Hence, from here on, we will assume all stages have multiplications in all the inner loops. Thus, applying the above techniques, we get

$$\begin{aligned} V(X_4) &= 2^4(6\Delta^2) + 2^4\delta^2 + 2^3(6\cdot 4\Delta^2) + 2^2(6\cdot 4^2\Delta^2) \\ &\quad + 2(6\cdot 4^3\Delta^2) + 6\cdot 4^4\Delta^2 \\ &\quad + 2^2K\Delta^2 + 2^3K\Delta^2 + 4^3\Delta^2 + 4^4\Delta^2 \end{aligned} \tag{20}$$

and, generally, if M is the last stage

$$\begin{aligned} V(X_M) &= 2^M(6\Delta^2) + 2^M\delta^2 + 2^{M-1}(6\cdot 4\Delta^2) + \cdots \\ &\quad + 2(6\cdot 4^{M-1}\Delta^2) + 2^{M-2}K\Delta^2 + (M-3)2^{M-1}K\Delta^2 \\ &\quad + 2^{M-4}(4^3\Delta^2) + 2^{M-4}(4^4\Delta^2) + \cdots + (4^M\Delta^2) \\ &= (1.5)2^{M+2}\Delta^2(1 + 2 + \cdots + 2^{M-1}) + 2^M\delta^2 \\ &\quad + (M-2.5)2^{M-1}K\Delta^2 \\ &\quad + 2^{M+2}\Delta^2 + 2^{M+4}(1 + 2 \cdots + 2^{M-4})\Delta^2 \end{aligned} \tag{21}$$

or

$$\begin{aligned} V(X_M) &\approx (1.5)2^{2M+2}\Delta^2 + 2^M\delta^2 + (M-2.5)2^{M-1}K\Delta^2 \\ &\quad + 2^{M+2}\Delta^2 + 2^{2M+1}\Delta^2 \\ &\approx 2^{2M+3}\Delta^2 + 2^M\delta^2 + (M-2.5)2^{M-1}K\Delta^2 + 2^{M+2}\Delta^2. \end{aligned} \tag{22}$$

K is the average of the square of the absolute values of the initial complex array. Hence, applying Parseval's theorem (3), the average of the square of the absolute values of the final array will be 2^MK. What is most meaningful in this case, however, is the mean square of the real numbers, which is $2^MK/2$. Hence we have

$$\frac{V(X_M)}{2^MK/2} \approx \frac{2^{M+3}\Delta^2}{K/2} + \frac{2\delta^2}{K/2} + \frac{(M-2.5)\Delta^2/2}{1/2} + \frac{2^2\Delta^2}{K/2}, \tag{23}$$

and, finally, for large M,

$$\frac{\text{rms (error)}}{\text{rms (result)}} \approx \frac{2^{(M+3)/2}\Delta}{\sqrt{K/2}} \approx \frac{2^{(M+3)/2}2^{-B}(0.3)}{\text{rms (initial array)}}. \tag{24}$$

Equation (24) gives an approximate upper bound for the ratio of the rms of the error to the rms of the answer. Notice that this bound increases as the $\sqrt{N}$ or $\frac{1}{2}$ bit per stage.

C. Lower Bound Analysis

We will now obtain an approximate lower bound for the ratio of the rms of the error to the rms of the answer. We obtain this lower bound by assuming that there are no overflows in the calculation and, hence, no shifts of the array. In this case,

$$\begin{aligned} V(X_0) &= \delta^2 \\ V(X_1) &= 2\delta^2 \\ V(X_2) &= 2^2\delta^2. \end{aligned} \tag{25}$$

In the third stage, half of the inner loops involve a multiplication and, hence,

$$V(X_3) = (1/2)(2^2K\Delta^2) + 1/2(\Delta^2) + 2^3\delta^2. \tag{26}$$

This can be seen by considering the first term of (17). The first term of (26) comes from the second term of the first of equations (17). The second term of (26) is caused by the rounding to B bits. Now, as before,

$$\begin{aligned} V(X_4) &= 2V(X_3) + 2^3K\Delta^2 + \Delta^2 \\ &= 2^2K\Delta^2 + 2^3K\Delta^2 + \Delta^2 + \Delta^2 + 2^4\delta^2. \end{aligned} \quad (27)$$

Finally,

$$\begin{aligned} &V(X_M) \\ &= 2^{M-2}K\Delta^2 + (M-3)2^{M-1}K\Delta^2 + 2^{M-3}\Delta^2 + 2^{M-5}\Delta^2 \\ &\quad + 2^{M-6}\Delta^2 + \cdots + \Delta^2 + 2^M\delta^2 \\ &= (M-2.5)2^{M-1}K\Delta^2 + 2^{M-3}\Delta^2 \\ &\quad + (1 + \cdots + 2^{M-5})\Delta^2 + 2^M\delta^2 \\ &\approx (M-2.5)2^{M-1}K\Delta^2 + 2^{M-3}\Delta^2 + 2^{M-4}\Delta^2 + 2^M\delta^2. \end{aligned} \quad (28)$$

As in Section IV-B, the mean square of the final sequence of real numbers is $2^M \cdot K/2$. Hence, we have

$$\frac{V(X_M)}{2^MK/2} \approx (M-2.5)\Delta^2 + \frac{\Delta^2/8}{K/2} + \frac{\Delta^2/6}{K/2} + \frac{\delta^2}{K/2} \cdot \quad (29)$$

Now one has to be careful in interpreting (29) to obtain an approximate lower bound. In actuality, the only way to have a situation in which there are no shifts is to have a small K and, in fact, one which approaches zero as N (or M) becomes large. However, if we assume that the word size expands to the left as necessary rather than overflowing, then this analysis does provide a lower bound to the error. With this interpretation, as M becomes large, we have

$$\frac{\text{rms (error)}}{\text{rms (result)}} \approx (M - 2.5)^{1/2}(.3)2^{-B}. \quad (30)$$

The lower bound increases as $M^{1/2} = \frac{1}{2}\log_2 N$. This is the rate of increase which has been observed for the floating-point calculation [5], [6].

D. Some Experimental Results

An IBM 7094 program was written to perform a fixed-point calculation using the fast Fourier transform algorithm, as described above. The program was capable of simulating a fixed-point machine of any word size up to 35 bits plus a sign. Experiments were run with fixed-point numbers of 17 bits plus a sign. This corresponds to $B=17$ in the analysis of Section IV-B and C.

We will now describe some experimental results. In these experiments we did not consider the propagation of the error present in the original sequence. Thus we considered the case where $\delta^2 = 0$. The experiments were performed as follows. Floating-point input was fixed to 17 bits plus a sign. This fixed input was then transformed with the fixed-point program. The fixed-point output was then floated. Next, the fixed-point 17-bit input was floated and a floating-point transform taken. Since this floating-point transform uses a floating-point word with a 27-bit mantissa, it was considered the correct answer. Finally, the rms of the difference between the fixed-point and floating-point answers was taken. We also obtained the maximum absolute error and average error.

Fig. 2 contains the result of transforming random numbers which lie between zero and one (placed in both the real and imaginary parts). In this and subsequent tests, three runs were made for every power of two from 8 to 2048. Since these random numbers have a dc component of one-half, the fixed-point program must rescale at least $N-1$ times. Hence, one would expect the error to lie close to the theoretical upper bound as given by (24). This theoretical upper bound is also plotted in Fig. 2 and the results are seen to lie slightly above it. The rms of the original array, $\sqrt{K/2}$, is approximately 0.58.

Fig. 3 contains the results of transforming three sine waves plus random numbers between zero and one-half in the real part and all zeros in the imaginary part. Specifically,

$$\begin{aligned} \text{Re}\,\{X(j)\} &= 1/2[Y(j) + (1/2)\sin(2\pi 8j/N) \\ &\quad + (1/4)\sin(2\pi 4j/N) \\ &\quad + (1/4)\sin(2\pi 8j/N)] \\ \text{Im}\,\{X(j)\} &= 0 \end{aligned}$$

where the $Y(j)$ are random numbers between zero and one. Again, there is a dc component of magnitude one-fourth and the array must be rescaled at least $N-2$ times. Thus, one would expect these results to be lower relative to the theoretical upper bound than the case depicted in Fig. 2. From Fig. 3 one can see that this is in fact the case. The rms of the original array $\sqrt{K/2}$ is, in this case, approximately 0.35. This is the reason the upper bound curve is higher than that of Fig. 2.

Fig. 4 contains the results of transforming random numbers from minus one to one (in both real and imaginary parts). In this case, the dc component is zero and there is no other strong component. The number of shifts should be approximately $(\log_2 N)/2$ or one-half shift per stage. Hence, one would expect the error curve to lie well below the theoretical upper bound, as is the case. In this case, $\sqrt{K/2} = 0.58$.

Fig. 5 contains the results of an experiment identical to that used for Fig. 3, except that the random numbers are between $\pm\frac{1}{2}$. The results are as expected. In this case, $\sqrt{k/2}$-0.35.

Finally, Fig. 6 contains the results of transforming a sine wave in the real part and zero in the imaginary part. The sine wave was $\sin(2\pi j/8)$. Although in this case the array must be rescaled in at least $N-2$ times, the error is well below the upper bound. Here, $\sqrt{K/2} = 0.5$.

In all these calculations the bias, as reflected by the average error, was negligible compared with the rms error. Furthermore, the maximum error was of the same order of magnitude as the rms error and hence the error was not due to the effect of a few, highly inaccurate terms.

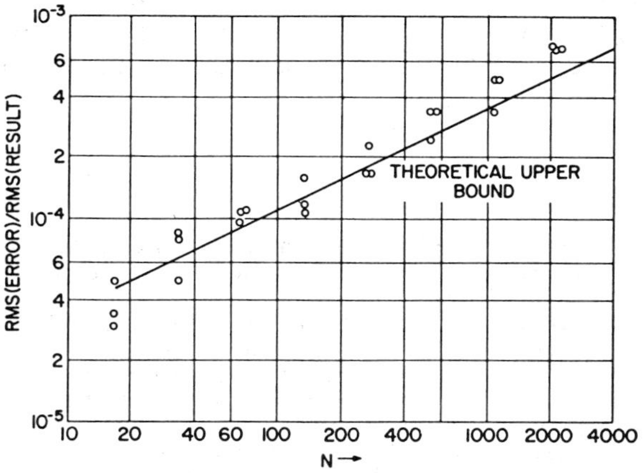

Fig. 2. Experimental error results: random numbers between 0 and 1; $B=17$.

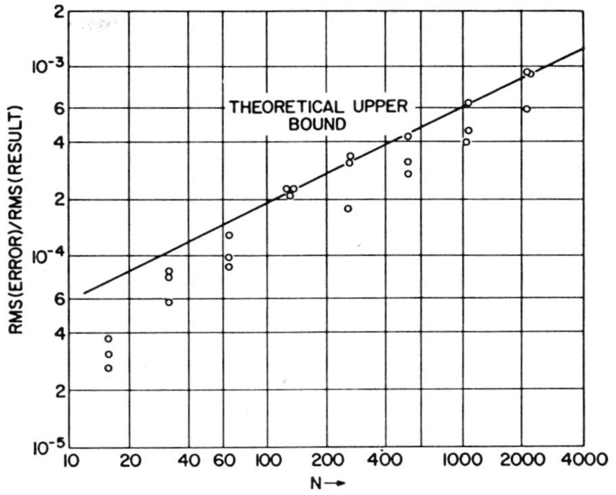

Fig. 3. Experimental error results: random numbers plus 3 sine waves; $0<$random numbers$<\frac{1}{2}$; $B=17$.

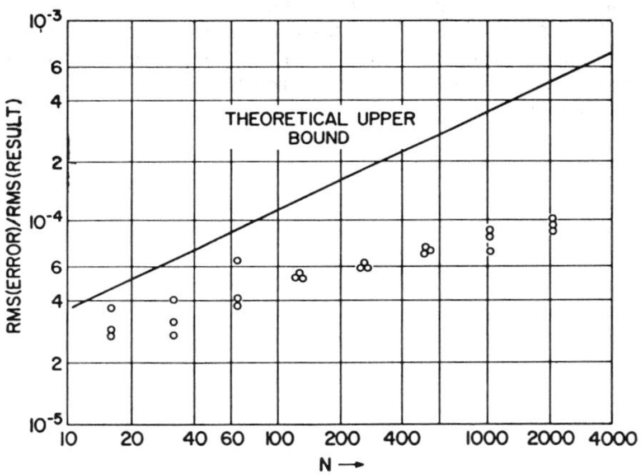

Fig. 4. Experimental error results: random numbers between -1 and 1; $B=17$.

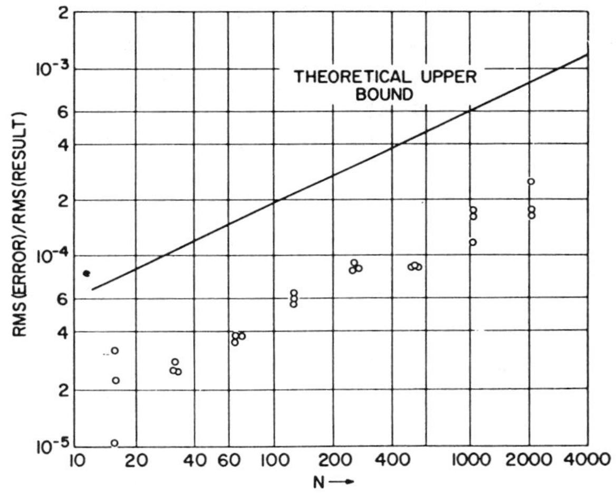

Fig. 5. Experimental error results: random numbers plus 3 sine waves; $-\frac{1}{2}<$random numbers$<\frac{1}{2}$; $B=17$.

Fig. 6. Experimental error results: single sine wave; $B=17$.

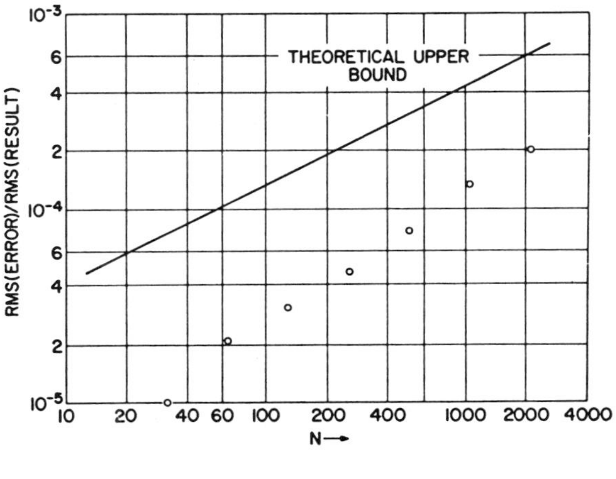

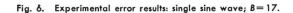

E. Conclusions and Additional Comments

The upper bound obtained in Section IV-B is of the form

$$\frac{\text{rms (error)}}{\text{rms (result)}} \leq \frac{2^{(M+3)/2}2^{-B}C}{\text{rms (initial sequence)}} \tag{31}$$

where $C=0.3$. On the basis of the experimental results we would recommend a bound with $C=0.4$.

We also carried through the analysis for a sign magnitude machine with truncation rather than rounding. In this case, the analytical upper bound was of the form given by (31) but with $C=0.4$. However, the experimental results were again higher and we would recommend a bound with $C=0.6$. The case of a twos-complement machine with truncation was not analyzed as analysis became exceedingly complex. However, experimental results indicated a bound of the form given by (31) with $C=0.9$.

It should be pointed out that if we are taking the transform to estimate spectra then we will be either averaging over frequency in a single periodogram or over time in a sequence of periodograms and this averaging will decrease the error discussed here as well as the usual statistical error. Finally, if we are taking a transform and then its inverse, Oppenheim and Weinstein have shown [7] that the errors in the two transforms are not independent.

Acknowledgment

The author would like to thank R. Ascher for assistance in programming the fixed-point calculations. He would also like to thank the referee for a number of corrections and valuable suggestions.

References

[1] J. W. Cooley and J. W. Tukey, "An algorithm for machine calculation of complex Fourier series," *Math. Comp.*, vol. 19, pp. 297–301, April 1965.

[2] J. W. Cooley, "Finite complex Fourier transform," SHARE Program Library: PK FORT, October 6, 1966.

[3] R. R. Shively, "A digital processor to generate spectra in real time," *1st Ann. IEEE Computer Conf., Digest of Papers*, pp. 21–24, 1967.

[4] "Experimental signal processing system," IBM Corp., 3rd Quart. Tech. Rept., under contract with the Directorate of Planning and Technology, Electronic Systems Div., AFSC, USAF, Hanscom Field, Bedford, Mass., Contract F19628-67-C-0198.

[5] J. W. Cooley, P. A. W. Lewis, and P. D. Welch, "The fast Fourier transform algorithm and its applications," IBM Corp., Res. Rept. RC 1743, February 9, 1967.

[6] W. M. Gentleman and G. Sande, "Fast Fourier transforms for fun and profit," *1966 Fall Joint Computer Conf., AFIPS Proc.*, vol. 29. Washington, D.C.: Spartan, 1966, pp. 563–578.

[7] A. V. Oppenheim and C. Weinstein, "A bound on the output of a circular convolution with application to digital filtering," this issue, pp. 120–124.

Roundoff Noise in Floating Point Fast Fourier Transform Computation

CLIFFORD J. WEINSTEIN, Student Member, IEEE

Lincoln Laboratory, Massachusetts Institute of Technology
Lexington, Mass.

Abstract

A statistical model for roundoff errors is used to predict output noise-to-signal ratio when a fast Fourier transform is computed using floating point arithmetic. The result, derived for the case of white input signal, is that the ratio of mean-squared output noise to mean-squared output signal varies essentially as $\nu = \log_2 N$ where N is the number of points transformed. This predicted result is significantly lower than bounds previously derived on mean-squared output noise-to-signal ratio, which are proportional to ν^2. The predictions are verified experimentally, with excellent agreement. The model applies to rounded arithmetic, and it is found experimentally that if one truncates, rather than rounds, the results of floating point additions and multiplications, the output noise increases significantly (for a given ν). Also, for truncation, a greater than linear increase with ν of the output noise-to-signal ratio is observed; the empirical results seem to be proportional to ν^2, rather than to ν.

Introduction

Recently, there has been a great deal of interest in the fast Fourier transform (FFT) algorithm and its application [1]. Of obvious practical importance is the issue of what accuracy is to be expected when the FFT is implemented on a finite-word-length computer. This note studies the effect of roundoff errors when the FFT is implemented using floating point arithmetic. Rather than deriving an upper bound on the roundoff noise, as Gentleman and Sande [2] have done, a statistical model for roundoff errors is used to predict the output noise variance. The statistical approach is similar to one used previously [3], [4] to predict output noise variance in digital filters implemented via difference equations. The predictions are tested experimentally, with excellent agreement.

Manuscript received July 11, 1969.

This work was sponsored by the U. S. Department of the Air Force. This work is based on the author's Ph.D. dissertation submitted to the M.I.T. Department of Electrical Engineering, July 1969.

The FFT Algorithm for $N=2^\nu$

The discrete Fourier transform (DFT) of the complex N point sequence $x(n)$ is defined as

$$X(k) = \sum_{n=0}^{N-1} x(n) W^{-nk} \qquad k = 0, 1, \cdots, N-1 \tag{1}$$

where $W = e^{j2\pi/N}$. For large N, the FFT offers considerable time savings over direct computation of (1). We restrict attention to radix 2 FFT algorithms; thus we consider $N=2^\nu$, where $\nu = \log_2 N$ is an integer. Here the DFT is computed in ν stages. At each stage, the algorithm passes through the entire array of N complex numbers, two at a time, generating a new N number array. The νth computed array contains the desired DFT. The basic numerical computation operates on a pair of numbers in the mth array, to generate a pair of numbers in the $(m+1)$st array. This computation, referred to as a "butterfly," is defined by

$$X_{m+1}(i) = X_m(i) + \tilde{W} X_m(j) \tag{2a}$$

$$X_{m+1}(j) = X_m(i) - \tilde{W} X_m(j). \tag{2b}$$

Here $X_m(i)$, $X_m(j)$ represent a pair of numbers in the mth array, and $\tilde{W}$ is some appropriate integer power of W, that is

$$\tilde{W} = W^p = e^{j2\pi p/N}.$$

At each stage, $N/2$ separate butterfly computations like (2) are carried out to produce the next array. The integer p varies with i, j, and m in a complicated way, which depends on the specific form of the FFT algorithm which is used. Fortunately, the specific way in which p varies is not important for our analysis. Also, the specific relationship between i, j, and m. which determines how we index through the mth array, is not important for our analysis. Our derived results will be valid for both decimation in time and decimation in frequency FFT algorithms [1], except in the section entitled "modified output noise analysis," where we specialize to the decimation in time case.

Reprinted from *IEEE Trans. Audio Electroacoust.*, vol. AU-17, pp. 209–215, Sept. 1969.

Propagation of Signals and Noise in the FFT

In the error analysis to be presented, we will need some results governing the propagation of signals and noise in the FFT. These results, specialized to correspond to the statistical model of signal and roundoff noise which we will use, are given in this section.

We assume a simple statistical model for the signal being processed by the FFT. Specifically, we consider the case where the signal $X_m(i)$ present at the mth array is white, in the sense that all $2N$ real random variables composing the array of N complex numbers are mutually uncorrelated, with zero means, and equal variances. More formally, we specify for $i=0, 1, \cdots, N-1$ that

$$\begin{aligned} &\mathcal{E}[\text{Re } X_m(i)] = \mathcal{E}[\text{Im } X_m(i)] = 0 \\ &\mathcal{E}\{[\text{Re } X_m(i)]^2\} = \mathcal{E}\{[\text{Im } X_m(i)]^2\} = \tfrac{1}{2}\mathcal{E}[\,|X_m(i)|^2] \\ &\qquad = \text{const.} = \tfrac{1}{2}\sigma_{X_m}{}^2 \\ &\mathcal{E}\{[\text{Re } X_m(i)][\text{Im } X_m(i)]\} = 0 \\ &\mathcal{E}\{[\text{Re } X_m(i)][\text{Re } X_m(j)]\} \\ &\quad = \mathcal{E}\{[\text{Im } X_m(i)][\text{Im } X_m(j)]\} \\ &\quad = \mathcal{E}\{[\text{Re } X_m(i)][\text{Im } X_m(j)]\} = 0 \qquad j \neq i \end{aligned} \tag{3}$$

($\mathcal{E}$ = expected value of).

Given this model for the statistics of the mth array, we can use (2), and the fact that

$$|\tilde{W}|^2 = [\text{Re } \tilde{W}]^2 + [\text{Im } \tilde{W}]^2 = 1, \tag{4}$$

to deduce the statistics of the $(m+1)$st array. First, the signal at the $(m+1)$st array is also white; that is, (3) all remain valid if we replace m by $m+1$. In verifying this fact, it is helpful to write out (2) in terms of real and imaginary parts. Secondly, the expected value of the squared magnitude of the signal at the $(m+1)$st array is just double that at the mth array, or

$$\mathcal{E}|X_{m+1}(i)|^2 = 2\mathcal{E}|X_m(i)|^2 \qquad i = 0, 1, \cdots, N-1. \tag{5}$$

This relationship between the statistics at the mth and $(m+1)$st array allows us to deduce two additional results, which will be useful below. First, if the initial signal array $X_0(i)$ is white, then the mth array $X_m(i)$ is also white, and

$$\mathcal{E}|X_m(i)|^2 = 2^m\mathcal{E}|X_0(i)|^2 \qquad i = 0, 1, \cdots, N-1. \tag{6}$$

Finally, let us assume that we add, to the signal present at the $(m+1)$st array, a signal-independent, white noise sequence $E_m(i)$ (which might be produced by roundoff errors) having properties as described in (3). This noise sequence will propagate to the νth, or output, array, independently of the signal, producing at the νth array white noise with variance

$$\mathcal{E}|E_\nu(i)|^2 = 2^{\nu-m-1}\mathcal{E}|E_m(i)|^2 \qquad i = 0, 1, \cdots, N-1. \tag{7}$$

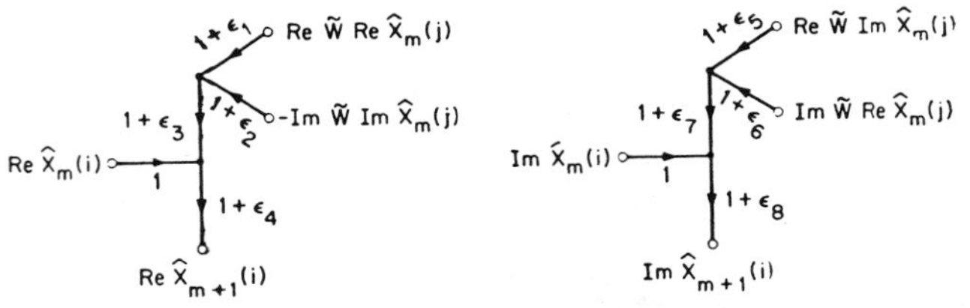

Fig. 1. Flow graphs for noisy butterfly computation.

Butterfly Error Analysis

To begin our FFT error analysis, we first analyze how roundoff noise is generated in, and propagated through, the basic butterfly computation. For reference, let the variables in (2) represent the results of perfectly accurate computation. Actually, however, roundoffs through the mth stage will cause the mth array to consist of the inaccurate results

$$\hat{X}_m(i) = X_m(i) + E_m(i) \qquad i = 0, 1, \cdots, N-1. \tag{8a}$$

and these previous errors, together with the roundoff errors incurred in computing (2), will cause us to obtain

$$\hat{X}_{m+1}(i) = X_{m+1}(i) + E_{m+1}(i). \tag{8b}$$

To analyze roundoff errors, we first express (2) in terms of real arithmetic. Thus (2a) becomes

$$\begin{aligned} \text{Re } [X_{m+1}(i)] &= \text{Re } [X_m(i)] + \text{Re } \tilde{W} \text{ Re } [X_m(j)] \\ &\quad - \text{Im } \tilde{W} \text{ Im } [X_m(j)] \end{aligned} \tag{9a}$$

$$\begin{aligned} \text{Im } [X_{m+1}(i)] &= \text{Im } [X_m(i)] + \text{Re } \tilde{W} \text{ Im } [X_m(j)] \\ &\quad + \text{Im } \tilde{W} \text{ Re } [X_m(j)], \end{aligned} \tag{9b}$$

and a similar pair of equations results for (2b). Let $fl(\cdot)$ represent the result of a floating point computation. We can write [5] that

$$fl(x+y) = (x+y)(1+\epsilon), \tag{10}$$

with $|\epsilon| \leq 2^{-t}$, where t is the number of bits retained in the mantissa. Also

$$fl(xy) = xy(1+\epsilon), \tag{11}$$

with again $|\epsilon| \leq 2^{-t}$. Thus, one could represent the actual floating point computation corresponding to (9) by the flow graphs of Fig. 1, or by the equations

$$\begin{aligned} &\text{Re } [\hat{X}_{m+1}(i)] \\ &\quad = \text{Re } [\hat{X}_m(i)](1+\epsilon_4) \\ &\qquad + \text{Re } \tilde{W} \text{ Re } [\hat{X}_m(j)](1+\epsilon_1)(1+\epsilon_3)(1+\epsilon_4) \\ &\qquad - \text{Im } \tilde{W} \text{ Im } [\hat{X}_m(j)](1+\epsilon_2)(1+\epsilon_3)(1+\epsilon_4) \end{aligned} \tag{12a}$$

$$\begin{aligned} &\text{Im } [\hat{X}_{m+1}(i)] \\ &\quad = \text{Im } [\hat{X}_m(i)](1+\epsilon_8) \\ &\qquad + \text{Re } \tilde{W} \text{ Im } [\hat{X}_m(j)](1+\epsilon_5)(1+\epsilon_7)(1+\epsilon_8) \\ &\qquad + \text{Im } \tilde{W} \text{ Re } [\hat{X}_m(j)](1+\epsilon_6)(1+\epsilon_7)(1+\epsilon_8). \end{aligned} \tag{12b}$$

Now we subtract (9) from (12) to obtain (using (8)) an equation governing noise generation and propagation in the butterfly. Neglecting terms of second or higher order in the ϵ_i and E_m, we obtain

$$E_{m+1}(i) = E_m(i) + \tilde{W}E_m(j) + U_m(i) \quad (13)$$

where

$$\begin{aligned} U_m(i) = \text{Re}\ &[X_m(i)](\epsilon_4) \\ &+ \text{Re}\ \tilde{W}\ \text{Re}\ [X_m(j)](\epsilon_1 + \epsilon_3 + \epsilon_4) \\ &- \text{Im}\ W\ \text{Im}\ [X_m(j)](\epsilon_2 + \epsilon_3 + \epsilon_4) \\ &+ j\{\text{Im}[X_m(i)](\epsilon_8) \\ &+ \text{Re}\ \tilde{W}\ \text{Im}\ [X_m(j)](\epsilon_5 + \epsilon_7 + \epsilon_8) \\ &+ \text{Im}\ \tilde{W}\ \text{Re}\ [X_m(j)](\epsilon_6 + \epsilon_7 + \epsilon_8)\}. \end{aligned} \quad (14)$$

Equations similar to (13) and (14) can be derived for $E_{m+1}(j)$. Equation (13) is the basic equation governing noise generation and propagation in the FFT. Comparing (13) and (2a), we see that the noise E_m already present at the mth array propagates through the butterfly, as if it were signal, to the next array. But also we have additional roundoff noise, represented by U_m, introduced due to errors in computing the $(m+1)$st array from the mth array. Note that the noise source $U_m(i)$ is a function of the signal, as well as of the roundoff variables ϵ_i. This signal dependence of roundoff errors is inherent in the floating point computation, and requires that we assume a statistical model for the signal, as well as for the ϵ_i, in order to obtain statistical predictions for the noise. We should note that the validity of the neglect of second order error terms in obtaining (13) and (14) needs to be verified experimentally.

Now we introduce a statistical model for the roundoff variables ϵ_i, and for the signal, which will allow us to derive the statistics of U_m and eventually predict output noise variance. We assume that the random variables ϵ_i are uncorrelated with each other and with the signal, and have zero mean and equal variances, which we call σ_ϵ^2. We also assume, for simplicity of analysis, that the signal $x(n)$ to be transformed is white, in the sense described above (see (3)). Thus, we have that all $2N$ real random variables (in the N point complex sequence) are mutually uncorrelated, with zero means, and equal variances, which we call $\frac{1}{2}\sigma_x^2$ so that

$$\mathcal{E}\,|\,x(n)\,|^2 = \mathcal{E}\,|\,X_0(n)\,|^2 = \sigma_x^2. \quad (15)$$

Given these assumptions, one can derive that

$$\mathcal{E}\,|\,U_m(i)\,|^2 \equiv \sigma_{u_m}^2 = 4\sigma_\epsilon^2\mathcal{E}\,|\,X_m(i)\,|^2. \quad (16)$$

In obtaining (16), one must take note of (4), and of the fact (see discussion preceding (6)) that the whiteness assumed for the initial signal array $X_0(n)$ implies whiteness for the mth array, so that Re $X_m(i)$, Im $X_m(i)$, Re $X_m(j)$, and Im $X_m(j)$ are mutually uncorrelated, with equal variance. One can use (6) to express the variance at the mth array in terms of the initial signal variance as

$$\mathcal{E}\,|\,X_m(n)\,|^2 = 2^m\mathcal{E}\,|\,X_0(n)\,|^2 = 2^m\sigma_x^2, \quad (17)$$

so that the variance of each noise source introduced in computing the $(m+1)$st array from mth array becomes

$$\sigma_{u_m}^2 = 2^{m+2}\sigma_\epsilon^2\sigma_x^2. \quad (18)$$

The argument leading to (18) implies that all the noise sources $U_m(i)$ in a particular array have equal variance. A slight refinement of this argument would include the fact that a reduced noise variance is introduced when $\tilde{W}=1$ or $\tilde{W}=j$, but this refinement is neglected for the moment. As indicated in (18), the noise variance depends on the signal variance. However, due to the fact that the roundoff variables ϵ_i are signal independent, the noise samples $U_m(i)$ are uncorrelated with the signal. Thus we can assume, in deriving output noise, that the roundoff noise propagates independently of the signal.

Output Noise Variance for FFT

In this section, our basic result for output noise-to-signal ratio in the FFT is derived. Because we are assuming that all butterflies (including where $\tilde{W}=1$ and $\tilde{W}=j$) are equally noisy, the analysis is valid for both decimation in time and decimation in frequency algorithms. Later we will refine the model for the decimation in time case, to take into account the reduced butterfly noise variance introduced when $\tilde{W}=1$ or $\tilde{W}=j$. But the quantitative change in the results produced by this modification is very slight.

Given the assumptions of independent roundoff errors and white signal, the variance of the noise at an FFT output point can be obtained by adding the variances due to all the (independent) noise sources introduced in the butterfly computations leading to that particular output point.

Consider the contribution to the variance of the noise $E_\nu(i)$ at a particular point in the νth, or output array, from just the noise sources $U_m(i)$ introduced in computing the $(m+1)$st array. These noise sources $U_m(i)$ enter as additive noise of variance $\sigma_{u_m}^2$ at the $(m+1)$st array, which (as implied by (13)) propagates to the output array as if it were signal. One can deduce (see (7)) that the resulting output noise variance is[1]

$$[\mathcal{E}\,|\,E_\nu(i)\,|^2]_m = 2^{\nu-m-1}\sigma_{u_m}^2 \quad i = 0, 1, \cdots, N-1, \quad (19)$$

or using (18),

$$[\mathcal{E}\,|\,E_\nu(i)\,|^2]_m = 2^{\nu+1}\sigma_\epsilon^2\sigma_x^2. \quad (20)$$

(20) states that the output noise variance, due to the mth array of noise sources, does not depend on m. This results

[1] Note that it is not quite true that the noise sequence $U_m(i)$, which is added to the signal at the $(m+1)$st array, is white, for in computing the two outputs of a butterfly, the same multiplications are carried out, and thus the same roundoff errors are committed. Thus, the pair of noise sources $U_m(i)$, $U_m(j)$ associated with each particular butterfly, will be correlated. However, all the noise sources $U_m(i)$ which affect a particular output point, are uncorrelated, since (as one could verify from an FFT flow-graph) noise sources introduced at the top and bottom outputs of the butterfly never affect the same point in the output array.

from the opposing effects (18), and (19). By (18), the noise source variance $\sigma_{u_m}^2$ increases as 2^m, as we go from stage to stage; this is due to the increase in signal variance, and the fact that the variance of floating point roundoff errors is proportional to signal variance. But (19) states that the amplification which $\sigma_{u_m}^2$ goes through in propagating to the output has a 2^{-m} dependence, that is the later a noise source is introduced, the less gain it will go through.

To obtain the output noise variance, we sum (20) over m to include the variance due to the computation of each array. Since ν arrays are computed, we obtain

$$[\mathcal{E} \mid E_\nu(i)^2]_{\text{total}} \equiv \sigma_E^2 = \nu 2^{\nu+1}\sigma_\epsilon^2\sigma_x^2. \tag{21}$$

We can recast (21) in terms of output noise-to-signal ratio, noting that (6) implies that

$$\mathcal{E} \mid X_\nu(k) \mid^2 = \sigma_X^2 = 2^\nu \sigma_x^2, \tag{22}$$

so that

$$\frac{\sigma_E^2}{\sigma_X^2} = 2\sigma_\epsilon^2 \nu. \tag{23}$$

Note the linear dependence on $\nu = \log_\nu N$ in the expression (23) for expected output mean-squared noise-to-signal ratio. For comparison, the bounding argument of Gentleman and Sande [2] led to a bound on output mean-squared noise-to-signal ratio which increased as ν^2 rather than as ν. (Actually, they obtained a bound on rms noise-to-signal ratio, which increased as ν.) Certainly, the fact that the bound on output signal-to-noise ratio is much higher than its expected value, is not surprising; since in obtaining a bound one must assume that the noises from the different stages add up in the worst possible way, rather than as uncorrelated random variables.

To express (23) quantitatively in terms of the register length used in the computation, we need an expression for σ_ϵ^2. Recall that σ_ϵ^2 characterizes the error due to rounding a floating point multiplication or addition (see (5) and (6)). Rather than assume [3] that ϵ is uniformly distributed in $(-2^{-t}, 2^{-t})$ with variance $\sigma_\epsilon^2 = \frac{1}{3}2^{-2t}$, σ_ϵ^2 was measured experimentally, and it was found that

$$\sigma_\epsilon^2 = (0.21)2^{-2t} \tag{24}$$

matched more closely the experimental results. Actually, σ_ϵ^2 for an addition was found to be slightly different from that for a multiplication, and σ_ϵ^2 for multiplication was found to vary slightly as the constant coefficient (Re $\tilde{W}$ or Im $\tilde{W}$) in the multiplication was changed. Equation (24) represents essentially an empirical average σ_ϵ^2 for all the multiplications and additions used in computing the FFT of white noise inputs.

Equations (23) and (24) summarize explicitly our predictions thus far for output noise-to-signal ratio. In the next section, the argument leading to (23) is refined to include the reduced butterfly error variance introduced when $\tilde{W}=1$ or $\tilde{W}=j$. We should remark again that the modification is slight, and that the essential argument and essential character of the results have been already given in this section.

Modified Output Noise Analysis

As mentioned above, we have so far not considered in our analysis the reduced error variance introduced by butterfly computations involving $\tilde{W}=1$ or $\tilde{W}=j$. To take these cases into account, we first need an equation corresponding to (16) for the butterfly error variance when $\tilde{W}=1$ or $\tilde{W}=j$. Observe that for $\tilde{W}=1$, we have in Fig. 1 (or (7)) that $\epsilon_1=\epsilon_2=\epsilon_3=\epsilon_5=\epsilon_6=\epsilon_7=0$, since multiplication by 1 or 0, or adding a number to 0, is accomplished noiselessly. Thus (14) becomes

$$\begin{aligned} U_m'(i) &= \epsilon_4\{\text{Re}\,[X_m(i)] + \text{Re}\,[X_m(j)]\} \\ &\quad + j\epsilon_8\{\text{Im}\,[X_m(i)] + \text{Im}\,[X_m(j)]\} \end{aligned} \tag{14'}$$

and (16) becomes

$$\sigma_{u_m'}^2 = 2\sigma_\epsilon^2 \mid X_m(i) \mid^2 = \tfrac{1}{2}\sigma_{u_m}^2, \tag{16'}$$

so that when $\tilde{W}=1$, the butterfly error variance is half the variance introduced when $\tilde{W}\neq 1$ and $\tilde{W}\neq j$. One can easily verify that the variance in (16′) is valid for $\tilde{W}=j$, also.

Now, not all the noise sources introduced in computing the $(m+1)$st array from the mth array will have equal variance. However, if $F(m)$ represents the fraction of the mth array of butterflies which involve either $\tilde{W}=1$ or $\tilde{W}=j$, then one can express the average noise variance for all butterflies used in this array of computations as

$$\begin{aligned} \sigma_{u_m}^2{}_{,\,\text{ave}} &= [1 - F(m)]\sigma_{u_m}^2 - F(m)\sigma_{u_m'}^2 \\ &= [1 - F(m)]\sigma_{u_m}^2 + F(m)\sigma_{u_m}^2/2 \\ &= [1 - F(m)/2]\sigma_{u_m}^2. \end{aligned} \tag{25}$$

The dependence of $F(m)$ on m depends on the form of the FFT algorithm which is used. We will consider the case of a decimation in time algorithm. For this case, only $\tilde{W}=1$ is used in the first array of computations, so $F(0)=1$. Only $\tilde{W}=1$ and $\tilde{W}=j$ are used in computing array 2 from array 1, so $F(1)=1$. In computing the array 3 from array 2, half the butterflies involve $\tilde{W}=1$ or $\tilde{W}=j$, in the next array $\frac{1}{4}$ of the butterflies involve $\tilde{W}=1$ or $\tilde{W}=j$, and so on. Summarizing, we have

$$F(m) = \begin{cases} 1 & m = 0 \\ (\frac{1}{2})^{m-1} & m = 1, 2, \cdots, \nu - 1 \end{cases} \tag{26}$$

and combining (25) and (26) we obtain

$$\sigma_{u_m}^2{}_{,\,\text{ave}} = \begin{cases} \frac{1}{2}\sigma_{u_m}^2 & m=0 \\ [1-(\frac{1}{2})^m]\sigma_{u_m}^2 & m=1, 2, \cdots, \nu-1, \end{cases} \tag{27}$$

where $\sigma_{u_m}^2$ is given in (18).

To derive our modified expression for output noise-to-signal ratio, we carry through the argument corresponding to (19) through (23), but replace $\sigma_{u_m}^2$ in (19) by $\sigma_{u_m}^2{}_{,\text{ave}}$. Two observations will be made before stating the result. First, the right hand side of (20) will now de-

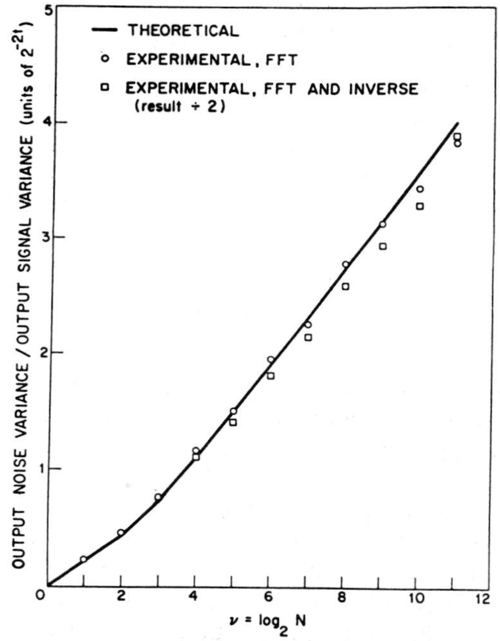

Fig. 2. Theoretical and experimental output noise-to-signal ratios for floating point FFT computations.

pend on m, and must actually be summed over m to obtain the equation corresponding to (21). Secondly, the fact that, in general, not all butterflies in a given array introduce the same roundoff noise variance, implies that there will be a slight variation of noise variance over the output array. Our result, which is thus to be interpreted as an average (over the output array) ratio of noise variance to signal variance, is

$$\sigma_E^2/\sigma_X^2 = 2\sigma_\epsilon^2[\nu - \tfrac{3}{2} + (\tfrac{1}{2})^{\nu-1}]. \qquad (28)$$

As ν becomes moderately large (say $\nu \geq 6$), one sees that (28) and (23) predict essentially the same linear rate of increase of σ_E^2/σ_X^2 with ν.

One further result, which can be derived using our model, is an expression for the final expected output noise-to-signal ratio which results after performing an FFT and an inverse FFT on a white signal $x(n)$. The inverse FFT introduces just as much roundoff noise as the FFT itself, and thus one can convince oneself that the resulting output noise-to-signal ratio is

$$\sigma_E^2/\sigma_x^2 = 4\sigma_\epsilon^2[\nu - \tfrac{3}{2} + (\tfrac{1}{2})^{\nu-1}], \qquad (29)$$

or just double the result in (28).

FFT versus Direct Fourier Transform Accuracy

An issue of some interest is the question of how the accuracy obtained in an FFT computation compares with that obtained in computing a DFT by "straightforward" means, i.e., direct summing of products as implied by (1). Gentleman and Sande [2] treated this problem by obtaining bounds on the mean-squared error for both methods, and the bound they obtain on the error in computing a DFT directly, increases much faster with N than their corresponding bound for the FFT. In deriving their bound for the direct Fourier transform case, they assume that the summing of products required by (1) is done in a cumulative way, that is the $(n+1)$st product is added to the sum of the first n products, and so on. However a slightly different, more error free, technique could be imagined for summing products.

Suppose instead, one were to sum (1) in a treelike fashion, that is the N products were summed in pairs, the $N/2$ results summed in pairs, and so on. If one were to derive, for this treelike summation, a strict bound on output mean-squared noise-to-signal ratio, the bound would increase more slowly with N than the corresponding bound for cumulative summation, but still would increase more rapidly with N than the corresponding bound for the FFT. However, a statistical analysis of roundoff errors incurred in implementing the treelike summation technique, predicts a linear dependence of the output noise-to-signal ratio on ν, similar to the result (23) for the FFT. Thus, on a statistical basis, the treelike summation makes the accuracy for a direct Fourier transform essentially the same as for an FFT. An intuitive explanation for this is that the FFT actually sums the Fourier series in a treelike fashion.

We should note that, compared with cumulative summing, the treelike summation requires some additional memory to store partial sums, and perhaps more inconvenient indexing. However, for any reasonably large N, this issue is academic, since for reasons of increased speed one would use the FFT, and neither cumulative nor treelike summing of products to perform the computation.

Experimental Verification

The results of the above analysis of FFT roundoff noise, as summarized in (28), (29), and (24), have been verified experimentally with excellent agreement. To check (28), a white noise sequence (composed of uniformly distributed random variables) was generated and transformed twice, once using rounded arithmetic with a short (e.g. 12 bit) mantissa, and once using a much longer (27 bit) mantissa. A decimation in time FFT algorithm was used. The results were subtracted, squared, and averaged to estimate the noise variance. For each $N=2^\nu$, this process was repeated for several white noise inputs to obtain a stable estimate of roundoff noise variance. The results, as a function of ν, are represented by the small circles on Fig. 2, which also displays the theoretical curve of (28).

To check (29), white noise sequences were put through an FFT and inverse, and the mean-squared difference between the initial and final sequences was taken. The results of this experiment (divided by a factor of 2 since (29) is twice (28)) are also plotted on Fig. 2.

To clarify the experimental procedure used, we should define carefully the convention used to round the results of floating point additions and multiplications. The results were rounded to the closest (t-bit mantissa) machine number, and if a result (say of an addition) lay midway between two machine numbers, a random choice was made as to whether to round up or down. If one, for example, merely truncates the results to t bits, the experimental noise-to-signal ratios have been observed to be significantly higher than in Fig. 2, and to increase more than linearly with ν. Sample results (to be compared with Fig. 2) of performing the first of the experiments described above, using truncation rather than rounding, are as follows: for $\nu=7$, 8, 9, 10, and 11.

$$\sigma_E^2/2^{-2t}\sigma_X^2 = 46,\ 63.5,\ 80,\ 101, \text{ and } 128, \text{ respectively.}$$

For $\nu=11$, for example, this represents an increase by a factor of 32 over the result obtained using rounding. This increased output noise can be partially explained by the fact that truncation introduces a correlation between signal and noise, in that the sign of the truncation error depends on the sign of the signal being truncated. However, a detailed theoretical analysis for the case of truncation, seems quite difficult.

Our empirical results for $\sigma_E^2/2^{-2t}\sigma_X^2$ for the case of truncation can be fitted closely with a curve of the form $a\nu^2$. This dependence on ν^2 has also been observed experimentally by Gentleman and Sande [2], for the case of truncation. For some of their experiments, however, rounding was used and a ν^2 dependence was still observed, in apparent disagreement with our experimental results. However, our rounding procedure included a random decision as to whether to round up or down, when a computed result lay midway between two machine numbers. It was found that when this random rule was not used, but was replaced by an upward round of the mantissa in this midway situation, the experimental noise-to-signal ratios increased significantly, and could indeed be better fitted with a quadratic than with a linear curve. This interesting observation is a possible explanation for the apparent discrepancy between our experimental results and those of these previous workers. The midway situation occurs quite frequently in floating point computation, especially for addition of numbers of the same order of magnitude, where the unrounded matissas are very often only one bit longer than the rounded mantissas. Always rounding the mantissa up in this situation introduces correlation between roundoff error and signal sign, which contradicts the assumption that roundoff errors are signal independent. This slight correlation seems to be enough to cause the output noise-to-signal ratio to increase significantly faster than ν.

Some experimental investigation has been carried out as to whether the prediction of (28) and (29) are anywhere near valid when the signal is nonwhite. Specifically, sinusoidal signals of several frequencies were put through the experiment corresponding to (28), for $\nu=8$, 9, 10, and 11. Our "randomized" rounding procedure was used. The results, averaged over the input frequencies used, were within 15 percent of those predicted by (28).

Comment on Register Length Considerations

A linear scale is chosen for the vertical axis of Fig. 2, in order to display the essentially linear dependence of output noise-to-signal ration on $\nu=\log_2 N$. To evaluate how many bits of noise are actually represented by the curve of Fig. 2, or equivalently by (28), one can use the expression

$$\left.\frac{\sigma_E^2}{\sigma_X^2\sigma_\epsilon^2}\right]_{\text{bits}} = \tfrac{1}{2}\log_2\left\{2\left[\nu-\tfrac{3}{2}+\left(\tfrac{1}{2}\right)^{\nu-1}\right]\right\} \tag{30}$$

to represent the number of bits by which the rms noise-to-signal ratio increases in passing through a floating point FFT. For example, for $\nu=8$, this represents 1.89 bits, and for $\nu=11$, 2.12 bits. One can use (30) to decide on a suitable register length for performing the computation.

According to (30), the number of bits of rms noise-to-signal ratio increases essentially as $\log_2(\log_2 N)$, so that doubling the number of points in the FFT produces a very mild increase in output noise, significantly less than the $\frac{1}{2}$ bit per stage increase predicted and observed by Welch [6] for fixed point computation. In fact, to obtain a $\frac{1}{2}$ bit increase in the result (30), one would essentially have to double $\nu=\log_2 N$, or square N.

Summary and Discussion

A statistical model has been used to predict output noise-to-signal ratio in a floating point FFT computation, and the result has been verified experimentally. The essential result is [see (23) and (24)]

$$\frac{\sigma_E^2}{\sigma_X^2} = (0.21)2^{-2t}\nu, \tag{31}$$

that is the ratio of output noise variance to output signal variance is proportional to $\nu=\log_2 N$; actually a slightly modified result was used for comparison with experiment.

In order to carry out the analysis, it was necessary to assume very simple (i.e. white) statistics for the signal. A question of importance is whether our result gives reasonable prediction of output noise when the signal is not white. A few experiments with sinusoidal signals seem to indicate that it does, but further work along these lines would be useful.

It was found that the analysis, and in particular the linear dependence on ν in (31), checked closely with experiment only when rounded arithmetic was used. Some results for truncated arithmetic, showing the greater than linear increase of σ_E^2/σ_X^2 with ν, have been given. In rounding, it was found to be important that a random choice be made as to whether to round up or down, when an unrounded result lay equally between two machine numbers. When, for example, results were simply rounded up in this midway situation, a greater than linear increase of σ_E^2/σ_X^2 with ν, was observed. Such a rounding procedure, it seems, introduces enough correlation between roundoff noise and signal, to make the experimental results deviate noticeably from the predictions of our model, which assumed signal and noise to be uncorrelated.

Acknowledgment

Discussions with Professor A. V. Oppenheim of M.I.T. contributed significantly to this work. Dr. W. M. Gentleman supplied the author with some valuable comments.

References

[1] W. T. Cochran *et al.*, "What is the fast Fourier transform," *Proc. IEEE*, vol. 55, pp. 1664–1674, October 1967.
[2] W. M. Gentleman and G. Sande, "Fast Fourier transforms—for fun and profit," *Proc. Fall Joint Computer Conf.*, pp. 563–578, 1966.
[3] T. Kaneko and B. Liu, "Round-off error of floating-point digital filters," presented at the Sixth Annual Allerton Conf. on Circuit and System Theory, October 2–4, 1968. To be published in the Proceedings of the Conference.
[4] C. Weinstein and A. V. Oppenheim, "A Comparison of round-off noise in floating point and fixed point digital filter realizations," *Proc. IEEE* (Letters), pp. 1181–1182, June 1969.
[5] J. H. Wilkinson, Rounding Errors in Algebraic Processes. Englewood Cliffs, N. J.: Prentice-Hall, 1963.
[6] P. D. Welch, "A fixed point fast Fourier transform error analysis," *IEEE Trans. Audio and Electroacoustics*, vol. AU-17, pp. 151–157, June 1969.

A-D AND D-A CONVERTERS:
THEIR EFFECT ON DIGITAL AUDIO FIDELITY

by

Thomas G. Stockham, Jr.

The potential for absolutely superb audio fidelity offered by the use of digital methods can be completely lost in the A-D and/or the D-A converters employed unless some basic but not so obvious performance specifications are met. This paper attempts to present a clear but thorough explanation of the issues involved.

The process of making a digital audio recording obviously involves the use of an analog-digital converter, a digital storage medium and a digital to analog converter. What is not so clear is that additional equipment is essential and that use of all equipment involved must be carefully controlled according to basic, but straightforward principles, which are easily forgotten or overlooked. In addition, the conventional analog equipment preceding the A-D converter and following the D-A converter, must be of the highest performance capability, in order that the full quality offered by the digital techniques be retained.

The additional equipments required to allow digital audio recording are: sharp cut-off, low-pass filters for conditioning the signal before and after digitization; precision, highly stable, low-jitter oscillators for triggering the A-D and D-A converters; and a frequency compensating filter to counteract the moderate high-frequency losses which are a basic part of the common method used to produce analog waveforms by means of D-A converters.

The distortions which can be produced by the improper use of this group of equipment range the gamut from subtle to obvious. What makes life difficult is that the subtle ones are insidious and the obvious ones not always easily explained. The following is a list of errors or imperfections which can contribute to these distortions in which the corresponding corrective action, degree of distortion, and difficulty of complete theoretical explanation is listed. The list is ordered in a manner corresponding to the flow of the audio signal. This list of errors or imperfections does not include hardward failures within the A-D or D-A converters or the digital storage medium; all of which are assumed to be operating according to manufacturer's specification.

This research was supported in part by the University of Utah, Computer Science Division and by the Advanced Research Projects Agency of the Department of Defense, monitored by Rome Air Development Center, Griffiss Air Force Base, New York 13440, under contract F30602-70-C-0300 ARPA order #829.

Reprinted with permission from *Oct. 5, 1971 Proc. 41st Conv. Audio Eng. Soc.*, New York, N.Y.

Error or Imperfection	Corrective Action	Degree of Distortion	Difficulty of Complete Theoretical Explanation
1. Frequency aliasing	Use pre-processing low-pass filter with appropriate cut-off frequency.	Subtle for audio signals except in extreme cases.	Difficult
2. Degraded signal to noise ratio performance or unwanted modulation on signal	Employ a precision oscillator to trigger the A-D converter whose pulse to pulse jitter is controlled to be less than 10 nanoseconds, preferably less than 5 nanoseconds.	Subtle	Fairly simple
3. Distortions in rapidly varying signals, especially loud, high-pitched sounds	Employ sample and hold circuit in A-D converter.	Moderate to high	Simple
4. Unacceptable signal to noise ratio	Employ A-D and D-A converters possessing a sufficient number of binary digits to resolve the dynamic range desired.	Subtle to moderate	Simple
5. Degraded signal to noise ratio performance or unwanted modulation on signal	Employ a precision oscillator to trigger the D-A converter whose pulse to pulse jitter is controlled to be less than 10 nanoseconds, preferably less than 5 nanoseconds.	Subtle	Fairly simple

Error or Imperfection	Corrective Action	Degree of Distortion	Difficulty of Complete Theoretical Explanation
6. Unexpected harmonic distortion and/or noise in playback signal	Employ an analog gate at the output of the D-A converter to key-out the transients in the output waveform which appear between individual digital samples or employ a D-A converter, the basic design of which guarantees that these distortions are small.	Subtle to moderate but easily measured	Moderate to difficult
7. Unwanted frequencies above the highest frequency to be recorded present in the output of the D-A converter	Employ a sharp cut-off low-pass post-processing filter.	High - comparable to the output of an amplitude modulation detector which is unfiltered.	Medium
8. Minor losses in the high-frequency output of the D-A converter	Use compensation filter.	Subtle	Difficult

Unless all of these errors are given appropriate attention, the quality of digital recording methods can be seriously degraded. We will now amplify each area of imperfection, one at a time, and clarify its cause. In certain cases we will give a simplified explanation instead of the complete theoretical one.

Aliasing distortion can be present in any sampling mechanism.

The effect is analogous to the wagon wheel rotating backwards in a motion picture sequence. Oscillations and vibrations which occur more rapidly than once every two samples, cannot be distinguished from those which occur less frequently. The only way to avoid problems with them is to filter them out before the A-D converter, so that they are simply not there. This is easily done with a high-quality, sharp cut-off, low-pass filter; the cut-off frequency of which is approximately 40% to 45% of the sampling frequency. For example: if a bandwidth of fifteen Khz is to be employed, the low-pass filter must have a cut-off frequency of fifteen Khz, which is 40% of 37.5 Khz; therefore, a sampling frequency of **37.5** Khz must be used. Alternately, if a sampling rate of forty Khz is to be used, the highest frequency which can be sampled without aliasing is twenty Khz and the sharp cut-off, low-pass filter should have a cut-off frequency of approximately sixteen Khz. The sharper the cut-off of the low-pass filter, the closer its cut-off frequency may approach half the sampling frequency. It should be noticed that because of the downward sloping character of typical spectra for audio signals, aliasing is not as serious a problem as it can be in digital instrumentation recording. Putting it more simply, the flatter the spectrum of the signal to be digitized, the more problem one will have with aliasing and the more complete one's low-pass filtering will have to be. Of course, the low-pass filter used will have to have distortion and noise characteristics commensurate with the rest of the system.

Unexpectedly low signal to noise ratio or unwanted modulation effects.

When the oscillator which determines the exact times at which the audio waveform is sampled is not jitter-free or when the sampling circuit, which it triggers, exhibits some uncertainty in the time at which it effects a sample, the resulting recorded digital waveform can act as if it had noise added to it. If the jitter follows a regular pattern (e.g. there may be a 60 hertz pattern to the jitter), the recorded digital waveform may appear to have been modulated by the waveform of this jitter pattern. This distortion is most severe when the signal being recorded contains high frequencies of high amplitudes, such that the slope of the waveform being recorded is high at the moment of sampling. Fortunately, most musical waveforms tend to be smoother than worst case full-amplitude waveforms of the same bandwidth. Jitters as large as 5 to 10 nanoseconds are required for 80 to 90 db dynamic range when recording worst case signals at 20 Khz bandwidth. Greater jitter can be tolerated when smaller bandwidths are employed and smoother signals, such as typical musical passages, are encountered. A worst case musical sound might be an extremely loud cymbal crash. The distortion caused by sampling jitter can be explained qualitatively through the observation that if you sample at the wrong time, you get the wrong sample value. The more rapidly the waveform changes, the bigger the error.

Distortions in rapidly varying signals, especially loud, high-pitched sounds.

Some A-D converters are manufactured and designed for use with signals which are naturally or deliberately held constant during the short interval of time in which the converter establishes its digital output. A-D converters of this type are often not supplied with analog input circuits employing a sample and hold amplifier. Unfortunately, audio waveforms almost never stand still. Even though the time interval required for a digital conversion seems relatively short, it is nowhere near short enough for the audio waveform to be considered as constant during that interval. Depending on the design of the A-D converter, the errors which can be induced by failure to use a sample and hold circuit at the input can be quite large and they will always result in substantial system performance degradation. It is not possible to construct a high-fidelity digital recording system without employing a sample and hold circuit or its equivalent at the input to the A-D converter. This distortion is caused by the fact that high-precision A-D converter design requires that the digital output be computed by the converter in a series of successive steps and that for each of these steps the same correct and constant analog input must be available.

Unacceptably low signal to noise ratio.

If the A-D converter, digital storage mechanism and D-A converter employed in the digital recording system do not use enough binary digits to represent the corresponding audio waveform with sufficient precision, the resulting distortion almost always appears as an additive noise in the playback signal, similar to tape hiss. A rule of thumb which is not exact, but works quite well, is that every bit employed adds six decibels to the signal to noise ratio obtainable with the recording system (e.g. a twelve-bit system will yield 72 db dynamic range, a fourteen-bit system will yield 84 db dynamic range, etc.). The signal to noise ratio obtained cannot exceed those allowed by the number of bits employed. Of course, pre- and post-emphasis techniques could be used to increase the subjective signal to noise ratio as is done in ordinary recording techniques, both disk and tape. This idea is an important one, but as yet is an unresolved issue and beyond the scope of this paper. The fact that the use of an insufficient number of binary digits should cause a noise-like signal to be added to the recording is fairly simply explained. Since each sample of the audio waveform is represented by a finite number of binary digits, the true analog value of the waveform must be rounded before the final digital sample can be stored. Each sample, being more or less independent of the last, will be subject to a completely new rounding error. Thus, the difference between the digitally recorded waveform and the analog waveform presented to the input of the recording device is a random signal with a peak to peak amplitude of one least significant bit and appears as a random hissing sound, if it is of sufficiently high-level.

Unexpectedly low signal to noise ratio or unwanted modulation effects.

We have already discussed this problem as it applies to the sampling of the input analog waveform. The problem appears again at the output of a digital recording system if similar jitter occurs in the clock which triggers the D-A converter. The reason is that such jitter gives unequal weights to the samples of the analog waveform being formed at the output; thus distorting the true values intended to be produced.

Unexpected harmonic distortion and/or noise in the playback signal

A digital converter is, in fact, a set of electronic switches which can sum a set of reference voltages to form an analog output. When the switches are set to a specific binary numerical value, the output then assumes the corresponding analog value. Unfortunately, it is impossible to design electronic switches which activate instantaneously. As a result, during a short interval of time after the precision oscillator which triggers the D-A converter has acted, one may find the switches in an apparent state of chaos as they change from one numerical value to the other. If a large percentage of the switches are being changed, and especially if a group of lesser significant bits are being exchanged for one more significant bit, an unwanted pulse (or "glitch" as it is called) will appear at the leading edge of the new analog sample value. If these unwanted pulses are large enough and/or last for a sufficiently long time, they can contribute components to the output audio waveform which appear as an audible distortion at the output. The amount of distortion depends on the sampling rate used. The higher the sampling rate, the greater the distortion; in spite of the fact that the size of the actual pulses themselves does not change. Surprisingly brief or small pulses can cause dramatic performance degradation in high performance situations. A smoothing filter, although it may eliminate the pulse on an oscilloscopic trace, will in no way improve the distortion situation associated with this kind of error! Either these pulses must be made sufficiently short by appropriate D-A converter design, or they must be gated out of the waveform produced by the D-A converter by a circuit similar in principle to the sample and hold used at the input of the A-D converter. The importance of this form of distortion cannot be over-emphasized. It is the most subtle form to be discussed here. The author is familiar with a 12-bit D-A converter installation theoretically capable of 72 db dynamic range and correspondingly low harmonic distortion performance, which was degraded to a performance commensurate with a 5-bit system due to the presence of this error alone.

Unwanted frequencies above the highest frequency to be recorded.

These may appear in the output of a digital recording system because they are basic to the harmonic structure of the typical staircase-like outputs that are obtained from D-A converters. Generally speaking, the output of a digital-analog converter cannot be fed directly to an amplifier-loud speaker system unless frequencies higher than half the sampling frequency are rejected by a suitable sharp cut-off, low-pass filter. The situation is similar, but not identical to the aliasing situation. For high-fidelity audio work, the frequencies in question would usually be beyond the range of human hearing, so the requirements for this filter might seem tenuous. However, generally it is advisable to smooth the square wave-like output of a D-A converter before presenting such a signal to conventional analog equipment which might be damaged or degraded by receiving such frequencies at its input. The fact that the frequencies in question can exist can best be envisioned by imagining that the output of the D-A converter is a sinewave of, let us say, 1 Khz frequency. Of course, the signal won't exactly be a sinewave, but rather a staircase approximation thereto, with each step in the staircase having a period corresponding to the sampling frequency. The output waveform then is really a sinewave with a bunch of little square jiggles on it which occur at the sampling frequency. If one considers the effect that these little jiggles would have upon a spectrum analyzer, one can readily envision how they contain frequencies of substantial amplitudes which are greater than one-half the sampling frequency; and, in fact, are clustered around the sampling frequency itself and multiples thereof.

Minor losses in the high-frequency output of the D-A converter.

The common method used to produce an analog waveform by means of a D-A converter involves creating a staircase-like wave, each step of which assumes an analog value proportional to the digital sample producing it. As explained above, this waveform contains unwanted frequencies higher than half the sampling frequency. These frequencies are easily removed, but unfortunately, desirable frequencies contained in the staircase-like wave are attenuated a small amount varying from zero db at very low frequencies to approximately 4 db at frequencies approaching half the sampling frequency. There is no very simple explanation for this phenomenon, but perhaps it is enough to say that this attenuation corresponds exactly to the frequency spectrum associated with a rectangular pulse whose width is equal to the sampling period. In essence, the D-A converter acts as if it were a circuit capable of producing such pulses acting in synchronism with a train of numerical commands.[1] The amplitude losses indicated above can be easily compensated to within standard audio specifications by any one of a number of simple analog filter designs. Of course, a different filter must be constructed for every sampling frequency anticipated. An additional solution is to modify the digital data at playback time, just before it is passed to the D-A converter by means of a compensating digital filter. This method has the advantage that the digital filter is the same for any sampling frequency which may be used, but has the disadvantage of requiring real time computations at moderately fast rates at the time of playback.

[1] The ideal D-A converter, which cannot be constructed for practical reasons, would have as its output a train of impulses, each of which would possess an area equal to the corresponding numerical sample. For such a D-A converter, there would be no frequency losses like we are describing here. Nonetheless, a practical D-A converter has an output identical to that which would be obtained if the ideal converter were followed by a linear filter whose impulse response were the rectangular pulse being described. The frequency response for such a linear filter is exactly the Fourier transform of that pulse. This frequency response has the analytical form sin x/x and its first zero appears at the sampling frequency. Sin x/x is unity at low frequencies and assumes a value of $2/\pi$ at half the sampling frequency. $2/\pi$ is approximately 0.64 which corresponds to an attenuation of approximately 4 db.

With one exception, all of the above difficulties are unique to the real time processes which take place at record and playback time. For a properly designed and maintained system, no degradations can be induced once a recording is made, regardless of the number of copies which may be desired from an original or the time period over which the recording has been stored. The one exception has to do with the number of binary digits used to resolve the audio waveform while it is in the digital state. Under certain circumstances, it may be practical to reduce the number of digits used for playback, copying, or digital computations performed on high-precision digital audio recordings. When properly carried out, the number of binary digits used in representing an audio waveform may be reduced without incurring any penalty other than the decreased signal to noise ratio associated with the number of digits remaining after the reduction. After certain forms of digital computation have been performed, it may, in fact, be meaningless to retain the original number of digits which were available in the input signal. Digital signal processing is, after all, nothing more than a numerical manipulation on data. As such, it is subject to round-off error just the same as any other numerical manipulation. Generally speaking, round-off errors appear as an additive noise in the output signal, the amplitude of which is commensurate with the number of significant digits remaining in the result of the computation.

There can be little doubt that digital recording systems will find their way into audio recording industries as the demand for and cost of this high-quality technique allow. It should not be overlooked that quality is not the only benefit of the digital recording method. We have already pointed out the error-free archival and copying properties which are made possible. It must also be remembered that exciting new developments in signal processing techniques will, in the forseeable technology, be largely dependent upon digital methods. Moreover, and perhaps most interesting of all, the extreme dynamic range capability and the built-in self-checking nature of digital recording methods assures a reduction of the occurrence of bad takes during expensive recording sessions due to set-up errors, variations in subject material, and undetected degradation of equipment performance.

APPENDIX

The following figures are illustrations of material in this paper.

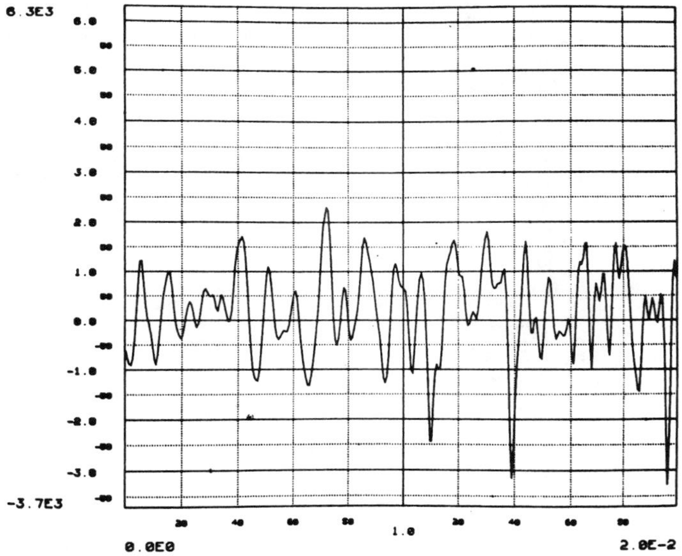

TIME FUNCTION

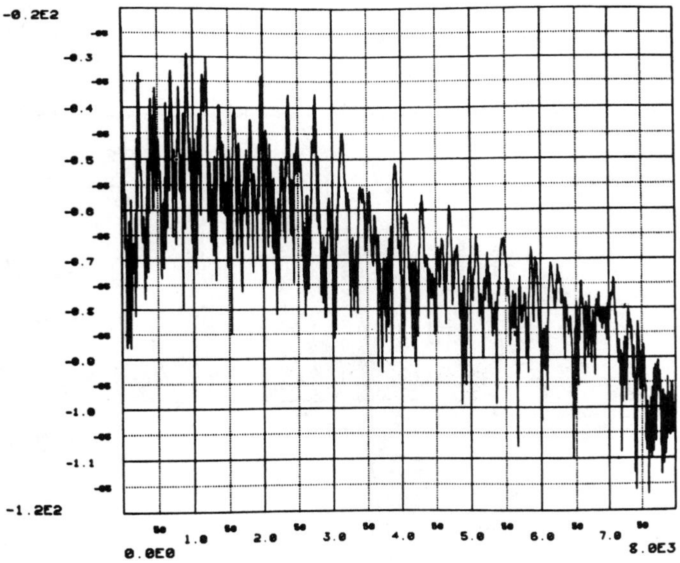

LOG SPECTRUM

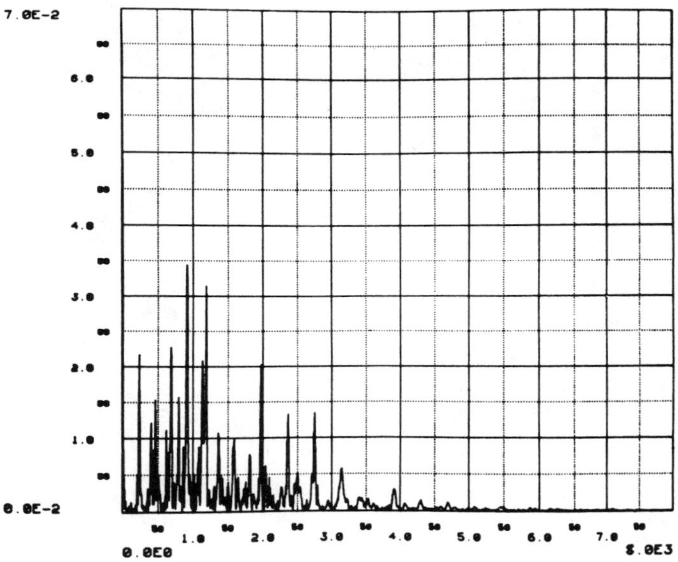

MAGNITUDE SPECTRUM

INPUT BEFORE LOW PASS FILTERING

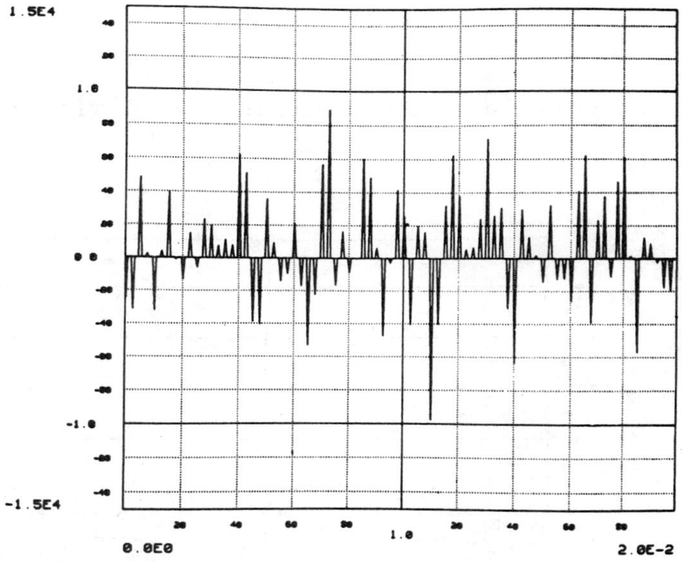

TIME FUNCTION

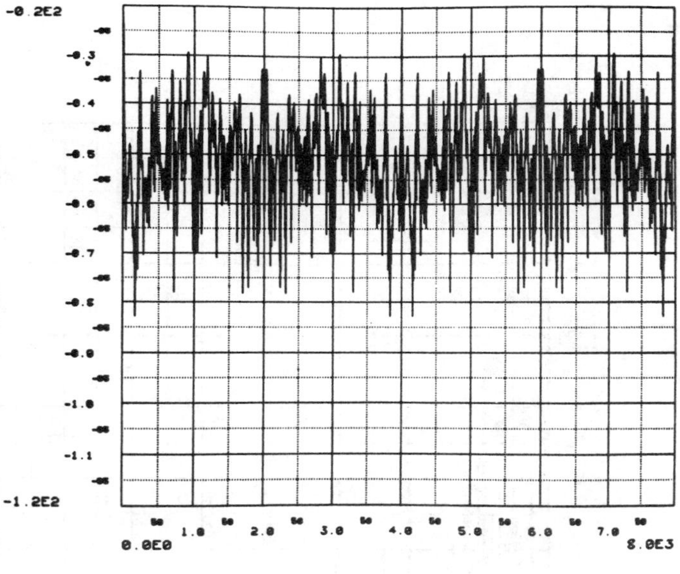

LOG SPECTRUM

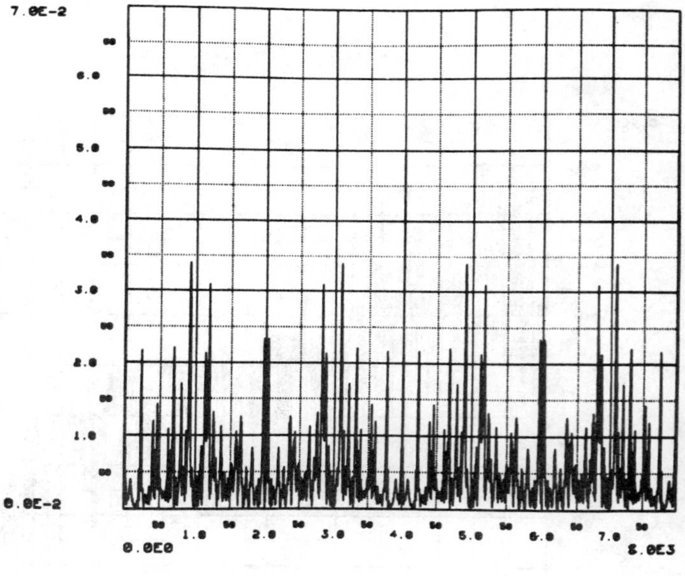

MAGNITUDE SPECTRUM

SAMPLES OF UNFILTERED INPUT SHOWING ALIASING

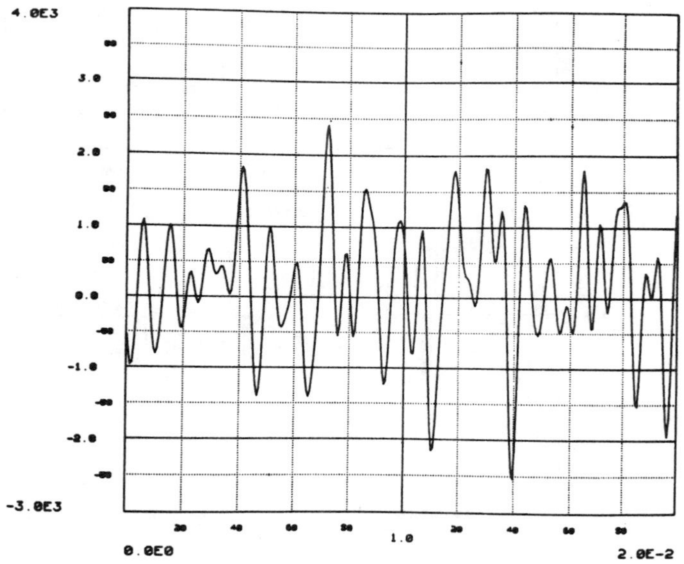

TIME FUNCTION

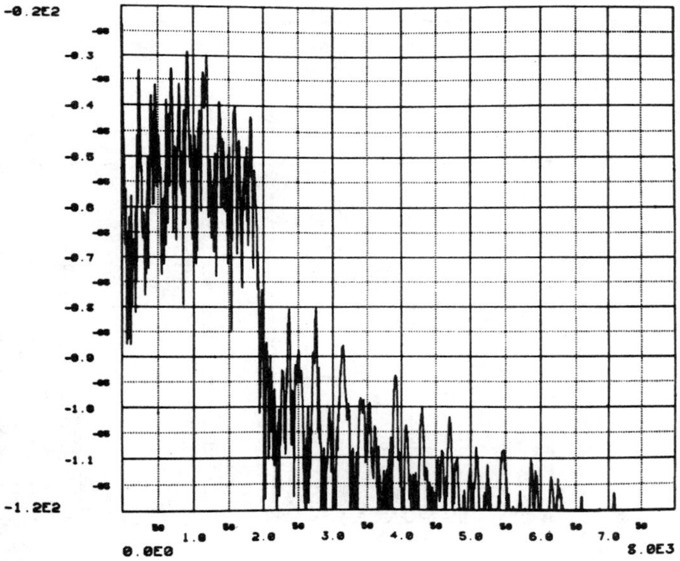

LOG SPECTRUM

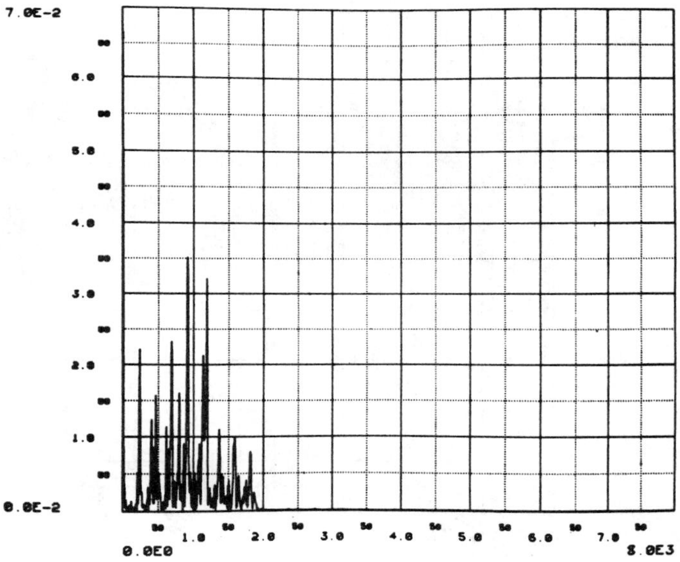

MAGNITUDE SPECTRUM

INPUT AFTER LOW PASS FILTERING

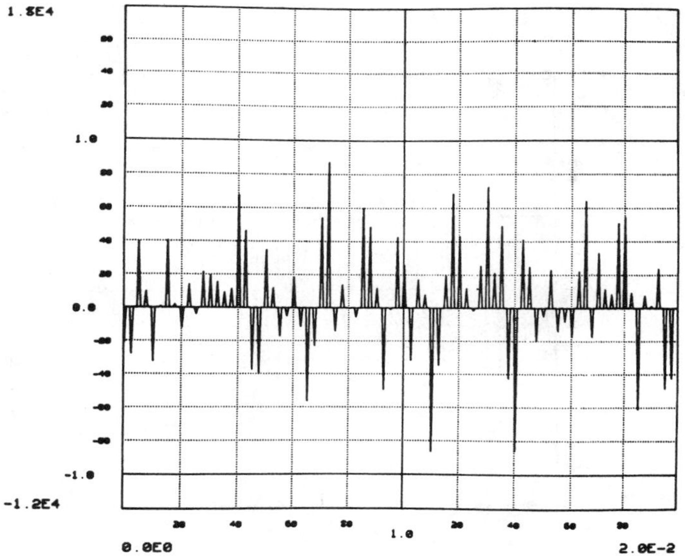

TIME FUNCTION

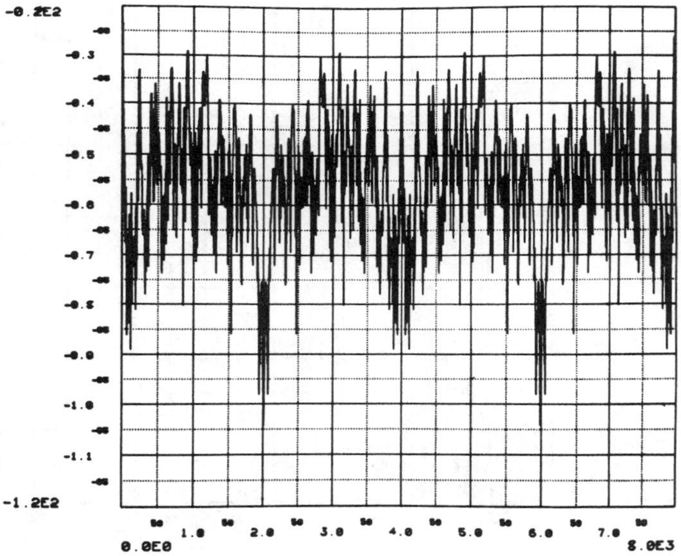

LOG SPECTRUM

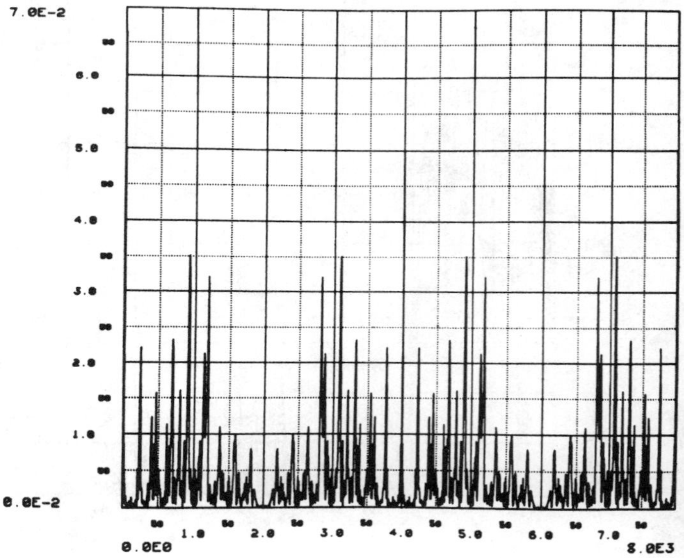

MAGNITUDE SPECTRUM

SAMPLES OF FILTERED INPUT

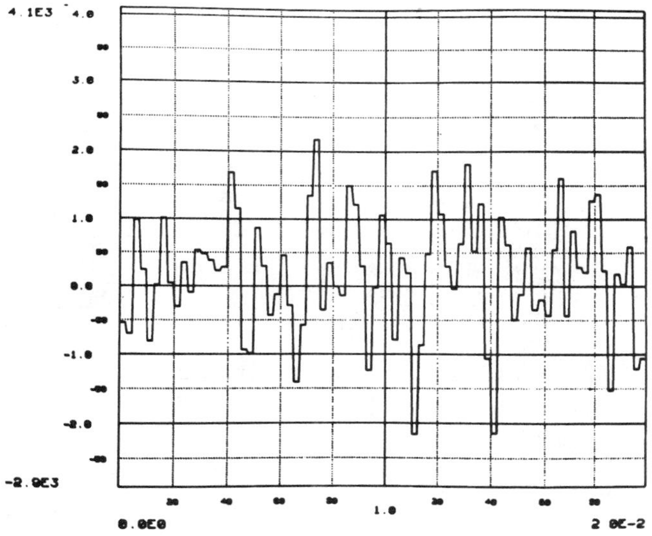

TIME FUNCTION

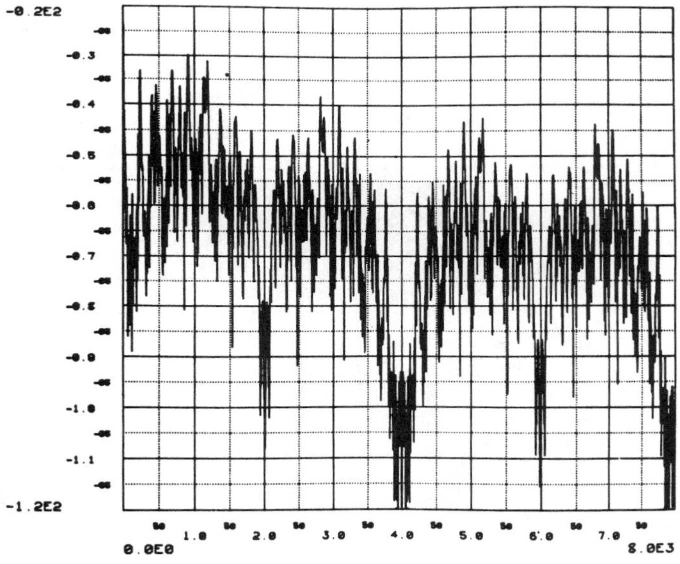

LOG SPECTRUM

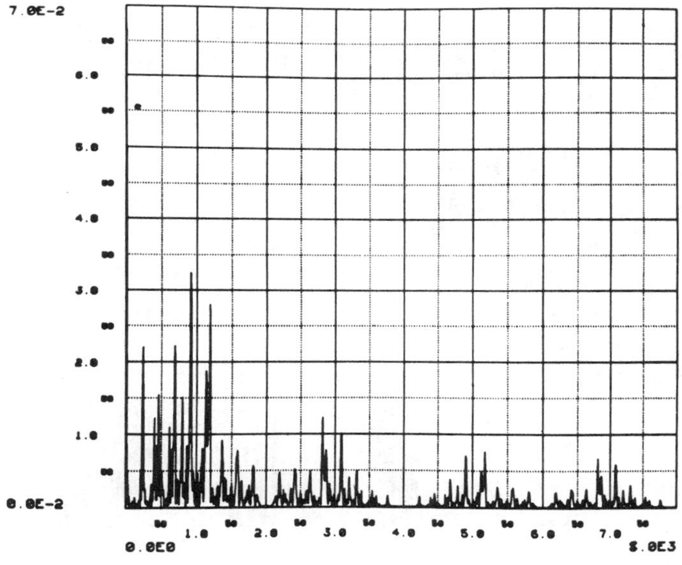

MAGNITUDE SPECTRUM

OUTPUT BEFORE LOW PASS FILTERING SHOWING IMAGES

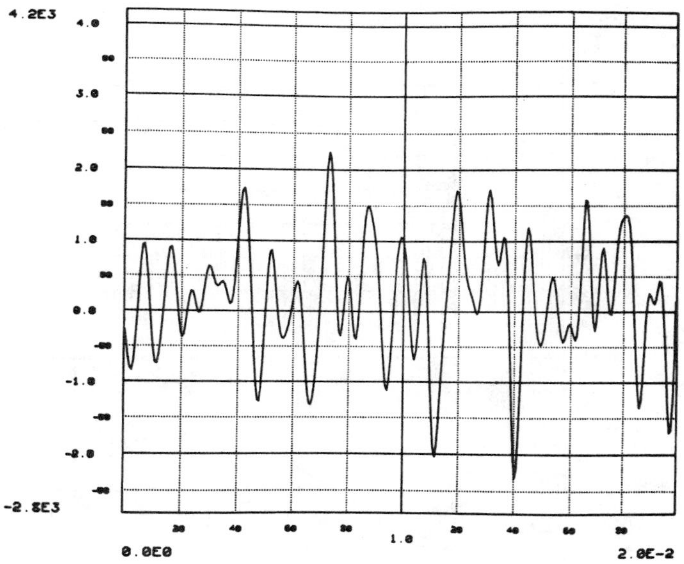

TIME FUNCTION

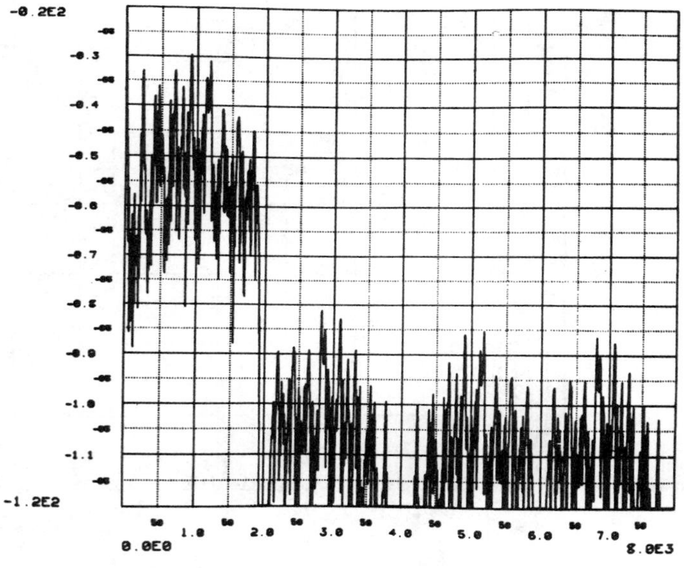

LOG SPECTRUM

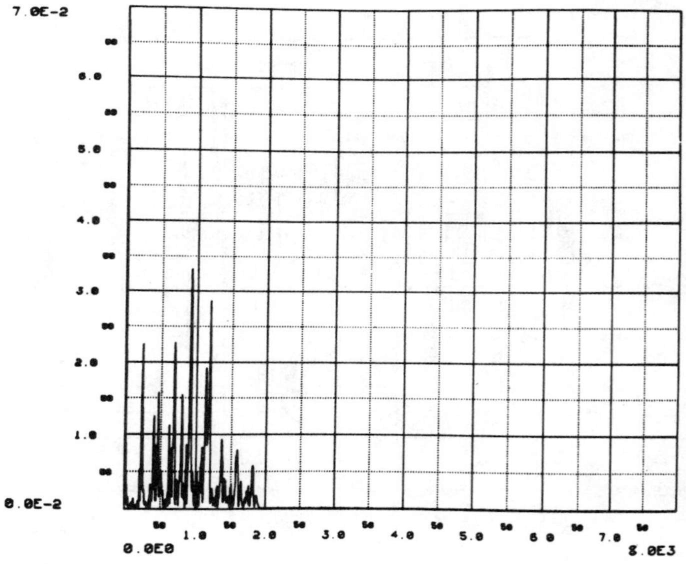

MAGNITUDE SPECTRUM

OUTPUT AFTER LOW PASS FILTERING

A Digital Frequency Synthesizer

JOSEPH TIERNEY, Member, IEEE
CHARLES M. RADER, Member, IEEE
BERNARD GOLD, Senior Member, IEEE
M.I.T. Lincoln Laboratory
Lexington, Mass. 02173

Abstract

A digital frequency synthesizer has been designed and constructed based on generating digital samples of exp $[j(2\pi nk/N)]$ at time nT. The real and imaginary parts of this exponential form samples of quadrature sinusoids where the frequency index k is allowed to vary $(-N/4) \leq K < (N/4)$. The digital samples drive digital to analog converters followed by low-pass interpolating filters to produce analog sinusoids. The method is superior to digital difference equations with poles on the unit circle since the noise or numerical inaccuracy remains bounded.

The digital technique used consists of factoring the exponential into two table look-ups from an efficiently organized small READ-ONLY memory table and performing a complex multiply to produce the real and imaginary components. A small array multiplier efficiently organized performs the multiplications.

The technique lends itself to the production of phase coherent or phase controlled sinusoids because of the indexing arrangement used. In addition finer frequency steps than the READ-ONLY memory allows are available by expanding the indexing register at no increase in inaccuracy.

Introduction

The generation of many different frequencies from a single stable source frequency is commonly achieved with analog circuits [1], [2] or combinations of analog and digital circuits [3], [4]. All of these approaches [5] generate a large set of frequencies from a single source in the analog or continuous sense by division, phase lock, mixing, or a combination of such techniques. An attractive alternate approach is the generation of a set of sampled sinusoids by digital computation of some kind (including simple table look-up) at the sampling times. Fig. 1 indicates the sample trains produced for two different frequencies. The return to analog or continuous sinusoids is accomplished with simple realizable smoothing filters.

One can consider the conventional frequency synthesizer as processing a single source frequency to produce a large set of new frequencies. The digital frequency synthesizer uses a single frequency to establish a stable sampling time at which sample values are computed. The difference in approach is one of using the source as a frequency directly or as a time reference.

Manuscript received July 22, 1970.

This work was sponsored by the Department of the Air Force.

The Digital Approach

Given the problem of computing samples of sinusoids the most obvious choice is a digital recursion, a difference equation whose Z transform has poles on the unit circle. By starting such a recursion with the proper initial conditions one can produce sinusoidal samples. There are at least two problems with this approach (Fig. 2). The frequency of the sampled sinusoid produced by such a typical recursion is not linearly related to a settable coefficient but related by the function cos WT, where T is the sampling interval, W is the produced frequency, and 2 cos WT is the actual coefficient of the recursion. The noise produced by such a recursion is in general worse than can be obtained by other methods and may in some cases be a function of the number of iterations [6].

The chosen approach is simply a direct computation of the samples,

$$\cos(\omega nT + \phi), \qquad \sin(\omega nT + \phi) \qquad n = \text{time index}.$$

Consider values of

$$\cos 2\pi fnT = \text{Re}\,(e^{j2\pi fnT})$$
$$\sin 2\pi fnT = \text{Im}\,(e^{j2\pi fnT})$$

where $f = kfo$, fo = lowest computed frequency. That is, we can compute multiples of some lowest frequency,

$$f_o = \frac{1}{NT}$$

where N is a design parameter. Then the exponential becomes

$$e^{j2\pi fnT} = \exp\left[j\frac{2\pi}{N}nk\right].$$

Reprinted from *IEEE Trans. Audio Electroacoust.*, vol. AU-19, pp. 48–56, Mar. 1971.

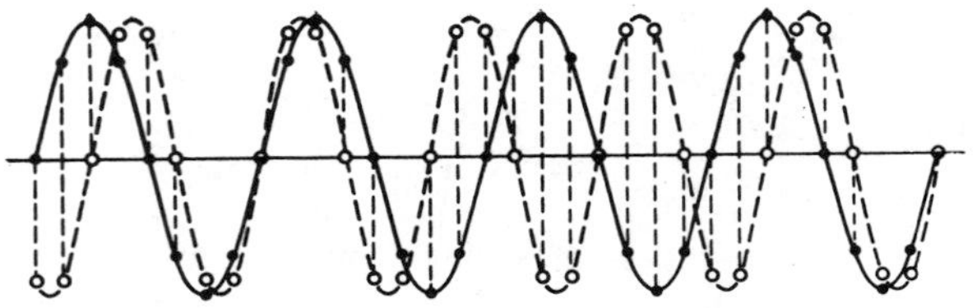

Fig. 1. Digital frequency determination.

Fig. 2. Pole position of the digital recursion for sinusoidal samples.

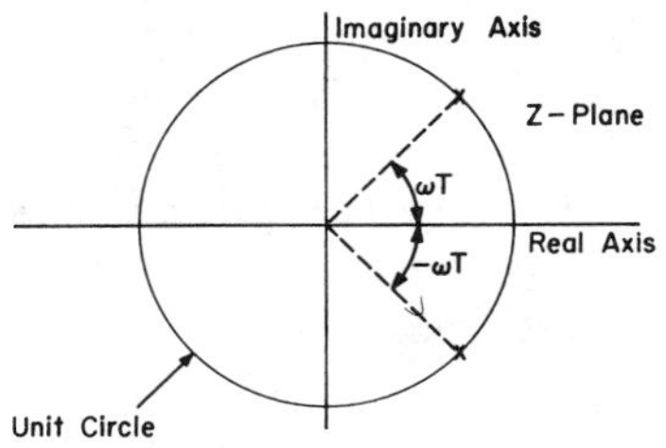

$$y_n = (2 \cos \omega T)\, y_{n-1} - y_{n-2}$$
$$y_0 = \cos \phi$$
$$y_{-1} = \cos [\phi - \omega T].$$

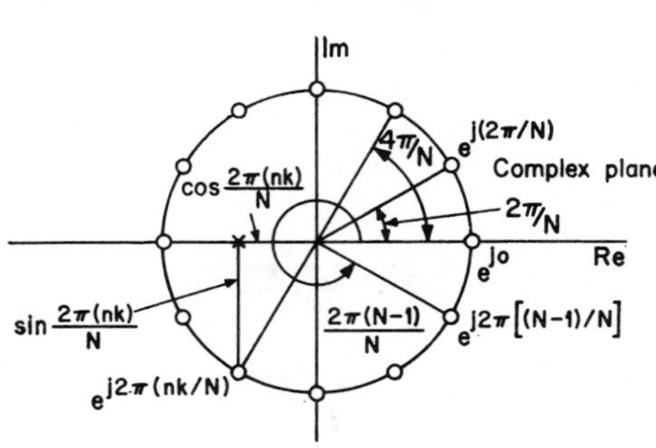

Fig. 3. Equispaced samples on the unit circle.

Samples of this complex exponential are equivalent to values of sine and cosine of the argument $2\pi f n T$ with

$$f = \frac{k}{NT}.$$

Computing samples of this complex exponential indexed on a frequency index k, and a time index n, is equivalent to computing coordinates of one of N equispaced points around the unit circle (Fig. 3) in the complex plane described by

$$\exp [jo], \quad \exp\left[j\frac{2\pi}{N}\right],$$
$$\exp\left[j\frac{2\pi}{N}2\right], \cdots, \exp\left[j\frac{2\pi}{N}(N-1)\right].$$

For a particular frequency index k, the argument of the exponential varies in increments of $(2\pi/N)k$ in successive time indices. The product nk is treated modulo N since $\exp[(j(2\pi/N)[X+N)] = \exp[j(2\pi/N)X]$ for any X. The generation of samples of a complex sinusoid consists then of accumulating multiples of k (i.e., nk at time n, $[n+1]k$ at time $n+1$), and using the accumulated value to calculate $\exp[j(2\pi/N)nk]$.[1]

[1] If in accumulating multiples of k, the accumulator is not initially zero but contains some constant C, the argument of the exponential becomes $(2\pi/N)(nk+C)$ which affects the phase of the result, but not the frequency. This can be useful in phase control.

The Calculation

To determine the value $\exp[j(2\pi/N)[nk+C]] = \exp[j(2\pi/N)Y]$ consider the simplest case, namely, a table storing N values from $\exp[j(2\pi 0/N)]$ to $\exp[j(2\pi(N-1)/N)]$. The computation then consists of using the value in the accumulator Y to index the table of N values producing a single complex sample [7]. As Y increases, the table is scanned producing sinusoidal samples. For a larger k value the table is covered faster (the interval between successive Ys, or nks is larger). Such an approach is suitable for small values of N, but more often we are interested in large N (a large set of frequencies) so that a single table look-up becomes impractical because of the size of the table.

For large N and $0 \leq Y \leq N-1$, Y may be broken into a sum of several words each of which represents a part of Y. If $Y = q+r+s$, then $\exp[j(2\pi Y/N)] = \exp[j(2\pi[q+r+s]/N)] = \exp[j(2\pi q/N)]\exp[j(2\pi r/N)]\exp[j(2\pi s/N)]$ where each factor takes on many fewer than N values and the overall storage has been reduced. For example, for N, a power of 2, say 2^b, then $0 \leq Y \leq 2^b - 1$ and can be represented as a binary number b digits long. $Y = \alpha_0 2^0 + \alpha_1 2^1 + \alpha_2 2^2 + \cdots + \alpha_{b-1} 2^{b-1}$. We can factor the exponential into b factors each of which is of the form $\exp[j(2\pi\alpha_i 2^i/N)]$ with $\alpha_i = 0$ or 1. Thus the table has been reduced from 2^b complex entries to b complex entries ($\log_2 2^b$) and we need $b-1$ complex multiplications to obtain the value $\exp[j(2\pi Y/N)] = \exp[j(2\pi/N)[nk+C]]$. Obviously the number of factors or the number of complex multiplies is one of the design parameters in this approach to frequency synthesis.

For our purposes factoring the complex exponential into two terms allows for a very efficient use of READ-ONLY memory. In addition we may take advantage of approximations to the sine and cosine of small angles as well as symmetries in the functions to effect further savings in READ-ONLY storage.

The Complete Synthesizer

The block diagram of a digital synthesizer which produces quadrature outputs is shown in Fig. 4. An input frequency control word k is stored in a register and used

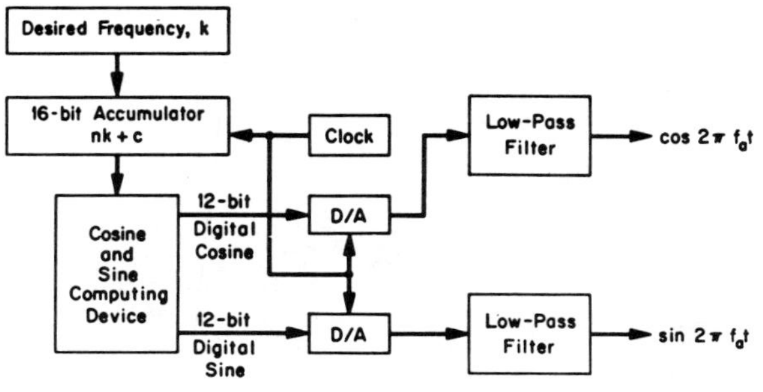

Fig. 4. Synthesizer block diagram.

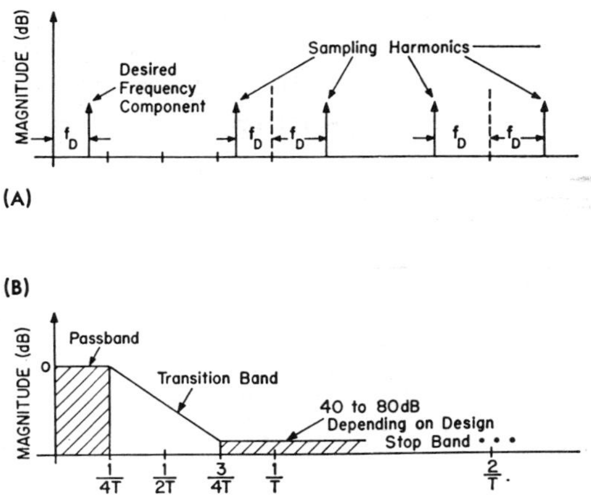

Fig. 5. (A) Sampled sinusoid spectrum. (B) Smoothing filter response.

to update an accumulator every T seconds. Each time the accumulator is changed, the value $nk+C$ is used to compute the real and imaginary parts of $\exp[j(2\pi/N)Y]$ by one of the methods proposed in the previous section. The computed values of $\sin(2\pi/N)Y$, $\cos(2\pi/N)Y$ are used to drive a pair of D–A converters of the proper word length to produce analog samples. These samples are interpolated by the output smoothing filters. At a sampling interval of T seconds the output spectrum before smoothing of a sampled sine or cosine would look like that shown in Fig. 5. The Nyquist condition would allow us to produce frequencies of up to but less than $1/2T$ which we could recover with ideal LP filters with cutoff at $1/2T$. However, for ease of filtering consider using only 1/4 of the sampling frequency as the band limit. Then we would like an output smoothing filter which passes all frequencies up to $1/4T$ with some design ripple, to have a transition band in the interval from $1/4T$ to $3/4T$, and to have some out of band attenuation depending on the allowed sampling harmonics. Such a filter is shown in Fig. 5(B). For an accumulator which overflows at some N and a sampling interval T, a digital frequency synthesizer as shown in Fig. 4 would produce a lowest frequency of $1/NT$, and a highest frequency of $1/4T$ for a total of $N/4$ different frequencies. If we take advantage of the quadrature outputs frum this realization we can double the bandwidth of $1/4T$ since we can modulate a carrier to $\pm 1/4T$ for a total of $N/2$ frequencies. Note that a single output synthesizer can be implemented using only one D–A converter if so desired. There is no constraint to produce quadrature outputs although they may in some cases be desirable, as shown above.

A Design Example

Up to this point the discussion of index accumulation and calculation of sine and cosine has been general. We could use any arithmetic and any size N. Consider now a design example using binary arithmetic and standard binary logic. The design specifications are: 1) 2^{16} frequencies; 2) 409.6 kHz bandwidth; 3) 12.5 Hz frequency spacing (or lowest frequency); 4) 70 dB spectral purity (ratio of power in desired frequency to power in any other 100 Hz band). If we assume the quadrature outputs will be used to obtain a bandwidth of $1/2T$, then $T=1/2\times 409.6$ kHz$\cong 1.22$ μs. In addition $N/2=2^{15}$ or $N=2^{16}$ so that the accumulator will be a 16-bit binary register, and the frequency spacing of $1/NT=12.5$ Hz. If quadrature outputs are desired, the synthesizer must produce values of $\exp[j(2\pi/2^{16})Y]$, $0\leq Y\leq 2^{16}-1$, that is, one of 2^{16} points equispaced around the unit circle. Since the highest frequency allowed is $1/4T$, four points around the circle, the largest k value corresponding to this value of frequency is $2^{16}/4$ or 2^{14}. A two's complement negative frequency input will cause samples to occur in the opposite sense around the unit circle, which is meaningful as a negative frequency only if quadrature outputs are available.

The remaining problem is the computation of the samples. To meet the design requirement of 70 dB purity the computation must be carried to about 12 bits. That is, if a sample out of the computation consists of a sign and 11 binary digits and the accuracy is $\pm$ the last digit or $\pm 2^{-11}$, a worst case harmonic caused by such an error will be 66 dB down from the output. As shown in a later section this bound is quite pessimistic and 70 dB is more likely. Using this 12-bit arithmetic the following procedure, which takes considerable advantage of the symmetries of the sine and cosine, is used.

First note that neglecting the least significant bit in the 16-bit accumulator will cause an amplitude error of no more than $\sin(2\pi/2^{16})$ which is roughly 2^{-12}. In fact one could extend the accumulator register and the input frequency word on the low significance end, use these extra bits for accumulation, but ignore them for the computation of sine and cosine, and still be bounded by $\sin(2\pi/2^{16})$ as an amplitude error. This implies generating finer and finer frequency steps by adding only to the accumulator, a feature which is unique to this type of synthesis and efficient in terms of memory use. Computing sine and cosine of multiples of $2\pi/2^{15}$ (ignoring bit 16) is solved by table look-up and interpolation as follows. Consider Y represented in binary form as

$$\begin{aligned} Y &= 2^0d_0 + 2^1d_1 + 2^2d_2 + 2^3d_3 + 2^4d_4 + 2^5d_5 + 2^6d_6 \\ &\quad + 2^7(d_7 + 2^1d_8 + 2^2d_9 + 2^3d_{10} + 2^4d_{11} \\ &\quad + 2^5d_{12} + 2^6d_{13} + 2^7d_{14}) \\ &= 2^0d_0 + 2^1d_1 + 2^2d_2 + 2^3d_3 + 2^4d_4 + 2^5d_5 - 2^6d_6 \\ &\quad + 2^7(d_6 + d_7 + 2^1d_8 + 2^2d_9 + 2^3d_{10} + 2^4d_{11} \\ &\qquad + 2^5d_{12} + 2^6d_{13} + 2^7d_{14}) \end{aligned}$$

or

$$Y = f + 2^7e$$

where

$$f = 2^0d_0 + 2^1d_1 + \cdots + 2^5d_5 - 2^6d_6$$
$$e = (d_6 + \cdots + 2^7d_{14})$$

so that the exponential can be factored as

$$\left(\exp\left[j\frac{2\pi}{2^{15}}f\right]\right)\left(\exp\left[j\frac{2\pi}{2^{15}}2^7e\right]\right) = \left(\exp\left[j\frac{2\pi}{2^{15}}f\right]\right)\left(\exp\left[j\frac{2\pi}{2^8}e\right]\right).$$

The computation of the complex exponential is then reduced to two table look-ups $\exp[j(2\pi/2^{15})f]$, $\exp[j(2\pi/2^8)e]$ and a complex multiply. The index e consists of the eight high-order bits of the accumulator rounded by the bit d_6, while the index f consists of the six lower bits d^5 through d_0 if d_6 is zero, or these bits two's complemented if d_6 is one. From the point of view of computing the value of a point at a particular angle $(2\pi/N)Y$, around the unit circle, the value of e determines which of 2^8 equally coarsely spaced points is nearest the desired point, and the value of f determines which of 64 possible angular corrections should be added to or subtracted from the coarse point to get the desired value. The angular correction is, of course, a complex multiplication. For this two factor approach the complex multiply is implementing the trigonometric identities

$$\sin(x + y) = \sin x \cos y + \cos x \sin y$$
$$\cos(x + y) = \cos x \cos y - \sin x \sin y$$

with

$$x = (2\pi/2^8)e, \qquad y = (2\pi/2^{15})f$$

where f may be positive or negative (according to bit d_6). Fig. 6 represents these operations on the unit circles.

To compute the value at X in Fig. 6, the eight high-order bits would be augmented by one (since $d_6 = 1$), giving e, and the value of f would be the two's complement of the distance above the center of the coarse interval. In other words we would go to a larger angle and decrement because $d_6 = 1$. If $d_6 = 0$ we would start at the lower angle and increment.

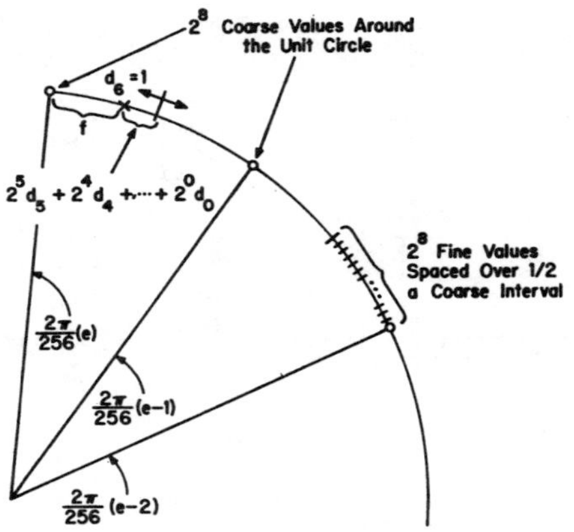

Fig. 6.

The bit d_6 is used to break the coarse intervals in half. If this bit is one, it means the computation needs the coarse value just bigger than the desired value. The fine correction is then clockwise. If d_6 is a zero, the coarse value just below the desired value will do, and the correction is counterclockwise.

Now we note that the cosine component of $\exp[j2\pi f/32\,768]$ is between 1.0 and 0.9999247, a difference in the 14th bit of its binary representation. Therefore we will approximate it by 1. The sine component is so small that its six most significant bits, not counting the sign bit, are equal to the sign bit, which is in turn equal to d_6. Thus the READ-ONLY memory indexed by f need only save the five least significant bits of the 11-bit plus sign representation of the sine corresponding to each of the 64 positive values of f; the sines corresponding to negative values of f can be found by taking the two's complement of f and changing the sign of the result. It is easier to take the one's complement, and this also results in an insignificant error.

The value of $\exp[j2\pi e/256]$ can also be found by look-up in a table with only 64 values corresponding to $\frac{1}{4}$ cycle of a sine wave. The sine and cosine components are addressed with the six least significant bits of e, and its two's complement, and the two most significant bits of e are used for exchanging the components and complementing either or both if necessary; mathematically, if the six least significant bits of e are g and the two most significant bits are h, we look up $\sin 2\pi(64-g)/256+j \sin 2\pi g/256$ and multiply the result by j^h.

The final operation is the multiplication of the coarse estimate $\exp[j2\pi e/256]$ by the fine corrector $(1+j \sin 2\pi f/32\,768)$. This requires two 8 by 5 bit multiplications and two additions. Two answers are kept to an accuracy of 11 bits plus sign and fed to the D–A converters.

The READ-ONLY memory requirements are 64 words of 5 bits fine angle and 11 bits coarse angle which can be combined into a 64 word by 16-bit memory which is accessed three times in a computation. The detailed block diagram of this synthesizer is shown in Fig. 7.

The array multiplier indicated in the block diagram is used to perform the 5 by 8 multiplication (5×8 rather than 5×11 because of the accuracy required). It can be implemented in several ways using the 4-bit TTL adder packages currently available. Using only the bits needed for the final accuracy of $\pm 2^{-12}$ and using a tree-like interconnection between partial sums the multiplication time is about 160 ns.

The digital-to-analog converters driven by the final 12-bit sine and cosine outputs are updated every 1.22 μs and hold the data words between transfer times. This sample and hold operation provides smoothing and filtering in addition to that provided by the output low-pass filters.

A synthesizer designed and built according to the above discussion requires about 85 TTL logic packages, and dissipates about 12 W. If quadrature outputs are not needed, a still simpler device will suffice.

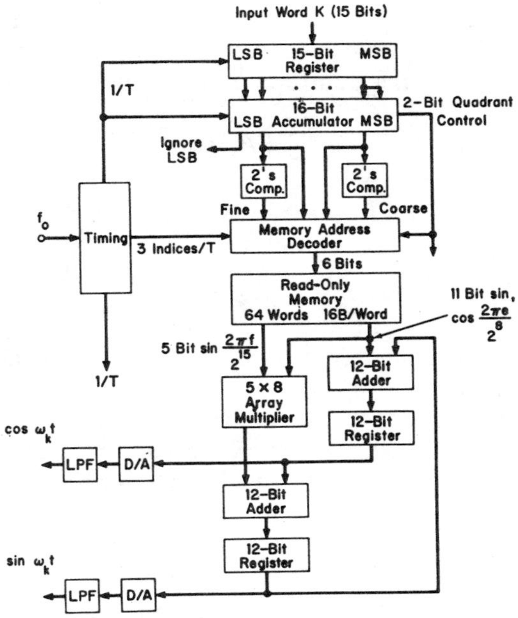

Fig. 7.

Output Noise and Time Response

The synthesizer noise output consists of three separate effects. The first contribution results from truncation or roundoff in computation of the sample value. The second effect is from the output digital to analog converter. Sampling harmonics passing through the smoothing filter contribute the third effect.

Truncation Noise

Since the calculation for a particular sample (that is, a particular value in the index accumulator) is always the same, the output samples are those of an exact sinusoid with some deterministic noise sample train added to it due to truncation. For an arbitrary generated frequency the truncation noise will have a period equal to NT, the largest period possible so that the truncation noise consists of a line spectrum with frequency spacing $f_0=1/NT$. If the generated frequency k/NT has one or more factors of 2 in k, or $k=2^l k^*$, then the truncation noise harmonics are multiples of $2^l f_0$. This is a consequence of N being a power of 2. The limiting case occurs for $k=2^a$, some power of 2. In that case the truncation harmonics are harmonics of the generated frequency. These cases are demonstrated for 16 equispaced samples around a unit circle in Fig. 8. Each sample has associated with it a truncation error ϵ_i and each line below represents the sequence of sample errors generated as the index k increases. The three cases are shown. Lines 1, 2, 4 are cases of a power of 2 times lowest frequency generated. Notice that the error period is the same as the generated period. Lines 3, 5, 7 represent noise periods of $16T$, the longest possible because the frequency index K has no factors of 2. Line 6 shows one factor of 2. As a consequence of our particular arithmetic implementation the error waveform only contains odd harmonics, that is, it has the property $\epsilon(n)=-\epsilon[n+(P/2)]$ where P is the period.

Fig. 8.

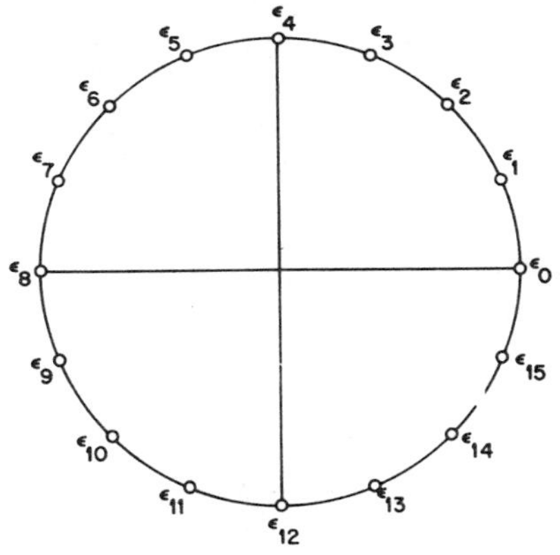

		Frequency			Noise Period
(1)	Lowest	1/16T	Noise period same as generated period	=	16T
(2)		1/8T	Noise period same as generated period	=	8T
(3)	•	3/16T	0 3 6 9 12 15 2 5 8 11 14 1 4 7 10 13	=	16T
(4)	•	T/4	Noise period same as generated period	=	4T
(5)	•	5/16T	0 5 10 15 4 9 14 3 8 13 2 7 12 1 6 11	=	16T
(6)		6/16T	0 6 12 2 8 14 1 10 0	=	8T
(7)	Highest	7/16T	0 7 14 5 12 3 10 1 8 15 6 13 4 11 2 9	=	16T

To bound the noise contributed by the truncation error, consider a generated frequency which is an odd factor k

times $2\pi/NT$. In this case the error waveform is of period NT or N samples long. A Parseval's relation for a discrete Fourier transform over N samples can be written as

$$\sum_{i=0}^{N-1} |F_i|^2 = N \sum_{n=0}^{N-1} \epsilon_n^2$$

where F_i is a frequency amplitude defined by

$$F_i = \sum_{l=0}^{N-1} \epsilon_l \exp\left(-j\frac{2\pi}{N}il\right)$$

and ϵ_n is the error associated with a particular sample. The ϵ_n is hopefully a pseudorandom variable uniformly distributed over the interval -2^{-11} to $+2^{-11}$ because of the truncation. We may assume all of the error energy in one frequency to obtain a bound. For a particular

$$|F|^2 = N \sum_{n=0}^{N-1} \epsilon_n^2$$

and assuming worst case conditions on ϵ_n,

$$|F|^2 = N^2(2^{-11})^2 \quad \text{or} \quad |F| = N2^{-11}.$$

The desired generated frequency will have an amplitude of N in the discrete transform, so that the ratio of noise amplitude to signal amplitude is 2^{-11}, the case we referred to earlier. A more realistic "bound" for the cases when the noise period includes many samples should be

$$\sum_{i=0}^{N-1} |F_i|^2 = N^2\left(\frac{1}{N}\sum_{n=0}^{N-1} \epsilon_n^2\right)$$

where the quantity in parenthesis is the error variance. Since the variance is $[(2^{-11})^2/3]$ for uniformly distributed noise, we expect

$$\sum_{i=0}^{N-1} |F_i|^2 = N^2\frac{(2^{-11})^2}{3}.$$

So the bound obtained assuming all the energy in one harmonic is $2^{-11}/\sqrt{3}$ noise amplitude to signal amplitude ratio or about -71 dB. If one assumes that the noise waveform is white in one period and using the fact that it contains only odd harmonics we have

$$\frac{N}{2}|F|^2 = N^2\frac{(2^{-11})^2}{3}$$

so that noise to signal ratio $=\sqrt{2/3N}\ 2^{-11}$ which is very small for large N.

Digital to Analog Converter Noise

Even in the case of perfectly calculated samples driving the digital to analog converter, the output analog samples will produce noise from switching time disparities between bits, and differences in on and off switching. These so called "glitches" which occur at the transitions between sample outputs depend on the initial and final words upon which the converter is acting. A transition that changes many bits such as 10000 to 01111 will tend to produce more noise energy than a transition of only a few bits. This noise has the same periodicity properties that truncation noise has since it depends on transitions which are periodic if the sample train is periodic. However, all harmonics of the basic frequency are present since the "glitches" do not have odd symmetry. If one assumes a "glitch" amplitude can be as high as one-half of full scale out of the converter, then such a noise pulse of width Δ occurring even only once in each period will produce a noise-to-signal rate of approximately Δ/NT. For the higher generated frequencies such a ratio can be large. For example if $N=8$ (a frequency generated of 2^{13} $(2\pi/NT)$) and we want to maintain a noise-to-signal ratio of 10^{-3} (60 dB), the "glitch" duration must be about 8 ns or less. Such a requirement is difficult to achieve even with current techniques. Basically then the time of nonlinearity must be small compared to the roughly 1-μs sample interval for low noise effects. In our design example a 12-bit converter with settling time of about 300 ns (presumably "glitches" of any consequence occupy only a small portion of this overall transient) in an interval of 1.22 μs produced a distortion curve as shown in Fig. 9. Since this curve represents only second-harmonic distortion it is not a result of truncation, but strictly converter distortion. Higher harmonics were smaller or equal to those shown on the curve. For generated frequencies which were not powers of 2 as given on the graph, the distortion products were less, falling between the two cases given. The large disparity between "good" and "bad" phase for the higher generated frequencies is a consequence of the small number of samples per period so that no averaging takes place. Either a good set of samples does or does not occur so that the noise is high or low.

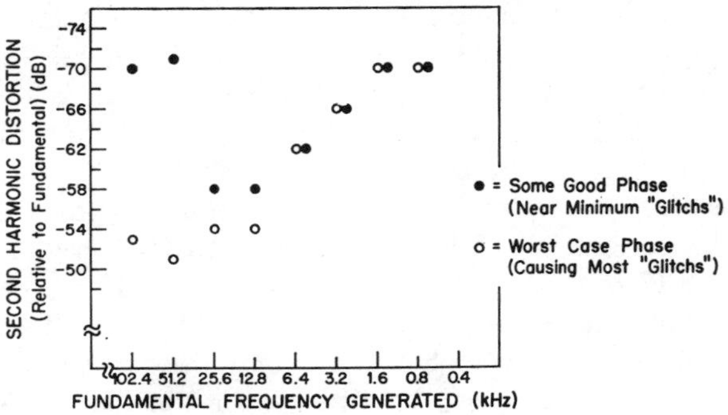

Fig. 9.

Even for odd generated frequencies the predominant noise tends to be harmonics of the generated frequency rather than harmonics of the lowest frequency. This indicates that certain more significant bits of the D–A converter are delayed more or less than others, and differences between on and off switch times are significant. In order to reduce these transition effects, one is forced to gating or sampling devices on the output of the converter. However, for times of less than 1 μs and linearities extending into the greater than 60 dB range, such circuits require careful design. Using such a gate for shorting the converter output to ground during transition time, the design presented earlier produced a worst case harmonic of 55 dB, an improvement upon the curve presented for the converter alone.

Smoothing Filter Time and Frequency Response

The third source of noise contributions to the synthesizer output is sampling harmonics of the generated frequency. As mentioned earlier a final smoothing filter is needed to interpolate computed sinusoidal samples or smooth the sampled spectrum. Since this smoothing filter is the energy storage device in the synthesizer the problem of time response arises. From Fig. 5(B) as previously seen, a filter was needed to pass the band up to 204.8 kHz and reject the band above 614.4 kHz. with a transition region between. The first tradeoff encountered is that between rejection or attenuation at 614.4 kHz and the time response of the filter. Consider a low-pass filter of fifth order (five poles) and examine its attenuation at 614.4 kHz as well as the time response. For several filter types the following table compares attenuation and step response. (Sample and hold response will add 10 dB to these figures.)

Type	Attenuation at 614.4 kHz	Step Response
Bessel (max flat phase)	~10 dB	~1 μs
Butterworth (max flat amp)	~48 dB	~4 μs
Chebyshev (1.0 dB ripple)	~65 dB	~7 μs
Cauer (elliptic function filter) 1.0 dB ripple	~85 dB	~9 μs

For this comparison the filters are 1.0 dB down at 204.8 and the step response is measured from 10 percent of final value to a ±10 percent window around final value. In other words overshoots must settle down to within a ±10 percent window. The obvious point made by this table is the tradeoff just mentioned. In addition, as the filter type moves from Bessel down, the phase or envelope delay characteristics become poorer. A similar trade can be effected between width of passband and step responses for a fixed attenuation at 614.4 kHz. If a certain attenuation is desired at that frequency a Bessel filter has a smaller passband than does a Butterworth, and so on down the list. For a fixed order of filter, and a fixed attenuation to be achieved at 614.4 kHz, the widest passband is obtained by using the sharpest filter and this in turn has the poorest time response.

The response times used in the above discussion have been step responses to dc rather than steps of sinusoidal input. It is clear that the time response of a step transition from one frequency to another consists of the response to two frequency steps; one to cancel the original frequency the other to start the new frequency. The amplitude response to each of these frequency steps reflects the natural frequencies of the driven filter. Therefore, the dc step response is still a measure of the time for amplitude response to settle down.

However, it is often the instantaneous frequency out of the synthesizer that is of importance, and the time this measure takes to settle down to some meaningful value. Obviously the instantaneous frequency must be influenced by the natural frequencies of the smoothing filter and must become some steady state value after the filter response time [8].

If an exact response for the instantaneous frequency is necessary, careful computation is necessary. If a rough response time is adequate, the table values will suffice. It is true that the smaller the frequency change, the smaller the frequency and amplitude perturbation.

It is also possible to reduce switching effects in both amplitude and phase response while still using a sharp cutoff filter such as an elliptic filter. This is accomplished by extending its cutoff and therefore its poor phase response into the transition region where no synthesizer outputs occur. In this way poorer properties of the filter are never manifested and some advantage is still taken of the sharp cutoff. Given any set of amplitude or phase criteria, there are a large set of filters to chose from satisfying smoothing requirements.

Producing RF Frequencies

Since the basic digital synthesizer cannot produce RF frequencies directly because of D–A speed limitations rather than logic problems, other methods must be used to produce an RF synthesizer output.

The method of single sideband modulation as mentioned earlier is useful if quadrature outputs are available. This is shown in Fig. 10 and requires a quadrature carrier as well. To move from upper to lower sideband the sign of one of the synthesizer outputs must change, and this is easily done. Since this technique requires a balanced subtraction at the output it is difficult to achieve high suppression of the opposite side frequency of more than 40–50 dB over a wide range.

A second approach to modulating the synthesizer output involves a single synthesizer output rather than quadrature. If the device need not produce quadrature outputs, it can work at a higher sampling or computation rate and produce a wider band of output frequencies ($\sim 1/4T$). Then the lower generated frequency is limited to produce a band from $1/4T$ down to some f_{low} which requires a modest filtering after modulation as shown in Fig. 11.

A third approach which is implemented at the digital level may be useful. This approach consists of computing quadrature outputs but at different sampling times (same

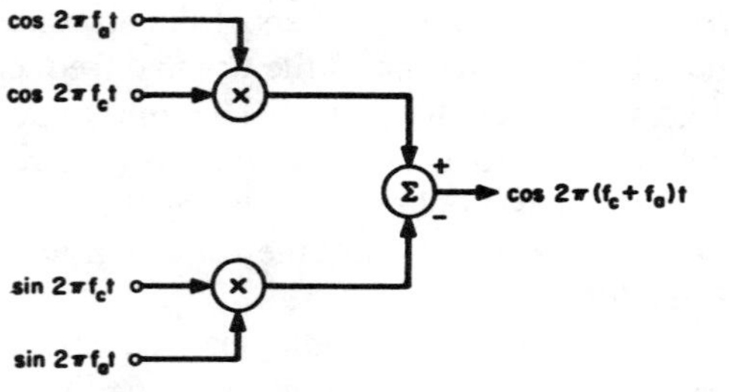

Fig. 10. Single sideband bandwidth doubling and frequency translation.

Fig. 11.

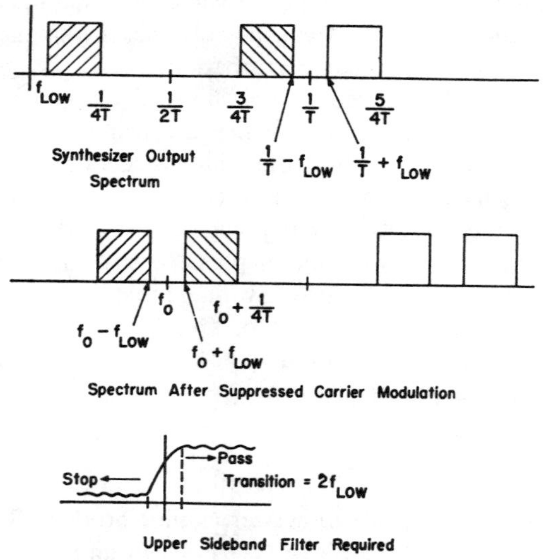

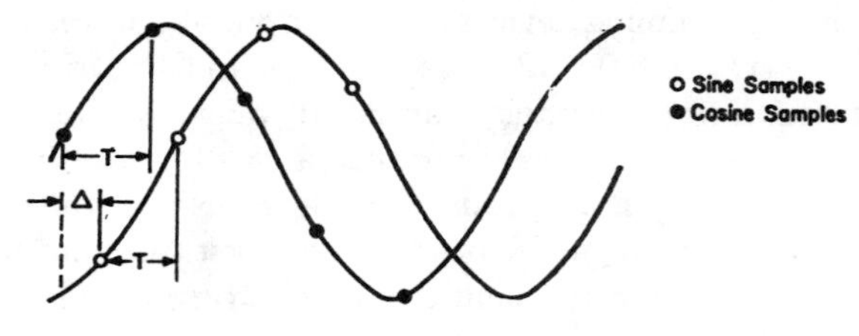

Computed Sampling

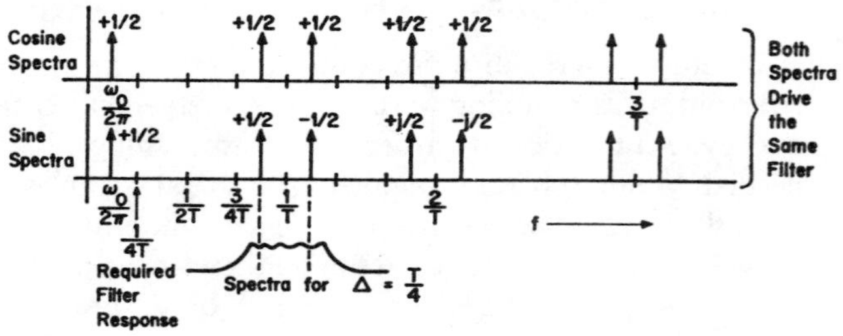

Fig. 12.

sampling interval) in order to cancel an upper or lower side frequency around some sampling harmonic (see Fig. 12). That is, if the cosine is computed at sampling times nT, then the sine must be computed at times $nT+\Delta$. To cancel the upper side frequency around the nth sampling harmonic $n2\pi/T$, Δ must be equal to $T/4n$, and the two computed samples are summed into the same smoothing filter (at their respective times). If the lower side frequency is to be eliminated, the difference between the two sample trains is used to drive the smoothing filter. This technique follows from the expressions for the spectra of sampled cosine computed at nT and sampled sine computed at $nT+\Delta$ as follows:

$$\text{cosine spectrum} = \frac{1}{2}\sum_{n=-\infty}^{\infty}\left[\delta\left(\omega - \omega o - \frac{n2\pi}{T}\right) + \delta\left(\omega - \omega o - \frac{n2\pi}{T}\right)\right]$$

$$\text{sine spectrum} = \frac{1}{2}\sum_{n=-\infty}^{\infty}\left[\delta\left(\omega - \omega o - \frac{n2\pi}{T}\right) - \delta\left(\omega + \omega o - \frac{n2\pi}{T}\right)\right] \exp\left[-j\frac{n2\pi\Delta}{T}\right].$$

This technique depends on generating narrow enough pulses out of the digital-to-analog converter to produce energy at some $n2\pi/T$ sampling harmonic.

Phase Control and Other Applications

The phase of the synthesizer output frequency at any computation time depends only on the stored value in the accumulator of Fig. 4 (ignoring any phase distortions introduced by the output smoothing filter). This allows for considerable phase control. In the normal mode of operation a change in frequency would be accomplished by changing only the input frequency word k and leaving unchanged the previous accumulator value. In this way the new frequency would be produced with no phase discontinuity since samples of the new frequency sinusoid would include the last sample of old frequency sinusoid. If on the other hand an arbitrary phase were desired at each change in frequency this means both a new frequency word k and a new accumulator value initial state. An example of such frequency switching would be resetting the accumulator (and phase) to zero whenever a new frequency is generated.

The ease with which phase can be controlled suggests many possible applications. The ability to frequency hop under phase control allows the implementation of new kinds of coherent communication systems if the channel phase is well behaved.

The synthesizer in the phase continuous mode can be used as a very linear frequency modulator. The modulating signal would drive an analog-to-digital converter (unless it is digital to begin with) whose output would be the digital frequency control word. The use of a large encoding range and a large encoding word would reduce distortion and noise. If the converter worked synchronous to the frequency synthesizer clock, no phase discontinuities could occur and a large deviation signal could be obtained with very high linearity.

Another class of applications is generation of sweep signals of various kinds. If the input frequency control word is incremented at certain fixed time intervals, a continuous stepped frequency waveform is generated. In the limit, this becomes a continuous frequency sweep so long as the sweep signal spectrum is not distorted by the sampling. The use of wired or stored frequency sequences could produce very complex sweep patterns.

Finally, the algorithm itself may be used in certain computational environments to produce digital samples of sine and cosine or complex coefficients for use in discrete Fourier transforms.

Some Design Considerations

A synthesizer of the type described here has several parameters which lead to design specifications. One is the time to produce a sample. This is composed of several elements, the accumulator time, the table look-up time, the multiplication time, and the D–A converter time. These times may be added, in some realizations, or they may be overlapped somewhat depending on the logic implementation. In our limited experience, the D–A time has been the largest element, and it is easily overlapped with other times, so that the time to produce a sample is the D–A time.

The reciprocal of the time to produce a sample is the sampling rate. This must be at least twice the highest frequency produced (or the bandwidth when a synthesizer design is intended to produce intermediate frequency sine waves), but it should be higher in order to simplify the requirements on the output smoothing filter. As discussed earlier there is, for any fixed sampling rate, a tradeoff between the complexity of the output filter required and the maximum frequency obtainable. This tradeoff is only interesting over a range of about 2:1 in maximum frequency, further simplification in the output filter as the sampling rate is increased becomes very small. The output filter also affects the switching time of the synthesizer output. Insofar as the sampling rate chosen affects the filter requirement, it also affects the obtainable switching time. This is an area where further work would be useful.

Another important set of parameters is the number of complex multiplications allowed in composing a sample. This is related to the number of subtables of constants required. With N words of storage, broken into $(n+1)$ subtables and combined by n multiplications (words, storage, multiplies all complex), the number of different complex exponentials (proportional to the number of frequencies which can be obtained) can be shown to be $(N/(n+1))^{n+1}$. By choosing n to maximize the number of exponentials, one can obtain a very large number indeed. For example, if $N=32$, it is quite straightforward to obtain 2^{16} exponentials with seven multiplies. In practice, we expect that it will not be desirable to use so many multiplications and probably one multiply and two subtables will be the most common choice.

The arithmetic precision needed, both in the stored constants and in the complex multiplications, will typically be determined by the capabilities of the output D–A converter. If the D–A converter is capable of resolving k bits, the computation should not attempt to produce numbers accurate to very much more than k bits. Note that the number of different frequencies obtainable is still unlimited, since the technique of using nearest samples is still available. If the sample used differs from the exact sample by less than 2^{-k}, the technique of using the nearest sample will not have any important error associated with it. This error is clearly bounded by $\sin 2\pi 2^{-k} \approx 2\pi 2^{-k}$. This implies using about $(k+3)$ bits of the accumulator in the computation, and using the lower bits only for accumulation to produce finer frequency increments. Many of these considerations become straightforward for particular design requirements.

Conclusion

A unique approach to the problem of frequency synthesis has been presented based on a sampled data realization. The simple factoring algorithm explained allows for very efficient use of storage and simple digital implementation. Finally, the phase properties of the synthesized signal and the ease of phase control allow the device to be used for generation of signals more complex than simple sinusoids.

References

[1] V. E. Van Duzer, "A 0–50 MHz frequency synthesizer with excellent stability, fast switching and fine resolution," *Hewlett-Packard J.*, vol. 15, May 1964, pp. 1–8.

[2] A. Noyes, Jr., "Coherent decade frequency synthesizers," *The Experimenter*, vol. 38, no. 9, Sept. 1964.

[3] E. Renschler and B. Welling, "An integrated circuit phase-locked loop digital frequency synthesizer," Motorola Semiconductor Products, Inc., Application Note 463.

[4] G. C. Gillete, "The digiphase synthesizer," *Frequency Technol.*, Aug. 1969, pp. 25–29.

[5] J. Noordanus, "Frequency synthesizers—A survey of techniques," *IEEE Trans. Commun. Technol.*, vol. COM-17, Apr. 1969, pp. 257–271.

[6] B. Gold and C. M. Rader, *Digital Processing of Signals.* New York: McGraw-Hill, 1969.

[7] A. W. Crooke, "A flexible digital waveform generator for use in matched filtering applications," presented at Arden House Workshop, Jan. 1970.

[8] H. Salinger, "Transients in frequency modulation," *Proc. IRE*, vol. 30, Aug. 1942, pp. 378–383.

Bibliography

M. R. Aaron and J. F. Kaiser, "On the calculation of transient response," *Proc. IEEE*, vol. 53, Sept. 1965, pp. 1269.

M. R. Aaron, R. A. McDonald, and E. N. Protonotarios, "Entropy power loss in linear sampled data filters," *Proc. IEEE*, vol. 55, June 1967, pp. 1093–1094.

T. A. Abele, "Transmission factors with Chebyshev-type approximations of constant group delay," *Arch. Elek. Ubertragung*, vol. 16, pp. 9–18, Jan. 1962.

V. R. Algazi, "Useful approximations to optimum quantization," *IEEE Trans. Commun. Technol.*, vol. COM-14, pp. 297–301, June 1966.

A. S. Alivi and G. M. Jenkins, "An example of digital filtering," *Appl. Statistics*, M116-AD, pp. 70–74.

L. E. Alsop and A. A. Nowroozi, "Fast Fourier analysis," *J. Geophys. Res.*, vol. 71, pp. 5482–5483, Nov. 15, 1966.

S. B. Alterman, "Discrete-time least squares, minimum mean square error, and minimax estimation," *IEEE Trans. Commun. Technol.*, vol. COM-14, pp. 302–308, June 1968.

E. B. Anders, "An error bound for a numerical filtering technique," *J. Ass. Comput. Mach.*, vol. 12, pp. 136–140, Jan. 1965.

E. B. Anders et. al., "Digital filters," NASA Contractor Rep. CR-136, Dec. 1964.

H. Andrews, "A high-speed algorithm for the computer generation of Fourier transforms," *IEEE Trans. Comput.* (Short note), vol. C-17, pp. 373–375, Apr. 1968.

C. B. Archambeau et. al., "Data processing techniques for the detection and interpretation of teleseismic signals," *Proc. IEEE*, vol. 53, Dec. 1965, pp. 1860–1884.

C. R. Arnold, "Laguerre functions and the Laguerre Network—their properties and digital simulation," Lincoln Lab., Lexington, Mass., Rep. 1966-28, May 4, 1966.

J. A. Athanassopoulos and A. D. Warren, "Design of discrete-time systems by mathematical programming," in *Proc. 1968 Hawaii Int. Conf. System Sci.* Honolulu: Univ. Hawaii Press, 1968, pp. 224–227.

——, "Time-domain synthesis by nonlinear programming," in *Proc. 4th Allerton Conf.*, 1968, pp. 766–775.

M. L. Attansoro D'atri, and T. Cianciolo, "Pole sensitivity to coefficient rounding in digital filters with multiple shift sequences," *Electron. Lett.*, vol. 7, pp. 29–31, Jan. 1971.

E. Avenhaus and W. Schüssler, "On the approximation problem in the design of digital filters with limited wordlength," *Arch. Elek. Übertragung*, vol. 24, pp. 571–572, 1970.

J. S. Bailey, "A fast Fourier transform without multiplications," in *Proc. Polytechnic Inst. Brooklyn Symp. Comput. Process. Commun.*, 1969.

D. C. Baxter, "The digital simulation of transfer functions," Nat. Res. Labs., Ottawa, Canada, DME Rep. MK-13, Apr. 1964.

——, "Digital simulation using approximate methods," Nat. Res. Council, Ottawa, Canada, Rep. MK-15, July 1965.

G. A. Bekey, "Sensitivity of discrete systems to variation of sampling interval," *IEEE Trans. Automat. Contr.*, vol. AC-11, pp. 284–287, Apr. 1966.

V. A. Benignus, "Estimation of the coherence spectrum and its confidence interval using the fast Fourier transform," *IEEE Trans. Audio Electroacoust.*, vol. AU-17, pp. 145–150, June 1969.

A. Bennett and A. Sage, "Discrete system sensitivity and variable increment sampling," *Proc. Joint Automat. Contr. Conf.*, pp. 603–612, 1967.

W. R. Bennett, "Spectra of quantized signals," *Bell Syst. Tech. J.*, vol. 27, pp. 446–472, July 1948.

G. D. Bergland and R. Klahn, "Digital processor for calculating Fourier coefficients," U.S. Patent 3 544 775, Dec. 29, 1966.

G. D. Bergland, "The fast Fourier transform recursive equations for arbitrary length records," *Math. Comput.*, vol. 21, pp. 236–238, 1967.

——, "A fast Fourier transform algorithm using base 8 iterations," *Math. Comput.*, vol. 22, pp. 275–279, Apr. 1968.

——, "A guided tour of the fast Fourier transform," *IEEE Spectrum*, vol. 6, pp. 41–52, July 1969.

——, "Fast Fourier transform hardware implementations—a survey," *IEEE Trans. Audio Electroacoust.*, vol. AU-17, pp. 109–119, June 1969.

——, "A radix-eight fast Fourier transform subroutine for real-valued series," *IEEE Trans. Audio Electroacoust.*, vol. AU-17, pp. 138–144, June 1969.

——, "Fast Fourier transform hardware implementations—an overview," *IEEE Trans. Audio Electroacoust.*, vol. AU-17, pp. 104–108, June 1969.

G. D. Bergland and H. W. Hale, "Digital real-time spectral analysis," *IEEE Trans. Electron. Comput.*, vol. EC-16, pp. 180–185, Apr. 1967.

G. D. Bergland and D. E. Wilson, "An FFT algorithm for a global, highly parallel processor," *IEEE Trans. Audio Electroacoust.*, vol. AU-17, pp. 125–127, June 1969.

J. E. Bertram, "The effect of quantization in sampled-feedback systems," *AIEE Trans.*, vol. 77, p. 177, 1958.

S. Bertram, "Frequency analysis using the discrete Fourier transform," *IEEE Trans. Audio Electroacoust.*, vol. AU-18, pp. 495–500, Dec. 1970.

——, "On the derivation of the fast Fourier transform," *IEEE Trans. Audio Electroacoust.*, vol. AU-18, pp. 55–58, Mar. 1970.

Bibliography, *Geophysics*, vol. 32, pp. 522–525, June 1967.

C. Bingham, M. D. Godfrey, and J. W. Tukey, "Modern techniques of power spectrum estimation," *IEEE Trans. Audio Electroacoust.*, vol. AU-15, pp. 56–66, June 1967.

L. I. Bluestein, "A linear filtering approach to the computation of discrete Fourier transform," *IEEE Trans. Audio Electroacoust.*, vol. AU-18, pp. 451–455, Dec. 1970.

M. Blum, "On exponential digital filters," *J. Ass. Comput. Mach.*, vol. 6, pp. 283–304, Apr. 1959.

B. Bogert, M. Healy, and J. Tukey, "The quefrency alanysis of time series for echoes," in *Proc. Symp. Time Series Analysis*, M. Rosenblatt, Ed. New York: Wiley, 1963, pp. 209–243.

B. Bogert and E. Parzen, "Informal comments on the uses of power spectrum analysis," *IEEE Trans. Audio Electroacoust.*, vol. AU-15, pp. 74–76, June 1967.

R. E. Bogner, "Frequency sampling filters—Hilbert transformers and resonators," *Bell Syst. Tech. J.*, vol. 48, pp. 501–510, Mar. 1969.

J. Boothroyd, "Complex Fourier series," *Comput. J.*, vol. 10, pp. 414–416, Feb. 1968.

E. M. Boughton, "Definition and synthesis of optimum-smoothing processes in filter terms," *IRE Trans. Instrum.*, vol. I-7, pp. 82–90, Mar. 1958.

R. Boxer, "Frequency analysis of computer systems," *Proc. IRE* (Corresp.), vol. 43, Feb. 1955, pp. 228–229.

——. "A note on numerical transform calculus," *Proc. IRE*, vol. 45, Oct. 1957, pp. 1401–1406.

R. Boxer and S. Thaler, "A simplified method of solving linear and nonlinear systems," *Proc. IRE*, vol. 44, Jan. 1956, pp. 89–101.

N. M. Brenner, "Fast Fourier transform of externally stored data," *IEEE Trans. Audio Electroacoust.*, vol. AU-17, pp. 128–132, June 1969.

——, "Three Fortran programs that perform the Cooley-Tukey Fourier transform," Lincoln Lab., Massachusetts Inst. Technol., Lexington, Tech. Note 1967-2, July 28, 1967.

E. O. Brigham and R. E. Morrow, "The fast Fourier transform," *IEEE Spectrum*, vol. 4, pp. 63–70, Dec. 1967.

H. W. Briscoe and P. L. Fleck, "A real-time computing system for LASA," in *1966 Spring Joint Computer Conf., AFIPS Conf. Proc.*, vol. 28. Washington, D.C.: Spartan, 1966.

J. Brogan, "Filters for sampled signals," *Proc. Symp. Networks Polytechnic Inst. Brooklyn*, Apr. 12–14, 1959, pp. 71–83.

P. W. Broome, "Discrete orthonormal sequences," *J. Ass. Comput. Mach.*, vol. 12, pp. 151–168, Apr. 1965.

P. W. Broome and W. C. Dean, "Seismic applications of orthogonal expansions," *Proc. IEEE*, vol. 53, Dec. 1965, pp. 1865–1869.

P. W. Broome, "A frequency transformation for numerical filters," *Proc. IEEE*, vol. 54, Feb. 1966, pp. 326–327.

J. D. Bruce, "Digital signal processing concepts," *IEEE Trans. Audio Electroacoust.*, vol. AU-18, pp. 344–353, Dec. 1970.

A. Budak and P. Aronhime, "Maximally flat low-pass filters with steeper slopes at cutoff," *IEEE Trans. Audio Electroacoust.*, vol. AU-18, pp. 63–66, Mar. 1970.

H. L. Buijs, "Fast Fourier transformation of large arrays of data," *Appl. Opt.*, vol. 8, pp. 211–212, Jan. 1969.

C. S. Burrus and T. W. Parks, "Time domain design of recursive digital filters," *IEEE Trans. Audio Electroacoust.*, vol. AU-18, pp. 137–141, June 1970.

A. M. Bush and D. C. Fielder, "An alternative derivation of the z-transform," *Amer. Math. Mon.*, vol. 70, pp. 281-284, Mar. 1963.
V. Cappellini, "Design of some digital filters with application to spectral estimation and data compression," in *Proc. Polytechnic Inst. Brooklyn Symp. Comput. Process. Commun.*, 1969.
——, "Digital filtering with sampled signal spectrum frequency shift," *Proc. IEEE* (Lett.), vol. 57, Feb. 1969, pp. 241-242.
V. Cappellini and T. D'Amico, "Some numerical filters obtained from the evaluation of the convolution integral and their application to spectral analysis," *Alta Freq.*, vol. 36, pp. 835-840, 1967.
C. C. Carroll and R. White, "Discrete compensation of control systems with integrated circuits," *IEEE Trans. Automat. Contr.*, vol. AC-12, pp. 579-582, Oct. 1967.
R. K. Cavin, C. H. Ray, and V. T. Rhyne, "The design of optimal convolutional filters via linear programming," *IEEE Trans. Geosci. Electron.*, vol. GE-7, pp. 142-145, July 1969.
R. K. Cavin and M. C. Budge, Jr., "A note on multirate z-transforms," *Proc. IEEE* (Lett.), vol. 58, Nov. 1970, pp. 1840-1841.
D. Chanoux, "Synthesis of recursive digital filters using the FFT," *IEEE Trans. Audio Electroacoust.*, vol. AU-18, pp. 211-212, June 1970.
E. W. Cheney and H. L. Loeb, "Generalized rational approximation," *SIAM J. Num. Analysis*, vol. 1, pp. 11-25, 1964.
V. Cizek, "Discrete Hilbert transform," *IEEE Trans. Audio Electroacoust.*, vol. AU-18, pp. 340-343, Dec. 1970.
——, "Numerische Hilbert-transformation," *Proc. Inst. Radio Eng. Electron.*, Czechoslovak Academy of Sciences, no. 11, 1961.
J. F. Claerbout and E. A. Robinson, "The error in least-squares inverse filtering," *Geophysics*, vol. 29, pp. 118-120, 1964.
W. T. Cochran et. al., "What is the fast Fourier transform?", *IEEE Trans. Audio Electroacoust.*, vol. AU-15, pp. 45-55, June 1967.
A. G. Constantinides, "Design of bandpass digital filters," *Proc. IEEE*, vol. 57, June 1969, pp. 1229-1231.
——, "Digital filters with equiripple passbands," *IEEE Trans. Circuit Theory* (Corresp.), vol. CT-16, pp. 535-538, Nov. 1969.
——, "Elliptic digital filters," *Electron. Lett.*, vol. 3, pp. 255-256, June 1967.
——, "Frequency transformations for digital filters," *Electron. Lett.* vol. 3, pp. 487-489, Nov. 1967.
——, "Spectral transformation for digital filters," *Proc. Inst. Elec. Eng.*, vol. 117, 1970, pp. 1585-1590.
——, "Synthesis of Chebychev digital filters," *Electron Lett.*, vol. 3, pp. 124-127, Mar. 1967.
J. W. Cooley, "Complex finite Fourier transform subroutine," Share Doc. 3465, Sept. 8, 1966.
——, "Application of the fast Fourier transform method," in *Proc. IBM Sci. Comput. Symp.*, 1966.
——, "Harmonic analysis of complex Fourier series," Share Program Library No. SDA 3425, Feb. 7, 1966.
J. W. Cooley et. al., "The 1968 Arden house workshop on fast Fourier transform processing," *IEEE Trans. Audio Electroacoust.*, vol. AU-17, pp. 66-76, June 1969.
J. W. Cooley, P. A. Lewis, and P. D. Welch, "Application of the fast Fourier transform to computation of Fourier integrals, Fourier series, and convolution integrals," *IEEE Trans. Audio Electroacoust.*, vol. AU-15, pp. 79-84, June 1967.
——, "The fast Fourier transform and its applications," IBM Res. Paper RC-1743, Feb. 9, 1967.
——, "The fast Fourier transform algorithm: programming considerations in the calculation of sine, cosine, and Laplace transforms," *J. Sound. Vib.*, vol. 12, pp. 315-337, 1970.
——, "The finite Fourier transform," *IEEE Trans. Audio Electroacoust.*, vol. AU-17, pp. 77-86, June 1969.
——, "Historical notes on the fast Fourier transform," *IEEE Trans. Audio Electroacoust.*, vol. AU-15, pp. 76-79, June 1967.
——, "The use of the fast Fourier transform algorithm for the estimation of spectra and cross spectra," in *Proc. Polytechnic Inst. Brooklyn Symp. Comput. Process. Commun.*, 1969.
J. W. Cooley and J. W. Tukey, "An algorithm for the machine computation of complex Fourier series," Share Doc. 3465, Sept. 8, 1966.
M. J. Corinthios, "A fast Fourier transform for high-speed signal processing," *IEEE Trans. Comput.*, vol. C-20, pp. 843-846, Aug. 1971.
——, "A time-series analyzer," in *Proc. Polytechnic Inst. Booklyn Symp. Comput. Process. Commun.*, 1969.
——, "The design of a class of fast Fourier transform computers," *IEEE Trans. Comput.*, vol. C-20, pp. 617-623, June 1971.
T. H. Crystal and L. Ehrman, "The design and applications of digital filters with complex coefficients," *IEEE Trans. Audio Electroacoust.*, vol. AU-16, pp. 315-321, Sept. 1968.
E. E. Curry, "The analysis of round-off and truncation errors in a hybrid control system," *IEEE Trans. Automat. Contr.*, vol. AC-12, pp. 601-604, Oct. 1967.
G. C. Danielson and C. Lanczos, "Some improvements in practical Fourier analysis and their application to x-ray scattering from liquids," *J. Franklin Inst.*, pp. 365-380, pp. 435-452, Apr.-May 1942.
J. A. D'Appolito, "A simple algorithm for discretizing linear stationary continuous time systems," *Proc. IEEE*, vol. 54, Dec. 1966, pp. 2010-2011.
A. C. Davies, "Digital filtering of binary sequences," *Electron. Lett.*, vol. 3, pp. 318-319, July 1967.
L. D. Divieti, C. M. Rossi, R. M. Schmid, and A. E. Vereschkin, "A note on computing quantization errors in digital control systems," *IEEE Trans. Automat. Contr.*, vol. AC-12, pp. 622-623, Oct. 1967.
C. L. Dolph, "A current distribution for broadside arrays which optimizes the relationship between beamwidth and side-lobe level," *Proc. IRE*, vol. 34, June 1946, pp. 335-348.
C. J. Drane, "Directivity and beamwidth approximations for large scanning Dolph-Chebyshev arrays," AFCRL Physical Sci. Res. Paper 117, AFCRL-65-472, June 1965.
G. Dumermuth and H. Fluhler, "Some modern aspects in numerical spectrum analysis of multichannel electroencephalographic data," *Med. Elec. Biol. Eng.*, vol. 5, pp. 319-331, 1967.
S. C. Dutta Roy, "On maximally flat sharp cutoff low-pass filters," *IEEE Trans. Audio Electroacoust.*, vol. AU-19, pp. 58-63, Mar. 1971.
P. M. Ebert, J. E. Mazo, and M. G. Taylor, "Overflow oscillations in digital filters," *Bell Syst. Tech. J.*, vol. 48, pp: 2999-3020, Nov. 1969.
R. Edwards and A. Bradley, "Design of digital filters by computers," *Int. J. Numer. Methods Eng.*, vol. 2, pp. 311-333, 1970.
R. Edwards, J. Bradley, and J. Knowles, "Comparison of noise performances of programming methods in the realization of digital filters," in *Proc. Polytechnic Inst. Brooklyn Symp. Comput. Process. Commun.*, 1969.
G. Epstein, "Recursive fast Fourier transforms," in *1968 Fall Joint Computer Conf. AFIPS Conf. Proc.*, vol. 33. Washington D.C.: Thompson, 1968, pp. 141-143.
M. J. Ferguson and P. E. Mantey, "Automatic frequency control via digital filtering," *IEEE Trans. Audio Electroacoust.*, vol. AU-16, pp. 392-398, Sept. 1968.
A. Fettweis, "A general theorem for signal-flow networks, with applications," *Arch. Elek. Übertragung*, vol. 25, pp. 557-561, Dec. 1971.
——, "Digital filter structures related to classical filter networks," *Arch. Elek. Übertragung*, vol. 25, pp. 79-89, 1971.
A. A. Filippini, "Synthesis of cascaded digital filters to achieve desired transfer characteristics," Tech. Rep. AD-699 529, Sept. 1969.
P. E. Fleischer, "Digital realization of complex transfer functions," *Simulation*, vol. 6, pp. 171-180, Mar. 1966.
R. Fletcher and M. J. Powell, "A rapidly convergent descent method for minimization," *Comput. J.*, vol. 6, pp. 163-168, 1963.
M. L. Forman, "Fast Fourier transform technique and its application to Fourier spectroscopy," *J. Opt. Soc. Am.*, vol. 56, pp. 978-997, July 1966.
M. E. Fowler, "A new numerical method for simulation," *Simulation*, vol. 4, pp. 324-330, May 1965.
P. A. Franaszek, "On sampled-data and time varying systems," Commun. Lab., Dept. Elec. Eng., Princeton Univ., Princeton, N.J., Tech. Rep. 11, Oct. 1965.
L. E. Franks, "Power spectral density of random facsimile signal," *Proc. IEEE* (Corresp.), vol. 52, Apr. 1964, pp. 431-432.
D. Fraser, "Associative parallel processing," in *1967 Spring Joint Computer Conf., AFIPS CONF. Proc.*, vol. 30. Washington, D.C.: Thompson, 1967, pp. 471-475.
R. Galpin, "Variable electronic allpass delay network," *Electron. Lett.*, vol. 4, pp. 137-139, 1968.
A. Gelb and P. Palosky, "Generating discrete colored noise from discrete white noise," *IEEE Trans. Automat. Contr.*, vol. AC-11,

pp. 148-149, Jan. 1966.
W. M. Gentleman and G. Sande, "Fast Fourier transforms—for fun and profit," in *1966 Fall Joint Computer Conf., AFIPS Conf. Proc.*, vol. 29. Washington, D.C.: Spartan, 1966, pp. 563-578.
W. M. Gentleman, "Matrix multiplication and fast Fourier transforms," *Bell Syst. Tech. J.*, vol. 47, pp. 1099-1103, July-Aug. 1968.
A. J. Gibbs, "An introduction to digital filters," *Aust. Telecommun. Res.*, vol. 3, pp. 3-14, Nov. 1969.
——, "The design of digital filters," *Aust. Telecommun. Res.*, vol. 4, pp. 29-34, 1970.
G. C. Gillete, "The digiphase synthésizer," *Frequency Technol.*, pp. 25-29, Aug. 1969.
M. Gilmartin, Jr. and R. R. Shively, "Digital processor for performing fast Fourier transforms," U.S. Patent 3 517 173, Dec. 29, 1966.
J. A. Glassman, "A generalization of the fast Fourier transform," *IEEE Trans. Comput.*, vol. C-19, pp. 105-116, Feb. 1970.
T. H. Glisson and A. P. Sage, "On discrete and complex representation of real signals," in *Proc. 12th Midwest Symp. Circuit Theory*, Apr. 1969.
T. H. Glisson, C. I. Black, and A. P. Sage, "The digital computation of discrete spectra using the fast Fourier transform," *IEEE Trans. Audio Electroacoust.*, vol. AU-18, pp. 271-287, Sept. 1970.
G. Goertzel, "An algorithm for the evaluation of finite trigonometric series," *Amer. Math. Mon.*, vol. 65, pp. 34-35, Jan. 1958.
B. Gold and K. Jordan, "A direct search procedure for designing finite duration impulse response filters," *IEEE Trans. Audio Electroacoust.*, vol. AU-17, pp. 33-36, Mar. 1969.
——, "A note on digital filter synthesis," *Proc. IEEE*, vol. 56, Oct. 1968, pp. 1717-1718.
B. Gold and L. R. Rabiner, "Analysis of digital and analog formant synthesizers," *IEEE Trans. Audio Electroacoust.*, vol. AU-16, pp. 81-94, Mar. 1968.
B. Gold and C. M. Rader, "Effects of quantization noise in digital filters," in *1966 Spring Joint Computer Conf., AFIPS Conf. Proc.*, vol. 28. Washington, D.C.: Spartan, 1966, pp. 213-219.
B. Gold, A. V. Oppenheim, and C. M. Rader, "Theory and implementation of the discrete Hilbert transform," in *Proc. Polytechnic Inst. Booklyn Symp. Comput. Process. Commun.*, 1969.
R. M. Golden, "Digital computer simulation of sampled-data communication systems using block diagram compiler: BLODIB," *Bell Syst. Tech. J.*, vol. 45, pp. 344-358, Mar. 1966.
——, "Digital computer simulation of sampled-data voice-excited vocoder," *J. Acoust. Soc. Am.*, vol. 35, pp. 1358-1366, 1963.
——, "Digital filter synthesis by sampled-data transformation," *IEEE Trans. Audio Electroacoust.*, vol. AU-16, pp. 321-329, Sept. 1968.
R. M. Golden and J. F. Kaiser, "Design of wideband sampled-data filters," *Bell Syst. Tech. J.*, vol. 43, pp. 1533-1546, July 1964.
——, "Root and delay parameters for normalized Bessel and Butterworth low-pass transfer functions," *IEEE Trans. Audio Electroacoust.*, vol. AU-19, pp. 64-71, Mar. 1971.
R. M. Golden and S. A. White, "A holding technique to reduce the number of bits in digital transfer functions," *IEEE Trans. Audio Electroacoust.*, vol. AU-16, pp. 433-437, Sept. 1968.
I. J. Good, "The interaction algorithm and practical Fourier series," *J. Roy Statist. Soc., Ser. B.*, vol. 20, pp. 361-372, 1958; Addendum, vol. 22, pp. 372-375, 1960.
——, "The relationship between two fast Fourier transforms," *IEEE Trans. Comput.*, vol. C-20, pp. 310-317, Mar. 1971.
D. J. Goodman, "Optimum digital filters for the estimation of continuous signals in noise," *Proc. Polytechnic Inst. Brooklyn Symp. Comput. Process. Commun.*, 1969.
L. M. Goodman and P. R. Drouilhet, Jr., "Asymptotically optimum pre-emphasis and de-emphasis networks for sampling and quantizing," *Proc. IEEE*, vol. 54, May 1966, pp. 795-796.
O. D. Grace, "Two finite Fourier transforms for bandpass signals," *IEEE Trans. Audio Electroacoust.* (Corresp.), vol. AU-18, pp. 501-502, Dec. 1970.
O. D. Grace and S. P. Pitt, "Quadrature sampling of high-frequency waveforms," *J. Acoust. Soc. Am.*, vol. 44, pp. 1453-1454, Nov. 1968.
R. J. Graham, "Determination and analysis of numerical smoothing elements," NASA Tech. Rep. TR-R-179, Dec. 1963.
A. Grassi and G. Strini, "Errors in the reconstruction of quantized and sampled random signals," *Alta Freq.*, vol. 33, pp. 547-555, Aug. 1964.
C. J. Greaves and J. A. Cadzow, "The optimal discrete filter corresponding to a given analog filter," *IEEE Trans. Automat. Contr.*, vol. AC-12, pp. 304-307, June 1967.
B. F. Green, J. E. Smith, and L. Klem, "Empirical tests of an additive random number generator," *J. Ass. Comput. Mach.*, vol. 6, pp. 527-537, Oct. 1959.
H. L. Groginsky and G. A. Works, "A pipeline fast Fourier transform," *IEEE Trans. Comput.*, vol. C-19, pp. 1015-1019, Nov. 1970.
S. C. Gupta and W. W. Happ, "Flowgraph approach to z-transforms and application to discrete systems," *Int. J. Contr.*, vol. 2, pp. 211-220, Sept. 1965.
C. A. Halijak, "Digital approximation of differential equations by trapezoidal convolution," Kansas State Univ. Bulletin, vol. 45, Rep. 10, pp. 83-94, July 1961.
S. R. Harrison and B. J. Leon, "Digital filters," Tech. Rep. N70-20579, Sept. 1969.
W. T. Hartwell and R. A. Smith, "Apparatus for performing complex wave analysis," U.S. Patent 3 544 894, July 10, 1967.
H. S. Heaps and W. Willcock, "The use of quantizing techniques in real time Fourier analysis," *Radio Electron. Eng.*, vol. 29, pp. 143-148, Mar. 1965.
D. Helman, "Tchebycheff approximations for amplitude and delay with rational functions," in *Proc. Polytechnic Inst. Brooklyn Symp. Modern Network Theory*, pp. 385-402.
H. D. Helms, "Digital filters with equiripple or minimax responses," *IEEE Trans. Audio Electroacoust.*, vol. AU-19, pp. 87-94, Mar. 1971.
——, "Fast Fourier transform method of computing difference equations and simulating filters," *IEEE Trans. Audio Electroacoust.*, vol. AU-15, pp. 85-90, June 1967.
——, "Nonrecursive digital filters: design methods for achieving specifications on frequency response," *IEEE Trans. Audio Electroacoust.*, vol. AU-16, pp. 336-342, Sept. 1968.
R. M. Hendrickson, "Frequency response functions of certain numerical filters," Space Technol. Labs., Redondo Beach, Calif., GM-00-4330-00281, Apr. 6, 1959.
O. Herrmann, "Design of nonrecursive digital filters with linear phase," *Electron. Lett.*, vol. 6, pp. 328-329, 1970.
——, "On the approximation problem in nonrecursive digital filter design," *IEEE Trans. Circuit Theory*, vol. CT-18, pp. 411-413, 1971.
O. Herrmann and W. Schüssler, "Design of nonrecursive digital filters with minimum phase," *Electron. Lett.*, vol. 6, 1970.
——, "On the accuracy problem in the design of nonrecursive digital filters," *Arch. Elek. Übertragung*, vol. 24, pp. 525-526, 1970.
G. E. Heyliger, "The scanning function approach to the design of numerical filters," Martin Comp., Denver, Colo., Rep. R-63-2, Apr. 1963.
——, "Simple design parameters for Chebyshev arrays and filters," *IEEE Trans. Audio Electroacoust.* (Corresp.), vol. AU-18, pp. 502-503, Dec. 1970.
F. B. Hills, "A study of incremental computation by difference equation," Servomechanisms Lab., Massachusetts Inst. Technol., Cambridge, Mass., Rep. 7849-R-1, May 1958.
M. J. Hinich and C. S. Clay, "The application of the discrete Fourier transform in estimation of power spectra, coherence and bispectra of geophysical data," *Rev. Geophys.*, vol. 6, pp. 347-363, Aug. 1968.
E. Hofstetter, A. V. Oppenheim, and J. Siegel, "A new technique for the design of nonrecursive digital filters," in *Proc. 5th Annu. Princeton Conf. Inform. Sci. Syst.*, 1971, pp. 64-72.
——, "On optimum nonrecursive digital filters," in *Proc. 9th Annu. Allerton Conf. Circuit System Theory*, Oct. 6-8, 1971.
H. Holtz and C. T. Leondes, "The synthesis of recursive filters," *J. Ass. Comput. Mach.*, vol. 13, pp. 262-280, Apr. 1966.
T. C. Hsia, "On synthesis of optimal digital filters," in *Proc. 1st Asilomar Conf. Circuit System Theory*, Nov. 1-3, 1967, pp. 473-480.
W. H. Huggins, "Signal Theory," *IRE Trans. Circuit Theory*, vol. CT-3, pp. 210-216, Dec. 1956.
L. B. Jackson, "An analysis of limit cycles due to multiplication rounding in recursive digital filters," in *Proc. 7th Annu. Allerton Conf. Circuit System Theory*, 1969, pp. 69-78.
——, "On the interaction of roundoff noise and dynamic range in digital filters," *Bell Syst. Tech. J.*, vol. 49, pp. 159-184, 1970.
——, "Roundoff-noise analysis for fixed-point digital filters realized in cascade or parallel form," *IEEE Trans. Audio Electroacoust.*, vol. AU-18, pp. 107-122, June 1970.

L. B. Jackson, J. F. Kaiser, and H. S. McDonald, "An approach to the implementation of digital filters," *IEEE Trans. Audio Electroacoust.*, vol. AU-16, pp. 413–421, Sept. 1968.

L. B. Jackson and H. S. McDonald, "Digital filtering," U.S. Patent 3 522 546, Aug. 4, 1970.

G. W. Johnson, D. P. Libdorff, and C. G. Nordling, "Extension of continuous-data system design techniques to sampled-data control systems," *AIEE Trans. Appl. Ind.*, vol. 74, part 2, pp. 252–263, Sept. 1955.

N. B. Jones, "Lowpass filters with approximately equal ripple modulus error," *Electron. Lett.*, vol. 3, pp. 516–517, Nov. 1967.

E. I. Jury, "A general z-transform formula for sampled-data systems," *IEEE Trans. Automat. Contr.*, vol. AC-12, pp. 606–608, Oct. 1967.

D. K. Kahaner, "Matrix description of the fast Fourier transform," *IEEE Trans. Audio Electroacoust.*, vol. AU-18, pp. 442–450, Dec. 1970.

R. E. Kahn and B. Liu, "Sampling with time jitter," Commun. Lab., Dept. Elec. Eng., Princeton Univ., Princeton, N.J., Tech. Rep. 8, June 1964.

J. F. Kaiser, "Computer aided design of classical continuous system transfer functions," in *Proc. Hawaii Int. Conf. System Sci.* Honolulu: Univ. Hawaii Press, pp. 197–200, 1968.

——, "Design methods for sampled-data filters," in *Proc. 1st Annu. Allerton Conf. Circuit System Theory*, 1963, pp. 221–236.

——, "Digital filters," in *System Analysis by Digital Computer*, F. F. Kuo and J. F. Kaiser, Ed. New York: Wiley, 1966.

——, "Some practical considerations in the realization of linear digital filters," in *Proc. 3rd Annu. Allerton Conf. Circuit System Theory*, 1963, pp. 621–633.

S. C. Kak, "The discrete Hilbert transform," *Proc. IEEE* (Lett.), vol. 58, Apr. 1970, pp. 585–586.

H. E. Kallmann, "Transversal filters," *Proc. IRE*, vol. 28, July 1940, pp. 302–310.

R. E. Kalman and J. E. Bertram, "A unified approach to the theory of sampled systems," *J. Franklin Inst.*, vol. 267, pp. 405–436, May 1959.

T. Kaneko and B. Liu, "Roundoff error of floating-point digital filters," in *Proc. 6th Annu. Allerton Conf. Circuit System Theory*, 1968, pp. 219–227.

——, "Accumulation of round-off errors in fast Fourier transform," *J. Ass. Comput. Mach.*, vol. 17, pp. 637–654, Oct. 1970.

B. J. Karafin, "The new block diagram compiler for simulations of sampled-data systems," in *1965 Fall Joint Computer Conf., AFIPS Conf. Proc.*, vol. 27. Washington, D.C.: Spartan, 1965, pp. 55–61.

J. Katzenelson, "On errors introduced by combined sampling and quantization," *IRE Trans. Automat. Contr.*, vol. 7, pp. 58–68, Apr. 1962.

——, "A note on errors introduced by combined sampling and quantization," Electron. Systems Lab., Massachusetts Inst. Technol., Cambridge, Mass., ESL-TM-101, Mar. 1961.

W. H. Kautz, "Transient synthesis in the time domain," *IRE Trans. Circuit Theory*, vol. CT-1, pp. 29–39, Sept. 1954.

J. E. Kelley, Jr., "An application of linear programming to curve fitting," *J. Soc. Ind. Appl. Math.*, vol. 6, pp. 15–22, 1968.

W. C. Kellogg, "Information rates in sampling and quantizing," *IEEE Trans. Inform. Theory*, vol. IT-13, pp. 506–511, July 1967.

J. L. Kelly, C. L. Lochbaum, V. A. Vyssotsky, "A block diagram compiler," *Bell Syst. Tech. J.*, vol. 40, pp. 669–676, 1961.

L. C. Kelly and J. N. Holmes, "Computer processing of signals with particular reference to simulation of electric filter networks," Eng. GPO Dept., Post Office Res. Station, London, Eng., Res. Rep. 21072, Feb. 15, 1965.

R. Klahn and R. R. Shively, "FFT—shortcut to Fourier analysis," *Electronics* vol. 41, pp. 124–129, Apr. 15, 1968.

R. Klahn, R. R. Shively, E. Gomez, and M. J. Gilmartin, "The timesaver-FFT hardware," *Electronics,* vol. 41, pp. 92–97, June 24, 1968.

J. B. Knowles and R. Edwards, "Aspects of subrate digital control systems," *Proc. Inst. Elec. Eng.*, vol. 113, Nov. 1966, pp. 1885–1892.

——, "Complex cascade programming and associated computational errors," *Electron. Lett.*, vol. 1, pp. 160–161, Aug. 1965.

——, "Computational error effects in a direct digital control system," *Automatica*, vol. 4, pp. 7–29, 1966.

——, "Effects of a finite word length computer in a sampled-data feedback system," *Proc. Inst. Elec. Eng.*, vol. 112, June 1965.

——, "Finite word-length effects in a multirate direct digital control system," *Proc. Inst. Elec. Eng.*, vol. 112, Dec. 1965, pp. 2376–2384.

J. B. Knowles and E. M. Olcayto, "Coefficient accuracy and digital filter response," *IEEE Trans. Circuit Theory*, vol. CT-15, pp. 31–41, Mar. 1968.

A. Kohlenberg, "Exact interpolation of band-limited functions," *J. Appl. Phys.*, vol. 24, pp. 1432–1436, Dec. 1953.

G. A. Korn, "Hybrid-computer techniques for measuring statistics from quantized data," *Simulation*, vol. 4, pp. 229–239, Apr. 1965.

E. Korngold, "The periodic analysis of sampled data," Lincoln Lab., Massachusetts Inst. Technol., Lexington, Mass., Group Rep. 1964–32, June 15, 1964.

A. A. Kosyakin, "The statistical theory of amplitude quantization," *Avtomat. Telemekh.*, vol. 22, p. 722, 1961.

W. Kuntz, "A new sample-and-hold device and its application to the realization of digital filters," *Proc. IEEE*, vol. 56, Nov. 1968, pp. 2092–2093.

W. Kuntz and H. W. Schüssler, "The numerical calculation of the time response networks with the aid of the z-transformation," *J. Nachrichtentech Z.*, vol. 5, pp. 121–124, 1967.

G. N. Lack, "Comments on upper bound on dynamic quantization error in digital control systems via the direct method of Liapunov," *IEEE Trans. Automat. Contr.* (Corresp.), vol. AC-11, pp. 331–333, Apr. 1966.

I. M. Langenthal, "Coefficient sensitivity and generalized digital filter synthesis," *1968 Eascon Rec.*, pp. 386–392, 1968.

——, "The synthesis of symmetrical bandpass digital filters," in *Proc. Polytechnic Inst. Brooklyn Symp. Comput. Process. Commun.*, 1969.

I. M. Langenthal and S. Gowrinathan, "Advanced digital processing techniques," Tech. Rep. AD 708736, 1970.

A. W. Langill, "Digital filters," *Frequency*, vol. 6, Nov.–Dec. 1964.

A. G. Larson and R. C. Singleton, "Real-time spectral analysis on a small general-purpose computer," *1967 Fall Joint Computer Conf., AFIPS Conf. Proc.*, vol. 31. Washington, D.C.: Thompson, 1967, pp. 665–674.

R. M. Lerner, "Band-pass filters with linear phase," *Proc. IEEE*, vol. 52, Mar. 1964, pp. 249–268.

L. B. Lesem, P. M. Hirsch, J. A. Jordan, Jr., "Computer synthesis of holograms for 3-D display," *Commun. Ass. Comput. Mach.*, vol. 11, pp. 661–674, Oct. 1968.

M. J. Levin, "Estimation of a system pulse transfer function in the presence of noise," *IEEE Trans. Automat. Contr.*, vol. AC-9, pp. 229–235, July 1964.

——, "Generation of a sampled Gaussian time series having a specified correlation function," *IRE Trans. Inform. Theory*, vol. IT-6, pp. 545–548, Dec. 1960.

P. M. Lewis, "Synthesis of sampled signal networks," *IRE Trans. Circuit Theory*, vol. CT-5, pp. 74–77, Mar. 1958.

R. N. Linebarger, "Precision-sample rate tradeoffs in quantized sampled-data systems," Systems Res. Center, Case Inst. Technol., Cleveland, Ohio, Rep. SRC-46-C-64-17, 1964.

W. K. Linvill, "Sampled-data control systems studied through comparison of sampling with amplitude modulation," *AIEE Trans.*, vol. 70, part 2, pp. 1779–1788, 1951.

B. Liu, "Effect of finite word length on the accuracy of digital filters—a review," *IEEE Trans. Circuit Theory*, vol. CT-18, pp. 670–677, Nov. 1971.

B. Liu and P. Franaszek, "A class of time-varying digital filters," *IEEE Trans. Circuit Theory*, vol. CT-16, pp. 467–471, Nov. 1969.

B. Liu and J. B. Thomas, "Error problems in sampling representations, part I, type I errors," Commun. Lab., Dept. Elec. Eng., Princeton Univ., Princeton, N.J., Tech. Rep. 4, Z08250, Apr. 1964.

——, "Error problems in the reconstruction of signals from sampled data," *Proc. Nec.*, vol. 23, pp. 803–807, 1967.

P. A. Lynn, "Economic linear-phase recursive digital filters," *Electron. Lett.*, vol. 6, pp. 143–145, Mar. 1970.

M. D. MacLaren and G. Marsaglia, "Uniform random number generators," *J. Ass. Comput. Mach.*, vol. 12, pp. 83–89, 1965.

C. E. Maley, "The effect of parameters on the roots of an equation system, *Comput. J.*, vol. 4, pp. 62–63, 1961–1962.

C. G. Maling, Jr., W. T. Morrey, and W. W. Lang, "Digital determination of third-octave and full-octave spectra of acoustical noise," *IEEE Trans. Audio Electroacoust.*, vol. AU-15, pp. 98–104, June 1967.

R. Manasse, "Tapped delay line realizations of frequency periodic filters and their application to linear FM pulse compression," Mitre Corp., Bedford, Mass., Tech. Doc. Rep. ESD-TDR-63-232, May 1963.

M. Mansour, "Instability criteria of linear discrete systems," *Auto-*

matica, vol. 2, pp. 167–178, Jan. 1965.

P. E. Mantey, "Convergent automatic-synthesis procedures for sampled data networks with feedback," Electron. Labs., Stanford Univ., Stanford, Calif., Tech. Rep. 6773-1, SU-SEL-64-112, Oct. 1964.

P. E. Mantey and G. F. Franklin, "Digital filter design techniques in the frequency domain," *Proc. IEEE*, vol. 55, Dec. 1967, pp. 2196–2197.

P. E. Mantey, "Eigenvalue sensitivity and state-variable selection," *IEEE Trans. Automat. Contr.*, vol. AC-13, pp. 263–269, June 1968.

G. Marsaglia, "A note on the construction of a multivarate normal sample," *IRE Trans. Inform. Theory*, vol. IT-3, p. 149, June 1957.

M. A. Martin, "Frequency domain applications to data processing," *IRE Trans. Space Electron. Telem.*, vol. SET-5, pp. 33–41, Mar. 1959.

——, "Digital filters for data processing," General Electric Comp., Missile Space Div., Tech. Inform., Series Rep. 62-SD484, 1962.

J. Max, "A new Fourier technique for frequency-domain synthesis of delay-line filters," in *Proc. 7th Midwest Symp. Circuit Theory*, May 4–5, 1964, pp. 165–170.

D. W. McCowan, "Finite Fourier transform theory and its application to the computation of convolutions, correlations, and spectra," Teledyne Industries, Inc., Earth Sciences Div., Oct., 1966.

R. N. McDonough, "Comment on 'z-transform' technique," *Proc. IEEE*, vol. 54, Nov. 1966, pp. 1616–1617.

R. N. McDonough and W. H. Huggins, "Best least-squares representation of signals by exponentials," *IEEE Trans. Automat. Contr.*, vol. AC-13, pp. 408–412, Aug. 1968.

B. T. McKeever, "The associative memory structure," in *1965 Fall Joint Computer Conf., AFIPS Conf. Proc.*, vol. 27. Washington, D.C.: Spartan, 1965, pp. 371–388.

T. H. McKinney, "A digital spectrum channel analyzer," in *Conf. Speech Commun. Process. Reprints*, pp. 442–444, Nov. 1967.

A. R. Memon, "Lowpass digital filters with linear phase," *Electron. Lett.*, vol. 6, pp. 253–254, Apr. 1970.

S. A. Miller, "A PDP-9 assembly-language program for the fast Fourier transform," Analog/Hybrid Computer Lab., Dept. Elec. Eng., Univ. Arizona, Tucson, ACL Memo. 157, Apr. 1968.

H. T. Nagle, Jr. and C. C. Carroll, "Organizing a special-purpose computer to realize digital filters for sampled-data systems," *IEEE Trans. Audio Electroacoust.*, vol. AU-16, pp. 398–413, Sept. 1968.

C. D. Negron, "Digital one-third octave spectral analysis," *J. Ass. Comput. Mach.*, vol. 13, pp. 605–614, Oct. 1966.

J. Noordanus, "Frequency synthesizers—a survey of techniques," *IEEE Trans. Commun. Technol.*, vol. COM-17, pp. 257–271, Apr. 1969.

D. J. Nowak and P. E. Schmid, "A nonrecursive digital filter for data transmission," *IEEE Trans. Audio Electroacoust.*, vol. AU-16, pp. 343–350, Sept. 1968.

A. Noyes, Jr., "Coherent decade frequency synthesizers," *Experimenter*, vol. 38, Sept. 1964.

G. C. O'Leary, "Nonrecursive digital filtering using cascade fast Fourier transformers," *IEEE Trans. Audio Electroacoust.*, vol. AU-18, pp. 177–183, June 1970.

A. V. Oppenheim, "Superposition in a class of nonlinear systems," Res. Lab. Electron., M.I.T., Cambridge, Mass., RLE Tech. Rep. 432, Mar. 31, 1965.

——, "Nonlinear filtering of convolved signals," Res. Lab. Electron., M.I.T., Cambridge, Mass., Quart. Prog. Rep. 80, pp. 168–175, Jan. 15, 1966.

——, "Realization of digital filters using block-floating-point arithmetic," *IEEE Trans. Audio Electroacoust.*, vol. AU-18, pp. 130–136, June 1970.

A. V. Oppenheim and R. W. Schafer, "Homomorphic analysis of speech," *IEEE Trans. Audio Electroacoust.*, vol. AU-16, pp. 221–226, June 1968.

A. V. Oppenheim, R. W. Schafer, and T. Stockham, "The nonlinear filtering of multiplied and convolved signals," *Proc. IEEE*, vol. 56, Aug. 1968, pp. 1264–1291.

A. V. Oppenheim and C. Weinstein, "A bound on the output of a circular convolution with application to digital filtering," *IEEE Trans. Audio Electroacoust.*, vol. AU-17, pp. 120–124, June 1969.

A. V. Oppenheim, D. Johnson, and K. Steiglitz, "Computation of spectra with unequal resolution using the FFT," *Proc. IEEE*, vol. 59, 1971, pp. 299–301.

H. J. Orchard, "The roots of the maximally flat-delay polynomials," *IEEE Trans. Circuit Theory* (Corresp.), vol. CT-12, pp. 452–454, Sept. 1965.

——, "Maximally flat approximation techniques," *Proc. IEEE*, vol. 56, Jan. 1968, pp. 65–66.

J. F. A. Ormsby, "Design of numerical filters with applications to missile data processing," *J. Ass. Comput. Mach.*, vol. 8, pp. 440–466, July 1961.

R. K. Otnes, "An elementary design procedure for digital filters," *IEEE Trans. Audio Electroacoust.*, vol. AU-16, pp. 330–336, Sept. 1964.

R. K. Otnes and L. P. McNamee, "Instability thresholds in digital filters due to coefficient rounding," *IEEE Trans. Audio Electroacoust.*, vol. AU-18, pp. 456–463, Dec. 1970.

——, "Exact second-order bandpass digital filters," *IEEE Trans. Audio Electroacoust.*, vol. AU-19, pp. 104–105 (Corresp.), Mar. 1971.

A. Papoulis, "On the approximation problem in filter design," *IRE Conv. Rec.*, part 2, pp. 175–185, 1957.

——, "Error analysis in sampling theory," *Proc. IEEE*, vol. 54, July 1966, pp. 947–955.

E. Parzen, "Notes on Fourier analysis and spectral windows," *Appl. Math. Statist. Labs.*, Stanford Univ., Stanford, Calif., Tech. Rep. 48, May 15, 1963.

——, "Statistical spectral analysis (single channel case) in 1968," Dept. Statist., Stanford Univ., Stanford, Calif., Tech. Rep. 11, ONR contract NONR-225 (80) (NR-042-234), June 10, 1968.

M. C. Pease and J. Goldberg, "Feasibility study of a special-purpose digital computer for on-line Fourier analysis," Adv. Res. Proj. Agency, Order 989, May 1967.

——, "Investigation of a special-purpose digital computer for on-line Fourier analysis," Stanford Res. Inst., Menlo Park, Calif., Special Tech. Rep. 1, Project 6557, Apr. 1967 (available from U.S. Army Missile Command, Redstone Arsenal, Ala., att: AMSMI-RNS).

M. C. Pease, "An adaptation of the fast Fourier transform for parallel processing," *J. Ass. Comput. Mach.*, vol. 15, pp. 252–264, Apr. 1968.

Y. Peless and T. Murakami, "Analysis and synthesis of transitional Butterworth-Thomson filters and bandpass amplifiers," *RCA Rev.*, vol. 18, pp. 60–94, Mar. 1957.

D. P. Petersen, "Smoothing and differential operators for digital processing of sampled-field data," *Nerem Record*, pp. 170–171, 1963.

S. E. A. Pinnell, "Design of a digital notch filter with tracking requirements," *IEEE Trans. Space Electron. Telem.* (Comment), vol. SET-10, p. 84, 1964.

M. J. Piovoso and L. P. Bolgiano, Jr., "Digital simulation using Poisson transform sequences," in *Proc. Polytechnic Inst. Brooklyn Symp. Comput. Process. Commun.*, 1969.

C. Pottle, "On the partial-fraction expansion of a rational function with multiple poles by a digital computer," *IEEE Trans. Circuit Theory*, vol. CT-11, pp. 161–162, Mar. 1964.

——, "Rapid computer time response for systems with arbitrary input signals," in *Proc. 5th Annu. Allerton Conf. Circuit System Theory*, pp. 523–533, 1967.

L. R. Rabiner, R. W. Schafer, and C. M. Rader, "The chirp z-transform algorithm and its application," *Bell Syst. Tech. J.*, vol. 48, pp. 1249–1292, May–June 1969.

——, "The chirp z-transform algorithm," *IEEE Trans. Audio Electroacoust.*, vol. AU-17, pp. 86–92, June 1969.

L. R. Rabiner, B. Gold and C. A. McGonegal, "An approach to the approximation problem for nonrecursive digital filters," *IEEE Trans. Audio Electroacoust.*, vol. AU-18, pp. 83–106, June 1970.

L. R. Rabiner and K. Steiglitz, "The design of wide-band recursive and nonrecursive digital differentiators," *IEEE Trans. Audio Electroacoust.*, vol. AU-18, pp. 204–209, June 1970.

L. R. Rabiner, "Techniques for designing finite-duration impulse-response digital filters," *IEEE Trans. Commun. Technol.*, vol. COM-19, pp. 188–195, Apr. 1971.

L. R. Rabiner and R. W. Schafer, "Recursive and nonrecursive realizations of digital filters designed by frequency sampling techniques," *IEEE Trans. Audio Electroacoust.*, vol. AU-19, pp. 200–207, Sept. 1971.

L. R. Rabiner, L. B. Jackson, R. W. Schafer, and C. H. Coker, "A hardware realization of a digital formant speech synthesizer," *IEEE Trans. Commun. Technol.*, vol. COM-19, pp. 1016–1020, Dec. 1971.

P. Rabinowitz, "Applications of linear programming to numerical analysis," *SIAM Rev.*, vol. 10, pp. 121–159, 1959.

C. M. Rader, "Speech compression simulation compiler," *J. Acoust. Soc. Am.*, vol. 37, p. 1199, June 1965.

——, "Discrete Fourier transforms when the number of data samples is prime," *Proc. IEEE*, vol. 56, June 1968, pp. 1107–1108.

——, "An improved algorithm for high-speed autocorrelation with applications to spectral estimation," *IEEE Trans. Audio Electroacoust.*

vol. AU-18, pp. 439-441, Dec. 1970.
C. Rader and B. Gold, "Digital filter design techniques in the frequency domain," *Proc. IEEE*, vol. 55, Feb. 1967, pp. 149-171.
——, "Effects of parameter quantization on the poles of a digital filter," *Proc. IEEE*, vol. 55, May 1967, pp. 688-689.
C. M. Rader, L. R. Rabiner and R. W. Schafer, "A fast method of generating digital random numbers," *Bell Syst. Tech. J.*, vol. 49, pp. 2303-2310, Nov. 1970.
C. M. Rader et al., "On digital filtering," *IEEE Trans. Audio Electroacoust.* vol. AU-16, pp. 303-315, Sept. 1968.
Q. I. Rahman, "The influence of coefficients on the zeros of polynomials," *J. London Math. Soc.*, vol. 36, part 1, pp. 57-64, Jan. 1961.
G. U. Ramos, "Roundoff error analysis of the fast Fourier transform," *Math. Comput.*, vol. 25, pp. 757-768, Oct. 1971.
R. R. Reed, "A method of computing the fast Fourier transform," M.A. thesis, Dept. Elec. Eng., Rice Univ., Houston, Tex., May 1968.
E. Renschler and B. Welling, "An integrated circuit phase-locked loop digital frequency synthesizer," Motorola Semiconductor Prod., Application Note 463.
A. A. G. Requicha and H. B. Voelcker, "Design of nonrecursive filters by specification of frequency-domain zeros," *IEEE Trans. Audio Electroacoust.*, vol. AU-18, pp. 464-471, Dec. 1970.
R. P. Rich and H. Shaw, Jr., "An application of digital filtering," *API Tech. Dig.*, vol. 4, pp. 13-18, Jan.-Feb. 1965.
J. Richalet, "Les systemes discrets," *Onde Elec.*, vol. 44, pp. 1011-1020, Oct. 1964.
R. A. Roberts and J. Tooley, "Signal processing with limited memory," in *Proc. Polytechnic Inst. Brooklyn Symp. Comput. Process. Commun.*, 1969.
H. H. Robertson, "Approximate design of digital filters," *Technometrics*, vol. 7, pp. 387-403, Aug. 1965.
E. A. Robinson and S. Treitel, "Principles of digital filtering," *Geophysics*, vol. 29, pp. 395-404, June 1964.
——, "Dispersive digital filters," *Rev. Geophys.*, vol. 3, pp. 433-461, Nov. 1965.
——, "Principles of digital Wiener filtering," Pan Am. Petroleum Corp., Tulsa, Okla., 1967.
A. E. Rogers and K. Steiglitz, "Maximum likelihood estimation of rational transfer function parameters," *IEEE Trans. Automat. Contr.*, vol. AC-12, pp. 594-597, Oct. 1967.
D. T. Ross, "Improved computational techniques for Fourier transformation," Servomechanisms Lab., M.I.T., Cambridge, Mass., Report 713-R-5, June 25, 1954.
C. Rossi, "Window functions for nonrecursive digital filters," *Electron. Lett.*, vol. 3, pp. 559-561, Dec. 1967.
E. N. Rozenvasser, "Stability criteria of nonlinear discrete systems," *Automat. Telemekh.*, vol. 27, pp. 58-66, Dec. 1966.
P. Rudnick, "Note on the calculation of Fourier series," *Math. Comput.*, vol. 20, pp. 429-430, July 1966.
M. Sablatash, "A Tellegen's theorem for digital filters," *IEEE Trans. Circuit Theory*, (Corresp.), vol. CT-18, pp. 201-203, Jan. 1971.
A. P. Sage and R. W. Burt, "Optimum design and error analysis of digital integrators for discrete system simulation," *Fall Joint Comput. Conf. AFIPS Conf. Proc.*, vol. 27. Washington, D.C.: Spartan, 1965, pp. 903-914.
A. P. Sage, "Discretization schemes and the optimal control of distributed parameter systems," in *Proc. 1st Asilomar Conf. Circuit System Theory*, 1967, pp. 191-200.
D. J. Sakrison, W. T. Ford, and J. H. Hearne, "The z-transform of a realizable time function," *IEEE Trans. Geosci. Electron.*, vol. GE-5, pp. 33-41, Sept. 1967.
I. W. Sandberg, "Floating-point-roundoff accumulation in digital filter realization," *Bell Syst. Tech. J.*, vol. 46, pp. 1775-1791, Oct. 1967.
R. W. Schafer, "Echo removal by generalized linear filtering," *Nerem Rec.*, pp. 118-119, 1967.
——, "Echo removal by discrete generalized linear filtering," Ph.D. dissertation, Dept. Elec. Eng., M.I.T., Cambridge, Mass., Feb. 1968.
R. W. Schafer and L. R. Rabiner, "Design of digital filter banks for speech analysis," *Bell Syst. Tech. J.*, vol. 50, pp. 3097-3115, Dec. 1971.
S. A. Schelkunoff, "A mathematical theory of linear arrays," *Bell Syst. Tech. J.*, vol. 22, pp. 80-107, Jan. 1943.
I. J. Schoenberg, "The finite Fourier series and elementary geometry," *Am. Math. Mon.*, vol. 57, Jun.-July 1950.
W. Schüssler and W. Winkelnkemper, "Variable digital filters," *Arch. Elek. Übertragung*, vol. 24, pp. 524-525, 1970.
W. Schüssler, "On the approximation problem in the design of digital filters," in *Proc. 5th Annu. Princeton Conf. Inform. Sci. Systems*, pp. 54-63, Mar. 1971.
A. Sekey, "A computer simulation study of real-zero interpolation," *IEEE Trans. Audio Electroacoust.*, vol. AU-18, pp. 43-54, Mar. 1970.
——, "Simulation of real-zero interpolation by the BLODIB compiler," in *Proc. Polytechnic Inst. Brooklyn Symp. Comput. Process. Commun.*, 1969.
J. L. Shanks, "Recursion filters for digital processing," *Geophysics*, vol. 23, pp. 33-51, Feb. 1967.
——, "Two planar digital filtering algorithms," in *Proc. 5th Annu. Princeton Conf. Inform. Sci. Systems*, Mar. 1971, pp. 48-53.
J. L. Shanks and T. W. Cairns, "Use of a digital convolution device to perform recursive filtering and the Cooley-Tukey algorithm," *IEEE Trans. Comput.*, vol. C-17, pp. 943-949, Oct. 1968.
J. P. Shelton, "Fast Fourier transforms and Butler matrices," *Proc. Elec. Eng.*, Australia, vol. 56, p. 350.
R. R. Shively, "A digital processor to perform a fast Fourier transform," in *Proc. 1st IEEE Comput. Conf.*, Sept. 1967, pp. 21-24.
——, "A digital processor to generate spectra in real time," *IEEE Trans. Comput.*, vol. C-17, pp. 485-491, May 1968.
R. C. Singleton, "A method for computing the fast Fourier transform with auxiliary memory and limited high-speed storage," *IEEE Trans. Audio Electroacoust.*, vol. AU-15, pp. 91-98, June 1967.
——, "On computing the fast Fourier transform," *Commun. Ass. Comput. Mach.*, vol. 10, pp. 647-654, Oct. 1967.
——, "An algol procedure for the fast Fourier transform with arbitrary factors," *Commun. Ass. Comput. Mach.*, Algorithm 339, vol. 11, pp. 776-779, Nov. 1968.
——, "Algol procedures for the fast Fourier transform," *Commun. Ass. Comput. Mach.*, Algorithm 338, vol. 11, p. 338, Nov. 1968.
——, "An algorithm for computing the mixed radix fast Fourier transform," *IEEE Trans. Audio Electroacoust.*, vol. AU-17, pp. 93-103, June 1969.
R. C. Singleton and T. C. Poulter, "Spectral analysis of the call of the male killer whale," *IEEE Trans. Audio Electroacoust.*, vol. AU-15, pp. 104-113, June 1967; also Comments by W. A. Watkins and authors' reply, vol. AU-16, p. 523, Dec. 1968.
J. B. Slaughter, "Quantization errors in digital control systems," *IEEE Trans. Automat. Contr.*, vol. AC-9, pp. 70-74, Jan. 1964.
J. M. Slazer, "Frequency analysis of digital computers operating in real time," *Proc. IRE*, vol. 42, Feb. 1954, pp. 457-466.
D. Slepian, H. O. Pollak, and H. J. Landau, "Prolate spheroidal wave functions, Fourier analysis and uncertainty," *Bell Syst. Tech. J.*, vol. 40, pp. 43-85, Jan. 1961.
E. A. Sloane, "Comparison of linearly and quadratically modified spectral estimates of Gaussian signals," *IEEE Trans. Audio Electroacoust.*, vol. AU-17, pp. 133-137, June 1969.
O. Sornmoonpin, "Investigation of quantization errors," M.S. Thesis, Univ. Manchester, England, 1966.
R. J. Stegen, "Excitation coefficients and beamwidths of Tschebyscheff arrays." *Proc. IRE*, vol. 41, Nov. 1953, pp. 1671-1674.
K. Steiglitz, "The approximation problem for digital filters," Dept. Elec. Eng., N.Y. Univ., Tech. Rep. 400-56, 1962.
——, "The general theory of digital filters with applications to spectral analysis," Ph.D. dissertation, N.Y. Univ., AFOSR rep. 64-1664, 1963.
——, "The equivalence of digital and analog signal processing," *Inform. Contr.*, vol. 8, pp. 455-467, Oct. 1965.
——, "Computer-aided design of recursive digital filters," *IEEE Trans. Audio Electroacoust.*, vol. AU-18, pp. 123-129, June 1970.
T. G. Stockham, Jr., "High-speed convolution and correlation," in *1966 Spring Joint Computer Conf., AFIPS Proc.*, vol. 28. Washington, D.C.: Spartan, 1966, pp. 229-233.
——, "The application of generalized linearity to automatic gain control," *IEEE Trans. Audio Electroacoust.*, vol. AU-16, pp. 267-270, June 1968.
——, "A-D and D-A converters: their effect on digital audio fidelity," in 41st meeting Audio Eng. Soc., N.Y.C., Oct. 5-8, 1971.
J. Stoer, "A direct method for Chebyshev approximation by rational functions," *J. Ass. Comput. Mach.*, vol. 11, pp. 59-69, Jan. 1964.
D. J. Storey and R. F. Donne, "Synthesis of the Hilbert transform of a train of rectangular pulses," *Electron. Lett.*, vol. 3, pp. 126-128, Mar. 1967.
J. I. Soliman and A. Al-Shaikh, "Sampled-data controls and the bilinear transformation," *Automatica*, vol. 2, pp. 235-242, July 1965.
J. I. Soliman and H. Kwoh, "Bilinear transformation for sampled-data

systems," *IEEE Trans. Automat. Contr.*, vol. AC-11, pp. 329-330, Apr. 1966.
D. A. Swick, "Discrete finite Fourier transforms: a tutorial approach," Naval Res. Labs., Washington, D.C., NRL Rep. 6557, June 1967.
G. C. Temes and D. A. Calahan, "Computer-aided network design—the state of the art," *Proc. IEEE*, vol. 55, Nov. 1967, pp. 1832-1863.
M. S. Thelliez and J. P. Gouyet, "Introduction a l'analyse des systemes asservis a information pulsee," *Ann. Radioelec.*, vol. 16, pp. 9-68, Jan. 1961.
F. Theilheimer, "A matrix version of the fast Fourier transform," *IEEE Trans. Audio Electroacoust.*, vol. AU-17, pp. 158-161, June 1969.
D. J. Thomson, "Generation of Gegenbauer prewhitening filters by fast Fourier transforming," in *Proc. Polytechnic Inst. Brooklyn Symp. Comput. Process. Commun.*, 1969.
J. Tierney, C. M. Rader, and B. Gold, "A digital frequency synthesizer," *IEEE Trans. Audio Electroacoust.*, vol. AU-19, pp. 48-58, Mar. 1971.
S. Treitel and E. A. Robinson, "The design of high resolution digital filters," *IEEE Trans. Geosci. Electron.*, vol. GE-4, pp. 25-39, June 1966.
S. Treitel, J. L. Shanks, and C. W. Frasier, "Some aspects of fan filtering," *Geophysics*, vol. 32, pp. 789-800, Oct. 1967.
W. F. Trench, "A general class of discrete time-invariant filters," *J. Soc. Ind. Appl. Math.*, vol. 9, pp. 406-421, Sept. 1961.
S. A. Tretter, "Some problems in the reconstruction and processing of sampled-data," Dept. Elec. Eng., Princeton Univ., Princeton, N.J., Ph.D. dissertation, pp. 105-122, Dec. 1965.
——, "Pulse-transfer-function identification using discrete orthonormal sequences," *IEEE Trans. Audio Electroacoust.*, vol. AU-18, pp. 184-187, June 1970.
Y. Z. Tsypkin, "Estimating the effect of quantization by level on the processes in automatic digital systems," *Automat. Telemekh.*, vol. 21, pp. 281-285, Mar. 1960.
——, "An estimate of the influence of amplitude quantization on processes in digital automatic control systems," *Automat. Telemekh.*, vol. 21, p. 195, 1960.
D. W. Tufts, H. S. Hersey, and W. E. Mosier, "Effects of FFT coefficient quantization on bin frequency response," *Proc. IEEE* (Lett.), vol. 60, Jan. 1972, pp. 146-147.
M. Tsu-Han Ma, "A new mathematical approach for linear array analysis and synthesis," Ph.D. dissertation, Syracuse Univ., Syracuse, N.Y., Univ. Microfilms, 62-3040, 1961.
A. Tustin, "A method of analyzing the behavior of linear systems in terms of time series," *Proc. Inst. Elec. Eng.*, Australia, vol. 94, part IIA, May 1947, pp. 130-142.
M. L. Uhrich, "Fast Fourier transforms without sorting," *IEEE Trans. Audio Electroacoust.*, (Corresp.), vol. AU-17, pp. 170-172, June 1969.
E. Ulbrich and H. Piloty, "The design of allpass, lowpass, and bandpass filters with Chebyshev-type approximations of constant group delay," *Arch. Elek. Übertragung,* vol. 14, pp. 451-467, Oct. 1960.
H. Urkowitz, "Analysis and synthesis of delay line periodic filters," *IRE Trans. Circuit Theory,* pp. 41-53, June 1957.
R. Vich, "Selective properties of digital filters obtained by convolution approximation," *Electron. Lett.*, vol. 4, pp. 1-2, Jan. 1968.
A. J. Villasenor, "Digital spectral analysis," NASA, Washington, D.C., Tech. Note D-4510, 1968.
H. B. Voelcker, "Toward a unified theory of modulation," *Proc. IEEE*, vol. 54, pt. 1, Mar. 1966, pp. 340-353; pt. 2, May 1966, pp. 735-755.
H. B. Voelcker and E. E. Hartquist, "Digital filtering via block recursion," *IEEE Trans. Audio Electroacoust.*, vol. AU-18, pp. 169-176, June 1970.
P. J. Walsh, "A study of digital filters," Tech. Rep. Ad 71-0381, Dec. 1969.
W. Wasow, "Discrete approximations to the Laplace transformation," *Z. Angew. Math. Phys.*, vol. 8, pp. 401-417, 1957.
D. G. Watts, "A general theory of amplitude quantization with applications to correlation determination," *Proc. Inst. Elec. Eng.,* Australia, vol. 109C, 1962, p. 209.
——, "Optimal windows for power spectra estimation," Math. Res. Center, Univ. Wisconsin, MRC-TCR-506, Sept. 1964.
C. S. Weaver, P. E. Mantey, R. W. Lawrence, and C. A. Cole, "Digital spectrum analyzers," Stanford Electron. Labs., Stanford, Calif., Rep. SEL 66-059 (Tr-109-1/1810-1), June 1966.
C. S. Weaver et al., "Digital filtering with applications to electrocardiogram processing," *IEEE Trans. Audio Electroacoust.*, vol. AU-16, pp. 350-392, Sept. 1968.
C. J. Weinstein, "Quantization effects in digital filters," Lincoln Lab. Tech. Rep. 468, Nov. 21, 1969.
C. Weinstein, "Quantization effects in frequency sampling filters," *Nerem Record,* p. 222, 1968.
——, "Quantization effects in digital filters," Ph.D. dissertation, Dept. Elec. Eng., M.I.T., Cambridge, Mass., July 1969.
——, "Roundoff noise in floating point fast Fourier transform computation," *IEEE Trans. Audio Electroacoust.*, vol. AU-17, pp. 209-215, Sept. 1969.
C. Weinstein and A. V. Oppenheim, "A comparison of roundoff noise in floating-point and fixed point digital filter realizations," *Proc. IEEE* (Lett.), vol. 57, June 1969, pp. 1181-1183.
P. D. Welch, "A direct digital method of power spectrum estimation," *IBM J. Res. Develop.*, vol. 5, pp. 141-156, Apr. 1961.
——, "The use of the FFT for estimation of power spectra: a method based on averaging over short, modified periodograms," *IEEE Trans. Audio Electroacoust.*, vol. AU-15, pp. 70-73, June 1967.
——, "A fixed-point fast Fourier transform error analysis," *IEEE Trans. Audio Electroacoust.*, vol. AU-17, pp. 151-157, June 1969.
M. A. Wesley, "Associative parallel processing for the fast Fourier transform," *IEEE Trans. Audio Electroacoust.*, vol. AU-17, pp. 162-165, June 1969.
J. E. Whelchel, Jr. and D. F. Guinn, "The fast Fourier-Hadamard transform and its use in signal representation and classification," *1968 Eascon Conv. Rec.*, pp. 561-571.
——, "FFT organizations for high-speed digital filtering," *IEEE Trans. Audio Electroacoust.*, vol. AU-18, pp. 159-168, June 1970.
W. D. White and A. E. Ruvin, "Recent advances in the synthesis of comb filters," *IRE Nat. Conv. Rec.*, pp. 186-199, 1957.
D. E. Whitney, "Computation errors in inertial navigation systems which employ digital computers," Instrum. Lab., Cambridge, Mass., T-409, Feb. 1965.
J. R. B. Whittlesey, "A rapid method for digital filtering," *Commun. Ass. Comput. Mach.*, vol. 7, pp. 552-556, Sept. 1964.
B. Widrow, "A study of rough amplitude quantization by means of Nyquist sampling theory," *IRE Trans. Circuit Theory*, vol. CT-3, pp. 266-276, Dec. 1956.
——, "Statistical analysis of amplitude-quantized sampled-data systems," *AIEE Trans.* (Appl. Ind.), vol. 79, part 2, pp. 555-568, 1961.
W. Winkelnkemper, "Unsymmetrical bandpass and bandstop digital filters," *Electron. Lett.*, vol. 5, Nov. 13, 1969.
J. H. Wilkinson, "Error analysis of floating-point comparison," *Numer. Math.*, vol. 2, pp. 319-340, 1960.
J. C. Wilson, "Computer calculation of discrete Fourier transforms using the fast Fourier transform," Center Naval Analyses, Arlington, Va., OEG Res. Contrib. 81, June 1968.
T. Y. Young, "Representation and analysis of signals, part X. Signal theory and electrocardiography," Dept. Elec. Eng., Johns Hopkins Univ., Baltimore, Md., May 1962.
——, "Binomial-weighted orthogonal polynomials," *J. Ass. Comput. Mach.*, vol. 14, pp. 120-127, Jan. 1967.
A. I. Zverev, "Digital MTI radar filters," *IEEE Trans. Audio Electroacoust.*, vol. AU-16, pp. 422-432, Sept. 1968.

Author Index

Subject Index

Editors' Biographies

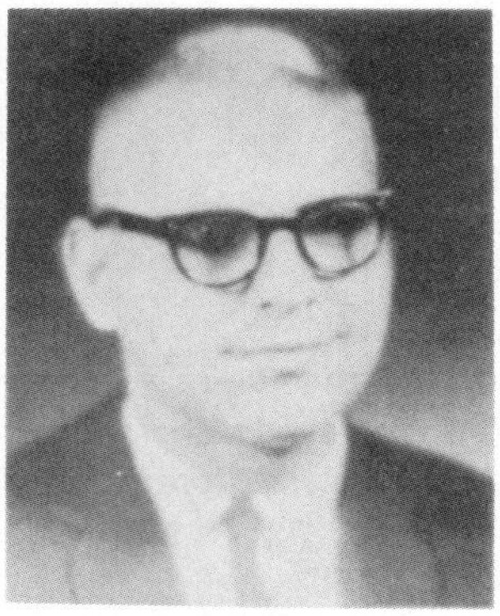

Lawrence R. Rabiner (S'62–M'62) was born in Brooklyn, N.Y., on September 28, 1943. He received the S.B. and S.M. degrees simultaneously in June 1964, and the Ph.D. degree in electrical engineering in June 1967, all from the Massachusetts Institute of Technology, Cambridge.

From 1962 through 1964 he participated in the cooperative plan in electrical engineering at Bell Telephone Laboratories, Inc., Whippany and Murray Hill, N.J. He worked on digital circuitry, military communications problems, and problems in binaural hearing. Presently he is engaged in research on speech communications and digital signal processing techniques at Bell Telephone Laboratories, Murray Hill.

Dr. Rabiner is a member of Eta Kappa Nu, Sigma Xi, Tau Beta Pi, and the Acoustical Society of America.

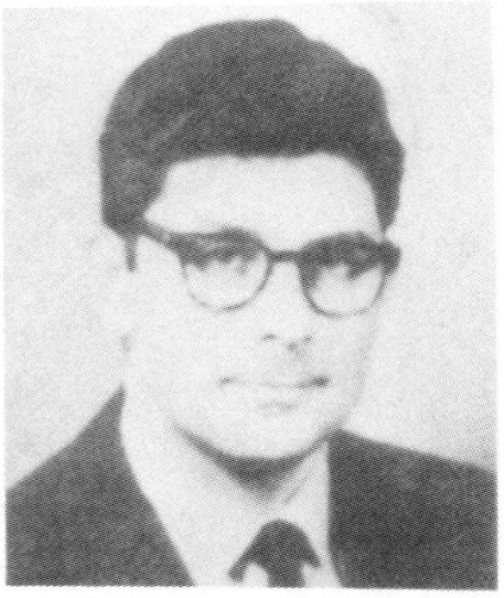

Charles M. Rader (S'59–M'62) was born in Brooklyn, N. Y., on June 20, 1939. He received the B. E. E. and M. E. E. degrees from the Polytechnic Institute of Brooklyn, Brooklyn, N. Y., in 1960 and 1961, respectively.

He has been with the M.I.T. Lincoln Laboratory, Lexington, Mass., since 1961, where he has worked on techniques for speech bandwidth compression and computer simulation.

Mr. Rader is a member of Eta Kappa Nu, Tau Beta Pi, and the Acoustical Society of America.